The Elements

Name	Symbol	Atomic Number	Atomic Mass[a]	Name	Symbol	Atomic Number	Atomic Mass[a]	Name	Symbol	Atomic Number	Atomic Mass[a]
Actinium	Ac	89	(227)[b]	Hassium	Hs	108	(265)[b]	Praseodymium	Pr	59	140.9
Aluminum	Al	13	26.98	Helium	He	2	4.003	Promethium	Pm	61	(145)[b]
Americium	Am	95	(243)[b]	Holmium	Ho	67	164.9	Protactinium	Pa	91	(231)[b]
Antimony	Sb	51	121.8	Hydrogen	H	1	1.008	Radium	Ra	88	(226)[b]
Argon	Ar	18	39.95	Indium	In	49	114.8	Radon	Rn	86	(222)[b]
Arsenic	As	33	74.92	Iodine	I	53	126.9	Rhenium	Re	75	186.2
Astatine	At	85	(210)[b]	Iridium	Ir	77	192.2	Rhodium	Rh	45	102.9
Barium	Ba	56	137.3	Iron	Fe	26	55.85	Rubidium	Rb	37	85.47
Berkelium	Bk	97	(247)[b]	Krypton	Kr	36	83.80	Ruthenium	Ru	44	101.1
Beryllium	Be	4	9.012	Lanthanum	La	57	138.9	Samarium	Sm	62	150.4
Bismuth	Bi	83	209.0	Lawrencium	Lr	103	(262)[b]	Scandium	Sc	21	44.96
Boron	B	5	10.81	Lead	Pb	82	207.2	Selenium	Se	34	78.96
Bromine	Br	35	79.90	Lithium	Li	3	6.941	Silicon	Si	14	28.09
Cadmium	Cd	48	112.4	Lutetium	Lu	71	175.0	Silver	Ag	47	107.9
Calcium	Ca	20	40.08	Magnesium	Mg	12	24.31	Sodium	Na	11	22.99
Californium	Cf	98	(251)[b]	Manganese	Mn	25	54.94	Strontium	Sr	38	87.62
Carbon	C	6	12.01	Mendelevium	Md	101	(258)[b]	Sulfur	S	16	32.07
Cerium	Ce	58	140.1	Meitnerium	Mt	109	(267)[b]	Tantalum	Ta	73	180.9
Cesium	Cs	55	132.9	Mercury	Hg	80	200.6	Technetium	Tc	43	(98)[b]
Chlorine	Cl	17	35.45	Molybdenum	Mo	42	95.94	Tellurium	Te	52	127.6
Chromium	Cr	24	52.00	Neilsborium	Ns	107	(262)[b]	Terbium	Tb	65	158.9
Cobalt	Co	27	58.93	Neodymium	Nd	60	144.2	Thallium	Tl	81	204.4
Copper	Cu	29	63.55	Neon	Ne	10	20.18	Thorium	Th	90	232.0
Curium	Cm	96	(247)[b]	Neptunium	Np	93	(237)[b]	Thulium	Tm	69	168.9
Dysprosium	Dy	66	162.5	Nickel	Ni	28	58.69	Tin	Sn	50	118.7
Einsteinium	Es	99	(252)[b]	Niobium	Nb	41	92.91	Titanium	Ti	22	47.88
Erbium	Er	68	167.3	Nitrogen	N	7	14.01	Tungsten	W	74	18[illegible]
Europium	Eu	63	152.0	Nobelium	No	102	(259)[b]	Unnilhexium	Unh	106	(2[illegible]
Fermium	Fm	100	(257)[b]	Osmium	Os	76	190.2	Unnilpentium	Unp	105	(2[illegible]
Fluorine	F	9	19.00	Oxygen	O	8	16.00	Unnilquadium	Unq	104	(2[illegible]
Francium	Fr	87	(223)[b]	Palladium	Pd	46	106.4	Uranium	U	92	[illegible]
Gadolinium	Gd	64	157.3	Phosphorus	P	15	30.97	Vanadium	V	23	[illegible]
Gallium	Ga	31	69.72	Platinum	Pt	78	195.1	Xenon	Xe	54	[illegible]
Germanium	Ge	32	72.61	Plutonium	Pu	94	(244)[b]	Ytterbium	Yb	70	[illegible]
Gold	Au	79	197.0	Polonium	Po	84	(209)[b]	Yttrium	Y	39	[illegible]
Hafnium	Hf	72	178.5	Potassium	K	19	39.10	Zinc	Zn	30	[illegible]
								Zirconium	Zr	40	[illegible]

[a]Atomic masses are 1987 IUPAC values based on carbon-12 = 12 u rounded to four significant figures.
[b]Radioactive element. The mass given in parentheses is the mass of the most stable isotope.

Understanding Chemistry

AN INTRODUCTION

Fred M. Dewey
Metropolitan State College of Denver

West Publishing Company
Minneapolis/St. Paul New York Los Angeles San Francisco

DEDICATION

To my father, ROBERT A. DEWEY, a writer and editor, and in his tenth decade of life, still a man of perseverance.

Production Credits

Copyediting Yvonne Howell
Design Paula Goldstein, Bookman Productions
Artwork Precision Graphics
Composition ColorType, San Diego
Page layout Judith Levinson, Suzanne Montazer, Renee Deprey, Wendy Goldberg, Madelaine Muir
Production management Hal Lockwood, Bookman Productions
Cover image (detail on spine and title page) Photomicrograph of metol/hydroquinone by Joseph Barabe/McCrone Associates
Cover design Roslyn Stendahl/Dapper Design
Index James Minkin
Prepress, printing, and binding West Publishing Company

West's Commitment to the Environment

In 1906, West Publishing Company began recycling materials left over from the production of books. This began a tradition of efficient and responsible use of resources. Today, up to 95 percent of our legal books and 70 percent of our college and school texts are printed on recycled, acid-free stock. West also recycles nearly 22 million pounds of scrap paper annually—the equivalent of 181,717 trees. Since the 1960s, West has devised ways to capture and recycle waste inks, solvents, oils, and vapors created in the printing process. We also recycle plastics of all kinds, wood, glass, corrugated cardboard, and batteries, and have eliminated the use of Styrofoam book packaging. We at West are proud of the longevity and the scope of our commitment to the environment.

Printed in the United States of America

01 00 99 98 97 96 95 94 8 7 6 5 4 3 2 1 0

Library of Congress Cataloging-in-Publication Data

Dewey, Fred M.
Understanding chemistry : an introduction / Fred M. Dewey.
p. cm.
Includes index.
ISBN 0-314-02825-0
1. Chemistry. I. Title.
QD33.D465 1994
540—dc20 93-33034
CIP

Brief Contents

Note: This text is available in two versions. *Understanding Chemistry: A Brief Introduction* is the shorter version of the book, which ends with Chapter 17 on oxidation–reduction reactions. *Understanding Chemistry: An Introduction* is the longer version, and includes three additional chapters covering nuclear chemistry, organic chemistry, and biochemistry.

Contents

4 Problem Solving 73

5 The Structure of Atoms. The Periodic Table 101

6 Electron Structure and the Periodic Table 129

7

Composition and Formulas of Compounds 167

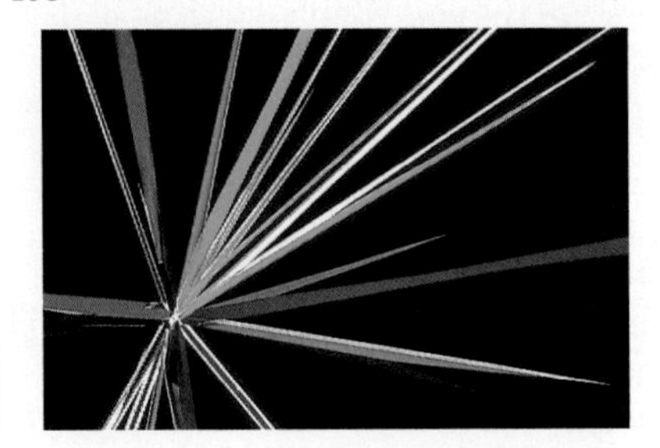

8

The Structure of Compounds. Chemical Bonds 197

9

Names and Formulas of Inorganic Compounds 229

10 Chemical Equations 267

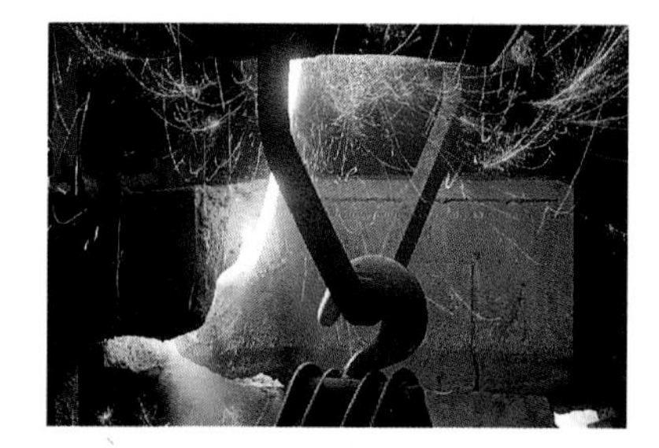

11 Calculations Involving Chemical Equations 299

12 The Gaseous State 323

13 Liquids, Solids, and Changes of State 365

14 Solutions 401

15 Acids and Bases 433

16 Reaction Rates and Chemical Equilibrium 461

17 Oxidation–Reduction Reactions 495

18

Radioactivity and Nuclear Energy 525

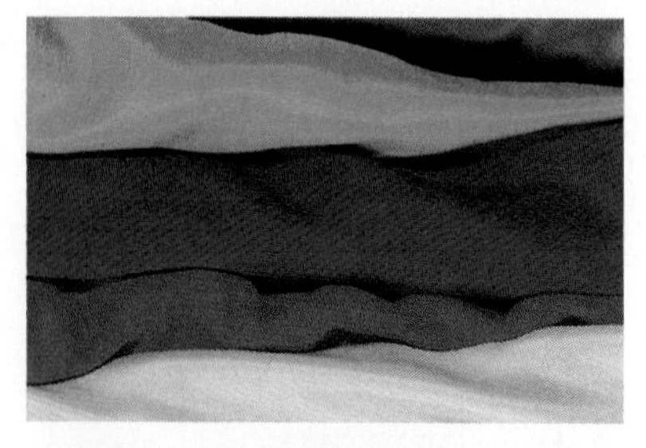

19

Introduction to Organic Chemistry 555

20

Biochemistry: The Chemistry of Life 599

Preface

Audience

Understanding Chemistry is intended to serve beginning chemistry students, those who have had little or no previous study of chemistry, and especially those who are preparing themselves for further study. The book is particularly appropriate for students preparing to take a full-year general chemistry course, or for those students of the health professions who may later take a semester course in organic chemistry and biochemistry. This book is available in two versions to provide options for use in either a one-semester or full-year course.

Philosophy and Approach

To communicate successfully, an author must consider not only the goals of the student audience, but also how students learn. Today's students are conditioned by media that are visually appealing and entertaining. Although this text is not intended to entertain, it is intended to meet students on their own terms, and to draw them into the experience of learning chemistry. Twenty years of teaching beginning chemistry students has made it clear that, with help, most students can succeed at mastering chemical concepts and applying those concepts in solving chemical problems.

This book provides support to students by acknowledging the importance of some underlying concerns. In particular, it recognizes that *students need confidence and motivation* to learn challenging new concepts and skills. The tone of the writing is friendly and encouraging, personal but not condescending. Topics are introduced at beginning levels and gradually built upon as ability and confidence increase. Underlying concepts and skills are not simply assumed; explanations and examples don't leave out steps or implicitly combine them. Chemical language (nomenclature, formulas, and terminology) is presented as needed and reviewed frequently.

Focus is given to *key concepts and basic reasoning strategies,* so that students have a basis of understanding to build on, and so they are less likely to try to solve problems by blindly following examples. A real understanding of what one is doing builds both confidence and competence. Students really can enjoy learning and gain satisfaction from reaching a level of understanding of a challenging subject.

Motivation is important, especially when understanding requires both learning new concepts and applying familiar ones in new ways. It is particularly important for students who may begin the course convinced that chemistry is difficult or not relevant to their lives and goals. A major theme of this book is that *the chemical world is constantly at work all around us.* Each chapter begins with an introduction that helps students recognize the chemistry common to their daily lives and experience. Descriptive chemistry is often used to introduce or to support new concepts, and essays on the chemical properties and behaviors of familiar objects appear in each chapter. For the "visual learners" of today, the book provides many photographs of chemistry in action and conceptual art to help students visualize the molecular and atomic world.

Organization

The order in which topics are presented supports the book's fundamental goal of *maximizing students' understanding of principles and minimizing the need for rote memorization.* For example, an overview of the science of chemistry and some basic concepts and terminology are presented in the first two chapters, so that students will have something to do in the following chapters on measurement and problem solving (and in the laboratory). Similarly, after emphasizing in Chapter 4 the need to approach problem solving as a process of thinking and reasoning, Chapters 5 and 6 present some atomic theory and ideas of electron structure and periodicity, so students will have something to think about and reason with. Chapter 7 then develops problem-solving strategies further with the mole concept and formulas of compounds.

A secondary concern in the organization of topics is *to provide the instructor with realistic options to reorganize* chapters or parts of chapters with minimum inconvenience. The chapters just mentioned above provide several examples. For instance, in the longer of the two versions of *Understanding Chemistry,* radioactive decay, nuclear equations, and other topics may be moved up from Chapter 18 to enrich the discussion of atomic structure in Chapter 5. With only modest adaptation, the problem-solving strategies and techniques in Chapter 7 may be brought forward to follow the introduction to problem solving in Chapter 4. In Chapter 9, the names and formulas of compounds could be covered prior to the discussion of electron structure and bonding, but students will have to rely upon rote memorization of the names of formulas of ions and compounds.

There are many other examples throughout the book of organizational flexibility among and within chapters, which provides instructors with realistic opportunities to modify the sequence, pace, or emphasis of a course. These options are discussed in the *Instructor's Manual* to accompany this text, along with other detailed information about the content and organization of the book.

At the most general level, *Understanding Chemistry* offers flexibility to instructors by offering two versions of the text. *Understanding Chemistry: A Brief Introduction* is the shorter version of the book. It provides coverage of chemical concepts, skills, and applications through oxidation–reduction reactions (Chapter 17). *Understanding Chemistry: An Introduction* expands the coverage to 20 chapters, with the addition of chapters on nuclear chemistry, organic chemistry, and biochemistry.

Problem Solving

A major emphasis of this book is on helping students develop their ability to solve problems by *an organized strategy based on reasoning.* The problem-solving approach is designed to maximize students' understanding of what they are doing and why. Chapters 4, 7, and 11 are particularly focused on problem solving. These chapters are purposely placed in the text, with simpler aspects of the material treated first, so that students are not overwhelmed and so that they have an opportunity to apply what they have learned about problem solving as they consider chemical principles and information in the subsequent chapters.

The unifying emphasis in all presentations of problem solving is on developing a strategy that will *enable students to attack any chemistry problem.* Dimensional analysis is emphasized, using an outline that helps students focus clearly on what they are trying to find (the unknown), what is known, what conversion factors are needed, and why an answer is reasonable or not. Algebra is presented as a strategy that must be used when conversion factors are not available but when there is an equation that defines the relationship between the unknown and known information.

The *dimensional-analysis strategy* presented here is slightly different from many texts in that the equation is set up beginning with the unknown rather than with the given information. This approach focuses on the goal of problem solving—the answer to the problem. More importantly, this strategy provides students with a basis for deciding what information is essential (and what is not) and in what sequence to put information in the equation. Stepwise, the equation starts with the symbol or word for the unknown (1), with the units in parentheses, on the left-hand side of the equation. The first term of the right side of the equation (2) is a conversion factor chosen by matching its units (those in the numerator) with the units of the unknown. Succeeding terms (3) are chosen so that their units cancel with unwanted units of preceding terms, leaving the units of the answer. For example, calculating mass from density would look like this:

$$\underset{(1)}{\text{mass (g)}} = \underset{(2)}{\left(\frac{1.02\text{ g}}{\cancel{\text{mL}}}\right)} \underset{(3)}{(35.0\ \cancel{\text{mL}})} = 35.7\text{ g}$$

(See Chapter 4 for numerous examples.) In the "check" step of the calculation, the terms are arranged in reverse order, starting with the given information. Many students will be familiar with this common practice of checking a calculation by repeating it in reverse order.

Often students try to memorize example problems, hoping they will be able to match a problem on a test with one they have memorized. They make little progress in developing their problem-solving abilities, and minor changes or the inclusion of extraneous information in the presentation of problems thoroughly confuses them. The many example problems in this book consistently use the problem-solving strategy outlined above. *This strategy emphasizes structured reasoning,* and example problems and the discussion of them frequently comment on the thinking process and the need to assess the reasonableness of answers. With this dimensional-analysis strategy, students always know where to begin; they are not confused by nonpertinent information, and they have a logical sequence of steps to follow for setting up an equation to calculate the unknown. The result is a framework within which they can think about the problem and obtain the confidence necessary to become a successful problem solver.

Pedagogy

The pedagogical framework of the text is designed to support the basic objectives of providing motivation, building confidence, and emphasizing the need for understanding and reasoning, not memorization or blind mechanics.

Each chapter includes these student aids:

Chapter Contents are listed for easy reference within each chapter.

"You Probably Know . . ." sections at the start of each chapter help give students confidence by reminding them of familiar objects or experiences that relate to the subject of the chapter. This section points out concrete examples of the chemical concepts they will be learning.

"Why Should We Study _____?" sections directly address students' need for relevance and give them an opportunity to see how the chemical concepts they are learning relate to each other, or to students' academic and career goals.

Learning Objectives focus on the key goals of each section.

Marginal Notes highlight important ideas and information, and remind students of relevant concepts studied earlier.

Key Terms in boldface, defined in the text and listed at the end of the chapter by section, are included in the Glossary at the end of the book.

Figures (mostly art) help students gain insight into the concepts presented and organize information, while photographs illustrate applications of the chemical concepts and help affirm that chemistry is part of daily life.

Worked Examples show how to apply chemical concepts and reasoning skills to solve quantitative problems.

Practice Problems follow Examples and provide opportunities to practice problem-solving skills as soon as they are presented.

Chemical World essays have been researched and prepared by Ronald DeLorenzo (Middle Georgia College). Topics have been carefully selected to illustrate how the principles and reactions presented in the text apply to our daily world.

The **Chapter Review** promotes thinking and active learning by requiring students to fill in missing terms. (Answers are provided on the last page of each chapter.)

Study Objectives provide a list by section of what students should understand and be able to do by the end of the chapter.

Questions and Problems are matched pairs of end-of-chapter exercises that are organized by section. Each exercise set, however, ends with a section of Additional Problems that are *not* sorted according to the section of the chapter they refer to, and may combine skills and concepts from among those covered in the chapter. Answers (not solutions) to items with blue numbers appear in Appendix C. The most difficult problems are marked by asterisks.

Solutions to Practice Problems are detailed step-by-step solutions to the in-chapter problems, using the same problem-solving strategies described in the chapter and shown in the Examples.

Appendices at the back of the text include instructions for the use of a scientific calculator and a review of relevant mathematics, followed by the Glossary.

■ Supplements

This text is supported by a complete package of ancillary materials designed to help meet the needs of the student and instructor:

Study Guide. The *Student Study Guide* by Leslie N. Kinsland (University of Southwestern Louisiana) summarizes and reviews the main themes of each chapter and section in the text. Over 200 example problems are provided, and students can test their understanding through Self-Test questions provided in both multiple-choice and open-ended formats.

Student's Solutions Manual. Prepared by Juliette A. Bryson (Las Positas College), the *Student's Solutions Manual* provides fully-worked-out solutions to all of the odd-numbered end-of-chapter exercises in the text. All solutions are carefully illustrated using the problem-solving methodology of the text.

Instructor's Manual. The *Instructor's Manual* by Fred M. Dewey and Juliette A. Bryson provides sequencing suggestions, teaching goals, and answers and solutions to all of the even-numbered problems in the text.

Laboratory Manual. *Experiments in Basic Chemistry* by Cynthia Hahn (Bristol Community College) provides 27 experiments in which students can further their understanding of chemical principles through guided laboratory exploration. The manual contains sections on laboratory safety and lab equipment and is supplemented by an *Instructor's Guide* and by spreadsheet files on disk that can be used for pooling and analyzing class data for selected experiments.

Test Bank. Prepared by Mitchel Fedak (Community College of Allegheny County), the *Test Bank* includes 1400 multiple-choice and short-answer questions that are listed in a print version or in the computerized Westest 3.0 format (available in DOS, Windows, or Macintosh versions).

Transparency Acetates. A set of over 100 full-color acetates of key figures and tables from the text is available for classroom use.

Interactive Software. *Concentrated Chemical Concepts* by Trinity Software (for IBM-compatible computers) is a computer-aided student tutorial covering numerous chemical principles, and is highly interactive, with instant feedback and reinforcement.

Videotapes. *Graphing Scientific Information* and *Calculators and Chemical Calculations* (both by Michael Clay, College of San Mateo) supply tutorial instruction on two key skills that every introductory chemistry student must thoroughly master in order to be successful.

Videodisk. West's *General Chemistry Videodisk* contains demonstrations that are difficult, expensive, or dangerous to perform in class. Most of the experiment footage is original. For most major topics, the videodisk presents figures (from the text and other sources), experiments, and animations.

■ Acknowledgments

This project would not have been possible without the support and encouragement of my Editor, Richard Mixter. The guidance and insight he has contributed during these years of hard work have been invaluable in shaping this book from early ideas to its ultimate form. Keith Dodson, the Developmental Editor, has also contributed a great deal toward the success of this project, particularly in developing much of the supplement package and working with accuracy checkers. I am also grateful to the West publishing team in Eagan, Minnesota, especially to Barbara Fuller for her outstanding work in the production of this book.

I gratefully acknowledge the assistance of several colleague chemists, including Richard Anderson, who did significant work on the biochemistry chapter; Ronald DeLorenzo, who wrote many of the Chemical World essays; and Jeffrey Hurlbut, who helped with proofreading.

Last, but certainly not least, I express my sincere gratitude to all of the reviewers who offered their various perspectives and insights on how to make this text work for students. These reviewers are:

Hugh Akers, Lamar University; Gilbert Albelo, Mount Hood Community College; Frank Andrews, University of California at Santa Cruz; Richard Angerer, Essex Community College; Jose Asire, Cuesta College; Caroline Ayers, East Carolina University; Edith Bartley, Tarrant County Junior College; Robert W. Batch, Canada College; Jon M. Bellama, University of Maryland; Larry Bennett, San Diego State University; Donna Bogner, Wichita State University; Michael E. Clay, College of San Mateo; John V. Clevenger, Truckee Meadows Community College; Ellene Tratras Contis, Eastern Michigan University; Jerry Driscoll, University of Utah; Vicky D. Ellis, Gulf Coast Community College; Lucy Pryde Eubanks, Clemson University; Gordon J. Ewing, New Mexico State University; Mitchel Fedak, Community College of Allegheny County; Elmer Foldvary, Youngstown State University; David Frank, Ferris State University; Patrick Garvey, Des Moines Area Community College; Lynn Geiger, University of Colorado at Boulder; Don Glover, Bradley University; Stan Grenda, University of Nevada, Las Vegas; Sue Griffin, Boston University; Howard Guyer, Fullerton College; William Hausler, Madison Area Technical College; Wayne Hiller, California State University, Chico; Richard Hoffman, Illinois Central College; Larry Houk, Memphis State University; Kirk Hunter, Texas State Technical Institute; Jeffrey A. Hurlbut, Metropolitan State College of Denver; Rebecca Jacobs, Ferris State University; Y. C. Jean, University of Missouri; Gerard F. Judd, Phoenix College; Floyd Kelly, Casper College; Christine S. Kerr, Montgomery College; Leslie N. Kinsland, University of Southwestern Louisiana; Willem R. Leenstra, University of Vermont; Leslie J. Lovett, West Virginia University; Jerome Maas, Oakton Community College; Felipe Macias, Broome Community College; Douglas L. Magnus, Saint Cloud State University; Joseph Maguire, Henry Ford Community College; Darwin Mayfield, California State University, Long Beach; Michael J. Millam, Phoenix College; Kenneth Miller, Milwaukee Area Technical College; Patricia Milliken-Wilde, Triton College; Richard S. Mitchell, Arkansas State University; Robin Monroe, Southeast Community College; John Ohlsson, University of Colorado at Boulder; Gordon A. Parker, University of Michigan—Dearborn; Robert M. Perrone, Community College of Philadelphia; James A. Petrich, San Antonio College; Robert Pool, Spokane Community College; Barbara Rainard, Community College of Allegheny County; Edith Rand, East Carolina University; Fred Redmore, Highland Community College; Erwin Richter, University of Northern Iowa; Keith Ries, Spokane Community College; Stephen P. Ruis, American River College; William D. Schulz, Eastern Kentucky University; Martha W. Sellers, Northern Virginia Community College; Ruth Sherman, Los Angeles Community College; Donald Slavin, Community College of Philadelphia; Leo H. Spinar, South Dakota State University; Vernon J. Thielmann, Southwest Missouri State University; Frina S. Toby, Rutgers University; Kathleen M. Trahanovsky, Iowa State University; Mary Vennos, Essex Community College; William J. Wasserman, University of Washington; Karen C. Weaver, University of Central Arkansas; John Weyh, Western Washington University; Hewitt G. Wight, California Polytechnic State University, San Luis Obispo.

I welcome your contributions to this book also. If you should find any errors, I would be grateful if you would let me know so that I can correct them. Your comments and suggestions about any part of the text or its supplements will be appreciated.

Fred M. Dewey
Department of Chemistry
Metropolitan State College of Denver
Denver, CO 80217

About the Author

Fred Dewey earned his doctorate in chemistry at the University of Colorado in Boulder, Colorado. Dr. Dewey has taught for more than 25 years at Metropolitan State College of Denver. He was a Fulbright Fellow in Manila, Philippines, in 1984–1985, where he taught at De La Salle University. In 1989, he was a member of an international delegation of environmental scientists that visited China. He also has worked as a research chemist and an environmental scientist in government and private laboratories. The Deweys, who have ten children and a growing number of grandchildren, live in Lakewood, Colorado.

1

Chemistry—The Central Science

Contents

The oceans, land, and air are made up of chemical substances.

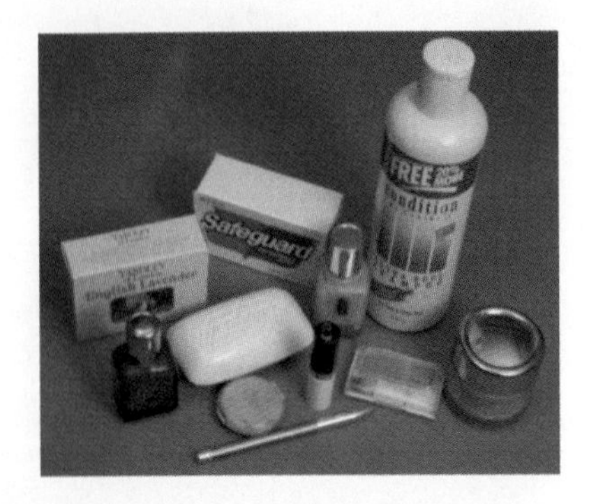

Chemical substances in our bodies act together to sustain life.

You Probably Know . . .

- Everything in the material world—the earth, the oceans, lakes and streams, and the atmosphere—is composed of chemical substances.
- The challenge of preserving our environment involves chemistry. Maintaining acceptable air and water quality includes evaluating and controlling pollutants—toxic chemical substances.
- Many things we use every day and take for granted are products of chemistry. These include soaps, shampoos, detergents, toothpaste, cosmetics, plastic articles such as toothbrushes, combs and brushes, pens, buttons, and many household articles. The list is practically endless.
- Medicines, from aspirin to cancer drugs, are chemical products. The lives of many people literally depend upon products of chemistry.
- Our bodies are an enormously complex mixture of chemical substances, which act together to sustain life.

Why Should We Study Chemistry?

Because you have chosen to study chemistry, you must think science will be important to you. Recognizing that science seeks to understand the natural world, perhaps you want to satisfy your curiosity about what you see around you. You may be concerned about the quality of the environment and expect that studying chemistry will help you to understand environmental issues. Perhaps you are planning a career in one of the medical sciences such as nursing or physical therapy, or one of the engineering sciences, or biology or forestry, or even one of the branches of chemistry. An understanding of chemistry is essential for all sciences. Furthermore, much of what we buy—foods, household products, medicines, personal items such as cosmetics, soaps, and shampoos, and fuels and automobile products—is composed of chemical substances, either natural or synthetic. We will be wiser consumers knowing something about chemistry.

1.1 What Is Chemistry?

Learning Objectives

- Explain what is involved in the study of chemistry.
- Name the principal physical and biological sciences.
- Identify and briefly describe the major divisions of chemistry.
- Describe what environmental chemists do.

What comes to mind when you hear the word "chemistry"? Perhaps you think of atoms and molecules. You may picture someone wearing a lab coat and watching over an experiment. You may think of some of the large companies that manufacture chemicals and chemical products. Chemistry may remind you of a person who is special to you because "the chemistry is right." Every college-age person has some notion about chemistry. But what is chemistry? Let's begin our study by giving attention to this question.

Chemistry is the study of our material world. In chemistry, we study the composition of materials and how their structure affects their properties and behavior. As we begin our study of chemistry, we will study the nature of the materials (matter) that make up our world. **Matter** is anything that has mass and occupies space. We find matter in the form of solids, liquids, and gases. Everything on the earth and in its atmosphere is composed of matter. To understand chemistry, we must have a good understanding of the fundamental nature of matter. Then, with this understanding, we will be ready to study changes that we observe going on around us every day.

A chemist in her laboratory.

Matter has mass and occupies space. Mass refers to the amount of matter.

Chemistry is one of the natural sciences. Natural sciences can be organized under the two broad headings of physical sciences and biological sciences, as shown in Figure 1.1. A **physical science** is concerned with the materials in our world, and in the entire universe, and how energy affects these materials. **Biological sciences** involve the study of life forms.

Chemistry is sometimes called the central science because it is an essential part of the other sciences. For example, a knowledge of the principles of chemistry is necessary to understand earth and atmospheric sciences. Astrochemistry has developed from astronomy. Materials scientists use their knowledge of both chemistry and physics to develop new space-age materials. Chemical and pharmaceutical research together have resulted in the development of antibiotics and other drugs to control diseases. Furthermore, chemistry is necessary to understand the functioning of cells, the basic units of life.

Identify the forms of matter you see here.

FIGURE 1.1 Organizational chart for the natural sciences. Chemistry is the central science.

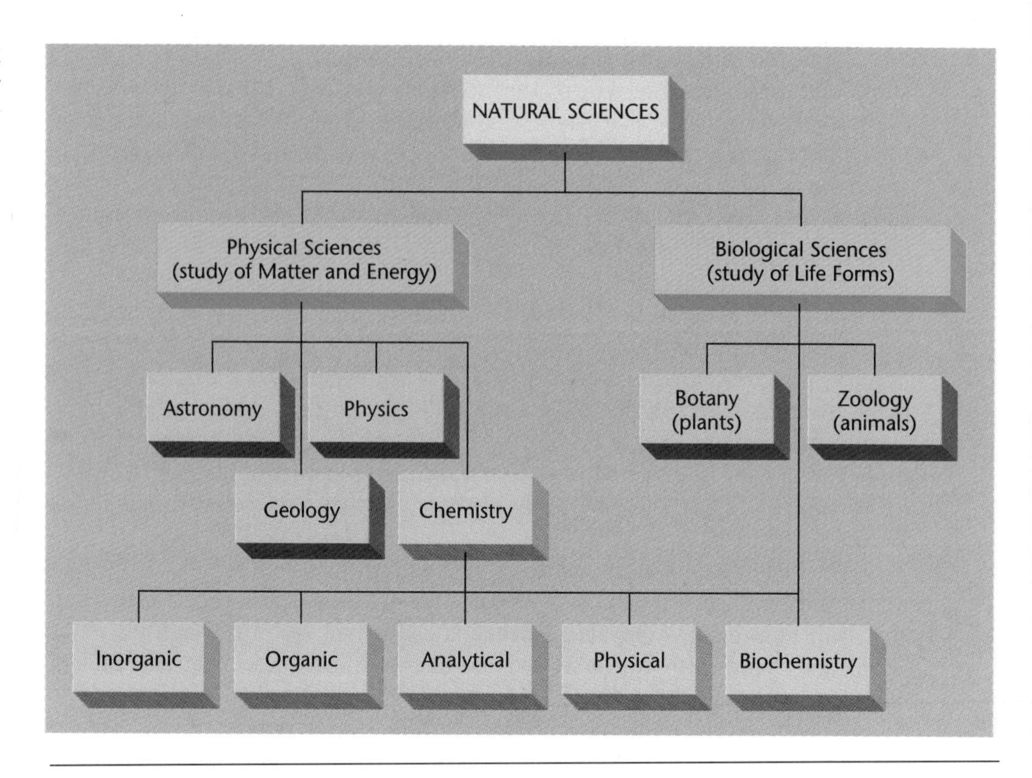

Because chemistry is such a broad field of study, it has been divided into a number of smaller fields. The main divisions of chemistry are shown in Figure 1.1. There is significant overlap of these divisions, as well as overlap with other sciences, and they should not be viewed as being totally separate and distinct. Furthermore, other subdivisions have been formed by scientists whose interests are more specialized. The introduction to the main divisions of chemistry given here is brief, and you will gain more understanding of each as you read on in this book.

Inorganic chemistry is the study of substances that are primarily of mineral origin. **Organic chemistry** is the study of carbon compounds. The word "organic" comes from "organism," a living being. Chemists once believed that organic compounds were found only in plants and animals or in substances left behind when they died. However, in 1828, when Friedrich Wöhler converted ammonium cyanate (inorganic) to urea (organic) in his laboratory, the door to the synthesis of organic compounds was opened. In recent years countless numbers of organic compounds have been synthetically prepared.

Analytical chemistry is the study of matter to determine the identity and quantity of its components. Chemical analysis is an important part of all the divisions of chemistry. **Physical chemistry** is concerned with the fundamental structure of matter and the influence of energy upon it. **Biochemistry** is the study of life processes and is the part of chemistry most essential to the medical sciences. Biochemistry is now considered by many to be a separate science, not just a division of chemistry.

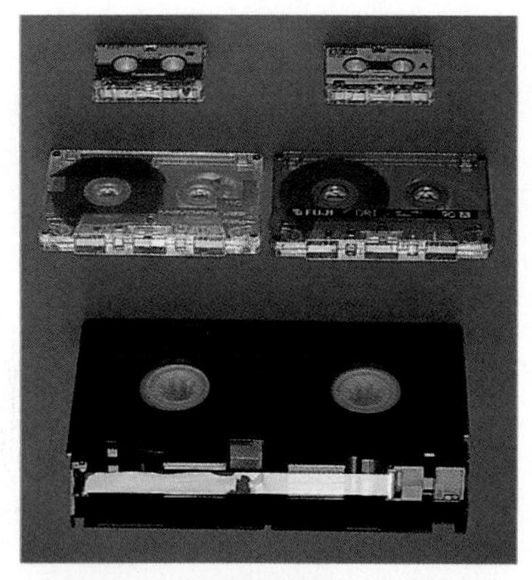

Mylar, a synthetic polymer, is used to make recording tape.

As chemistry has been commercialized for our benefit, a number of specialized commercial fields have developed. We now have polymer chemistry, agricultural and food chemistry, paint and coatings chemistry, medicinal chemistry, and so on. The chemical industry and related industries employ large numbers of people and make up a major part of our economy.

Another of the specialized areas of chemistry to develop in recent decades is environmental chemistry. Environmental chemists identify chemical pollutants and determine what quantities are present in soil, water, and air. Other environmental scientists join with chemists to determine the sources of pollutants and to evaluate their effects on plant, animal, and human life. This information is used by state and federal governments to establish policies to protect and improve the environment for the welfare and pleasure of the people who inhabit it. Ultimately, it is the responsibility of all citizens—all of us—to be informed and to elect responsible representatives who will make the laws that govern us.

History of Chemistry

Learning Objectives

- Identify the oldest chemical art.
- Briefly trace the history of chemistry from ancient civilizations to the nineteenth century.
- Identify the two main pursuits of the alchemists and their principal contributions to science.
- Identify the main scientific contributions of Aristotle, Democritus, Robert Boyle, Joseph Priestley, Antoine Lavoisier, and John Dalton.

Someone has said that history gives us a perspective about the present. As we begin our study of chemistry, let's look briefly at its history to see how chemistry has developed in becoming a modern science.

The roots of chemistry can be traced to ancient civilizations. Gold was likely the first metal discovered and has long been considered a measure of wealth. Copper utensils found in Egypt have been dated at about 3000 B.C. Bronze, a copper–tin alloy, was used for weapons and armor as early as 2000 B.C. Iron, when first discovered, was considered too precious to be used in making weapons. However, the secret of smelting iron from iron ore was eventually discovered, and by the tenth century B.C. the Assyrian army was equipped with iron weapons. Metal-tipped weapons gave some armies a great advantage over others that did not have them. The ancients sought weapons of war that would give them superiority over their neighbors, much as nations do today with their conventional weapons and nuclear stockpiles.

Gold and copper were used by ancient civilizations.

After the discovery of metals, people began to develop skills in working with other materials obtained from the earth, from plants, and from animals. The chemical arts emerged and produced materials that changed people's lifestyles. Pottery has been called the oldest of the arts. The development of pottery gave people waterpots and dishes for everyday life. Over the years pottery-making led to the science that we now call ceramics. Dyes were used by ancient peoples to color pottery, cloth, and other materials. The chemical arts influenced religious rituals as well. For example, the Egyptians used an embalming process, along with

A Chinese white porcelain "Yu" vase made during the T'ang dynasty.

Alchemists developed laboratory equipment and techniques in their search for the means of making gold from other metals.

perfumes, in preparing their dead for burial. Perfumes have long been used in religious ceremonies. Incense, a kind of perfume, was used in Old Testament times, and its use continues today. Alcoholic beverages, prepared by fermenting fruits and grains, have been known for thousands of years.

The chemical arts progressed by means of valuable but unexpected discoveries. There were no theories that could lead artisans to develop better products. Development was slow in ancient times, and hundreds of years often passed between the discovery of a material and the period of its widespread use.

We can trace the beginnings of chemical theory back to about 600 B.C. Greek philosophers left us records of early speculations about the nature of matter. One of these speculations was the idea that the universe was made up of four basic elements: air, earth, fire, and water. Aristotle (384–322 B.C.) accepted and expanded this theory to include a fifth element that he called *ether* (from the word meaning "to glow"). He thought the heavenly bodies must be composed of ether because they were luminous. Even prior to Aristotle, Democritus (about 470–380 B.C.)

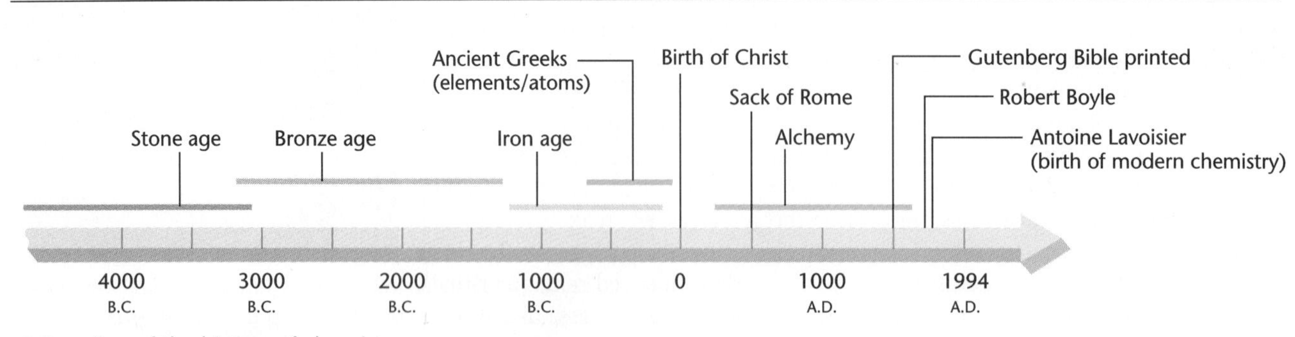

A time line of the history of chemistry.

suggested the idea of tiny indivisible particles of matter called *atoms*. However, his idea was not supported by Aristotle and was not widely accepted. In fact, it was more than 2000 years later that John Dalton (1766–1844) proposed his atomic theory based upon the thoughts of Democritus. (See Chapter 5.)

Robert Boyle (1627–1691) was the first to record experimental data. The results of his work were published for others to read and scrutinize.

The alchemists (about A.D. 500–1600) were forerunners of the people who later were leaders in the development of chemistry as a science. Two factors motivated the alchemists: the search for a means of changing base metals into gold and the search for an elixir of life that would lead to immortality. So, for centuries alchemy followed two paths, one involving mineralogy and the other medicine. Although the alchemists failed in both pursuits, they developed laboratory equipment and techniques that were of value to scientists who followed them.

The science of chemistry began to develop in the seventeenth and eighteenth centuries. Although alchemists were captivated by the search for wealth and immortality, chemistry was advanced during this period by people with an intense curiosity about the fundamental nature of the material world. Solids and liquids such as metals and water were the focus of attention of the ancient peoples and the alchemists, but it was with the investigation of gases that science began to progress.

Antoine Lavoisier (1743–1794) is sometimes called the father of modern chemistry. He met an untimely and tragic death by guillotine during the French Revolution in 1794.

For centuries people were puzzled about the nature of air and the mystery of combustion. Robert Boyle (1627–1691) studied air and the effect of pressure on the volume of air. (Boyle's law is discussed in Chapter 12.) Boyle's experiments were very systematic. He was the first to carefully record experimental observations, and the results of his experiments were published for others to read and scrutinize. Many other scientists also made significant contributions. One of these was Joseph Priestley (1733–1804), an English clergyman who came to America because of religious persecution. Priestley is remembered primarily for his discovery that oxygen is a component of air and that it supports combustion. He also studied carbon dioxide and other gases. The French chemist Antoine Lavoisier (1743–1794) used a balance to make careful mass measurements that served to establish quantitative (numerical) relationships among materials used in his experiments. He is sometimes called the father of modern chemistry because his work established chemistry as a quantitative science. Lavoisier's work led to recognition of the law of conservation of matter: Matter can neither be created nor destroyed. Furthermore, his work stimulated many other scientists in their investigations.

John Dalton proposed his atomic theory during the period 1803–1808. Since that time, chemistry has made great advances. The combined outstanding achievements of chemists and the parallel development of chemical technology have touched almost every area of life of people who live in the modern world.

The Scientific Method

Learning Objectives

- Outline the key elements of the scientific method.
- Distinguish among hypothesis, theory, and law.

As we begin our study of modern chemistry, let's give some thought to some characteristics of science in general. Science observes, collects information, organizes

Polluted air over New York City. Some of the buildings are barely visible through the heavy haze. Air pollution is a serious health threat.

it, and tries to explain what has been observed. Armed with information, science poses questions and identifies problems. Science is active in answering its own questions. Experimental research is the active part of science. Through this means, science solves problems. It is important for us to examine the approach used by scientists in problem solving.

To understand how scientists approach problems, let's look at one of the major problems confronting us today: air pollution. Many people have asked, "What can we do about air pollution?" The air over many cities looks dirty, smells bad, and is sometimes irritating to our eyes and lungs. People with respiratory diseases can become sick on days when the air is very polluted. Lung cancer rates in large metropolitan areas are higher than in rural areas. Environmental scientists are continuing to study air pollution. Everyone agrees that this is a major problem. How do scientists tackle a problem of this magnitude?

In solving problems, scientists often follow a logical, systematic process called the scientific method. Although there is no strict order to the elements of the scientific method, the key elements can be condensed and summarized in logical sequence:

1. State the problem precisely.
2. Identify what is known about the problem.
3. Propose a hypothesis (a tentative answer) to the problem or an explanation for what has been observed.
4. Conduct experiments to test the hypothesis. Resulting observations must be carefully recorded.
5. State a conclusion based upon the experimental observations.

The question asked earlier was, "What can we do about air pollution?" This question identifies a problem as "air pollution" and seems to involve the first element of the scientific method. However, this statement of the problem is too broad and too vague to be very useful to a scientist. The problem must be subdivided into smaller problems so that the specialized knowledge and experience of individual scientists can be effectively utilized to address each of them. For example, a scientist might ask, "What role is played by dust in the air over New York City on high pollution days?" This question identifies a small part of the problem of air pollution and is a more precisely worded question.

You may have heard about one of the "think tanks" that have been formed in industry and government. These groups of respected problem solvers tackle some of the country's most challenging problems. One of the *first* tasks they face is to divide a major problem into precisely defined smaller problems. From their experience, they know that a precisely worded question or statement of a problem helps scientists choose the direction to follow in solving the problem.

The *second* element of the process is to determine what is already known about the problem. Scientists rely on the results and conclusions of the work of other scientists. It would be a waste of resources and time to ignore available information and to repeat the work of others. Information can be sought in scientific journals, government publications, and books, and may be available at scientific meetings. In some cases, vital information may be common knowledge. For example, a survey of residents of inner city neighborhoods may reveal that a heavy layer of dust settles on their cars and in their homes on high pollution days in New York City.

Armed with what is known about a problem, a scientist moves on to the *third* element of the scientific method and proposes a hypothesis. As mentioned before, a **hypothesis** is simply a tentative answer to a problem or an explanation of what has been observed. A hypothesis is not necessarily supported by a large amount of experimental evidence. For the air pollution problem, a hypothesis might be, "Dust is a major component of the air over New York City on high pollution days."

An environmental scientist can determine dust levels in air using a particulate sampler.

A hypothesis leads a scientist to consider what experiments are needed to test the hypothesis. Experiments are the *fourth* element of the scientific process. During experiments, observations are made from which we learn *empirical facts,* so called because we discover them by *observing* the properties and behaviors of materials. Empirical facts are called *data.* For example, measurements of the quantity of dust per cubic meter of air would provide data that might determine the validity of the hypothesis about polluted air over New York City. An environmental scientist might place an air sampler on the roof of one of the buildings in the city. The sampler collects dust from the air and measures the volume of air sampled, thereby giving data needed to test the hypothesis. Following Robert Boyle's example, today's scientists recognize the importance of carefully recording data. Experimental results are too easily forgotten and lost in the absence of careful documentation.

The *fifth* element of the scientific method is the scientist's conclusion based on data obtained in experiments. Recording the conclusion is an essential element of the problem-solving process. This part of the process is just as important as writing the answer to an arithmetic problem you have worked. Failing to record a conclusion denies others access to information that may benefit them. In writing a conclusion, a scientist interprets data and explains what the data mean. If the data indicate that the hypothesis is incorrect, another hypothesis can be proposed. Then additional experiments can be suggested that test the new hypothesis. The process continues in a cycle until there is a satisfactory conclusion for the problem (Figure 1.2).

Perhaps scientists have wondered if dust is a major component of polluted air in other parts of New York City, or in other cities. They may continue to test and refine the hypothesis by performing more experiments and collecting additional data. If the weight of the evidence in support of a hypothesis is great, a hypothesis may be elevated to the status of a **theory** (Figure 1.3). *We must remember that*

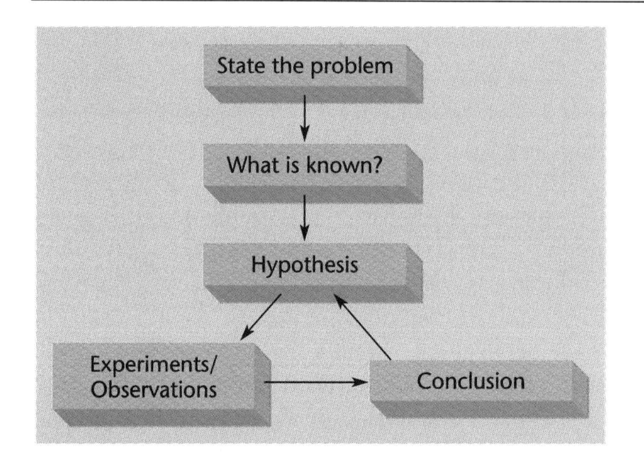

FIGURE 1.2 Elements of the scientific method.

Hypothesis: A tentative answer to a question or an explanation of observations supported by a limited amount of data.

Theory: A good explanation of observed phenomena supported by a significant amount of data.

Law: A statement of natural events that occur with consistency under the same conditions.

FIGURE 1.3 Hypothesis, Theory, and Law.

CHEMICAL WORLD

Air Pollution—Indoors

The effects of industrial and household chemicals on the environment—air, water, and soil—have concerned chemists and everyone else for several decades. Recently, some people have realized that our homes, schools, and workplaces are the environments in which we spend 90% of our lives, and that these environments can be more polluted than our smoggy streets.

Uranium-bearing soils, which lie under many buildings in some regions, exude a radioactive gas, radon, that collects in cellars. We build our houses with plywood and adhesives and insulate them with plastic foam, all of which contain volatile formaldehyde. Then we paint, carpet, and tile our buildings with materials that require organic solvents in their use or manufacture, and scrub them with household cleaners that contain ammonia (a biological poison) and phosphoric acid. In summer we control pests with insecticides and in winter we heat with furnaces, which, when they get old, give off carbon monoxide. And all year we cook over unvented stoves, taking in more carbon monoxide along with some nitrogen oxides. After dinner, we dress for an evening out, using antiperspirant sprays and various cosmetics. Does anyone in your house smoke? Add nicotine and tar. An Environmental Protection Agency study showed that typical indoor concentrations of all target chemicals exceed outdoor levels by factors of up to 100.

Of course, some of these pollutants have been present indoors since we first lived in buildings. Our modern problems stem from trying to save energy by using tight-fitting windows and doors, which trap pollution inside.

As buildings have become more airtight, a new illness called "sick building syndrome" has appeared with increasing frequency. Symptoms are similar to those suffered by people living in contaminated air: hoarseness, stinging eyes, headaches, nausea, and lethargy. But there may also be a problem with "mind pollution." According to one study, people who dislike their jobs are much more likely to suffer from sick building syndrome than people who like their jobs. Sick building syndrome probably reflects a combination of psychological and chemical factors.

Until solutions are found for these problems, we will need literally to go out for a breath of fresh air and find jobs that we can enjoy.

Some workers suffer from "sick building syndrome."

hypotheses and theories are both explanations for what we observe. They should not be considered to be facts. In the case of polluted air, it is possible that dust is not a major component of all polluted city air. Hypotheses and theories are usually examined critically by the scientific community. Such scrutiny leads to further experimentation, which leads to eventual refinement of a theory.

Law of conservation of matter: Matter can neither be created nor destroyed (Section 2.5).

You should note the difference between a law and a theory. A **law** is a simple statement of observed natural events that occur with consistency under the same conditions. A law tells us *what* happens. A theory is our explanation about *how* events occur. You may already be familiar with some of the basic laws, such as the law of gravity and the law of conservation of matter.

What Are Scientists Like?

1.4

Learning Objective

- Describe several important characteristics of scientists.

As you continue your study of science, you will also be learning what scientists are like. Scientists are curious, inquiring, and careful observers. They are trained to make measurements and to be careful and thorough in recording and evaluating data. They are creative in designing experiments that will provide data needed to test hypotheses and provide answers to important questions and problems. Scientists are learners. They are committed to searching for truth. Intellectual honesty is an important personal characteristic of scientists that enables them to avoid personal bias in forming their conclusions. They are also people who can communicate the results of their research to other scientists and to the public for the welfare of others. So, as you study science, remember that progress in science depends on the capabilities and character of individual scientists.

Perhaps you will choose to pursue a career in science. If you do, you will find it is a challenging and noble journey.

Chapter Review

Directions: From the list of choices at the end of the Chapter Review, choose the word or term that best fits in each blank. Use each choice only once. The answers appear at the end of the chapter.

Section 1.1 Chemistry is the study of the __________ (1) of materials, their properties, and their behavior. Matter is anything that has __________ (2) and occupies space.

Chemistry is a physical science. A physical science is concerned with matter and how __________ (3) affects changes in matter. __________ (4) sciences involve the study of life forms. The main divisions of chemistry are inorganic, __________ (5), analytical, physical, and __________ (6). Inorganic substances are primarily of __________ (7) origin. Organic chemistry is a study of __________ (8) compounds. Analytical chemists determine the identity and quantity of the components of matter. Physical chemists study the __________ (9) of matter and the effect of energy upon it. Biochemistry is the study of __________ (10) processes. Other specialized areas of chemistry have developed to address specific kinds of problems. One of these is environmental chemistry which is concerned with the identity and quantity of chemical __________ (11) in soil, water, and air.

Section 1.2 Chemical arts were known to ancient civilizations. The first metal discovered was probably __________ (12), followed by copper, bronze, and iron. The oldest of the arts is __________ (13). From this art came dishes and waterpots needed for everyday life. The chemical arts progressed by valuable but unexpected __________ (14). There were no theories that could lead the artisans to develop better products.

A concept of the atom can be traced to ______ (15) philosophers about 400 B.C. However, atomic theory was not generally accepted until after Dalton's theory was proposed in the nineteenth century. Alchemists, in their pursuit of __________ (16) and immortality, developed techniques and laboratory equipment that were of value to scientists who followed them. However, chemistry advanced little until the seventeenth and eighteenth centuries. __________ __________ (17) studied air and the effect of pressure on the volume of air. He carefully recorded his observations and published his results. Antoine Lavoisier used a __________ (18) to make careful mass measurements and established chemistry as a quantitative science. His work led to the recognition of the law of __________ __________ __________ (19).

Section 1.3 In solving problems, scientists use a systematic approach called the __________ __________ (20). Proposing an explanation of what is observed or a tentative answer to a problem—a __________(21)—is an important step. A theory is a good explanation of observed phenomena that is supported by a significant amount of data. A statement summarizing natural events that occur with consistency under the same conditions is called a __________ (22).

Section 1.4 Intellectual __________ (23) is an important personal characteristic of scientists.

Choices: balance, biochemistry, biological, carbon, composition, conservation of matter, discoveries, energy, gold, Greek, honesty, hypothesis, law, life, mass, mineral, organic, pollutants, pottery, Robert Boyle, scientific method, structure, wealth

Study Objectives

After studying Chapter 1, you should be able to:

Section 1.1

1. Explain what is involved in the study of chemistry.
2. Name the principal physical and biological sciences.
3. Identify and briefly describe the main divisions of chemistry.
4. Describe what environmental chemists do.

Section 1.2

5. Identify the oldest chemical art.
6. Briefly trace the history of chemistry from ancient civilizations to the nineteenth century.
7. Identify the two main pursuits of the alchemists and their principal contributions to science.
8. Identify the main scientific contributions of Aristotle, Democritus, Robert Boyle, Joseph Priestley, Antoine Lavoisier, and John Dalton.

Section 1.3

9. Outline the key elements of the scientific method.
10. Distinguish among hypothesis, theory, and law.

Section 1.4

11. Describe several important characteristics of scientists.

Key Terms

Review the definition of each of the terms listed here by chapter section number. You may use the glossary if necessary.

1.1 analytical chemistry, biochemistry, biological science, chemistry, inorganic chemistry, matter, organic chemistry, physical chemistry, physical science

1.3 hypothesis, law, scientific method, theory

Questions

Answers to questions with blue numbers appear in Appendix C.

1. Which of the following are not classified as matter?
   a. salt
   b. sunlight
   c. gasoline
   d. electricity
   e. water
   f. air
   g. heat
   h. love
   i. fear
2. Which of the following are physical sciences?
   a. chemistry
   b. physics
   c. psychology
   d. history
   e. algebra
   f. biology
   g. English
   h. sociology
   i. astronomy
3. A physician needed to know the level of theophyllin (an asthma drug) in a patient's blood. What kind of chemist could provide the information?
4. Which division of chemistry is most important to the medical sciences?
5. Which division of chemistry would be involved with air and water pollution studies?
6. Identify the following as either organic or inorganic:
   a. sugar from sugar cane
   b. table salt
   c. a piece of wood
   d. human hair
   e. sand
   f. water

7. Match each person with his contribution to science.

1. Aristotle	a. idea of atoms
2. Democritus	b. discovered oxygen
3. alchemists	c. atomic theory
4. Robert Boyle	d. law of conservation of matter
5. Joseph Priestley	e. theory of five elements
6. Antoine Lavoisier	f. systematic study of gases
7. John Dalton	g. laboratory techniques

8. What were the two principal pursuits of the alchemists? Compare the pursuits of the alchemists with those of modern science and technology.
9. Outline the key elements of the scientific method.
10. Why is a precise statement of a problem important to a scientist?
11. Identify each of the following as a hypothesis, theory, or law.
    a. Humpty Dumpty's shell broke because it was weak due to a dietary deficiency.
    b. Energy can neither be created nor destroyed.
    c. The pressure of a gas is inversely proportional to its volume.
    d. A weight of evidence supports the idea that the pressure of a gas in a balloon is due to gas molecules colliding with the inside surface of the balloon.
    e. A man's hair turned gray because of many problems he had with his children.
    f. Cars rust faster during the winter months because of salt that is spread on slippery streets and highways.
    g. Life originated as a result of a series of chance events that occurred over a period of millions of years.
12. Think of one or more scientists that you know personally, perhaps some of your professors. What characteristics do they have in common? Do they fit preconceived stereotypes (if any) that you have had about scientists? Which of their characteristics are important in their practice of science?

Answers to Chapter Review

1. composition
2. mass
3. energy
4. biological
5. organic
6. biochemistry
7. mineral
8. carbon
9. structure
10. life
11. pollutants
12. gold
13. pottery
14. discoveries
15. Greek
16. wealth
17. Robert Boyle
18. balance
19. conservation of matter
20. scientific method
21. hypothesis
22. law
23. honesty

2

Matter and Energy

Contents

Icebergs from Portage Glacier in Alaska float in Glacier Bay. Water is shown here as a solid, a liquid, and a gas in the air.

Gold nuggets.

Burning candles give off heat and light.

You Probably Know . . .

- Matter exists in three states. For example, water is commonly found as a solid, liquid, or gas.
- Mixtures such as air, soil, and gasoline are common, but pure substances such as gold and diamonds are rare.
- Matter is made up of tiny particles that are invisible to the eye. These particles, called atoms, were first proposed by the Greeks several centuries before the time of Christ (Section 1.2).
- Changes we observe in various materials generally involve energy. For example, heating water causes it to boil and cooking (heating) our food changes its texture and flavor. Furthermore, burning a fuel such as candle wax gives off light and heat.

Why Should We Study Matter and Energy?

Chemistry is concerned with the materials that make up the universe and how energy affects these materials. Many of these materials are very complex, both those that are inorganic (of mineral origin) and those that are organic. To begin to understand our environment, the various consumer products we use, and the processes of living organisms, including those of our own bodies, we need to understand the materials involved. In this chapter, we will begin to classify and examine the nature of these materials (matter). We will also begin to study the role energy plays in the changes we observe in matter.

2.1 States of Matter

Learning Objectives

- Describe the characteristics of a solid, a liquid, and a gas in terms of visible properties and motion of particles.
- Describe the kinetic theory of matter.

Matter is anything that has mass and occupies space (Section 1.1). Matter includes the bricks, cement, wood, and glass used for buildings. It is the steel used for cars and trucks and the flesh and bone of our bodies. It includes the earth, the water in lakes and streams, the air we breathe, and the food we eat. Drugs, plastics, natural and synthetic fibers, gasoline and other fuels are all included. Matter also includes the substances that pollute our air and water. However, matter does not include energy such as heat, light, or electricity. Emotions such as love, hate, and fear are likewise excluded by the definition.

Matter has three physical states: solid, liquid, and gas.

As we observe various materials around us every day, we see them in different forms or states. The food we eat, the water we drink and use for bathing, and the air we breathe are examples of matter in different physical states called *solids, liquids,* and *gases.* We identify the physical states of substances almost unconsciously because their characteristics are so familiar.

A **solid** has a definite volume and a shape that does not depend on the shape of its container. The shape of a coin in a piggy bank is the same as its shape in a purse. Likewise, seashells have the same shape lying on the beach as they have as part of a shell collection in a jar. Numerous materials, both natural and synthetic, are solids. Included are many of the raw materials taken from the earth such as gravel, coal, and various minerals.

A **liquid** has a definite volume and takes the shape of its container to the level that it reaches. It has no definite shape of its own and differs from a solid in this respect. Liquids familiar to all of us are water, gasoline, and vegetable oil. When spilled, a liquid spreads over a surface, or flows. A material that flows is called a **fluid.** Because liquids such as water and gasoline flow, they can be pumped through pipes from a location where they are stored to the place where they are used.

A **gas** has no definite shape or volume. A gas completely fills its container, so its shape and volume are determined by the container. Gases, like liquids, are fluids and can be moved through pipes from one location to another. For example, natural gas is usually transported from a gas well to our homes through pipelines. Another example is the movement of air through ventilation ducts in schools, office buildings, and factories to maintain a comfortable and healthful work environment.

Kinetic theory: Matter is made up of tiny particles that are in constant, random motion.

The differences we observe among solids, liquids, and gases can be explained in terms of the **kinetic theory** of matter. (Kinetic refers to motion.) The kinetic theory proposes that all matter is made of tiny particles that are in constant, random motion. The velocity of this motion increases with increasing temperature. Even the particles of a solid substance are moving to some extent.

The kinetic theory explains what happens when warm, moist air cools. When the temperature drops, the water molecules slow down. Because of their mutual

Erupting Kilauea and lava flow in Hawaii. As a liquid, lava (melted rock) is a fluid.

Fog is formed when warm, moist air cools.

attractions, the water molecules move close enough to each other to become liquid. Tiny drops of water form and remain suspended in the air, forming fog. When the tiny drops in the fog combine into larger drops, they become too heavy to remain suspended in the air. Then it begins to rain. If rain falls into a bucket, the water assumes the bucket's shape.

As the temperature continues to drop, the speed of the particles in the liquid is reduced further and a solid is formed. The particles in the solid can no longer move easily past each other. Their movement is limited to small vibrations in an almost fixed position.

When rapid cooling occurs in a cloud in the sky, sleet (frozen rain) or hail (balls of ice) may form. The shape of the sleet crystals is almost spherical, like the shape of the raindrops before they were frozen. When the rainwater in the bottom of a bucket cools and freezes solid, the ice takes on the shape of the bottom of the bucket. If there is a design on the bottom of the bucket, it becomes part of the shape of the piece of ice. The shape of the ice remains unchanged when it is removed from the bucket. Having a definite shape is one of the characteristics of a solid.

Figure 2.1 summarizes the properties of solids, liquids, and gases. The characteristics of water in each of its three physical states are typical of those of other substances.

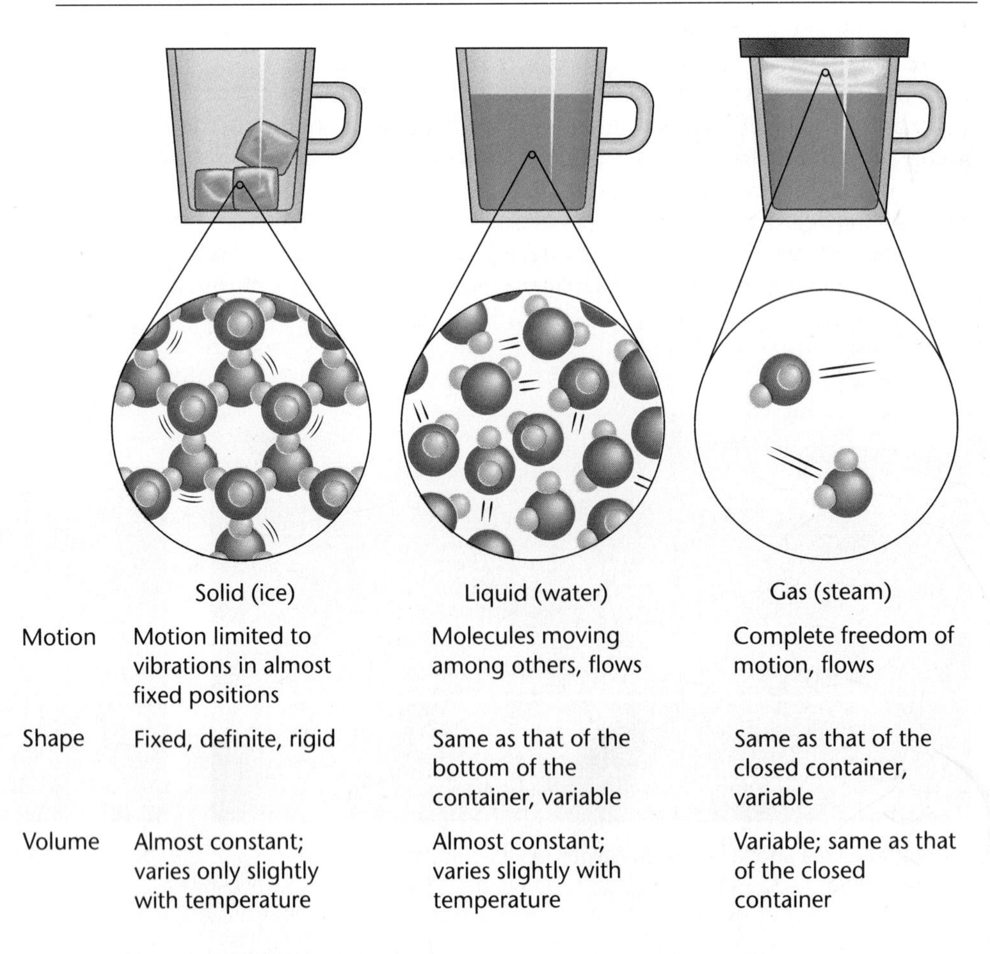

Solid (ice)
Liquid (water)
Gas (steam)
Motion
Motion limited to vibrations in almost fixed positions
Molecules moving among others, flows
Complete freedom of motion, flows
Shape
Fixed, definite, rigid
Same as that of the bottom of the container, variable
Same as that of the closed container, variable
Volume
Almost constant; varies only slightly with temperature
Almost constant; varies slightly with temperature
Variable; same as that of the closed container

FIGURE 2.1 Water in three physical states

Example 2.1

Describe the shape of each of the following:

a. a nickel
b. air in a hollow ball
c. gasoline in a half-filled spherical tank
d. a marble
e. orange juice in a can ■

Solutions

a. round
b. sphere
c. hemisphere
d. sphere
e. cylinder

Pure Substances and Mixtures

2.2

Learning Objectives

- Distinguish between a pure substance and a mixture.
- Distinguish between homogeneous and heterogeneous mixtures.

All materials are either single substances or mixtures. A **substance** is a particular type of matter with a definite composition. A substance, sometimes called a pure substance, is either an element or a compound (Section 2.3). You are already familiar with elements such as carbon, gold, and silver, and compounds such as water, carbon dioxide, and sugar.

Most of the materials that we see around us are mixtures of two or more substances. The air we breathe is a mixture of oxygen, nitrogen, and small amounts of other gases and dust. Our water, even after it has been treated to make it safe to drink, contains small amounts of many contaminants. We rarely find a pure substance. When we do find one, such as a gold nugget or a diamond (a form of carbon), it is often very valuable.

Pure substances are rare.

The outline in Figure 2.2 will help us begin to classify our material world.

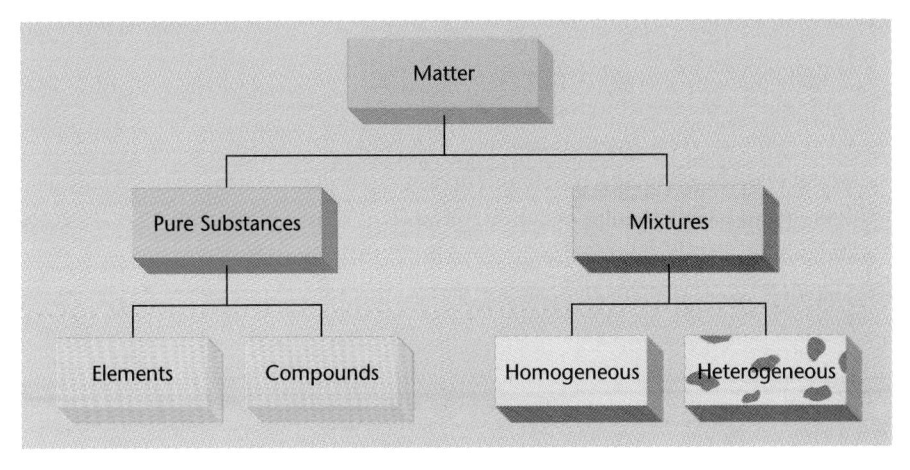

FIGURE 2.2 Classification of matter

A solution is a homogeneous mixture.

Mixtures are composed of two or more substances and can be classified as either homogeneous or heterogeneous. A **homogeneous** mixture is uniform in composition, properties, and appearance. It may be a solid, liquid, or gas. We often call a homogeneous mixture a **solution** (Chapter 14). Examples include a solution of two metals (called an alloy), solutions of salt or sugar in water, and air. In contrast, **heterogeneous** mixtures are nonuniform. Mud, for example, is a heterogeneous mixture of dirt and water. Vegetable oil and water are two liquids that form a nonuniform mixture. Although substances are usually homogeneous, you should note that they are heterogeneous when two physical states are mixed together. Ice water is an example.

The composition* of a mixture can vary widely, and this is one characteristic that distinguishes it from a compound (Section 2.6). For example, sugar solutions may contain varying amounts of sugar and water. We decide how much sugar to put in our coffee or tea depending upon our taste.

Example 2.2

Identify each of the following as a mixture or a pure substance.

a. sugar
b. sugar-coated puffed rice
c. shaving cream
d. bricks
e. paint in a can ■

Solutions

a. Sugar is a compound—a pure substance.
b. Sugar-coated puffed rice is a mixture.
c. Shaving cream is a mixture of water, soap, lubricants, a scent, and other ingredients.
d. Bricks are a mixture of clay, dyes, and other ingredients.
e. Paint is a mixture of compounds such as water, a pigment, and organic compounds.

2.3 Elements and Compounds

Learning Objectives

- Distinguish between elements and compounds.
- Distinguish between compounds and mixtures of elements.
- Write symbols and names for common elements.
- Explain the meaning of symbols and numbers (subscripts) in a chemical formula.
- Write names and formulas of binary compounds of two nonmetals.
- Write names of ionic binary compounds from their formulas.
- State the law of definite composition.

*Composition can be expressed as mass percentages of the components.

A pure substance can be either an element or a compound. An **element** is a substance that cannot be broken down into simpler substances by chemical means. Gold, silver, and iron are elements. A **compound** is a substance that is made up of two or more elements joined together and with a definite composition (definite mass percentages of the elements). Water, sugar, and table salt (sodium chloride) are compounds. A compound can be broken down by chemical means (chemists say decomposed) into its elements. For example, water can be decomposed by electrolysis into hydrogen and oxygen. This is illustrated in Figure 2.3.

Electrolysis: a chemical reaction caused by an electric current.

More than 100 elements are known, and new ones are being made by nuclear physicists at a rate of one every few years. However, only 88 elements are known to occur in nature, although trace amounts of several others have been found in the environment. In some ways the elements are like the 26 letters of the English alphabet, which are used to form hundreds of thousands of words. The millions of compounds on the earth and in its atmosphere are composed of these relatively few elements. Some common elements are listed in Table 2.1. A periodic table of the elements is found on the inside front cover of this book.

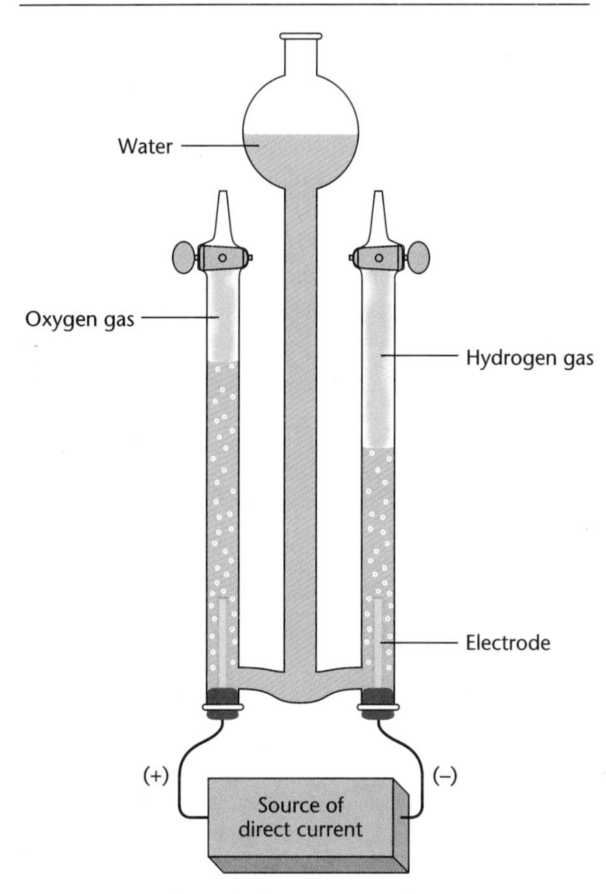

FIGURE 2.3 Electrolysis of water to form hydrogen and oxygen:

$$\text{water} \xrightarrow{\text{electric current}} \text{hydrogen} + \text{oxygen}$$

TABLE 2.1 Common Elements

Element	*Symbol*
aluminum	Al
argon	Ar
barium	Ba
beryllium	Be
boron	B
bromine	Br
calcium	Ca
carbon	C
chlorine	Cl
chromium	Cr
cobalt	Co
copper	Cu
fluorine	F
gold	Au
helium	He
hydrogen	H
iodine	I
iron	Fe
lead	Pb
lithium	Li
magnesium	Mg
nitrogen	N
oxygen	O
phosphorus	P
potassium	K
silicon	Si
silver	Ag
sodium	Na
sulfur	S
tin	Sn

Each element is represented by its symbol. The symbols of the elements are the basis for the chemical language, just as the letters of the alphabet are the basis for the English language. As you began to learn to read and write, you had to learn the 26 letters that are used to form all the words. Now, as you are beginning to learn the chemical language, you must learn the symbols for the elements that combine into compounds. The symbols for most of the elements have one or two letters. Only a few of the more recently synthesized elements have three letters. In writing chemical symbols, the first letter is always capitalized and the others are lower case. For example, Co is the symbol for cobalt, a metal used in making high speed cutting tools, not to be confused with carbon monoxide, CO, a toxic gas composed of carbon and oxygen. You can see how someone could be confused if you are not careful in writing the symbols for the elements.

Only the first letter of a chemical symbol is capitalized.

The smallest particle of an element that can combine with other elements to form compounds is an **atom.** Atoms are extremely small, and even a thumbtack contains 4×10^{21} or 4 000 000 000 000 000 000 000 iron atoms! Chemists often use the symbol of an element to represent one atom. For example, Au may represent one atom of gold, and Ca may indicate one atom of calcium. The structure of atoms is considered in Chapters 5 and 6.

4×10^{21} is scientific notation for a very large number (Section 3.3).

A **molecule** is a neutral particle composed of two or more atoms joined together. Some elements exist naturally in the form of molecules. For example, one form of phosphorus is composed of molecules that have four phosphorus atoms. There is a form of sulfur that is composed of molecules made up of eight sulfur atoms. The molecules of some elements have only two atoms. These are said to be **diatomic.** Nitrogen and oxygen, the main components of air, and hydrogen exist naturally as diatomic molecules. These are represented by the formulas N_2, O_2, and H_2. A number written as a subscript shows how many of that atom are present in a molecule. Thus, a chemical **formula** includes the symbols of the elements that make up a molecule and the numbers, as subscripts, that indicate the number of atoms of each element in a molecule (Figure 2.4).

Phosphorus (P_4), sulfur (S_8), and oxygen (O_2) are some of the molecular elements.

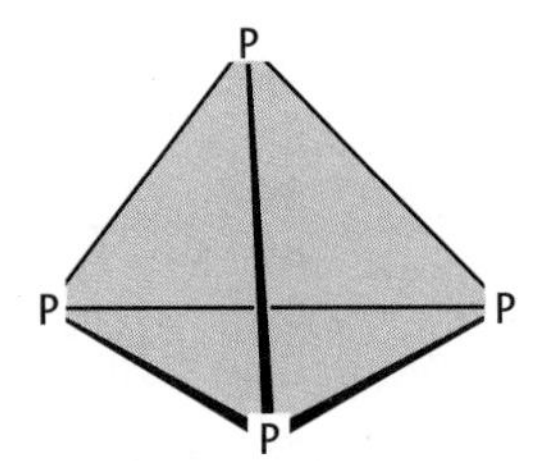

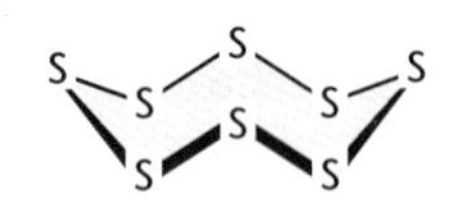

Example 2.3

Fluorine, chlorine, bromine, and iodine are diatomic elements. Write formulas for these elements. ■

Solution The formula of a diatomic molecule is written using the subscript 2: F_2, Cl_2, Br_2, I_2.

Many compounds are also molecular. Familiar ones are water, sugar, and carbon dioxide. A water molecule, represented by H_2O, is the smallest particle of water. One molecule of water is made up of two hydrogen atoms and one oxygen atom as indicated by the subscripts in the formula. Note that a subscript of "1" is understood when there is no subscript number.

Represents hydrogen ↷ ↶ Represents oxygen

H_2O_1

Means 2 hydrogen atoms ↗ ↖ "1" is understood

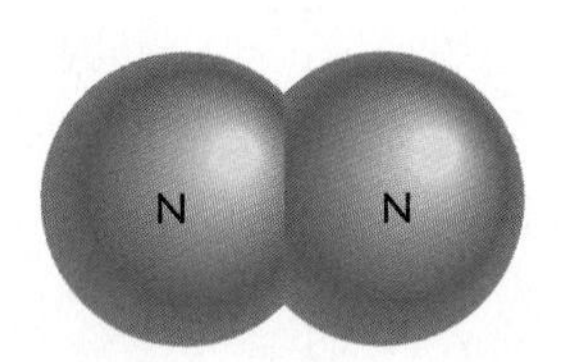

Models of diatomic molecules

The molecules of many compounds are much larger than water. For example, a sugar (sucrose) molecule contains 12 carbon atoms, 22 hydrogen atoms, and 11 oxygen atoms. As in other formulas, the subscripts in the formula of sucrose, $C_{12}H_{22}O_{11}$, indicate the number of atoms of each element in a sucrose molecule.

The formula for carbon dioxide, CO_2, is apparent from its name. Counting prefixes such as *mono-, di-,* and *tri-* are used to indicate the numbers of atoms in a molecule of a binary compound (two elements). Thus, carbon monoxide is CO, and sulfur trioxide is SO_3. Note that *mono-* is seldom used except to distinguish between different compounds of the same elements such as carbon dioxide (CO_2) and carbon monoxide (CO). Note also that the names of binary compounds end with *-ide.*

Represents a nitrogen atom → N_2 ← Means two atoms in the molecule

FIGURE 2.4 Formulas for molecules

Counting Prefixes

Number	Prefix
1	mono-
2	di-
3	tri-
4	tetra-
5	penta-
6	hexa-
7	hepta-
8	octa-
9	nona-
10	deca-

Some compounds are not molecular. For example, sodium chloride is composed of sodium ions (Na^+) and chloride ions (Cl^-) rather than molecules. An **ion** is a positively or negatively charged atom or group of atoms. A positive ion (e.g., Na^+) is called a **cation,** and a negative ion (e.g., Cl^-) is called an **anion.** In a grain of salt, the ions are held together by attractive forces that exist between positive and negative charges.

Cation: a positive ion.

Anion: a negative ion.

Ionic compounds are generally composed of a metal and a nonmetal, whereas molecular compounds are usually made up of two or more nonmetals. In the periodic table (inside front cover), metals are found on the left side, whereas nonmetals are on the right side. Thus, magnesium oxide, MgO, and potassium iodide, KI, are ionic compounds.

Binary ionic compounds are named by naming the cation and then the anion. The name of a cation is usually the same as the name of the metal, although there are some exceptions that will be covered later. The name of a monatomic anion, such as Cl^-, ends with *-ide* (the chloride ion). To name a monatomic anion, add *-ide* to the root of the name of the nonmetal. The following examples illustrate this pattern.

Ionic compound: a metal and a nonmetal.

Molecular compound: two nonmetals.

$CaBr_2$	calcium bromide
K_2O	potassium oxide

Thus, the names of monatomic anions and their binary compounds both end with *-ide.* We do not use prefixes such as *di-* in the names of ionic compounds because the numbers of ions are known from the charges on the cation and anion. Charges for some common ions are given in Table 2.2. Naming compounds is covered in detail in Chapter 9, including those involving complex ions composed of several atoms.

Counting prefixes are not used in the names of ionic compounds.

TABLE 2.2 Some Common Ions

Cations			*Anions*		
1+	*2+*	*3+*	*3−*	*2−*	*1−*
Li^+	Mg^{2+}	Al^{3+}	N^{3-}	O^{2-}	F^-
Na^+	Ca^{2+}		P^{3-}	S^{2-}	Cl^-
K^+	Ba^{2+}				Br^-
					I^-

You must be careful to distinguish between a compound made up of several elements and a mixture of those elements. There is a very important difference. This is stated by the **law of definite composition,** which says that a compound always has a definite and fixed composition by mass. In other words, the percentage of each element in a particular compound is always the same. For example, pure water is always 88.8% oxygen and 11.2% hydrogen by mass. On the other hand, a mixture has variable composition. Mixtures of oxygen and hydrogen gases can have any ratio of the two gases.

Naming Binary Compounds (Suffix: *-ide*)

Ionic (metal + nonmetal)

1. Cation: name of the metal ion
2. Anion: root of the name of the nonmetal with *-ide* suffix

Example: CaI_2, calcium iodide (Ca^{2+} and I^-)

Molecular (two nonmetals)

1. Prefixes to indicate numbers of each kind of atom
2. *-ide* suffix

Example: N_2O, dinitrogen oxide (2N and 1O)

Example 2.4

Explain the difference between the formulas CS_2 and Cs_2. Why is it important to capitalize only the first letter of the symbol for an element? ■

Solution CS_2 represents a molecule composed of 1 carbon atom and 2 sulfur atoms. Cs_2 represents a molecule of 2 cesium (Cs) atoms; it is not known to exist.

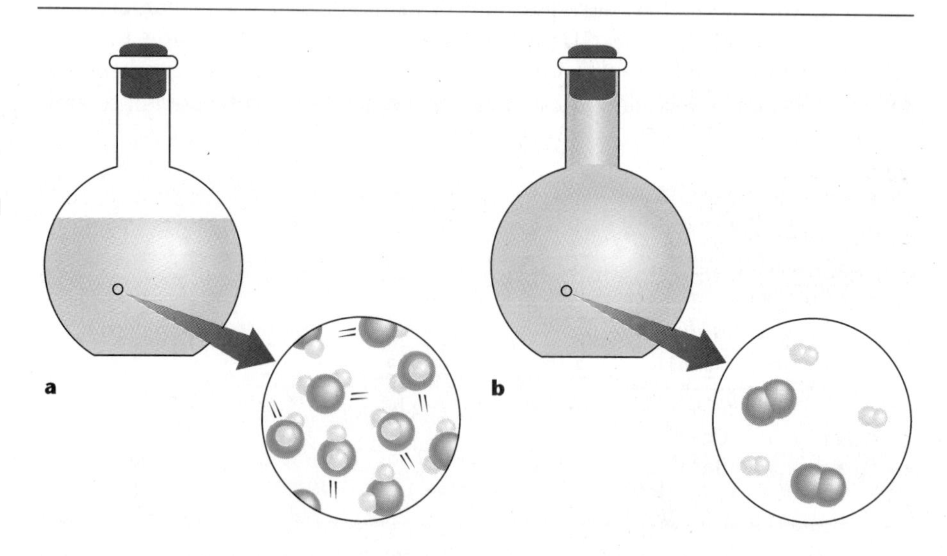

FIGURE 2.5 (a) Water, a compound, is composed of hydrogen and oxygen atoms in a fixed ratio as shown by its formula, H_2O. (b) Mixtures of hydrogen and oxygen can have various ratios of hydrogen and oxygen.

Example 2.5

Which of the following are elements and which are compounds?

a. iron
b. alcohol
c. caffeine
d. nitrogen ■

Solutions Refer to the list of elements inside the front cover of the book.
elements: **a, d** compounds: **b, c**

Example 2.6

Name the following binary compounds:

a. SO_2
b. PCl_3
c. CaS ■

Solutions To name a binary compound, use the suffix *-ide*. When naming compounds of two nonmetals, use counting prefixes to indicate the number of each kind of atom.

a. sulfur dioxide
b. phosphorus trichloride
c. calcium sulfide

Example 2.7

How many atoms of each element are in each of the following molecules?

a. SO_2
b. PCl_3
c. $C_{10}H_{14}N_2$ (nicotine) ■

Solutions Remember, a subscript of 1 is understood when there is no subscript number.

a. 1S and 2O atoms
b. 1P and 3Cl atoms
c. 10C, 14H, and 2N atoms

Example 2.8

Write formulas for the following:

a. carbon disulfide
b. phosphorus pentachloride
c. nitrogen triiodide ■

Solutions A counting prefix indicates the number of that kind of atom.

a. CS_2
b. PCl_5
c. NI_3

Practice Problem 2.1 Name the following compounds:
a. N_2O_3 **b.** KBr **c.** $MgCl_2$
(Solutions to practice problems appear at the end of the chapter.)

Practice Problem 2.2 Write formulas for the following compounds:
a. phosphorus tribromide **b.** nitrogen dioxide **c.** carbon tetrachloride

2.4 Physical and Chemical Properties

Learning Objectives

- Distinguish between physical and chemical properties.
- Distinguish between physical and chemical changes.
- Label the reactants and products of a chemical reaction.

How do you distinguish one substance from another? The answer is that every substance has its own unique set of characteristics or properties. Properties are classified as either physical or chemical. A **physical property** is a characteristic of a substance that can be determined without changing its composition. For example, if you were asked to describe water, you might say it is a colorless, odorless liquid. You also might mention that its boiling point is 100°C and its freezing point is 0°C. Common physical properties of a substance include its physical state, color, odor, boiling point, and melting (or freezing) point. All of these properties can be determined without changing the composition of a substance. Physical properties can be used to distinguish one substance from others and to identify a substance.

Some of the physical properties of some common substances are shown in Table 2.3. We can find the physical properties of the elements and many compounds in references such as a handbook of chemistry and physics.

TABLE 2.3 Physical Properties of Common Substances

Substance	*Physical State, 20°C*	*Melting Point, °C*	*Boiling Point, °C*
iron	solid	1535	2800
copper	solid	1083	2582
mercury	liquid	−39	357
ethanol (ethyl alcohol)	liquid	−117	78
water	liquid	0	100
octane	liquid	−57	126
oxygen	gas	−219	−183
nitrogen	gas	−210	−196
chlorine	gas	−101	−35

TABLE 2.4 Examples of Chemical and Physical Changes

Chemical Changes	*Physical Changes*
burning natural gas	melting ice
cooking an egg	boiling water
bleaching a stained shirt	dissolving sugar in water
burning a marshmallow	breaking a dish
digesting a meal	writing on a chalkboard
a rusting nail	sawing a log

Dissolving sugar in water is a physical change.

In a chemical reaction, new substances are formed.

A change in a substance that alters its physical properties without changing its identity (chemical composition) is called a **physical change.** Examples are listed in Table 2.4. Melting an ice cube is an example of a physical change. The form of the water has changed, but it has retained its identity as water. Boiling water is also a physical change. Dissolving sugar in water is another example. The form of the sugar has changed but not its identity. Furthermore, the water can be evaporated to recover the sugar—another physical change.

The conversion of one substance into another is a **chemical change.** A chemical change is also called a **chemical reaction.** For example, when a piece of paper is burned, it is converted into carbon dioxide, water, and ash (which is primarily of mineral content). The new substances formed or produced in a reaction (water, carbon dioxide, and ash) are called the reaction **products.** The products have different compositions and different physical properties from the original substances (paper and oxygen), which are called **reactants.**

A **chemical property** of a substance is described by one of its chemical reactions (see Table 2.4). Because natural gas (mostly methane) burns, calling it a fuel describes one of its chemical properties. Together with its composition, the chemical properties of a substance establish its chemical identity and distinguish it from other substances.

You should note that the physical and chemical properties of a compound differ from those of its elements. For example, water is a liquid at room temperature, whereas its elements, hydrogen and oxygen, are both gases. Hydrogen burns, and oxygen supports combustion, whereas water neither burns nor supports combustion. Also consider sodium chloride, ordinary table salt, a colorless solid that melts at 801°C. The physical properties of its elements are very different from those of the compound. Sodium is a solid metal that melts at 98°C, and chlorine is a greenish-yellow gas. There are also significant differences in their chemical properties. Sodium chloride is unreactive and nontoxic relative to sodium, which reacts vigorously with nonmetals, and relative to chlorine, which reacts vigorously with metals. Chlorine is a very toxic substance and was used as a war gas in World War I. It is now used in water treatment in swimming pools and municipal water treatment plants. Accidental exposure to chlorine can cause serious damage to the respiratory tract and even death.

This gas light burns natural gas, a fuel.

A chemical reaction can be represented by a **chemical equation** in which the symbols or formulas of reactants are written on the left side of an arrow and those of the products are written on the right side. When we read the equation, the arrow is read using words such as "yield" or "react to give."

$$\text{reactants} \longrightarrow \text{products}$$

To illustrate how a chemical equation is written for a chemical reaction, consider a familiar example. You may have used charcoal at a barbeque. Charcoal is mostly carbon. When carbon burns, it combines with oxygen in the air to form carbon dioxide, a new substance. This reaction can be represented by a word equation:

$$\text{carbon} + \text{oxygen} \longrightarrow \text{carbon dioxide}$$

This word equation is read "carbon and oxygen react to give carbon dioxide." A chemical equation can be written from the word equation. The elements and compounds are represented by their symbols and formulas:

$$C + O_2 \longrightarrow CO_2$$

Remember that oxygen is diatomic. The chemical equation concisely describes the chemical reaction. Carbon dioxide (CO_2) is the new substance formed when carbon (C) and oxygen (O_2) combine.

2.5 Law of Conservation of Matter

Learning Objective

- State the law of conservation of matter (mass).

The law of conservation of matter is also called the law of conservation of mass.

Combustion, or burning, puzzled early chemists. When wood or charcoal burned, it appeared that some of the matter ceased to exist. However, Antoine Lavoisier and others who followed him demonstrated in carefully controlled experiments that the mass of the products formed in a reaction is equal to the mass of the reactants. This was later summarized in the **law of conservation of matter,** which states: *in a chemical reaction, matter is neither created nor destroyed.* Because mass is a measure of matter, this law is sometimes stated as the **law of conservation of mass:** *in a chemical reaction, mass is conserved.* There is no measurable change in mass.

Chemical equations must be balanced to give equal numbers of atoms of each element on both sides.

Thus, to correctly describe a reaction, a chemical equation must show what substances are reacting and being formed and that matter is conserved. To show that matter is conserved, *the same number of atoms of each element must appear on both sides of the equation.* When this is done, the equation is said to be balanced. Inspection of the equation for the burning of charcoal confirms that the equation is balanced.

$$C + O_2 \longrightarrow CO_2$$

One atom of carbon and two atoms of oxygen appear on both sides of the equation.

Often, just writing the symbols for the reactants and the products of a reaction gives us an equation that is not balanced. For example, writing formulas for the reactant and products of the decomposition of water (Figure 2.3) gives an unbalanced equation:

$$\begin{array}{ccc} H_2O & \longrightarrow & H_2 + O_2 \\ \text{2H, 1O} & & \text{2H} \quad \text{2O} \end{array}$$

But we can balance the equation by inspection. Because the oxygen atoms are not balanced, two water molecules are needed on the left side.

2O are needed

$$2H_2O \longrightarrow H_2 + O_2$$

However, this gives 4H on the left side, and 4H are needed on the right side to balance the equation.

$$2H_2O \longrightarrow 2H_2 + O_2$$

$2 \times 2H = 4H$

4H are needed

We will study chemical equations in greater detail in Chapter 10.

Example 2.9

Balance the following equation:

$$CH_4 + Cl_2 \rightarrow CCl_4 + HCl$$ ■

Solution Inspect the equation, beginning with the first formula. Carbon is balanced, and to balance hydrogen, four hydrogen atoms are needed on the right side.

4H are needed

$$CH_4 + Cl_2 \rightarrow CCl_4 + 4HCl$$

To balance chlorine, eight chlorine atoms are needed on the left side.

$$CH_4 + 4Cl_2 \rightarrow CCl_4 + 4HCl$$

$4Cl + 4 \times 1Cl = 8Cl$

8Cl are needed

Check to make sure all atoms are balanced.

Practice Problem 2.3 Balance the following equations:
a. $CO + H_2 \rightarrow CH_4O$
b. $P_4 + Cl_2 \rightarrow PCl_3$

2.6 Separation of Mixtures

Learning Objectives

- Distinguish between the properties of a mixture and the properties of a compound.
- Briefly describe filtration, distillation, crystallization, and chromatography.

We see mixtures of all kinds of substances every day. The foods we eat, such as orange juice, cereal, bacon and eggs, and a five-course dinner, are mixtures. When we are on a beach, we see a mixture of water and sand in the shallow water. Sea water is a mixture of salt and water. The water piped to our homes contains small amounts of minerals and other contaminants, so it is a mixture also. Air is a mixture mostly of oxygen, nitrogen, and water vapor, and it also contains varying amounts of pollutants.

The principal differences between mixtures and compounds are summarized in Table 2.5. The individual substances of a mixture (its components) have different physical properties, and they are attracted to one another only by weak forces. This allows them to be separated by physical methods, which do not involve chemical changes. However, the elements making up a compound can be separated only by chemical decomposition of the compound.

Using physical methods, the components of a mixture can be separated without altering their compositions or chemical identities. Some separations are easy and others are difficult. Separation of the components of a mixture is easiest when they have very different physical properties.

A number of methods for separating mixtures are used by chemists in their laboratories. These include *filtration, distillation, crystallization,* and *chromatography.* To decide which method to use, you must identify differences in one or more physical properties for the components of the mixture.

For example, sand and water form a heterogeneous mixture. One component is a solid and the other is a liquid. The mixture can be separated simply by placing the wet sand on a piece of cloth and allowing the water to drain. Because cloth is porous, it allows the liquid to pass through as the solid is collected. The cloth is called a **filter,** and we call this process **filtration.** You may have used a coffee filter in

TABLE 2.5 Differences Between Mixtures and Compounds

Mixture	*Compound*
variable composition	fixed composition
properties are a blend of the properties of its components	properties are different from the properties of its components
components are only weakly attracted to one another	components are tightly bonded together
components can be separated by physical methods	components cannot be separated by physical methods

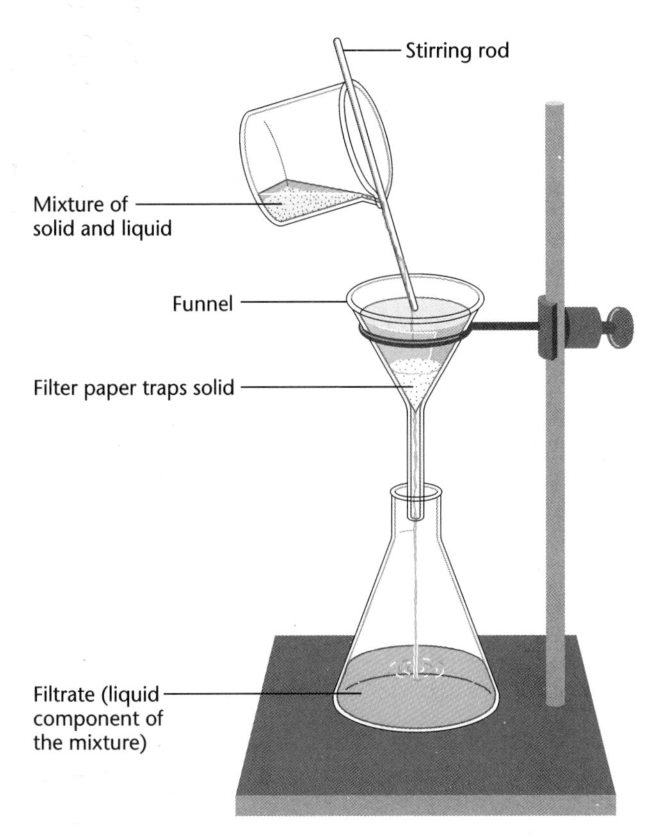

FIGURE 2.6 Filtration is used to separate a solid from a liquid in a heterogeneous mixture.

brewing coffee. Many types of filter paper are available for laboratory use. Filtration is commonly used to separate a solid from a liquid in a heterogeneous mixture (Figure 2.6).

Distillation is a very useful technique for separating the components of a mixture when they differ in their volatility (how easily they vaporize). Distillation can be used to separate a solid–liquid solution such as sea water (Figure 2.7). This is done by boiling the water and then condensing the steam to water again. The solid is left behind as a residue. Generally, liquid–liquid solutions can also be separated by distillation if the liquids boil at different temperatures.

You may have observed the separation of solid sugar from pancake syrup. The separation of a solid from a solution is called **crystallization** (Figure 2.8). Crystallization of sugar may have occurred either when some of the water in the syrup was evaporated or when the syrup was cooled in a refrigerator. Evaporation of water from a solution reduces the amount of sugar that can remain dissolved. Furthermore, cooling a solution reduces the solubility of sugar in water. Once crystals have formed, they can be separated from the liquid by filtration. Separation by crystallization combines some of the features of distillation (evaporation) with filtration.

Solubility of sugar refers to the amount that can be dissolved in a certain quantity of water at a given temperature (Chapter 14).

Chromatography is a separation method that depends on differences in the ability of the components of a mixture to stick (or **adsorb**) to surfaces. Paper chro-

Adsorb means to stick to a surface.

FIGURE 2.7 When salt water is distilled, water is boiled. The resulting steam is cooled in the condenser, and pure water is collected in the receiving flask.

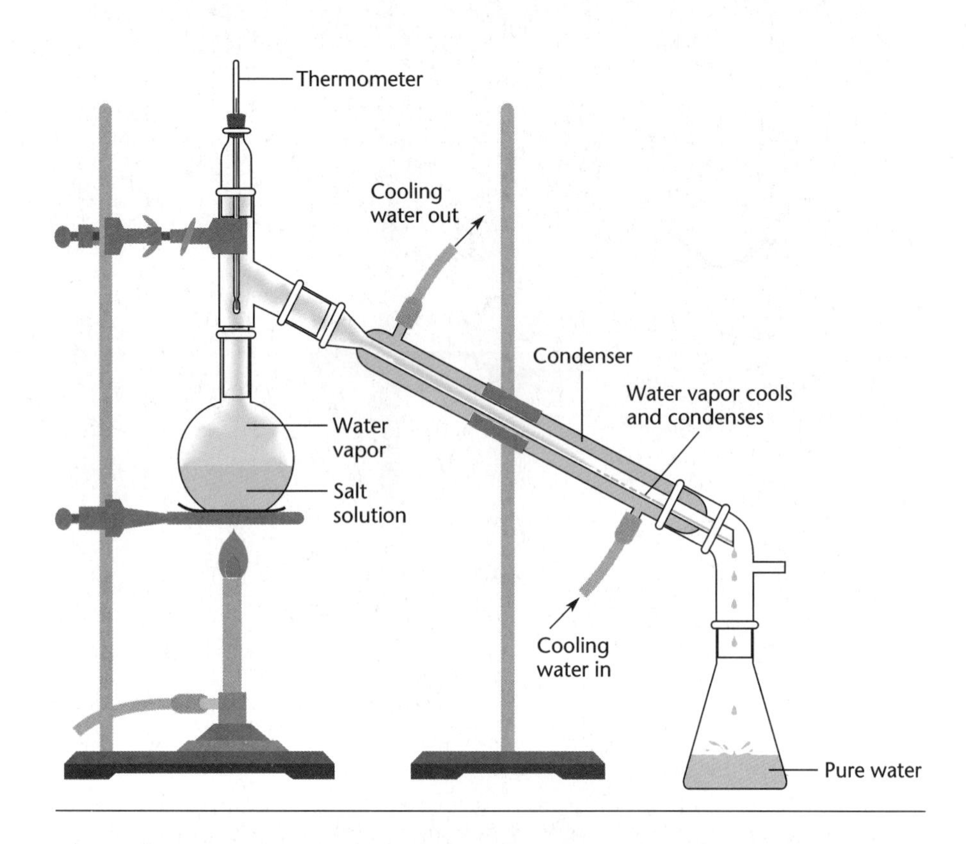

FIGURE 2.8 *Left:* Crystals of alum form by cooling a hot, concentrated solution of alum (used in some baking powders). *Right:* A large crystal of alum.

CHEMICAL WORLD

Chromatography

Chromatography (*chromato,* colored; *graphy,* writing) is used to detect and analyze air and water pollutants. Very low levels of pollutants have been detected in what had been thought to be clean air and water. Flammable liquids used in arson can be so precisely analyzed that the brand of a gasoline or other flammable liquid can be determined. In cases of homicide, drugs in body fluids can be detected and identified, and in more ordinary matters, minute contaminants in foods can be detected.

Components of a mixture differ in their affinities for various adsorbent materials, called *adsorbents* in chromatography. You may have noticed the coffee color climbing a paper filter. The original technique, paper chromatography, used filter paper to separate mixtures of plant pigments. It is now used for many other separations. A sheet of filter paper (called the stationary phase) is spotted with the mixture to be analyzed and suspended vertically with its bottom edge immersed in a solvent (the moving phase). As the solvent climbs the paper it carries the various pigments various distances, depending on their solubility in the solvent and their affinity for the paper. Because solubility and affinity for a solvent and paper differ for most components, the mixture is separated, and each component can be identified by the distance it travels.

Modern chromatography separates colorless liquids and even gases.

Thin-layer chromatography (TLC) is like paper chromatography except that a glass plate or plastic sheet, coated with a solid adsorbent such as alumina or silica gel, replaces the paper in the stationary phase. In gas chromatography (GC), also called vapor-phase chromatography (VPC), helium or some other inert gas serves as the moving phase. As a vapor mixture passes through a long column containing an adsorbent, the components are separated according to their affinity for the adsorbent. In high-performance liquid chromatography (HPLC), the mixture is forced under pressure through a column that is tightly packed with adsorbent. The components of a mixture can be identified and the quantities present determined by comparing the resulting chromatogram with those of known mixtures.

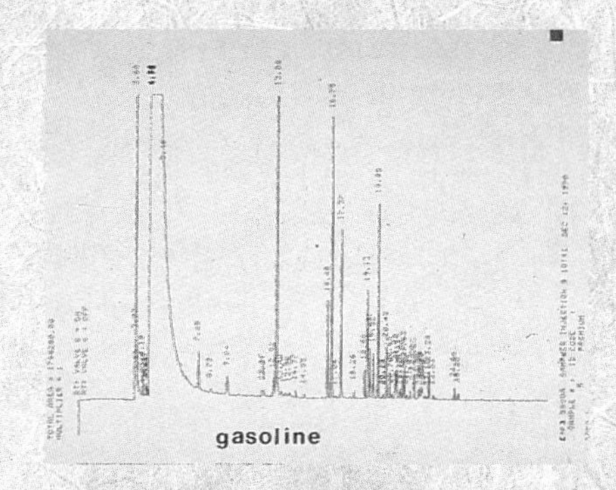

Above: A gas chromatograph is a powerful tool for analyzing mixtures. *Below:* A sample of gasoline is made up of more than 200 compounds. Each compound is represented by a peak on the chromatogram. Data can be stored in a computer for later evaluation.

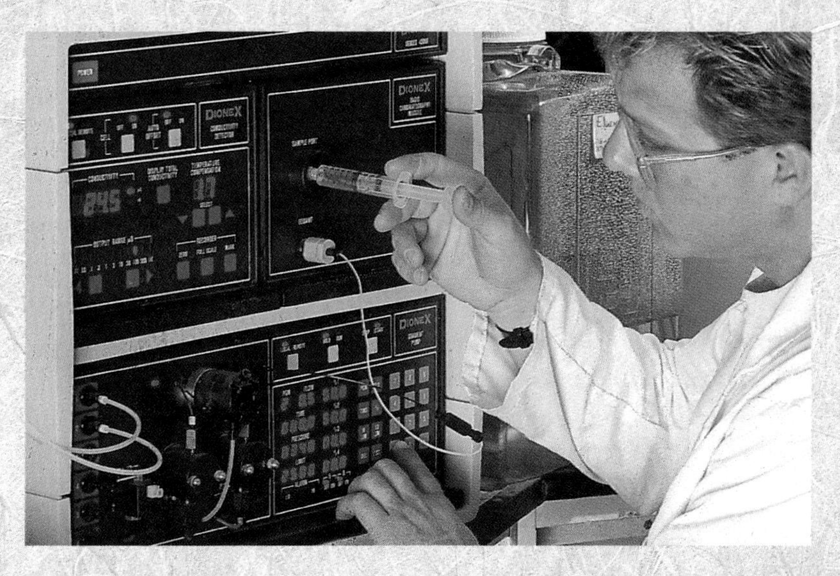

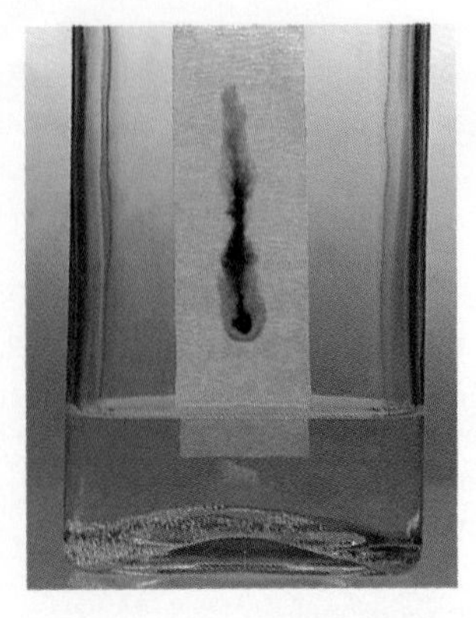

FIGURE 2.9 Paper chromatography separates a mixture of dyes, which show up on the paper as colored spots.

matography is the simplest of the various forms of chromatography (Figure 2.9). With this method a solution containing a mixture is washed along a piece of paper. Compounds that adsorb strongly move short distances, and those that are weakly adsorbed move farther. If the compounds are colored, spots of color at various places on the paper indicate the compounds in the mixture. Other forms of chromatography, including gas chromatography (GC) and high performance liquid chromatography (HPLC), have been automated and are extremely valuable analytical techniques.

Example 2.10

What method could be used to separate the components of each of the following mixtures:

a. sand in gasoline
b. acetone (nail polish remover) and water
c. salt water ▪

Solutions

a. A heterogeneous solid–liquid mixture is best separated by filtration.
b and **c.** Homogeneous liquid mixtures can be separated by distillation.

2.7 Electrical Nature of Matter

Learning Objective

- Describe the effects of the electrostatic forces between like charges and between opposite charges.

You may have had the experience of walking across a carpet and getting a shock when you touched a lamp or a light switch. You may have rubbed a balloon in your hair and then stuck it on the wall of the room. These experiences illustrate that matter has electrical properties.

Like charges repel each other; opposite charges attract each other.

Attractive and repulsive forces exist between objects that have electrical charges. When a hard rubber comb or rod is rubbed with fur, it develops a "negative" charge. When a glass rod is rubbed with a piece of silk, it takes on a "positive"

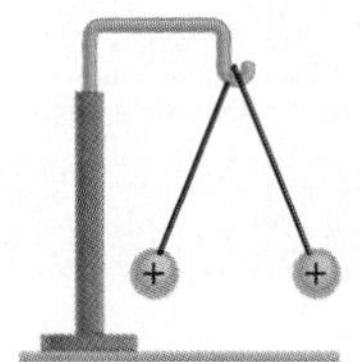

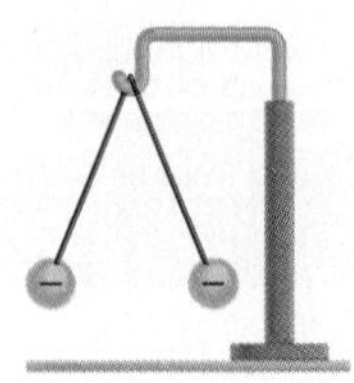

a

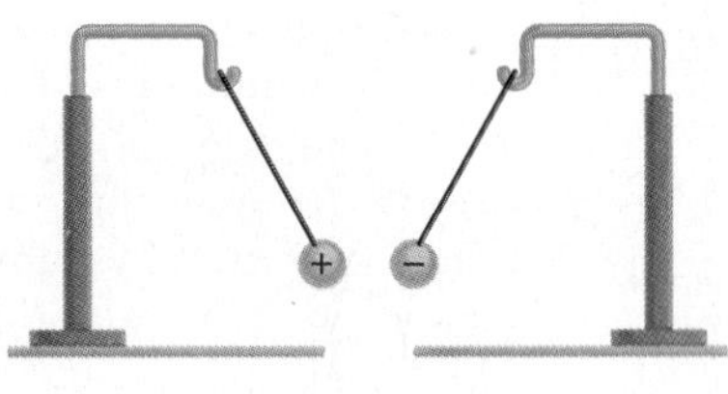

b

FIGURE 2.10 Repulsive and attractive forces between charged objects. (a) When the two balls have the same charge, they repel each other. (b) Balls having opposite charges are attracted to one another.

charge. The nature of the forces between positive and negative charges can be illustrated by some very simple experiments.

Suppose you make a tiny ball of fibers (or pith) found in a cornstalk and hang it by a thread (Figure 2.10). The pith ball can be positively charged by touching it with a positively charged rod. Now that the pith ball has the same charge as the rod, it will fly away when the rod is moved close to it. Similarly, two positively charged pith balls repel each other. Two pith balls with negative charges also repel each other. However, when the pith balls have opposite charges, they attract each other.

Thus, we see that the electrical properties of matter result in **electrostatic forces** of attraction and repulsion. Atoms, molecules, and ions interact with each other much like the balls, and as we examine their natures we will begin to understand how these forces are responsible for energy changes observed in chemical reactions.

Role of Energy in Physical and Chemical Changes

Learning Objectives

- Distinguish between kinetic and potential energy.
- Distinguish between exothermic and endothermic changes.
- State the law of conservation of energy.

Energy is the ability to do work. Energy causes objects to move. For example, someone may move a chair, carry books, or push a lawnmower. All of these tasks involve **work,** moving an object through a distance. Energy is different from matter. Unlike objects, we cannot touch energy or pick it up. However, we all have felt the effects of energy. Heat and light are part of our everyday lives. Sunlight, solar energy, not only makes plants grow but heats homes and powers hand-held calculators and satellites. Electrical energy is used to light and heat our homes and to operate appliances, tools, and machinery (Figure 2.11).

FIGURE 2.11 A modern kitchen uses electrical energy for lighting and cooking and for operating kitchen appliances.

Heat flows spontaneously from a hot burner to a cooler pan.

Energy causes both physical and chemical changes. You have seen the wind blow leaves off a tree. A flooding river can cause great destruction to buildings and the environment. Heat and sunlight melt snow and ice. These are physical changes and do not involve a change in composition of the materials involved.

Heat and microwave energy can cook our food. Sunlight causes plants to grow. In our bodies food is digested, giving the energy needed for bodily functions and releasing carbon dioxide, water, and other waste products. These are chemical changes and illustrate that energy is active in producing new substances.

Of the various forms of energy, heat will concern us the most in our study of chemistry. **Heat** is a form of energy that flows spontaneously from a warm object to a cool object. For example, when charcoal burns, energy is released by the chemical reaction to the surroundings. Energy flows from the hot charcoal to the cooler surrounding air. Heat is associated with the rapidly moving molecules formed in the reaction. These rapid molecules collide with others in the surroundings, causing them to move faster. As a result, they have greater energy of motion, or **kinetic energy,** making them warmer. A chemical reaction that releases heat to the surroundings is called an **exothermic** reaction. Physical changes can be exothermic as well. For example, steam condensing to liquid water or liquid water freezing to ice are exothermic processes.

Exothermic change: Heat is released to the surroundings.

Endothermic change: Heat is absorbed from the surroundings.

Melting ice is an example of an **endothermic** process, one that absorbs heat from the surroundings. Melting ice and boiling water are examples of endothermic physical changes. Many chemical changes are also endothermic, such as cooking food and the decomposition of water.

As was noted, energy and work are closely related. Because work involves moving a mass, a swimmer climbing a ladder to a diving board is doing work. Moving a book from a table to a high shelf is doing work. The book's potential to do work is increased by moving it against the force of gravity, which attracts it. The **potential** energy of the book, or its energy of position, is greater on the shelf than on the table. Potential energy can be thought of as stored energy.

Chemical energy is an important form of potential energy. It is energy stored in chemical substances. When a fuel burns, some of its chemical energy is released as kinetic energy in the form of heat. Potential energy is converted into kinetic energy. As a tree is growing, energy from the sun is being converted to chemical energy stored in the trunk, branches, and leaves of the tree. This energy is released as heat when a log cut from the tree is burned. Thus, solar energy can be converted into chemical energy, and chemical energy can be converted into heat.

Energy is conserved in chemical and physical changes.

It has been shown by careful studies of energy conversions that energy is conserved. The results of such studies can be summarized in the **law of conservation of energy.** This law states that *energy can neither be created nor destroyed in an ordinary chemical or physical change. Energy is conserved.* During endothermic reactions, heat is absorbed from the surroundings, and kinetic energy is converted to potential energy (chemical energy), the energy stored in the products of the reaction. During exothermic reactions, some of the potential energy (chemical energy) of the reactants is converted to kinetic energy of the products, and heat is released to the surroundings.

Our understanding of chemical energy will depend upon an understanding of the electrostatic forces involved in molecular and ionic compounds. These will be discussed in detail in Chapter 8.

Potential energy is energy of position, or stored energy. The potential energy of a book is greater on a high shelf than on a table.

Chapter Review

Directions: From the list of choices at the end of the Chapter Review, choose the word or term that best fits in each blank. Use each choice only once. The answers appear at the end of the chapter.

Section 2.1 All the materials that make up the earth and its atmosphere are called __________ (1). These materials may exist in three states called __________ (2), __________ (3), __________ and gases. A solid has a definite volume and __________ (4) that does not depend upon the shape of its container. A liquid takes the shape of the part of the container it occupies. A __________ (5) has no definite shape or volume; these are determined by its container. Gases and liquids flow and are called __________ (6). The differences between the three states of matter can be explained in terms of the __________ __________ (7) of matter.

Section 2.2 Matter can be considered to be either pure substances or __________ (8). A pure substance has a definite, fixed __________ (9). Pure substances are either elements or __________ (10). Mixtures can be either homogeneous or __________ (11). Homogeneous mixtures are uniform in properties and appearance.

Section 2.3 An __________ (12) is a substance that cannot be broken down into simpler substances by chemical means. A compound is a substance with a definite composition and is made up of two or more elements. Chemists represent each element by its __________ (13). The smallest particle of an element that can combine with other elements is called an __________ (14). A __________ (15) is a neutral particle composed of two or more atoms joined together. Both elements and compounds can be molecular. Molecules can be represented by formulas such as __________ (16) for nitrogen and __________ (17) for water. Some compounds are not molecular and are made up of charged atoms called __________ (18). A positively charged ion is called a __________ (19), and a negative ion is called an __________ (20). The name of a binary compound ends with __________ (21). The law of __________ __________ (22) distinguishes between a compound and a mixture of elements.

Section 2.4 Each substance has its own unique set of physical and chemical properties. __________ (23) properties are those characteristics of a substance that can be determined without changing its composition. A physical change results in a change in physical properties without changing the composition of the substance.

A __________ (24) change, also called a chemical __________ (25), always results in the formation of new substances. The substances entering into the reaction are called the __________ (26), and the substances formed in the reaction are called the __________ (27). A chemical __________ (28) of a substance is described by one of its reactions. Chemists represent a chemical reaction by writing a __________ __________ (29).

Section 2.5 The law of __________ __________ __________ (30) states that matter can neither be created nor destroyed in a chemical reaction. This law is sometimes called the law of conservation of __________ (31). In recognition of this law, chemical equations must be __________ (32).

Section 2.6 Mixtures have __________ (33) composition. The components of a mixture are only __________ (34) attracted to one another. Separation of the components of a mixture is easiest when their physical properties are very __________ (35). The method usually used to separate a solid from a liquid making up a heterogeneous mixture is __________ (36). The method used to separate a solid–liquid solution such as sea water, as well as many liquid–liquid solutions, is __________ (37). A separation method that depends upon differences in the ability of compounds to adsorb on surfaces is __________ (38).

Section 2.7 Matter has electrical properties. A hard rubber or glass rod can be __________ (39) by rubbing it with a piece of fur or silk. Like charges __________ (40) each other; opposite charges __________ (41) each other. The __________ (42) forces of attraction and repulsion help us to understand energy changes for chemical reactions.

Section 2.8 __________ (43) is the ability to do work. Energy can cause both physical and chemical changes. __________ (44) is a form of energy that flows from a warm object to a cool object. Heat is related to the kinetic energy of molecules, or energy of __________ (45). Changes in which heat is evolved are called __________ (46) changes. A change is said to be endothermic when heat is __________ (47) from the surroundings. Melting ice is an __________ (48) change.

Stored energy, or energy of position, is called __________ (49) energy. When a fuel is burned, some of its chemical energy (a form of potential energy) is converted to heat (a form of __________ (50) energy). During endothermic reactions, heat is converted into chemical energy of the products.

Choices: absorbed, anion, atom, attract, balanced, cation, charged, chemical, chemical equation, chromatography, composition, compounds, conservation of matter, definite composition, different, distillation, electrostatic, element, energy, endothermic, exothermic, filtration, fluids, gas, H_2O, heat, heterogeneous, -ide, ions, kinetic, kinetic theory, liquids, mass, matter, mixtures, molecule, motion, N_2, physical, potential, products, property, reactants, reaction, repel, shape, solids, symbol, variable, weakly

Study Objectives

After studying Chapter 2, you should be able to:

Section 2.1

1. Describe the characteristics of a solid, a liquid, and a gas in terms of visible properties and motion of particles.
2. Describe the kinetic theory of matter.

Section 2.2

3. Distinguish between a pure substance and a mixture.
4. Distinguish between homogeneous and heterogeneous mixtures.

Section 2.3

5. Distinguish between elements and compounds.
6. Distinguish between compounds and mixtures of elements.
7. Write symbols and names for common elements.
8. Explain the meaning of symbols and numbers (subscripts) in a chemical formula.
9. Write names and formulas of binary compounds of two nonmetals.
10. Write names of ionic binary compounds from their formulas.
11. State the law of definite composition.

Section 2.4

12. Distinguish between physical and chemical properties.
13. Distinguish between physical and chemical changes.
14. Label the reactants and products of a chemical reaction.

Section 2.5

15. State the law of conservation of matter (mass).

Section 2.6

16. Distinguish between the properties of a mixture and the properties of a compound.
17. Briefly describe filtration, distillation, crystallization, and chromatography.

Section 2.7

18. Describe the effects of the electrostatic forces between like charges and between opposite charges.

Section 2.8

19. Distinguish between kinetic and potential energy.
20. Distinguish between exothermic and endothermic changes.
21. State the law of conservation of energy.

Key Terms

Review the definition of each of the terms listed here by chapter section number. You may use the glossary if necessary.

2.1 fluid, gas, kinetic, liquid, matter, solid

2.2 heterogeneous, homogeneous, mixture, substance

2.3 anion, atom, cation, compound, diatomic, element, formula, ion, law of definite composition, molecule

2.4 chemical change, chemical equation, chemical property, chemical reaction, physical change, physical property, product, property, reactant

2.5 law of conservation of matter (mass)

2.6 adsorb, chromatography, crystallization, distillation, filter, filtration

2.7 electrostatic forces

2.8 chemical energy, endothermic, energy, exothermic, heat, kinetic energy, law of conservation of energy, potential energy, work

Questions and Problems

Answers to questions and problems with blue numbers appear in Appendix C.

States of Matter (Section 2.1)

1. Compare the shape and volume of a substance such as water in the solid, liquid, and gaseous states.
2. Compare the motion of the particles of a substance in the gaseous, liquid, and solid states.
3. Describe the kinetic theory of matter.
4. In terms of the kinetic theory of matter, explain what happens when a gas such as steam is cooled and becomes a liquid. Explain what happens when a liquid such as water is cooled and becomes a solid (ice).
5. Which of the following are fluids?
   a. gravel
   b. air
   c. motor oil
   d. sugar
   e. gasoline
   f. bricks

Pure Substances and Mixtures (Section 2.2)

6. Identify each of the following as a pure substance or a mixture:
   a. gold
   b. carbon
   c. sea water
   d. silver
   e. ashes
   f. carbon dioxide
7. Identify each of the following as either homogeneous or heterogeneous:
   a. salt water
   b. Italian salad dressing
   c. iced tea
   d. orange juice
   e. gasoline
   f. vegetable oil
   g. grape jam
   h. concrete
8. Think of things (other than food) where you live or work. Can you identify six things that are heterogeneous? Identify six things that are homogeneous.

Elements and Compounds (Section 2.3)

9. Which of the following are elements?
   a. hydrogen
   b. $C_5H_{10}O_5$
   c. aluminum
   d. gasoline
   e. salt
   f. air
   g. CO_2
   h. oxygen
   i. water
10. Which of the following are compounds?
    a. silver
    b. CO
    c. C_2H_6O
    d. sulfur
    e. helium
    f. NaCl
11. Rubbing alcohol is a colorless liquid with a noticeable odor. A small quantity was poured onto a dry dish and ignited. After the flame went out, a colorless, odorless liquid remained. Do these observations suggest that rubbing alcohol is a compound or a mixture? Explain.
12. So-called white vinegar is a colorless liquid with a strong odor and a distinct taste. Careful distillation of white vinegar gives an odorless liquid that boils at 100°C and a second liquid that has a sharp irritating odor and boils at 118°C. Is white vinegar a single substance or a mixture? Explain.
13. Baking soda is a solid with a definite composition of 27.4% sodium, 1.20% hydrogen, 14.3% carbon, and 57.1% oxygen. Is baking soda a single substance or a mixture? Explain.
14. Write the symbols for the following elements. Use the periodic table as necessary.
    a. carbon
    b. sodium
    c. aluminum
    d. iron
    e. fluorine
    f. magnesium
    g. nickel
    h. sulfur
15. Name the elements represented by the following symbols:
    a. N
    b. P
    c. Si
    d. Li
    e. Cl
    f. He
16. Which of the following are diatomic elements?
    a. H_2
    b. O_2
    c. Cl_2
    d. CO
17. How many atoms of oxygen are in each molecule of the following compounds?
    a. CO_2 b. SO_3 c. $C_6H_{12}O_6$
18. Write formulas for the following:
    a. phosphorus pentabromide
    b. oxygen difluoride
    c. carbon tetrabromide
19. Name the following compounds:
    a. NaH b. CCl_4 c. CaS
20. Write formulas for the following compounds:
    a. silicon tetrachloride
    b. sulfur dichloride
    c. dinitrogen pentoxide
21. Name the following compounds:
    a. Mg_3N_2 b. LiCl c. Al_2O_3

Physical and Chemical Properties (Section 2.4)

22. Which of the following are physical properties and which are chemical properties?
    a. Alcohol is a colorless liquid.
    b. The boiling point of water is 100°C.
    c. Water decomposes to give hydrogen and oxygen gases.
    d. Apple juice turns sour.
    e. Sugar dissolves in water.
23. Which of the following involve a chemical change?
    a. breaking an egg
    b. exploding a firecracker
    c. melting snow
    d. tearing paper
    e. drying a wet towel
    f. mixing salt and water
    g. lighting a fire
    h. cutting hair
    i. slicing bread

24. Identify the reactants and products in each of the following reactions. Remember that oxygen is a reactant when a substance is burned (combustion).
    a. Propane burns to give carbon dioxide and water.
    b. Iron combines with oxygen to give iron oxide (rust).
    c. Electrolysis of water gives hydrogen and oxygen.
    d. Acetone (used in nail polish remover) burns to give carbon dioxide and water.
    e. Baking soda (sodium bicarbonate) reacts with vinegar (acetic acid) to give carbon dioxide, sodium acetate, and water.
25. Adding hydrogen to oxygen at room temperature gives a gas that is visibly no different from the individual components. Is the resulting gas a compound or a mixture? Did a chemical change take place upon mixing? When this gas is ignited, a liquid results. Did a chemical change occur? What is the liquid?

Law of Conservation of Matter (Section 2.5)

26. When charcoal is burned, the resulting ash weighs much less than the charcoal. Explain. How is this understood in terms of the law of conservation of matter?
27. Which of the following equations are *not* balanced?
    a. $HCl + NaOH \longrightarrow H_2O + NaCl$
    b. $Li + O_2 \longrightarrow Li_2O$
    c. $CH_4 + O_2 \longrightarrow CO_2 + 2H_2O$
    d. $CO_2 + H_2O \longrightarrow CH_2O + O_2$
    e. $N_2 + O_2 \longrightarrow 2NO_2$

Separation of Mixtures (Section 2.6)

28. Which of the following can be separated into its components by physical methods?
    a. water
    b. mixture of sand and water
    c. carbon dioxide
    d. sugar
    e. sea water
    f. table salt
    g. mixture of sugar and sand
29. Which of the following can be separated by filtration?
    a. a solution of baking soda and water
    b. noodles and water in a pan
    c. water and sawdust
    d. mixture of oil and water
    e. mixture of broken glass and water
30. Which of the following mixtures can be separated by distillation?
    a. salt and water
    b. sugar and water
    c. mixture of nail polish remover (acetone) and water
31. What separation method depends upon a difference in the ability of compounds to adsorb on a surface such as paper?

Electrical Nature of Matter (Section 2.7)

32. Do the following attract or repel each other?
    a. two negatively charged balloons
    b. a negatively charged balloon and a positively charged rod
    c. a Na^+ ion and a K^+ ion
    d. a Na^+ ion and a Cl^- ion
33. A diagram for an electrolysis experiment showed Cl^- ions in a solution were attracted toward a wire marked "anode" that was connected to a battery. What was the charge on the anode?

Energy in Physical and Chemical Changes (Section 2.8)

34. Which of the following changes are exothermic and which are endothermic?
    a. evaporation of water from your skin
    b. melting snow
    c. steam changing to water
    d. clothes drying
    e. freezing water
35. Which of the following are exothermic and which are endothermic changes?
    a. a burning candle
    b. pancakes cooking on a griddle
    c. digesting a meal
    d. a roasting marshmallow
    e. a burning marshmallow
36. What term is described by each of the following?
    a. energy of position
    b. energy of motion
    c. energy stored in chemical substances
    d. energy due to the moving molecules of a substance
37. When an apple falls out of a tree to the ground, does its potential energy increase or decrease? Does its kinetic energy increase or decrease as it is falling? Does its total energy change when it falls? Explain.
38. When a candle burns, wax disappears, smoke (a mixture of carbon dioxide, water vapor, and carbon) is observed, and heat is released. Explain the energy changes in terms of the law of conservation of energy.

Additional Problems

39. Using a test tube and a beaker of water, explain how you could show that air is matter.
40. Which of the following are compounds and which are elements?
    a. platinum
    b. sulfur dioxide
    c. aspirin
    d. uranium
41. Classify each of the following properties as chemical or physical. Explain your reasoning in each case.
    a. Cooking oil burns.
    b. Gold is yellow in color.
    c. A black coating forms on copper when it is heated in air.

42. Write formulas for each of the following:
    a. dinitrogen tetroxide
    b. boron trifluoride
    c. hydrogen bromide

43. Name the following compounds:
    a. CaF_2 b. N_2O c. XeF_4

44. Balance the following equations:
    a. $H_2 + Br_2 \longrightarrow HBr$
    b. $N_2 + H_2 \longrightarrow NH_3$
    c. $C_2H_4 + O_2 \longrightarrow CO_2 + H_2O$

45. Which method, filtration or distillation, could be used to separate each of the following mixtures?
    a. cooked carrots and water
    b. a solution of paraffin wax and gasoline
    c. ice and vinegar
    d. cereal and milk

46. Which of the following changes are endothermic and which are exothermic?
    a. melting candle wax
    b. an exploding firecracker
    c. evaporation of alcohol
    d. burning charcoal

Solutions to Practice Problems

PP 2.1

To name a binary compound, use the suffix *-ide*. When naming compounds of two nonmetals, use counting prefixes to indicate the number of each kind of atom.
a. dinitrogen trioxide
b. potassium bromide
c. magnesium chloride

PP 2.2

With a compound of two nonmetals, a counting prefix indicates the number of that kind of atom. Remember, because the prefix *mono-* is generally not used, the absence of a prefix indicates one of that kind of atom.
a. PBr_3
b. NO_2
c. CCl_4

PP 2.3

a. Inspection shows that C and O are balanced. To balance H, 4H are needed on the left side.

4H are needed

$$CO + 2H_2 \rightarrow CH_4O$$
$$2 \times 2H = 4H$$

b. To balance phosphorus, 4P are needed on the right side.

4P are needed

$$P_4 + 6Cl_2 \rightarrow 4PCl_3$$
$$4 \times 3Cl = 12Cl$$

12Cl are needed

To balance Cl, 12Cl are needed on the left side.

Answers to Chapter Review

1. matter
2. solids
3. liquids
4. shape
5. gas
6. fluids
7. kinetic theory
8. mixtures
9. composition
10. compounds
11. heterogeneous
12. element
13. symbol
14. atom
15. molecule
16. N_2
17. H_2O
18. ions
19. cation
20. anion
21. -ide
22. definite composition
23. physical
24. chemical
25. reaction
26. reactants
27. products
28. property
29. chemical equation
30. conservation of matter
31. mass
32. balanced
33. variable
34. weakly
35. different
36. filtration
37. distillation
38. chromatography
39. charged
40. repel
41. attract
42. electrostatic
43. energy
44. heat
45. motion
46. exothermic
47. absorbed
48. endothermic
49. potential
50. kinetic

3

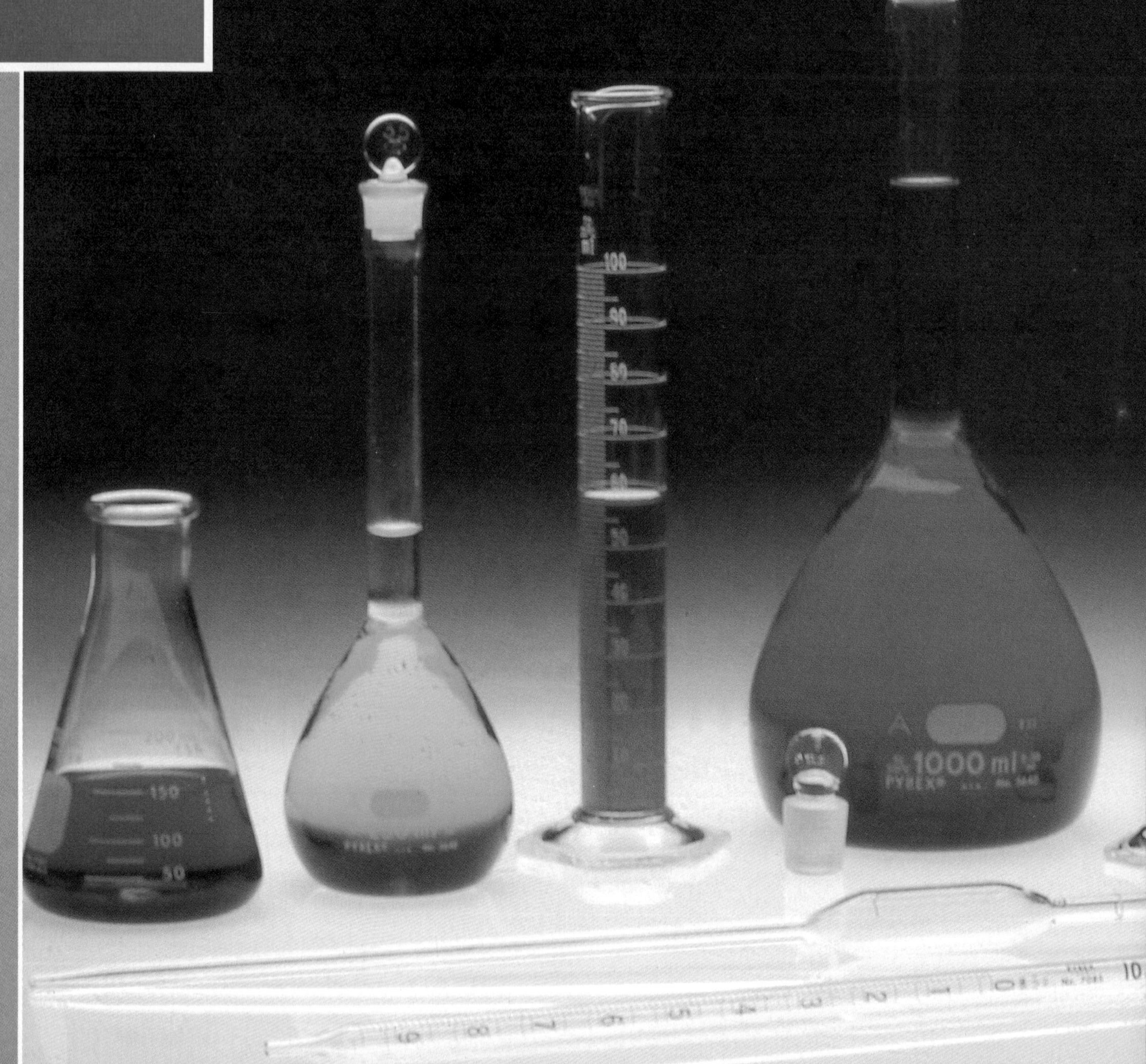

Measurements

Contents

Various kinds of glassware are used for measuring volumes of liquids. *Left to right:* 250 mL Erlenmeyer flask, 250 mL volumetric flask, 100 mL graduated cylinder, 1000 mL volumetric flask, 25 mL graduated cylinder, and 250 mL breaker. *In front:* 20 mL pipet and 10 mL pipet.

Gasoline is measured in gallons.

Metric measurements are used in most countries.

You Probably Know . . .

- We are constantly observing, and we describe what we see in various ways. For example, we could say that it is snowing hard, or we could tell others that snow is falling at a rate of one inch per hour—if we measured the snowfall.
- When we measure something, the kind of measurement determines how it is expressed: gasoline in gallons, soft drinks in liters, height in feet and inches, land in acres, oven temperature in degrees Fahrenheit. The numbers of measurements are meaningless without their units.
- The United States is gradually changing to the metric system of measurement. People in most other countries, as well as scientists everywhere, already use the metric system.

Why Should We Study Measurements?

Whether you are cooking, sewing, building with wood, or working in a laboratory, you make measurements. To bake a cake that has the desired texture and flavor requires careful measurements. Careful measurements are also necessary to make a dress that fits.

In a laboratory investigation, quantities must be measured carefully and the measured values recorded properly. Frequently, the measured values are used in calculations to obtain an answer to a problem being studied. Measurements of various kinds are an essential part of any scientific investigation.

3.1 Observations

Learning Objective

- Distinguish between quantitative and qualitative observations.

Problem solving is an important part of chemistry and other physical sciences. Earlier we noted that in solving problems, scientists usually follow a logical, systematic procedure called the scientific method. Making observations is one of the key elements in this procedure.

Quantitative observations result from measurements.

Observing is something we do every day. Some of our observations are qualitative, and some may be quantitative. We may observe that it is a cold morning (qualitative), or that the thermometer at 6:30 A.M. reads 10°F (quantitative). The air may feel humid (qualitative), or the relative humidity may have been measured as 95% (quantitative). The air over our city may smell polluted (qualitative), or the sulfur dioxide concentration may have been measured to be 0.5 ppm (quantitative).*

Quantitative observations are expressed numerically and result from making measurements of various kinds. This distinguishes them from **qualitative observations,** which are not numerical and refer to what, what kind, or to a quality observed. In preparing to be problem solvers, we must learn how to make measurements, how to properly record measured values, and how to express their units.

A **unit** is a defining quantity for a measurement. It tells us what scale was used for the measurement. You have learned many units such as seconds, minutes, and hours for time; inches, feet, and yards for length; and miles per hour for speed. Units that are used with scientific measurements must be memorized as well. A measurement is meaningless without its unit. For example, if a friend told you the distance to the stadium was 5.3, you would wonder 5.3 *what?* The stadium might be 5.3 miles or 5.3 kilometers away.

An outdoor thermometer tells us the temperature on a cold morning.

The Metric System and SI Units

Learning Objectives

- Give the SI units for mass, length, temperature, and time.
- Define the prefixes commonly used with metric units.
- Use prefixes to derive smaller or larger units from base units.

The metric system was first developed in the latter part of the eighteenth century by the French Academy of Sciences. Modifications in the metric system were made in 1960, and the new system is called the **International System of Units,** or **SI.** In spite of the new name, we often refer to these units as metric units. The seven SI base units are shown in Table 3.1. The first five base units are used in chemistry and are the ones of interest to us.

TABLE 3.1 SI Base Units of Measurement

Quantity	*Unit*	*Abbreviation*
length	meter	m
mass	kilogram	kg
temperature	kelvin	K
time	second	s
amount of substance	mole	mol
electric current	ampere	A
luminous intensity	candela	cd

*Sulfur dioxide is a major air pollutant. Most people can smell it at a concentration of 0.5 ppm (parts per million).

TABLE 3.2 Prefixes Used in the Metric System

Prefix	*Symbol*	*Multiple*	
tera	T	10^{12}	= 1 000 000 000 000
giga	G	10^{9}	= 1 000 000 000
mega	M	10^{6}	= 1 000 000
kilo	k	10^{3}	= 1 000
hecto	h	10^{2}	= 100
deka	da	10^{1}	= 10
		basic unit (e.g., gram, liter)	1
deci	d	10^{-1}	= 0.1
centi	c	10^{-2}	= 0.01
milli	m	10^{-3}	= 0.001
micro	μ	10^{-6}	= 0.000 001
nano	n	10^{-9}	= 0.000 000 001
pico	p	10^{-12}	= 0.000 000 000 001

The metric system of units is a decimal system. A quantity can be expressed in a larger or smaller unit simply by moving the decimal point. Thus, larger units differ from smaller units by multiples of 10. A prefix is added to a base unit (not always the SI base unit) to give a new unit that is larger or smaller than the base unit. Table 3.2 shows the names, symbols, and values of many of the prefixes.

Let's illustrate with some of the more common prefixes:

1 kilometer, km = 1000 meters

1 centimeter, cm = 0.01 meter

1 millimeter, mm = 0.001 meter

1 micrometer, μm = 0.000 001 meter (or 10^{-6} meter)

Example 3.1

Write the following times in seconds:

a. 1 nanosecond, ns **c.** 1 millisecond, ms
b. 1 kilosecond, ks **d.** 1 picosecond, ps ■

Solutions Refer to Table 3.2 for the definitions of the prefixes.

a. 0.000 000 001 s **c.** 0.001 s
b. 1000 s **d.** 0.000 000 000 001 s

The units derived with these prefixes may be very large or very small compared with the base unit. A large or small number can be simplified and conveniently expressed by using a power (or exponent) of 10 as described in the next section.

Exponential Notation

3.3

Learning Objectives

- Write decimal numbers in exponential (scientific) notation and vice versa.
- Carry out addition, subtraction, multiplication, and division of exponential numbers.

When you work with large or small units that are multiples of a base unit, it is convenient to write a number as a power of 10 rather than using common decimal notation. Writing a number as a power of 10 is **exponential notation.** Exponential notation is a concise expression of a large or small number. With this notation, we don't have to write many zeros. The exponent tells us how many times a base number, usually 10, is repeated as a factor. For example, 10^6 means 10 is repeated six times as a factor:

$$10^6 = 10 \times 10 \times 10 \times 10 \times 10 \times 10 = 1\ 000\ 000$$

From this example we can see that a positive exponent indicates a large number (>10). The exponent tells us the magnitude of the number and how many zeros must be included when the number is written in nonexponential form.

Remember: A positive exponent means a large number. A negative exponent means a small number.

Negative exponents are used to express small numbers. A negative exponent tells us how many times 1/10 is repeated as a factor. For example:

$$10^{-3} = 1/10 \times 1/10 \times 1/10 = 1/10^3 = 1/1000 = 0.001$$

A negative exponent indicates a small number (<1) and tells us the magnitude of the number. It tells us how many places the decimal point must be moved to the left when the number is written in decimal form. To convert to decimal notation:

$$\begin{aligned} 10^{-3} &= 1 \times 10^{-3} && \text{(multiply by 1)} \\ &= 0.001 && \text{(move the decimal point 3 places)} \end{aligned}$$

Note that multiplying 10^{-3} by 1 does not change the value of the number.

Example 3.2

Write the following in decimal or nonexponential form.

a. 10^{-5}

b. 10^4 ■

Solutions

a. 10^{-5} is equivalent to 1×10^{-5}. An exponent of -5 tells us to move the decimal point five places to the left. The answer is

$$0.000\ 01$$

b. 10^4 is equivalent to 1×10^4. An exponent of 4 tells us to move the decimal point four places to the right. The answer is

$$10\ 000$$

Example 3.3

Write the following in exponential notation.

a. 0.0001
b. 10 000 000 ■

Solutions Remember that a negative exponent means a small number and a positive exponent means a large number.

a. 10^{-4}
b. 10^{7}

Practice Problem 3.1 Express the following in decimal or nonexponential form.
a. 10^{9}
b. 10^{-6}
(Solutions to practice problems appear at the end of the chapter.)

Practice Problem 3.2 Write the following in exponential notation.
a. 100 000
b. 0.000 000 001

Scientific notation is a form of exponential notation. With scientific notation, the number is written in two parts: (1) a number between 1 and 10, and (2) a power of 10. The first part is multiplied by the power of 10. For example, the moon at its nearest distance is 221 600 miles from the earth. This distance can be expressed as 2.216×10^{5} miles.

Practice using scientific notation by working the following examples.

Example 3.4

Write each number in scientific notation.

a. 4 543 211
b. 5206
c. 0.0046
d. 0.0513 ■

Solutions The number of places you move the decimal point determines the exponent.

a. $4.543\ 211 \times 10^{6}$
b. 5.206×10^{3}
c. 4.6×10^{-3}
d. 5.13×10^{-2}

Example 3.5

Write each of the following numbers in decimal notation or nonexponential notation.

a. 2.1×10^{-3}

b. 7.3411×10^{4}

c. 4.78×10^{-5}

d. $1.633\ 32 \times 10^{5}$ ■

Solutions The exponent tells you the number of places to move the decimal point. Remember, a negative exponent means a small number (smaller than 1), and a positive exponent means a number larger than 10.

a. 0.0021

b. 73 411

c. 0.000 0478

d. 163 332

Practice Problem 3.3 Write each of the following in scientific notation.

a. 0.0358

b. 32 306

c. 12.01

Practice Problem 3.4 Express each of the following in decimal or nonexponential notation.

a. 1.56×10^{-3}

b. 7.93×10^{5}

c. 4.982×10^{3}

■ Addition and Subtraction of Exponential Numbers

To add or subtract exponential numbers by hand, they must be adjusted so that all exponents are the same. For example, to add $1.02 \times 10^{2} + 4.6751 \times 10^{4}$, rewrite one of the numbers so that the power of 10 agrees with the power of 10 of the other numbers and then add:

$$\begin{array}{r} .0102 \times 10^{4} \\ +\ 4.6751 \times 10^{4} \\ \hline 4.6853 \times 10^{4} \end{array}$$

Another way of adding the numbers is to rewrite them in nonexponential notation before adding. Your calculator automatically makes this adjustment when the numbers are entered (Appendix A).

$$\begin{array}{r} 102 \\ +\ 46\ 751 \\ \hline 46\ 853 \end{array}$$

Example 3.6

Add or subtract the following numbers by hand. Check your work with a calculator.

a. $1.042 \times 10^{-3} + 3.9 \times 10^{-5}$
b. $2.651 \times 10^{3} - 4.55 \times 10^{2}$
c. $8.36 \times 10^{2} + 3.8 \times 10^{1}$ ■

Solutions Adjust one of the numbers so its exponent agrees with the exponent of the other, then add or subtract.

a.
$$\begin{array}{r} 1.042 \times 10^{-3} \\ +\ 0.039 \times 10^{-3} \\ \hline 1.081 \times 10^{-3} \end{array} \quad \text{or} \quad 0.001\ 081$$

b.
$$\begin{array}{r} 2.651 \times 10^{3} \\ +\ 0.455 \times 10^{3} \\ \hline 2.196 \times 10^{3} \end{array} \quad \text{or} \quad 2196$$

c.
$$\begin{array}{r} 8.36 \times 10^{2} \\ +\ 0.38 \times 10^{2} \\ \hline 8.74 \times 10^{2} \end{array} \quad \text{or} \quad 874$$

■ Multiplication and Division of Exponential Numbers

To multiply, add exponents:
$10^x \times 10^y = 10^{x+y}$
To divide, subtract exponents:
$10^x/10^y = 10^{x-y}$

To multiply two exponential numbers, the exponents are added. To divide, the exponent of the divisor (denominator) is subtracted from the exponent of the dividend (numerator). For example:

$$10^3 \times 10^5 = 10^8$$

$$10^4 \times 10^{-2} = 10^2$$

$$10^5/10^2 = 10^3$$

$$10^{-3}/10^{-1} = 10^{-2} \qquad \text{Note: } -3 - (-1) = -3 + 1 = -2$$

Grouping together the nonexponentials and the exponentials before multiplying or dividing may avoid mistakes. An alternative is to enter the entire number into your calculator and let it do the work (Appendix A). However, mental multiplication and division of exponentials enables you to check the magnitude of the answer to a problem. These operations are illustrated in the following examples. The answers to these problems have been rounded off to three digits (Section 3.9).

Example 3.7

Multiply 3.57×10^2 times 1.43×10^3. ■

Solution Perform the operations on the nonexponentials and exponentials separately:

$$(3.57 \times 10^2)\,(1.43 \times 10^3) = 5.11 \times 10^5$$

$$3.57 \times 1.43 = 5.11$$

$$10^2 \times 10^3 = 10^5$$

Example 3.8

Divide 7.85×10^2 by 2.48×10^5. ■

Solution

$$\frac{(7.85 \times 10^2)}{(2.48 \times 10^5)} = 3.17 \times 10^{-3}$$

$7.85 \div 2.48 = 3.17$

$10^2 \div 10^5 = 10^{-3}$

Practice Problem 3.5 Carry out the following operations:

a. Multiply 1.31×10^{-3} times 2.73×10^4.

b. Divide 9.68×10^{-4} by 3.33×10^{-6}.

Mass and Weight

Learning Objective

- Distinguish between mass and weight.

Mass expresses the quantity of matter in a sample or an object. The mass of an object is the same regardless of where it is measured. **Weight** is the gravitational attractive force exerted upon an object. Weight is proportional to mass, but the weight of an object differs on the earth, the moon, and in space where it is far from any celestial body. If you weigh 100 lb on earth, you would weigh only 17 lb on the moon. Thus, the weight of an object may vary, but its mass does not (Figure 3.1).

Gravitational force is not uniform even on earth. It is greater at the north and south poles than at the equator because the earth's diameter is slightly greater at the equator. This causes small differences in the weight of an object depending upon where it is weighed. For example, a bag of potatoes that weighs 10 lb at the north pole would weigh about 9 lb 15 oz at the equator. However, differences in weights measured at various locations within the continental United States are insignificant for most purposes.

In spite of the distinction between weight and mass, we often use the words interchangeably. Sometimes this is confusing, but it is common practice and usually causes few difficulties.

Weight is usually measured with spring scales such as your bathroom scale or those in grocery stores for weighing fruits and vegetables, as in Figure 3.2. How far the spring is stretched depends on the gravitational force exerted on the fruit in the pan.

Mass is measured using a balance. Laboratory balances express mass measurements in grams (abbreviated g). A penny has a mass of about 3 g, so you can see that a gram is a small mass. You may have seen the quantity of corn flakes printed in grams on its box.

FIGURE 3.1 An object may be weightless in space, but its mass is the same in space as it is on earth.

a

Unit	Abbreviation	Gram Equivalent		
kilogram	kg	1000 g	or	10^3 g
gram	g	1 g		10^0 g
decigram	dg	0.1 g		10^{-1} g
centigram	cg	0.01 g		10^{-2} g
milligram	mg	0.001 g		10^{-3} g
microgram	μg	0.000 001 g		10^{-6} g
nanogram	ng	0.000 000 001 g		10^{-9} g

TABLE 3.3 Units of Mass

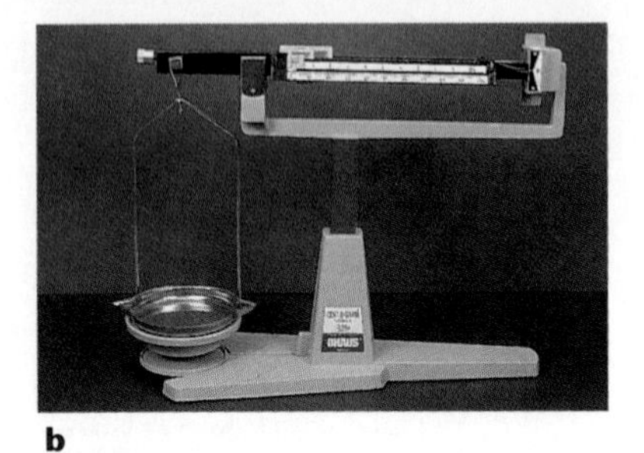

b

FIGURE 3.2 (a) A scale for weighing fruits and vegetables. (b) A laboratory centigram balance.

The SI unit of mass is the *kilogram,* kg (Table 3.1). One kilogram is defined as the mass of a platinum block kept at the International Bureau of Weights and Measures at Sevres, France. One kilogram is equal to 2.205 lb. One pound is equivalent to 453.6 g.

Table 3.3 shows the use of prefixes to express units of mass larger and smaller than 1 g. Using the information in the table, you can see that

$$1 \text{ kg} = 1000 \text{ g or } 10^3 \text{ g}$$

$$1 \text{ g} = 1000 \text{ mg or } 10^3 \text{ mg}$$

$$1 \text{ g} = 1\ 000\ 000\ \mu\text{g or } 10^6\ \mu\text{g}$$

3.5 Length, Area, and Volume

Learning Objective

- Give the SI and other common units of length, area, and volume.

The SI unit of length is the *meter,* m (Table 3.1). A meter is defined as the distance light travels in a vacuum in 1/299 792 458 second. A very precise standard for the unit of length is needed to calculate accurate distances between points on earth and between locations on earth and locations in space.

One meter is equivalent to 1.09 yard and is a convenient measure of length for everyday use. For example, most people are between 1.5 and 2 m tall. Many cars are between 4 and 5 m long. However, most laboratory samples are much smaller than people and cars and have dimensions around 0.01 to 0.1 m. A convenient unit of length for laboratory work is the *centimeter,* cm. The width of a person's little fingernail is about 1 cm. Comparing metric and American units of length:

1 in. = 2.54 cm (exactly)

$$1 \text{ in.} = 2.54 \text{ cm}$$

A centimeter ruler is shown in Figure 3.3.

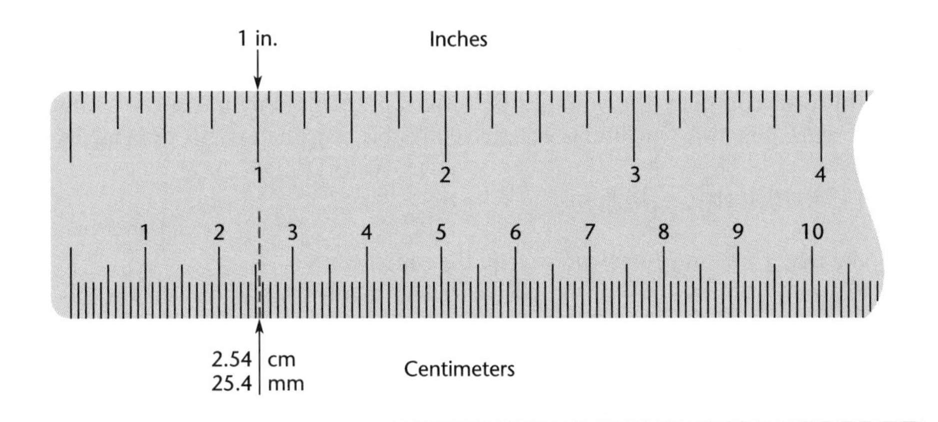

FIGURE 3.3 A centimeter ruler. A comparison of metric and American length measurement.

Area

Area is two dimensional (m^2, cm^2, and so on).

Area is a measure of a surface bounded by a set of lines. The area (A) of the floor in a rectangular room is its length (L) multiplied by its width (W). A room that measures 7.0 m by 6.0 m has an area of 42 m^2 (read as "42 square meters").

$$A = L \times W = 7.0 \text{ m} \times 6.0 \text{ m} = 42 \text{ m}^2$$

Because two dimensions (length and width for a rectangle) are needed in the calculation, area is considered to be two dimensional and is expressed in "square units." Common American units are in^2, ft^2, and yd^2. Using the metric system, we express area in m^2, cm^2, and so on.

Volume

Volume is the amount of space occupied by matter. The volume (V) of a box is its length (L) times its width (W) times its height (H). A box that measures 2.5 ft by 2.5 ft and is 3.0 ft high has a volume of 19 ft^3 (read as "19 cubic feet").

$$V = L \times W \times H = 2.5 \text{ ft} \times 2.5 \text{ ft} \times 3.0 \text{ ft}$$
$$= 18.75 \text{ ft}^3$$
$$= 19 \text{ ft}^3 \quad \text{(rounded off to two digits)}$$

Because three dimensions are needed in the calculation, volume is said to be three dimensional.

The SI unit of volume is the *cubic meter,* m^3. This is a large volume, equivalent to a cube 1 m on a side. It is useful for measuring volumes of air moving in a ventilation system or for measuring quantities of dirt removed from an excavation site. However, the *cubic centimeter,* cm^3, is more practical for expressing volumes of samples handled in the laboratory. One teaspoon is about 5 cm^3.

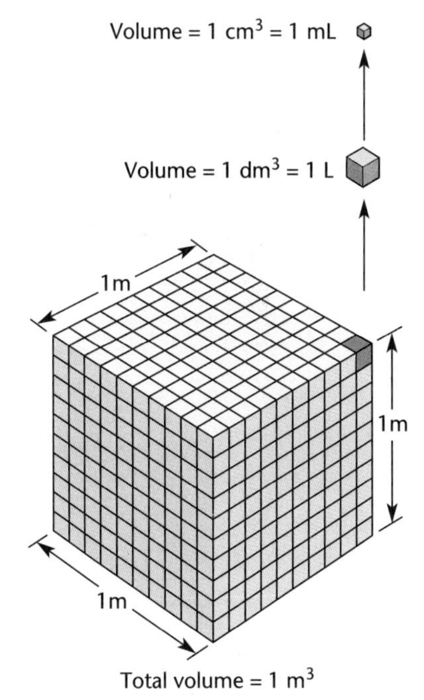

Volume is three dimensional. A volume of 1 m^3 (the large cube) is equivalent to 1 000 000 cm^3 (small cube).

A 2-liter soft drink bottle and a bottle of liquid medicine

Most people are familiar with the *liter,* L, for measuring the volumes of consumer products such as soft drinks and gasoline. One liter is equivalent to 1.0567 quart. In most countries the volumes of liquids are commonly measured using liters or *milliliters,* mL. One liter is exactly 1000 cubic centimeters. So, remember:

$$1\ \text{L} = 1000\ \text{cm}^3 = 1000\ \text{mL}$$

Also note that 1 mL and 1 cm^3 are exactly the same volume.

Figure 3.4 shows some of the glassware commonly used in measuring volumes of liquids in the laboratory.

3.6 Density

Learning Objectives

- Write the defining equation for density.
- Compare the densities of gases with the densities of liquids and solids.

Have you ever picked up a bag of cement? Did you notice it was heavier than a pillow of about the same size? You know from experience that different substances with the same volume have different masses. **Density** is the ratio of the mass of a substance to its volume. It can be expressed in equation form as follows:

$$\text{density} = \frac{\text{mass}}{\text{volume}}$$

Using letter abbreviations for the words gives an algebraic equation:

$$D = \frac{m}{V}$$

This is called a *defining equation** because it is a definition of density in equation form. When the density is given for a solid or liquid, the mass is usually expressed in grams and the volume in milliliters or cubic centimeters.

$$D = \frac{m}{V} = \frac{\text{g}}{\text{mL}} = \frac{\text{g}}{\text{cm}^3}$$

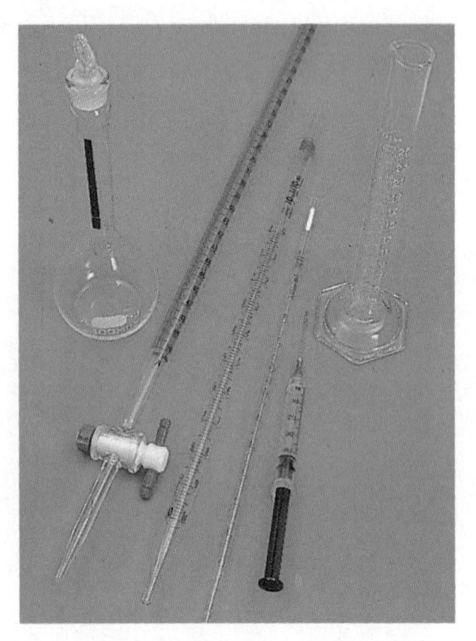

FIGURE 3.4 Laboratory glassware for measuring volumes. *Left to right:* volumetric flask, buret, graduated pipet, volumetric pipet, syringe, and graduated cylinder.

Because the volume of a substance changes when its temperature changes, density is temperature dependent. For example, the volume of exactly one gram of

*This term is used by E. I. Peters, *Introduction to Chemical Principles,* Fifth Ed. (Philadelphia: Saunders College Publishing, 1990), p. 55.

water at 4°C is 1.0000 mL, resulting in a density of 1.0000 g/mL. As the temperature is raised, the volume of one gram of water increases, resulting in a lower density. The density of water is one of its characteristic physical properties. (See Chapter 13.)

The density of a gas is much less than the density of a liquid or a solid. For example, the density of air at 20°C is about 0.0012 g/mL or 1.2 g/L. The density of water is about 1000 times greater.

The densities of solids vary widely. Some solids, such as wood, are less dense than water. Water is used as a reference liquid when comparing densities of various substances. Whether a substance sinks or floats on water classifies it as more or less dense than water. For example, sand and other mineral substances are more dense than water. The density of gold (Table 3.4) is among the highest of all substances and is much greater than the densities of the sand and mud commonly found with gold in stream beds. Thus, in panning for gold, sand and mud are washed away from the more dense gold particles left in the pan.

Because the densities of solids and liquids are much greater than the densities of gases, they are referred to as condensed phases. The densities of some common substances are given in Table 3.4. The density of water is included for comparison.

Density calculations are covered in Chapter 4 where problem-solving strategies are considered.

TABLE 3.4 Densities of Some Common Substances

	Substance	*Density at 20°C, 1 atm pressure*
Gaseous	hydrogen (least dense gas)	0.0840 g/L
	helium	0.168
	methane	0.672
	air	1.2
	carbon dioxide	1.85
	chlorine	2.95
Liquid	ethyl alcohol	0.789 g/cm^3
	kerosene	0.80
	soybean oil	0.924
	ethylene glycol	1.116
	water	**0.998**
	trichloroethylene (TCE)	1.456
	sulfuric acid	1.84
	mercury	13.55
Solid	sugar	1.59 g/cm^3
	salt	2.16
	aluminum	2.70
	lead	11.34
	gold	19.3
	osmium (element of greatest density)	22.48

3.7 Heat and Temperature

Learning Objectives

- Distinguish between heat and temperature.
- Give the SI units of heat and temperature.
- Write equations for the conversion of Celsius to Kelvin temperatures, for the conversion of Fahrenheit temperatures to Celsius, and vice versa.
- Write the defining equation for specific heat.
- Give the units of specific heat.

In the previous chapter (Section 2.8) you learned that heat is a form of kinetic energy of small particles of matter. Heat spontaneously flows from a hot object such as the burner of a stove to a cooler object such as a pan on the burner. The temperatures of the objects determine the direction of the flow of heat. It is important to remember that heat is a form of energy and is quite different from temperature.

Now think for a moment about your experience with heat. You probably know that more heat is needed to boil a quart of water than a cup of water. This also means that when a quart of boiling water cools, more heat is released than from a cup of boiling water. In other words, the heat absorbed or released by water, or any substance, depends upon both the temperature change and the mass of the sample.

Heat is related to the total *kinetic energy of the particles in a sample.*

1 cal = 4.184 J

Imagine that you could determine the kinetic energy of each of the molecules in a quart of water. The sum of these would give the total kinetic energy of the sample. A quart of boiling water contains more molecules than a cup of boiling water and therefore has a higher total kinetic energy. Upon cooling to room temperature, the quart of boiling water loses more kinetic energy in the form of heat.

Scientists express heat in calories or joules. A **calorie,** cal, is the quantity of energy required to raise the temperature of one gram of water one Celsius degree. One calorie is equivalent to 4.184 **joules,** J.* Because a calorie and a joule (SI energy unit) are small quantities of energy, kcal and kJ are often used.

$$1 \text{ kcal} = 1000 \text{ cal} \qquad \text{and} \qquad 1 \text{ kJ} = 1000 \text{ J}$$

The Calorie, Cal, is used to express the energy value of food.

1 Cal = 1000 cal = 1 kcal

The Calorie, Cal (capital "C"), used by nutritionists tell us the energy-producing value of food and is actually 1 kcal. This is sometimes called the "large calorie" and should not be confused with the calorie (small "c") used in scientific work.

$$1 \text{ Cal} = 1000 \text{ cal} = 1 \text{ kcal}$$

Temperature depends on the average *kinetic energy of the particles in a sample.*

Temperature is a measure of intensity of heat. Temperature depends on the *average* kinetic energy of the particles making up a sample. A quart of water and a cup of water at the same temperature have molecules with the same average kinetic energy. Be careful to note the distinction between temperature and heat. Although at the same temperature, more heat is available from a quart than from a cup of boiling water.

---

*One *joule* is defined as the kinetic energy of a 2-kg mass moving with a velocity of 1 m/s: $KE = \frac{1}{2}mv^2 = \frac{1}{2}(2 \text{ kg})(1 \text{ m/s})^2 = 1 \text{ kg} \cdot \text{m}^2/\text{s}^2$

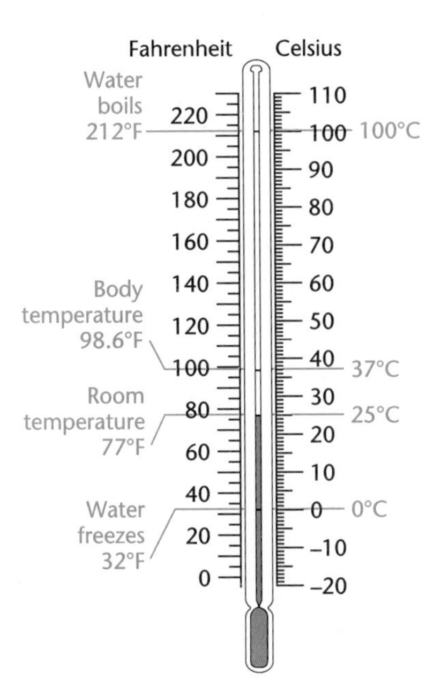

FIGURE 3.5 A comparison of Fahrenheit and Celsius temperatures. The freezing point (32°F, 0°C) and the boiling point (212°F, 100°C) of water are reference temperatures.

If you have grown up in the United States, you are probably used to measuring temperatures in degrees Fahrenheit. However, most people in the world, including all scientists, measure temperatures in degrees Celsius. Figure 3.5 shows the relationship between Fahrenheit and Celsius temperatures.

The Fahrenheit and Celsius scales use the freezing point and boiling point of water as reference points. The freezing point of water is assigned a value of 0°C, and the boiling point is 100°C. On the Fahrenheit scale these temperatures are 32°F and 212°F. Comparison of these scales leads to the following equation, which can be used to convert from one scale to the other:

$$T_F = 1.8T_C + 32$$

where T_F is the Fahrenheit temperature and T_C is the Celsius temperature.

The SI unit of temperature is the **kelvin,** K. The Kelvin scale is based on **absolute zero,** theoretically the lowest possible temperature. Absolute zero is equivalent to -273.15°C. In our calculations, we will generally round off to the nearest degree. Therefore, the two scales are related by the following equation:

$$T_K = T_C + 273$$

0 K = −273°C

where T_K is the Kelvin temperature.

Note that the degree sign is not used with Kelvin temperatures. We will not use Kelvin temperatures until we study gases (Chapter 12).

Examples of conversions from one temperature scale to another are found in Chapter 4 where problem-solving strategies are considered.

CHEMICAL WORLD

Why Does Oatmeal Stick to Your Ribs?

The adage "Oatmeal sticks to your ribs" simply means that a breakfast that includes a bowl of hot oatmeal satisfies your appetite longer than a presweetened cold cereal or a doughnut. Why?

Oatmeal is a complex carbohydrate, and complex carbohydrates are broken down by your body more slowly than simple carbohydrates such as the sugar in ready-to-eat presweetened cereals (more than 50% of the total content of some of these). Simple carbohydrates satisfy your hunger immediately by raising your blood sugar, but your body digests and burns them so quickly that your blood sugar falls soon after you eat them. Then you're hungry again. Because your body digests complex carbohydrates more slowly, your blood sugar stays elevated for longer, and you retain a feeling of being full for a longer time after eating oatmeal.

Another reason that oatmeal sticks to your ribs has to do with specific heat. Consider what happens when you eat a bowl of cold cereal with cold milk. To keep this illustration simple, we'll assume that the specific heat of skim milk is about the same as water, that is, 1.000 cal/g·C° (Table 3.5). This is a pretty fair assumption because skim milk is mostly water. We'll assume that you have 500 g (5.00×10^2 g) of cold milk (one cup on your cereal and another cup to drink). If your body temperature is 37.0°C and the milk is 4.0°C, your body must burn food calories to produce enough heat to raise the temperature of the 500 g of milk from 4.0°C to 37.0°C (Think about how long it takes to heat a cup of milk on the stove!). We can calculate the number of calories your body must burn by solving the specific heat equation,

$$SH = \frac{q}{m \times \Delta T}$$

and for q(cal),

$$q = SH \times m \times \Delta T$$

Substituting our values for specific heat, mass, and temperature change into this second equation, we have

$$q = (1.000\ \text{cal}/\cancel{\text{g}}\cdot\cancel{\text{C}^\circ})\ (500\ \cancel{\text{g}})\ (33.0\cancel{\text{C}^\circ})$$
$$= 16\ 500\ \text{cal}$$

Thus, your body must burn 16 500 calories' worth of your breakfast just to warm the milk you consumed with your cereal.* If you had eaten hot oatmeal, you would have conserved this energy, and the energy provided by the hot oatmeal would have taken you that much farther into your day.

So, if you're one of the many people who get hungry way before lunch time, try a bowl of hot oatmeal for breakfast. Prove to yourself that the combination of the cereal's warmth and its complex carbohydrates really do stick to your ribs.

---

*Remember that 1000 calories (cal) is 1 food Calorie (Cal).

TABLE 3.5 Specific Heats of Some Common Substances

	Specific Heat	
Substance	*J/g·C°*	*cal/g·C°*
water	4.184	1.000
ice	2.06	0.492
aluminum	0.900	0.215
copper	0.385	0.0921
iron	0.473	0.113
mercury	0.138	0.0330
sand	0.80	0.19

Specific Heat

The **specific heat** of a substance is the quantity of heat required to raise the temperature of one gram of the substance one Celsius degree. This definition may be expressed in equation form:

$$\text{specific heat} = \frac{\text{calorie}}{\text{gram} \cdot \text{degree}} = \frac{\text{heat}}{\text{mass} \times \text{temperature change}}$$

By using the symbols SH (specific heat), q (heat), m (mass), and ΔT (temperature change) and substituting, we get a defining equation for specific heat:

$$SH = \frac{q}{m \times \Delta T} \quad \text{or} \quad q = SH \times m \times \Delta T$$

To understand the meaning of specific heat, think of times that you worked in the kitchen. You may have noticed that an empty metal pan gets hot very quickly when placed on a burner. However, water equal in mass to the pan takes much longer to get hot. Because metals have much smaller specific heats than water, less heat is required for a metal object of the same mass as the water to reach a given temperature. In other words, when the specific heat of a substance is small, a small quantity of heat is needed to raise the temperature of a given mass of the substance. Table 3.5 gives the specific heats of some common substances. Calculations involving specific heats are discussed in Chapter 4.

Uncertainty in Measurements

3.8

Learning Objectives

- Identify exact numbers.
- Distinguish between accuracy and precision.
- Identify the number of significant figures in a measurement.

As a child you learned that 12 make a dozen, 10 pennies make a dime, and a minute is 60 seconds. These numbers are definitions of quantities. They are **exact numbers.** Numbers obtained by counting are also exact numbers. For example, there may be

Exact numbers: Definitions of quantities, or small numbers obtained by counting.

FIGURE 3.6 An illustration of precision and accuracy. The group of darts on target (a) was highly precise and accurate, (b) shows shots that are precise but not accurate, and (c) was neither precise nor accurate.

five spoons on the table. Your car has four wheels. These are all examples of exact numbers and have no uncertainty.

You also may have learned that weighing yourself on a bathroom scale does not give your exact weight. You may not observe the same weight when you weigh yourself several times in succession. Numbers from measurements may agree fairly closely, but they are not exact. They are always uncertain to some degree, and the repeatability or precision depends on the limitations of the measuring instrument and how carefully a person uses the instrument. **Precision** refers to how closely repeated measurements agree.

Precision: The agreement among repeated measurements.

Accuracy: How close a measured value is to a true value.

Repeatable measurements are not necessarily accurate. **Accuracy** refers to how close a measured value is to the true value. Suppose a person was weighed five times using a bathroom scale to give weights of 126, 127, 127, 128, and 127 pounds. The average is 127 pounds, and because the weights agree closely the uncertainty is small. However, if the person's actual weight is 133 pounds, the weights obtained using the bathroom scale are not very accurate. To determine the accuracy of a measurement, the instrument used must be compared with a standard instrument and adjusted to agree with the standard. This process is called **calibration** of the instrument. It must be remembered that the accuracy of a measurement determined with an uncalibrated instrument is unknown.

The distinction between precision and accuracy is illustrated in Figure 3.6.

■ Significant Figures

Measurements usually involve reading a scale. For example, to measure the length of a small object you use a ruler (Figure 3.7). The length of the object often falls between marks on the scale. In everyday life, people simply read the nearest mark on the scale. Using the bottom ruler in the figure, they would say the length of the object is 9.6 cm. However, in scientific work, the length of the object would be measured by estimating between the marks on the scale. Because the edge of the object is estimated to be seven-tenths of the way between the marks, the length is read as 9.57 cm. Because the last digit is estimated, the scale might also be read one time as 9.58 cm and another time as 9.56 cm. Note that the estimated digit is uncertain. In scientific work, *the uncertain or estimated digit should always be recorded.* In a given measurement, all the digits that are known plus the estimated (uncertain) digit are called **significant figures.**

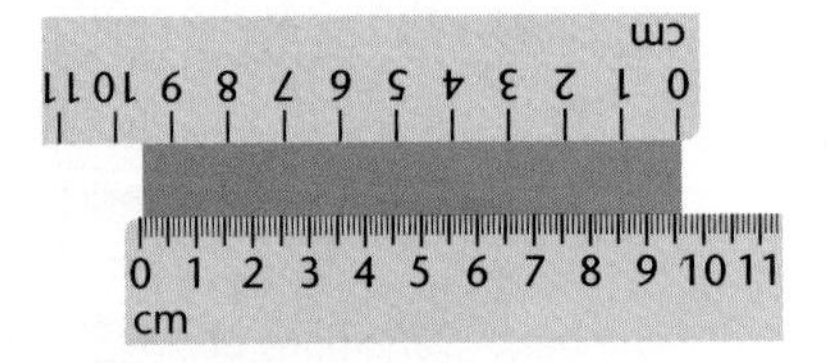

FIGURE 3.7 The last digit recorded is the estimated digit.

Note that the number of significant figures recorded for a measurement depends on the instrument scale. Using the top ruler in Figure 3.7 and recording the known and estimated digits gives us 9.6 cm (two significant figures). Using the bottom ruler, and again recording the known and estimated digits, gives us 9.57 cm (three significant figures). A length of 9.6 cm was measured with a ruler subdivided into centimeters, and a length of 9.57 cm was measured with a ruler subdivided into tenths of centimeters. Regardless of the instrument, the last digit recorded is always the estimated digit.

0 1 2 3 4 5 6 7 8 9 10 11
cm

The length of the rectangle is 10.00 cm.

Remember that the estimated digit must always be recorded to properly express the precision of the measurement. Furthermore, when a measurement falls on a mark on the scale, a zero must be recorded as the estimated digit. For example, the length of the rectangle as measured with the centimeter ruler is 10.00 cm. The zeros after the decimal point are necessary to properly express the precision of the measurement.

At this point, it should be emphasized that a measurement includes three pieces of information: quantity, units, and precision. The meaning of a measurement is not totally clear unless all three, including precision, are expressed correctly.

A measurement includes quantity, units, *and* precision.

It is important to record measurements carefully, using the proper number of significant figures so that whoever reads the data will recognize the uncertainty in the measurement. You will recognize the importance of this when you are reading measurements recorded by others and are using these numbers in calculations.

Counting Significant Figures

As you prepare to use measured quantities in calculations, you must learn to count the number of significant figures in a number resulting from someone else's measurement. Counting digits is not difficult, but it is easy to make mistakes. Here are some things to remember.

When counting significant figures, begin counting with the first nonzero digit on the left and end with the last digit written (estimated digit). Look at the data in Table 3.6 and see if you agree with the number of significant figures for each measurement. Pay particular attention to the zeros in the numbers. Use the following rules to determine which zeros should be counted:

Counting significant figures: Begin counting with the first nonzero digit on the left and end with the estimated digit.

1. Zeros between nonzero numbers are significant. For example, the zeros in 3.002 are significant, and this number has four significant figures.

TABLE 3.6 Winning Track Times

Race, meters	*Time, seconds*	*Number of Significant Figures*
40	5.6	2
100	12.3	3
200	26.0	3
400	60.2	3
800	130.0	4

What was the winning time?

2. Zeros at the end of a number to the right of a decimal point (trailing zeros) are significant. For example, the zero in 4.30 is significant, and this number has three significant figures.
3. Zeros at the left of nonzero numbers (leading zeros) are not significant. For example, 0.024 has two significant figures. The zeros simply place the decimal point. They are not shown when the number is written in scientific notation:

$$0.024 = 2.4 \times 10^{-2}$$

In other words, *zeros that simply locate a decimal point are not counted as significant figures.*

The track times (Table 3.6) illustrate these rules. The winning time for the 200-meter race has three significant figures. The zero is the estimated digit and is counted. The time for the 400-meter race has three significant figures. A zero between two nonzero digits is counted. The time for the 800-meter race has four significant figures. The last zero is the estimated digit and both zeros are counted.

When a measurement is written with no decimal point, the number of significant figures may be ambiguous when the number ends with zeros. For example, if the distance between two cities was reported to be 100 miles, we would not be sure how many significant figures are intended. If the distance was actually estimated to the nearest mile, then the measured distance should be written showing three significant figures. We can show the proper number of significant figures using scientific notation and writing the distance as 1.00×10^2 miles. An alternative and simpler notation is to include the decimal point in the number to indicate that the trailing zero is significant:

100. miles (three significant figures)

Significant Figures in Measurements

1. The number of significant figures in a measurement is the number of known digits plus the estimated (uncertain) digit. The estimated digit, even a zero, must be recorded to show the uncertainty in the measurement.
2. Count significant figures from left to right beginning with the first nonzero digit and ending with the estimated digit.
3. The rules for zeros are as follows:
   a. Zeros between nonzero numbers are significant.
   b. Zeros at the end of a number to the right of a decimal point (trailing zeros) are significant.
   c. Zeros at the left of nonzero numbers (leading zeros) are *not* significant.
4. When no decimal point is written, zeros to the right of the last nonzero digit may or may not be significant. The number can be written in scientific notation to show the number of significant figures, or a decimal point can be added to indicate that the zeros are significant. Otherwise, we do not count these zeros as significant figures.

Example 3.9

Count the significant figures in each of the following:

a. 3.5 m
b. 10 people
c. 0.0250 g
d. 25.20 s
e. 0.10 mile
f. 37.8°C
g. 730 ft
h. 730. ft ■

Solutions Remember that numbers obtained by counting or that define a quantity are exact numbers and have no uncertainty.

a. two
b. an exact number
c. three
d. four
e. two
f. three
g. two
h. three

Practice Problem 3.6 How many significant figures are shown in each of the following quantities?

a. 120.0 mile
b. 21 bottles
c. 3.25×10^{-3} cm
d. 6.02×10^{23} atoms
e. 0.0089 g
f. 20.5 hr

Significant Figures in Calculations

3.9

Learning Objective

- When performing calculations using measured values, round off the answer to the correct number of significant figures.

When using measured quantities in calculations, the answer can be no more precise than the least-precise quantity. Therefore, the results of calculations should always be rounded off to show the correct number of significant figures.

■ Rounding Off Numbers

When measured quantities are used in calculations, most calculators show more digits for an answer than are significant. When this occurs, you must *round off* the answer.

We will use the following rules for rounding off numbers:*

1. If the first digit to be dropped is less than 5, the preceding digit is left unchanged.

*Some people favor following a third rule when the first digit to be dropped is exactly 5. We will use only the two rules stated here.

2. If the first digit to be dropped is 5 or greater, the preceding digit is increased by 1.

For example, to round off 69.374 to four significant figures, the first digit to be dropped (4) is less than 5, so the preceding digit (7) is left unchanged to give 69.37. To round off this number to three significant figures, the first digit to be dropped (7) is greater than 5, so the preceding digit is increased by 1 to give 69.4.

Example 3.10

Round off each of the following numbers to three significant figures.

a. 88.016
b. 334.3
c. 0.076 65
d. 1.007 35
e. 14.15
f. 10.55 ■

Solutions Give attention to the *first* digit to be dropped. In **a** and **b,** the digit to be dropped is less than 5, and the preceding digit is not changed when the number is rounded to three significant figures. In **c, d, e,** and **f,** the digit to be dropped is 5 or greater, and the preceding digit is increased by 1 when the number is rounded.

a. 88.0
b. 334
c. 0.0767
d. 1.01
e. 14.2
f. 10.6

Example 3.11

Round off each of the following numbers to four significant figures.

a. 10 234
b. 11.235
c. 5.7867×10^5
d. 1.0670×10^{-3}
e. 0.245 116 ■

Solutions Scientific notation can be used when expressing significant figures.

a. 1.023×10^4
b. 11.24
c. 5.787×10^5
d. 1.067×10^{-3}
e. 0.2451

Most of the calculations in this book will involve multiplication, division, addition, or subtraction or a combination of these operations. Two rules tell us how to write the answers to problems using the proper number of significant figures.

Multiplication and Division Calculations

The answer is rounded off to agree with the quantity that has the least number of significant figures. For example, the area (A) of the floor in a rectangular room can be found by multiplying the length (L) times the width (W) of the room:

$$A = L \times W$$

If the floor measured 8.5 m by 10.2 m, multiplying using a calculator gives

$$A = (10.2\text{ m})\,(8.5\text{ m}) = 86.7\text{ m}^2$$
$$= 87\text{ m}^2 \qquad \text{(two significant figures)}$$

The answer is rounded off to agree with the number of significant figures in the width, 8.5 m (two significant figures).

Exact numbers used in a calculation do not limit the number of significant figures in the answer. For example, the total of the lengths of three boards each 2.25 m long is:

$$3 \times 2.25 = 6.75\text{ m} \qquad \text{(three significant figures)}$$

The number of boards (3) is an exact number and does not limit the number of significant figures in the total length.

Example 3.12

Express the answers to the following problems using the proper number of significant figures.

a. $11.6 \times 3.4 =$
b. $213.0 \times 0.12 =$
c. $0.4237 \times 13.0 =$
d. $879/43.70 =$
e. $40.1/1.2 =$
f. $0.0210/2.100 =$ ■

Solutions Round off the answer to agree with the quantity that has the least number of significant figures.

a. 39
b. 26
c. 5.51
d. 20.1
e. 33
f. 0.0100

Addition and Subtraction Calculations

The answer to an addition or subtraction calculation should be rounded off to agree with the last decimal place of the number having the least number of decimal places.

For example, add the lengths of three boards measuring 4.2 m, 3.25 m, and 2 m.

$$\begin{array}{l} 4.2\ \text{m} \\ 3.25\ \text{m} \\ \underline{2\quad\ \ \text{m}} \\ 9.45\ \text{m} \end{array}$$

The answer is rounded off to 9 m to agree with the place value of "2 m" (one's place).

Sometimes significant figures are lost in subtraction. For example, when performing an experiment, if the temperature rose from 22.6°C to 25.3°C, the rise in temperature would be

$$\begin{array}{r} 25.3°\text{C} \\ \underline{-\ 22.6°\text{C}} \\ 2.7°\text{C} \end{array}$$

The temperatures have *three* significant figures. The difference has only *two* significant figures.

Example 3.13

Express the answers to the following problems using the proper number of significant figures.

a. 23.467 − 2.6 =
b. 0.352 + 0.02 =
c. 3.78 − 3.2 =
d. (1.65) (100.2 − 23) =
e. (2.010 − 1.3)/(8.73) =
f. 1.6 + 0.345 + 10.2 + 0.0510 = ■

Solutions When adding or subtracting, round the answer to agree with the last decimal place of the number having the least number of decimal places.

a. 20.9
b. 0.37
c. 0.6
d. 130
e. 0.08
f. 12.2

In part d, note that the result of the subtraction has only two significant figures, limiting the number expressed in the answer. In part e, the result of the subtraction has only one significant figure.

Practice Problem 3.7 Express the answers to the following problems using the proper number of significant figures.

a. 12.3 (28.4 − 22.6) =
b. 5.62 (1.206 + 0.7) =
c. (2.43) (1.53)/(5.71 + 0.082) =

■ Chapter Review

Directions: From the choices listed at the end of the Chapter Review, choose the word or term that best fits in each blank. Use each choice only once. The answers appear at the end of the chapter.

Sections 3.1, 3.2 One of the key elements in problem solving is making __________ (1). Measurements result in __________ (2) observations or numerical data. Scientific measurements are expressed using metric or __________ (3) units. A __________ (4) is a defining quantity for a measurement. Base units are modified by using prefixes to give units that are larger or smaller than a base unit.

Section 3.3 __________ (5) notation is used when we want to write either very large or very small numbers and is helpful in defining the prefixes used in the metric system.

Section 3.4 The quantity of matter in a sample is indicated by its __________ (6). The weight of an object is the __________ (7) force acting on the object. __________ (8) is proportional to mass but varies depending on the gravitational attraction of the celestial body where it is found. Mass is measured using a __________ (9). The SI unit of mass is the __________ (10).

Section 3.5 The SI unit of length is the __________ (11), but the __________ (12) is more convenient for laboratory measurements. __________ (13) is two dimensional and is expressed in "square units." The amount of space occupied by matter is its __________

(14), which is __________ (15) dimensional. It is expressed in units such as m^3 and __________ (16). The __________ (17), equal to 1.0567 qt, is used for measuring volumes of consumer products such as soft drinks and gasoline. One liter is exactly 1000 cm^3; 1 cm^3 is equivalent to 1 __________ (18).

Section 3.6 The ratio of the mass of a substance to the volume it occupies is its __________ (19). Density is temperature dependent. The density of a __________ (20) is much less than the density of a liquid or a solid. The least-dense gas is __________ (21). The defining equation for density is $D =$ __________ (22). This equation can be used to calculate either density, mass, or volume if the other two quantities are given.

Section 3.7 Heat is expressed in calories or __________ (23). One __________ (24) is the quantity of heat required to raise the temperature of one gram of __________ (25) one Celsius degree. One calorie = 4.184 joules. The Calorie, used by nutritionists, tells us the energy-producing value of food. One Calorie = __________ (26) cal.

__________ (27) is a measure of intensity of heat and depends upon the average __________ (28) energy of the particles making up a sample. Most people in the world, including all scientists, measure temperatures in degrees __________ (29). The SI unit of temperature is the __________ (30). __________ (31) zero is equivalent to $-273°C$.

The quantity of heat required to raise the temperature of one __________ (32) of a substance one Celsius degree is its __________ __________ (33). The defining equation for specific heat is $SH =$ __________ (34). This equation can be used to calculate specific heat, __________ (35), mass, or a temperature change when the other three quantities are given.

Section 3.8 Numbers that have no uncertainty, and differ from measured values in that respect, are called __________ (36) numbers. The degree of uncertainty in a measured value depends upon the limitations of the measuring instrument and how carefully a person uses it.

__________ (37) refers to how close a measured value is to the true value. The accuracy of a measurement is unknown unless the instrument has been ______ (38).

In a measured value, the digits that are known plus the first one that is estimated or uncertain are called __________ __________ (39). When counting significant figures, begin counting with the first __________ (40) digit and end with the estimated digit.

Section 3.9 Many calculators show more digits for an answer than are significant. When this occurs, you must __________ __________ (41) the answer to show the proper number of significant figures. In a multiplication and division problem, the answer is rounded off to agree with the quantity having the __________ (42) number of significant figures. In addition and subtraction calculations, the answer is rounded off to agree with the last __________ __________ (43) of the number having the least number of decimal places.

Choices: absolute, accuracy, area, balance, calibrated, calorie, Celsius, centimeter, cm^3, decimal place, density, exact, exponential, gas, gram, gravitational, heat, hydrogen, joules, kelvin, kilogram, kinetic, least, liter, m/V, mass, meter, mL, nonzero, observations, 1000, quantitative, round off, SI, significant figures, specific heat, temperature, three, unit, volume, water, weight, $q/(m \times \Delta T)$

Study Objectives

After studying Chapter 3, you should be able to:

Section 3.1

1. Distinguish between quantitative and qualitative observations.

Section 3.2

2. Give the SI units for mass, length, temperature, and time.
3. Define the prefixes commonly used with metric units.
4. Use prefixes to derive smaller or larger units from base units.

Section 3.3

5. Write decimal numbers in exponential (scientific) notation and vice versa.
6. Carry out addition, subtraction, multiplication, and division of exponential numbers.

Section 3.4

7. Distinguish between mass and weight.

Section 3.5

8. Give the SI and other common units of length, area, and volume.

Section 3.6

9. Write the defining equation for density.
10. Compare the densities of gases with the densities of liquids and solids.

Section 3.7

11. Distinguish between heat and temperature.
12. Give the SI units of heat and temperature.
13. Write equations for the conversion of Celsius to Kelvin temperatures, for the conversion of Fahrenheit temperatures to Celsius, and vice versa.
14. Write the defining equation for specific heat.
15. Give the units of specific heat.

Section 3.8

16. Identify exact numbers.
17. Distinguish between accuracy and precision.
18. Identify the number of significant figures in a measurement.

Section 3.9

19. When performing calculations using measured values, round off the answer to the correct number of significant figures.

Key Terms

Review the definition of each of the terms listed here by chapter section number. You may use the glossary if necessary.

3.1 qualitative observation, quantitative observation, unit

3.2 SI

3.3 exponential notation, scientific notation

3.4 mass, weight

3.5 area, volume

3.6 density

3.7 absolute zero, calorie, joule, specific heat, temperature

3.8 accuracy, calibration, exact number, precision, significant figures

Questions and Problems

Answers to questions and problems with blue numbers appear in Appendix C.

Observations (Section 3.1)

1. Which of the following observations are qualitative and which are quantitative?
   a. red hair
   b. a long rope
   c. a ten-story building
2. Which of the following observations are quantitative?
   a. a heavy truck
   b. a three-pound fish
   c. a 250-mile trip

SI Units (Section 3.2)

3. Which SI unit would you use in the following measurements?
   a. The distance from your home to your school.
   b. The winning time in a 100-meter race.
   c. The outside temperature at twelve o'clock.
   d. The quantity of tomatoes picked by a farm worker in one hour.
4. Express each of the following in meters.
   a. 1 Mm
   b. 1 mm
   c. 1 cm
   d. 1 dm
   e. 1 nm
   f. 1 km
5. Using the appropriate metric prefix, rewrite each of the following quantities as 1 of a derived unit. For example, 1000 m = 1 km.
   a. 1 000 000 m
   b. 0.001 m
   c. 100 m
   d. 0.000 000 001 m
   e. 0.01 m
   f. 0.000 001 m

Exponential Notation (Section 3.3)

6. Express the numbers in problem 5 in scientific notation.
7. Write the following numbers in scientific notation.
   a. 0.003 45
   b. 103.2
   c. 45 623
8. Write the following numbers in decimal notation.
   a. 1.23×10^{-4}
   b. 6.3×10^{5}
   c. 9.75×10^{-7}
9. Twelve grams of carbon contains 6.02×10^{23} atoms. Write this number in decimal form.
10. Express the following in scientific notation.
   a. The city of Manila in the Philippines has 75 000 homeless children.
   b. Light travels at a speed of 186 000 miles per second.
   c. One gram of carbon contains 50 200 000 000 000 000 000 000 carbon atoms.

d. Every day, 42 000 children die from malnutrition and preventable diseases.
e. The maximum contaminant level for lead in drinking water is 0.000 005 g/L.

11. Add the following numbers, expressing the answers in scientific notation.
a. $(7.32 \times 10^{-3}) + (4.67 \times 10^{-2})$
b. $(7.8 \times 10^{11}) + (8.67 \times 10^{12})$

12. Carry out the following operations. Express the answers in scientific notation.
a. $(9.42 \times 10^{9}) - (7.6 \times 10^{8})$
b. $(2.19 \times 10^{-6}) - (7.64 \times 10^{-7})$

13. Complete the following operations. Express the answers in scientific notation.
a. $(4.74 \times 10^{4})(7.32 \times 10^{7})$
b. $(3.66 \times 10^{3})(6.59 \times 10^{6})$
c. $(9.12 \times 10^{-3})(1.07 \times 10^{-5})$

14. Express the answers to the following in scientific notation.
a. $\dfrac{8.37 \times 10^{7}}{3.08 \times 10^{5}}$
b. $\dfrac{2.37 \times 10^{17}}{3.45 \times 10^{-3}}$
c. $\dfrac{7.42 \times 10^{-12}}{6.18 \times 10^{-25}}$

15. Express the answers to the following problems using scientific notation.
a. $\dfrac{(1.25 \times 10^{8})(3.78 \times 10^{-3})}{5.11 \times 10^{4}}$
b. $\dfrac{8.73 \times 10^{-5}}{(2.31 \times 10^{-7})(3.68 \times 10^{6})}$

16. Write the answers to the following problems in scientific notation.
a. $\dfrac{(1.01 \times 10^{2}) + (7.32 \times 10^{3})}{2.77 \times 10^{-4}}$
b. $\dfrac{(2.67 \times 10^{-3}) - (8.79 \times 10^{-4})}{3.48 \times 10^{-5}}$

Mass and Weight (Section 3.4)

17. Distinguish between mass and weight. Give an example of how the weight of an object can change even though its mass is constant.
18. What is the SI unit of mass? What is its equivalent in pounds?
19. Express each of the following in grams:
a. 1 Mg
b. 1 mg
c. 1 cg
d. 1 dag
e. 1 ng
f. 1 kg

20. Using the appropriate metric prefix, rewrite each of the following quantitites as 1 of a derived unit. For example, 1000 g = 1 kg.
a. 1 000 000 g
b. 0.001 g
c. 1000 g
d. 0.000 000 001 g
e. 0.01 g
f. 0.000 001 g

Length, Area, and Volume (Section 3.5)

21. What are meant by "two dimensional" and "three dimensional" quantities? Give examples of each.
22. What units are commonly used in the laboratory for expressing length, area, and volume? Give the abbreviations for each.
23. Which of the following measures volumes with the best precision: a pipet, a beaker, or a graduated cylinder?

Density (Section 3.6)

24. Define density in a sentence and with a defining equation.
25. What units of density are commonly used for liquids? For solids? For gases?
26. Compare the densities of gases with the densities of liquids and solids.
27. Explain why more exertion is required to run through waste-deep water than to run on land (through air).
28. Why does a piece of wood float on water whereas a piece of iron sinks?

Heat and Temperature (Section 3.7)

29. Define or explain each of the following:
a. calorie
b. the food Calorie
c. absolute zero
30. Distinguish between heat and temperature.
31. All of the following are at 50°C. Which can release more heat to its surroundings in cooling to 20°C?
a. a pound of gold or an ounce of gold
b. a quart of water or a gallon of water
c. ten bricks or one brick
32. In each pair in problem 31, which is hotter at 50°C?
33. What is the SI unit of temperature?
34. What are the reference temperatures on the Fahrenheit, Celsius, and Kelvin scales?
35. Write equations that show the relationship between
a. Fahrenheit and Celsius temperatures.
b. Celsius and Kelvin temperatures.
36. Which is the higher temperature?
a. 0°C or 20°F
b. 0°C or 0 K
c. 100 K or 0°C
37. Which is the lower temperature?
a. 200 K or 100°F
b. 212°F or 90°C

Precision and Accuracy (Sections 3.8 and 3.9)

38. What are exact numbers? Give some examples.

39. Explain how numbers from measurements differ from exact numbers.

40. What is meant by the precision of measured values?

41. Explain what is meant by the accuracy of a measurement. Distinguish between the accuracy and the precision of a measurement.

42. What is meant by calibration of an instrument? Why is the accuracy of measurements often unknown?

43. When recording a measurement, explain how to determine the estimated digit.

44. Explain how to count the number of significant figures in a measured value.

45. Which of the following kinds of zeros are counted as significant figures?
    a. zeros to the left of digits
    b. zeros to the right of a decimal point
    c. a zero between two nonzero digits

46. Give an example of a measured value for which the number of significant figures is ambiguous. How could the number of significant figures be expressed clearly by a person recording a measurement?

47. How many significant figures are expressed in each of the following?
    a. 62.5 lb corn
    b. 450°C
    c. 0.02 mile
    d. 12 mg salt
    e. 2254 m
    f. 40.0 cm thread

48. How many significant figures are expressed in each of the following?
    a. 24.5 s
    b. 2.5×10^{-6} g DDT
    c. 200. g sodium chloride
    d. 0.0040 m wire

49. Round off each of the following numbers to three significant figures.
    a. 0.102 36
    b. 10.840
    c. 213.43
    d. 13 402

50. Round off each of the following numbers to two significant figures.
    a. 2983
    b. 0.004 57
    c. 1.009
    d. 30.7

51. Round off each of the following numbers to three significant figures and express the numbers in scientific notation.
    a. 36.083
    b. 203 088
    c. 0.000 4398
    d. 765.3

52. Express the answers to the following problems using the proper number of significant figures.
    a. $67.3 \times 0.001\,30 =$
    b. $4.670/10.0 =$
    c. $123 \times 3.2 =$
    d. $0.20\,(1.00 \times 10^{2}) =$

53. Express the answers to the following problems using the proper number of significant figures.
    a. $8.5 \text{ m} \times 9.3 \text{ m} =$
    b. 0.324 g aspirin tablet $\times$ 5 tablets $=$
    c. 546 miles/9.5 hr $=$

54. Express the answers to the following problems using the proper number of significant figures.
    a. $4.41 \text{ cm} + 2.300 \text{ cm} + 102 \text{ cm} =$
    b. $5.24 \text{ g} - 2.1 \text{ g} - 1.03 \text{ g} =$
    c. $8.6 \text{ s} + 12.5 \text{ s} + 22.8 \text{ s} + 103 \text{ s} =$

55. A student took 10 tablets of acetaminophen (Tylenol) over a two-day period. Each tablet contained 0.500 g of the pain reliever. What mass of acetaminophen did he take?

56. What was the average weight of the cookies in a package if the individual cookies weighed 0.12, 0.13, 0.10, 0.09, 0.15, and 0.14 lb?

57. A package delivery service shipped boxes weighing 123 lb, 34.5 lb, 5.8 lb, and 18 lb. What was the total weight of the boxes shipped?

58. Overseas first class postage is \$0.50 per ½ oz or fraction of ½ oz. How much postage would be required to mail fifty 0.0352-oz vitamin tablets, including 1.5 oz of packaging material, to Calcutta, India?

59. A bar of soap, originally weighing 4.75 oz, was used by two people for exactly two weeks. At the end of this period, the bar of soap weighed 2.1 oz. What weight of soap was used per person per day?

Additional Problems

60. Write the answers to the following problems in scientific notation.

    a. $$\frac{(2.13 \times 10^{2}) + (3.33 \times 10^{3})}{(3.27 \times 10^{-4})\,(1.12 \times 10^{-2})}$$

    b. $$\frac{(6.42 \times 10^{-3})\,(1.23 \times 10^{-5})}{(2.67 \times 10^{-3}) - (8.79 \times 10^{-4})}$$

61. After a successful hunting trip, a friend suggested that the meat should be frozen at a temperature of 0° or below. Realizing that the meat should be solidly frozen, should you check the freezer with a Celsius or a Fahrenheit thermometer? Explain.

62. Express each of the following in liters:
    a. 1 ML
    b. 1 mL
    c. 1 cL
    d. 1 µL
    e. 1 nL
    f. 1 kL

63. How many significant figures are expressed in each of the following?
    a. 23.60 g silver
    b. 5.00×10^{6} lb sugar
    c. 0.1205 g dye
    d. 5 doz eggs

64. An electronic analytical balance will detect and display mass measurements to 0.1 mg. What is its detection limit in grams?

65. Record the following measured quantities using the correct number of significant figures:
   a. Using a graduated cylinder with the scale marked in milliliters, the meniscus of a solution was resting on the 16 mL mark.
   b. Using a Celsius thermometer with the scale marked at one degree intervals, the mercury was halfway between 36°C and 37°C.
   c. Using a millimeter ruler with a scale marked in millimeters, what is the length (in cm) of a piece of wire having its end resting on the 12-centimeter mark?

66. Using a scale in the school locker room after a vigorous workout, a person's weight was 146, 145, and 147 lb in three successive measurements. Two days later, using a certified scale in a doctor's office, the person's weight was 151.6 lb, compared with 155.5 lb three months earlier. Which measurements are more accurate? Explain your answer.

Solutions to Practice Problems

PP 3.1

a. 10^9 is equivalent to 1×10^9. An exponent of 9 tells us to move the decimal point nine places to the right. The answer is

1 000 000 000

b. 10^{-6} is equivalent to 1×10^{-6}. An exponent of -6 tells us to move the decimal point six places to the left. The answer is

0.000 001

PP 3.2

Count the number of places the decimal point must be moved. Remember, a number larger than 10 is written exponentially with a positive exponent. A number smaller than 1 has a negative exponent.

a. 10^5 b. 10^{-9}

PP 3.3

Move the decimal point to give a number between 1 and 10. The number of places the decimal point was moved tells you the power of 10.

a. 3.58×10^{-2} b. 3.2306×10^4 c. 1.201×10^1

PP 3.4

The power of 10 tells you the number of places to move the decimal point.

a. 0.001 56 b. 793 000 c. 4982

PP 3.5

You can perform the operations on the nonexponentials and the exponentials separately. Remember to add exponents when multiplying and to subtract exponents when dividing.

a. $(1.31 \times 10^{-3})(2.73 \times 10^4) = 3.58 \times 10^1$

$1.31 \times 2.73 = 3.58$

$10^{-3} \times 10^4 = 10^1$

b. $\dfrac{(9.68 \times 10^{-4})}{(3.33 \times 10^{-6})} = 2.91 \times 10^2$

$9.68 \div 3.33 = 2.91$

$10^{-4} \div 10^{-6} = 10^2$

PP 3.6

a. Four. Trailing zeros to the right of a decimal point are counted.
b. The bottles were counted, and this is an exact number.
c. three
d. three
e. Two. Leading zeros are not counted.
f. three

PP 3.7

a. 71 Subtraction gave two significant figures.
b. 11 Addition gave two significant figures.
c. 0.642 Addition gave three significant figures.

Answers to Chapter Review

1. observations
2. quantitative
3. SI
4. unit
5. exponential
6. mass
7. gravitational
8. weight
9. balance
10. kilogram
11. meter
12. centimeter
13. area
14. volume
15. three
16. cm^3
17. liter
18. mL
19. density
20. gas
21. hydrogen
22. m/V
23. joules
24. calorie
25. water
26. 1000
27. temperature
28. kinetic
29. Celsius
30. kelvin
31. absolute
32. gram
33. specific heat
34. $\dfrac{q}{m \times \Delta T}$
35. heat
36. exact
37. accuracy
38. calibrated
39. significant figures
40. nonzero
41. round off
42. least
43. decimal place

4

Problem Solving

Contents

Chemistry problems often involve experimental data.

Problem solving is a major part of a beginning chemistry course.

You Probably Know . . .

- To find answers to problems, scientists conduct experiments in which they make measurements and collect numerical data (Section 3.1). The data are often used in calculations that give answers to the problems being investigated.
- Solving chemistry problems is a major part of a beginning chemistry course, and you will be working many problems as you study this chapter and the chapters that follow.
- Mathematical problem solving is part of many occupations. Improving your problem-solving skills will help you to prepare for whatever you choose to do in the future.

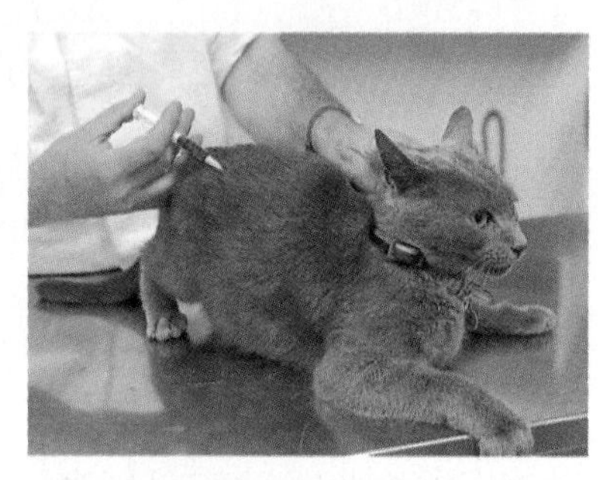

A veterinarian must solve problems when treating animals.

Why Should We Study Problem Solving?

Mathematical problem solving has long been an essential part of scientific investigation. Chemists encounter a variety of problems in their work. Some problems are simple, and some are complex and difficult. As you study chemistry, you must learn how to perform various calculations and how to solve various kinds of applied math problems. Regardless of the kind of problem you are working, you will need a working knowledge of certain skills and problem-solving strategies.

Furthermore, problem solving has become a necessary skill in coping with the complexities of our society. One of the benefits of studying chemistry is learning to apply mathematical skills that will help us to succeed in this competitive world.

4.1 Skills and Strategies

Learning Objective

- Apply the steps involved in problem solving.

To become a successful problem solver, you need basic math skills. Most of the problems in this book can be solved by using the basic operations of addition, subtraction, multiplication, and division. However, some problems will require basic algebra skills. A review of basic math skills and the fundamentals of algebra is presented in Appendix B. Master this material as rapidly as you can. It is there to help you get a fast start.

With almost all problems, you will improve the speed and accuracy of your computations by using a calculator. Easy-to-use calculators can be purchased for a reasonable price. One that uses exponential notation is recommended. Be sure you understand how to use your calculator and have it with you whenever you work chemistry problems.

Right: A simple hand calculator can perform basic mathematical operations and is satisfactory for most problems in this book. *Left:* A scientific calculator can handle exponential notation and can be used for basic operations as well as finding a reciprocal, square root, square, and logarithm of a number. Other functions, used mostly in engineering calculations, are also included.

■ Strategies

It is common for people to solve various simple math problems when they are shopping, working in the kitchen, sewing, making crafts, and working at fix-it projects around their homes. If you frequently work in the kitchen, you may find that the calculations used in doubling a recipe are easy for you. Problems that you work frequently may require little conscious thought. You may use the same problem-solving method or strategy so often that you are not consciously aware of the steps you follow.

However, when you are faced with an unfamiliar problem, you may not even know how to begin. As you continue to study chemistry, you will be faced with many new kinds of problems. To be successful, it is important that you have a strategy for solving problems.

In developing strategies for solving problems, we can learn from others who are experienced and successful problem solvers. Experienced problem solvers have identified several important elements in problem solving:

Organizing information. Outlining a problem helps you to get started and aids in checking your work.

Sorting information. You must be able to identify what information is essential to the problem and what is unimportant.

Setting up a calculation. Sequencing, or arranging information in a logical order, is an important part of setting up a calculation. You need strategies to help you arrange information in logical order as you set up a calculation for a problem.

Dividing a complex problem into several simpler problems. Solving several simpler problems is usually easier than solving one complex problem.

Problem-solving skills are necessary to be a wise shopper.

These elements of problem solving are summarized in the problem-solving steps shown in Figure 4.1. The last of the preceding elements is included in the second step and is one of the strategies that will be used in setting up the calculation. This strategy will be discussed later as we encounter complex problems.

Step 1	Organize the information
Step 2	Set up the calculation
Step 3	Perform the calculation
Step 4	Check your work

FIGURE 4.1 Problem solving

A warning is in order. The purpose of worked examples in the text is to illustrate problem-solving methods. *Resist the temptation to memorize how particular example problems are worked.* A minor variation in the wording of a problem can derail a "memorize the example" approach. Instead, put forth the effort *now* to learn good strategies that can be used with a variety of problems. Then practice these strategies by working as many problems as possible. Using the strategies suggested in this chapter and later in the book will help you become a successful problem solver.

Many problems can be worked using more than one method.

Many problems can be solved by more than one method. Dimensional analysis (Section 4.4) is an effective and often-used method that uses the units of given and known quantities to set up a calculation. Most problems in this book can be worked with dimensional analysis, and this strategy will be emphasized. However, rearrangement of algebraic expressions (Section 4.5) is an important strategy for some problems and should not be ignored.

Most problems can be solved using dimensional analysis.

4.2 Outlining a Problem

Learning Objective

- Outline word problems.

Be organized! Outline each problem.

Organizing information is the crucial first step in getting started in solving a problem. Furthermore, an outline is needed so you can check your work after finishing the calculation. Even good problem solvers make mistakes. However, *good problem solvers are able to find their mistakes and correct them.* Organize your work and keep it neat so you can check your work easily.

Now, consider a simple numerical problem. Suppose you have decided to write a detailed plan for using your time each week. To get started with your plan, you need to know how many hours are in one week. This is not a difficult problem, so you may have already figured out how to work it. You want to calculate the number of hours in seven days. You know there are 24 hours in one day, so you reason that you must multiply to find the answer:

$$\text{time (hours)} = 24 \text{ hours/day} \times 7 \text{ days} = 168 \text{ hours}$$

Note that the slash symbol / is read *per* (or "in one") and is a division sign.

Good work! That is the correct answer. Now let's use this problem to illustrate our problem-solving strategy.

We begin our strategy with an outline of the problem. First, we identify what we want to find. The parts of the outline look like this:

Find: time (hr) = ?

Given: 7 da

Known: 24 hr = 1 da or 24 hr/1 da

Solution: time (hr) = ?

Always *write the units with each number.*

Two things should be noted about the outline. The first is the format. Although the exact format is not crucial, we will find this four-part format useful for many problems. The second thing to note is that we have been careful to *express the units* or dimensions (using abbreviations) for the unknown, the given, and the known information. Remember, *a unit is a defining quantity for a measurement.* It tells us what scale was used for the measurement. A number is meaningless without its unit. Knowledge of units and relationships among units is essential to be able to use dimensional analysis in working problems.

Imagine that as you were outlining a problem you wrote "known" but you couldn't think of *anything* to write down! Don't panic. If you can't find "known" information in your memory, you must look elsewhere. Simply move on to the "solution" step. The strategy for the "solution" step may help you identify what "known" information to look for. You may need to look in a reference such as a table of data in the chapter, in an appendix at the end of the book, or in a book from the library.

Conversion Factors

4.3

Learning Objectives

- Write equivalence statements in equation form.
- Write two conversion factors for each equivalence statement.

A **conversion factor** is a ratio that expresses the relationship of one quantity to another. Conversion factors are used in the dimensional analysis approach to problem solving. So, when outlining a problem, it will be helpful to express given and known information as conversion factors whenever possible. You can learn how this is done by looking at some familiar conversion factors.

Conversion factor: A ratio that expresses the relationship of one quantity to another.

We all have learned how to express common measurements in various units. For example, we can convert minutes to hours or hours to days because we know that 60 minutes are equivalent to one hour and 24 hours are equivalent to one day. We might call these equivalence statements. We write them in equation form like this:

$$60 \text{ min} = 1 \text{ hr} \quad \text{and} \quad 24 \text{ hr} = 1 \text{ da}$$

An equivalence statement can be changed into a conversion factor by dividing both sides of the equation by one of the quantities. This is how it is done:

$$\frac{60 \text{ min}}{1 \text{ hr}} = \frac{1 \text{ hr}}{1 \text{ hr}} = 1 \quad \text{and} \quad \frac{24 \text{ hr}}{1 \text{ da}} = \frac{1 \text{ da}}{1 \text{ da}} = 1$$

Thus, the equivalence statements have been changed into conversion factors:

$$60 \text{ minutes per hour} \left(60 \text{ min/hr, or } \frac{60 \text{ min}}{1 \text{ hr}}\right)$$

and

$$24 \text{ hours per day,} \left(24 \text{ hr/da or } \frac{24 \text{ hr}}{1 \text{ da}}\right)$$

You should remember two things about conversion factors. First, *a conversion factor from an equivalence statement always has a value of 1.* Multiplying a quantity by 1 does not change its value. Although multiplying a quantity by a conversion factor changes the number and its units, the *value* of the quantity remains the same. As an example, consider the previous conversion of days to hours:

A conversion factor from an equivalence statement always has a value of 1.

$$\text{time (hr)} = 24 \text{ hr/da} \times 7 \text{ da} = 168 \text{ hr}$$

Note that 7 da is the same amount of time as 168 hr, only the number and units have changed.

The second thing to remember is that *conversion factors always come in pairs.* One is always the inverse of the other. Simply invert one to write the other one. So, with hours and days we have two conversion factors:

$$\frac{24 \text{ hr}}{1 \text{ da}} \quad \text{and} \quad \frac{1 \text{ da}}{24 \text{ hr}}$$

Conversion factors come in pairs.

$$\frac{24 \text{ hr}}{1 \text{ da}} \quad \text{and} \quad \frac{1 \text{ da}}{24 \text{ hr}}$$

$$\frac{60 \text{ min}}{1 \text{ hr}} \quad \text{and} \quad \frac{1 \text{ hr}}{60 \text{ min}}$$

Example 4.1

Write a pair of conversion factors from each of the following:

a. 5280 ft = 1 mi
b. 1 mi = 1.609 km
c. 1 lb = 453.6 g
d. 1 L = 1.057 qt ■

Solutions

a. $\frac{5280 \text{ ft}}{1 \text{ mi}}$ $\frac{1 \text{ mi}}{5280 \text{ ft}}$

b. $\frac{1 \text{ mi}}{1.609 \text{ km}}$ $\frac{1.609 \text{ km}}{1 \text{ mi}}$

c. $\frac{453.6 \text{ g}}{1 \text{ lb}}$ $\frac{1 \text{ lb}}{453.6 \text{ g}}$

d. $\frac{1 \text{ L}}{1.057 \text{ qt}}$ $\frac{1.507 \text{ qt}}{1 \text{ L}}$

4.4 Solution Strategy I: Dimensional Analysis

Learning Objectives

- Use dimensional analysis to set up a calculation to solve a problem.
- Using a calculator, calculate the answer to a problem from the data in an equation, and express the answer using the correct number of significant figures.

After the unknown, given, and known information are identified and outlined, you are ready to begin the solution process. This involves setting up and performing a calculation. To set up the calculation, you must build a mathematical expression that you can use to calculate the answer. Sorting information and arranging it in logical order (sequencing) are the tricky parts of problem solving. You need a good strategy for this part.

Remember, most problems can be solved using dimensional analysis. **Dimensional analysis** is a method of writing an equation by examining the units (dimensions) of given and known quantities in a problem. It is a valuable tool for building an equation needed to solve a problem. *We can use dimensional analysis when a conversion factor is known that relates the unknown to known or given quantities.*

Let's look again at the problem we worked earlier, finding the number of hours in 7 days. Remember, the outline for the problem looks like this:

Find: time (hr) = ?
Given: 7 da
Known: 24 hr = 1 da or $\frac{24 \text{ hr}}{1 \text{ da}}$ (a conversion factor)
Solution: time (hr) =

Our strategy for setting up a calculation is simple, but it works. Like marching, the first step is important. If you get started on the correct foot, the next step is easy. The equation for this problem is:

$$\text{time (hr)} = \left(\frac{24\text{ hr}}{1\cancel{\text{da}}}\right)(7\cancel{\text{da}}) = 168\text{ hr}$$

Note that the quantities in parentheses are multiplied. Units in the numerator and denominator can be canceled.

This equation was set up following these steps:

Step 1 *Begin with the unknown, including its units.* Then write an equal sign.

Step 2 First term of the equation: Choose a conversion factor by *matching its units* (those in the numerator) with the units of the unknown.

Step 3 Succeeding terms of the equation: Choose a conversion factor or given quantity with units that will *cancel unwanted units* in a preceding term.

Summarizing these steps, we have:

$$\underset{\text{unknown (units)}}{(1)} = \underset{\substack{\textit{match units}\text{ with}\\ \text{unknown}}}{\overset{(2)}{\text{(conversion factor)}}}\ \underset{\substack{\textit{cancel units}\text{ of}\\ \text{preceding term}}}{\overset{(3)}{\text{(given quantity)}}}$$

For our problem, the three steps are indicated by numbers above the equation, as follows:

$$\overset{(1)}{\text{time (hr)}} = \overset{(2)}{\left(\frac{24\text{ hr}}{1\cancel{\text{da}}}\right)}\overset{(3)}{(7\cancel{\text{da}})} = 168\text{ hr}$$

In Step 2, remember to match the units in the *numerator* with the units of the unknown.

After performing the calculation, be sure to *check your answer.* Have the units canceled to give appropriate units for the unknown? Yes, hr is a time unit. Have you expressed the correct number of significant figures? Yes, both numbers used in the calculation are exact numbers, so the answer is also an exact number. Is the answer reasonable? Consider which unit has the larger number associated with it—da or hr:

Check your work!

Units can be canceled (treated algebraically) just as numbers can.

1 da = 24 hr (more hours than days)

Thus, the length of time in hours should be a larger number.

Also *check the calculation* to see that the numerical answer is reasonable. One way to check the calculation is to reverse the order of the terms, beginning with the information given.

Check: $(7\cancel{\text{da}})\left(\frac{24\text{ hr}}{1\cancel{\text{da}}}\right) = 168\text{ hr}$ ✔

As was mentioned earlier, different solution methods may work equally well for many problems. For example, in working the previous problem it makes little difference whether you set up the calculation starting with the unknown or starting with the given quantity, as shown in the check. Both give the same result. However, with some problems, there is a definite advantage in starting with the unknown. (See Examples 4.8 and 4.12.) In this book, the unknown-first approach will be emphasized, but examples using the given-first approach will be shown periodically.

The unknown-first approach can be more effective with problems that include more information than is needed with multistep problems. These are characteristic of many real-world problems, making them different from textbook problems, which are often simplified for instructional purposes. In particular, there are two advantages to the unknown-first approach:

1. When unneeded information is given in a problem, we have a strategy for choosing which given information to use: matching units with the unknown (Step 2) and then canceling unwanted units (Step 3). Note that the given-first approach can be confusing. How will we decide which given information to use first? Which information is essential and which is unimportant?
2. In a multistep problem, the unknown-first approach outlined here gives us a strategy for getting started (start with the unknown) and for arranging information in logical order as we set up the calculation in steps.

Take careful note of the use of these strategies as you study the example problems that follow. Then use these strategies as you work the practice problems and those at the end of the chapter.

Example 4.2

How many seconds are in 3 days? ■

Find: time (s) = ?

Given: 3 da

Known: $\dfrac{60\text{ s}}{1\text{ min}}, \dfrac{60\text{ min}}{1\text{ hr}}, \dfrac{24\text{ hr}}{1\text{ da}}$

Solution: The steps in setting up the calculation are indicated by numbers:

$$\underset{(1)}{\text{time (s)}} = \underset{(2)}{\left(\frac{60\text{ s}}{1\,\cancel{\text{min}}}\right)} \underset{(3)}{\left(\frac{60\,\cancel{\text{min}}}{1\,\cancel{\text{hr}}}\right)} \left(\frac{24\,\cancel{\text{hr}}}{1\,\cancel{\text{da}}}\right)(3\,\cancel{\text{da}}) = 2.592 \times 10^5\text{ s}$$

All numbers in the calculation are exact numbers, so the answer is an exact number. The units cancel properly to give units of s. The answer is reasonable because the time in seconds (a small unit of time) should be a large number. To check the calculation, reverse the order of the factors.

Is the answer reasonable?

Check: $(3\,\cancel{\text{da}})\left(\dfrac{24\,\cancel{\text{hr}}}{1\,\cancel{\text{da}}}\right)\left(\dfrac{60\,\cancel{\text{min}}}{1\,\cancel{\text{hr}}}\right)\left(\dfrac{60\text{ s}}{1\,\cancel{\text{min}}}\right) = 2.592 \times 10^5\text{ s}$

1. Outline the problem. Begin by identifying the unknown. Always write units with numbers. Rewrite known information as conversion factors.
2. Set up the calculation.
   a. If there is a conversion factor that relates the unknown to a given quantity, use dimensional analysis.
      i. Begin writing the equation with the unknown, including its units.
      ii. For the first term in the equation, choose a conversion factor by *matching its units* with the units of the unknown.
      iii. For each succeeding term of the equation, choose a given quantity or conversion factor with units that *cancel unwanted units* in a preceding term, thereby leaving the units of the unknown.
   b. If there is no conversion factor, and the unknown is part of an algebraic equation, use algebra. (See Section 4.5.)
3. Calculate the answer.
4. Check your work. Check the *units* of the answer. Check *significant figures.* Inspect the numbers in the calculation to see if the answer is *reasonable.*

FIGURE 4.2 Steps in problem solving.

Remember, sorting information and arranging it in logical order are important elements in problem solving. Our strategy of *beginning the equation with the unknown* helps you with these tricky parts of problem solving. Later (Section 4.5), you will see that this is consistent with the algebra strategy, which also begins with the unknown.

Figure 4.2 summarizes our problem-solving strategy. This simple strategy works because of the following:

1. An outline of the problem gets you started by organizing the information.
2. You have a strategy for sorting information and identifying what is essential and what is not needed.
3. You always know how to begin to set up the calculation. You have a strategy for arranging information in logical order as you build an equation.

Use this strategy as you work the example problems.

Example 4.3

A farm family has 25 laying hens. If 20 eggs are collected each day from the hen house, how many dozen eggs will the hens have laid in one year? ■

Find: $\frac{\text{\# eggs (doz)}}{1 \text{ yr}} = ?$

Given: 25 hens, $\frac{20 \text{ eggs}}{1 \text{ da}}$

Known: $\frac{12 \text{ eggs}}{1 \text{ doz}}$ and $365 \text{ da} = 1 \text{ yr}$ or $\frac{365 \text{ da}}{1 \text{ yr}}$

Solution: $\dfrac{\#\text{ eggs (doz)}}{1\text{ yr}} = \left(\dfrac{1\text{ doz}}{12\text{ }\cancel{\text{eggs}}}\right)\left(\dfrac{20\text{ }\cancel{\text{eggs}}}{1\text{ }\cancel{\text{da}}}\right)\left(\dfrac{365\text{ }\cancel{\text{da}}}{1\text{ yr}}\right)$

$= \dfrac{608\text{ doz}}{1\text{ yr}}$

Check: $\left(\dfrac{20\text{ }\cancel{\text{eggs}}}{1\text{ }\cancel{\text{da}}}\right)\left(\dfrac{365\text{ }\cancel{\text{da}}}{1\text{ yr}}\right)\left(\dfrac{1\text{ doz}}{12\text{ }\cancel{\text{eggs}}}\right) = \dfrac{608\text{ doz}}{1\text{ yr}}$ 

Note that "25 hens" was unnecessary information and was not used in the problem. With the unknown-first approach, you can easily decide which given quantity to use.

Example 4.4

Pike's Peak near Colorado Springs is 14 110 feet above sea level. What is the elevation of the mountain in meters? 1 m = 3.281 ft. ■

Find: elev (m) = ?

Given: 14 110 ft

Known: 1 m = 3.281 ft or $\dfrac{3.281\text{ ft}}{1\text{ m}}$ or $\dfrac{1\text{ m}}{3.281\text{ ft}}$

Solution: $\text{elev (m)} = \left(\dfrac{1\text{ m}}{3.281\text{ }\cancel{\text{ft}}}\right)(14\ 110\text{ }\cancel{\text{ft}}) = 4301\text{ m}$

Check the answer for correct units!

Check: $(14\ 110\text{ }\cancel{\text{ft}})\left(\dfrac{1\text{ m}}{3.281\text{ }\cancel{\text{ft}}}\right) = 4301\text{ m}$ (4 significant figures) 

Example 4.5

A sample of iron powder was found to have a mass of 0.434 g. What was the mass of iron in milligrams? ■

Find: mass (mg) = ?

Given: 0.434 g Fe

Known: 1 mg = 10^{-3} g (by definition) or $\frac{1 \text{ mg}}{10^{-3} \text{ g}} = \frac{10^3 \text{ mg}}{1 \text{ g}}$

Solution: $\text{mass (mg)} = \left(\frac{10^3 \text{ mg}}{1 \not{g}}\right)(0.434 \text{ g}) = 434 \text{ mg}$

Check: $(0.434 \text{ g})\left(\frac{10^3 \text{ mg}}{1 \not{g}}\right) = 434 \text{ mg}$ 

Example 4.6

A student measured 125 mL of water in a graduated cylinder. What was the volume of water in microliters? ■

Find: V (μL) = ?

Given: 125 mL water

Known: 1 μL = 10^{-6} L (by definition), 1 L = 10^3 mL

or $\frac{1 \text{ μL}}{10^{-6} \text{ L}} = \frac{10^6 \text{ μL}}{1 \text{ L}}$ $\frac{1 \text{ L}}{10^3 \text{ mL}}$

Solution: $V(\text{μL}) = \left(\frac{10^6 \text{ μL}}{1 \not{L}}\right)\left(\frac{1 \not{L}}{10^3 \not{mL}}\right)(125 \not{mL})$

$= 1.25 \times 10^5 \text{ μL}$

Check: $(125 \not{mL})\left(\frac{1 \not{L}}{10^3 \not{mL}}\right)\left(\frac{10^6 \text{ μL}}{1 \not{L}}\right) = 1.25 \times 10^5 \text{ μL}$ 

Example 4.7

How many liters of gasoline would fill a car's 10.5-gal tank? 1 L = 1.057 qt. ■

Find: V (L) = ?

Given: 10.5 gal

Known: $\frac{1.057 \text{ qt}}{1 \text{ L}}, \frac{4 \text{ qt}}{1 \text{ gal}}$

Solution: $V\,(\text{L}) = \left(\dfrac{1\ \text{L}}{1.057\ \cancel{\text{qt}}}\right)\left(\dfrac{4\ \cancel{\text{qt}}}{1\ \cancel{\text{gal}}}\right)(10.5\ \cancel{\text{gal}})$

$= 39.7\ \text{L}$

Check: $(10.5\ \cancel{\text{gal}})\left(\dfrac{4\ \cancel{\text{qt}}}{1\ \cancel{\text{gal}}}\right)\left(\dfrac{1\ \text{L}}{1.057\ \cancel{\text{qt}}}\right) = 39.7\ \text{L}$

Practice Problem 4.1 What is the distance in cm for a 400.-m race? (Solutions to practice problems appear at the end of the chapter.)

Practice Problem 4.2 Find the volume in liters equivalent to 1 m^3.

Practice Problem 4.3 A certain brand of maple flavoring contains 7% alcohol by volume (7 mL alcohol/100 mL flavoring). What is the volume of alcohol in 59 mL of maple flavoring?

Air pollution was mentioned in Chapter 1 as one area of study in which scientists apply problem-solving strategies. One of the instruments used by environmental scientists in air-pollution studies is a particulate air sampler. This kind of air sampler is used to determine the level of dust in the air. The following example illustrates the kind of problem encountered in air-pollution studies.

Example 4.8

A particulate air sampler mounted on the roof of a city building was equipped with a filter having a mass of 3.2381 g. The sampler was operated for 24.0 hr at a flow rate of 4.50 L/min. At the end of the sampling period, the filter was removed and taken to a laboratory where its mass was measured using an analytical balance (Figure 4.3). The mass of the filter and dust was found to be 3.3562 g. What was the concentration of dust in the air in mg/m^3? ■

FIGURE 4.3 The mass of an air filter is determined using an analytical balance.

Find: conc (mg/m^3) = ?

Given: 24.0 hr, 4.50 L/min
mass of filter: 3.2381 g
mass of filter and dust: 3.3562 g

Known: 1000 L = 1 m^3 or $\dfrac{1000\ \text{L}}{1\ \text{m}^3}$

Solution: conc (mg/m^3) = ?

This problem is more complex than previous ones. However, it can be simplified by dividing it into smaller parts. This problem illustrates a multistep approach to the solution process.

Divide and conquer the problem!

Examining the units of the unknown tells us that two quantities are needed to calculate the answer: (1) mass (mg) of dust and (2) volume (m^3) of air. Thus, we must calculate both quantities and then divide mass (mg) of dust by volume (m^3) of air.

1. Mass of dust in milligram units (mg):

$$\text{mass (mg)} = \left(\frac{1000 \text{ mg}}{1 \text{ g}}\right) (? \text{ g})$$

Mass of dust in gram units (g):

mass of filter and dust	3.3562 g
− mass of filter	− 3.2381 g
	= 0.1181 g

Substituting in the first equation:

$$\text{mass (mg)} = \left(\frac{1000 \text{ mg}}{1 \cancel{\text{g}}}\right) (0.1181 \text{ g}) = 118.1 \text{ mg}$$

Check: $(0.1181 \cancel{\text{g}}) \left(\frac{1000 \text{ mg}}{1 \cancel{\text{g}}}\right) = 118.1 \text{ mg}$

2. Volume of air (V) in cubic meter units (m^3):

$$V (\text{m}^3) = \left(\frac{1 \text{ m}^3}{1000 \cancel{\text{L}}}\right) \left(\frac{4.50 \cancel{\text{L}}}{1 \cancel{\text{min}}}\right) \left(\frac{60 \cancel{\text{min}}}{1 \cancel{\text{hr}}}\right) (24 \cancel{\text{hr}}) = 6.48 \text{ m}^3$$

Be sure the unwanted units cancel to give volume units (m^3).

Finally, dividing (1) by (2) gives

$$\text{conc (mg/m}^3) = \frac{118.1 \text{ mg}}{6.48 \text{ m}^3} = 18.2 \text{ mg/m}^3 \qquad \text{(3 significant figures)}$$

Density

Density (Section 3.6) is the mass-volume equivalence for a substance. For example, because the density of gold is 19.3 g/cm^3, 19.3 g of gold occupies a volume of 1.00 cm^3. The density of a substance can be used as a conversion factor to find its mass from its volume or vice versa. Dimensional analysis can be applied to problems involving the density of a substance.

Example 4.9

What is the density of glycerin if 65.0 mL has a mass of 81.90 g? ■

Find: D (g/mL) = ?
Given: 65.0 mL, 81.90 g

Known: density $(D) = \frac{\text{mass (g)}}{\text{volume (mL)}}$

Solution: The units tell us that density can be calculated by dividing mass (g) by volume (mL):

$$D(\text{g/mL}) = \frac{81.90 \text{ g}}{65.0 \text{ mL}} = 1.26 \text{ g/mL}$$

Note that beginning with the unknown with its units tells us how to set up the calculation.

Example 4.10

The density of soybean oil is 0.924 g/cm^3. Calculate the mass of 234 mL of oil. ■

Find: mass (g) = ?

Given: $V = 234$ mL, $D = 0.924$ g/cm^3

Known: density $(D) = \frac{\text{mass (g)}}{\text{volume (mL)}}$

Solution: Because the density of soybean oil (0.924 g/cm^3) is a conversion factor, use dimensional analysis. Note that 1 cm^3 = 1 mL, so 234 mL = 234 cm^3. Follow the steps:

(1) (2) (3)

$$\text{mass (g)} = \left(\frac{0.924 \text{ g}}{1 \cancel{\text{cm}^3}}\right) (234 \cancel{\text{cm}^3}) = 216 \text{ g}$$

Check: $(234 \cancel{\text{cm}^3}) \left(\frac{0.924 \text{ g}}{1 \cancel{\text{cm}^3}}\right) = 216 \text{ g}$

Practice Problem 4.4 Find the mass of 1.2 mL of mercury that has a density of 13.6 g/mL.

Practice Problem 4.5 What is the volume of an aluminum nail that has a mass of 7.64 g? The density of aluminum is 2.70 g/cm^3.

■ Specific Heat

Specific heat (Section 3.7), expressed in units of J/g · C° or cal/g · C°, can be thought of as a conversion factor that relates heat (J or cal) to a mass (g) and to a temperature change (C°)* for a substance. Accordingly, dimensional analysis can be used to solve problems involving specific heat.

Example 4.11

Calculate the heat (joules) needed to raise the temperature of 425 g of water from 20.°C to 55°C. ■

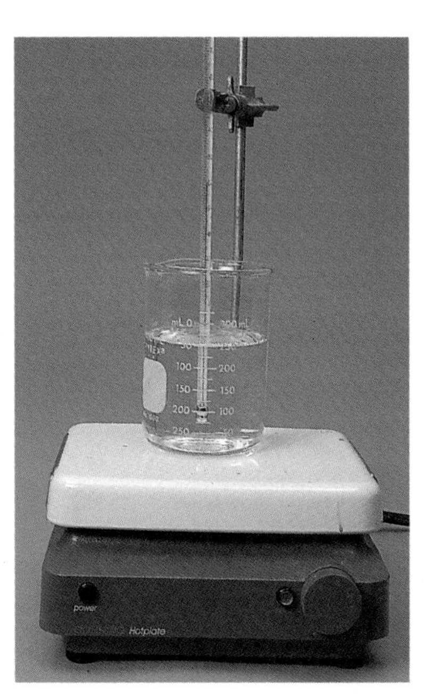

Begin by outlining the problem. A common symbol for heat is q.

Find: $q(\text{J}) = ?$

Given: $m = 425$ g, initial temperature $T_i = 20.°\text{C}$, final temperature $T_f = 55°\text{C}$, $\Delta T = 55°\text{C} - 20.°\text{C} = 35\text{C}°$

Known: specific heat, $SH(H_2O) = 4.184$ J/g · C° (Table 3.5)

Solution: Set up the equation by following the three steps of the dimensional analysis strategy:

$$\underset{(1)}{q(\text{J})} = \underset{(2)}{\left(\frac{4.184\ \text{J}}{1\ \cancel{\text{g}}\cdot\cancel{\text{C}°}}\right)} \underset{(3)}{(425\ \cancel{\text{g}})} \underset{(3)}{(35\ \cancel{\text{C}°})} = 62\ 000\ \text{J}$$

$$= 6.2 \times 10^4\ \text{J} \qquad \text{(2 significant figures)}$$

The answer can also be expressed as 62 kJ. Check the units of the answer to make sure they are correct units for heat.

Check: $(425\ \cancel{\text{g}}) \left(\frac{4.184\ \text{J}}{1\ \cancel{\text{g}}\cdot\cancel{\text{C}°}}\right) (35\ \cancel{\text{C}°}) = 62\ 000\ \text{J}$ 

Example 4.12

When a 25.8 g sample of copper at a temperature of 65.2°C was placed in cold water, it lost 79.0 cal and its temperature dropped to 31.6°C. From these data, calculate the specific heat of copper. ■

Find: specific heat, SH (cal/g · C°) = ?

Given: $m = 25.8$ g, $T_i = 65.2°\text{C}$, $T_f = 31.6°\text{C}$,
$\Delta T = 65.2°\text{C} - 31.6\text{C}° = 33.6\text{C}°$, $q = 79.0$ cal

*The unit of temperature change (C°) should not be confused with the unit for a Celsius temperature (°C).

Solution: Examining the units of specific heat, we see that we must divide heat (cal) by both mass (g) and the temperature change (C°):

$$SH \left(\frac{\text{cal}}{\text{g} \cdot \text{C°}}\right) = \frac{(79.0 \text{ cal})}{(25.8 \text{ g})\ (33.6\text{C°})}$$

$$= 0.0911 \text{ cal/g} \cdot \text{C°}$$

Check the units of the answer to make sure they are correct units for specific heat. Three significant figures are justified for the answer based upon the data (given information).

Note that the units of the unknown can guide us in setting up the calculation.

Practice Problem 4.6 How much heat is lost from 150 g of water as it cools from 95°C to 47°C?

Practice Problem 4.7 What is the mass of an aluminum object that absorbed 855 J of heat as its temperature was raised from 25°C to 56°C? For aluminum, $SH = 0.900$ J/g·C°.

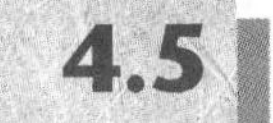

4.5 Solution Strategy II: Using Algebra

Learning Objectives

- Use algebra to solve an equation for the unknown in a problem.
- Use the defining equation for density to calculate density, mass, or volume when the other two quantities are given.
- Use the equations relating the Fahrenheit, Celsius, and Kelvin temperature scales to convert from one temperature scale to the others.
- Use the defining equation for specific heat to calculate specific heat, heat, mass, or a temperature change when the other three quantities are given.

If no conversion factor relates the unknown to a given quantity, dimensional analysis cannot be used as a problem-solving strategy. However, if the unknown is part of an algebraic equation, these problems can be solved by using algebra. You may need to review the basic algebraic operations in Appendix B before you begin this section. We will illustrate the use of algebra with problems involving density, temperature conversions, and heat.

Temperature Conversions

The following equations relating the Fahrenheit, Celsius, and Kelvin temperature scales were given in Chapter 3. These equations can be used to convert temperatures from one scale to another.

$T_F = 1.8T_C + 32$

$T_K = T_C + 273$

Note that the reference points on the Fahrenheit scale (32°F and 212°F) and Celsius scale (0°C and 100°C) are exact numbers. Because 1.8 (equivalent to 9/5) is derived from these reference points, it is also an exact number and does not limit the uncertainty in the calculated temperature. On the other hand, 273 is not an exact number (Section 3.7).

The simplest conversion is from a Celsius to a Kelvin temperature or vice versa. The example problems that follow illustrate the use of these equations.

$T_F = 1.8T_C + 32$ — 1.8 and 32: Exact numbers

Example 4.13

Normal body temperature is 37°C. What is this temperature on the Kelvin scale? ■

For very short problems, there is little benefit in following the four-part outline. Here, we will simplify the outline:

Find: $T_K = ?\ \text{K}$

$T_K = T_C + 273$

$= 37 + 273 = 310.\ \text{K}$

Example 4.14

Oxygen boils at 90. K. What is the boiling point of oxygen on the Celsius scale? ■

Find: $T_C = ?\ °\text{C}$

Known: $T_K = T_C + 273$

Solution: Rearrange the equation to solve for T_C:

$T_K - 273 = T_C + 273 - 273$ — Subtract 273 from both sides

$T_C = T_K - 273$ — Rearrange terms and substitute

$= 90. - 273 = -183°\text{C}$

It is better to solve the equation for the unknown before making substitutions. Note that problem solving with algebra is also an unknown-first method.

In preparation for a launch, space shuttle Discovery is loaded with liquid hydrogen and liquid oxygen.

Example 4.15

Alcohol boils at 78.3°C. What is the boiling point of alcohol in °F? ■

Find: $T_F = ?\ °F$

$$T_F = 1.8T_C + 32$$
$$= (1.8)\,(78.3) + 32 = 173°F$$

Example 4.16

If the temperature of your food freezer was 0°F, what was the temperature in °C? ■

Find: $T_C = ?\ °C$

Known: $T_F = 1.8T_C + 32$

Solution: Solve the equation for the unknown, T_C:

$T_F - 32 = 1.8T_C + 32 - 32$ Subtract 32 from both sides

$T_F - 32 = 1.8T_C$ Simplify

$\frac{(T_F - 32)}{(1.8)} = \frac{\cancel{1.8}\,T_C}{\cancel{1.8}}$ Divide by 1.8

$T_C = \frac{(T_F - 32)}{(1.8)}$ Rearrange

After substituting, perform the subtraction inside the parentheses first. Then divide to get the answer.

$$T_C = \frac{(0 - 32)}{(1.8)} = \frac{(-32)}{(1.8)} = -18°C$$

Practice Problem 4.8 On a July day in Denver, the temperature reached 104°F. What was the temperature in °C?

Practice Problem 4.9 The temperature on a cold winter day was −12°C. What was the temperature in K?

■ Density

In Chapter 3, the density of a substance was defined as:

$$\text{density} = \frac{\text{mass}}{\text{volume}} \quad \text{or} \quad D = \frac{m}{V}$$

Although we usually use dimensional analysis to set up density problems, we can also use algebra to solve the density equation for the unknown in the problem. This is illustrated in the following example.

Example 4.17

The density of gold is 19.3 g/cm³. What is the volume of a gold ring having a mass of 9.50 g? ■

The volume of a gold ring can be determined from its mass and its density.

Find: $V\ (\text{cm}^3) = ?$

Given: $m = 9.50\ \text{g}$

Known: $D = \dfrac{m}{V} = 19.3\ \text{g/cm}^3$

Solution: Solve the density equation for V:

$$V \times D = \frac{m}{\cancel{V}} \times \cancel{V} \qquad \text{Multiply both sides by } V$$

$$V \times \frac{\cancel{D}}{\cancel{D}} = \frac{m}{D} \qquad \text{Divide both sides by } D$$

or

$$V = \frac{m}{D}$$

Substituting and dividing gives the answer.

$$V = \frac{(9.50\ \text{g})}{(19.3\ \text{g/cm}^3)} = 0.492\ \cancel{\text{g}} \left(\frac{\text{cm}^3}{\cancel{\text{g}}}\right) = 0.492\ \text{cm}^3$$

Check the units to verify that the answer has units of volume (cm³). Note that it was necessary to divide grams (g) by density units (g/cm³), which are in fraction form. Remember, to divide by a fraction, you must invert the fraction and multiply. This allows you to cancel grams (g) from the calculation leaving units of volume (cm³) for the answer.

■ Specific Heat

Earlier, we learned that specific heat is defined by the equation:

$$SH = \frac{q}{m \times \Delta T} \qquad \text{or} \qquad q = (SH)\,(m)\,(\Delta T)$$

We have already shown that dimensional analysis can be used to set up specific-heat problems. Example 4.18 shows how algebra can also be used.

Example 4.18

An aluminum pan having a mass of 875 g and at a temperature of 22°C was heated on a stove. What was the final temperature, T_f, of the pan after it had absorbed 10.0 kJ of heat? (Note that T_f is a symbol for the final temperature in °C and is different from T_F, which stands for a Fahrenheit temperature.) ■

Find: $T_f = ?\ ^\circ\text{C}$

Given: $m = 875\ \text{g},\ T_i = 22^\circ\text{C},\ q = 10.0\ \text{kJ} = 1.00 \times 10^4\ \text{J}$

CHEMICAL WORLD

Quick Approximations

If you travel outside the United States, you'll be confronted by metric road signs giving distances in kilometers and metric thermometers giving temperatures in degrees Celsius. The best way to adapt to these situations is to develop a feeling for the units so you won't have to solve algebraic equations to decide whether to walk to a restaurant or catch a bus and whether to wear a coat. However, if your visit is to be brief, you'll be relieved to know that there are alternatives. Here are two shortcuts for determining temperatures and distances in English units without the need of pencil, paper, and calculator.

Temperature

To convert from Celsius to Fahrenheit, merely double the Celsius temperature and add 30°. This doesn't give you an exact answer, but it does give a quick and reasonably accurate approximation. To illustrate, consider this. You're about to go walking in the streets of Paris when from your hotel window you see a thermometer outside reading 20°C. You quickly double the 20° and add 30. Voilà, it's 70°F and you don't need a coat! But if you're walking to a colleague's laboratoire, you'd better take your calculator because the exact temperature involved in her experiment is T_F in the equation $T_F = 1.8T_C + 32$.

In most countries, distances are measured in kilometers.

A quick look at the T_F formula tells us why this shortcut works. Doubling T_C gives a bit more than multiplying it by 1.8, and adding 30° compensates for that by being a bit less than 32°. Thus, a useful approximation.

Distance

Your friend's laboratoire is 7 kilometers from your hotel. Paris is a great walking city, but do you want to walk that far? Or had you better take a cab? All you have to do to find out is to add 10% of the metric distance to one-half of the metric distance: 0.7 + 3.5 = about 4 miles. This is a long walk, so you'd better take a cab. And take your calculator in case your friend asks you how tall you are.

Why does this approximation work? A kilometer is 0.62 137 of a mile, very close to 60%—which is 50%, easy to figure, plus 10%, which is also easy to figure. How many miles are there in 642 kilometers?

General Approach

You can simplify almost any conversion equation or formula by making such approximations. Begin by rounding off all numbers or fractions to one-significant-digit equivalents. Then experiment a little. You'll find that with some trial and error and a little practice, you can become proficient in making these mental conversion approximations.

But it's still best to get a feel for metric!

Known: $SH = \dfrac{q}{m \times \Delta T}$, SH (Al) = 0.900 J/g·C° (Table 3.5)

Solution: The equation for specific heat includes ΔT, not T_f. However, T_f is related to ΔT:

$$\Delta T = T_f - T_i \tag{1}$$

Rearranging terms gives

$$T_f = T_i + \Delta T$$
$$= 22°C + \Delta T$$

Therefore, you must first find the temperature change, ΔT. Rearrange the equation for ΔT and substitute:

$$\Delta T = \frac{q}{m \times SH} = \frac{(1.00 \times 10^4 \cancel{J})}{(875 \cancel{g})(0.900)\cancel{J}/\cancel{g} \cdot C°)} \qquad (2)$$

$$= 12.7C°$$

Then, going back to equation (1):

$$T_f = 22 + 12.7 = 34.7°C$$

$$= 35°C \qquad \text{(2 significant figures)}$$

4.6 Summary of Problem-Solving Strategies

Our problem-solving strategy begins with an outline of the unknown, given, and known information. It is important that you write the proper units with all numerical data. Most problems can be solved by using dimensional analysis. However, algebra is an important strategy that is used when a needed conversion factor is not available but an equation relates the unknown to a given quantity. Figure 4.2 summarized the steps of our problem-solving strategy using dimensional analysis. The flow diagram in Figure 4.4 adds the algebraic option.

Remember that some problems (such as density problems) can be solved with either dimensional analysis or algebra. In solving those problems choose the solution strategy that is easier for you.

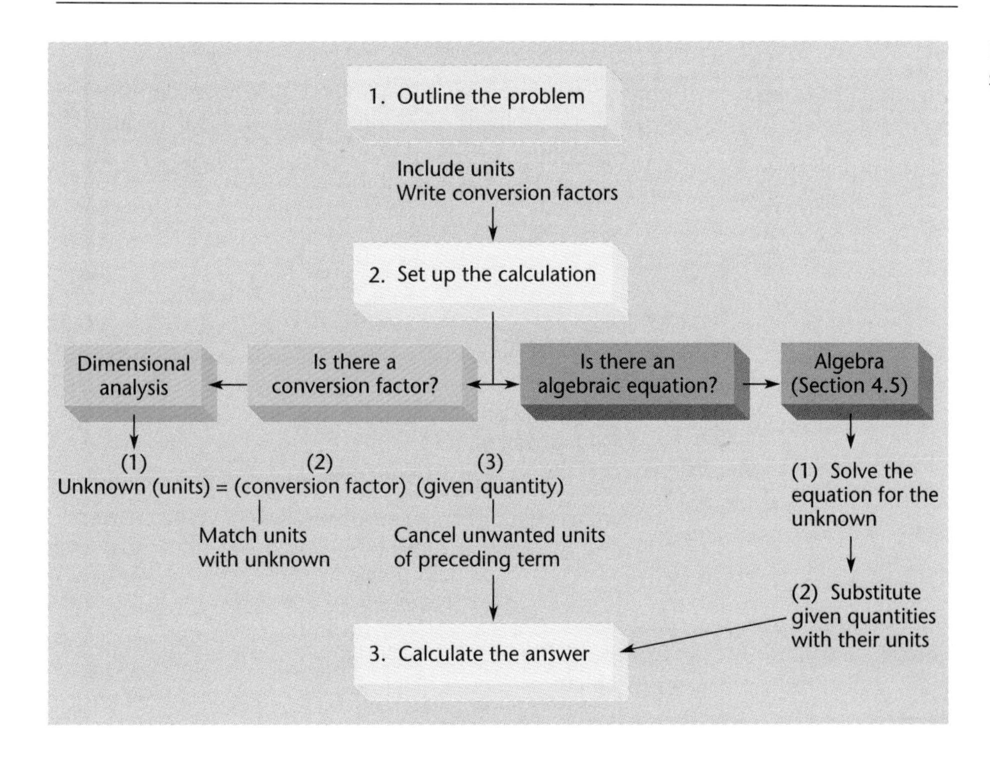

FIGURE 4.4 Problem-solving strategy.

Chapter Review

Directions: From the choices listed at the end of the Chapter Review, choose the word or term that best fits in each blank. Use each choice only once. The answers appear at the end of the chapter.

Section 4.1 As you study chemistry you must learn to solve problems. You will need to use math skills and problem-solving strategies. Most of the problems in this book can be solved by addition, subtraction, multiplication, and ____________ (1).

Experienced problem solvers have identified several important elements in problem solving, including ____________ (2) information, ____________ (3) information, and setting up a calculation. ____________ (4), or arranging information in a logical order, is an important part of setting up a calculation. Most problems can be solved using ____________ ____________ (5). However, using ____________ (6) is also an important solution strategy.

Section 4.2 The problem-solving strategy used in this book begins with an ____________ (7) of the problem. As you outline the unknown, given, and known information, it is important that you include the ____________ (8) with the numbers. A unit is a ____________ (9) quantity for a measurement; a number is ambiguous without its unit.

Section 4.3 Given and known information should be expressed in the form of ____________ ____________ (10) whenever appropriate. These always come in pairs where one is the ____________ (11) of the other.

Section 4.4 Dimensional analysis is a valuable tool for setting up a calculation. In writing an equation, remember to begin with the ____________ (12) with its units followed by an equal sign. The first term of the equation is chosen by ____________ (13) its units with the units of the unknown. The second and succeeding terms are chosen to cancel ____________ (14) units, thereby leaving the units of the unknown. After setting up an equation, use your calculator to find the answer.

Section 4.5 When the unknown is part of an algebraic ____________ (15), algebra can be used to solve for the unknown in the problem. For example, the defining equation for density is $D =$ ____________ (16). This equation can be used to calculate either density, mass, or volume if the other two quantities are given.

Equations relating the Fahrenheit, ____________ (17), and Kelvin scales can be used to convert from one scale to another.

The defining equation for specific heat is $SH =$ ____________ (18). This equation can be used to calculate specific heat, ____________ (19), mass, or a temperature change when the other three quantities are given or known.

Section 4.6 In summary, our problem-solving strategy involves four steps: (1) Outline the problem. (2) Set up the calculation using either dimensional analysis or algebra. (3) Calculate the answer using your calculator. (4) Check the answer. Check the units, significant figures, and the calculation.

Choices: algebra, Celsius, conversion factors, defining, dimensional analysis, division, equation, heat, inverse, m/V, matching, organizing, outline, $(q)/(m \times \Delta T)$, sequencing, sorting, units, unknown, unwanted.

Study Objectives

After studying Chapter 2, you should be able to:

Section 4.1

1. Apply the steps involved in problem solving.

Section 4.2

2. Outline word problems.

Section 4.3

3. Write equivalence statements in equation form.
4. Write conversion factors from equivalence statements.

Section 4.4

5. Using dimensional analysis, set up a calculation to solve a problem.
6. Using a calculator, calculate the answer to a problem from the data in an equation, and express the answer using the correct number of significant figures.

Section 4.5

7. Use algebra to solve an equation for the unknown in a problem.

8. Use the defining equation for density to calculate density, mass, or volume when the other two quantities are given.
9. Use the equations relating the Fahrenheit, Celsius, and Kelvin temperature scales to convert from one temperature scale to the others.
10. Use the defining equation for specific heat to calculate specific heat, heat, mass, or a temperature change when the other three quantities are given.

Key Terms

Review the definition of each of the terms listed here. For reference, they are listed here by chapter section number. You may use the glossary as necessary.

4.3 conversion factor

4.4 dimensional analysis

Questions and Problems

Answers to questions and problems with blue numbers appear in Appendix C. The most difficult problems are marked with an asterisk (*).

Problem-Solving Strategies (Sections 4.1, 4.2)

1. Identify the math skills you will need to solve the problems in this book.
2. List four steps involved in problem solving.
3. What are two good reasons for outlining a problem?
4. What are two reasons for writing units for the unknown, given, and known information?
5. What are the two methods used to solve most problems?
6. In using dimensional analysis, what is an advantage of the unknown-first approach over the given-first approach?

Conversion Factors (Section 4.3)

7. What is meant by a conversion factor?
8. Express the following as conversion factors. Remember to write a pair of conversion factors for each.
   a. 12 doz = 1 gross
   b. 39.37 in = 1 m
   c. 365 da = 1 yr
   d. 1 hr = 3600 s
   e. 1 case = 24 cans
9. Write conversion factors for each of the following:
   a. 1 lb = 16 oz
   b. 2.20 lb = 1 kg
   c. 10^6 μg = 1 g
   d. 1 tsp = 5 mL
   e. 1760 yd = 1 mi

Dimensional Analysis Problems (Section 4.4)

10. A delivery truck makes 31 trips in an average day for a total of 247 mi. What is the average length of a trip?
11. Calculate the distance in mm for a 1500-m race.
12. How many hairs are in a wig having a mass of 365 g if each hair weighs 12 mg?
13. A box of detergent weighs 2 lb 10 oz. What is its equivalent in grams? 1 lb = 454 g, 1 lb = 16 oz.
14. A 12-oz can of corned beef costs $1.59. What is the mass of the corned beef in kg? 1 kg = 2.20 lb.
15. A package of 50 paper cups weighed 6.4 oz. What was the mass in grams of each cup? 1 lb = 454 g, 1 lb = 16 oz.
16. A package of raisins weighed 8 oz and contained 14 mini-snack packs. How many grams of raisins were in each of the small packs?
17. One penny weighs about 3.4 g. What is the mass of $20.00 of pennies?
18. How many microliters are in 1 mL of liquid?
19. How many liters of gasoline will fill a 12-gal automobile gas tank? 1.000 L = 1.0567 qt.
20. Gasohol is 10% ethanol by volume. How many liters of ethanol are in a 15-gal tank filled with gasohol?
21. A berry juice drink was sold in packages of three 8½ oz boxes. How many liters of drink were in a package? 1 qt = 32 oz.
22. Suppose you are planning a party for eight of your friends. You want enough soft drinks so each person (you included) will have four 10-oz cups of beverage. How many 2-liter bottles of soft drinks should you buy?
23. A child's dose of a liquid antihistamine medicine was 2 teaspoons (tsp). How many doses are in a 250-mL bottle of the medicine? 1 tsp = 5 mL.
24. A spray steam iron had a capacity of 75 mL of water. Starting with the iron full of water, it took 40 sprays to empty the iron. How many mL of water will remain in the iron after 15 sprays?

25. How many 8-oz cups of lemonade can be served from a completely full 2-gal insulated jug? 1 qt = 32 oz, 1 gal = 4 qt.
26. A peck (pk) basket was found to hold 21 apples. How many apples would be in 8 bushels (bu)? 1 bu = 4 pk.
27. An acre is equivalent to 43 560 sq ft. You have decided to buy a lot having dimensions of 80. ft $\times$ 105 ft on which to build a new home. The price of the lot is $4 500. What is the price of the land per acre?
28. You are preparing to pour a new concrete floor in the basement of your summer home. The floor is going to be 0.50 ft thick. The dimensions of the basement are 24 ft $\times$ 36 ft. How many cubic yards of concrete should you order?
29. A package of 5 razor cartridges costs $1.69. Dad uses each cartridge for 7 shaves. What is the cost per shave for a razor cartridge?
30. A package of 50 paper cups costs $0.89. How much would it cost for cups to serve punch to a class of 23 students?
31. The population of China is estimated to be 1.1 billion people. Beijing is estimated to have 10. million people and 7.0 million bicycles. If 10.% of these are actually tricycles, how many bicycle and tricycle wheels are in the city of Beijing?
32. The Yangzi River, 3915 mi long, flows easterly and divides China into North and South China. Approximately 410 million people live in the Yangzi River basin. Calculate the number of people per foot of river frontage.
33. The Great Wall of China, extending for 3728 mi across northern China, is the only man-made structure visible from space. If the wall averages 25 ft high and 12 ft thick, what is the total volume of the wall?
34. The yuan, equal to 100 fen, is the currency of China. While visiting Shanghai, your Chinese host paid 12 yuan and 40 fen for your two dinners. If $1.00 = 5.74 yuan, what was the cost of the dinners in dollars?
35. Japan is made up of almost 4000 islands having a total land area of 145 823 sq mi. The population of Japan is estimated to be 131 million people. Tokyo has about 12.2 million people living in an area of about 825 sq mi. Find the population densities of Tokyo and of Japan.
36. One cup of detergent is required to wash one load of laundry in a washing machine. A cup of detergent weighs 3⅓ oz. If a 2 lb 10 oz box of detergent costs $2.28, what is the cost of the detergent needed for one load of laundry?
37. The Doud's cow gives an average of 5 gal of milk per day. One gallon of milk weighs about 9 lb. What weight of milk does the Doud's cow give per week?
38. What volume (in liters) of vaccine would be used by a nurse in vaccinating 350 students if each injection had a volume of 2.5 mL?
39. What minimum volume (m^3) of earth would need to be excavated to bury a 4.00×10^4 L gasoline tank for a new service station? 1 m^3 = 1000 L.
40. Calculate the volume (L) of air in a room 5.0 m $\times$ 12 m $\times$ 3.0 m. 1 m^3 = 1000 L.
41. A case of juice contains eight 48-oz bottles. One quart is equivalent to 32 oz. How many quarts of juice are in one case?
42. One box of business envelopes contains 500 envelopes. A case contains eight boxes. How many envelopes are in four cases?
43. A box of six computer printer ribbons costs $16.60. One ribbon will print an average of 265 pages. What is the ribbon cost per page printed?
44. A concrete truck is delivering 9 yd^3 of concrete for a construction project. The concrete must be moved by wheelbarrow from the truck to the concrete forms. The wheelbarrow can hold 6 ft^3 of concrete. How many trips with the wheelbarrow will be needed to move the truckload of concrete?
45. One quart of sea water contains about 1 oz of salt. How many gallons of sea water must be evaporated to yield 50 lb of salt? 1 gal = 4 qt, 1 lb = 16 oz.
46. A typical female salmon lays 10 000 eggs during a spawning season. For every 5000 eggs laid, only about 50 hatch. How many spawning female salmon are needed to hatch 1000 minnows?
47. A baseball player was paid $440 000 per year. If he played 145 games last season and each game lasted an average of 2 hr and 45 min, what was his pay per hour of playing time?
*48. In some third-world countries, a child can be fed a nutritious meal for as little as $0.15. How many children could be fed two meals per day for one year with a baseball player's annual salary of $440 000?

Algebra Problems (Section 4.5)

Density

49. An apple was found to have a mass of 88.6 g and a volume of 102 cm^3. What was the density of the apple?
50. Oak has a density of 0.82 g/cm^3. What is the mass of an oak block having dimensions of 15.2 cm $\times$ 5.5 cm $\times$ 9.2 cm?
51. The mercury recovered from a broken thermometer was found to have a mass of 4.82 g. What was the volume of the mercury? The density of mercury is 13.6 g/mL.
52. Five large marbles had a mass of 174 g. When the marbles were added to a large graduated cylinder containing water, the water level rose from 125 mL to 194 mL. What was the density of the marbles? Note that the volume of the water displaced is equal to the volume of the marbles.

53. What is the mass of 50.0 mL of mercury? Density = 13.6 g/mL.

54. The density of lead is 11.3 g/cm^3. What mass of lead is needed to make 50 lead bullets each having a volume of 1.34 cm^3?

55. The density of acetone (an ingredient of fingernail polish remover) is 0.791 g/mL. What is the mass of 250. mL acetone?

56. A sample of gasoline had a density of 0.745 g/mL. What was the volume of 5.0 kg of gasoline?

57. The density of platinum is 21.5 g/cm^3. What mass of platinum will occupy the same volume as 1.0 tsp of water? 1 tsp = 5.0 cm^3.

58. A copper candlestick displaced 78.5 mL of water. The density of copper is 8.96 g/cm^3. What was the mass of the candlestick?

59. A vegetable oil has a density of 0.832 g/mL. A recipe called for ¼ cup of vegetable oil. What was the mass of the oil needed for the recipe? 1 qt = 4 cups, 1 L = 1.0567 qt.

Temperature Conversions

60. Water in mile-high Denver boils at 95.5°C. What is the boiling point of water in Denver in K? In °F?

61. A nurse in Bangkok, Thailand, measured a child's temperature to be 39.4°C. What was the child's temperature in °F?

62. Dry ice, solid carbon dioxide, vaporizes (sublimes) at −78°C. What is its sublimation temperature in K?

63. On a summer day on the Arabian desert the temperature was 128°F. What was the temperature in °C?

Specific Heat

64. A 33.4-g sterling silver spoon at 60.0°C was placed in a container of cold water, resulting in a final temperature of 18.8°C. The water was found to have absorbed 78.9 J of heat from the spoon. Calculate the specific heat of silver.

65. Calculate the quantity of heat needed to raise the temperature of 40.0 g of aluminum from 25°C to 45°C. *SH*(Al) = 0.900 J/g·C°.

66. How much heat would be needed to raise the temperature of a 44.8-g copper cup from 22°C to 83°C? *SH*(Cu) = 0.385 J/g·C°.

67. How much water could be heated from 25°C to 95°C upon absorption of 4.22 kcal of heat?

*68. Borosilicate glass, used to make laboratory glassware, has a specific heat of 0.20 cal/g·C°. A 118-g glass beaker at 25°C absorbed 1.25 kcal of heat during an experiment. What was the final temperature of the beaker?

Additional Problems

69. For underground coal miners, the permissible exposure limit to carbon monoxide is 55 mg/m^3 of air. At this concentration, how many grams of carbon monoxide would a miner inhale with 0.25 m^3 of air?

70. A big-league pitcher can throw a fast ball at a speed of 98 mph. The distance traveled by the ball from the pitcher's hand to home plate is 58 ft. How many seconds would it take a fast ball to reach home plate?

71. If placed side by side, how many dust particles, 2.5 μm in diameter, would make a "dust trail" 1.0 m long?

*72. A bottle contained 8.0 oz of cough syrup. If a child's dose is 2 tsp, how many doses are in the bottle? 1 tsp = 5.0 mL, 1 qt = 32 oz, 1 L = 1.0567 qt.

73. Ethylene glycol, the common automobile antifreeze, boils at 192°C. What is its boiling point in K? In °F?

*74. A 155-lb man displaced 18 gal of water when he was submerged in a swimming pool. What was the man's density in g/cm^3? 1 L = 1.0567 qt, 1 lb = 454 g.

*75. After a rainstorm, water standing on a flat roof was 3.0 cm deep. The dimensions of the roof were 62 m × 86 m. The density of water is 0.998 g/mL. What was the mass of the water standing on the roof?

*76. How long does it take a family to travel by car from Denver to Chicago, a distance of 1025 mi, at an average speed of 62 mph? Include in the trip time a 15-min rest stop every 2 hr, two 1-hr and 15-min breaks for meals, and 10. min for each refueling stop. They begin the trip with a full tank of fuel, and one tankful takes them an average of 300. mi.

*77. A large dump trunk can hold 12 tons of coal. An 86-car freight train had 15 boxcars of lumber. The remaining cars were loaded coal cars. Each coal car had a capacity of 110 tons of coal. How many dump trucks could be loaded with the coal on the freight train?

*78. An electric clothes dryer was used an average of 3.5 hr per day, 3 days per week for 1 yr. The dryer was rated at 4.5 kilowatts. At a cost of $0.075 per kilowatt-hour, what did it cost to operate the dryer for the year? Note that an appliance rated at one kilowatt operated for one hour would use one kilowatt-hour of electricity.

*79. On a recent holiday, a carnival booth had 839 contestants trying to win a prize. During the day's festivities, seven teddy bears were awarded as first prizes, each costing the proprietor $7.50. One hundred and three balloons, costing $0.20 each, were given as second prizes. The remaining contestants were given suckers costing $2.25 per gross (12 doz) as consolation prizes. If each contestant paid $0.50, how much did the proprietor net for the day's activities?

*80. A piece of aluminum at 78.6°C was immersed in 50.0 g of water at 22.3°C in an insulated vessel. The final temperature of the water and the aluminum was 31.5°C. What was the mass of the aluminum? *SH*(Al) = 0.900 J/g·C°.

Solutions to Practice Problems

PP 4.1

Find: distance (cm) = ?

Given: 400. m

Known: 100 cm = 1 m, or $\frac{100\ \text{cm}}{1\ \text{m}}$

Solution: $\text{distance (cm)} = \left(\frac{100\ \text{cm}}{\cancel{\text{m}}}\right)(400.\ \cancel{\text{m}})$

$= 4.00 \times 10^4\ \text{cm}$

PP 4.2

Find: volume (L) = ?

Given: 1 m^3

Known: 1 L = 1000 cm^3 or $\frac{1\ \text{L}}{1000\ \text{cm}^3}$

1 m = 100 cm or $\frac{100\ \text{cm}}{1\ \text{m}}$

Solution: volume (L) =

$$\left(\frac{1\ \text{L}}{1000\ \cancel{\text{cm}^3}}\right)\left(\frac{100\ \cancel{\text{cm}}}{1\ \cancel{\text{m}}}\right)\left(\frac{100\ \cancel{\text{cm}}}{1\ \cancel{\text{m}}}\right)\left(\frac{100\ \cancel{\text{cm}}}{1\ \cancel{\text{m}}}\right)(1\ \cancel{\text{m}^3})$$

$= 1000\ \text{L}$

1000 L is equivalent to 1 m^3

PP 4.3

Find: V (mL) alcohol = ?

Given: 59 mL maple flavoring

7% alcohol (by volume), or $\frac{7\ \text{mL alcohol}}{100\ \text{mL flavoring}}$

Solution: V (mL) alcohol =

$$\left(\frac{7\ \text{mL alcohol}}{100\ \cancel{\text{mL flavoring}}}\right)(59\ \cancel{\text{mL flavoring}})$$

$= 4\ \text{mL alcohol}$

(one significant figure)

PP 4.4

Find: mass (g) = ?

Given: 1.2 mL of mercury, D = 13.6 g/mL

Solution: $\text{mass (g)} = \left(\frac{13.6\ \text{g}}{\cancel{\text{mL}}}\right)(1.2\ \cancel{\text{mL}}) = 16.3\ \text{g}$

$= 16\ \text{g}$ (two significant figures)

PP 4.5

Find: volume (cm^3) = ?

Given: mass = 7.64 g, density = 2.70 g/cm^3

Known: $\text{density} = \left(\frac{\text{mass}}{\text{volume}}\right)\left(\frac{\text{g}}{\text{cm}^3}\right)$

Solution: $\text{volume (cm}^3) = \left(\frac{1\ \text{cm}^3}{2.70\ \cancel{\text{g}}}\right)(7.64\ \cancel{\text{g}}) = 2.83\ \text{cm}^3$

PP 4.6

Find: q (J) = ?

Given: 150 g water

$T_i = 95°\text{C}$, $T_f = 47°\text{C}$

$\Delta T = 95°\text{C} - 47°\text{C} = 48\text{C}°$

Known: SH (water) = 4.184 J/g·C°

Solution: $q\ (\text{J}) = \left(\frac{4.184\ \text{J}}{\cancel{\text{g}}\cdot\cancel{\text{C}°}}\right)(150\ \cancel{\text{g}})(48\ \cancel{\text{C}°}) = 3.0 \times 10^4\ \text{J}$

(two significant figures)

PP 4.7

Find: mass (g) = ?

Given: q = 855 J

$\Delta T = 56°\text{C} - 25°\text{C} = 31\text{C}°$

$SH\ (\text{Al}) = 0.900\ \frac{\text{J}}{\text{g}\cdot\text{C}°}$

Solution: $\text{mass (g)} = \left(\frac{1\ \text{g}\cdot\cancel{\text{C}°}}{0.900\ \cancel{\text{J}}}\right)\left(\frac{1}{31\ \cancel{\text{C}°}}\right)(855\ \cancel{\text{J}}) = 31\ \text{g}$

(two significant figures)

PP 4.8

Solution: See Example 4.16.

$$T_C = \frac{(T_F - 32)}{(1.8)} = \frac{(104 - 32)}{(1.8)} = 40.°\text{C}$$

Note that 32 and 1.8 are exact numbers. One significant figure was lost in subtraction, and the answer has two significant figures.

Check: Would a reasonable answer in °C be greater than or less than 104°F?

PP 4.9

$T_K = T_C + 273 = -12 + 273 = 261\ \text{K}$

Answers to Chapter Review

1. division
2. organizing
3. sorting
4. sequencing
5. dimensional analysis
6. algebra
7. outline
8. units
9. defining
10. conversion factors
11. inverse
12. unknown
13. matching
14. unwanted
15. equation
16. m/V
17. Celsius
18. $\frac{q}{m \times \Delta T}$
19. heat

5

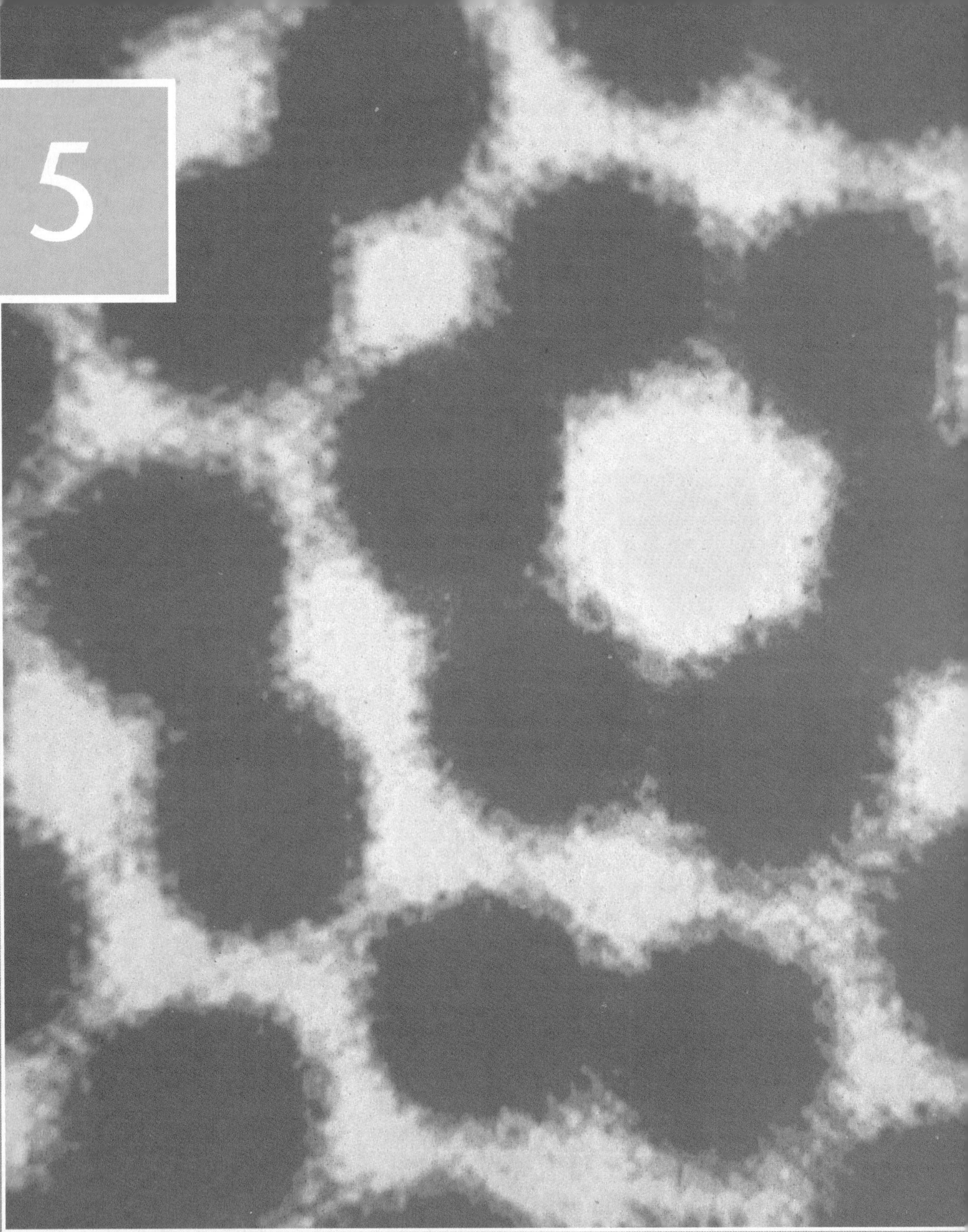

The Structure of Atoms. The Periodic Table

Contents

Silicon atoms as seen through a tunneling electron microscope.

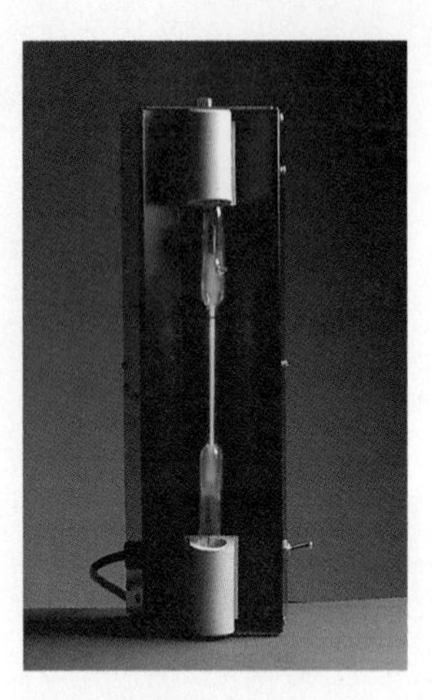

A cathode ray tube.

You Probably Know . . .

- An atom is so small that nobody has ever seen one with the naked eye. Only with powerful instruments such as an electronic microscope have images of atoms been photographed.
- Although atoms are extremely small, they are made up of particles that are even smaller—mainly protons, neutrons, and electrons. These particles were discovered by scientists in the late nineteenth and early twentieth century in experiments using devices such as a cathode ray tube (CRT), which is similar to a television picture tube.
- Metals are elements that have characteristic physical properties. Nonmetals, such as the neon used in neon signs, are elements that typically have properties different from those of metals.
- Two elements can combine to form a compound (Section 2.3). An example is burning magnesium, which is the combination of magnesium and oxygen to form magnesium oxide.

Burning magnesium.

Neon signs create a bright street scene at night.

Why Should We Study the Structure of Atoms?

All matter is composed of various combinations of elements. About 100 elements are known, each of which has atoms that are different from those of the other elements. How do the atoms of one element differ from the atoms of another?

Some elements are solids, some are liquids, and others are gases at room temperature. Some elements, such as metals, have similar characteristic properties that are very different from those of other elements, such as oxygen, nitrogen, and other gaseous elements. Why do certain elements have similar physical properties that are radically different from others?

Some elements, such as oxygen, are very reactive with other substances. Other elements, such as gold (a solid) and neon (a gas) are not. How do we account for the differences in chemical properties?

Answers to these questions require an understanding of the structure of atoms. In this chapter we begin our study of atomic structure and the periodic table. In Chapter 6 we will take a close look at electron structure of atoms. Our study of atomic structure will prepare us to understand the physical and chemical properties of elements and compounds.

Atomic Theory

Learning Objectives

- Summarize Dalton's atomic theory.
- State the law of multiple proportions.

Have you ever looked at a grain of sand and wondered about the atoms in that small speck? What is the nature of those tiny unseen particles of which matter is composed? It was just that sort of curiosity that led the Greeks, many centuries ago, to propose the concept of an indivisible particle of matter, which they called an atom.

The Greek ideas about atoms were practically forgotten for more than 2000 years. They contributed little to our understanding of the nature of matter until they were revived by John Dalton in the early nineteenth century.

John Dalton (1776–1844), an English schoolteacher and chemist, based his atomic theory on observations made by scientists during the previous century. He built upon the ancient Greek ideas to explain these observations. Dalton's *atomic theory* can be summarized as follows:

1. All matter is made up of tiny particles called atoms.
2. Atoms are indivisible and indestructible.
3. The atoms of an element are identical. They all have the same mass and are the same size.
4. Atoms of one element can combine with the atoms of another to form compounds. Combinations of these atoms occur in simple numerical ratios such as 1:1, 1:2, 2:3, and so on.

John Dalton (1776–1844) proposed an atomic theory to explain many scientific observations of the previous century.

Although Dalton's theory was not immediately widely accepted, some of its key features have withstood the test of time and are valid today. One important example is now stated as the **law of multiple proportions.** It says that if two elements combine in different proportions to form more than one compound, the masses of one element that will combine with a given mass of the other element are in a ratio of small whole numbers. To illustrate, consider two compounds of carbon and oxygen: carbon dioxide, CO_2, and carbon monoxide, CO. The ratio of the masses of oxygen combined with a given mass of carbon in the two compounds is 2:1, a small, whole-number ratio.

Dalton's theory presents a picture of an atom as a tiny particle, too small to be seen even with the most powerful conventional microscope. However, atoms of some of the elements have been photographed with a tunneling electron microscope (Figure 5.1).

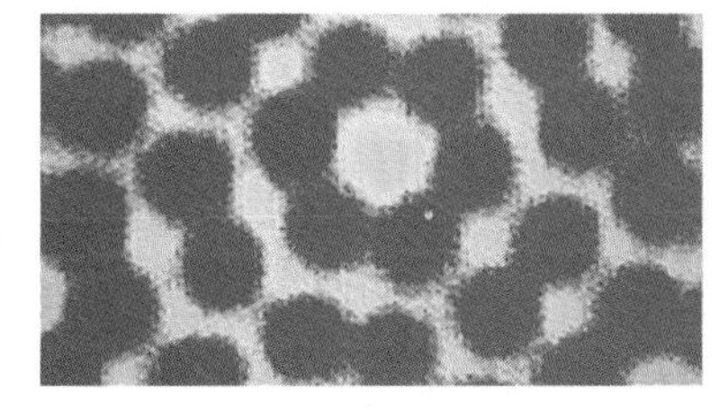

FIGURE 5.1 Surface atoms on a crystal of silicon are shown in this photograph taken through a tunneling electron microscope.

Experiments conducted after Dalton's time by chemists and atomic physicists have led to two revisions of his theory. First, discovery of the existence of isotopes showed that not all atoms of an element have the same mass. We will discuss isotopes later in this chapter. The second revision, arising from the discovery of subatomic particles, led to changes in the course of history as we entered the atomic age. Atoms are not indivisible, but can be split into smaller atoms and particles with the release of tremendous amounts of energy. Atomic energy can be used for peaceful purposes such as to generate electricity, but it also has great destructive potential.

5.2 Subatomic Particles

Learning Objective

- Specify the relative masses and charges of the proton, electron, and neutron.

Atoms are made up of electrons, protons, and neutrons.

Atoms are no longer thought to be the smallest particles of matter. We now understand that they are made up principally of electrons, protons, and neutrons. Because these particles are smaller than atoms, they are referred to as **subatomic particles.**

Dalton's idea of the indivisible atom was revised after the discovery of natural radioactivity by Henri Becquerel in 1896. Becquerel observed that some natural materials, such as uranium and radium ores, spontaneously emit penetrating radiations.* This observation suggested that atoms have some kind of substructure. The discovery and description of the electron by J. J. Thomson (1856–1940) in 1897 provided more evidence for an atomic substructure. This encouraged the search for other subatomic particles, resulting in the discovery of the proton and the neutron in the years that followed.

Mass of a proton = mass of a neutron = 1 amu
Mass of an electron = ~0 amu

The properties of the electron, proton, and neutron are summarized in Table 5.1. The **electron** has been assigned an electrical charge of -1. Its mass, 9.109×10^{-28} g, is extremely small. The **proton** was identified by Ernest Rutherford in 1919. It has an electrical charge of $+1$, equal to the charge on an electron but opposite in sign. The mass of a proton is about 1837 times greater than the mass of an electron. The third subatomic particle is the **neutron,** discovered by James Chadwick in 1932. The mass of a neutron is approximately the same as the mass of a proton, but it has no electrical charge, as its name implies.

Recognizing the existence of subatomic particles was a major step toward unravelling the mysteries of the atom.

TABLE 5.1 Properties of Subatomic Particles

Particle	*Symbol*	*Charge*[a]	*Mass, g*	*Mass, amu*[b]
electron	e^-	-1	9.109×10^{-28}	~0
proton	p	$+1$	1.673×10^{-24}	~1
neutron	n	0	1.675×10^{-24}	~1

[a]Relative electrical charge
[b]An atomic mass unit, or amu, is a small unit of mass used to express the masses of atoms and subatomic particles. See Section 5.5.

*Marie Curie, a contemporary of Becquerel, was the first to isolate the element radium. Natural radioactivity will be discussed in Chapter 18.

The Nucleus and Atomic Number

5.3

Learning Objectives

- Briefly describe and interpret Rutherford's experiments.
- Identify the number of protons and electrons from the atomic number of an atom.

At this point, you may wonder: if electrons, protons, and neutrons are building blocks of an atom, how do these blocks fit together? Part of the answer to that question was provided in 1911 by the experiments of Ernest Rutherford (1871–1937), a New Zealander who carried on his research in England. Rutherford and his students conducted a series of experiments in which a beam of **alpha particles** (helium ions, He^{2+}) was directed at a thin sheet of metal foil. These experiments are described in Figure 5.2.

Interpretation of Rutherford's experiments (Figure 5.3) led to some very important conclusions about atoms:

1. At the center of every atom is a tiny, very dense region called a **nucleus.** All of the positive charge and most of the mass of the atom is concentrated in the nucleus.
2. Outside the nucleus is a very large region of almost empty space. The electrons are found here, so this is a region of negative charge. The total negative charge exactly balances the positive charge of the nucleus.

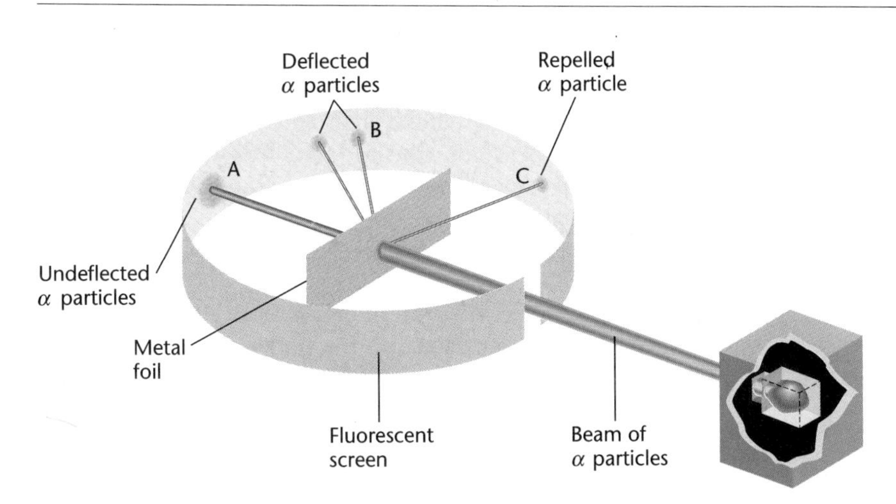

FIGURE 5.2 Rutherford's experiments. A beam of alpha particles from a natural source of radioactivity was directed at a sheet of very thin metal foil. A fluorescent screen almost completely surrounded the foil. Most of the alpha particles were observed to pass through the foil undeflected. They caused flashes of light where they hit the screen behind the foil at (A). Some of the particles were deflected, hitting the screen at random points (B). A small number of alpha particles were repelled by the foil and hit the screen at (C).

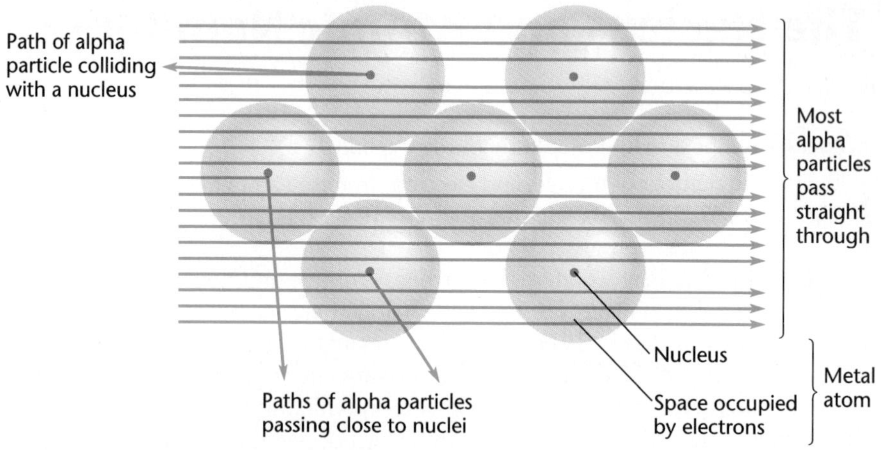

FIGURE 5.3 Interpretation of Rutherford's experiments. Atoms are pictured as being mostly empty space. At the center of each atom is its nucleus, a tiny region of concentrated mass and positive charge. The space surrounding the nucleus is extremely large by comparison and is sparsely populated by electrons. Most of Rutherford's alpha particles were able to penetrate the metal foil, easily passing through the almost empty space between the nuclei. A very small number of alpha particles passed close to nuclei and were deflected by their positive charge. Occasionally, an alpha particle collided with a nucleus and was repelled back toward the source.

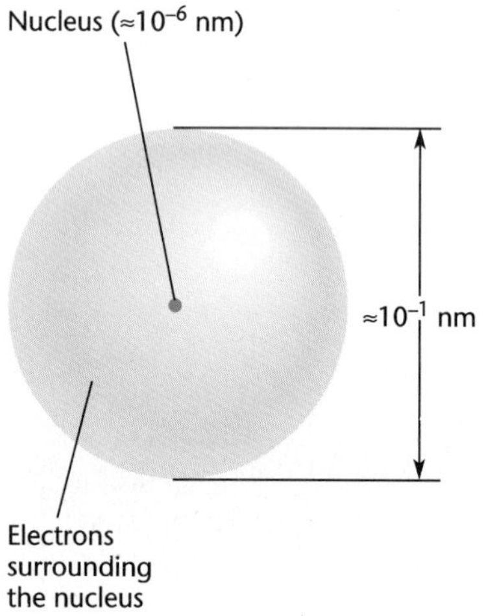

A picture of the atom is now beginning to take shape. At the center of the atom is its nucleus—positively charged and having very high mass density. Surrounding the nucleus is almost empty space, sparsely populated by electrons. The nucleus is extremely tiny, and the space around it is extremely large by comparison (Figure 5.4). To visualize these dimensions, consider a nucleus whose diameter is the width of your little finger nail (about 1 cm). The diameter of the atom would be about 1000 m, more than one-half mile!

Nucleus ($\approx 10^{-6}$ nm)

$\approx 10^{-1}$ nm

Electrons surrounding the nucleus

FIGURE 5.4 Dimensions of an atom. The diameter of an atom is approximately 0.1 nm, about 100 000 times the diameter of its nucleus.

The **atomic number** of an atom tells us the number of protons in its nucleus. Because a proton has a charge of +1, the atomic number, represented by the letter *Z*, also tells us the charge on the nucleus. Furthermore, because atoms are electrically neutral, the number of electrons is equal to the number of protons for an atom. Thus, for an atom:

atomic number = number of protons = number of electrons

In the periodic table (inside front cover), the atomic number is shown with the symbol for the element as a whole number. For example, the atomic number of carbon is 6, and the atomic number of uranium is 92. *Note that the atomic number of an element determines its identity.* All the atoms of a given element have the same atomic number and the same number of protons. Atomic number 6 always refers to carbon and never to boron (number 5) or nitrogen (number 7). All carbon atoms have 6 protons, all boron atoms have 5 protons, all nitrogen atoms have 7 protons, and so on.

The other number shown with the symbol for each element in the periodic table is the atomic mass of that element. However, before we discuss atomic mass, we need to consider the role of neutrons in atomic structure.

An element's atomic number determines its identity.

Isotopes and Mass Numbers

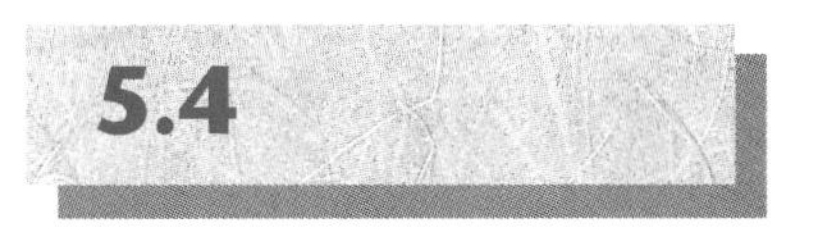

Learning Objectives

- Write the nuclear symbol for a given isotope of an element.
- Identify the number of protons and the number of neutrons for an atom from its nuclear symbol.

Dalton's atomic theory proposed that all atoms of an element are identical. However, since then evidence has shown that, for many elements, not all atoms have the same mass. Atoms of an element that have different masses are called **isotopes.** Most naturally occurring elements exist as mixtures of isotopes. Isotopes have different masses because they contain different numbers of neutrons. However, because isotopes are atoms of the same element, they have the same number of protons (that is, the same atomic number).

Isotopes are atoms of an element that have different numbers of neutrons and hence different masses.

One isotope can be distinguished from others by its **nuclear symbol.** The nuclear symbol for a nucleus includes the symbol for the element, its atomic number, and its mass number (Figure 5.5). The **mass number** is the sum of the protons and neutrons in the nucleus.* Let's illustrate by considering the isotopes of hydrogen. Hydrogen has an atomic number of 1, so every hydrogen atom has one proton in its nucleus. Ordinary hydrogen (sometimes called protium) does not have any neutrons, so its mass number is also 1. Thus, the nuclear symbol for this isotope of hydrogen is

Mass number = number of protons + number of neutrons.

Protium, deuterium, and tritium are isotopes of hydrogen. They have the same atomic number but different mass numbers.

Only isotopes of hydrogen have specific names.

$${}^{1}_{1}\mathrm{H}$$

Another isotope of hydrogen has one neutron. Thus, the sum of the protons and neutrons is 2, and this isotope has a mass number of 2. It is called *deuterium,* and its nuclear symbol is

$${}^{2}_{1}\mathrm{H}$$

There is a third isotope of hydrogen called *tritium.* Tritium has two neutrons, so its mass number is 3. (Remember, "tri" means "3.") The nuclear symbol for tritium is

$${}^{3}_{1}\mathrm{H}$$

Because a name or symbol is sufficient to identify an element, the atomic number is sometimes omitted. We could refer to hydrogen-3 and understand that the number refers to the mass number for this isotope of hydrogen. Thus, an isotope is named by writing its mass number after the name of the element.

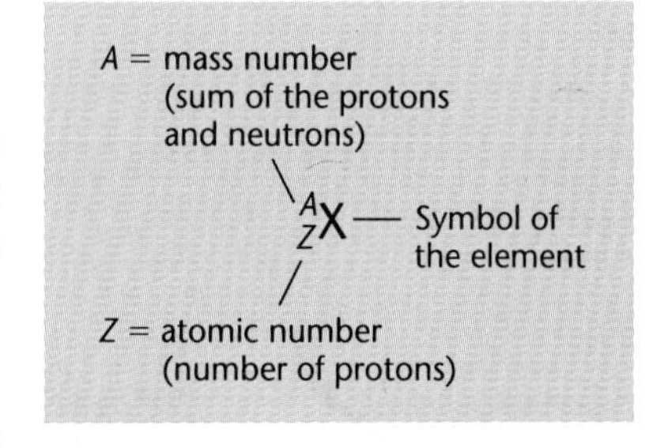

FIGURE 5.5 A nuclear symbol

---

*The mass number is a whole number. The exact isotopic mass is slightly different but rounds off to the same whole number. Thus, the mass number of an atom is approximately equal to its mass in amu.

Carbon-12, carbon-13, and carbon-14 are isotopes of carbon. Because all carbon atoms have atomic number 6, the nuclear symbols are written as follows:

$$^{12}_{6}C \qquad ^{13}_{6}C \qquad ^{14}_{6}C$$

Abbreviated symbols can be written without the atomic number as ^{12}C, ^{13}C, and ^{14}C.

Remember, the mass number is the sum of the protons and neutrons and is approximately the mass of the atom in amu. It can be calculated from the equation:

$$\text{mass number} = \text{number of protons} + \text{number of neutrons}$$

$$A = Z + \text{number of neutrons}$$

From the nuclear symbol of an atom you can determine the number of protons and neutrons in its nucleus. Rearranging the equation gives

$$\text{number of neutrons} = \text{mass number} - \text{number of protons}$$

$$= A - Z$$

For carbon-14:

$$\text{number of neutrons} = 14 - 6 = 8$$

Atomic number of an atom
= number of protons
= number of electrons

Table 5.2 summarizes the atomic data for selected isotopes. Note again that the number of electrons and the number of protons in a neutral atom of an element are always equal.

Example 5.1

How many neutrons are in each of the following atoms?

a. carbon-13
b. nitrogen-15
c. phosphorus-31 ■

Solutions Number of neutrons = mass number − atomic number.

a. for carbon-13, number of neutrons = 13 − 6 = 7
b. for nitrogen-15, 8 neutrons
c. for phosphorus-31, 16 neutrons

TABLE 5.2 Atomic Data for Selected Isotopes

Element	*Atomic Number*	*Mass Number*	*Nuclear Symbol*	*Number of* *Protons*	*Number of* *Electrons*	*Number of* *Neutrons*
boron	5	11	$^{11}_{5}B$	5	5	6
fluorine	9	19	$^{19}_{9}F$	9	9	10
sodium	11	23	$^{23}_{11}Na$	11	11	12
neon	10	20	$^{20}_{10}Ne$	10	10	10
iron	26	57	$^{57}_{26}Fe$	26	26	31

CHEMICAL WORLD

Biological Clocks, Isotopes, and Scientific Prediction

You live in Connecticut, and it's deep winter when you win a week's stay in sunny California. Just picture it: sleeping late every morning, brunching, surfing in the afternoons.

But in California you're always dead tired. Every morning you wake at 4 A.M. wired and ready to go and can't get back to sleep for several hours, sometimes missing brunch. It's the same every morning until the day before you have to go home. What's wrong with you? Don't worry, it's only that your biological clock is not made for such quick trips. You're still waking up for your chemistry class, just as you did every morning in Connecticut at 7 A.M. Eastern Standard Time.

The ability to keep time is found in many animals and plants, even in one-celled organisms. Biologists can predict within 15 minutes when hamsters will get on and off their exercise wheels, even if these hamsters are kept in total darkness, perfectly isolated from all external clues of time passage. Potatoes consume more oxygen in the daytime even when they are kept in lightproof containers that are held at constant temperature and pressure. Also consider the surprising behavior of the one-celled algae that live on Cape Cod, Massachusetts. These tiny plants bury themselves in the sand just before the Cape Cod tide comes in and resurface just after the tide goes out. This habit keeps them safe from rushing water. When the algae are moved to a laboratory miles from the beach and are placed in an environment that has constant light and constant temperature, they still bury themselves and resurface in synchrony with the tide at Cape Cod. Similar observations have been made of various organisms sent into outer space, far from every terrestrial time clue, even the earth's rotation.

To understand how life-forms can keep time within minutes per day, we must find out what all living things have in common that recurs and that takes place at a constant speed. It may startle you to learn that the answer is not astronomical or geological, but biochemical. For years scientists have speculated that our metabolism (the chemical processes of building up and breaking down molecules that living things carry on to maintain life) may be governed by an *internal* clock. To test this idea, some scientists tinkered with the metabolism of a group of deer mice. To understand these experiments, we have to know how isotopes of an element are similar to each other and how they differ.

Isotopes of an element are chemically identical but physically different. Deuterium (^{2}H), which was used in the deer mice experiments, is an isotope of hydrogen (^{1}H). When reacted with fluorine, hydrogen produces hydrogen fluoride and deuterium produces deuterium fluoride. The reaction between hydrogen and fluorine takes place so rapidly that it is explosive, but the reaction between deuterium and fluorine is nonexplosive. Deuterium reacts more slowly than hydrogen because it is twice as massive.

The deer mice were given mixtures of ordinary water ($^{1}H_2O$) and "heavy" water ($^{2}H_2O$) to drink. The two forms of water appear (and taste) identical, but heavy water reacts more slowly in metabolic processes. The experimenters predicted that the timekeeping of the deer mice's biological clocks would vary according to the concentration of heavy water they consumed. This is exactly what happened. The greater the heavy-water concentration, the slower their biological clocks ran.

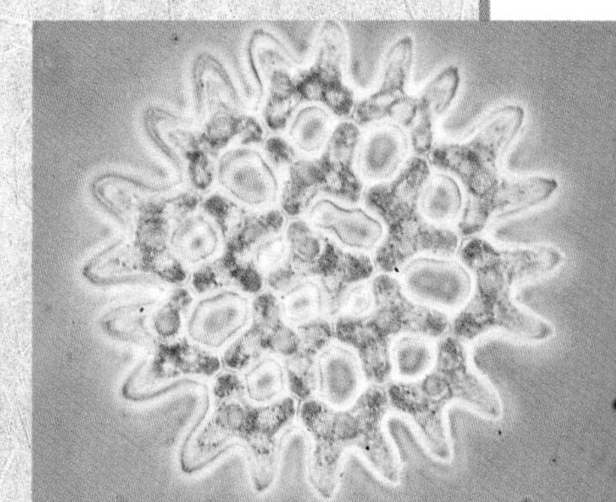

Even one-celled organisms, like this pediastrum, can keep time.

Today we think that the biological clock may be associated with the production of a certain protein. When the concentration of this protein is high, the organism decreases its production. Then, when the protein concentration falls below a preestablished level, the organism increases its production. (Such feedback loops are a common type of biological control.) The concentration of this protein rises and falls with fair regularity every 24 hours. This is probably the recurring, constant-rate event by which living cells tell time.

Scientists are currently linking problems such as your sleep disorder in California, and the more serious sleep disorders of aging people, with malfunctions of our biological clocks. They also are hopeful that study of these clocks will lead to a better understanding of some of our mental processes.

Example 5.2

Write nuclear symbols for the following:

a. radon-222
b. oxygen-18
c. uranium-238 ■

Solutions Find the symbol and atomic number of each element in the periodic table or in the table of elements in the front of the book.

a. $^{222}_{86}Rn$
b. $^{18}_{8}O$
c. $^{238}_{92}U$

Practice Problem 5.1. Determine the number of neutrons in each of the following atoms. (Solutions to practice problems appear at the end of the chapter.)

a. oxygen-18
b. $^{35}_{17}Cl$
c. $^{235}_{92}U$

Practice Problem 5.2 Write a nuclear symbol for each of the following:

a. helium-3
b. sulfur-35
c. iodine-123

5.5 Atomic Mass

Learning Objectives

- Write the atomic number and atomic mass for any element using information in the periodic table of the elements.
- Calculate the atomic mass of an element from the masses and abundances of its isotopes.

When Dalton proposed his atomic theory, the mass of an atom could not be measured directly. Even today the mass of a single large atom such as uranium-238 is far too small to be measured with even the most sensitive laboratory balance. Only recently have the masses of atoms been determined directly with an instrument

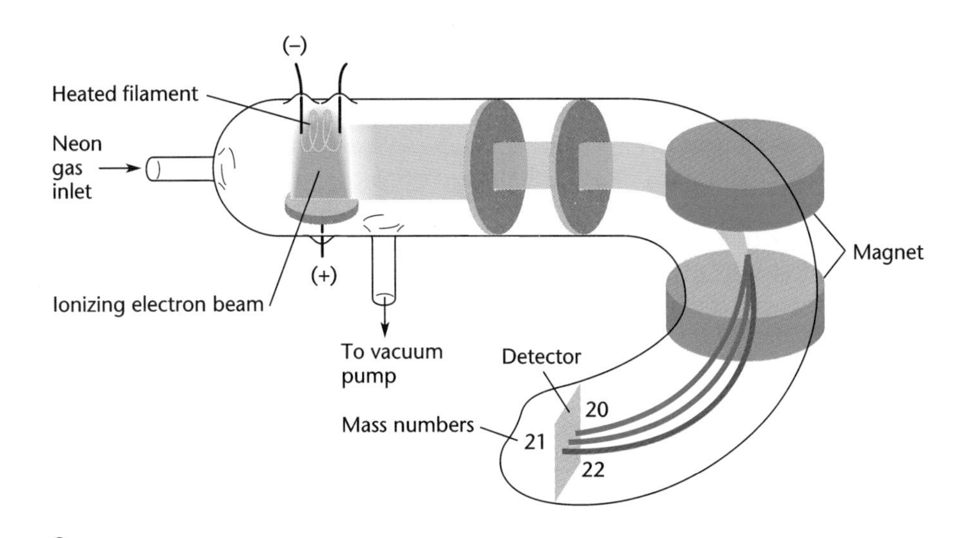

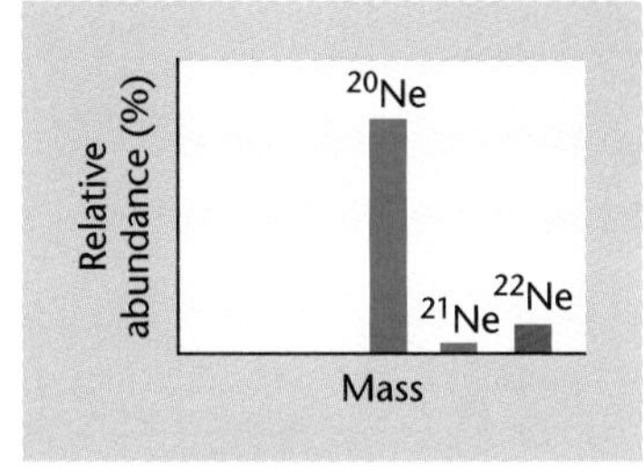

FIGURE 5.6 (a) The masses of the atoms in a sample of an element and their abundances can be determined with a mass spectrometer. After the sample is introduced into the instrument, the atoms are converted into a beam of positive ions. The positive ions are separated according to their masses. (b) The mass spectrum of neon shows three isotopes. Neon-20 is the most abundant isotope.

called a mass spectrometer (Figure 5.6).* For example, the mass of a single carbon-12 atom is 1.9926×10^{-23} g.

In comparing the masses of various atoms, it would be inconvenient to work with these very small numbers. It is much more convenient to have a unit that results in simpler numerical values, so atomic masses are often expressed in **atomic mass units,** or amu. One **amu** is defined as 1/12 the mass of a carbon-12 atom and is equal to a mass of 1.6605×10^{-24} g.†

The masses of a proton and a neutron are both very close to 1.0 amu. The mass of the electron is much smaller, only 0.000 549 amu. For most calculations, we can consider the electron's mass to be approximately 0 amu (Table 5.1).

You have already learned that most elements occur naturally as mixtures of isotopes. The abundance of each isotope of an element can be expressed as the percentage of the total number of atoms making up a sample of the element. For example, carbon-12 is the most abundant isotope of carbon. Its natural abundance can be expressed as follows:

$$\text{abundance of carbon-12 (\%)} = \frac{\text{number of atoms of carbon-12}}{\text{total number of carbon atoms in the sample}} \times 100$$

A mass spectrometer can determine the abundance of the isotopes of an element as well as their masses. The natural abundance of carbon-12 is 98.89%. The natural abundance of other isotopes of carbon can be expressed in the same way. Only 1.11% of carbon is carbon-13, and only a trace of carbon-14 occurs naturally.

*A mass spectrometer is a powerful analytical instrument. Today, mass spectrometers are widely used in the analysis of environmental samples and drugs, and in a wide variety of research problems.

†The SI abbreviation is "u," but "amu" is commonly used.

The **atomic mass** (*AM*) of an element is an average mass of its atoms calculated from the masses of the isotopes and their abundances in a naturally occurring sample.* For example, a sample of chlorine gas is made up of 75.77% chlorine-35 (mass = 34.97 amu) and 24.23% chlorine-37 (mass = 36.97 amu). The atomic mass of chlorine can be calculated to be 35.45 amu, as illustrated in the following example. Note that no single chlorine atom has a mass of 35.45 amu. The atomic mass, 35.45 amu, is an average value based on the masses of the individual isotopes and their natural abundances.

Example 5.3

Calculate the atomic mass of chlorine from the mass and abundance of each isotope. ■

Solution The data needed for this calculation are given in the previous paragraph. The atomic mass is calculated by multiplying the fraction of each isotope (from its percentage) times its mass and adding the resulting masses.

$$AM = \left(\frac{75.77}{100}\right)(34.97\text{ amu}) + \left(\frac{24.23}{100}\right)(36.97\text{ amu}) = 35.45\text{ amu}$$

Example 5.4

A sample of copper was found to be made up of 69.09% copper-63 (mass = 62.93 amu) and 30.91% copper-65 (mass = 64.93 amu). Calculate the atomic mass of copper. ■

Solution

$$AM = \left(\frac{69.09}{100}\right)(62.93\text{ amu}) + \left(\frac{30.91}{100}\right)(64.93\text{ amu}) = 63.55\text{ amu}$$

Practice Problem 5.3 Calculate the atomic mass of magnesium from the following isotopic composition data:
magnesium-24 (mass = 23.99 amu), 78.70%
magnesium-25 (mass = 24.99 amu), 10.13%
magnesium-26 (mass = 25.98 amu), 11.17%

***Atomic mass* is a better term, but **atomic weight** has been used for many years and is still commonly used.

As you look at the periodic table of the elements, you will notice that the atomic masses of most of the elements are not whole numbers. Now you can see why. We have just said that the atomic mass of an element is the average of the masses of its isotopes. You will use atomic masses frequently in calculations as you learn more about the composition of chemical substances and their chemical reactions.

The periodic table gives the atomic number and the atomic mass of each element.

6	— Atomic number
C	
12.01	— Atomic mass

The Periodic Table

5.6

Learning Objectives

- Describe the organization of the elements in the periodic table.
- Explain what is meant by the periodic law.

Imagine walking into a large library to look for a biography of Abraham Lincoln. Approaching the information desk, you ask where you might find Lincoln's biography. The librarian responds that the library has biographies of Lincoln but does not know where they are. You are told that this library has over 100 000 volumes covering many different topics. However, the books are shelved wherever people happened to put them. The librarian suggests that you just browse through the books to find what you need. You immediately realize that you may never find a biography of Lincoln in such a disorganized library. For information to be accessible and useful, it must be organized.

A wealth of information about elements and compounds has been gathered by scientists over the years. This information has been presented at conferences and published in books and scientific journals. As with books in a library, to be accessible and useful, this information must be organized. The *periodic table* of the elements is an organized and very abbreviated compilation of a vast amount of information that has proved to be highly useful to anyone wanting to understand chemistry. You have previously referred to the periodic table for certain information such as atomic numbers and atomic masses. Now, let's take a closer look at it and see how we can make use of still more of the information it provides in understanding chemistry.

By the nineteenth century, chemists had learned enough about the elements to be able to recognize similarities in the properties of groups of elements. Recognition of these similarities led to the publication of various tables for different groups of elements showing their similar properties. Then in 1869, the Russian chemist Dmitri Ivanovitch Mendeleev (1834–1907) and Lothar Meyer (1830–1895) of Germany independently published periodic tables of the elements. Later revision of these tables led to the modern periodic table. Although the original tables were organized according to atomic masses of the elements, the elements in the modern table are arranged by atomic numbers (Figure 5.7).

The word "periodic" means recurring at regular intervals. As we study the periodic table, we will find that similar chemical and physical properties of the elements recur periodically when the elements are arranged according to their atomic numbers. This statement is called the **periodic law.** As you continue your

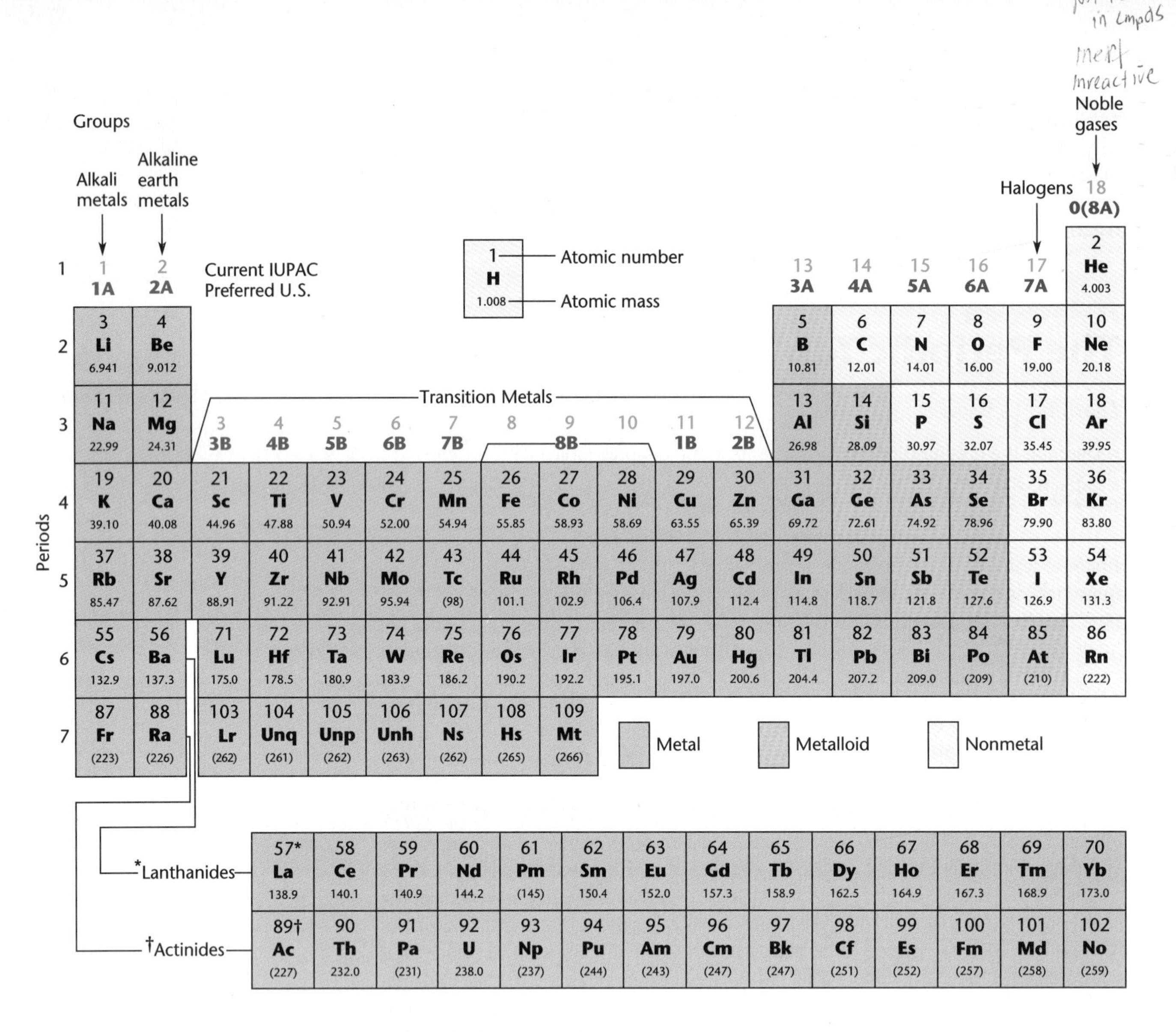

Note: The atomic masses shown in this table are the 1987 IUPAC atomic masses, which are based on carbon-12 = 12 u, rounded to four significant figures (if four are available). Masses in parentheses are masses of the most stable isotope of radioactive elements.

FIGURE 5.7 The periodic table of the elements.

study, you will learn how the position of an element in the periodic table tells us some of its important chemical and physical properties. We will begin by noting several broad classifications of the elements based upon their positions in the table.

Periodic law: Similar chemical and physical properties of the elements recur periodically when the elements are arranged according to their atomic numbers.

Language of the Periodic Table

5.7

Learning Objectives

- Identify the representative elements, transition elements, actinides, and lanthanides.
- Name Groups 1A, 2A, 7A, and 8A in the periodic table.

In a modern periodic table (Figure 5.7) the elements are arranged horizontally in numerical order, by atomic number. The resulting rows of elements are called **periods.** Elements are also arranged in columns called **groups** or families of elements. Elements in a group have similar properties; elements in a period generally have dissimilar properties. Periods are often numbered 1, 2, 3, and so on. In the system used in the United States, each group of elements has a number-and-letter designation. Either Arabic numbers or Roman numerals may be used. The International Union of Pure and Applied Chemistry (IUPAC) has approved a system in which the groups are numbered from left to right with numbers 1 through 18.

Elements in groups designated 1A–8A are called the **representative elements,** or **main-group elements.** There are two blocks of these elements. The block on the left (Groups 1A and 2A) is separated from the block on the right (Groups 3A–8A) by the **transition elements.** The number of each group of transition elements is labeled with the letter B. A fourth block, the **lanthanides** and **actinides,** are offset from the other elements to reduce the width of the table. The significance of the grouping of elements in blocks is discussed further in Section 6.6.

Some of the groups of representative elements have names. The elements of Group 1A are called the **alkali metals.** Group 2A elements are the **alkaline earth metals.** The **halogens** are the elements in Group 7A. The elements of Group 0 (often numbered 8A) are called the **noble gases.** These names are widely used when referring to the elements in these groups, so you should learn them. The other groups of representative elements are generally referred to by their respective group number.

More will be said about the properties of the various groups of elements in Chapter 6 and in chapters that follow.

Metals, Nonmetals, and Metalloids

5.8

Learning Objectives

- Identify elements as metals, nonmetals, and metalloids.
- Write formulas for cations formed from elements in Groups 1A, 2A, and 3A.
- Write formulas for monatomic anions from elements in Groups 5A, 6A, and 7A.
- Write formulas for binary ionic compounds.
- Write a balanced chemical equation for the reaction of a metal with a nonmetal.

Elements are classified as metals, nonmetals, and metalloids. The majority of the elements are metals. Remember, metals are found toward the left side of the periodic table (inside front cover), and nonmetals are on the right side. Elements between the metals and the nonmetals in the periodic table are called metalloids because their properties are in between those of the metals and the nonmetals. Some of the properties of metals, nonmetals, and metalloids are shown in Table 5.3.

FIGURE 5.8 Some familiar metals: aluminum foil, copper wire, mercury (in the dish), a steel spoon, and an iron cylinder.

Metals

Based upon your experience with metals, you may be able to recall some of their typical physical properties. Thinking of a stainless steel knife or spoon, or gold or silver jewelry, you may think of metals as elements that can be polished to a shiny luster (Figure 5.8). Or, you may think of a pan that you use in cooking and remember that a metal is a substance that is a good heat conductor. You may also think of the electrical wiring in your home and realize that a metal can be a good conductor of electricity. Metals can also be formed or bent into a variety of shapes and drawn into wires. In other words, metals are malleable and ductile. These are all physical properties of metals (see Table 5.3).

Cation: A positively charged atom (Section 2.3).

However, the chemical properties of metals are of greater interest to chemists than their physical properties. In describing the chemical properties of a metal, a chemist would say a **metal** is an element that has a tendency to lose one or more electrons in a chemical reaction to form a positive ion (cation). In general, the tendency to lose electrons is greatest for the elements on the left side of the periodic table.

We can often identify the charge on a metal ion from the position of the metal in the periodic table in accordance with the periodic law. For example, sodium (number 11) is found in Group 1A, and the formula for a sodium ion is Na^+. All the metals in Group 1A form ions having a charge of +1, such as Li^+ and K^+. The metals in Group 2A all form ions having a charge of +2, such as Mg^{2+}, Ca^{2+}, and Ba^{2+}. Aluminum (Group 3A) forms the Al^{3+} ion. In showing charges for ions, it is customary to write the + or − sign after the number (a superscript), although the number "1" is not written. This correlation of ionic charge with group number is explained in Section 6.8.

TABLE 5.3 Properties of Metals, Nonmetals, and Metalloids

Element Class	*Physical Properties*	*Chemical Properties*
metals	lustrous malleable, ductile electrical conductors thermal conductors most are solids	lose electrons
nonmetals	not lustrous poor electrical conductors poor thermal conductors brittle if solid many are gases	gain or share electrons
metalloids	poor electrical conductors poor thermal conductors most are solids	gain or share electrons

Remember that a cation has fewer electrons than its corresponding metal atom. For example, a sodium atom (atomic number 11) has 11 protons (+11) and 11 electrons (−11), resulting in no net charge. However, a sodium ion, Na^{+}, has 11 protons (+11) and 10 electrons (−10), resulting in a +1 ionic charge. Note that when an ion forms, the nucleus remains unchanged. Only the number of electrons changes.

Some metals found in other A-groups and in the B-groups form two cations of different charges. For these metals, the charge on a cation is not necessarily the same as the group number. Some of these cations are considered in Section 9.5.

Metal cations (Groups 1A, 2A, 3A):

charge = +(group number)

Nonmetals

In general, nonmetals have physical properties quite different from those of metals (see Table 5.3). About two-thirds of the nonmetals are gases. Only bromine is a liquid. Those that are solids are generally poor electrical and thermal conductors, are brittle, and are not malleable or ductile. For example, carbon in its common forms, such as charcoal and soot, bears little resemblance to a metal,* nor does sulfur, a yellow solid that melts at a low temperature.

The chemical behavior of nonmetals is also different from that of metals. In general, a **nonmetal** has a tendency to gain one or more electrons in a chemical reaction. This is a general chemical property of nonmetals. Note, however, that the elements in Group 8A (noble gases) do not exhibit a tendency to gain electrons like other nonmetals and are not chemically active. In general, the tendency to gain electrons is greatest for the elements near the top of Groups 6A and 7A.

A metal reacts with a nonmetal to give an ionic compound made up of cations formed from the metal and anions (negative ions) formed from the nonmetal.† As

Nonmetals are
gases: noble gases (helium, neon, argon, krypton, xenon, radon)
halogens (fluorine, chlorine)
others (hydrogen, nitrogen, oxygen)
solids: carbon, phosphorus, sulfur, selenium, iodine, astatine
liquid: bromine

Carbon and sulfur are typical nonmetals.

a

b

c

Halogens. (a) Chlorine gas with liquid chlorine (inner tube) under pressure. (b) Liquid bromine (bottom of the bottle), with bromine vapor above it. (c) Iodine crystals hanging from a glass surface with iodine vapor filling a beaker.

*However, note that one form of carbon, called graphite, is the only natural nonmetallic substance that conducts electricity.

†When the elements reacting are both nonmetals, a molecular rather than an ionic compound is formed (Section 2.3). For example:

$$C + O_2 \longrightarrow CO_2$$

The differences between ionic and molecular compounds are discussed in some detail in Section 8.3.

we did with cations, we can determine the charge on the anion from the position of the nonmetal in the periodic table. For example, chlorine is found in Group 7A, and the formula for a chloride ion is Cl^-. All the elements in Group 7A form monatomic ions (made up of only one atom) having a charge of -1. The nonmetals in Group 6A form monatomic ions having a charge of -2, such as O^{2-} and S^{2-}. Thus, the charge on a monatomic anion is determined from the group number for the nonmetal:

Anions formed from the halogens are called halide *ions, such as Br^-, bromide ion (Section 2.3).*

$$\text{charge on anion} = -(8 - \text{group number})$$

Monatomic anion: charge = $-(8$ − group number).

Remember, the $-$ sign follows the number in the formula of the anion. This relationship of ionic charge with group number is explained in Section 6.8.

We should emphasize that an anion has a negative charge because it has more electrons than its corresponding nonmetal atom. For example, a chlorine atom (atomic number 17) has 17 protons ($+17$) and 17 electrons (-17) and is neutral. However, a chloride ion, Cl^-, has 17 protons ($+17$) and 18 electrons (-18), resulting in an ionic charge of -1. Also note that when an ion forms, the nucleus remains unchanged. Only the number of electrons changes.

Formation of ions by a loss or gain of electrons is covered in Section 6.8 and Chapter 17.

As was noted earlier (Section 2.3), the names of the monatomic anions and their compounds end with *-ide*. For example:

O^{2-}, oxide ion CaO, calcium oxide

S^{2-}, sulfide ion Na_2S, sodium sulfide

Monatomic anion: Use -ide *suffix (Section 2.3).*

This discussion illustrates something important about the position of elements in the periodic table. *Elements in the same group have similar chemical properties.* Elements in different groups have dissimilar chemical properties. Many more examples of this point will be presented in later chapters.

As was mentioned earlier (Section 2.4), chemical reactions can be described using chemical equations. For example, the chemical equation for the reaction of sodium with chlorine to form sodium chloride is

$$2Na + Cl_2 \rightarrow 2NaCl$$

When writing the formula for chlorine, we must remember that it is one of the diatomic elements. Note that although sodium chloride is ionic, the numbers of Na^+ ions and Cl^- ions are equal, and hence the compound is neutral. The charges on the ions are not shown in its formula. The coefficients—"2" for Na and for NaCl—are needed to balance the equation (law of conservation of matter, Section 2.5).

When the charges on the cation and anion are not numerically equal, subscripts must be written in the formula of the compound to balance the charges. For example, calcium chloride is made up of Ca^{2+} and Cl^- ions, and its formula is $CaCl_2$. The subscript 2 (Section 2.3) means that there are two Cl^- ions for each Ca^{2+} ion. This 2:1 ratio of -1 anions to $+2$ cations is necessary to have a neutral compound:

$$CaCl_2 \quad \text{means} \quad Ca^{2+} + \begin{matrix} Cl^- \\ Cl^- \end{matrix}$$

Writing formulas of ionic compounds is covered in greater detail in Section 9.5.

Example 5.5

Write formulas for the following ions:

a. magnesium ion
b. rubidium ion
c. fluoride ion
d. sulfide ion ■

Solutions

a. Mg^{2+} (Group 2A)
b. Rb^{+} (Group 1A)
c. F^{-} [ionic charge = −(8 − group number)]
d. S^{2-} [ionic charge = −(8 − group number)]

Example 5.6

Give the number of protons and electrons for each of the ions in Example 5.5. ■

Solutions

a. Number of p's = atomic number = 12
ionic charge = sum of + charges (protons) and − charges (electrons)
+2 = 12+ (protons) + 10− (electrons)
Number of e^-'s = 10

b. Number of p's = 37
ionic charge = +1 = +37 (protons) + −36 (electrons)
Number of e^-'s = 36

c. Number of p's = 9
ionic charge = −1 = +9 (protons) + −10 (electrons)
Number of e^-'s = 10

d. Number of p's = 16
ionic charge = −2 = +16 (protons) + −18 (electrons)
Number of e^-'s = 18

Example 5.7

Write formulas for the ions that make up the following compounds, then write the formula for the compound.

a. potassium chloride
b. lithium oxide
c. sodium sulfide
d. magnesium bromide ■

Solutions

a. The ions are K^+ and Cl^-. A 1:1 ratio of the cation and anion is needed to balance the charges. Thus, the formula of the compound is KCl.

b. Li^+ and O^{2-}. A 2:1 ratio of the ions is needed to balance the charges: Li_2O

c. Na^+ and S^{2-}. A 2:1 ratio of the ions is needed to balance the charges:

$$\begin{matrix} Na^+ \\ Na^+ \end{matrix} + S^{2-} \text{ resulting in the formula } Na_2S$$

d. Mg^{2+} and Br^-. A 1:2 ratio of the ions is needed to balance the charges:

$$Mg^{2+} + \begin{matrix} Br^- \\ Br^- \end{matrix} \text{ resulting in the formula } MgBr_2$$

Example 5.8

Write chemical equations for the following reactions. Be careful to write correct formulas for diatomic elements and for the compounds that are formed in the reactions.

a. Sodium reacts with bromine to form sodium bromide.
b. Magnesium reacts with oxygen to form magnesium oxide.
c. Barium reacts with chlorine to form barium chloride.
d. Potassium reacts with sulfur to form potassium sulfide. (Consider sulfur to be monatomic for this exercise.) ■

Solutions

a. $2Na + Br_2 \longrightarrow 2NaBr$ (made up of Na^+ and Br^-)
b. $2Mg + O_2 \longrightarrow 2MgO$ (made up of Mg^{2+} and O^{2-})
c. $Ba + Cl_2 \longrightarrow BaCl_2$ (made up of Ba^{2+} and $2Cl^-$)
d. $2K + S \longrightarrow K_2S$ (made up of $2K^+$ and S^{2+})

Practice Problem 5.4 Write formulas for the following ions:

a. bromide ion
b. barium ion
c. cesium ion
d. iodide ion

Practice Problem 5.5 Write chemical equations for the following reactions. Show correct formulas for diatomic elements and for the compounds formed in the reactions.

a. Calcium reacts with bromine to form calcium bromide.
b. Magnesium reacts with sulfur to form magnesium sulfide.

Practice Problem 5.6 How many protons and electrons make up each of the ions in Practice Problem 5.4?

■ Metalloids

The **metalloids** are positioned in the periodic table (inside front cover) between the metals and the nonmetals. In general, metalloids have properties intermediate between those of metals and nonmetals. Most metalloids have nonmetallic chemical properties but have physical properties that border on being metallic (see Table

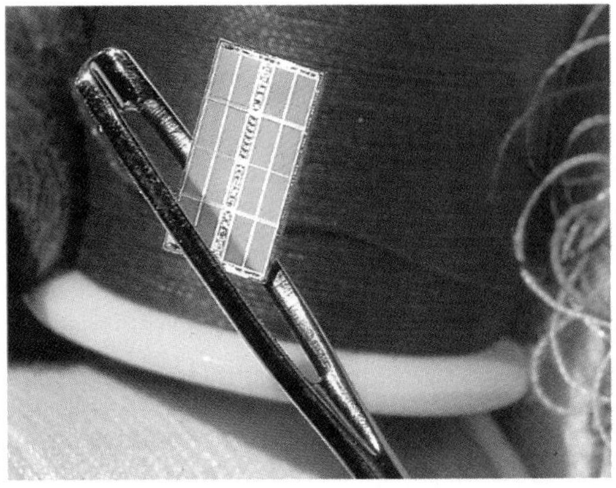

FIGURE 5.9 The electronics industry uses semiconductors in the manufacture of its products. *Left:* A computer. *Above:* A micro chip.

5.3). For example, antimony, Sb, has a bluish-white luster similar to many metals, but it is quite brittle like nonmetallic solids. Silicon and germanium are the principal raw materials for semiconductors, which are essential for the worldwide electronics industry (Figure 5.9).

Chapter Review

Directions: From the choices listed at the end of the Chapter Review, choose the word or term that best fits in each blank. Use each choice only once. Answers appear at the end of the chapter.

Section 5.1 John Dalton proposed his atomic theory in the early nineteenth century. He said that all matter is composed of tiny particles called __________ (1). He considered atoms to be indivisible. He also thought atoms of an element all had the same __________ (2).

Section 5.2 We now understand that atoms are not the smallest particles of matter. They are made up of subatomic particles called __________ (3), protons, and __________ (4). The __________ (5) and the neutron have essentially the same mass. The proton has a __________ (6) charge. The particle having a −1 charge is the __________ (7). Compared with the neutron and proton, its mass is essentially zero.

Section 5.3 __________ __________ (8), conducted experiments in which a beam of __________ __________ (9) was directed at a thin sheet of metal foil. From the results of these experiments, he concluded that the atom is mostly empty space surrounding a tiny __________ (10) which is positively charged and in which is concentrated most of the mass of the atom. The number of protons in an atom is its __________ __________ (11). The atomic number tells us the number of protons and the number of electrons making up an atom. The atomic number is the same for all atoms of an __________ (12).

Section 5.4 Atoms of an element that have different masses are called __________ (13). Most of the naturally occurring elements exist as __________ (14) of isotopes. Isotopes have different masses because they have __________ (15) numbers of neutrons. The nucleus of an atom may be represented by its __________ (16) symbol. This symbol includes the symbol for the element, its __________ (17) number, and its mass number, which is the __________ (18) of the protons and neutrons in the nucleus. Protium, __________ (19), and __________ (20) are names for the isotopes of hydrogen. The number of neutrons can be calculated by subtracting the atomic number from the __________ __________ (21).

Section 5.5 The __________ __________ (22) of an element is the average mass of the isotopes in a naturally occurring sample, taking into consideration the abundance of each. Masses of atoms and subatomic particles are expressed using units of (abbreviated) __________ (23).

Section 5.6 A __________ __________ (24) of the elements is an abbreviated and very useful compilation of information. The first tables were developed independently by __________ (25), a Russian, and Meyer, a German, in 1869. In the original tables, the elements were arranged according to their atomic __________ (26), but in modern tables elements are listed by atomic __________ (27). The variation of the chemical and physical properties of the elements when they are arranged according to their atomic numbers is called the __________ __________ (28).

Section 5.7 A __________ (29) is a horizontal row of elements in the periodic table. Elements are also found in columns called __________ (30) or families, which are identified by a number and letter. Elements in Groups 1A through 8A are called the __________ (31) elements or main-group elements. Those in the B-groups are called the __________ (32) elements. The elements called the lanthanides and actinides are offset from the others in the table.

Some of the groups of representative elements have group names. The elements in Group 1A are called __________ (33) metals. Those in Group 2A are the __________ __________ (34) metals. The __________ (35) are found in Group 7A, and the elements in Group 8A are called noble gases.

Section 5.8 Elements are classified as metals, __________ (36), and metalloids. The majority of the elements are __________ (37). In general, a metal is an element that has a tendency to __________ (38) one or more electrons to form a positive ion called a __________ (39). The charge on a metal ion of an element in Group 1A or 2A can be determined from its group number. For example, ions of Group 2A elements have a charge of __________ (40).

When a nonmetal reacts with a metal, an __________ (41) compound forms. The charge on a monatomic __________ (42) is $-(8 -$ __________ __________$)$ (43). Although made up of ions, an ionic compound has no net __________ (44). Its formula is written without showing the charges on the ions. When writing a chemical equation, remember to write correct formulas for the elements and compounds involved and to __________ (45) the equation.

Silicon and germanium, two of the __________ (46), are the principal raw materials for the semiconductors used in the electronics industry.

Choices: alkali, alkaline earth, alpha particles, amu, anion, atomic, atomic number, atomic mass, atoms, balance, cation, charge, deuterium, different, electron, electrons, element, Ernest Rutherford, group number, groups, halogens, ionic, isotopes, lose, masses, mass number, Mendeleev, metals, metalloids, mixtures, neutrons, nonmetals, nuclear, nucleus, numbers, period, periodic table, periodic law, +1, +2, proton, representative, sum, transition, tritium

■ Study Objectives

After studying Chapter 5, you should be able to:

Section 5.1

1. Summarize Dalton's atomic theory.
2. State the law of multiple proportions.

Section 5.2

3. Specify the relative masses and charges of the proton, electron, and neutron.

Section 5.3

4. Briefly describe and interpret Rutherford's experiments.
5. Identify the number of protons and electrons from the atomic number of an atom.

Section 5.4

6. Write the nuclear symbol for a given isotope of an element.
7. Identify the number of protons and the number of neutrons for an atom from its nuclear symbol.

Section 5.5

8. Write the atomic number and atomic mass for any element using information in the periodic table of the elements.
9. Calculate the atomic mass of an element from the masses and abundances of its isotopes.

Section 5.6

10. Describe the organization of the elements in the periodic table.
11. Explain what is meant by the periodic law.

Section 5.7

12. Identify the representative elements, transition elements, actinides, and lanthanides.
13. Name Groups 1A, 2A, 7A, and 8A in the periodic table.

Section 5.8

14. Identify elements as metals, nonmetals, and metalloids.
15. Write formulas for cations formed from elements in Groups 1A, 2A, and 3A.
16. Write formulas for monatomic anions from elements in Groups 5A, 6A, and 7A.
17. Write formulas for binary ionic compounds.
18. Write a balanced chemical equation for the reaction of a metal with a nonmetal.

Key Terms

Review the definition of each of the terms listed here by chapter section number. You may use the glossary if necessary.

5.1 law of multiple proportions

5.2 electron, neutron, proton, subatomic particle

5.3 alpha particle, atomic number, nucleus

5.4 isotopes, mass number, nuclear symbol

5.5 atomic mass unit (amu), atomic mass, atomic weight

5.6 periodic law

5.7 alkali metal, alkaline earth metal, group, halogen, main-group element, noble gas, period, representative element, transition element

5.8 metal, metalloid, nonmetal

Questions and Problems

The answers to questions and problems with blue numbers appear in Appendix C. Asterisks indicate the more challenging problems.

Atomic Theory (Section 5.1)

1. Summarize Dalton's atomic theory.
2. In what two respects did Dalton think the atoms of an element were identical?
3. State the law of multiple proportions.
4. What two revisions have been made in Dalton's atomic theory? Explain why these revisions were necessary.

Atomic Structure (Sections 5.2, 5.3, 5.4)

5. Summarize the properties of the three main subatomic particles by completing the following table.

Particle	Symbol	Mass, amu	Charge
proton	p		
		0	
			0

6. Describe the experiments of Ernest Rutherford.

7. Summarize the major conclusions from Rutherford's experiments.
8. What is an alpha particle? Write its nuclear symbol.
9. In what part of an atom is almost all of its mass found?
10. What subatomic particles make up the nucleus of an atom?
11. Compare the diameter of an atom with the diameter of its nucleus.
12. What is meant by the atomic number of an element?
13. What number tells us the charge on the nucleus of an atom?
14. What are isotopes?
15. Oxygen-16 and oxygen-18 are isotopes of oxygen. What do these numbers mean? What is the atomic number for these isotopes?
16. How many protons, electrons, and neutrons are found in each of the following atoms? Refer to the table of elements in the front of the book and to the periodic table as necessary.
    a. copper-63
    b. silver-109
    c. argon-40
17. Write nuclear symbols for the following. Refer to the table of elements in the front of the book and to the periodic table as necessary.
    a. helium-4
    b. carbon-14
    c. plutonium-244
    d. beryllium-10
    e. potassium-39
    f. chlorine-37
18. Write nuclear symbols for the following:
    a. deuterium
    b. tritium
19. Complete the following table:

Element	Nuclear Symbol	Atomic Number	Mass Number	Number of Protons	Number of Electrons	Number of Neutrons
helium			3			
sodium			22			
chlorine			35			
		29	65			
		26	57			
				92		143
			27	13		
			19	9		
arsenic						37
gold						118

Atomic Mass (Section 5.5)

20. Define *atomic mass unit*. Explain why amu's are used rather than grams to express the masses of atoms.
21. Explain why the atomic mass of an element is different from the mass number of the most abundant isotope of the element.
22. The two isotopes of chlorine are chlorine-35 (mass = 34.97 amu) and chlorine-37 (mass = 36.97 amu). The average of their masses is 35.97. Why is the atomic mass for chlorine in the periodic table 35.45 rather than 35.97, the average of the two masses?
23. Calculate the atomic mass of the element having the following isotopes: 51.82%, mass = 106.90 amu; 48.18%, mass = 108.90 amu. Identify the element by referring to the periodic table.
24. Sulfur is composed of 95.02% sulfur-32 (mass = 31.97 amu), 0.78% sulfur-33 (mass = 32.97 amu), and 4.20% sulfur-34 (mass = 33.97 amu). Calculate the atomic mass of sulfur.
25. Carbon consists of 98.89% carbon-12 (mass = 12 amu exactly) and 1.11% carbon-13 (mass = 13.003 amu). Calculate the atomic mass of carbon.
26. Refer to the periodic table and identify the atomic masses for the first ten elements.

*27. The mass of a carbon-12 atom has been determined to be 1.9926×10^{-23} g. Assuming the mass of the atom is the sum of the masses of its subatomic particles, calculate the mass of a carbon-12 atom from the data in Table 5.1. Can you suggest why the two masses for carbon-12 are different? (See Chapter 18.)

*28. From the mass of a carbon-12 atom given in problem 27, calculate the number of carbon atoms in 12.00 g of carbon.

*29. The mass of a sulfur-32 atom is 31.97 amu. From the mass of a carbon-12 atom given in problem 27, calculate the mass in grams of one sulfur-32 atom.

The Periodic Table (Sections 5.6, 5.7)

30. What number serves as the basis for the arrangement of the elements in the periodic table?
31. Who proposed the first periodic tables?
32. How were the elements arranged in the first periodic tables?
33. State the periodic law.
34. Explain the following terms:
    a. period
    b. alkali metals
    c. group
    d. representative elements
    e. transition elements
35. Where are each of the following found in the periodic table?
    a. halogens
    b. noble gases
    c. alkaline earth metals
    d. actinides
36. Which of the following are alkali metals?

    Na, Mg, Cu, Ag, N, P, K, Ca, Cs, Rn, Rb

37. Write the symbols for all the alkaline earth metals.

38. Which of the following are noble gases?

K, Kr, Rn, Rb, Xe, As, N, Ni, Ar, Ag, Au, H, He

39. Identify the halogens in the following list of elements:

C, Cl, Ca, Fe, F, I, In, S, Si, B, Br

40. Which of the following are transition metals?

Fe, Mg, Mn, Ca, Co, N, Ni, Pt, P

41. Give the group number for each of the following:
   a. halogens
   b. noble gases
   c. alkali metals
   d. alkaline earth metals

42. Identify the group number for each of the following elements:
   a. sulfur
   b. beryllium
   c. neon
   d. potassium
   e. lithium
   f. phosphorus

Metals, Nonmetals, and Metalloids (Section 5.8)

43. Describe some common physical properties of metals.

44. Describe the general chemical property of a metal.

45. Describe the general chemical property of a nonmetal.

46. Which of the following nonmetals is a liquid under ordinary conditions?

carbon, nitrogen, sulfur, bromine, iodine, neon

47. When a metal reacts with a nonmetal, what kind of compound results? Give an example.

48. What two elements are the principal raw materials for making semiconductors for the electronics industry?

49. Identify the nonmetal in each list of elements:
   a. K, Al, Ca, N, Ni
   b. Na, Mg, Ba, Br, Cu
   c. Rb, Cr, Fe, F, Mo

50. Identify the metal in each list of elements:
   a. S, Sr, Si, Se
   b. N, Ne, Ni, He
   c. Ca, C, Cl, Xe

51. Identify the metalloid in each list:
   a. Ge, Ga, Ca, Ba
   b. Sc, S, Sb, Sn
   c. Ar, Au, As, Ag

52. Which of the following elements form +1 ions?

K, P, Rb, Rn, Ne, Na

53. Which of the following elements form +2 ions?

As, C, Ca, B, Ba, Mg

54. Write formulas for the following ions:
   a. potassium ion
   b. calcium ion
   c. chloride ion
   d. iodide ion
   e. barium ion

55. Write formulas for the following ions:
   a. oxide ion
   b. nitride ion
   c. aluminum ion

56. Write formulas for the ions that make up the following compounds, then formulas for the compounds.
   a. sodium bromide
   b. magnesium sulfide
   c. potassium oxide
   d. calcium fluoride

57. Write a balanced equation for each of the following reactions. Give formulas of the ions that are formed.
   a. Potassium reacts with bromine to give potassium bromide.
   b. Lithium and oxygen react to form lithium oxide.
   c. Calcium reacts with sulfur to give calcium sulfide.

58. Which of the following pairs of elements have similar chemical properties?
   a. Na/K
   b. Ar/Kr
   c. C/Ca
   d. Rb/Rn

59. Identify from the following pairs of elements those that have *similar* chemical properties and those that have *different* chemical properties.
   a. N/Ne
   b. O/S
   c. P/As
   d. K/Kr

Additional Problems

60. Write formulas for the ions making up each compound, then the formula for the compound.
   a. barium chloride
   b. potassium iodide
   c. lithium bromide
   d. strontium fluoride

61. Write a balanced equation for each of the following reactions:
   a. Sodium and sulfur react to give sodium sulfide.
   b. Calcium and oxygen react to form calcium oxide.
   c. Magnesium and iodine react to form magnesium iodide.

62. Give the number of protons, electrons, and neutrons in each of the following atoms:
   a. zinc-70
   b. gold-197
   c. iron-57

63. Write the nuclear symbol for each of the atoms in problem 62.

64. Potassium exists as three isotopes: potassium-39 (mass = 38.964 amu, 93.11%), potassium-40 (mass = 39.974, 0.001180%), and potassium-41 (mass = 40.962, 6.88%). Calculate the atomic mass of potassium.

65. How many protons and electrons make up each of the following ions:
    a. O^{2-} c. Br^-
    b. Ca^{2+} d. Al^{3+}

66. From the following list of elements—Au, Rb, I, Kr, Sr, Cs, Li, Mn, F, Be, He—identify those that are
    a. alkali metals d. alkaline earth metals
    b. halogens e. noble gases
    c. transition elements

67. An element in Group 4A has a shiny, bright luster, is malleable, and is a good thermal conductor. Which of the elements could it possibly be?

68. An element in Group 5A is a colorless, odorless gas. Which element is it?

69. An element in Group 3A is a hard, brittle, black solid that has semiconductor properties. Identify the element.

Solutions to Practice Problems

PP 5.1

Number of neutrons = mass number − atomic number
a. For oxygen-18: number of neutrons = 18 − 8 = 10
b. 18
c. 143

PP 5.2

Find the symbol and atomic number of each element in the periodic table or the table of elements in the front of the book.
a. ${}^{3}_{2}He$ b. ${}^{35}_{16}S$ c. ${}^{123}_{53}I$

PP 5.3

Multiply each mass times its fractional abundance and add.

$$(23.99 \text{ amu})(0.7870) = 18.88 \text{ amu}$$
$$(24.99 \text{ amu})(0.1013) = 2.532 \text{ amu}$$
$$(25.98 \text{ amu})(0.1117) = 2.902 \text{ amu}$$
$$24.31 \text{ amu}$$

PP 5.4

Use the group number in the periodic table to determine the ionic charge.
a. Br^- b. Ba^{2+} c. Cs^+ d. I^-

PP 5.5

a. $Ca + Br_2 \longrightarrow CaBr_2$ (made up of Ca^{2+} and $2Br^-$)
b. $Mg + S \longrightarrow MgS$ (made up of Mg^{2+} and S^{2-})

PP 5.6

a. Number of p's = atomic number = 35
   ionic charge = sum of +charges (protons) and − charges (electrons)
   −1 = +35 (protons) + −36 (electrons)
   Number of e^-'s = 36
b. Number of p's = 56
   ionic charge = +2 = +56 (protons) + −54 (electrons)
   Number of e^-'s = 54
c. Number of p's = 55
   ionic charge = +1 = +55 (protons) + −54 (electrons)
   Number of e^-'s = 54
d. Number of p's = 53
   ionic charge = −1 = +53 (protons) + −54 (electrons)
   Number of e^-'s = 54

Answers to Chapter Review

1. atoms
2. mass
3. electrons
4. neutrons
5. proton
6. +1
7. electron
8. Ernest Rutherford
9. alpha particles
10. nucleus
11. atomic number
12. element
13. isotopes
14. mixtures
15. different
16. nuclear
17. atomic
18. sum
19. deuterium
20. tritium
21. mass number
22. atomic mass
23. amu
24. periodic table

25. Mendeleev
26. masses
27. numbers
28. periodic law
29. period
30. groups
31. representative
32. transition
33. alkali
34. alkaline earth
35. halogens
36. nonmetals
37. metals
38. lose
39. cation
40. +2
41. ionic
42. anion
43. group number
44. charge
45. balance
46. metalloids

6

Electron Structure and the Periodic Table

Contents

A rainbow forms when sunlight is refracted (bent) and dispersed into a spectrum of colors by rain or mist. The analysis of emission spectra of gaseous elements gives us information about their electron structures.

You Probably Know . . .

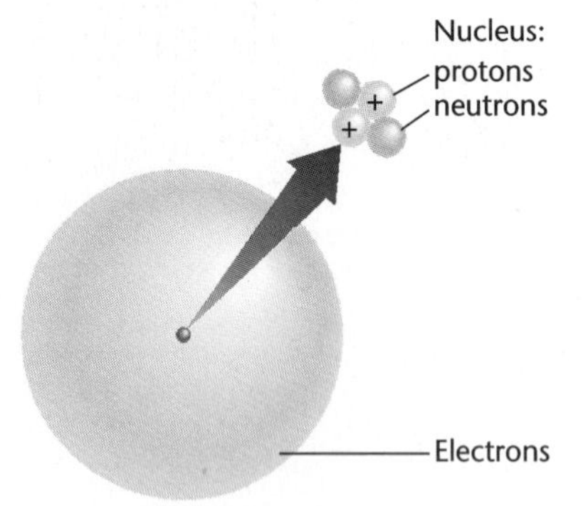

An atom has zero charge because the number of protons (positive charges) equals the number of electrons (negative charges).

- An atom, once thought to be the smallest particle of matter, has a substructure. At its center is a nucleus, made up of protons and neutrons. Electrons surround the nucleus. (Chapter 5)
- The number of electrons equals the number of protons in an atom. Consequently, atoms are neutral.
- Atoms can either lose or gain electrons to form charged particles (+ or −) called ions. (Section 5.8)

Why Should We Study Electron Structure?

According to Dalton's atomic theory, atoms of two elements can combine to form compounds. Chemists have found that certain elements (such as sodium and chlorine) combine readily to form a compound (sodium chloride), whereas others (such as calcium and magnesium) do not combine. Why is it that some elements react together and some do not?

Chemists have also observed that elements combine in different ratios. For example, sodium and chlorine combine in a 1:1 ratio of atoms, whereas magnesium and chlorine combine in a 1:2 ratio. Why are the combining ratios different?

Chemists have also discovered that metals have a tendency to lose electrons, thereby becoming positive ions (cations). Certain metals (such as sodium) always lose one electron and form cations with a +1 charge, whereas other metals (such as magnesium) always lose two electrons and so form cations with a +2 charge. Why do different metals lose different numbers of electrons?

Nonmetals are known to gain electrons, thereby becoming negative ions (anions). Certain nonmetal atoms (such as chlorine) gain one electron and form anions with a −1 charge, whereas others (such as oxygen) gain two electrons and form anions with a −2 charge. Why do different nonmetals gain different numbers of electrons?

All these questions (and more) can be answered from a knowledge of the electron structure of atoms. We have already seen that elements such as the alkali metals (Li, Na, K, etc.) are grouped together in the periodic table because of their chemical similarity. Also, the halogens (F, Cl, Br, and I), having similar chemical properties, are found together in one group. Knowledge of electron structure is the key to understanding the chemical properties of the elements.

6.1 Properties of Electrons

Learning Objective

- Describe the properties of electrons.

Electricity and some of the electrical properties of matter (Section 2.7) were known in the early 1800s when Dalton proposed his atomic theory (Section 5.1). During the nineteenth century, electricity was put to practical use in many ways with such inventions as the electric battery, the electric generator, and the electric lightbulb. However, it wasn't until the end of that century that the electron was discovered and characterized by J. J. Thomson. His experiments showed that an electron is an extremely small particle having a charge of −1. Its mass is only 0.000 549 atomic mass unit (amu), compared with a mass of about 1 amu for a proton or a neutron.

However, as scientists continued to study electrons, they discovered that electron behavior is more complex than this description implies. Although some experiments suggest that an electron has properties of a small particle, other experiments show that electrons have wave properties similar to light. This dual character of an electron, behaving both like a particle and like a wave, continues to puzzle scientists.

Our purpose, however, is not to gain a thorough understanding of the properties of electrons. Rather, we want to come to an understanding of the arrangement of electrons within atoms so we can understand how atoms take part in chemical reactions. We now consider how scientists have arrived at a modern description, or model, of the electron structure of atoms.

Benjamin Franklin showed the connection between lightning and electricity in 1752 by flying a kite during a thunderstorm. He received an electric shock from a key tied to the kite string.

The Bohr Model for Hydrogen

6.2

Learning Objectives

- Diagram and describe the Bohr model for a hydrogen atom.
- Distinguish between a continuous spectrum and a bright-line spectrum.

Niels Bohr (1885–1962) proposed a model for a hydrogen atom in 1913 (Figure 6.1). He imagined a hydrogen atom to have a central nucleus surrounded by a number of distinct paths, or **orbits,** along which an electron might travel. Bohr suggested that the electron could travel only in an orbit at a definite distance from the nucleus. It could never be found in the space between orbits. Bohr's concept was a revolutionary idea for his time. Let's try to understand how he arrived at his model.

Bohr's model was an interpretation of the bright-line emission spectrum for hydrogen. A light **spectrum** is a photograph or chart of the display resulting when light is separated into its various colors (Figure 6.2). You have seen the continuous rainbow of colors, or spectrum, that results when sunlight passes through raindrops in the sky. Now let's see what an emission spectrum for hydrogen looks like.

An emission spectrum for hydrogen can be observed by energizing a sample of hydrogen gas with a high voltage until the sample glows. When the light from the sample passes through a narrow slit and then through a glass or quartz prism, it separates into its various colors. The resulting bright-line emission spectrum for hydrogen is shown in Figure 6.3. It is different from the familiar rainbow because not all of the colors are present, and so the colors are lines that do not blend together.

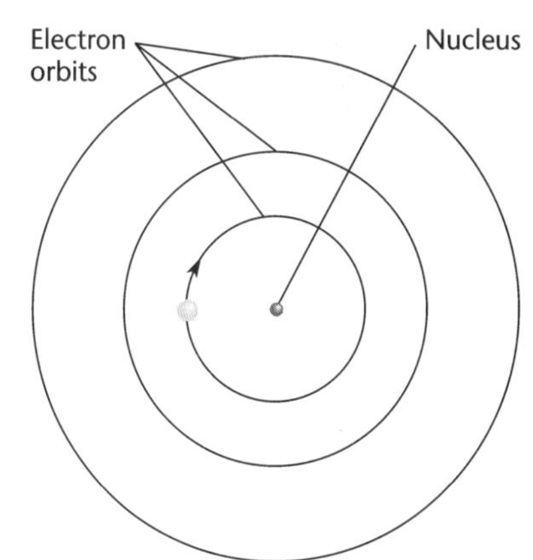

FIGURE 6.1 The Bohr model for a hydrogen atom. The electron is imagined to travel in an orbit around the nucleus, much like a planet around the sun.

FIGURE 6.2 A light spectrum. The color and energy of light are related.

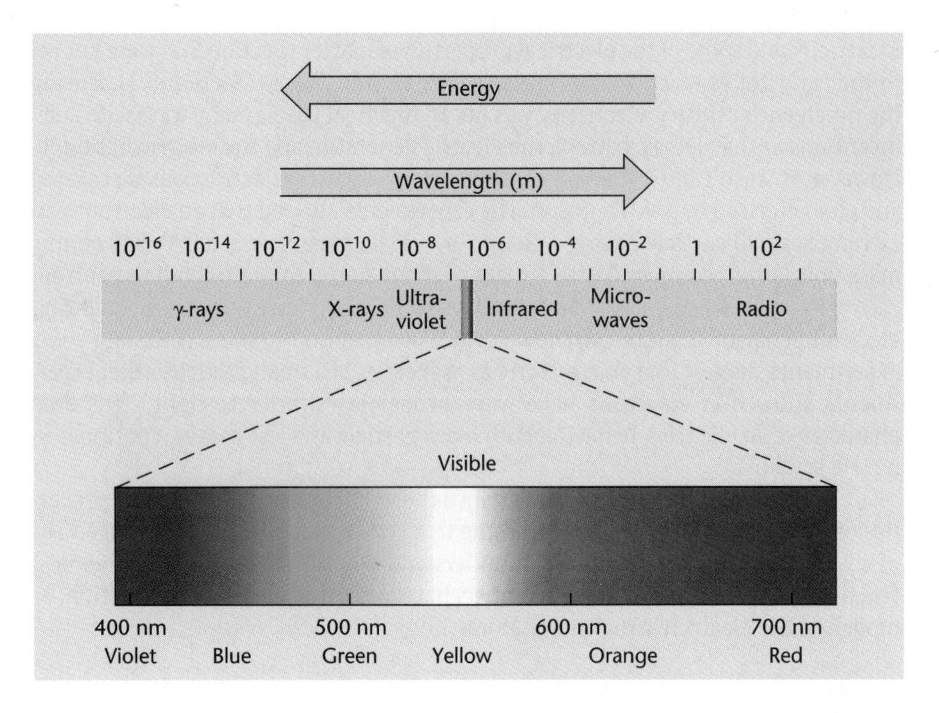

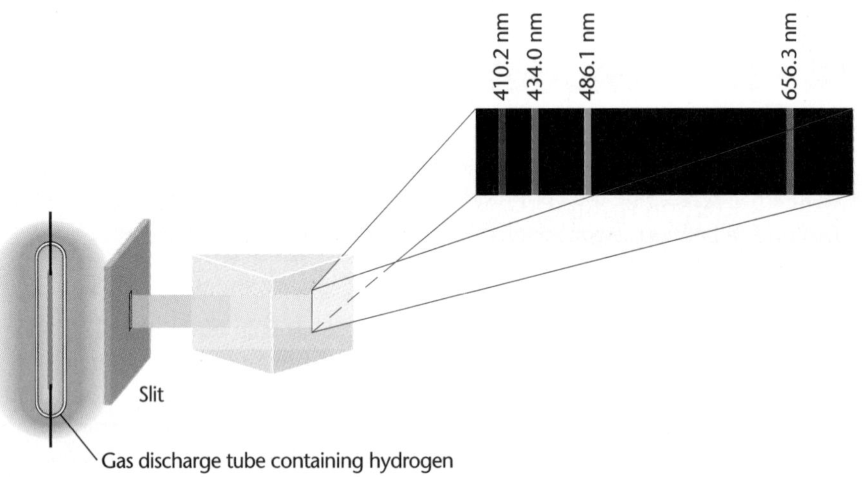

FIGURE 6.3 The emission spectrum for hydrogen is a bright-line spectrum. Red, green, blue, and violet lines are visible.

Figure 6.4 shows how Bohr correlated the bright-line spectrum for hydrogen with his model. Knowing that energy is required to separate objects that are attracted to each other, he reasoned that when a hydrogen atom absorbs energy, that energy causes its electron to move away from the positive nucleus to a higher-energy orbit. This so-called excited electron then drops to a lower-energy orbit with the emission of light. Each time the electron drops to a lower-energy orbit, a specific amount of energy is emitted in the form of light.* This definite amount of

*The energy (E) of light is related to its wavelength (λ) by the equation $E = hc/\lambda$, where h is a constant and c is the speed of light. Light of different wavelengths has different colors.

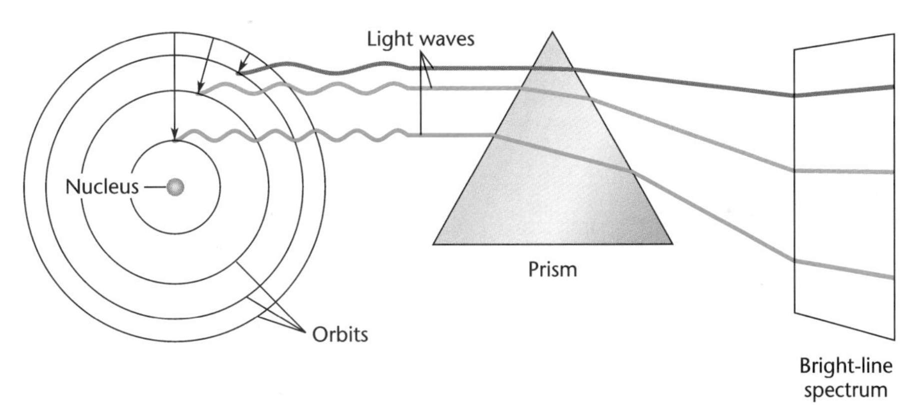

FIGURE 6.4 The formation of a bright-line spectrum as explained by Bohr. An excited electron drops to a lower-energy orbit. With each transition to a lower-energy orbit, a quantum of energy is released as light, thereby producing a bright line in the spectrum.

energy is called a **quantum of energy.** Each bright line in the spectrum corresponds to a quantum of energy emitted. From the colors of the lines he was able to calculate the energy of the light for each of the bright lines. From the energies of the lines, he determined the differences in energies of the orbits.

The orbits of the Bohr model are sometimes referred to as quantized energy levels. Let's illustrate this idea by considering a familiar example (Figure 6.5). You may have seen a child playing with a ball on a flight of steps. Think of the ball as being like an electron. When the ball is thrown to the top of the steps, it is like an excited electron. As the ball falls off the top step, it drops to the next step just like the electron drops to the next orbit of lower energy. The ball continues to bump down the steps, but it can stop only on each step. It is not possible for the ball to stop between steps. Every time it drops to the next lower step, a quantum of energy is released. If the energy released by the ball were in the form of light, you would see a bright-line spectrum. However, a ball rolling down a ramp would give a continuous spectrum similar to a rainbow. Thus, the Bohr model suggests that quantized atoms have electrons in orbits of fixed energy states.

Bohr's model for a hydrogen atom was revolutionary for his time and a tremendous step toward understanding the secrets of the atom. However, its limitations were significant. The most serious limitation of the model was that it did not satisfactorily explain the emission spectra of other elements.

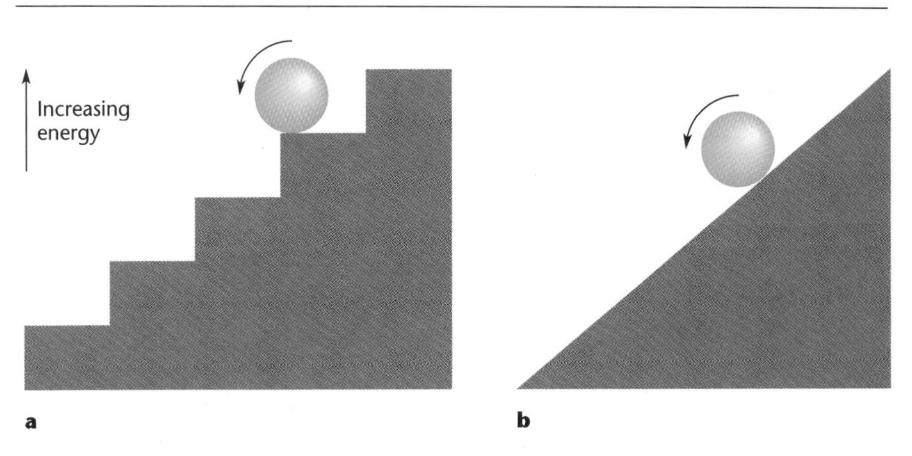

FIGURE 6.5 (a) A ball bumping down a flight of steps moves through quantized energy levels. (b) A ball rolling down a ramp moves through continuous energy levels.

Atomic physicists meet at the 1927 Solvay Conference in Brussels, Belgium. Front row: I. Langmuir, M. Planck, M. Curie, H. A. Lorentz, A. Einstein, P. Langevin, Ch. E. Guye, C. T. R. Wilson, O. W. Richardson. Middle row: P. Debye, M. Knudsen, W. L. Bragg, H. A. Kramers, P. A. M. Dirac, A. H. Compton, L. de Broglie, M. Born, N. Bohr. Back row: A. Piccard, E. Henriot, P. Ehrenfest, Ed. Herzen, Th. de Donder, E. Schrödinger, E. Verschaffelt, W. Pauli, W. Heisenberg, R. H. Fowler, L. Brillouin.

During the 1920s, another model of atoms was developed. This new model, which incorporates some of the features of Bohr's model, was the result of the work of a number of European scientists. It was the French physicist Louis de Broglie who proposed that electrons could have wave properties like light; and Erwin Schrödinger (1887–1961), an Austrian, used a new method of calculation that led to the development of quantum mechanics, or wave mechanics. The resulting quantum mechanical model (or wave mechanical model) of atoms has been the basis for research in atomic physics and chemistry from that time to the present. Let's examine the principal features of the quantum mechanical model of atoms.

6.3 Energy States for Electrons

Learning Objectives

- List the principal energy levels and their sublevels in order of increasing energy.
- Specify the number of sublevels in each principal energy level.

When Bohr calculated the energies of the orbits of the hydrogen atom, he assigned a number n, called a principal quantum number, to each of the orbits. The orbit, or energy level, closest to the nucleus was assigned a value of $n = 1$, and the numbers increased consecutively for the higher-energy levels farther from the nucleus.

Although Bohr's calculations did not work for atoms other than hydrogen, the idea that electrons are found in quantized energy levels agreed with the data from bright-line spectra. In the modern model of atoms, these are called **principal energy levels,** or **shells,** and they are numbered 1, 2, 3, and so on, just as the orbits

Principal energy levels (shells) are numbered 1, 2, 3, 4, and so on.

TABLE 6.1 Principal Energy Levels and Sublevels for Electrons

Principal Energy Level, n		*Sublevels*
	Increasing energy ↑	→
6		*s, p, d,* —[a]
5		*s, p, d, f*
4		*s, p, d, f*
3		*s, p, d*
2		*s, p*
1		*s*

[a]The 6*f* sublevel, although predicted by quantum theory, is not needed for any of the presently known elements.

of Bohr's model were numbered. The principal energy levels are divided into **sublevels,** or **subshells.**

The number of sublevels in each principal energy level is equal to its number n. These sublevels are identified by the letters ***s, p, d,*** and ***f.**** Within each principal energy level, the *s* sublevel has the lowest energy, and the *f* sublevel has the highest energy. In the first principal energy level, $n = 1$ and there is one sublevel, identified as 1*s*. For the second principal energy level, $n = 2$ and there are two sublevels, identified as 2*s* and 2*p*. The third principal energy level has three sublevels: 3*s*, 3*p*, and 3*d*. The fourth principal energy level has four sublevels: 4*s*, 4*p*, 4*d*, and 4*f*. Quantum theory predicts there should be sublevels beyond the *f* sublevel, but these are not needed to describe the electron arrangements for the presently known elements (Table 6.1).

The energy of an electron depends on the principal energy level and the sublevel it is in. Remember, higher n values indicate higher principal energy levels. The energy also increases within an n level for the sublevels in order from *s* to *f*. For example, energy increases from 3*s* to 3*p* to 3*d*. This will be examined in greater detail in the next section.

Atomic Orbitals

6.4

Learning Objectives

- Specify the number of orbitals in each sublevel.
- Specify the electron capacity of an orbital, sublevel, and principal energy level.
- Describe the shapes of *s* and *p* atomic orbitals.
- State the Heisenberg uncertainty principle and the Pauli exclusion principle.
- Describe the spins of two electrons in the same orbital.

Erwin Schrödinger's new method of calculation allowed him to describe a region around the nucleus of an atom where there would be a high probability of finding a

*The letters that identify the various sublevels originated with early spectroscopists who described the types of lines in a line spectrum as sharp (*s*), principal (*p*), diffuse (*d*), and fundamental (*f*).

FIGURE 6.6 Shapes of *s* and *p* atomic orbitals according to the quantum mechanical model.

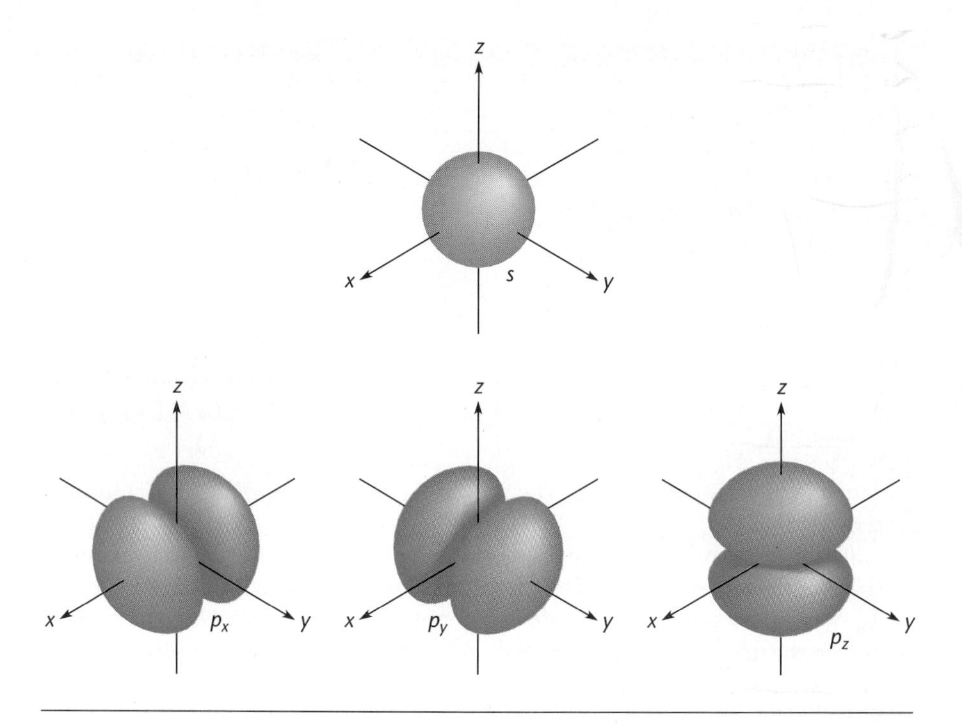

Atomic orbital: A region in space around the nucleus of an atom where there is a high probability of finding an electron.

particular electron. This region in space is called an **atomic orbital.** Note the difference between an *orbit,* a specific path for an electron, and an *orbital,* a space that is three-dimensional. Atomic orbitals cannot be seen, but they have been described mathematically using Schrödinger's calculations.

Each kind of sublevel has a specific number of atomic orbitals. An *s* sublevel has only one orbital, but there are three *p* orbitals in each *p* sublevel. All *d* sublevels have five *d* orbitals, and *f* sublevels have seven orbitals. So, for the *s, p, d,* and *f* sublevels there are 1, 3, 5, and 7 orbitals, respectively.

Because an atomic orbital is a region in space, we can describe its shape. The shapes of the *s* and *p* atomic orbitals are illustrated in Figure 6.6. The nucleus of the atom is located at the intersection of the *x, y,* and *z* axes. The *p* orbitals are designated p_x, p_y, and p_z because they lie along the *x, y,* and *z* axes. The shapes of the *d* and *f* orbitals are more complex; we will not be concerned with them.

Describing the location of an electron in terms of probability resulted from an idea of one of Bohr's students, Werner Heisenberg (1901–1975), a German physicist. He proposed what is now called the **Heisenberg uncertainty principle.** This principle states that it is impossible to precisely determine the exact position and speed of an electron at the same time. One or the other can be determined, but not both. The uncertainty principle suggested that Bohr's model of an electron traveling in a precise orbit is not valid. The best we can do is to describe its most probable location by specifying the orbital in which it is located. So, the idea of an orbital involving electron probabilities replaced the precise orbit of the Bohr model.

Wave properties of electrons gave atomic physicists another reason to replace orbits with orbitals in the modern model for atoms. We can think of the orbitals as electron waves. These waves are sometimes called standing waves because they do not spread out like water waves. Furthermore, these standing waves are three-dimensional. Often, atomic orbitals are said to be like electron clouds. The analogy of a cloud is a good one for an atomic orbital because a cloud occupies space (and has shape) but has almost no mass. Remember that an electron has practically no mass. Figure 6.7 compares the Bohr model with the modern, quantum mechanical model of a hydrogen atom.

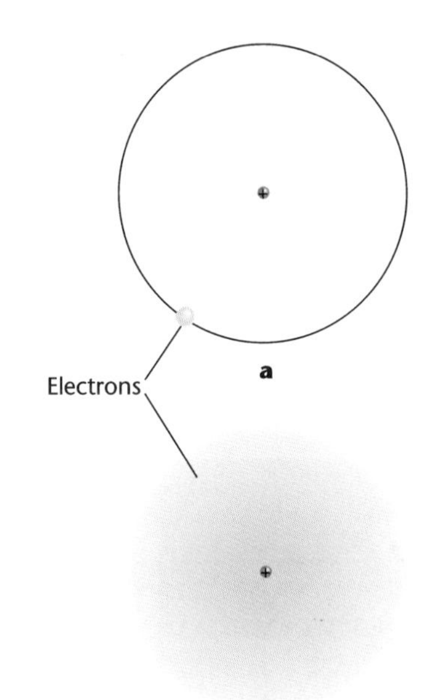

FIGURE 6.7 (a) The Bohr model and (b) the modern, quantum mechanical model for a hydrogen atom.

Orbital Occupancy

An atomic orbital can be occupied by 0, 1, or 2 electrons. Because each kind of sublevel has a specific number of orbitals, we can easily figure the electron capacity for each sublevel:

$$\text{number of electrons} = 2 \times \text{number of orbitals}$$

The electron capacity for each sublevel is given in Table 6.2.

The electron capacity for each principal energy level is also 2 × number of orbitals. The number of orbitals in a principal energy level equals n^2, and it follows that

$$\text{number of electrons} = 2n^2$$

For example, the first principal energy level can have a maximum of two electrons ($2n^2 = 2 \times 1^2 = 2$); the second principal energy level can have up to eight electrons ($2 \times 2^2 = 2 \times 4 = 8$). Table 6.3 gives the electron capacity for each of the first five principal energy levels.

An atomic orbital can contain no more than two electrons.

TABLE 6.2 Population of Electrons in Sublevels

Sublevel	*Number of Orbitals*	*Electron Capacity*
s	1	2
p	3	6
d	5	10
f	7	14

TABLE 6.3 Electron Capacity for Principal Energy Levels

Principal Energy Level	*Electron Capacity, $2n^2$*
1	$2 \times 1^2 = 2$
2	$2 \times 2^2 = 8$
3	$2 \times 3^2 = 18$
4	$2 \times 4^2 = 32$
5	$2 \times 5^2 = 50$[a]

[a]For the elements discovered to date, the fifth principal energy level is occupied by no more than 32 electrons.

CHEMICAL WORLD

Why Do Electrons Spin?

The best answer to the question of why electrons spin seems to be "Why not?" It's simply more probable. The universe is rich with spinning entities, including the gigantic, pinwheel-shaped galaxies comprised of hundreds of billions of stars and the huge roiling clouds of gas and dust in interstellar space. These large collections of matter do not sit quietly and regally in space. They are continuously acted upon by a multitude of forces: the infall of debris from elsewhere, the pressure of intense radiation, tugs of gravity, the explosions of dying stars, and sometimes a slow-motion (to us) collision with another galaxy or gas cloud. If these objects weren't spinning, it would mean that all the forces they feel have always been perfectly balanced. Such a balancing act is highly improbable, so we expect, and we can measure, that galaxies and gas clouds do spin.

A spinning top.

Stars also spin on their axes. Astronomers have measured the rotation of many stars in our own galaxy and have watched sunspots move across the face of the sun as it turns. Even closer to home, you and this book are spinning around with the earth at about 1000 miles per hour.

Constant motion, it seems, leads to frequent interactions, which often lead to spin. Head-on collisions do not happen often because there are so many ways for glancing collisions to occur. Picture two moving billiard balls (a useful analogy in science). If the balls collide, they will probably give each other a glancing blow. (Think about the one chance for a perfect head-on collision compared with the many chances for lesser collisions at all other angles.) Glancing collisions make both balls spin.

Atoms and molecules are in constant motion and are constantly colliding, or otherwise disturbing each other. Like the billiard balls, most close encounters are glancing, so we are not surprised to find in the laboratory that atoms and molecules have spin.

What about electron spin? Well, no one has ever seen electrons well enough to know firsthand if they spin. They certainly do not spin like billiard balls (or like tops either). They are quantum particles and obey other rules. But electrons also are in constant motion, like atoms and molecules, and they *act* like they're spinning. Electrons act so much like they're spinning that chemists and physicists have to allow for electron spin in some kinds of experiments. They have repeatedly found that only by assuming that electrons spin can they successfully predict the outcome of these experiments.

But do electrons *really* spin?

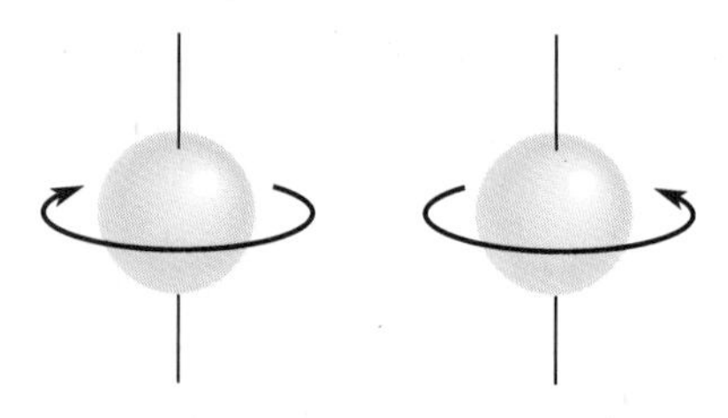

FIGURE 6.8 Electron spins: two possible directions.

Electron Spin

In 1925, Wolfgang Pauli (1900–1958) proposed that an electron behaves as if it is spinning on its axis like a top (Figure 6.8). Orbital occupancy is governed by the **Pauli exclusion principle:** that is, no more than two electrons can occupy a given orbital; and when two electrons occupy an orbital, their spins must be opposite. Arrows are sometimes used to represent electrons in orbitals. The spins of two electrons are *paired* if one is designated $\uparrow$ and the other is $\downarrow$. Paired spins are shown $\uparrow\downarrow$.

Energy Level Diagram for Electrons

6.5

Learning Objectives

- Draw an electron energy level diagram showing the first three principal energy levels and their sublevels.
- Draw pictures of the occupied orbitals of any of the first ten elements.
- State Hund's rule.

Before we proceed with the description of the quantum mechanical model of an atom, let's pause to review what we have discussed so far. Principal energy levels, sublevels, and orbitals are features of the energy states for electrons. Each principal energy level (shell) is identified by a number, $n = 1, 2, 3, 4$, and so on. Sublevels (subshells) are identified by the letters *s, p, d,* and *f* in order of increasing energy. The *s* sublevel is made up of one orbital, and the *p, d,* and *f* sublevels are made up of three, five, and seven orbitals, respectively. Table 6.4 summarizes these ideas.

Now, using these ideas, let's try to develop a mental picture of an atom. Imagine that you have been given a model atom kit as a gift. (This is sort of like a model airplane kit, but it was made especially for people with an inclination toward science.) After opening the box, you begin to read the instructions, which tell you how to build a model of an atom.

First, you must select one of the tiny balls in the kit to be the nucleus. These balls are marked +1, +2, +3, and so on to represent the nuclear charges of different atoms. You must choose the ball with the charge that matches the atomic number of the atom you are going to build. You have decided to build a model of a carbon atom, so you choose a tiny ball marked +6 to represent a nucleus that contains six protons.

Because atoms are neutral, you must place six electrons around the nucleus so that their negative charge will balance the positive charge of the protons in the

TABLE 6.4 Principal Energy Levels, Sublevels, and Orbitals in the Quantum Mechanical Model of an Atom

				Electron Capacity	
n	*Number of Sublevels*	*Sublevel*	*Number of Orbitals*	*Sublevel*	*Principal Energy Level*
4	4	*f*	7	14	32
		d	5	10	
		p	3	6	
		s	1	2	
3	3	*d*	5	10	18
		p	3	6	
		s	1	2	
2	2	*p*	3	6	8
		s	1	2	
1	1	*s*	1	2	2

nucleus. The net charge for the atom will then be zero. The instructions now tell you to continue building the atom by placing the first electrons in the lowest principal energy level, which is actually the 1*s* sublevel. When that level is filled, you are to place the remaining electrons in the next-higher principal energy level. Because the second level has two sublevels, 2*s* and 2*p*, you add electrons in the 2*s* sublevel until it is filled and then add the last two electrons to the 2*p* sublevel.

At this point, you may begin to be confused. However, an energy level diagram similar to Figure 6.9 has been provided with instructions to help you see how to proceed. To complete your model for a carbon atom, you must put two electrons in the 1*s* orbital and two electrons in the 2*s* orbital. The last two electrons go in the 2*p* sublevel but in different orbitals because of the repulsion of their negative charges. In general, electrons in a given sublevel are placed in different orbitals

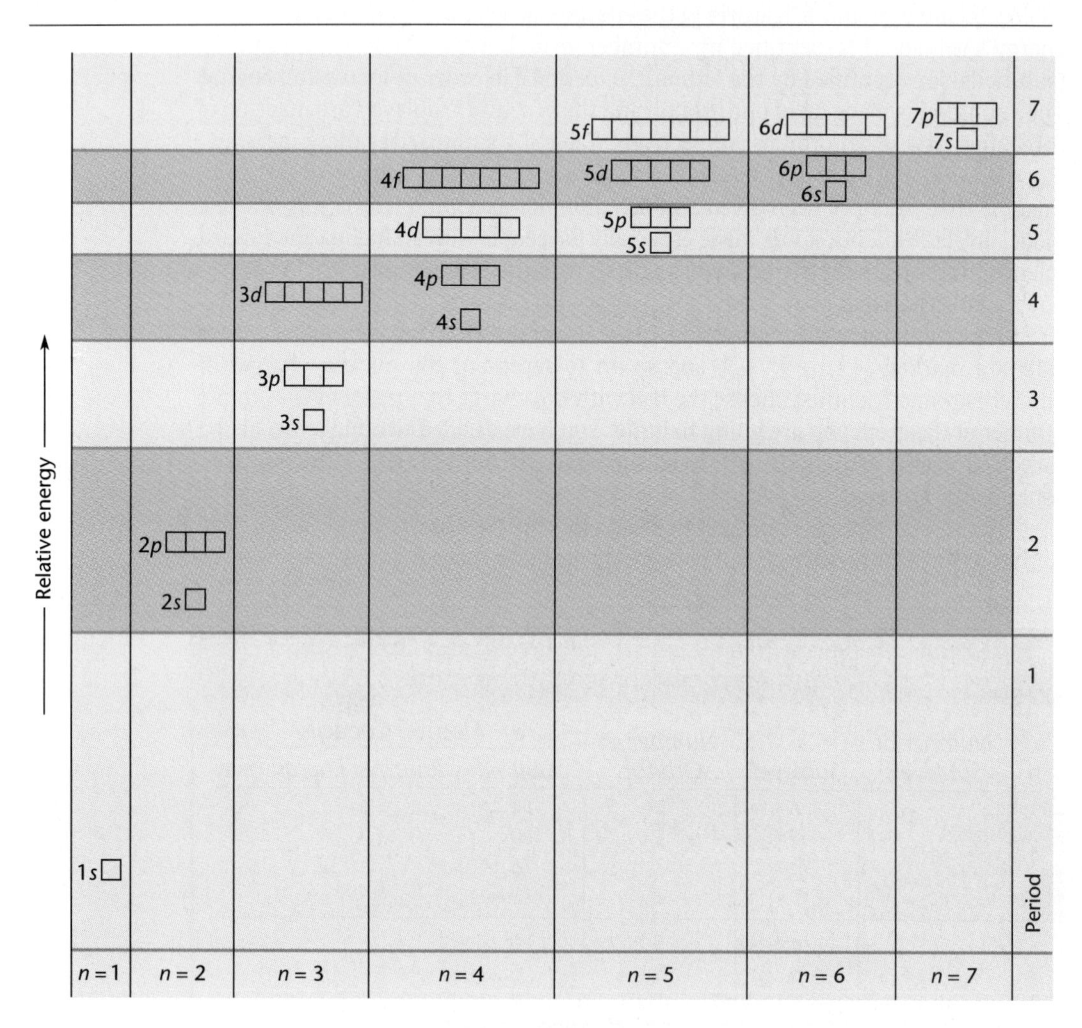

FIGURE 6.9 Energy level diagram showing principal energy levels, sublevels, and orbitals in a model of an atom.

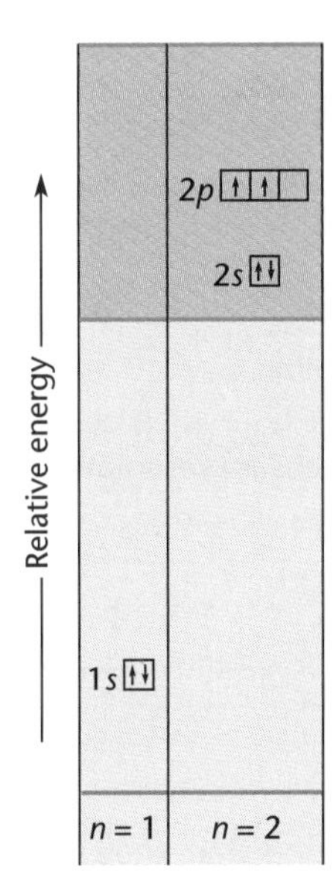

FIGURE 6.10 Energy level diagram for a carbon atom. Electrons are placed in lowest energy levels and sublevels. In the 2*p* sublevel, electrons are placed to minimize pairing.

FIGURE 6.11 Orbital picture for a carbon atom

until all orbitals of that sublevel have one electron (half filled). The instructions illustrated this with a diagram similar to Figure 6.10 and a picture of the orbitals involved like that of Figure 6.11. You will use plastic 1*s*, 2*s*, and 2*p* orbitals in the kit to finish your model. When you are done, your model will look like Figure 6.11.

Using your model atom kit, you have learned to build an orbital model of a six-electron atom. You can imagine how complicated this kind of model would be for an atom that has a large number of electrons. To help you imagine orbital models for atoms, you must learn to draw electron energy level diagrams. You can do this by following the diagram illustrated in Figure 6.9 and observing these rules:

1. Electrons will occupy the lowest energy levels and sublevels. Higher energy levels are occupied only when the lower energy levels are filled.
2. An orbital can contain no more than two electrons (often referred to as a *pair* of electrons).
3. Orbitals in a given sublevel are at the same energy level. Because like charges repel one another, an electron enters an empty orbital before entering an orbital already occupied by one electron. This process follows **Hund's rule:** electrons are evenly distributed among the orbitals of a subshell. (See Figure 6.10.)

Now you are ready to build the electron structure for other atoms. An oxygen atom (element 8) has eight electrons. Its energy level diagram is shown in Figure 6.12. You should be able to diagram larger atoms by following the preceding rules.

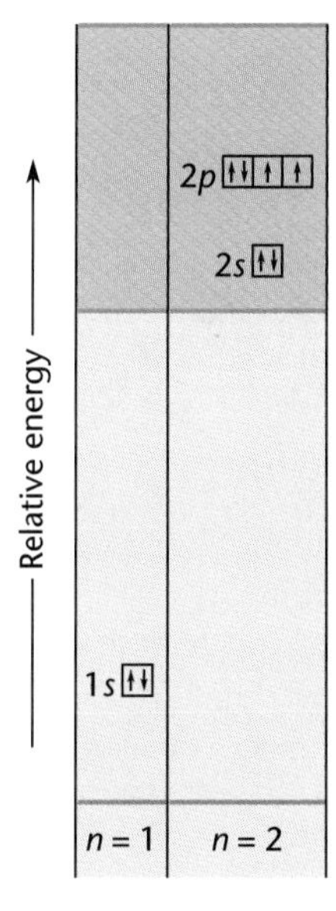

FIGURE 6.12 Energy level diagram for an oxygen atom

6.6 Electron Configurations and the Periodic Table

Learning Objectives

- Find the *s*-block, *p*-block, *d*-block, and *f*-block in the periodic table.
- Using the periodic table as a guide, write complete electron configurations and valence-shell configurations for atoms and monatomic ions of the representative elements.
- Using noble gas core abbreviations, write electron configurations for atoms of the representative elements.
- Specify the number of valence electrons for an atom of any representative element from its group number.

The **electron configuration** for an atom is a shorthand notation that shows the energy states of its electrons. This notation provides most of the information given in an energy level diagram in abbreviated form.

When writing electron configurations, we will only concern ourselves with the lowest energy state, or **ground state** of the atom. Let's illustrate by considering the eight electrons of an oxygen atom. The ground state for an oxygen atom was previously described in the energy level diagram in Figure 6.12. You can see that two electrons were placed in the 1*s* orbital, two in the 2*s* orbital, and four in the 2*p* sublevel. The electron configuration for an oxygen atom looks like this:

O: $1s^22s^22p^4$

Remember that the letters identify the sublevels and a superscript indicates the number of electrons in the sublevel. The sum of the superscripts tells us the number of electrons for the atom.

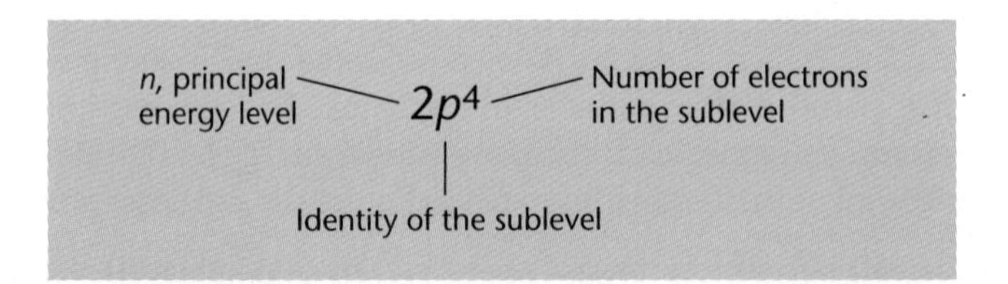

The electron configurations for the first ten elements are shown in Table 6.5.

In writing electron configurations, it is common practice to indicate only the sublevel in which the electrons are found, not the individual orbitals. However, the configuration can be expanded to show individual orbitals if greater detail is needed for some specific purpose. The expanded configuration for carbon is

C: $1s^22s^22p_x{}^12p_y{}^1$

The letters *x*, *y*, and *z* are used to distinguish the three orbitals in the *p* subshell. In a similar way, the expanded electron configuration for oxygen (element 8) is:

O: $1s^22s^22p_x{}^22p_y{}^12p_z{}^1$

TABLE 6.5 Electron Configurations of the First Ten Elements

Atomic Number[a]	*Element*	*Electron Configuration*
1	H	$1s^1$
2	He	$1s^2$
3	Li	$1s^22s^1$
4	Be	$1s^22s^2$
5	B	$1s^22s^22p^1$
6	C	$1s^22s^22p^2$
7	N	$1s^22s^22p^3$
8	O	$1s^22s^22p^4$
9	F	$1s^22s^22p^5$
10	Ne	$1s^22s^22p^6$

[a]Atomic number = number of electrons for a neutral atom
= sum of the superscripts

Example 6.1

Write the electron configuration for a sodium atom. ■

Solution Sodium is element 11, so a sodium atom has 11 electrons. Referring to an energy level diagram (Figure 6.9), you should be able to write:

$1s^22s^22p^63s^1$

Example 6.2

Write electron configurations for the following elements:

a. Li
b. Si
c. S ■

Solutions Find the atomic number of each element in the periodic table. Refer to Figure 6.9 as you write the electron configurations.

a. $1s^22s^1$
b. $1s^22s^22p^63s^23p^2$
c. $1s^22s^22p^63s^23p^4$

Example 6.3

Identify each element from its electron configuration:

a. $1s^22s^2$
b. $1s^22s^22p^3$ ■

Solutions Count superscripts (electrons), and look in the periodic table to find the symbol having that atomic number.

a. Be
b. N

Practice Problem 6.1 Write electron configurations for the following elements. (Solutions to practice problems appear at the end of the chapter.)

a. Mg
b. F
c. Cl

Practice Problem 6.2 Identify each element from its electron configuration:

a. $1s^22s^22p^63s^23p^3$
b. $1s^22s^22p^6$

Electron Configurations and the Periodic Table

To this point, we have used the energy level diagram (Figure 6.9) to help us remember the order of filling for principal energy levels and sublevels. We can also use the periodic table to determine the order of filling of electrons.

Figure 6.13 shows that the elements are grouped in the periodic table in blocks according to their outermost occupied sublevel. The elements of Groups 1A and 2A, plus hydrogen and helium, make up the *s-block*. Note that although hydrogen ($1s^1$) belongs in the *s*-block and appears in Group 1A in some periodic tables, it is not an alkali metal. Helium ($1s^2$) also belongs in the *s*-block but is usually placed in Group 8A with the other noble gases. The *p-block* includes the elements in Groups 3A through 8A (the noble gases), except helium. The *d-block* is made up of

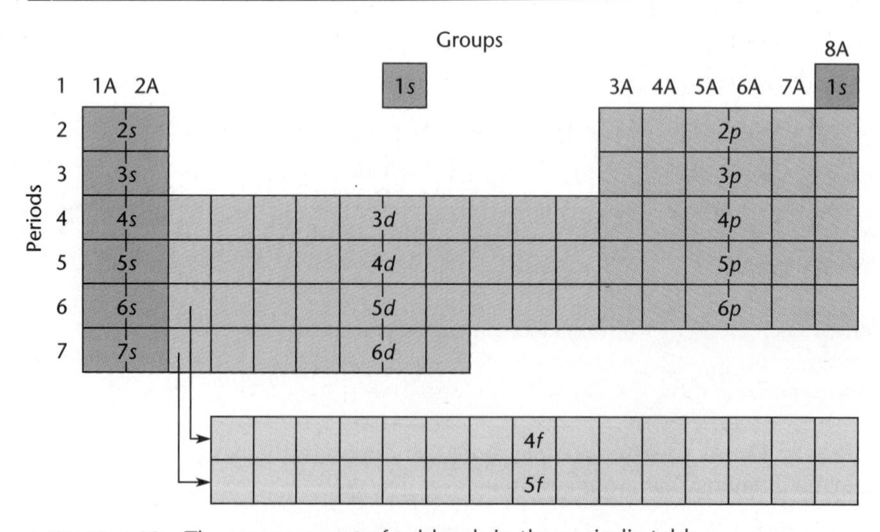

FIGURE 6.13 The arrangement of sublevels in the periodic table.

the transition elements, and the lanthanides and actinides are found in the *f-block.* We will use the periodic table to guide us in writing electron configurations.

Outer sublevel configuration for phosphorus:

Period number — $3p^3$ — Position of phosphorus in the *p*-block

Let's try writing the electron configuration for phosphorus (element 15). Looking at the periodic table, we find phosphorus in the *p*-block in period 3. The period number is the same as the number of the outermost occupied principal energy level, or **valence shell,** for the representative elements found in that period. Because phosphorus is the third element in the *p*-block in period 3, its outer sublevel configuration is $3p^3$. The complete electron configuration for phosphorus can be written by following the order of the sublevels shown in the periodic table. Reading the periodic table left to right, top down, we have

P $1s^2 2s^2 2p^6 3s^2 3p^3$

For nd sublevels, n = period number minus 1.

For nf sublevels, n = period number minus 2.

For the electrons in a *d* sublevel, the principal quantum number equals the period number minus 1. For electrons in an *f* subshell, the principal quantum number equals the period number minus 2. Thus, scandium (Sc, element 21), in period 4, has a configuration ending with $4s^2 3d^1$.

The sublevels are filled in order through the $3p$ sublevel, which is completed with argon (element 18). At this point, we might expect that the 19th electron for potassium (element 19) would be in the $3d$ sublevel. However, the position of potassium in the periodic table shows this electron in the $4s$ sublevel, in agreement with the order of sublevels shown in Figure 6.9. Based on the positions of potassium and calcium in the periodic table, their electron configurations are

K $1s^2 2s^2 2p^6 3s^2 3p^6 4s^1$

Ca $1s^2 2s^2 2p^6 3s^2 3p^6 4s^2$

Look again at the periodic table. You will notice that the outermost electrons of rubidium (Rb) and strontium (Sr) are in the $5s$ sublevel, and the $4d$ sublevel remains empty. For indium (In, element 49) through xenon (Xe, element 54), the $5p$ sublevel fills before the $4f$. There is a similar relationship for the $6s$ and $5d$ as well as the $6p$ and $5f$ sublevels. Although you may have difficulty remembering these details, the order of filling of the sublevels is easily determined from the periodic table.

Example 6.4

Using the periodic table as a guide to the order of filling of the sublevels, write electron configurations for the following elements:

a. Ga, element 31
b. Ar, element 18 ■

Solutions Gallium, Ga, is the first element in the *p*-block of period 4. Thus its electron configuration ends with $4p^1$. Argon, Ar, is the sixth element in the *p*-block of period 3, so its electron configuration ends with $3p^6$.

a. $1s^2 2s^2 2p^6 3s^2 3p^6 4s^2 3d^{10} 4p^1$
b. $1s^2 2s^2 2p^6 3s^2 3p^6$

Practice Problem 6.3 Write electron configurations for the following elements:

a. As, element 33

b. Al, element 13

Valence Electrons

Remember that elements in the same group have similar chemical properties. Most of the chemical properties of an element depend on the number of electrons in its outermost occupied principal energy level, or **valence shell.** The word *valence* refers to the combining capacity of an element when it reacts with other elements. Elements in the same group have the same number of valence electrons and therefore have similar combining capacities. Let's take another look at the representative elements, those in A-groups.

Number of valence electrons = group number (A-groups).

For the representative elements, the group (family) number identifies the number of valence electrons for atoms in that group. So, you can see that the elements in Group 1A all have one valence electron in an *s* sublevel. Lithium's valence electron is $2s^1$, sodium's is $3s^1$, and potassium's is $4s^1$. Those in Group 2A all have two valence electrons in an s sublevel. The elements in Group 3A have three valence electrons (ns^2np^1). The Group 4A elements have four valence electrons (ns^2np^2), and so on. The noble gases (Group 8A) with eight valence electrons (helium with two) ordinarily do not participate in chemical reactions. Elements in the other families have fewer than eight valence electrons and are chemically reactive.

Example 6.5

Specify the number of valence electrons for each of the following atoms:

a. F

b. C

c. K ■

Solutions The group number tells us the number of valence electrons:

a. 7

b. 4

c. 1

Example 6.6

Write valence-shell electron configurations for the following elements:

a. Cl

b. Si

c. Rb, element 37 ■

Solutions Refer to the periodic table. The group number tells the number of valence electrons. Find which block the element is in to determine the outermost occupied sublevel.

a. $3s^23p^5$
b. $3s^23p^2$
c. $5s^1$

Table 6.6 shows the valence electrons for the representative elements. Electron-dot symbols are discussed in the following section.

Example 6.7

Identify the following elements from their valence-shell configurations:

a. $3s^1$
b. $2s^22p^2$
c. $4s^2$
d. $3s^23p^3$ ■

Solutions The number of valence electrons tells us the group number. The period is determined from the principal quantum number (for example, for (**a**) we have Group 1A, period 3).

a. Na
b. C
c. Ca
d. P

TABLE 6.6 Valence Electrons for the Representative Elements

Group	*Number of Valence Electrons*	*Valence-Shell Configuration (Ground State)*	*Electron-Dot Symbol (Second-Period Element)*
1A (Li, Na, K, etc.)	1	ns^1	Li·
2A (Be, Mg, Ca, etc.)	2	ns^2	Be:
3A (B, Al, etc.)	3	ns^2np^1	·B:
4A (C, Si, etc.)	4	ns^2np^2	·Ċ:
5A (N, P, etc.)	5	ns^2np^3	·N: (with dot below)
6A (O, S, etc.)	6	ns^2np^4	:O: (with dots above and below)
7A (F, Cl, etc.)	7	ns^2np^5	:F: (with dots above and pair below)
8A (Ne, Ar, etc.)	8	ns^2np^6	:Ne: (with pairs above and below)

Practice Problem 6.4 How many valence electrons does each of the following have?

a. S
b. Ne
c. Ba

Practice Problem 6.5 Write valence-shell configurations for the following elements:

a. Ge, element 32
b. I, element 53
c. Cs, element 55

Noble Gas Core

It is common practice to abbreviate electron configurations using a symbol for a noble gas core followed by the additional subshells needed for the remaining electrons in the atom. A **noble gas core** is the electron configuration for one of the noble gases. For example, the electron configuration for neon, element 10, is:

$$1s^22s^22p^6 \quad \text{or} \quad [\text{Ne}]$$

These 10 electrons are the neon core, represented by the symbol [Ne]. The noble gas core symbols are shown in Table 6.7.

You can see that [Ar] stands for 18 electrons, [Kr] represents 36 electrons, and [Xe] represents 54 electrons.

TABLE 6.7 Noble Gas Core Symbols

Symbol	*Electron Configuration*
[He]	$1s^2$
[Ne]	$1s^22s^22p^6$
[Ar]	$1s^22s^22p^63s^23p^6$
[Kr]	$1s^22s^22p^63s^23p^64s^23d^{10}4p^6$
[Xe]	$1s^22s^22p^63s^23p^64s^23d^{10}4p^65s^24d^{10}5p^6$

Example 6.8

Using the noble gas core notation, write electron configurations for the following elements:

a. Sc
b. Rb
c. Ba
d. P
e. Be ■

Solutions Find the noble gas that appears before the element and write its core symbol. Add the remaining electrons to agree with the element's position in the table (group number and block).

a. [Ar]$4s^2 3d^1$
b. [Kr]$5s^1$
c. [Xe]$6s^2$
d. [Ne]$3s^2 3p^3$
e. [He]$2s^2$

Electron-Dot Symbols

6.7

Learning Objective

- Write electron-dot or Lewis symbols for atoms of the representative elements.

An electron-dot or **Lewis symbol** shows the symbol for the element surrounded by dots that represent its valence electrons. The electron-dot symbols for the elements in the second period are shown in Table 6.6. When two electrons occupy the same orbital, they are shown as a pair of dots; an unpaired electron is shown as one dot. Lewis symbols are useful because chemical reactions for most atoms are directly related to the number of their valence electrons.

A Lewis symbol shows valence electrons:

Unpaired electron — Paired electrons

·Al:

Symbol for the atom

Example 6.9

Write Lewis symbols for the following atoms. Note the number of unpaired electrons for each.

a. sodium
b. silicon
c. nitrogen
d. sulfur

Solutions To determine the number of valence electrons for each atom, find its group number in the periodic table. Remember to show only the valence electrons.

a. Na ·, 1 unpaired electron
b. :Si· (with one dot above), 2 unpaired electrons
c. :N· (with one dot above and one below), 3 unpaired electrons
d. :S: (with one dot below), 2 unpaired electrons

Practice Problem 6.6 Write Lewis symbols for the following atoms. Indicate the number of unpaired electrons for each.

a. calcium
b. argon
c. chlorine
d. boron

6.8 The Octet Rule and Formation of Ions

Learning Objectives

- Write Lewis (electron-dot) symbols for monatomic ions.
- Write chemical equations for half-reactions and balanced redox reactions in which ions (ionic compounds) are formed from atoms.
- Identify elements oxidized and reduced and oxidizing and reducing agents.

The chemical stability of the noble gases (Group 8A) is attributed to their stable electron configurations. Noble gas atoms have eight valence electrons, except for helium with two. A set of eight electrons is often referred to as an **octet** of electrons. An octet of outer electrons is thought to be a uniquely stable electron structure because the noble gases are so unreactive.

Although helium has only two valence electrons, it is stable because its first principal energy level is filled. It seems to be a completely unreactive element. Because of its stability and because it has the second lowest density of any gas, it is now used in weather balloons, dirigibles, and party balloons in place of hydrogen which is highly flammable.

Octet rule: When forming compounds, an atom has a tendency to lose, gain, or share electrons until it has eight electrons in its last occupied shell.

When a representative element (one in an A-group) combines with another element to form a compound, it loses, gains, or shares electrons so that it ends up with the same number of electrons as the noble gas closest to it (by atomic number). This is known as the **noble gas rule,** or **octet rule,** because the noble gases (except for helium) have eight valence electrons. The octet rule is observed for most of the representative elements as they lose, gain, or share electrons to form either ions or molecules. Exceptions to the octet rule occur with some of the molecular compounds of metalloids and nonmetals such as boron, phosphorus, and sulfur, and with some of the metals in the *p*-block. The concept of sharing electrons and examples of exceptions to the octet rule are found in Chapter 8.

Formation of Cations: Oxidation

Let's consider a reaction of sodium atoms with chlorine atoms to form sodium chloride, a compound made up of sodium ions and chloride ions. Sodium (element 11) is found in Group 1A in the periodic table. So a sodium atom, like all elements in this group, has one valence electron, and its Lewis symbol is:

$$\text{Na}\cdot$$

Remember that sodium, a metal, has a tendency to lose one electron. Looking at the periodic table, note that the noble gas nearest to sodium is neon (element 10). Therefore, you might expect sodium to lose one electron as it combines with chlorine. Then it would have 10 electrons, the same number as neon. The following equation shows a sodium atom losing one electron to form a sodium ion. Remember that a sodium ion is a *cation,* or positively charged ion (Section 2.3). The charge on an ion is determined by whether protons or electrons are in excess. Cations are always electron deficient, so protons are in excess resulting in a positive charge.

$$\begin{array}{lll} \text{Na}\cdot & \rightarrow & \text{Na}^+ \quad + \quad e^- \\ 11p, \ +11 & & 11p, \ +11 \\ 11e^- \ \underline{-11} & & 10e^- \ \underline{-10} \\ \quad 0 \text{ charge} & & \quad +1 \text{ charge} \end{array}$$

A Na^+ ion and a Ne atom have the same electron configuration:

Na^+ $1s^22s^22p^6$

Ne $1s^22s^22p^6$

Both Na^+ and Ne are very stable, whereas Na is vigorously reactive. Atoms and ions that have the same electron configuration are said to be **isoelectronic.** A cation formed from an A-group metal is isoelectronic with the nearest noble gas. Magnesium ion, Mg^{2+}, and aluminum ion, Al^{3+}, also are isoelectronic with neon. Potassium ion, K^+, and calcium ion, Ca^{2+}, both have 18 electrons and are isoelectronic with argon, Ar.

Atoms and ions having the same electron configuration are isoelectronic.

Now you can understand why the charge on a cation is the same as the group number for a metal in Group 1A, 2A, or 3A (Section 5.8). The group number identifies the number of valence electrons for an atom in that group. The charge on a cation is equal to the number of valence electrons lost by the metal atom.

The loss of an electron by a sodium atom is an example of oxidation, and we say that sodium was oxidized. **Oxidation** is a chemical change in which an element loses electrons. It frequently occurs when a substance reacts with oxygen, such as when charcoal or another fuel is burned, which accounts for the origin of the term. Actually, oxidation is only half the reaction between sodium and chlorine. However, writing an equation showing only the loss of electrons (the oxidation half-reaction) helps us to understand oxidation. We will see in a moment that something else must gain electrons. In other words, electrons lost in oxidation are actually transferred to another substance.

Loss of electrons is oxidation (LEO).

Formation of Anions: Reduction

Now, let's think about the chlorine atoms in the reaction of sodium with chlorine.* Chlorine is in Group 7A, so a chlorine atom has seven valence electrons. You could represent a chlorine atom with its Lewis symbol:

$$:\ddot{\underset{\cdot\cdot}{Cl}}\cdot$$

Because chlorine is a nonmetal, its properties are practically opposite those of metals. As it combines with sodium, you might expect a chlorine atom to gain one electron to give it an octet. Then it would be isoelectronic with an argon atom. The following equation shows chlorine gaining one electron to form a chloride ion. Note that chloride ion is an *anion,* or negatively charged ion (Section 2.3).

$$:\ddot{\underset{\cdot\cdot}{Cl}}\cdot + e^- \rightarrow \left[:\ddot{\underset{\cdot\cdot}{Cl}}:\right]^-$$

This is an example of **reduction,** a chemical change in which an element gains electrons. We may say that chlorine was reduced. This equation shows only the reduction part of the reaction, called the reduction half-reaction.

Gain of electrons is reduction (GER).

---

*Although chlorine is a diatomic element and exists as Cl_2 molecules, chlorine atoms are shown here to simplify the discussion.

The anion formed when a nonmetal atom gains electrons is isoelectronic with the nearest noble gas. A Cl^- ion has the same electron configuration as an Ar atom:

$$\text{Cl}^- \quad 1s^22s^22p^63s^23p^6$$

$$\text{Ar} \quad 1s^22s^22p^63s^23p^6$$

Other examples are fluoride ion, F^-, and oxide ion, O^{2-}, which are isoelectronic with neon.

Remember: LEO says GER.

It is important to note that an element cannot gain electrons (reduction) unless electrons are lost (oxidation) by another element. In other words, electrons are transferred in oxidation–reduction reactions. An element cannot be reduced without another element being oxidized at the same time.

The chemical equation for the reaction of sodium and chlorine is the sum of the equations for the two half-reactions. Electrons are not written in this final equation. When the equations for the half-reactions are added, the electrons must cancel.

$$\text{Na}\cdot \rightarrow \text{Na}^+ + \cancel{e^-}$$

$$:\underset{\cdot\cdot}{\overset{\cdot\cdot}{\text{Cl}}}\cdot + \cancel{e^-} \rightarrow \left[:\underset{\cdot\cdot}{\overset{\cdot\cdot}{\text{Cl}}}:\right]^-$$

$$\text{Na}\cdot + \cdot\underset{\cdot\cdot}{\overset{\cdot\cdot}{\text{Cl}}}: \rightarrow \text{Na}^+ + \left[:\underset{\cdot\cdot}{\overset{\cdot\cdot}{\text{Cl}}}:\right]^- \text{ or NaCl}$$

Thus, sodium chloride, made up of sodium ions and chloride ions, forms when electrons are transferred from sodium atoms to chlorine atoms.

Redox reactions are electron-transfer reactions.

Metals are reducing agents.

The preceding equation shows an example of an oxidation–reduction or **redox reaction.** Redox reactions are sometimes called electron-transfer reactions. A sodium atom lost an electron and was oxidized, and a chlorine atom gained an electron and was reduced. Chlorine is called the **oxidizing agent,*** an electron acceptor. An oxidizing agent causes another substance to be oxidized. Sodium is called a **reducing agent,** an electron donor. A reducing agent causes another substance to be reduced. Because of their tendency to lose electrons when they react, metals are always reducing agents. Because of their tendency to gain electrons, nonmetals are oxidizing agents when they react with metals. Fluorine, chlorine, and oxygen are examples of good oxidizing agents.

Oxidation and Reduction

Substance *oxidized* is the *reducing agent* (loses electons).	**Substance** *reduced* is the *oxidizing agent* (gains electrons).

*This term originated with oxygen, O_2, the oxidizing agent in the ordinary combustion of fuels.

Example 6.10

Write Lewis symbols for the following ions:

a. lithium ion
b. magnesium ion
c. fluoride ion ■

Solutions Remember that the number of valence electrons lost by a metal atom is equal to the positive charge on the cation that is formed. The numbers of electrons gained by a nonmetal atom (octet rule) is equal to the negative charge on its anion.

a. Li^{+}

b. Mg^{2+}

c. $[:\ddot{\underset{..}{F}}:]^{-}$

Example 6.11

Using Lewis symbols, write an equation for the reaction of potassium and bromine atoms. Identify the element oxidized and the element reduced. Which is the oxidizing agent? The reducing agent? ■

Solution Potassium, found in Group 1A in the periodic table, has one valence electron. The oxidation half-reaction is:

$$K\cdot \rightarrow K^{+} + e^{-}$$

Bromine is in Group 7A and has seven valence electrons. The reduction half-reaction is:

$$:\ddot{\underset{..}{Br}}\cdot + e^{-} \rightarrow [:\ddot{\underset{..}{Br}}:]^{-}$$

Adding the equations gives:

$$K\cdot + :\dot{\underset{..}{Br}}: \rightarrow K^{+} + [:\ddot{\underset{..}{Br}}:]^{-} \quad \text{or} \quad KBr$$

The product is potassium bromide and can be represented by the simple formula KBr. Potassium was oxidized and acted as a reducing agent (electron donor). Bromine was reduced and acted as an oxidizing agent (electron acceptor).

Practice Problem 6.7 Write Lewis symbols for the following ions:
a. sulfide ion
b. aluminum ion
c. nitride ion

Practice Problem 6.8 Using Lewis symbols, write the equation for the reaction of sodium and oxygen atoms. Identify the element oxidized and the element reduced. Identify the oxidizing and reducing agents.

6.9 Periodic Properties of the Elements

Learning Objectives

- Predict relative sizes for atoms and ions.
- Predict relative ionization energies for the elements.
- Identify elements that have high electron affinities and elements that have low electron affinities.

The periodic law (Section 5.6) tells us that when the elements are arranged according to their atomic numbers, similar chemical and physical properties recur at regular intervals. At this point, you are beginning to understand some of the chemical properties of the elements based upon their location in the periodic table. For example, a metal tends to lose electrons to form a positive ion. On the other hand, a nonmetal, found on the right side of the table, tends to gain electrons until it has the same number as the nearest noble gas. For the representative elements, the number of electrons lost, gained, or shared can usually be predicted from the number of valence electrons for the atom, using the octet rule.

Now let's consider three properties of elements and how these properties vary as the elements are arranged in the periodic table. Knowing the variation in these properties helps us to understand the differences in the chemical properties of the elements. We will consider sizes of atoms and ions, ionization energy, and electron affinity.

Sizes of Atoms and Ions

An atom can be thought of as a tiny, invisible particle that is spherical in shape.* Because it is so small, we cannot simply measure its diameter as if it were a marble or a baseball. Instead, x-ray techniques are used to estimate the dimensions of an atom. The atomic radii shown in Figure 6.14 were calculated from x-ray data.

Notice that the sizes of the atoms increase as you move down a group. For example, a cesium atom (Group 1A) is larger than a lithium atom. An iodine atom (Group 7A) is larger than a fluorine atom. This trend is consistent in all the A-groups.

As you look at the elements in a period, you see that atoms on the right are smaller than those on the left. For example, a fluorine atom is smaller than a lithium atom, and a chlorine atom is smaller than a sodium atom. This trend is

*In view of the shapes of atomic orbitals, many atoms probably are not spherical. However, we will assume a spherical shape for convenience in discussing size and other physical properties.

1A	2A	3A	4A	5A	6A	7A
Li 152	Be 111	B 80	C 77	N 74	O 74	F 71
Na 186	Mg 160	Al 143	Si 118	P 110	S 103	Cl 99
K 227	Ca 197	Ga 122	Ge 123	As 125	Se 116	Br 114
Rb 248	Sr 215	In 163	Sn 141	Sb 145	Te 143	I 133
Cs 265	Ba 217	Tl 170	Pb 175	Bi 155		

FIGURE 6.14 Atomic radii for selected elements in picometers (1 pm = 10^{-12} m). (Data are from T. Moeller, *Inorganic Chemistry: A Modern Introduction;* Wiley: New York, 1982; pp. 70–72.)

consistent for most of the representative elements in any period. However, there are some inconsistencies among the transition elements, the lanthanides, and the actinides.

Why does chlorine, with six more electrons than sodium, have a smaller atomic radius? The variation in atomic size is principally the result of two factors:

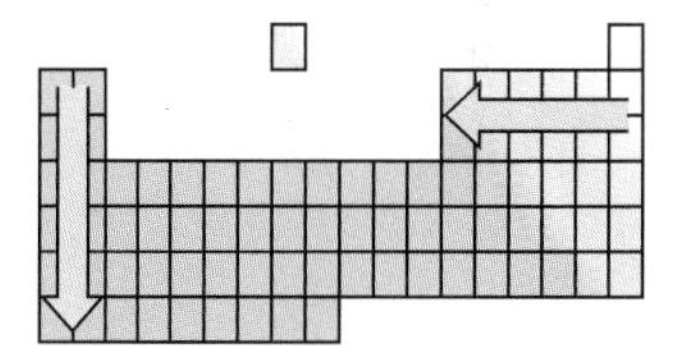

Variation in atomic size. In general, atomic radii increase from right to left across a period and from top to bottom down groups.

1. *The highest occupied principal energy level, n.* Within a given group, the atom having the highest occupied principal energy level (shell) is the largest. This explains why the atoms of elements at the bottom of a group are larger.
2. *Nuclear charge.* Moving from left to right in a period, the nuclear charge increases. A nucleus having a large positive charge strongly attracts outer electrons in a given shell, pulling them toward the nucleus. In a chlorine atom, outer electrons are in the same shell as those in a sodium atom. Thus, chlorine (element 17) is smaller than sodium (element 11).

Ions differ in size from the corresponding atoms because they have different numbers of electrons (Figure 6.15). For example, a sodium atom, Na·, is larger than a sodium ion, Na^+. A sodium atom has 11 electrons attracted by its nuclear charge of +11, but a sodium ion has only 10 electrons attracted by its +11 nuclear charge. In addition, the valence shell of the ion is vacant, and the atom has a higher occupied shell. Their electron configurations are:

sodium atom, Na· $1s^22s^22p^63s^1$

sodium ion, Na^+ $1s^22s^22p^6$

An anion, in contrast to a cation, is larger than its corresponding atom. For example, a fluoride ion is larger than a fluorine atom. Their electron configurations are:

FIGURE 6.15 Sizes of ions compared with atoms for some of the representative elements. Radii are in picometers. (Data are from T. Moeller, *Inorganic Chemistry: A Modern Introduction;* Wiley: New York, 1982; pp. 141–144.)

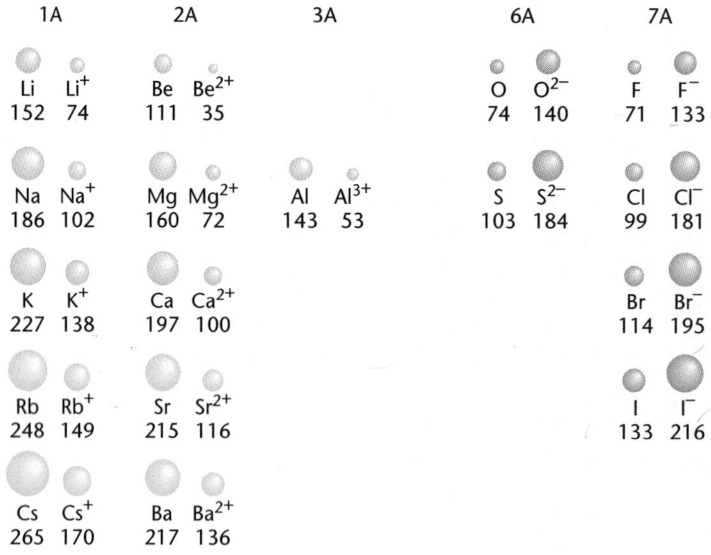

1A		2A		3A		6A		7A	
Li 152	Li^{+} 74	Be 111	Be^{2+} 35			O 74	O^{2-} 140	F 71	F^{-} 133
Na 186	Na^{+} 102	Mg 160	Mg^{2+} 72	Al 143	Al^{3+} 53	S 103	S^{2-} 184	Cl 99	Cl^{-} 181
K 227	K^{+} 138	Ca 197	Ca^{2+} 100					Br 114	Br^{-} 195
Rb 248	Rb^{+} 149	Sr 215	Sr^{2+} 116					I 133	I^{-} 216
Cs 265	Cs^{+} 170	Ba 217	Ba^{2+} 136						

fluorine atom, $:\ddot{\underset{..}{F}}\cdot$ $1s^2 2s^2 2p^5$

fluoride ion, $\left[:\ddot{\underset{..}{F}}:\right]^-$ $1s^2 2s^2 2p^6$

Cations are smaller than their related atoms; anions are larger than their related atoms.

Each has a nuclear charge of +9, but the fluoride ion has 10 electrons compared with 9 electrons for the fluorine atom. For 10 electrons, the effective nuclear charge attracting each electron is smaller. Thus the anion is larger.

Example 6.12

Which is larger:

a. Na atom or Na^+?
b. Cl atom or Cl^-?
c. Na^+ or Cl^-? (Refer to Figure 6.15.) ■

Solutions Remember that cations are smaller than their respective atoms, and anions are larger than their respective atoms.

a. Na atom
b. Cl^-
c. Cl^-

Practice Problem 6.9 Which is smaller:
a. Na^+ or K^+?
b. F^- or I^-?
c. Br^- or K^+? (Refer to Figure 6.15.)

■ Ionization Energy

The **ionization energy** for an element is the amount of energy required to remove its outermost electron. You know from experience that energy is required to move an object away from the force attracting it. Thus, removing a negatively charged electron from the attraction of the positive nuclear charge is an endothermic process. Atoms must absorb energy to lose their outer electrons. For sodium, this process can be represented by the equation:

$$\text{energy} + \text{Na}\cdot \longrightarrow \text{Na}^{+} + e^{-}$$

Ionization energies of elements are measured for gaseous atoms. The variation of ionization energy with atomic number is shown in Figure 6.16. You should take note of two trends. First, ionization energy increases as you move left to right across a period. Second, ionization energy increases from bottom to top in a group of elements. Note that the trends for ionization energy are opposite to the trends in atomic size. This is because the outermost electron of a large atom is farther from its nucleus, and so less energy is required to remove it than to remove the outermost electron of a small atom.

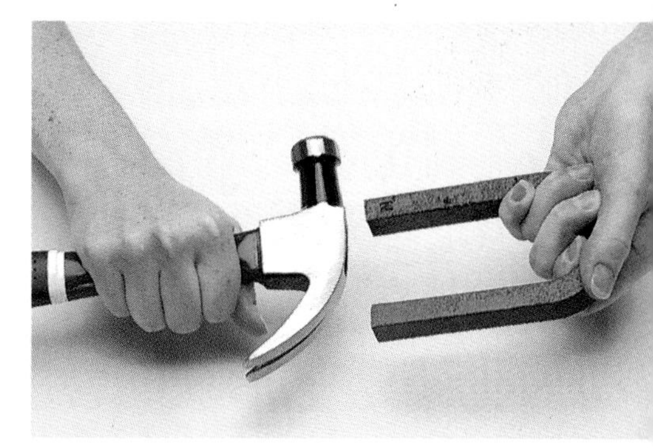

Energy is required to move an object away from the force attracting it.

Endothermic: Energy is absorbed from the surroundings (Section 2.8).

Example 6.13

Which has the larger ionization energy:

a. Na or Cs? **c.** Al or P?
b. Na or S? **d.** K or Br? ■

Solutions Remember the trends shown in Figure 6.16 and the marginal diagram.

a. Na **c.** P
b. S **d.** Br

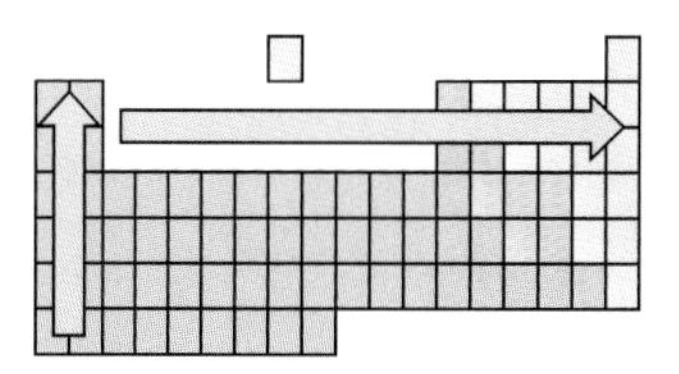

Variation in ionization energy. In general, ionization energies increase from bottom to top in main groups and from left to right across periods.

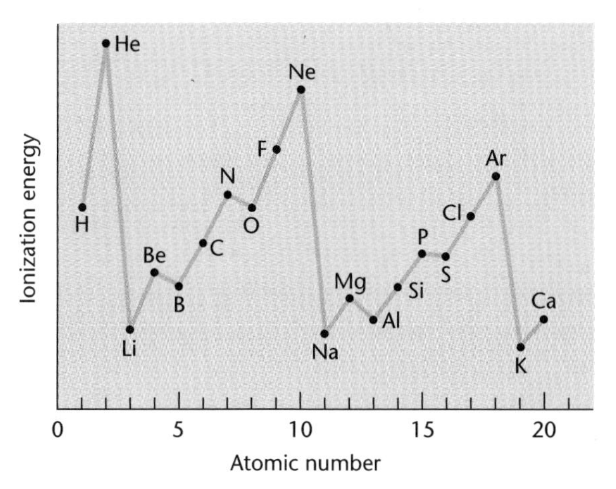

FIGURE 6.16 Variation of ionization energies with atomic number for elements 1–20.

Practice Problem 6.10 Which has the smaller ionization energy:
a. K or Br?
b. K or Li?
c. P or Cl?

Electron Affinity

To understand the chemical properties of nonmetals, we need to consider another energy quantity, the **electron affinity** for an element, which is the energy change resulting when gaseous atoms each gain an electron into their valence shell. Electron affinity could be called electron-gain energy. When chlorine atoms gain electrons, the process is exothermic, as is illustrated by the following equation:

Exothermic: Energy is released to the surroundings (Section 2.8).

$$:\ddot{\underset{\cdot\cdot}{Cl}}\cdot + e^- \rightarrow [:\ddot{\underset{\cdot\cdot}{Cl}}:]^- + \text{energy}$$

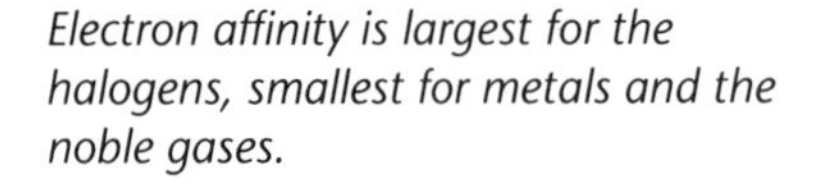

Electron affinity is largest for the halogens, smallest for metals and the noble gases.

As you might expect, the halogens (Group 7A) have the largest electron affinities (Figure 6.17); that is, the electron gain is highly exothermic. These elements have a tendency to gain electrons. For elements that have little tendency to gain electrons, the electron gain is endothermic or only weakly exothermic. Metals have a tendency to lose rather than gain electrons, so they would be expected to have small electron affinities. In general, electron affinity increases from left to right and from bottom to top in the periodic table. However, the noble gases, which are generally unreactive, have small electron affinities.

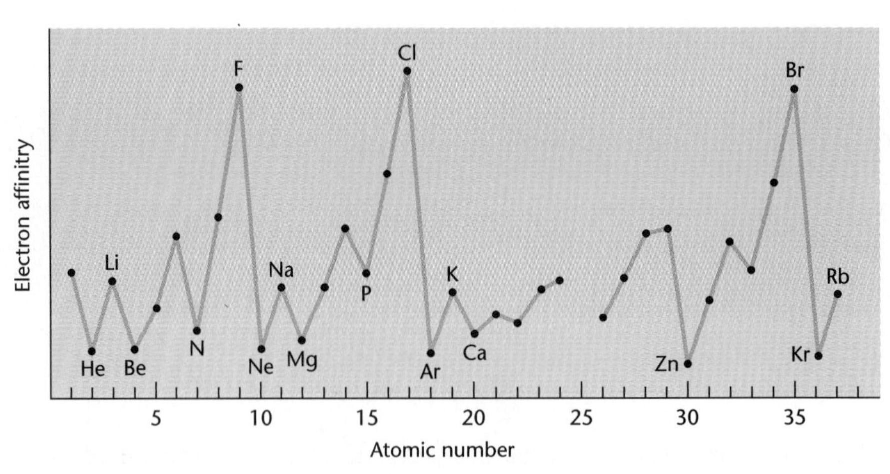

FIGURE 6.17 Variation of electron affinities with atomic number.

Example 6.14

Which of the following has the larger electron affinity:

a. Na or Cl?
b. K or O?
c. O or Ne? ■

Solutions Refer to Figure 6.17.

a. Cl
b. O
c. O

Practice Problem 6.11 Which has the smaller electron affinity:

a. S or Mg?
b. Cl or I?
c. Si or Cl?

Ionization energies and electron affinities explain what chemists repeatedly have observed about the chemical reactivity of the elements. Metals are elements that have a tendency to lose electrons, and thus form positive ions, because they have small ionization energies and electron affinities. The elements in Groups 1A and 2A are the most metallic. We now can say that cesium (element 55) is the most metallic element because it has the smallest ionization energy of all the elements.* Nonmetals show the opposite behavior because they have large ionization energies and large electron affinities. Noble gases have very large ionization energies and very small (actually endothermic) electron affinities, in agreement with their lack of reactivity.

The periodic table contains a wealth of information that can help you to understand chemistry. It is the most important chemistry reference available, and you will be using it often as you continue your study of chemistry.

■ Chapter Review

Directions: From the choices listed at the end of the Chapter Review, choose the word or term that best fits in each blank. Use each choice only once. The answers appear at the end of the chapter.

Section 6.1 An electron may be considered to be a very small particle having a charge of __________ (1) and a very small mass, approximately __________ (2) amu. An electron can behave both as a particle and as a __________ (3).

Section 6.2 In 1913, Niels Bohr proposed a model for a hydrogen atom. He suggested the nucleus was surrounded by a number of __________ (4) in which the electron could travel. Bohr's model was an interpretation of the bright-line __________ (5) for hydrogen. An electron could move from one orbit to another with the absorption or emission of a __________ (6) of energy. The inability of the Bohr model to explain the emission spectra of other elements was a serious limitation, but it

*Francium (Fr) is below cesium in Group 1A and would be expected to be the most metallic element. However, it is a very rare, radioactive element, and few of its properties have been determined.

laid the foundation for the __________ __________ (7) model, which is still used today.

Section 6.3 The quantum mechanical model for an atom shows electrons occupying principal energy levels, sublevels, and orbitals. The principal energy levels, or __________ (8) are numbered 1, 2, 3, and so on. The number of sublevels in each principal energy level is equal to its principal __________ __________ (9), *n*. The __________ (10) are designated with the letters *s, p, d,* and *f.*

Section 6.4 An atomic __________ (11) is a region in space around the nucleus of an atom in which there is a high probability of finding an electron. Orbitals have characteristic shapes. For example, the shape of an *s* orbital is __________ (12). The *s, p, d,* and *f* sublevels have 1, 3, __________ (13), and 7 orbitals, respectively. When you show the energy states for electrons in an atom, remember that electrons fill lowest-energy sublevels first.

The Heisenberg __________ (14) principle states that it is impossible to precisely determine the exact position and speed of an __________ (15) at the same time.

Each atomic orbital can be occupied by 0, 1, or __________ (16) electrons. Thus, the capacity of the sublevels is 2, 6, 10, and __________ (17) electrons, respectively. The electron capacity of the __________ (18) energy levels is given by the formula __________ (19).

Orbital occupancy is governed by the Pauli __________ (20) principle: When two electrons occupy an orbital, their __________ (21) must be opposite. Two electrons in the same orbital are said to be ______ (22).

Section 6.5 An __________ (23) level diagram shows the order of energy of the principal energy levels and sublevels of an atom. When completing an energy level diagram for an atom, remember three rules:

1. Electrons will occupy the __________ (24) energy levels and sublevels.
2. An __________ __________ (25) may contain no more than two electrons.
3. Electrons are evenly distributed among the orbitals of a sublevel according to __________ (26) rule.

Section 6.6 An electron __________ (27) is a shorthand notation that shows the energy states for the electrons in an atom. The elements in the periodic table are grouped in blocks according to their outermost occupied sublevel. The *s*-block includes Groups ______ (28) and __________ (29). Elements in Groups 3A through 8A make up the __________ (30). The *d*-block is made up of the transition elements, and the lanthanides and actinides are found in the __________ (31).

Electrons in the outermost principal energy level of an atom are called __________ (32) electrons. For the __________ (33) elements, the __________ (34) number identifies the number of valence electrons.

When writing electron configurations, you may abbreviate using a symbol for the __________ (35) gas core followed by the additional sublevels needed for the remaining electrons.

Section 6.7 An electron-dot or __________ (36) symbol shows the symbol for the element surrounded by dots that represent its valence electrons. Single dots represent __________ (37) electrons.

Section 6.8 The chemical stability of the noble gases is due to their __________ (38) electron configurations. Except for He, they all have an __________ (39) of electrons. Atoms of the representative elements tend to gain or lose electrons and thus become __________ (40) with the nearest noble gas. The tendency for elements to combine so as to have __________ (41) electrons in their valence shell is called the octet rule.

A chemical change in which an element loses electrons is called __________ (42). __________ (43) are the most easily oxidized elements. Reduction is a chemical change in which an element __________ (44) electrons. When an element is reduced, it acts as an __________ (45) agent. When an element is __________ (46), it acts as a reducing agent. Metals are always __________ (47) agents. Electron transfer reactions are often called __________ (48) reactions. Oxidation and __________ (49) must occur together.

Section 6.9 Sizes of atoms __________ (50) as you move down a group and as you move right to left in a period. Cations are __________ (51) than the corresponding atoms, and anions are __________ (52) than their related atoms.

The amount of energy needed to remove the outermost electron of an atom is called its __________ __________ (53). A large atom will have a relatively __________ (54) ionization energy and will tend to lose its outer electron easily. The most metallic element is __________ (55).

The energy change resulting when atoms of an element each gain an electron into their valence shell is called its __________ __________ (56). Electron affinity is highest for the __________ (57) and lowest for metals and the elements in Group __________ (58).

Choices: atomic orbital, cesium, configuration, 8A, eight, electron, electron affinity, energy, exclusion, *f*-block, 5, 14, gains, group, halogens, Hund's, increase, ionization energy, isoelectronic, larger, Lewis, low, lowest, metals, -1, noble, 1A, octet, orbital, orbits, oxidation, oxidized, oxidizing, *p*-block, paired, principal, quantum, quantum mechanical, quantum number, redox, reducing, reduction, representative, shells, small, smaller, spectrum, spherical, spins, stable, sublevels, 2, 2A, $2n^2$, uncertainty, unpaired, valence, wave, 0

Study Objectives

After studying Chapter 6, you should be able to:

Section 6.1

1. Describe the properties of electrons.

Section 6.2

2. Diagram and describe the Bohr model for a hydrogen atom.
3. Distinguish between a continuous spectrum and a bright-line spectrum.

Section 6.3

4. List the principal energy levels and their sublevels in order of increasing energy.
5. Specify the number of sublevels in each principal energy level.

Section 6.4

6. Specify the number of orbitals in each sublevel.
7. Specify the electron capacity of an orbital, sublevel, and principal energy level.
8. Describe the shapes of *s* and *p* atomic orbitals.
9. State the Heisenberg uncertainty principle and the Pauli exclusion principle.
10. Describe the spins of two electrons in the same orbital.

Section 6.5

11. Draw an electron energy level diagram showing the first three principal energy levels and their sublevels.
12. Draw pictures of the occupied orbitals of any of the first ten elements.
13. State Hund's rule.

Section 6.6

14. Find the *s*-block, *p*-block, *d*-block, and *f*-block in the periodic table.
15. Using the periodic table as a guide, write complete electron configurations and valence-shell configurations for atoms and monatomic ions of the representative elements.
16. Using noble gas core abbreviations, write electron configurations for atoms of the representative elements.
17. Specify the number of valence electrons for an atom of any representative element from its group number.

Section 6.7

18. Write Lewis (electron-dot) symbols for atoms of the representative elements.

Section 6.8

19. Write Lewis (electron-dot) symbols for monatomic ions.
20. Write chemical equations for half-reactions and balanced redox reactions in which ions (ionic compounds) are formed from atoms.
21. Identify elements oxidized and reduced and oxidizing and reducing agents.

Section 6.9

22. Predict relative sizes for atoms and ions.
23. Predict relative ionization energies for the elements.
24. Identify elements that have high electron affinities and elements that have low electron affinities.

Key Terms

Review the definition of each of the terms listed here by chapter section number. You may use the glossary if necessary.

6.2 orbit, spectrum, quantum of energy

6.3 principal energy level, shell, sublevel, subshell

6.4 atomic orbital, Heisenberg uncertainty principle, Pauli exclusion principle

6.5 Hund's rule

6.6. electron configuration, ground state, noble gas core, valence shell

6.7 Lewis (electron-dot) symbol

6.8 isoelectronic, noble gas rule, octet, octet rule, oxidation, oxidizing agent, redox reaction, reducing agent, reduction

6.9 electron affinity, ionization energy

Questions and Problems

Answers to questions and problems with blue numbers appear in Appendix C.

Bohr Model (Section 6.2)

1. Draw a diagram and describe the Bohr model for an atom.
2. Describe an orbit.
3. Explain the difference between a continuous spectrum like a rainbow and a bright-line emission spectrum for an element such as hydrogen.
4. What is a quantum of energy? Explain why quanta of energy are emitted from an excited sample of hydrogen atoms.
5. What are quantized energy levels for electrons?
6. What evidence has led physicists to propose quantized energy levels for the electrons within an atom?
7. Explain why the Bohr model for an atom is no longer used today.

Modern (Quantum Mechanical) Model (Sections 6.3, 6.4, 6.5)

8. What is a principal quantum number, n? Explain the significance of a small number for n and a large number for n.
9. What are principal energy levels (shells) for electrons?
10. What notation is used to designate principal energy levels for electrons?
11. What is the electron capacity for each of the first four principal energy levels (shells)?
12. What letters are used to label the sublevels?
13. How many sublevels make up each of the first four principal energy levels?
14. Arrange the sublevels of the 4th principal energy level according to increasing energy.
15. Construct an energy diagram for electrons in an atom, showing principal energy levels and sublevels through $4p$.
16. What is an atomic orbital? How is it different from an orbit?
17. What is the electron capacity of an atomic orbital?
18. What is the shape of an s orbital?
19. What is the shape of a p orbital? Sketch the three $2p$ orbitals.
20. How many orbitals are in s, p, d, and f sublevels?
21. What is the electron capacity for each of the sublevels?
22. What is the Heisenberg uncertainty principle?
23. What is the Pauli exclusion principle?
24. How many spin states are possible for an electron?
25. What is meant by paired electrons? When are electrons paired?
26. What is Hund's rule?
27. Draw an energy level diagram for a boron atom. Using the diagram, sketch the occupied atomic orbitals of a boron atom.
28. Draw an energy level diagram and an orbital picture for a nitrogen atom.
29. Draw energy level diagrams for the following atoms:
    a. fluorine
    b. argon
30. Draw energy level diagrams for the following atoms:
    a. potassium
    b. titanium

Electron Configurations (Sections 6.6, 6.7)

31. Write electron configurations for the first twelve elements.
32. Write electron configurations for the following atoms:
    a. aluminum
    b. phosphorus
    c. chlorine
33. Write expanded electron configurations for:
    a. phosphorus
    b. sulfur
    c. silicon
    How many unpaired electrons are found in the atoms of each of these elements?
34. Write the electron configurations for:
    a. lithium
    b. sodium
    c. potassium
    What similarity do you see in the configurations of these three elements? Where are these elements found in the periodic table?
35. Write the electron configurations for:
    a. fluorine
    b. chlorine
    c. bromine
    What similarity do you see in the configurations of these three elements? Where are these elements found in the periodic table?
36. Write the electron configurations for:
    a. magnesium
    b. calcium
    c. strontium

What similarity do you see in the configurations of these three elements? Where are these elements found in the periodic table?

37. What groups of elements make up the *s*-block in the periodic table? The *p*-block?

38. Which elements make up the *d*-block in the periodic table? The *f*-block?

39. How are the transition elements similar in their electron configurations?

40. How are the lanthanides and actinides similar in their electron configurations?

41. Write electron configurations for:
    a. indium, element 49
    b. bismuth, element 83

42. Write the valence-shell configurations for the following elements:
    a. Cl
    b. As
    c. Si
    d. Rb
    e. Sb

43. Using the noble gas core notation where appropriate, write the electron configurations for the following:
    a. Sr
    b. Ga
    c. Se
    d. I
    e. Sb

44. Identify the following elements from their valence-shell configurations:
    a. $5s^2$
    b. $2s^2 2p^1$
    c. $2s^2 2p^6$
    d. $6s^2 6p^3$
    e. $4s^2 4p^2$

45. Write the valence-shell configuration for each of the following elements:
    a. the third alkali metal
    b. the second halogen
    c. the first alkaline earth metal
    d. the fourth noble gas

46. Identify the number of valence electrons for each of the following elements:
    a. Mg
    b. K
    c. P
    d. Si
    e. S

47. Write Lewis (electron-dot) symbols for the following atoms:
    a. strontium
    b. potassium
    c. sulfur
    d. arsenic

48. Write Lewis symbols for the following atoms:
    a. phosphorus
    b. boron
    c. bromine
    d. selenium

49. Specify the number of unpaired electrons for each of the following atoms:
    a. selenium
    b. bromine
    c. phosphorus
    d. potassium
    e. calcium
    f. arsenic

Octet Rule and Formation of Ions (Section 6.8)

50. What is the octet rule? Which atoms have an octet of electrons?

51. Which noble gas does not have eight (an octet) valence electrons?

52. Why is the octet rule also called the noble gas rule?

53. Which elements generally follow the octet rule when they combine with other elements?

54. Specify the charge on each of the following ions:
    a. potassium ion
    b. magnesium ion
    c. fluoride ion
    d. oxide ion
    e. phosphide ion
    f. barium ion

55. What is the charge on an alkali metal ion?

56. What is the charge on an alkaline earth metal ion?

57. Write Lewis (electron-dot) symbols for the following ions:
    a. calcium ion
    b. chloride ion
    c. sulfide ion
    d. potassium ion
    e. aluminum ion

58. Write electron configurations for the ions in question 57.

59. Write Lewis symbols for the following ions:
    a. magnesium ion
    b. phosphide ion
    c. bromide ion
    d. rubidium ion
    e. hydride ion (from hydrogen)

60. Write electron configurations for the ions in question 59. For those with more than 10 electrons, use noble gas core abbreviations.

61. Which noble gas is isoelectronic with each of the following pairs of ions?
    a. Na^+ and F^-
    b. N^{3-} and O^{2-}
    c. K^+ and Ca^{2+}
    d. Ca^{2+} and Cl^-
    e. S^{2-} and Cl^-

62. What cations are isoelectronic with a neon atom?

63. What anions are isoelectronic with a neon atom?

64. What cations are isoelectronic with a Br^- ion?

65. What is oxidation?

66. Write an equation for the half-reaction showing the oxidation of lithium.

67. What is reduction?

68. Write an equation for the half-reaction for the reduction of chlorine atoms.

69. Using Lewis symbols, write equations for the following redox reactions. Identify the elements oxidized and reduced and the oxidizing and reducing agents.
    a. lithium and chlorine atoms react
    b. calcium and sulfur atoms react

70. Explain why redox reactions are called electron-transfer reactions.

Periodic Properties (Section 6.9)

71. What factors influence the size of an atom?

72. Explain why a sodium atom is larger than a lithium atom.

73. Why is a sodium atom larger than a fluorine atom?

74. Which is larger:
    a. a potassium atom or a cesium atom?
    b. a chlorine atom or an iodine atom?
    c. a barium atom or a magnesium atom?
    d. a potassium atom or a bromine atom?
    e. an oxygen atom or a beryllium atom?

75. Which is smaller:
    a. a sodium atom or a sodium ion?
    b. a magnesium atom or a magnesium ion?
    c. an oxygen atom or an oxide ion?

76. Which is larger:
    a. a chlorine atom or a chloride ion?
    b. a lithium ion or a chloride ion?
    c. a sulfide ion or an oxide ion?

77. Which is the smallest halogen? The smallest alkali metal?

78. What is meant by the ionization energy of an element?

79. How do ionization energies vary from top to bottom in Group 1A?

80. How do ionization energies vary from left to right in the third period of elements?

81. Which element in each pair has the larger ionization energy?
    a. Na or Cs
    b. Cs or I
    c. Mg or Cl
    d. Mg or Ca

82. Which element in each pair has a greater tendency to lose its outermost electron?
    a. Na or Cs
    b. Cs or Mg
    c. Mg or Cl
    d. Ca or Br

83. What is meant by the electron affinity of an element?

84. How do electron affinities vary from bottom to top in Group 6A?

85. How do electron affinities vary from left to right in the third period of elements?

86. Which element in each pair has the larger electron affinity?
    a. Li or F
    b. K or Cl
    c. Mg or O
    d. Ar or O
    e. Cl or Ne

87. Which element in each pair has a greater tendency to gain an electron into its outer shell?
    a. F or I
    b. Cl or S
    c. O or Na
    d. S or Al

Additional Problems

88. Draw energy level diagrams for the following atoms:
    a. magnesium
    b. silicon

89. Identify the number of valence electrons for each of the following elements:
    a. I
    b. As
    c. Al
    d. O
    e. N
    f. F

90. Which is larger:
    a. magnesium ion or barium ion?
    b. Sr atom or I atom?
    c. O^{2-} or Se^{2-}?

91. Which element of each pair has the smaller ionization energy?
    a. Ne or K
    b. Mg or Ar
    c. Al or S
    d. Ca or As

92. Which element of each pair has the larger electron affinity?
    a. S or Ar
    b. Cl or Si
    c. C or F

93. Using Lewis symbols, write equations for the redox reactions that occur when:
    a. magnesium and bromine atoms react
    b. potassium and oxygen atoms react
    Identify the elements oxidized and reduced and the oxidizing and reducing agents.

94. Write electron configurations for each of the following:
    a. Sn
    b. Rb^{+} ion
    c. Se^{2-} ion

95. Write the valence-shell configurations for:
    a. Ga
    b. Xe
    c. Sr
    d. Te

96. Identify the element described by each of the following:
    a. has one 3*s* electron
    b. has three 4*p* electrons
    c. has two paired and two unpaired 4*p* electrons

97. Krypton has 36 electrons. Give the number of filled
    a. *s* sublevels
    b. *p* sublevels
    c. *d* sublevels
    d. *f* sublevels

98. Sketch the occupied atomic orbitals of a nitrogen atom. Note that the differences in energy levels of the orbitals are related to the average distances of the orbitals from the nucleus.

99. According to the trends in periodic properties, which element in Group 6A should have:
    a. the smallest size?
    b. the largest ionization energy?
    c. the largest electron affinity?
    d. the greatest tendency to form a positive ion?
    e. the greatest tendency to form a negative ion?

Solutions to Practice Problems

PP 6.1

Add electrons to the energy level diagram (Figure 6.9) until you have the total number of electrons for the atom. Then write the electron configuration.

a. $1s^22s^22p^63s^2$
b. $1s^22s^22p^5$
c. $1s^22s^22p^63s^23p^5$

PP 6.2

Count superscripts (electrons), and look in the periodic table to find the symbol having that atomic number.

a. P b. Ne

PP 6.3

Use the periodic table as a guide.

a. $1s^22s^22p^63s^23p^64s^23d^{10}4p^3$ b. $1s^22s^22p^63s^23p^1$

PP 6.4

The number of valence electrons is the same as the group number.

a. 6 b. 8 c. 2

PP 6.5

Refer to the periodic table. The group number tells the number of valence electrons. Find which block the element is in to determine the outermost occupied sublevel.

a. $4s^24p^2$ b. $5s^25p^5$ c. $6s^1$

PP 6.6

Remember that a Lewis symbol shows only the valence electrons.

a. Ca:, no unpaired electrons
b. :Ar:, no unpaired electrons
c. :Cl·, 1 unpaired electron
d. :B·, 1 unpaired electron

PP 6.7

Show the valence electrons and the charge for the ion.

a. $[:\ddot{S}:]^{2-}$ b. Al^{3+} c. $[:\ddot{N}:]^{3-}$

PP 6.8

The oxidation half-reaction is

$Na\cdot \longrightarrow Na^+ + e^-$

An oxygen atom must gain two electrons to have an octet. The reduction half-reaction is

$:\dot{O}: + 2e^- \longrightarrow [:\ddot{O}:]^{2-}$

When the equations are added, the oxidation half-reaction must be multiplied by 2 to cancel electrons. The final balanced equation is

$2Na\cdot + :\dot{O}: \longrightarrow 2Na^+ + [:\ddot{O}:]^{2-}$ or Na_2O

The product is sodium oxide and can be represented by the simple formula Na_2O. Sodium was oxidized and was the reducing agent. Oxygen was reduced and was the oxidizing agent.

PP 6.9

An ion or atom low in a particular group is larger than one higher in that group.

a. Na^+ b. F^- c. K^+

PP 6.10

Ionization energy increases from bottom to top and from left to right in the periodic table.

a. K b. K c. P

PP 6.11

Electron affinity increases from left to right and from bottom to top in the periodic table.

a. Mg b. I c. Si

Answers to Chapter Review

1. -1
2. 0
3. wave
4. orbits
5. spectrum
6. quantum
7. quantum mechanical
8. shells
9. quantum number
10. sublevels
11. orbital
12. spherical
13. 5
14. uncertainty
15. electron
16. 2
17. 14
18. principal
19. $2n^2$
20. exclusion
21. spins
22. paired
23. energy
24. lowest
25. atomic orbital
26. Hund's
27. configuration
28. 1A
29. 2A
30. *p*-block
31. *f*-block
32. valence
33. representative
34. group
35. noble
36. Lewis
37. unpaired
38. stable
39. octet
40. isoelectronic
41. eight
42. oxidation
43. metals
44. gains
45. oxidizing
46. oxidized
47. reducing
48. redox
49. reduction
50. increase
51. smaller
52. larger
53. ionization energy
54. small
55. cesium
56. electron affinity
57. halogens
58. 8A

7

Composition and Formulas of Compounds

Contents

Every compound is made up of two or more elements and has a fixed composition. Shown here left to right are copper(II) nitrate, $Cu(NO_3)_2$, ammonium dichromate, $(NH_4)_2Cr_2O_7$, and magnetite, Fe_3O_4, an iron ore.

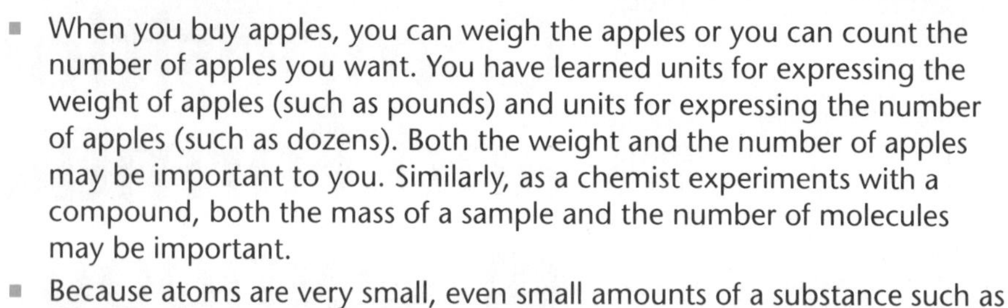

You Probably Know . . .

Both the weight and number of apples are important.

- When you buy apples, you can weigh the apples or you can count the number of apples you want. You have learned units for expressing the weight of apples (such as pounds) and units for expressing the number of apples (such as dozens). Both the weight and the number of apples may be important to you. Similarly, as a chemist experiments with a compound, both the mass of a sample and the number of molecules may be important.
- Because atoms are very small, even small amounts of a substance such as a drop of water or a grain of sand contain very large numbers of atoms.
- Certain compounds are of value primarily because of the content of only one element. For example, an iron ore (Fe_2O_3) is of value because of its iron content. The percentage of iron in the compound is important to those who are producing iron from iron ore.
- A chemist represents a compound by its formula. Some formulas are quite simple, such as H_2O and NaCl, whereas others, such as $C_{27}H_{46}O$ (cholesterol), are rather complex.

A grain of sand contains a very large number of atoms.

Hematite, Fe_2O_3.

Why Should We Study the Composition and Formulas of Compounds?

Chemists are continuing to discover ways to recover elements and compounds from natural mineral, plant, and animal materials. Furthermore, chemists have synthesized many useful new compounds—including plastics such as PVC and polystyrene, saran and mylar film, polyester and nylon fibers, synthetic rubber, pharmaceuticals such as aspirin and cancer drugs, food additives, and cosmetics.

You have learned that a pure substance has a definite composition—a feature, along with its characteristic physical properties, that distinguishes it from mixtures of substances. You also know that a compound is represented by a chemical formula, which you will soon see is a description of the composition of the compound. In preparing new chemical products, chemists must know the composition and the formulas of the substances they are working with. With this knowledge, the necessary quantities of reactants for a reaction can be calculated to produce certain quantities of products. Determining the composition of a new compound and its formula are necessary to identify it and to describe it to other chemists.

Early in your study of chemistry, you learned that the quantity of a substance is found by measuring its mass. The composition of a compound can be expressed in terms of the masses of the elements of which it is composed. However, as with apples, it is also important to know the number of particles (atoms, molecules, or ions) in a particular sample of a substance. In this chapter, we will consider a counting unit for atoms and molecules—a mole. We will see how mass composition and formulas are related by the numbers of moles of the elements making up a compound. Furthermore, we will see how to calculate formulas of compounds.

The Mole

7.1

Atoms and molecules are so tiny that they cannot be seen or handled as individual particles. Just one drop of water contains many trillions of water molecules. The unit needed for counting such large numbers of molecules must be an extremely large unit. Other counting units such as a dozen or a gross (12 dozen) would be much too small to be useful in counting such large numbers of molecules. The counting unit used by chemists for this purpose is called a *mole.*

By definition, a **mole** is the amount of substance that contains the same number of formula units (atoms, molecules, and so on) as there are carbon atoms in exactly 12 g of carbon-12. (You will recall from Section 5.5 that carbon-12 is the reference atom for atomic masses.) This number is **Avogadro's number,** named after an Italian scientist, Amadeo Avogadro (1776–1856). Avogadro's number has been determined experimentally by several methods to be 6.022×10^{23}.

A **formula unit** is the atom or group of atoms indicated by the formula of the substance. For example, C, Fe, O_2, H_2O, and NaCl represent formula units. We use this term in the definition of a mole to include all types of chemical substances. Remember, some elements are made up of atoms, and others such as oxygen and nitrogen are molecular. Some compounds such as water are made up of molecules, and others such as sodium chloride are made up of ions. Thus, a mole is Avogadro's number of formula units of any substance regardless of what type it is. If the substance is an element such as carbon, then one mole of carbon is 6.022×10^{23} carbon atoms. If the substance is a compound such as water, then one mole of water is 6.022×10^{23} water molecules. Remember, just as one dozen of anything is always 12 objects, one mole of any substance is always 6.022×10^{23} formula units of that substance.

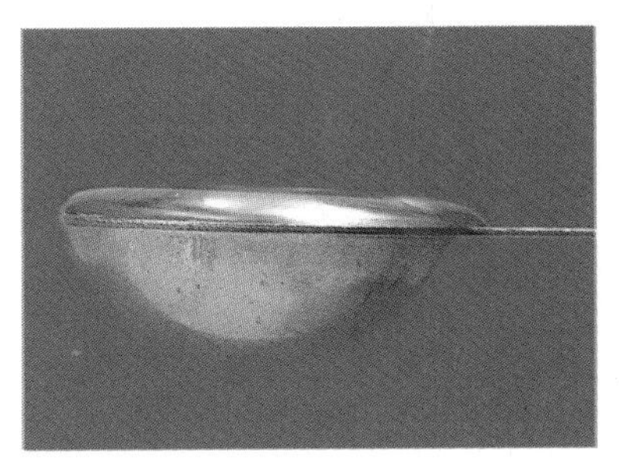

One mole of water is 602 200 000 000 000 000 000 000 water molecules, just a little more than one tablespoon of water!

Avogadro's number is an incredibly large number. To illustrate, it has been estimated that one mole of rice grains would cover the continental United States with rice one-half mile deep! So you should keep in mind that a mole, equivalent to 6.022×10^{23} objects, is a very large counting unit that is useful when dealing with very small particles such as atoms, molecules, and ions.

You will gain a better understanding of the importance of Avogadro's number and the mole as we continue our study of the composition of matter.

Molar Mass

Molar mass, *MM,* is the mass in grams of one mole of any substance. From the definition of the mole, the molar mass of carbon-12 is exactly 12 g of carbon-12. The molar mass of carbon-14 must be 14 g because each atom of this isotope of carbon is two amu's greater in mass than an atom of carbon-12. Because naturally occurring carbon is a mixture of isotopes with different mass numbers, its molar mass is equal to its atomic mass in grams (Section 5.5):

$AM\text{-C} = 12.01$

$MM\text{-C} = 12.01\ \text{g/mol}$

Molar mass (MM) of an element = mass of one mole = gram atomic mass.

It follows that *the molar mass of an element in atomic form is equal to its atomic mass expressed in grams.* Saying this another way, the molar mass of an element is the mass per mole of the element, or abbreviating:

The SI abbreviation for mole is mol.

$MM = \text{mass/mole} = \text{g/mol}$

Although the atomic masses given in the periodic table are dimensionless numbers, we use them most often in mole calculations with units of g/mol. Atomic masses expressed in grams are often called *gram atomic masses.* Thus, the molar mass (gram atomic mass) of oxygen can be expressed as 16.00 g/mol O, and the molar mass of magnesium is 24.31 g/mol Mg. Or, abbreviating:

Remember: MM = g/mol.

$MM\text{-O} = 16.00\ \text{g/mol O}$

$MM\text{-Mg} = 24.31\ \text{g/mol Mg}$

Writing molar masses with units of g/mol gives us conversion factors that will be very useful in problem solving.

You should now understand that the molar mass of any element (in atomic form) contains one mole, or 6.022×10^{23} atoms (Figure 7.1). For example:

Remember, any equivalence statement can be written as a conversion factor; for example,

$$\frac{12.01\ \text{g C}}{6.022 \times 10^{23}\ \text{C atoms}}.$$

12.01 g carbon = 1 mol carbon = 6.022×10^{23} carbon atoms

24.31 g magnesium = 1 mol magnesium
= 6.022×10^{23} magnesium atoms

39.10 g potassium = 1 mol potassium = 6.022×10^{23} potassium atoms

32.07 g sulfur = 1 mol sulfur = 6.022×10^{23} sulfur atoms

55.85 g iron = 1 mol iron = 6.022×10^{23} iron atoms

7.3 Molar Masses of Compounds

Learning Objective

- Calculate the molar mass of a compound.

As was mentioned earlier, the definition of a mole applies to compounds as well as elements. Water is represented by its formula, H_2O. From the previous discussion, you know that one mole of water is 6.022×10^{23} H_2O molecules. Because each molecule is made up of two hydrogen atoms and one oxygen atom, one mole of water molecules is a combination of two moles of hydrogen atoms and one mole of oxygen atoms. Thus, the molar mass (*MM*) of a compound can be calculated from the molar masses (or atomic masses) of the elements in the formula unit. The atomic mass (*AM*) of each element is multiplied by the number of atoms of the element in the formula. This is illustrated in the calculation of the molar mass of water:

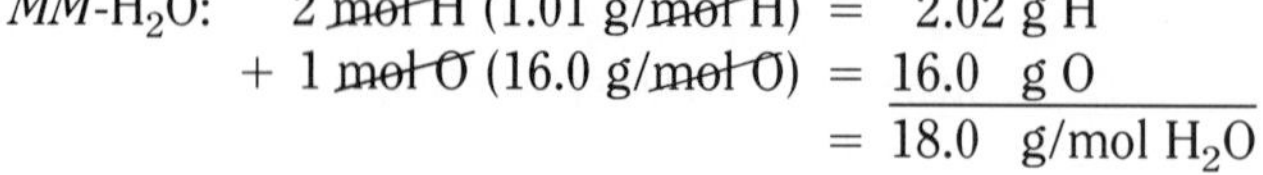

$$\begin{aligned} MM\text{-}H_2O: \quad & 2\ \text{mol H}\ (1.01\ \text{g/mol H}) = 2.02\ \text{g H} \\ + & 1\ \text{mol O}\ (16.0\ \text{g/mol O}) = 16.0\ \text{g O} \\ & = 18.0\ \text{g/mol } H_2O \end{aligned}$$

FIGURE 7.1 *Left:* One mole (32.07 g) of sulfur. *Right:* One mole (55.85 g) of iron. Each of these samples contains 6.022×10^{23} atoms.

The older terms **formula weight** (*FW*) and **molecular weight** (*MW*) continue to be used in place of molar mass. Although you may encounter these terms in your work, the term *molar mass* (*MM*) is currently favored. Remember to use units of g/mol for molar mass and the other equivalent terms (*FW, MW*). As with atomic masses, values with these units can be used as conversion factors in problem solving.

Other terms for molar mass (MM):

formula weight, FW = g/mol
molecular weight, MW = g/mol

As you review the calculation of the molar mass of water, you can see that you must multiply the subscript for each element by its atomic mass and then add the numbers. Let's see how this works for other compounds such as sulfuric acid (common battery acid) and sucrose (table sugar).

Example 7.1

Calculate the molar mass of sulfuric acid, H_2SO_4. ■

Solution

$$
\begin{aligned}
MM\text{-}H_2SO_4\text{:}\ & (2\ \cancel{\text{mol H}})\ (1.01\ \text{g}/\cancel{\text{mol H}}) = 2.02\ \text{g H}\\
& (1\ \cancel{\text{mol S}})\ (32.1\ \text{g}/\cancel{\text{mol S}}) = 32.1\ \text{g S}\\
& (4\ \cancel{\text{mol O}})\ (16.0\ \text{g}/\cancel{\text{mol O}}) = \underline{64.0\ \text{g O}}\\
& \qquad\qquad\qquad\qquad\qquad\qquad 98.1\ \text{g /mol } H_2SO_4
\end{aligned}
$$

Sulfuric acid is used in car batteries.

One molecule of sulfuric acid is far too small to weigh, even with the most sensitive balance known. However, the mass of one mole (6.022×10^{23} molecules) of sulfuric acid (98.1 g) can be determined easily with a common laboratory balance.

Example 7.2

Calculate the molar mass of sucrose, $C_{12}H_{22}O_{11}$. ■

Solution

$$
\begin{aligned}
MM\text{-}C_{12}H_{22}O_{11}\text{:}\ & (12\ \cancel{\text{mol C}})\ (12.0\ \text{g}/\cancel{\text{mol C}}) = 144\ \text{g C}\\
& (22\ \cancel{\text{mol H}})\ (1.01\ \text{g}/\cancel{\text{mol H}}) = 22.22\ \text{g H}\\
& (11\ \cancel{\text{mol O}})\ (16.0\ \text{g}/\cancel{\text{mol O}}) = \underline{176\ \text{g O}}\\
& \qquad\qquad\qquad\qquad\qquad\qquad 342\ \text{g/mol } C_{12}H_{22}O_{11}
\end{aligned}
$$

As a matter of convenience we'll use only three significant figures in our calculations of molar masses unless other measured values are clearly more precise. That is why the molar mass of sucrose in Example 7.2 was rounded to 342 g/mol.

Table sugar is sucrose.

Practice Problem 7.1 Calculate the molar mass of each of the following compounds. (Solutions to practice problems appear at the end of the chapter.)

a. CO_2
b. $MgSO_4$
c. $Fe_2(SO_4)_3$
d. NO_2
e. $CHCl_3$
f. $C_8H_{10}O$
g. NH_4Cl
h. K_3PO_4
i. $Ca(ClO_3)_2$

7.4 Mole Calculations

Learning Objectives

- Calculate the number of moles of an element or a compound from its mass.
- Calculate the mass of a substance needed to provide a certain number of moles.

Before you try some mole calculations, let's briefly review the key ideas of the mole concept.

1. The mole is a counting unit equal to 6.022×10^{23} particles (atoms, molecules, etc.). It is simply a word for a definite number, just as "pair" is a word for 2, "dozen" is a word for 12, and "gross" is a word for 144.
2. The mass of one mole (molar mass) of an element (or compound) is different from the mass of one mole of another because the atoms (or molecules, etc.) of each have different masses.
3. The molar mass of an element is its atomic mass expressed in grams. The molar mass of a compound is the sum of the molar masses of its elements, each multiplied by the number of atoms of the element in the formula. Remember the units of molar mass:

$$MM = \text{g/mol}$$

FIGURE 7.2 *Left to right:* One mole (12.0 g) of carbon. Two moles (24.0 g) of carbon. Three moles (36.0 g) of carbon.

Now you are ready to perform mole calculations. Some calculations can be done mentally without writing out the problem. For example, because 12.0 g of carbon is one mole of carbon, 24.0 g of carbon is two moles of carbon, and 36.0 g of carbon is three moles of carbon (Figure 7.2). Because 18.0 g of water is one mole of water, 9.00 g of water is ½ mole of water. However, many problems will involve calculations you cannot do mentally. For these problems you need to use our problem-solving strategy. This is illustrated in the following examples.

Example 7.3

How many moles of water are in 275 g of water? ■

In applying our strategy, let's first outline the problem.

Find: number of moles H_2O = ?
Given: 275 g H_2O
Known: MM-H_2O = 18.0 g/mol H_2O (calculated previously)
Solution: Using the dimensional analysis strategy to set up an equation, we start with the unknown:

$$\text{number of moles } H_2O = \left(\frac{1 \text{ mol } H_2O}{18.0 \not{g}}\right)(275 \not{g}\, H_2O)$$

$$= 15.3 \text{ mol } H_2O$$

Check: $(275 \not{g}\, H_2O)\left(\frac{1 \text{ mol } H_2O}{18.0 \not{g}}\right) = 15.3 \text{ mol } H_2O$

Example 7.4

Dry ice (solid carbon dioxide) is usually sold by weight. Find the number of moles of carbon dioxide in 25.0 lb of dry ice. 1.00 lb = 454 g. ■

Find: number of moles CO_2 = ?

Given: 25.0 lb CO_2

Known: 1.00 lb = 454 g, or $\frac{454 \text{ g}}{1 \text{ lb}}$

$$
\begin{aligned}
MM\text{-}CO_2:\ & (1 \cancel{\text{mol C}})(12.0 \text{ g}/\cancel{\text{mol C}}) &= 12.0 \text{ g C} \\
& (2 \cancel{\text{mol O}})(16.0 \text{ g}/\cancel{\text{mol O}}) &= \underline{32.0 \text{ g O}} \\
& &44.0 \text{ g/mol } CO_2
\end{aligned}
$$

Solution: Set up the calculation using dimensional analysis:

$$
\text{number of moles } CO_2 = \left(\frac{1 \text{ mol } CO_2}{44.0 \cancel{\text{g}}}\right)\left(\frac{454 \cancel{\text{g}}}{1 \cancel{\text{lb}}}\right)(25.0 \cancel{\text{lb}}) = 258 \text{ mol } CO_2
$$

Check: $(25.0 \cancel{\text{lb}})\left(\frac{454 \cancel{\text{g}}}{1 \cancel{\text{lb}}}\right)\left(\frac{1 \text{ mol } CO_2}{44.0 \cancel{\text{g}}}\right) = 258 \text{ mol } CO_2$

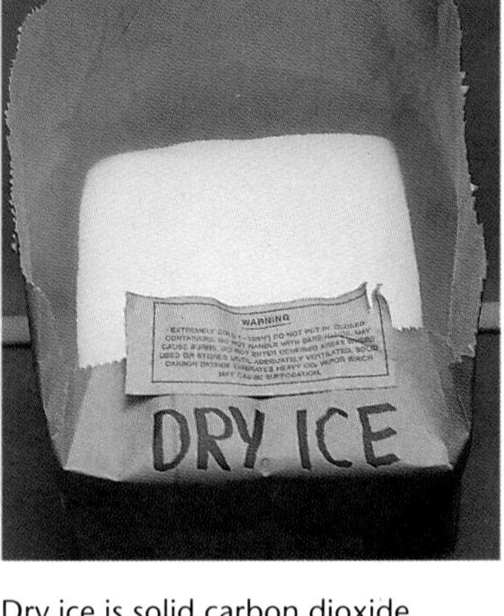

Dry ice is solid carbon dioxide.

Example 7.5

Calculate the mass of 3.50 moles of TNT, $C_7H_5(NO_2)_3$. ■

Find: mass (g) = ?

Given: 3.50 mol $C_7H_5(NO_2)_3$ (TNT)

Known:

$$
\begin{aligned}
MM\text{-TNT}:\ & (7 \cancel{\text{mol C}})(12.0 \text{ g}/\cancel{\text{mol C}}) &= 84.0 \text{ g C} \\
& (5 \cancel{\text{mol H}})(1.01 \text{ g}/\cancel{\text{mol H}}) &= 5.05 \text{ g H} \\
& (3 \cancel{\text{mol N}})(14.0 \text{ g}/\cancel{\text{mol N}}) &= 42.0 \text{ g N} \\
& (6 \cancel{\text{mol O}})(16.0 \text{ g}/\cancel{\text{mol O}}) &= \underline{96.0 \text{ g O}} \\
& &227 \text{ g/mol } C_7H_5(NO_2)_3
\end{aligned}
$$

Solution: $\text{mass (g)} = \left(\frac{227 \text{ g}}{1 \cancel{\text{mol}}}\right)(3.50 \cancel{\text{mol}}) = 795 \text{ g } C_7H_5(NO_2)_3$

Check: $(3.50 \cancel{\text{mol}})\left(\frac{227 \text{ g}}{1 \cancel{\text{mol}}}\right) = 795 \text{ g } C_7H_5(NO_2)_3$ 

Practice Problem 7.2 Find the number of moles of each substance in the following samples:

a. 74.9 g tin, Sn

b. 675 g iodine, I_2

c. 24.62 g $Al_2(SO_4)_3$

Practice Problem 7.3 Calculate the mass of each of the following:
a. 1.43 mol Fe
b. 2.55 mol Mg
c. 0.346 mol CH_3OH (methyl alcohol), wood alcohol
d. 0.346 mol $C_6H_4Cl_2$ (*p*-dichlorobenzene), moth crystals

7.5 Number of Atoms or Molecules in a Sample

Learning Objectives

- Calculate the number of atoms or molecules in a sample.
- Calculate the mass of a given number of atoms or molecules.
- Calculate the number of moles of a substance from the number of atoms or molecules.

Because atoms or molecules are too small to be seen, we cannot count them as we can visible objects. However, using your new counting unit (the mole), you can now understand how we can calculate the number of atoms or molecules in a sample when we know the mass of the sample. For example, to calculate the number of iron atoms in 74.9 g of iron:

Find: number of atoms Fe = ?
Given: 74.9 g Fe
Known: 6.02×10^{23} atoms/mol
MM-Fe = 55.9 g/mol Fe
Solution: Using dimensional analysis, set up an equation:

$$\text{number of atoms Fe} = \left(\frac{6.02 \times 10^{23}\ \text{atoms}}{1\ \cancel{\text{mol Fe}}}\right)\left(\frac{1\ \cancel{\text{mol Fe}}}{55.9\ \cancel{\text{g Fe}}}\right)(74.9\ \cancel{\text{g Fe}})$$

$$= 8.07 \times 10^{23}\ \text{atoms Fe}$$

Check: $(74.9\ \cancel{\text{g Fe}})\left(\frac{1\ \cancel{\text{mol Fe}}}{55.9\ \cancel{\text{g Fe}}}\right)\left(\frac{6.02 \times 10^{23}\ \text{atoms}}{1\ \cancel{\text{mol Fe}}}\right)$

$$= 8.07 \times 10^{23}\ \text{atoms Fe}$$

Example 7.6

Find the number of molecules of CO_2 in 1.00 lb (454 g) of dry ice. ■

Find: number of molecules CO_2 = ?
Given: 1.00 lb (454 g) CO_2

Known: 6.02×10^{23} molecules CO_2/mol CO_2

MM-CO_2 = 44.0 g/mol (from atomic masses)

Solution: Set up an equation using dimensional analysis:

$$\text{number of molecules } CO_2 = \left(\frac{6.02 \times 10^{23} \text{ molecules } CO_2}{1 \cancel{\text{mol } CO_2}}\right)\left(\frac{1 \cancel{\text{mol } CO_2}}{44.0 \cancel{g}}\right)(454 \cancel{g})$$

$$= 6.21 \times 10^{24} \text{ molecules } CO_2$$

Check: $(454 \text{ g})\left(\frac{1 \cancel{\text{mol } CO_2}}{44.0 \text{ g}}\right)\left(\frac{6.02 \times 10^{23} \text{ molecules } CO_2}{1 \cancel{\text{mol } CO_2}}\right) = 6.21 \times 10^{24} \text{ molecules } CO_2$ ✔

Note: A variation of the dimensional analysis strategy is to break the problem into smaller parts. First, set up an equation to calculate the number of molecules of CO_2:

$$\text{number of molecules } CO_2 = \left(\frac{6.02 \times 10^{23} \text{ molecules } CO_2}{1 \text{ mol } CO_2}\right)(? \text{ mol } CO_2) \quad (1)$$

Then, set up a second equation to calculate the number of moles of CO_2:

$$\text{number of moles } CO_2 = \left(\frac{1 \text{ mol } CO_2}{44.0 \cancel{g}}\right)(454 \cancel{g}) = 10.3 \text{ mol } CO_2 \quad (2)$$

Substituting into equation (1) gives

$$\text{number of molecules } CO_2 = \left(\frac{6.02 \times 10^{23} \text{ molecules } CO_2}{1 \cancel{\text{mol } CO_2}}\right)(10.3 \cancel{\text{mol } CO_2})$$

$$= 6.20 \times 10^{24} \text{ molecules } CO_2$$

Example 7.7

What is the mass of 1.00 million molecules of glucose ($C_6H_{12}O_6$), blood sugar? ■

Find: mass (g) = ?

Given: 1.00×10^6 molecules $C_6H_{12}O_6$

Known: 6.02×10^{23} molecules/mol $C_6H_{12}O_6$

MM-$C_6H_{12}O_6$:

$$\begin{aligned} (6 \cancel{\text{mol C}})(12.0 \text{ g}/\cancel{\text{mol C}}) &= 72.0 \text{ g C} \\ (12 \cancel{\text{mol H}})(1.01 \text{ g}/\cancel{\text{mol H}}) &= 12.1 \text{ g H} \\ (6 \cancel{\text{mol O}})(16.0 \text{ g}/\cancel{\text{mol O}}) &= 96.0 \text{ g O} \\ & \overline{180. \text{ g/mol}} \end{aligned}$$

Solution:

$$\text{mass (g)} = \left(\frac{180. \text{ g}}{1 \cancel{\text{mol}}}\right)\left(\frac{1 \cancel{\text{mol}}}{6.02 \times 10^{23} \cancel{\text{molecules}}}\right)1.00 \times 10^6 \cancel{\text{molecules}})$$

$$= 2.99 \times 10^{-16} \text{ g } C_6H_{12}O_6$$

Check: $(1.00 \times 10^6 \cancel{\text{molecules}})\left(\frac{1 \cancel{\text{mol}}}{6.02 \times 10^{23} \cancel{\text{molecules}}}\right)\left(\frac{180. \text{ g}}{1 \cancel{\text{mol}}}\right) = 2.99 \times 10^{-16} \text{ g } C_6H_{12}O_6$ ✔

Example 7.8

What is the mass of one molecule of nitrogen, N_2? ■

Find: mass (g) = ?

Given: 1 molecule N_2

Known: $MM\text{-}N_2 = (2 \text{ mol N})(14.0 \text{ g/mol N}) = 28.0 \text{ g/mol } N_2$

6.02×10^{23} molecules/mol N_2

Solution:

$$\text{mass (g)} = \left(\frac{28.0 \text{ g}}{1 \text{ mol}}\right)\left(\frac{1 \text{ mol}}{6.02 \times 10^{23} \text{ molecules}}\right)(1 \text{ molecule})$$

$$= 4.65 \times 10^{-23} \text{ g}$$

Check:

$$(1 \text{ molecule})\left(\frac{1 \text{ mol}}{6.02 \times 10^{23} \text{ molecules}}\right)\left(\frac{28.0 \text{ g}}{1 \text{ mol}}\right) = 4.65 \times 10^{-23} \text{ g}$$

Example 7.9

Find the number of moles of toluene (C_7H_8), a high-octane compound in gasoline, in a sample containing 1.23×10^8 molecules. ■

Find: number of moles = ?

Given: 1.23×10^8 molecules

Known: 6.02×10^{23} molecules/mol

Solution:

$$\text{number of moles} = \left(\frac{1 \text{ mol}}{6.02 \times 10^{23} \text{ molecules}}\right)(1.23 \times 10^8 \text{ molecules})$$

$$= 2.04 \times 10^{-16} \text{ mol toluene}$$

Practice Problem 7.4 How many atoms or molecules are in each of the following samples?

a. 10.3 g C_8H_{18} (octane), in gasoline

b. 0.346 g Zn

c. 0.0215 g P_4

Practice Problem 7.5 Find the mass in grams of one atom or molecule of each of the following:

a. uranium-235

b. Freon-12, CCl_2F_2

c. C_2HCl_3

Practice Problem 7.6 How many moles of each element are in each of the following samples?

a. 4.65×10^{45} copper atoms

b. 8.33×10^{24} magnesium atoms

Percentage Composition of Compounds

7.6

Learning Objective

- Calculate the percentage composition of a compound.

You have already learned that a compound is a substance made up of more than one element and has a definite composition. The composition of a compound may be expressed in terms of the mass percentage of each element in the compound. "Percent" means parts per hundred parts of a substance. When you read that a compound contains 36% chlorine, this means it contains 36 grams of chlorine per 100 grams of the compound. The **percentage composition** of a compound is the mass percentage of each element making up the compound.

Law of definite composition: Every pure substance has a definite and fixed composition (Section 2.3).

The percentage of an element in a compound can be calculated from its formula. Remember, a formula expresses the numbers of atoms making up a formula unit, not the masses. As an example, consider CCl_2F_2, the chlorofluorocarbon, or Freon, used in air conditioners and refrigerators (Figure 7.3). To calculate the percentage of chlorine in CCl_2F_2, you must find the fraction of the mass that is chlorine and then multiply by 100:

$$\begin{aligned} \%\ \text{Cl} &= \text{mass fraction} \times 100 \\ &= \frac{\text{mass of Cl/mol } CCl_2F_2}{\text{mass of 1 mol } CCl_2F_2} \times 100 \\ &= \frac{\text{mass of Cl/mol } CCl_2F_2}{MM\text{-}CCl_2F_2} \times 100 \end{aligned}$$

Note that there are three steps in this problem:

Step 1 Calculation of the molar mass of CCl_2F_2 (the denominator)

Step 2 Calculation of the mass of Cl in 1 mol of CCl_2F_2 (the numerator)

Step 3 Calculation of the percentage of chlorine in CCl_2F_2

Accordingly:

$$\begin{aligned} \textbf{1. } MM\text{-}CCl_2F_2\text{:}\ & (1\ \cancel{\text{mol C}})(12.0\ \text{g}/\cancel{\text{mol C}}) = 12.0\ \text{g C} \\ & (2\ \cancel{\text{mol Cl}})(35.5\ \text{g}/\cancel{\text{mol Cl}}) = 71.0\ \text{g Cl} \\ & (2\ \cancel{\text{mol F}})(19.0\ \text{g}/\cancel{\text{mol F}}) = \underline{38.0\ \text{g F}} \\ & \qquad\qquad\qquad\qquad\qquad\ \ 121\ \ \text{g/mol } CCl_2F_2 \end{aligned}$$

FIGURE 7.3 Freon (CCl_2F_2) is used in air conditioners and refrigerators. Freons, or chlorofluorocarbons (CFCs), are a threat to the ozone layer in the earth's atmosphere. Scientists are continuing to develop Freon substitutes that are environmentally safe.

2. Careful examination of the molar mass calculation shows us that in 1 mol (121 g) of CCl_2F_2 there are 2 mol Cl, or 71.0 g of Cl.

3. $$\% \text{ Cl} = \frac{2\ (MM\text{-Cl})}{MM\text{-CCl}_2\text{F}_2} \times 100 = \frac{71.0\ \text{g}}{121\ \text{g}} \times 100 = 58.7\%\ \text{Cl}$$

Example 7.10

Find the percentage of oxygen in sodium peroxide, Na_2O_2. ■

Solution

$$\% \text{ O} = \frac{\text{mass O/mol Na}_2\text{O}_2}{MM\text{-Na}_2\text{O}_2} \times 100$$

$$= \frac{2(MM\text{-O})}{MM\text{-Na}_2\text{O}_2} = \frac{2(16.0)}{2(23.0) + 2(16.0)} \times 100 = 41.0\%\ \text{oxygen}$$

Practice Problem 7.7 Calculate the percentage composition of each of the following compounds:

a. NH_4ClO_4, an ingredient in solid rocket propellants

b. $(NH_4)_3PO_4$, a fertilizer

7.7 Formulas for Compounds

Learning Objective

- Identify the principal kinds of chemical formulas.

In earlier chapters, we encountered many formulas, ranging from those for simple compounds such as NaCl, H_2O, and CO_2 to more complicated ones such as $C_{12}H_{22}O_{11}$ (sucrose). Remember, just as an element is represented by its symbol, a compound is represented by its formula. A formula shows the symbols of the elements making up the compound, followed by subscripts to show the number of atoms of each element.

An empirical formula shows the simplest mole ratio of the elements.

Several kinds of formulas are widely used by chemists. An **empirical formula** is the simplest. The subscripts in an empirical formula tell us the simplest mole ratio of the elements making up the compound. A familiar example is the formula for water, H_2O. This formula indicates that hydrogen is combined with oxygen in a 2:1 mole ratio. The ratios are generally in the form of small, whole numbers such as CO (1:1), N_2O_3 (2:3), and CCl_4 (1:4). However, for some compounds, such as $C_{12}H_{22}O_{11}$ (sucrose), the simplest mole ratios involve larger numbers.

A **molecular formula** tells us the actual number of each kind of atom making up a molecule. Sometimes the empirical formula and the molecular formula are identical. This is true for water, H_2O, because the ratio of numbers of hydrogen and oxygen atoms in a molecule is the same as the simplest mole ratio of hydrogen and oxygen in water. However, for many other compounds these formulas are different. For example, the molecular formula for hydrogen peroxide, H_2O_2, indicates that a hydrogen peroxide molecule is made up of two hydrogen atoms and two oxygen atoms. However, the simplest mole ratio is 1:1, and the empirical formula is HO. A stable molecule with the formula HO doesn't exist, although we will meet an important ion, OH^-, that does. Note that the number "1" is understood and is never written in formulas. Empirical formulas are generally used for ionic compounds. On the other hand, a molecular formula is preferred for molecular compounds.

A molecular formula shows the actual numbers of atoms in a molecule.

A third kind of formula, a **structural formula,** shows how the atoms are joined to each other in a molecule. When reading a structural formula, we must remember that it is a two-dimensional representation of a three-dimensional molecule, so the formula may not clearly show the shape of the molecule. You will learn how to write structural formulas (Lewis type) in Chapter 8. Figure 7.4 illustrates the three kinds of formulas.

A structural formula shows how atoms are joined together in a molecule.

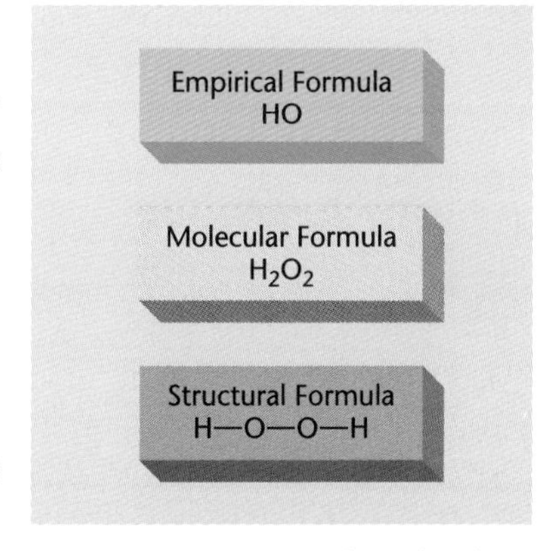

FIGURE 7.4 Empirical, molecular, and structural formulas for hydrogen peroxide.

Example 7.11

Which of the following are empirical formulas? What kind of formula must the others be?

a. CH_4O
b. C_2H_6
c. $C_3H_{10}N_2$
d. $C_6H_6O_2$ ■

Solutions Remember, an empirical formula shows the simplest mole ratio of the elements.

a. Empirical formula.
b. The 2:6 ratio can be reduced to 1:3, and the empirical formula is CH_3. The formula given must be a molecular formula.
c. Empirical formula.
d. The 6:6:2 ratio can be reduced to 3:3:1, and the empirical formula is C_3H_3O. The formula given must be a molecular formula.

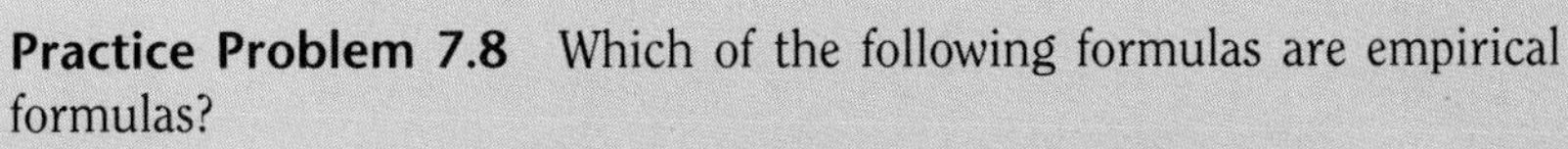

Practice Problem 7.8 Which of the following formulas are empirical formulas?
a. C_4H_5N
b. Al_2S_3
c. P_4O_{10}
d. N_2O_4

Hydrogen peroxide is widely used as an antiseptic.

7.8 Calculation of Empirical Formulas

Learning Objective

- Calculate the empirical formula of a compound.

A chemist who discovers or synthesizes a new compound must determine its empirical formula to be able to describe the compound to others. But to calculate the empirical formula of a compound, its composition must be known. Thus, the chemist first must do a qualitative analysis on the compound to determine which elements are present. Next, quantitative analysis is performed to determine the mass composition of the compound (Figure 7.5). Then the chemist uses these data to calculate the empirical formula.

Qualitative analysis determines which *substances are present.*

Quantitative analysis determines how much *of each substance is present.*

In calculating empirical formulas, remember that an empirical formula expresses the simplest whole-number *mole ratio* of the elements making up the compound. The mole ratio is expressed in the formula by the subscripts that follow the symbols of the elements making up the compound. For example, water is made up of hydrogen and oxygen in a 2:1 mole ratio. When writing the formula, remember that the "1" is not written.

The formula for water, H_2O, expresses a 2:1 mole ratio of hydrogen:oxygen.

Calculating an empirical formula involves four steps, and to avoid mistakes it is important that you follow the steps carefully. Remember, an empirical formula expresses the simplest whole-number mole ratio of the elements in the compound, and your objective in this calculation is to determine this mole ratio.

Step 1 Write the *mass (g) of each element* in the compound.

Step 2 Convert the mass of each element to *moles.*

Step 3 Calculate the *simplest whole-number mole ratio* of the elements. This is easily done by dividing the number of moles of each element by the smallest number obtained in Step 2. If the result is not a whole-number ratio, multiply all numbers by an integer (2, 3, etc.) to give a whole-number ratio.

Step 4 Write the *empirical formula.* For inorganic compounds, the symbol for the most positive element is written first. For organic compounds, the order is C, H, followed by other elements in alphabetical order.

The following examples illustrate the calculation of an empirical formula.

FIGURE 7.5 A modern combustion analyzer is used to determine the percentage of carbon, hydrogen, and nitrogen in organic compounds.

Example 7.12

Ethylene glycol, the common automobile antifreeze, contains 38.7% carbon, 9.70% hydrogen, and 51.6% oxygen. What is the empirical formula for the compound? ■

Solution Outline the four steps of the problem:

Element	*Step 1:* Mass, g	*Step 2:* Number of Moles	*Step 3:* Simplest Whole-Number Mole Ratio	*Step 4:* Formula
C	38.7	$\frac{38.7\ \text{g}}{12.0\ \text{g/mol}} = 3.23\ \text{mol}$	$\frac{3.23\ \cancel{\text{mol}}}{3.23\ \cancel{\text{mol}}} = 1.00$ or 1	
H	9.70	$\frac{9.70\ \text{g}}{1.01\ \text{g/mol}} = 9.60\ \text{mol}$	$\frac{9.60\ \cancel{\text{mol}}}{3.23\ \cancel{\text{mol}}} = 2.97$ or 3	CH_3O
O	51.6	$\frac{51.6\ \text{g}}{16.0\ \text{g/mol}} = 3.23\ \text{mol}$	$\frac{3.23\ \cancel{\text{mol}}}{3.23\ \cancel{\text{mol}}} = 1.00$ or 1	

Ethylene glycol is the common automobile antifreeze.

Now, let's go through the calculation one step at a time.

Step 1 To find the mass (g) of each element, note that the percentage of an element means parts per hundred, or grams of the element in 100 g of the compound. Thus:

$$X\% = \frac{X\ \text{g of element}}{100\ \text{g of compound}}$$

38.7% C means 38.7 g C in 100 g of the compound
9.70% H means 9.70 g H in 100 g of the compound
51.6% 0 means 51.6 g O in 100 g of the compound

Step 2 Use dimensional analysis to find the number of moles of each element:

$$\text{number of moles C} = \left(\frac{1\ \text{mol C}}{12.0\ \cancel{\text{g C}}}\right)(38.7\ \cancel{\text{g C}}) = 3.23\ \text{mol C}$$

The number of moles of hydrogen and the number of moles of oxygen are calculated in the same way.

Step 3 Inspect the number of moles calculated in Step 2. Divide the number of moles of each element by the smallest number of moles to obtain the simplest whole-number mole ratio. Because only whole numbers are used in formulas, some rounding off may be necessary because of the uncertainty in the experimental values used in the calculation.

For carbon: $\frac{3.23\ \cancel{\text{mol}}}{3.23\ \cancel{\text{mol}}} = 1.00$ or 1

For hydrogen: $\frac{9.60\ \cancel{\text{mol}}}{3.23\ \cancel{\text{mol}}} = 2.97$ or 3

For oxygen: $\frac{3.23\ \cancel{\text{mol}}}{3.23\ \cancel{\text{mol}}} = 1.00$ or 1

Step 4 The empirical formula is CH_3O.

Calculation of an Empirical Formula

Step 1 Find the masses of each element in a sample of the compound.

Step 2 Find the number of moles of each element from its mass and atomic mass.

Step 3 Find the simplest whole-number mole ratio of the elements.

Step 4 Write the formula of the compound.

Example 7.13

A 1.45-g sample of chromium oxide contains 0.990 g of chromium. The remaining mass is oxygen. What is the empirical formula of chromium oxide? ■

Solution Outline the problem as before, starting with the masses of the elements of chromium oxide (law of conservation of mass).

$$\underset{0.990\text{ g}}{\text{chromium}} + \underset{0.46\text{ g}}{\text{oxygen}} \longrightarrow \underset{1.45\text{ g}}{\text{chromium oxide}}$$

Element	*Step 1: Mass, g*	*Step 2: Number of Moles*	*Step 3: Simplest Whole-Number Mole Ratio*	*Step 4: Formula*
Cr	0.990	$\frac{0.990\text{ g}}{52.0\text{ g/mol}} = 0.0190\text{ mol}$	$\frac{0.0190\ \cancel{\text{mol}}}{0.0190\ \cancel{\text{mol}}} = 1.00$	
			$1.00 \times 2 = 2.00$ or 2	Cr_2O_3
O	0.46	$\frac{0.46\text{ g}}{1.6.0\text{ g/mol}} = 0.029\text{ mol}$	$\frac{0.029\ \cancel{\text{mol}}}{0.0190\ \cancel{\text{mol}}} = 1.5$	
			$1.5 \times 2 = 3.0$ or 3	

Remember that the simplest whole-number mole ratio (Step 3) is most easily obtained by dividing the smallest number of moles by itself and then dividing the number of moles of the other elements by that same number. However, this time the numbers expressing the mole ratio are *not close* to whole numbers, so 1.5 *should not be rounded* to 2. In this case, to find the whole-number ratio, multiply by 2, and then round off to the whole number:

Mole Ratio	*Whole-Number Ratio*
$1.00 \times 2 = 2.00$	2
$1.5 \times 2 = 3.0$	3

The whole-number ratio is used to write the formula, Cr_2O_3.

For some problems, the decision to round a number or multiply by an integer may not be clear. The following is a guide for working empirical formula problems in this book. If a number is within 0.1 of a whole number, round off to the whole number. In other words:

Mole Ratio	*Whole-Number Ratio*
1	1
1.07 round to	1

However, when a number is not within 0.1 of a whole number, multiply by the appropriate integer to give a whole-number ratio:

Mole Ratio	*Whole-Number Ratio*
$1 \times 4 = 4$	4
$1.25 \times 4 = 5.00$ or 5	5

Practice Problem 7.9 Find the empirical formula of each of the following compounds from the percentage composition data given:

a. 40.0% C, 6.70% H, 53.3% O (formaldehyde, a tissue preservative and embalming agent)

b. 20.2% Mg, 26.6% S, 53.2% O

Practice Problem 7.10 A reaction between zinc and sulfur produced 1.68 g of a compound that contained 1.13 g of zinc. Find the empirical formula of the compound.

Calculation of Molecular Formulas

7.9

Learning Objective

- Calculate the molecular formula of a compound.

The molecular formula of a compound is always a whole-number multiple of the empirical formula. For example, the molecular formula for hydrogen peroxide is H_2O_2, and its empirical formula is HO. Thus, the molecular formula is the empirical formula multiplied by 2.

$$H_2O_2 = (HO)_2 = (HO)_x \qquad \text{where } x = 2$$

When a molecular formula is not known, we can calculate it by reversing this thinking process. For example, if we didn't know the molecular formula for hydrogen peroxide, we could find it from its empirical formula (HO) and x, the number of

To find the molecular formula you must find x, the number of empirical formula units per molecule.

empirical formula units per molecule. The empirical formula can be determined from the mass composition of the compound as illustrated in Section 7.8. The factor, x, can be calculated by dividing the molar mass (MM) of the compound by the empirical formula mass (EFM).

$$x = \frac{MM}{EFM} \quad \begin{array}{l}\text{(determined experimentally)} \\ \text{(calculated from the empirical formula)}\end{array}$$

Thus, the molecular formula can be determined from the empirical formula and the molar mass of the compound.

To determine the molecular formula of a compound, you need its:

empirical formula
molar mass

The molar mass of a compound can be determined experimentally by several methods, including mass spectrometry (Section 5.5). A mass spectrometer is the most technologically advanced and most accurate instrument for determining molar masses.

Now, let's see how the number x can be calculated. For hydrogen peroxide:

$MM\text{-}H_2O_2 = 34.0$ g/mol (determined experimentally)

$EFM\text{-}HO = 17.0$ g/mol of formula units (from atomic masses)

Then,

$$x = \frac{MM}{EFM} = \frac{34.0}{17.0} = 2.00 \text{ or } 2 \text{ formula units/molecule}$$

And the molecular formula is

$$(HO)_x = (HO)_2 = H_2O_2$$

Calculation of a Molecular Formula

Step 1 Determine the empirical formula for the compound.

Step 2 Calculate the empirical formula mass.

Step 3 Find x, the number of formula units per molecule.

$$x = \frac{MM}{EFM}$$

Step 4 Write the molecular formula.

Example 7.14

The molar mass of ethylene glycol is 62.0 g/mol, and its empirical formula is CH_3O (Example 7.12). Find the molecular formula of ethylene glycol. ■

First, outline the problem.

Find: x in the formula $(CH_3O)_x$

Given: $MM = 62.0$ g/mol (of molecules)

CH_3O is the empirical formula

Known: $x = \frac{MM}{EFM}$

EFM = 31.0 g/mol of formula units (from the empirical formula, CH_3O, and atomic masses)

Solution: $x = \dfrac{MM}{EFM} = \dfrac{62.0}{31.0} = 2.00$ or 2 formula units/molecule

$(CH_3O)_x = C_2H_6O_2$

Practice Problem 7.11 The percentage composition of dextrose (glucose) is 40.0% carbon, 6.72% hydrogen, and 53.3% oxygen. Its molar mass is 180. g/mol. Find the molecular formula of dextrose.

Mass Relationships from Formulas

7.10

Learning Objectives

- Calculate the mass of an element in a given mass of a compound.
- Calculate the mass of a compound needed to provide a certain mass of one of its elements.

The formula of a compound gives us both qualitative and quantitative information about the compound. For example, the formula for carbon dioxide, CO_2, tells us the compound is made up of the elements carbon and oxygen. This is qualitative information. The formula tells us that each molecule is made up of one atom of carbon and two atoms of oxygen. Furthermore, the formula tells us there is a 1:2 mole ratio of carbon to oxygen. This is quantitative information.

Mole ratios can be expressed precisely as follows:

$$\frac{1 \text{ mol carbon}}{2 \text{ mol oxygen}} \quad \text{or} \quad \frac{2 \text{ mol oxygen}}{1 \text{ mol carbon}}$$

$$\frac{1 \text{ mol carbon}}{1 \text{ mol } CO_2} \quad \text{or} \quad \frac{1 \text{ mol } CO_2}{1 \text{ mol carbon}}$$

$$\frac{2 \text{ mol oxygen}}{1 \text{ mol } CO_2} \quad \text{or} \quad \frac{1 \text{ mol } CO_2}{2 \text{ mol oxygen}}$$

Note that mole ratios, like conversion factors in general, always come in pairs. One ratio is simply the inverse of the other.

The information provided by a chemical formula can be used to answer questions involving the masses of elements making up the compound. Let's illustrate with the following example problems.

The formation of rust (Fe_2O_3) is a costly problem.

Example 7.15

Corrosion of machinery and metal parts in the United States results in losses of hundreds of millions of dollars per year. Formation of rust is an oxidation–

reduction reaction in which iron is the reducing agent and oxygen (in the air) is the oxidizing agent. Calculate the mass of iron lost in the formation of 1.00 kg of rust, Fe_2O_3. ■

Find: mass (g) Fe = ?

Given: 1.00 kg (1.00×10^3 g) Fe_2O_3

Known: $\frac{2 \text{ mol Fe}}{1 \text{ mol Fe}_2\text{O}_3}$

MM-Fe = 55.9 g/mol Fe

MM-Fe_2O_3 = 160. g/mol (from atomic masses)

Solution: This, like many problems, can be solved by more than one approach. The solution shown here uses dimensional analysis and illustrates how a complex problem can be divided into smaller, simpler problems.

Remember the three steps of the solution strategy:

Step 1 Begin with the unknown, including its units.

Step 2 First term: Match its units with the unknown.

Step 3 Succeeding terms: Cancel unwanted units.

First, you want to find the mass (g) Fe:

$$\text{mass (g) Fe} = \left(\frac{55.9 \text{ g}}{1 \text{ mol Fe}}\right) (\text{? mol Fe}) \quad (1)$$

Second, because the number of moles of Fe was not given, this quantity must be calculated. Set up an equation for this calculation:

$$\text{number of moles Fe} = \left(\frac{2 \text{ mol Fe}}{1 \text{ mol Fe}_2\text{O}_3}\right) (\text{? mol Fe}_2\text{O}_3) \quad (2)$$

Third, because the number of moles of Fe_2O_3 was not given, this quantity must also be calculated:

$$\text{number of moles Fe}_2\text{O}_3 = \left(\frac{1 \text{ mol Fe}_2\text{O}_3}{160. \cancel{\text{g Fe}_2\text{O}_3}}\right) (1.00 \times 10^3 \cancel{\text{g Fe}_2\text{O}_3}) \quad (3)$$

$$= 6.25 \text{ mol Fe}_2\text{O}_3$$

Now, substitute in equation (2) to calculate the number of moles of Fe:

$$\text{number of moles Fe} = \left(\frac{2 \text{ mol Fe}}{1 \cancel{\text{mol Fe}_2\text{O}_3}}\right) (6.25 \cancel{\text{mol Fe}_2\text{O}_3})$$

$$= 12.5 \text{ mol Fe}$$

Substitute 12.5 mol Fe in equation (1) to calculate the mass of Fe:

$$\text{mass (g) Fe} = \left(\frac{55.9 \text{ g}}{1 \cancel{\text{mol Fe}}}\right) (12.5 \cancel{\text{mol Fe}})$$

$$= 699 \text{ g Fe}$$

The answer is 699 g Fe.

$$Check: \quad (1.00 \times 10^3\ \cancel{g\ Fe_2O_3}) \left(\frac{1\ \cancel{mol\ Fe_2O_3}}{160.\ \cancel{g\ Fe_2O_3}}\right) \left(\frac{2\ \cancel{mol\ Fe}}{1\ \cancel{mol\ Fe_2O_3}}\right) \left(\frac{55.9\ g\ Fe}{1\ \cancel{mol\ Fe}}\right)$$

$$= 699\ g\ Fe$$

Example 7.16

Magnesium is a lightweight metal used in the manufacture of airplanes. Magnesium is made by the electrolysis of molten magnesium chloride, $MgCl_2$. How much magnesium chloride is needed to make 10.0 kg magnesium? ■

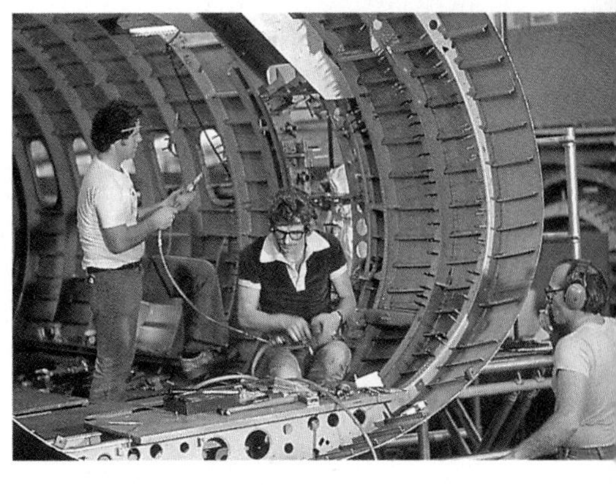

Lightweight metals such as magnesium are used in the manufacture of airplanes.

Find: mass (g) $MgCl_2$ = ?

Given: 10.0 kg (1.00×10^4 g) Mg

Known: *MM*-Mg = 24.3 g/mol, *MM*-$MgCl_2$ = 95.3 g/mol

$$\frac{1\ mol\ Mg}{1\ mol\ MgCl_2}$$

Solution: Using dimensional analysis, you first want to find:

$$\text{mass (g) } MgCl_2 = \left(\frac{95.3\ g}{1\ mol\ Mg}\right) (?\ mol\ MgCl_2) \qquad (1)$$

Second, because the number of moles $MgCl_2$ was not given or known, set up an equation to calculate this quantity:

$$\text{number of moles } MgCl_2 = \left(\frac{1\ mol\ MgCl_2}{1\ mol\ Mg}\right) (?\ mol\ Mg) \qquad (2)$$

Third, calculate the number of moles Mg:

$$\text{number of moles Mg} = \left(\frac{1\ mol\ Mg}{24.3\ \cancel{g\ Mg}}\right) (1.00 \times 10^4\ \cancel{g\ Mg}) \qquad (3)$$

$$= 412\ mol\ Mg$$

Substituting into equation (2) gives

$$\text{number of moles } MgCl_2 = \left(\frac{1\ mol\ MgCl_2}{1\ \cancel{mol\ Mg}}\right) (412\ \cancel{mol\ Mg})$$

$$= 412\ mol\ MgCl_2$$

Substituting into equation (1) gives

$$\text{mass (g) } MgCl_2 = \left(\frac{95.3\ g}{1\ \cancel{mol\ MgCl_2}}\right) (412\ \cancel{mol\ MgCl_2})$$

$$= 3.93 \times 10^4\ g\ MgCl_2 \quad \text{or} \quad 39.3\ kg\ MgCl_2$$

Check: Begin with the given information:

$$(1.00 \times 10^4\ \cancel{g\ Mg}) \left(\frac{1\ \cancel{mol\ Mg}}{24.3\ \cancel{g\ Mg}}\right) \left(\frac{1\ \cancel{mol\ MgCl_2}}{1\ \cancel{mol\ Mg}}\right) \left(\frac{95.3\ g\ MgCl_2}{1\ \cancel{mol\ MgCl_2}}\right)$$

$$= 3.92 \times 10^4\ g\ MgCl_2$$

Note that the difference in answers is due to rounding in equation (3).

FIGURE 7.6 Solution ladder for complex problems.

Planning the steps (writing the sequence of equations)

Equation (1)

Equation (2)

Equation (3)

Performing the calculations

Examples 7.15 and 7.16 illustrate an effective strategy for organizing the solution to a problem: dividing a complex problem into smaller, simpler problems. In setting up the steps of the solution, you must organize your work carefully to avoid confusion about what step comes next. *Planning* the sequence of steps (equations) of the solution is similar to drawing a ladder from the top down. After the ladder is drawn, the *sequence is reversed* in performing the calculations. The answer to the problem is calculated by "climbing the ladder" of equations. The last equation written, equation (3), is the first step of the calculation (first rung of the ladder). This quantity is then substituted into equation (2) (second rung) to obtain the quantity needed in equation (1) (top rung). The answer to the problem is calculated using equation (1). Figure 7.6 illustrates this process.

Practicing this solution process will enable you to tackle complex problems with speed and confidence. Remember that practice brings familiarity with the process. What at first may appear to be a long and complicated series of steps should become a rapid and familiar path to your goal. Good strategies and serious practice will help you to be a good problem solver.

Practice Problem 7.12 Calculate the mass of chlorine in 227 g $Ca(ClO)_2$, a compound used to treat water in swimming pools.

Chapter Review

Directions: From the list of choices at the end of the Chapter Review, choose the word or term that best fits in each blank. Use each choice only once. The answers appear at the end of the chapter.

Section 7.1 A __________ (1) is a counting unit equal to 6.022×10^{23} formula units of a substance. This number is called __________ (2) number. By definition, a mole is the amount of substance that contains the same number of __________ __________ (3) as there are carbon atoms in exactly 12 g of carbon-12.

Section 7.2 The mass of one mole of a substance is called its __________ __________ (4). The molar mass of an __________ (5) is equal to its gram atomic mass. The molar mass of a compound is calculated from the __________ (6) masses of the elements making up the compound. Molar mass may be expressed using units of __________ (7). These units are very useful in problem solving.

Section 7.6 The law of __________ __________ (8) says that every pure substance has a definite and fixed composition. The __________ (9) composition of a compound is the mass percentage of each element making up the compound. The percentage composition of a compound can be calculated from its formula and the atomic masses of elements making up the compound.

Sections 7.7, 7.8, 7.9, 7.10 Several kinds of formulas are widely used by chemists. The simplest mole ratio of the elements in a compound is expressed in its _______ (10) formula. A __________ (11) formula shows the number of each kind of atom making up a molecule. How atoms are joined together in a molecule is shown in its __________ (12) formula. Empirical formulas can be calculated from mass or percentage __________ (13) data. Molecular formulas can be calculated from the empirical formula and the __________ (14) mass of the compound. Empirical formulas can be used to answer questions involving the __________ (15) of elements in a specific mass of a compound.

Choices: atomic, Avogadro's, composition, definite composition, element, empirical, formula units, g/mol, masses, molar, molar mass, mole, molecular, percentage, structural

Study Objectives

After studying Chapter 7, you should be able to:

Section 7.3

1. Calculate the molar mass of a compound.

Section 7.4

2. Calculate the number of moles of an element or a compound from its mass.
3. Calculate the mass of a substance needed to provide a certain number of moles.

Section 7.5

4. Calculate the number of atoms or molecules in a sample.
5. Calculate the mass of a given number of atoms or molecules.
6. Calculate the number of moles of a substance from the number of atoms or molecules.

Section 7.6

7. Calculate the percentage composition of a compound.
8. Identify the principal kinds of chemical formulas.

Section 7.8

9. Calculate the empirical formula of a compound.

Section 7.9

10. Calculate the molecular formula of a compound.

Section 7.10

11. Calculate the mass of an element in a given mass of a compound.
12. Calculate the mass of a compound needed to provide a certain mass of one of its elements.

Key Terms

Review the definition of each term listed here by chapter section number. You may use the glossary if necessary.

7.1 mole, Avogadro's number, formula unit

7.2 molar mass

7.3 formula weight and molecular weight (see molar mass)

7.6 percentage composition

7.7 empirical formula, molecular formula, structural formula

Questions and Problems

The answers to questions and problems with blue numbers appear in Appendix C. Asterisks indicate the more challenging problems.

The Mole, Molar Mass (Sections 7.1, 7.2)

1. What is a mole of a substance?
2. What is Avogadro's number?
3. What is a formula unit of a substance?
4. What is the formula unit for each of the following substances?
   a. water
   b. sodium
   c. carbon dioxide
   d. nitrogen gas
   e. hydrogen gas
   f. sodium chloride
5. What is meant by the molar mass of a substance?

6. What is the molar mass (gram atomic mass) of each of the following elements?
   a. sodium
   b. iron
   c. arsenic
   d. zinc
   e. phosphorus
   f. calcium
7. What is the molar mass (gram atomic mass) of each of the following elements?
   a. copper
   b. sulfur
   c. potassium
   d. arsenic
   e. argon
   f. magnesium

Molar Masses of Compounds (Section 7.3)

8. Calculate the molar masses of the following compoun
   a. KI
   b. CaO, lime
   c. CO, a toxic gas
   d. $Sn(NO_3)_2$
9. What is the molar mass of each of the following?
   a. $Fe_3(PO_4)_2$
   b. PCl_3
   c. NaClO, dissolves in water to give liquid bleach
   d. SO_3
   e. $Mg(OH)_2$, milk of magnesia
10. Calculate the molar masses of the following organic compounds:
   a. C_6H_6, benzene
   b. $C_3H_8O_3$, glycerin
   c. $C_9H_8O_4$, aspirin
   d. $NaC_7H_5O_2$, sodium benzoate, a preservative in soft drinks
11. Find the molar mass of each of the following compounds:
   a. C_4H_{10}, butane, a fuel for lighters and campstoves
   b. C_2H_6O, ethyl alcohol
   c. H_3PO_4, used in ceramic tile cleaner
   d. $C_6H_5NO_2$, nitrobenzene
   e. P_4O_{10}
   f. N_2O, nitrous oxide, "laughing gas"
   g. C_3H_6O, acetone, nail polish remover

Mole Calculations (Section 7.4)

12. How many moles are in each of the following samples?
   a. 675 g water
   b. 49.2 g carbon
   c. 2.00 oz (56.8 g) gold
   d. 250. g H_2SO_4
   e. 75.88 g iron
13. Calculate the number of moles in each of the following samples:
   a. 126 g propane (C_3H_8), liquified petroleum gas
   b. 65.9 g Al_2O_3, alumina
   c. 12.56 g $KClO_3$
   d. 135.7 g NH_4Br
14. Find the number of moles of each substance:
   a. 1.00 kg sodium carbonate, Na_2CO_3
   b. 4.5 mg KI
   c. 3.8 μg $C_2H_4Br_2$, ethylene dibromide (EDB), a pesticide
   d. 1.25 lb glucose, $C_6H_{12}O_6$
15. What mass of each of the following is needed to provide the number of moles specified?
   a. 4.33 mol Cr
   b. 8.03 mol CO_2
   c. 2.54 mol Mg
   d. 0.234 mol $Al_2(SO_4)_3$
16. Calculate the mass of each of the following substances needed to provide the number of moles specified.
   a. 0.650 mol Br_2
   b. 2.004 mol $C_8H_{10}N_4O_2$, caffeine
   c. 5.40 mol $NaHCO_3$, baking soda
   d. 10.5 mol S
17. Calculate the mass of each of the following:
   a. 1.06 mol K_2CrO_4
   b. 1.50 mol NaBr
   c. 0.542 mol NaOH, lye
   d. 0.385 mol CH_4N_2O, urea, a natural fertilizer in mammalian excrement

Number of Atoms or Molecules in a Sample (Section 7.5)

18. How many atoms are in each of the following samples?
   a. 3.50 mol zinc
   b. 1.25 mol Na
   c. 0.345 mol Cu
   d. 1.20×10^{-3} mol Se, a trace element of nutritional importance
19. Calculate the number of molecules in each of the following samples:
   a. 0.0235 mol N_2
   b. 0.0050 mol O_3, ozone
   c. 0.65 mol $C_5H_{10}O_4$, deoxyribose, the sugar in DNA
   d. 1.25 mol NO_2, a pollutant in photochemical smog
20. Find the number of molecules in each of the following samples:
   a. 10.52 g SO_3
   b. 125 g C_7H_8, toluene, in gasoline (high octane rating)
   c. 0.011 g CH_4, methane, in natural gas, also formed by the decay of biowaste (sewage, garbage, etc.)
21. Calculate the number of molecules in each of the following:
   a. 1.00 kg of $HC_2H_3O_2$, acetic acid, in vinegar
   b. 6.20 g Br_2
   c. 2.9 mg vanillin, $C_8H_8O_3$, in vanilla extract
22. Calculate the total number of atoms in each of the following:
   a. 0.0120 mol tin
   b. 24.0 g Cl_2, used to treat drinking water
   c. 1.1 mol H_2S, odor of rotten eggs, and a toxic gas
   d. 8.668 g Co
23. Find the total number of atoms in each of the following substances:
   a. 1.65×10^{-4} mol Mg
   b. 1.0 μg ozone, O_3
   c. 3.2 mg silver
   d. 2.60 kg sulfur
24. Find the number of:
   a. nitrogen atoms in 10.0 kg urea, CH_4N_2O
   b. chlorine atoms in 5.0 g CCl_4
   c. oxygen atoms in 1.50 g ammonium perchlorate, NH_4ClO_4, an oxidizer in solid rocket propellants
   d. carbon atoms in 125 g octane, C_8H_{18}

25. Find the mass of one molecule of:
   a. H_2S
   b. CO_2
   c. propane, C_3H_8
   d. trichloroethane, $C_2H_3Cl_3$, a dry cleaning solvent
26. Calculate the mass of 10 molecules of:
   a. $HC_2H_3O_2$
   b. H_2
   c. DDT, $C_{14}H_9Cl_5$
   d. Br_2
   e. N_2
27. Calculate the mass of one atom of each of the following elements:
   a. calcium
   b. nickel
   c. tin
   d. lithium
   e. silicon
   f. arsenic
28. Find the mass of 25 atoms of each of the following elements:
   a. vanadium, V
   b. radon, Rn
   c. plutonium, Pu
   d. lead, Pb
29. How many moles of iron are 2.5×10^6 iron atoms?
30. A tablespoon of ethyl alcohol (grain alcohol) contains approximately 1.55×10^{23} molecules. How many moles of ethyl alcohol are in one tablespoon?

Percentage Composition (Section 7.6)

31. Calculate the mass percentage composition of each of the following compounds:
   a. NaCl
   b. CCl_4
   c. KI
   d. $NaHCO_3$
32. Calculate the mass percentage of:
   a. nitrogen in NH_3
   b. iron in Fe_2O_3
   c. sulfur in $Ca(HSO_3)_2$
   d. copper in $CuCl_2$
   e. aluminum in $Al_2(SO_4)_3$
33. Find the percentage of chlorine in each of the following compounds:
   a. Cl_2O
   b. NaClO, household bleaching agent
   c. CH_2Cl_2, methylene chloride, a solvent
   d. $C_{14}H_9Cl_5$, DDT, dichlorodiphenyltrichloroethane
   e. dioxin, $C_{12}H_4Cl_4O_2$, a toxic contaminant in agent orange, a defoliating agent used during the Vietnam war
34. Calculate the mass percentage of:
   a. oxygen in $KClO_3$
   b. chlorine in *para*-dichlorobenzene (moth crystals), $C_6H_4Cl_2$
   c. aluminum in $AlCl_3$
35. Fertilizers are rated by their nitrogen content. Calculate the percentage of nitrogen in urea, CH_4N_2O, a valuable fertilizer.
36. Calculate the percentage of phosphorus in each of the following compounds:
   a. Na_3PO_4
   b. H_3PO_4
   c. $Ca_3(PO_4)_2$
   d. P_2O_5
37. Calculate the percentage of the first element in each of the following compounds:
   a. Ag_2O
   b. FeS
   c. $KMnO_4$
   d. $CuBr_2$
   e. $CoSO_4$
   f. Cr_2O_3
38. Find the percentage of oxygen in each of the following compounds, all oxidizing agents:
   a. $KClO_3$
   b. $K_2Cr_2O_7$
   c. NH_4ClO_4
   d. $KMnO_4$
   e. H_2O_2

Empirical Formulas (Sections 7.7, 7.8)

39. Which of the following are empirical formulas?
   a. Na_2O_2
   b. $HC_2H_3O_2$
   c. C_2H_2
   d. $NaHCO_3$
   e. CH_2O
   f. C_4H_9
40. Identify which of the following are empirical formulas:
   a. Na_3PO_4
   b. KI
   c. $C_9H_8O_4$
   d. $CaCO_3$
   e. $Mg(OH)_2$
   f. $BaSO_4$
41. Which of the following are *not* empirical formulas? Write the empirical formula for each.
   a. H_2O
   b. H_2O_2
   c. CH_4
   d. $C_6H_{12}O_6$
   e. $C_{12}H_{22}O_{11}$
   f. C_3H_8
   g. $C_9H_8O_3$
42. Find the empirical formula of each of the following from the percentage composition data for the compound.
   a. 40.6% carbon, 5.1% hydrogen, 54.2% oxygen
   b. 93.7% carbon, 6.29% hydrogen
   c. 26.5% chromium, 24.5% sulfur, 49.0% oxygen
   d. 63.1% carbon, 11.9% hydrogen, 25.0% fluorine
   e. 49.0% carbon, 2.72% hydrogen, 48.3% chlorine
43. Find the empirical formula of the compound formed when 3.65 g iron reacts completely with 3.15 g sulfur.
44. A 1.08-gram sample of titanium chloride contained 0.273 g titanium. What is the empirical formula of titanium chloride?
45. A 0.254-gram sample of a compound was found to contain 0.0412 g sodium and 0.0980 g manganese. The balance of the mass was oxygen. Find the empirical formula of the compound.
46. Adrenalin, an important hormone, contains 59.0% C, 7.10% H, 26.2% O, and 7.65% N. Calculate the empirical formula of adrenalin.
47. *p*-Aminobenzoic acid, or PABA, is a sunscreen agent. Analysis of PABA gave its composition as 61.3% C, 5.11% H, 23.4% O, and 10.2% N. What is the empirical formula of PABA?
48. Teflon™ is the Du Pont name for polytetrafluoroethylene, a nonstick plastic material. Analysis revealed teflon to be composed of 24.0% carbon and 76.0% fluorine. What is the empirical formula of teflon?

49. What is the empirical formula of an oxide of nitrogen containing 30.4% nitrogen?

50. What is the empirical formula of a compound that is 21.6% magnesium, 21.4% carbon, and 57.0% oxygen?

51. Calculate the empirical formula of the nitrogen oxide that is 36.8% nitrogen and 63.2% oxygen.

52. The composition of *cis*-1,2-dichloroethylene is 24.7% carbon, 2.06% hydrogen, and 73.2% chlorine. What is its empirical formula?

53. A sample of iron pyrite was composed of 0.730 g iron and 0.840 g sulfur. What is its empirical formula?

*54. Determine the formula of hydrated zinc sulfate ($ZnSO_4$) that was 43.9% water. The formula of the hydrate should be written $ZnSO_4 \cdot xH_2O$. (See Section 13.8.)

*55. A hydrocarbon (a compound that contains only hydrogen and carbon) was burned, and the combustion products were collected and weighed. After the combustion, all of the carbon in the hydrocarbon was present in 1.47 g CO_2. All of the hydrogen in the original compound was present in 0.891 g H_2O. Calculate the empirical formula of the hydrocarbon.

Molecular Formulas (Sections 7.7, 7.9)

56. Explain how a molecular formula can differ from an empirical formula. Some molecular formulas are also empirical formulas. Explain.

57. What information is needed to calculate the molecular formula of a compound?

58. Find the molecular formulas of the following compounds:
   a. CH_2O, *MM* = 180. g/mol (fructose, a sugar)
   b. CH, *MM* = 78.0 g/mol (benzene, a carcinogenic solvent)
   c. C_3H_2Cl, *MM* = 219 g/mol
   d. C_4H_5, *MM* = 106 g/mol (xylene, an organic solvent)
   e. NO_2, *MM* = 92.0 g/mol

59. The percentage composition of citric acid is 37.5% C, 4.20% H, and 58.3% O. Its molar mass is 192 g/mol. What is the molecular formula of citric acid?

60. Hydrazine, a liquid rocket fuel, is composed of 87.5% N and 12.5% H. Its molar mass is 32.0 g/mol. What is the molecular formula for hydrazine?

61. Putrescine, a compound responsible for the odor of decaying fish, has a molar mass of 88.0 g/mol. Its percentage composition is 54.5% C, 13.8% H, and 31.8% N. What is the molecular formula of putrescine?

62. Nicotine is a highly toxic compound found in tobacco. Its percentage composition is 74.1% C, 8.70% H, and 17.3% N, and its molar mass is 162 g/mol. What is the molecular formula of nicotine?

*63. The composition of naphthalene, used as moth crystals, is 93.7% carbon and 6.29% hydrogen. Its molar mass is 128 g/mol. What is the molecular formula of naphthalene?

*64. Ribose, a sugar, is composed of 40.0% carbon, 6.67% hydrogen, and 53.3% oxygen. The molar mass of ribose is 150. g/mol. Calculate the molecular formula of ribose.

Additional Problems

65. Calculate the mass of metal needed to make 1.00 mol of each of the following compounds:
   a. $Fe(NO_3)_3$
   b. $CuCl_2$
   c. AgCl
   d. $Mg(OH)_2$
   e. CrO_3
   f. $Mg(ClO_4)_2$

66. Find the mass of oxygen in 10.0 g of each of the following compounds:
   a. H_2O_2
   b. $KMnO_4$
   c. $K_2Cr_2O_7$
   d. Na_2O_2
   e. $KClO_3$
   f. CrO_3

67. Find the mass of chlorine in 1.00 g of each of the following compounds?
   a. Cl_2O
   b. CH_2Cl_2
   c. $C_{14}H_9Cl_5$
   d. dioxin, $C_{12}H_4Cl_4O_2$

68. Calculate the mass of phosphorus in 12.45 g of each of the following compounds:
   a. Na_3PO_4
   b. H_3PO_4
   c. $Ca_3(PO_4)_2$
   d. P_2O_5

*69. Calculate the empirical formula of a hydrate from the following analytical data:

Na: 12.1% Al: 14.2% Si: 22.1%
O: 42.1% H_2O: 9.48%

Write the formula as illustrated in problem 54.

70. Chlordane, $C_{12}H_6Cl_8$, is an insecticide that has been used to control termites. Calculate the mass of chlorine in 1.00 g of chlordane.

71. Carbon tetrachloride, CCl_4, is an organic solvent believed to be a human carcinogen. Calculate the mass of chlorine in 1.00 mol of carbon tetrachloride.

72. Asbestos is believed to be a threat to public health because of its toxicity and its widespread use as insulation in schools and other public buildings. Asbestos is represented by the formula $Mg_6(Si_4O_{11})(OH)_6$. How much magnesium is contained in 1.00×10^{-5} g of asbestos?

73. What is the mass of 5.00×10^{28} atoms of gold?

*74. A fire consumed 1.00 kg of charcoal in 135 min. Assuming the charcoal to be pure carbon, how many carbon atoms were consumed in the fire per minute?

75. Complete the following table:

Compound	Mass, g	Number of Moles	Number of Molecules
$C_{14}H_9Cl_5$		0.250	
$C_{10}H_8$	10.0		
NO_2			1.25×10^{24}
C_2H_6O			35

76. What mass of lead contains the same number of atoms as 0.100 g of gold?

77. What mass of carbon contains the same number of atoms as 4.00 g of sulfur?

*78. Toluene, a hydrocarbon with a high octane number, is found in gasoline. A sample of toluene that had a mass of 4.25 g, equivalent to 0.0462 mol, was composed of 3.88 g carbon and 0.370 g hydrogen. What is the molecular formula of toluene?

79. Calculate the percentage composition of each of the following compounds:
   a. $CaCO_3$, limestone
   b. $Mg(OH)_2$, milk of magnesia

80. Formaldehyde is a tissue preservative and embalming agent. Find the empirical formula of formaldehyde from its percentage composition data:
40.0% C, 6.70% H, 53.3% O

Solutions to Practice Problems

PP 7.1

When parentheses appear in a formula, remember to multiply the number of atoms inside the parentheses by the subscript that follows.

a. 44.0 g/mol CO_2
b. 120. g/mol $MgSO_4$
c. $MM\text{-}Fe_2(SO_4)_3$:

$$(2\ \cancel{\text{mol Fe}})(55.9\ \text{g}/\cancel{\text{mol Fe}}) = 112\ \text{g Fe}$$
$$(3\ \cancel{\text{mol S}})(32.1\ \text{g}/\cancel{\text{mol S}}) = 96.3\ \text{g S}$$
$$(12\ \cancel{\text{mol O}})(16.0\ \text{g}/\cancel{\text{mol O}}) = 192\ \text{g O}$$
$$400.\ \text{g/mol}\ Fe_2(SO_4)_3$$

d. 46.0 g/mol NO_2
e. 120 g/mol $CHCl_3$
f. 122 g/mol $C_8H_{10}O$
g. 53.5 g/mol NH_4Cl
h. 212 g/mol K_3PO_4
i. $MM\text{-}Ca(ClO_3)_2$:

$$(1\ \cancel{\text{mol Ca}})(40.1\ \text{g}/\cancel{\text{mol Ca}}) = 40.1\ \text{g Ca}$$
$$(2\ \cancel{\text{mol Cl}})(35.5\ \text{g}/\cancel{\text{mol Cl}}) = 71.0\ \text{g Cl}$$
$$(6\ \cancel{\text{mol O}})(16.0\ \text{g}/\cancel{\text{mol O}}) = 96.0\ \text{g O}$$
$$207\ \text{g/mol}\ Ca(ClO_3)_2$$

PP 7.2

Outline each problem and use dimensional analysis to set up the calculation:

a. *Find:* number of moles = ?
Given: 74.9 g Sn
Known: MM-Sn = 119 g/mol
Solution:
$$\text{number of moles} = \left(\frac{1\ \text{mol Sn}}{119\ \cancel{\text{g}}}\right)(74.9\ \cancel{\text{g}}) = 0.629\ \text{mol Sn}$$

b. *Find:* number of moles = ?
Given: 675 g I_2
Known:
$$MM\text{-}I_2 = (2\ \cancel{\text{mol I}})\left(\frac{127\ \text{g}}{1\ \cancel{\text{mol I}}}\right) = 254\ \text{g/mol}$$
Solution:
$$\text{number of moles} = \left(\frac{1\ \text{mol I}}{254\ \cancel{\text{g}}}\right)(675\ \cancel{\text{g}}) = 2.66\ \text{mol}\ I_2$$

c. *Find:* number of moles = ?
Given: 24.62 g $Al_2(SO_4)_3$
Known: $MM\text{-}Al_2(SO_4)_3$ = 342.2 g/mol (from *AM*'s)
Solution:
$$\text{number of moles} = \left(\frac{1\ \text{mol}\ Al_2(SO_4)_3}{342.2\ \cancel{\text{g}}}\right)(24.62\ \cancel{\text{g}}) = 0.07195\ \text{mol}\ Al_2(SO_4)_3$$

PP 7.3

Use dimensional analysis to set up each calculation.

a. *Find:* mass (g) = ?
Given: 1.43 mol Fe
Known: MM-Fe = 55.9 g/mol
Solution:
$$\text{mass (g)} = \left(\frac{55.9\ \text{g Fe}}{1\ \cancel{\text{mol}}}\right)(1.43\ \cancel{\text{mol}}) = 79.9\ \text{g Fe}$$

b. *Find:* mass (g) = ?
Given: 2.55 mol Mg
Known: MM-Mg = 24.3 g/mol
Solution:
$$\text{mass (g)} = \left(\frac{24.3\ \text{g Mg}}{1\ \cancel{\text{mol}}}\right)(2.55\ \cancel{\text{mol}}) = 62.0\ \text{g Mg}$$

c. *Find:* mass (g) = ?
Given: 0.346 mol CH_3OH
Known: $MM\text{-}CH_3OH$ = 32.0 g/mol (from *AM*'s)
Solution:
$$\text{mass (g)} = \left(\frac{32.0\ \text{g}\ CH_3OH}{1\ \cancel{\text{mol}}}\right)(0.346\ \cancel{\text{mol}}) = 11.1\ \text{g}\ CH_3OH$$

d. *Find:* mass (g) = ?
Given: 0.346 mol $C_6H_4Cl_2$
Known: $MM\text{-}C_6H_4Cl_2$ = 147 g/mol (from *AM*'s)
Solution:
$$\text{mass (g)} = \left(\frac{147\ \text{g}\ C_6H_4Cl_2}{1\ \cancel{\text{mol}}}\right)(0.346\ \cancel{\text{mol}}) = 50.9\ \text{g}\ C_6H_4Cl_2$$

PP 7.4

a. *Find:* number of molecules = ?
Given: 10.3 g C_8H_{18}
Known: $MM\text{-}C_8H_{18}$ = 114 g/mol (from *AM*'s)
6.02×10^{23} molecules/mol

Solution:

$$\text{number of molecules} = \left(\frac{6.02 \times 10^{23}\ \text{molecules}}{1\ \cancel{\text{mol}}}\right)\left(\frac{1\ \cancel{\text{mol}}}{114\ \cancel{\text{g}}}\right)(10.3\ \cancel{\text{g}})$$

$$= 5.44 \times 10^{22}\ \text{molecules}\ C_8H_{18}$$

b. *Find:* number of atoms = ?

Given: 0.346 g Zn

Known: *MM*-Zn = 65.4 g/mol

6.02×10^{23} atoms/mol

Solution:

$$\text{number of atoms} = \left(\frac{6.02 \times 10^{23}\ \text{atoms}}{1\ \cancel{\text{mol}}}\right)\left(\frac{1\ \cancel{\text{mol}}}{65.4\ \cancel{\text{g}}}\right)(0.346\ \cancel{\text{g}})$$

$$= 3.14 \times 10^{21}\ \text{atoms Zn}$$

c. *Find:* number of molecules = ?

Given: 0.0215 g P_4

Known: $MM\text{-}P_4 = (4\ \cancel{\text{mol P}})\left(\frac{31.0\ \text{g}}{1\ \cancel{\text{mol P}}}\right) = 124\ \text{g/mol}$

6.02×10^{23} molecules/mol

Solution:

$$\text{number of molecules} = \left(\frac{6.02 \times 10^{23}\ \text{molecules}}{1\ \cancel{\text{mol}}}\right)\left(\frac{1\ \cancel{\text{mol}}}{124\ \cancel{\text{g}}}\right)(0.0215\ \cancel{\text{g}})$$

$$= 1.04 \times 10^{20}\ \text{molecules}\ P_4$$

Note that the formula tells us there are 4 P atoms/molecule, so the number of *atoms* of P is 4 times the number of molecules.

$$\text{number of P atoms} = \left(\frac{4\ \text{atoms}}{1\ \cancel{\text{molecule}}}\right)(1.04 \times 10^{20}\ \cancel{\text{molecules}})$$

$$= 4.16 \times 10^{20}\ \text{P atoms}$$

PP 7.5

a. *Find:* mass (g) = ?

Given: 1 atom uranium-235 (exactly)

Known: *MM*-U-235 = 235 g/mol

6.02×10^{23} atoms/mol

Solution:

$$\text{mass (g)} = \left(\frac{235\ \text{g}}{1\ \cancel{\text{mol}}}\right)\left(\frac{1\ \cancel{\text{mol}}}{6.02 \times 10^{23}\ \cancel{\text{atoms}}}\right)(1\ \cancel{\text{atom}})$$

$$= 3.90 \times 10^{-22}\ \text{g}$$

b. *Find:* mass (g) = ?

Given: 1 molecule CCl_2F_2 (exactly)

Known: $MM\text{-}CCl_2F_2$ = 121 g/mol (from *AM*'s)

6.02×10^{23} molecules/mol

Solution:

$$\text{mass (g)} = \left(\frac{121\ \text{g}}{1\ \cancel{\text{mol}}}\right)\left(\frac{1\ \cancel{\text{mol}}}{6.02 \times 10^{23}\ \cancel{\text{molecules}}}\right)(1\ \cancel{\text{molecule}})$$

$$= 2.01 \times 10^{-22}\ \text{g}\ CCl_2F_2$$

c. *Find:* mass (g) = ?

Given: 1 molecule C_2HCl_3 (exactly)

Known: $MM\text{-}C_2HCl_3$ = 132 g/mol (from *AM*'s)

6.02×10^{23} molecules/mol

Solution:

$$\text{mass (g)} = \left(\frac{132\ \text{g}}{1\ \cancel{\text{mol}}}\right)\left(\frac{1\ \cancel{\text{mol}}}{6.02 \times 10^{23}\ \cancel{\text{molecules}}}\right)(1\ \cancel{\text{molecule}})$$

$$= 2.19 \times 10^{-22}\ \text{g}\ C_2HCl_3$$

PP 7.6

a. *Find:* number of moles = ?

Given: 4.65×10^{45} copper atoms

Known: 6.02×10^{23} molecules/mol

Solution: Set up the calculation using dimensional analysis:

$$\text{number of moles} = \left(\frac{1\ \text{mol}}{6.02 \times 10^{23}\ \cancel{\text{atoms}}}\right)(4.65 \times 10^{45}\ \cancel{\text{atoms}})$$

$$= 7.72 \times 10^{21}\ \text{mol copper}$$

b. *Find:* number of moles = ?

Given: 8.33×10^{24} magnesium atoms

Known: 6.02×10^{23} atoms/mol

Solution:

$$\text{number of moles} = \left(\frac{1\ \text{mol}}{6.02 \times 10^{23}\ \cancel{\text{atoms}}}\right)(8.33 \times 10^{24}\ \cancel{\text{atoms}})$$

$$= 13.8\ \text{mol magnesium}$$

PP 7.7

Follow the steps outlined in Section 7.6. Remember to consider the subscripts in the formula when you calculate the mass of each element (Step 2). Check your work by finding the sum of the percentages. The sum should be 100% (allowing for rounding numbers).

a. $MM\text{-}NH_4ClO_4$ = 118 g/mol
N: 11.9%, H: 3.42%, Cl: 30.1%, O: 54.2%

b. $MM\text{-}(NH_4)_3PO_4$ = 149 g/mol
N: 28.2%, H: 8.12%, P: 20.8%, O: 43.0%

PP 7.8

a. Empirical formula

b. Empirical formula

c. The 4:10 ratio can be reduced to 2:5, and the empirical formula is P_2O_5. The formula given, P_4O_{10}, must be a molecular formula.

d. The 2:4 ratio can be reduced to 1:2, and the empirical formula is NO_2. The formula given, N_2O_4, must be a molecular formula.

PP 7.9

a. Outline the four steps of the problem:

Element	*Step 1:* Mass, g	*Step 2:* Number of Moles	*Step 3:* Simplest Whole-Number Mole Ratio	*Step 4:* Formula
C	40.0	$\frac{40.0\ \text{g}}{12.0\ \text{g/mol}} = 3.33\ \text{mol}$	$\frac{3.33\ \cancel{\text{mol}}}{3.33\ \cancel{\text{mol}}} = 1.00$ or 1	CH_2O
H	6.70	$\frac{6.70\ \text{g}}{1.01\ \text{g/mol}} = 6.63\ \text{mol}$	$\frac{6.63\ \cancel{\text{mol}}}{33.3\ \cancel{\text{mol}}} = 1.99$ or 2	
O	53.3	$\frac{53.3\ \text{g}}{16.0\ \text{g/mol}} = 3.34\ \text{mol}$	$\frac{3.34\ \cancel{\text{mol}}}{3.33\ \cancel{\text{mol}}} = 1.00$ or 1	

The empirical formula is CH_2O.

b. Outline the four steps of the problem:

Element	Step 1: Mass, g	Step 2: Number of Moles	Step 3: Simplest Whole-Number Mole Ratio	Step 4: Formula
Mg	20.2	$\frac{20.2\text{ g}}{24.3\text{ g/mol}} = 0.831$ mol	$\frac{0.831\text{ mol}}{0.829\text{ mol}} = 1.00$ or 1	$MgSO_4$
S	26.6	$\frac{26.6\text{ g}}{32.1\text{ g/mol}} = 0.829$ mol	$\frac{0.829\text{ mol}}{0.829\text{ mol}} = 1.00$ or 1	
O	53.2	$\frac{53.2\text{ g}}{16.0\text{ g/mol}} = 3.33$ mol	$\frac{3.33\text{ mol}}{0.829\text{ mol}} = 4.02$ or 4	

The empirical formula is $MgSO_4$.

PP 7.10

Outline the problem as before, starting with the masses of the elements making up the compound. Note that the mass of sulfur combined with zinc can be calculated by subtracting the mass of zinc from the mass of the compound.

Element	Step 1: Mass, g	Step 2: Number of Moles	Step 3: Simplest Whole-Number Mole Ratio	Step 4: Formula
Zn	1.13	$\frac{1.13\text{ g}}{65.4\text{ g/mol}} = 0.0173$ mol	$\frac{0.0173\text{ mol}}{0.017\text{ mol}} = 1.0$ or 1	ZnS
S	0.55	$\frac{0.55\text{ g}}{32.1\text{ g/mol}} = 0.017$ mol	$\frac{0.017\text{ mol}}{0.017\text{ mol}} = 1.0$ or 1	

The empirical formula is ZnS.

PP 7.11

Find: molecular formula

Given: $MM = 180.$ g/mol

40.0% C, 6.72% H, 53.3% O

Known: $x = \frac{MM}{EFM}$, where *EFM* is the empirical formula mass for dextrose

Solution: Because the empirical formula is not given, it must be calculated from the percentage composition of dextrose.

Element	Step 1: Mass, g	Step 2: Number of Moles	Step 3: Simplest Whole-Number Mole Ratio	Step 4: Formula
C	40.0	$\frac{40.0\text{ g}}{12.0\text{ g/mol C}} = 3.33$ mol	$\frac{3.33\text{ mol}}{3.33\text{ mol}} = 1.00$ or 1	CH_2O
H	6.72	$\frac{6.72\text{ g}}{1.01\text{ g/mol H}} = 6.65$ mol	$\frac{6.65\text{ mol}}{3.33\text{ mol}} = 2.00$ or 2	
O	53.3	$\frac{53.3\text{ g}}{16.0\text{ g/mol O}} = 3.33$ mol	$\frac{3.33\text{ mol}}{3.33\text{ mol}} = 1.00$ or 1	

For CH_2O, $EFM = 30.0$ g/mol CH_2O (from *AM*'s). To calculate x, the number of formula units per molecule:

$$x = \frac{MM}{EFM} = \frac{180.}{30.0} = \text{6.00 or 6 formula units/molecule}$$

The molecular formula is $(CH_2O)_6$, or $C_6H_{12}O_6$.

PP 7.12

Find: mass (g) Cl = ?

Given: 227 g $Ca(ClO)_2$

Known: *AM*-Cl = 35.5 g/mol

MM-$Ca(ClO)_2$ = 143 g/mol (from *AM*'s)

Note from the formula that there are 2 mol Cl per mol of $Ca(ClO)_2$:

$$\frac{2\text{ mol Cl}}{1\text{ mol Ca(ClO)}_2}$$

Solution: Using dimensional analysis, first set up the calculation for the mass of Cl:

$$\text{mass (g) Cl} = \left(\frac{35.5\text{ g}}{1\text{ mol Cl}}\right)(\text{? mol Cl}) \quad (1)$$

Second, because the number of moles of Cl was not given or known, set up an equation to calculate this quantity:

$$\text{number of moles Cl} = \left(\frac{2\text{ mol Cl}}{1\text{ mol Ca(ClO)}_2}\right)(\text{? mol Ca(ClO)}_2) \quad (2)$$

Third, to calculate the number of mol $Ca(ClO)_2$, set up another equation:

$$\text{number of mol Ca(ClO)}_2 = \left(\frac{1\text{ mol Ca(ClO)}_2}{143\text{ g}}\right)(227\text{ g}) \quad (3)$$
$$= 1.59\text{ mol Ca(ClO)}_2$$

Substituting into equation (2) gives

$$\text{number of moles Cl} = \left(\frac{2\text{ mol Cl}}{1\text{ mol Ca(ClO)}_2}\right)(1.59\text{ mol Ca(ClO)}_2)$$
$$= 3.17\text{ mol Cl}$$

Substituting into equation (1) gives

$$\text{mass (g) Cl} = \left(\frac{35.5\text{ g}}{1\text{ mol Cl}}\right)(3.17\text{ mol Cl})$$
$$= 113\text{ g Cl}$$

Check: Begin with the given information to check your calculations:

$$(227\text{ g Ca(ClO)}_2)\left(\frac{1\text{ mol Ca(ClO)}_2}{143\text{ g Ca(ClO)}_2}\right)\left(\frac{2\text{ mol Cl}}{1\text{ mol Ca(ClO)}_2}\right)\left(\frac{35.5\text{ g Cl}}{1\text{ mol Cl}}\right)$$
$$= 113\text{ g Cl}$$

Answers to Chapter Review

1. mole
2. Avogadro's
3. formula units
4. molar mass
5. element
6. atomic
7. g/mol
8. definite composition
9. percentage
10. empirical
11. molecular
12. structural
13. composition
14. molar
15. masses

8

The Structure of Compounds. Chemical Bonds

Contents

A photomicrograph of crystalline aspartame, or NutraSweet®, a molecular compound.

A fertilizer is rated according to its composition, especially the percentage of nitrogen.

A structural formula for sucrose ($C_{12}H_{22}O_{11}$).

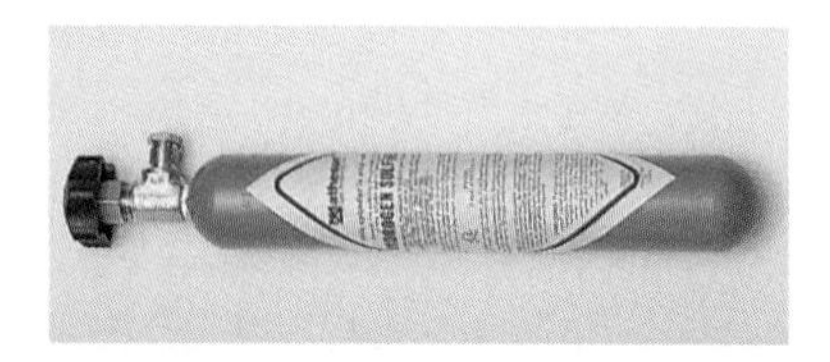

Hydrogen sulfide has the odor of rotten eggs.

You Probably Know . . .

- Some compounds, such as water, are composed of molecules whereas others, such as sodium chloride (table salt), are made up of ions (Section 2.3). Although both of these compounds are made up of just two elements, you are well aware that they have very different properties.
- Compounds have definite composition whereas mixtures have variable composition (Section 2.3).
- As with all forms of matter, compounds are three dimensional in their structure. Chemists use structural formulas to describe how atoms of a molecule are joined to each other (Section 7.7).
- Two molecular compounds that have similar formulas, such as water (H_2O) and hydrogen sulfide (H_2S), can have very different properties. Water is a colorless, odorless liquid whereas hydrogen sulfide is a poisonous gas that has the foul odor of rotten eggs.

Why Should We Study the Structure of Compounds?

Why are some compounds solids and others liquids or gases? Why do some compounds dissolve readily in water and others do not? Why do some solids, such as candle wax, melt at moderate temperatures whereas others, such as sand, melt only at very high temperatures? Why do some liquids evaporate readily and others do not? The answers to these questions, and our understanding of all physical properties of compounds, depend upon our understanding of their structures.

In earlier chapters, we began to see that the structure of atoms—the nucleus and electron arrangements—determines how elements combine to form compounds. As we continue our study, we'll also see that the structure of the molecules or ions of a compound determines how the compound reacts to form new substances. In this chapter, we consider the structure of molecular and ionic compounds as preparation for more extensive study in later chapters of their chemical reactions and physical properties.

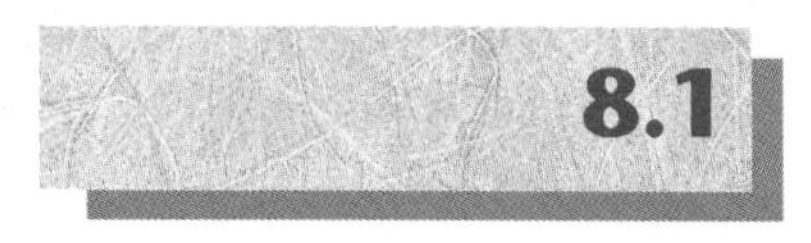

8.1 Chemical Bonds

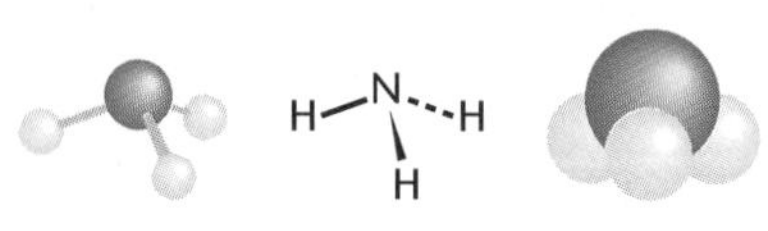

Ammonia (NH_3), a compound, has a definite composition.

Ammonia (NH_3) molecules contain one nitrogen atom and three hydrogen atoms. Like all other compounds, ammonia has a definite composition, a feature that distinguishes it from mixtures of nitrogen and hydrogen gases, which have variable composition.

What is responsible for maintaining the definite composition of a compound such as ammonia or water? What prevents the composition of water from changing when it is heated to its boiling point? The answers to these questions are based on our understanding of **chemical bonds,** attractive forces that keep the oppositely charged ions of an ionic compound together and hold the atoms of a molecule together. The two principal kinds of bonds are called *ionic* and *covalent* bonds.*

Chemical bonds: Attractive forces that keep the oppositely charged ions of an ionic compound together and hold the atoms of a molecule together.

Compounds are often classified according to the type of bonding involved—ionic or covalent. Earlier (Section 2.3), we described compounds as being made up of ions or molecules, and we referred to compounds as being either ionic or molecular. For example, sodium chloride (NaCl) is an ionic compound because it is made up of Na^+ and Cl^- ions. Water is a molecular compound because it is composed of H_2O molecules. Molecular compounds are often called covalent compounds because of the types of bonding involved.

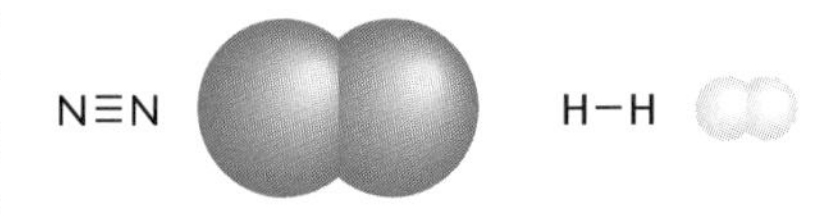

Mixtures of N_2 and H_2 molecules can vary in composition.

Ionic Bonding

8.2

Learning Objectives

- Identify compounds that are ionic.
- Describe the crystal structure of sodium chloride.

Sodium chloride forms in the reaction of sodium with chlorine (Section 6.8), as illustrated by the following equation:

$$\text{Na}\cdot + \cdot\ddot{\underset{..}{\text{Cl}}}: \rightarrow \text{Na}^+ + \left[:\ddot{\underset{..}{\text{Cl}}}:\right]^- \quad \text{or} \quad \text{NaCl}$$

This is an electron transfer (oxidation–reduction) reaction resulting in the formation of sodium ions (cations) and chloride ions (anions) in equal numbers as indicated by the formula NaCl. We recognize that opposite charges attract each other. The attractive force between these oppositely charged ions in a compound is called an **ionic bond.** Ionic bonding usually occurs when a metal reacts with a nonmetal.

Remember that NaCl is the empirical formula of sodium chloride and represents the smallest unit, or formula unit (Section 7.1), of the compound. Note that the formula unit of an ionic compound should not be called a molecule. On the contrary, the formula unit of an ionic compound gives us the simplest ratio of cations to anions.

The crystal structure for sodium chloride is illustrated in Figure 8.1. Notice that Na^+ ions are smaller than Cl^- ions. The packing arrangement of the cations and anions is determined by their sizes and by their relative charges. Here, each Na^+ ion is surrounded by six Cl^- ions, and each Cl^- ion is surrounded by six Na^+ ions. The result is a multidirectional network of attractive forces. This multidirectional feature of ionic bonding helps to distinguish it from a covalent bond, which is directed between two specific atoms. Sometimes when students look at the formula, NaCl, they make the mistake of thinking the ionic bond is an attractive force

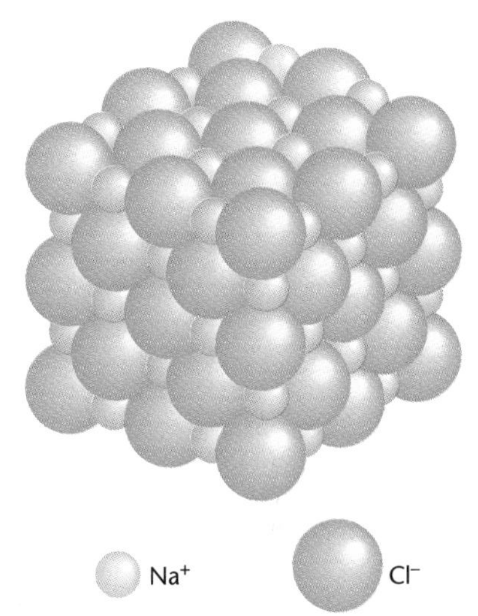

FIGURE 8.1 Crystal structure of sodium chloride. Each Na^+ ion (gray) is surrounded by six Cl^- ions (green), and each Cl^- ion is surrounded by six Na^+ ions. The crystal is neutral because the number of positive charges equals the number of negative charges.

*Other kinds of attractive forces, such as metallic bonding (Section 13.5) and intermolecular forces (Section 13.2), are discussed later.

between one sodium ion and one chloride ion. This is incorrect. No single Na^+ is bonded to any particular Cl^-. Remember, molecules of ionic compounds do not exist.

Having ions of opposite charge next to each other tends to keep ions of like charge apart. This maximizes attractive forces between ions of opposite charge (Na^+ and Cl^-) and minimizes repulsive forces between ions of like charge. The strength of ionic bonding depends upon the relative strengths of the attractive and repulsive forces. In general, ionic bonds are very strong. This is why ionic compounds have high melting points (see also Section 13.5) and are always solids at room temperature. Moreover, ionic solids are hard and rigid, and they shatter easily upon impact.

Calcium chloride, $CaCl_2$, forms in the reaction of calcium with chlorine:

$$\mathrm{Ca{:}} + 2\,\mathrm{{:}\ddot{C}l{\cdot}} \rightarrow \mathrm{Ca^{2+}} + 2\left[\mathrm{{:}\ddot{C}l{:}}\right]^- \quad \text{or} \quad \mathrm{CaCl_2}$$

This equation illustrates that cations and anions in different compounds are not always present in the same numerical ratio. Following the octet rule, a calcium atom loses two valence electrons to form a calcium ion, Ca^{2+}. However, a chlorine atom completes its octet of valence electrons by gaining only one electron to form a chloride ion, Cl^-. Therefore, two chlorine atoms are needed to accept the two electrons lost by one calcium atom. The 1:2 ratio of Ca^{2+} ions to Cl^- ions is shown in the formula for calcium chloride, $CaCl_2$. Depending upon the charges, the formulas of other compounds can involve various ratios of cations to anions to give a neutral compound. See Chapter 9 for a detailed discussion of the formulas and names of compounds.

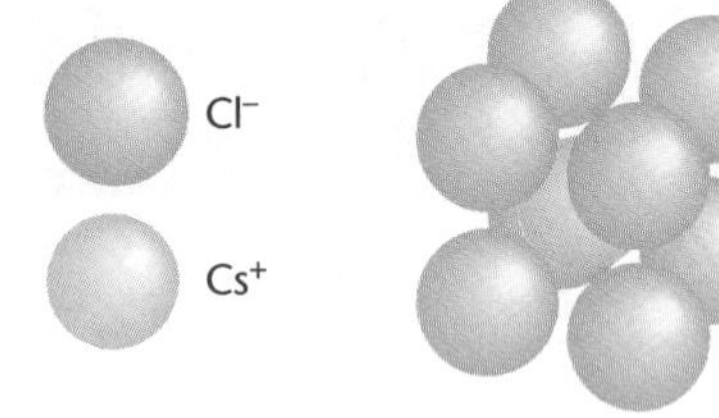

FIGURE 8.2 The structure of CsCl. Each Cs^+ ion (gray) is surrounded by eight Cl^- ions (green) and vice versa.

The crystal structure of $CaCl_2$ is different from the structure of sodium chloride because of the different cation:anion ratio and because a Ca^{2+} ion is smaller than a Na^+ ion. Even when a compound has a 1:1 ratio of cations to anions, the packing arrangement can differ from the sodium chloride structure described earlier. For example, in a crystal of cesium chloride, CsCl (Figure 8.2), each large Cs^+ ion is surrounded by eight Cl^- ions and vice versa. This structure is less common than the sodium chloride structure, but it illustrates how a different packing arrangement can accommodate ions of different sizes.

In the solid state, where ions are immobile, ionic compounds are nonconductors of electricity. However, when an ionic compound is dissolved in water, the cations and anions separate from one another (dissociate):

The abbreviation (aq) stands for "aqueous" and means "dissolved in water."

$$\mathrm{NaCl(s)} \xrightarrow{\text{water}} \mathrm{Na^+(aq)} + \mathrm{Cl^-(aq)}$$

They are then free to move independently of one another. Melting an ionic compound also mobilizes the ions. Because of the mobility of the charges, both the solution and the molten compound conduct an electric current. A compound, either as a liquid or dissolved in water, that is electrically conducting is called an **electrolyte.** Testing the electrical conductivity of an aqueous solution of a compound is a simple method of testing for ions.

Electrical conductivity: a test for ions.

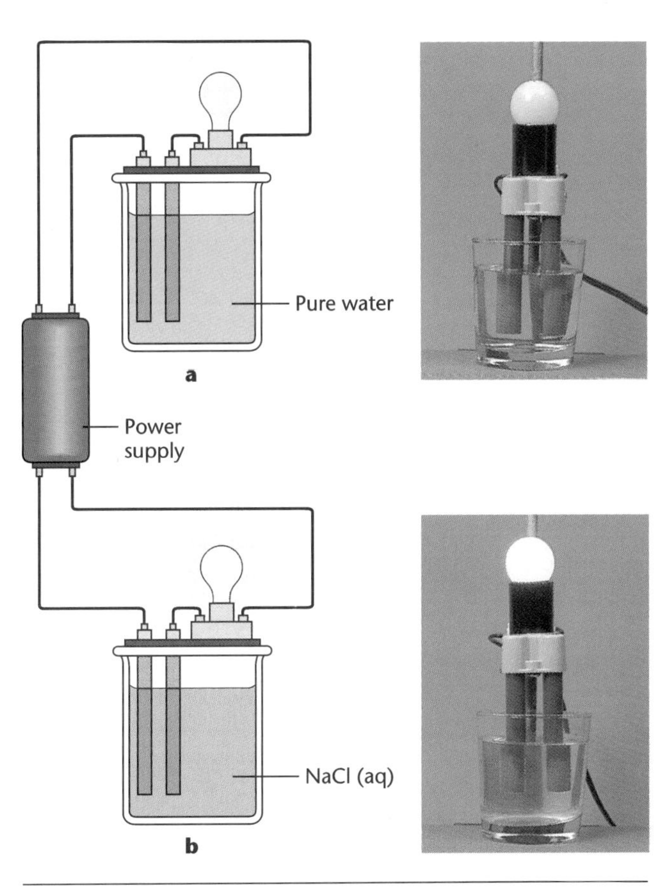

(a) Pure water does not conduct a detectable electric current, indicating the absence of ions. (b) Current flows through a sodium chloride solution, and the bulb lights.

Example 8.1

Which of the following compounds are ionic?

a. NaBr
b. CaF_2
c. CO_2
d. HCl
e. CuCl ▪

Solution Ionic compounds (metal + nonmetal) are a, b, and e.

Covalent Bonding

Learning Objectives

- Describe the bonding for a compound as ionic or covalent.
- Describe the kind of energy changes that occur when bonds are formed and when bonds are broken.
- Describe differences in bond length and in the strength of single, double, and triple bonds.

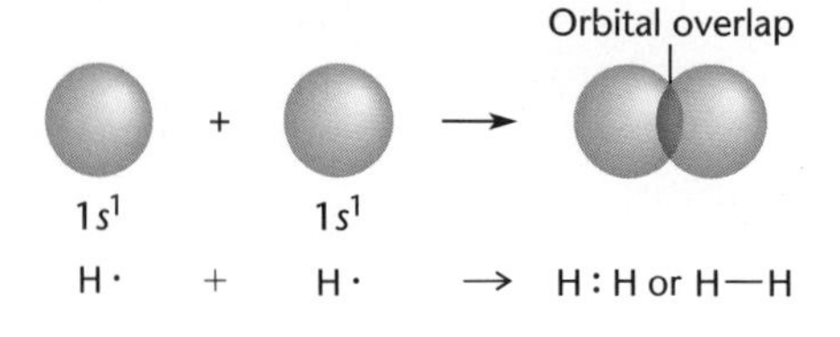

FIGURE 8.3 Formation of a hydrogen molecule from two hydrogen atoms. The covalent bond is shown as a line in the formula.

Covalent bond: Joins two nonmetal atoms of a molecule.

A **covalent bond** is an attractive force between the atoms of a molecule that is a result of their sharing a pair of electrons. Among the millions of known compounds, the covalent bond is the prevalent kind of chemical bond. Remember, also, that some elements such as nitrogen, oxygen, and chlorine are made up of molecules. So the covalent bond is the attractive force that holds together molecules of all kinds. Covalent bonds usually join atoms that are nonmetals.

To help us understand the covalent bond, let's look at a hydrogen molecule. We may think of a hydrogen molecule as being formed from two hydrogen atoms. A hydrogen atom (found in the *s*-block in the periodic table) has one electron in its 1*s* atomic orbital. This orbital has a capacity of two electrons, which also is the capacity of the first principal energy level. Figure 8.3 illustrates the overlapping of the 1*s* orbitals of two hydrogen atoms to form a hydrogen molecule. Two electrons are now shared by the two atoms, so both have a filled principal energy level. The pair of electrons is attracted by the positive nuclei of both atoms, resulting in a covalent bond, shown as a line in the formula.

Because of the strong attractive force joining the hydrogen atoms, the hydrogen molecule is more stable than two hydrogen atoms. Energy is released (an exothermic process) when the bond forms. On the other hand, because breaking a covalent bond is the reverse of bond formation, bond breaking is endothermic. These energy differences are shown in the diagram in Figure 8.4. The strength of the bond can be determined by the energy needed to break it. This is called the **bond dissociation energy** or just *bond energy*. Bond strengths vary widely as you can see from the bond dissociation energies listed in Table 8.1. Remember that bond breaking is endothermic, and this is the same quantity of energy that is released in bond formation.

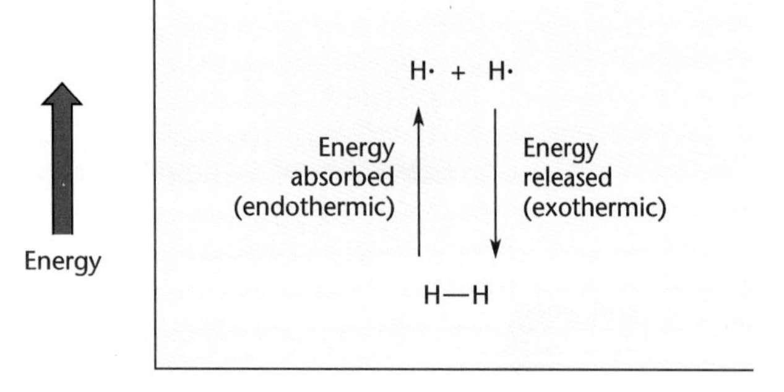

FIGURE 8.4 Energy changes in forming and in breaking a covalent bond.

TABLE 8.1 Bond Dissociation Energies for Diatomic Molecules

Molecule	*Energy, kJ/mol*
H_2	435
F_2	159
Cl_2	243
Br_2	192
I_2	151
HF	565
HCl	431
HBr	364

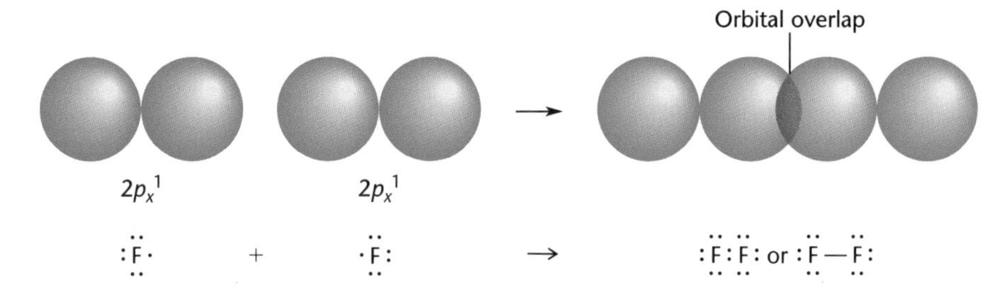

FIGURE 8.5 Formation of a fluorine molecule from two fluorine atoms.

A fluorine molecule (F_2) is also diatomic. Fluorine is in the *p*-block in the periodic table, and a pair of electrons is shared when a 2*p* atomic orbital (with one electron) overlaps a 2*p* orbital of the other atom. This gives each atom of the molecule eight valence electrons (octet rule, Section 6.8). The formation of the fluorine molecule is illustrated in Figure 8.5. Lewis formulas of molecules are discussed in detail in Section 8.4.

You should begin to see that you can use the periodic table and the octet rule to predict the number of covalent bonds expected for a nonmetal. For example, a chlorine atom (Group 7A) has seven valence electrons, like fluorine. When a chlorine atom forms a covalent bond by sharing one electron with another atom, it gains one electron to give it eight valence electrons (an octet). Because all halogen atoms have seven valence electrons, the octet rule predicts each will form one covalent bond. In similar fashion, the number of bonds typically formed by other nonmetals is

For a nonmetal:

number of bonds (predicted) = 8 − group number

$$\text{number of bonds (predicted)} = 8 - \text{group number}$$

For example, nitrogen (Group 5A) typically forms 3 bonds (8 − 5) as in ammonia, NH_3. Carbon (Group 4A) usually forms 4 bonds (8 − 4) as in methane, CH_4.

In each of the previous examples, two atoms shared one pair of electrons and were joined by a **single bond.** However, numerous compounds have atoms joined by multiple bonds, both double and triple bonds. Atoms joined by a **double bond** share two pairs of electrons, and atoms joined by a **triple bond** share three pairs of electrons. To illustrate, ethane has a carbon–carbon single bond, whereas the carbons of ethylene (used to make the plastic polyethylene) are joined by a double bond. The carbons of acetylene (the fuel for a welder's torch) are joined by a triple bond.

```
    H  H          H       H
    |  |           \     /
 H—C—C—H            C=C            H—C≡C—H
    |  |           /     \
    H  H          H       H

  Ethane          Ethylene          Acetylene
```

Double and triple bonds are stronger and shorter than single bonds.

Multiple bonds are *stronger* and *shorter* than single bonds. For example, the double bond (four electrons) in ethylene is a stronger attractive force than the carbon–carbon single bond of ethane. Moreover, the triple bond of acetylene is shorter than the double bond of ethylene, and the carbon–carbon single bond of ethane is the longest of the three bonds.

Example 8.2

Identify the following compounds as either ionic or covalent.

a. Fe_2O_3
b. CuS
c. PCl_3
d. NO_2
e. C_2H_6 ■

Solutions Ionic compounds (metal + nonmetal): a, b
Covalent compounds (two nonmetals): c, d, e

Example 8.3

Identify the following as exothermic or endothermic.

a. $Br_2 \rightarrow 2Br$
b. $CH_3 + Cl \rightarrow CH_3Cl$ ■

Solutions

a. The equation describes bond breaking, which is endothermic.
b. A bond is forming, so this is exothermic.

Example 8.4

Which bond is stronger? Which bond is shorter?

a. C—N or C=N?
b. C=N or C≡N? ■

Solutions Bond strength follows the order single < double < triple. Bond length follows the reverse order; single bonds are the longest.

a. C=N is the stronger and a shorter bond.
b. C≡N is the stronger and a shorter bond.

Practice Problem 8.1 Which of the following compounds are ionic and which are covalent? (Solutions to practice problems appear at the end of the chapter.)

a. Cr_2O_3
b. P_2O_5
c. CH_3I
d. $AgCl$
e. NiS

Practice Problem 8.2 Which bond is stronger? Which bond is shorter?

a. C—O or C=O?
b. N—O or N=O?

Lewis Formulas

8.4

Learning Objectives

- Draw Lewis structures for simple molecules and ionic compounds.
- Identify several substances that are exceptions to the octet rule.
- Draw resonance structures to represent simple molecules and ions.

In Chapter 7, you were introduced to several types of formulas. An empirical formula tells us the simplest mole ratio of the elements making up the compound. A molecular formula tells us the actual number of atoms making up a molecule. A structural formula shows how the atoms are joined together in a molecule. Structural formulas for H_2 and F_2 are shown in Figures 8.3 and 8.5.

The type of structural formula most often used to represent compounds is called a Lewis formula. This kind of formula was named for G. N. Lewis who first introduced the concept of covalent bonding. A **Lewis formula** uses a line to represent a covalent bond and dots for nonbonding valence electrons. Because a covalent bond is a pair of electrons that is shared by two atoms, *a line in the formula means two electrons.* As with electron-dot symbols for atoms (Section 6.7), a complete Lewis formula shows all valence electrons including the nonbonding valence electrons of each atom. Most atoms involved in covalent bonding have an octet of electrons like that of the nearest noble gas in the periodic table (Section 6.8). The Lewis formulas for hydrogen, chlorine, and hydrogen chloride, all diatomic molecules, are shown in Figure 8.6.

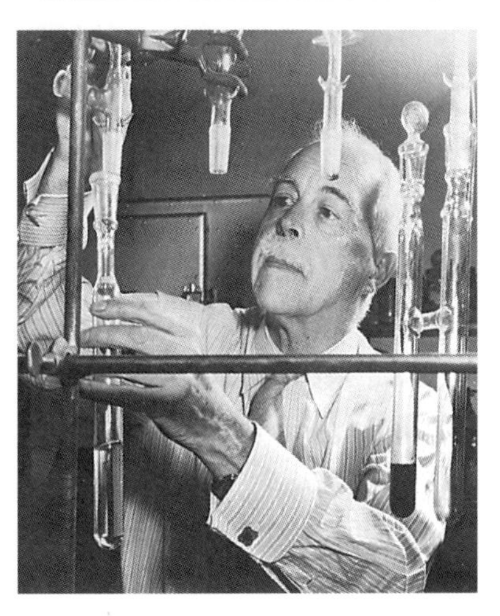

The covalent bond was proposed in 1916 by the American chemist G. N. Lewis (1875–1946).

Lewis formula: A line for each bond (two electrons), dots for nonbonding valence electrons.

Lewis (electron-dot) symbols show all valence electrons; for example,

·N̈: ·F̤̈: [:C̤̈l:]⁻ *(Section 6.7).*

The steps for writing Lewis formulas of covalent compounds are summarized in Figure 8.7. To illustrate, consider the Lewis formula of carbon dioxide, CO_2. The bonding is covalent because carbon and oxygen are nonmetals.

Step 1 One carbon and two oxygen atoms make up the formula. According to their position in the periodic table, we can calculate the number of valence electrons they would contribute as follows:

	number of valence electrons
carbon, from Group 4A	$4\ e^- \times 1$ atom $= 4\ e^-$
oxygen, from Group 6A	$6\ e^- \times 2$ atoms $= 12\ e^-$
	total $= 16\ e^-$

Step 2 Predict the number of bonds for carbon and oxygen.

carbon (Group 4A) $8 - 4 = 4$ bonds

oxygen (Group 6A) $8 - 6 = 2$ bonds

Step 3 Arrange the atoms for symmetry and connect with lines:

O—C—O

Step 4 To comply with the predicted number of bonds for both C and O (Step 2), add two bonds as shown:

O=C=O

Step 5 Add dots for nonbonding electrons (octet rule):

:Ö=C=Ö:

Hydrogen	H—H
Chlorine	:C̤̈l—C̤̈l:
Hydrogen chloride	H—C̤̈l:

FIGURE 8.6 Lewis formulas of simple molecules

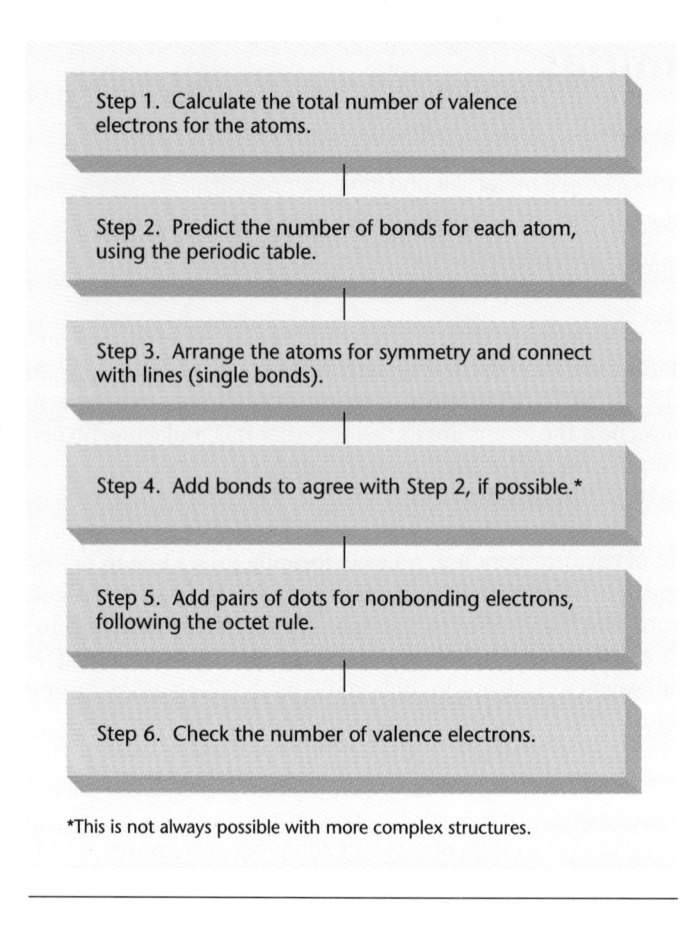

FIGURE 8.7 Writing Lewis formulas of molecules and polyatomic ions

Step 6 Check the valence electrons: 16 valence electrons are shown in the formula. The number of valence electrons checks. The Lewis formula shows that CO_2 has two double bonds.

Example 8.5

Write the Lewis formula for nitrogen, N_2, the principal component of air. ■

Solution

Step 1 Nitrogen (Group 5A) has 5 valence electrons.

$5\ e^- \times 2\ \text{atoms} = 10\ e^-$

Step 2 Predict 3 bonds for nitrogen (Group 5A).

Step 3 N—N

Step 4 Add 2 bonds to agree with Step 2.

N≡N

Step 5 Add one pair of dots to each atom (octet rule).

:N≡N:

Step 6 Check valence electrons. Because the formula shows 10 valence electrons, the formula checks.

The Lewis formula shows that N_2 has a triple bond.

Example 8.6

Write the Lewis formula for formaldehyde, CH_2O, a preservative for biology specimens. ■

Solution

Step 1

carbon (Group 4A)	$4\ e^- \times 1$ atom	$= 4\ e^-$
hydrogen (element 1)	$1\ e^- \times 2$ atoms	$= 2\ e^-$
oxygen (Group 6A)	$6\ e^- \times 1$ atom	$= 6\ e^-$
	total	$= 12\ e^-$

Step 2 Predict:

4 bonds for carbon (Group 4A)
2 bonds for oxygen (Group 6A)
1 bond for hydrogen (limited by the capacity of the first shell—two e^-)

Step 3 Carbon, with more bonds, is the central atom:

```
   O
   |
H—C—H
```

Step 4 Add a second bond for C—O to give oxygen 2 bonds and carbon 4 bonds (see Step 2).

```
   O
   ‖
H—C—H
```

Step 5 Add 2 pairs of dots to O (octet rule).

```
  :O:
   ‖
H—C—H
```

Step 6 Check valence electrons. Because the formula has $12\ e^-$, it checks.

Practice Problem 8.3 Write Lewis formulas for the following compounds:

a. NH_3
b. HCN

■ Lewis Formulas of More Complex Structures

Polyatomic: Composed of more than two atoms.

The Lewis formulas of certain molecules and polyatomic ions (Section 9.8) cannot be written showing the predicted number of bonds for each element. Sulfur trioxide, SO_3, is an example. The steps must be modified slightly, beginning with Step 4, as shown in Example 8.7.

Example 8.7

Write the Lewis formula of sulfur trioxide, SO_3. ■

Solution

Step 1 oxygen (Group 6A) $6\ e^- \times 3\ \text{atoms} = 18\ e^-$
sulfur (Group 6A) $6\ e^- \times 1\ \text{atom} = 6\ e^-$
total $= 24\ e^-$

Step 2 Predict 2 bonds for both sulfur and oxygen (Group 6A).

Step 3 Arrange the atoms for symmetry and connect with single bonds:

```
    O
    |
H—S—H
```

Step 4 Because the number of bonds for S in the formula is greater than predicted, no additional bonds are added even though the oxygen atoms do not have two bonds. Go to Step 5.

Step 5 Add pairs of dots to give all but the central atom (S in this case) an octet of electrons. Then, to give sulfur an octet, change a pair of dots to a bond.

```
      ..                        ..
     :O:       changes to      :O:
  ..  |  ..                 ..  |  ..
 :O—S—O:                   :O—S=O:
  ..     ..                 ..     ..
```

Step 6 Check valence electrons. The formula shows $24\ e^-$. It checks.

In compounds, O—O bonds are found only in peroxides such as hydrogen peroxide, H_2O_2.

This example points out something you should remember about oxygen. In compounds, oxygen atoms are rarely bonded together. Oxygen–oxygen bonds are found only in peroxides (e.g., H_2O_2), which are chemically unstable compounds.

■ Resonance

In writing the formula for SO_3, you may have drawn the double bond in a different position than shown in the preceding example. Actually, three Lewis formulas can be drawn for SO_3 that differ in the position of the double bond:

```
      ..               ..               ..
     :O:              :O:              :O:
  ..  |  ..        ..  |  ..        ..  ‖  ..
 :O—S=O:  ↔   :O=S—O:  ↔   :O—S—O:
  ..              ..     ..        ..     ..
```

This example illustrates resonance. **Resonance** exists if two or more Lewis formulas can be drawn for a substance that differ only in the position of electrons. The formulas used to show resonance are called **resonance structures.** Two-headed arrows are drawn between the formulas. The positions of the atoms must be the same in all resonance structures.

Resonance structures are Lewis formulas that differ only in the position of electrons.

Resonance is generally used to show equivalency of bonds when one Lewis formula is inadequate to do so. If we consider only one formula for sulfur trioxide, we would think that a SO_3 molecule had two single bonds and one double bond. Actually, *all the bonds are alike* and have a character that falls between a single bond and a double bond. We can visualize the formula for the actual SO_3 molecule if we imagine drawing each resonance structure on a transparent sheet, superimposing the sheets, and projecting the image on a screen. The image would show that all bonds are equivalent. Sometimes a formula called a *resonance hybrid* is used to show the equivalency of the bonds:

Resonance shows that all bonds in SO_3 are equivalent.

```
     O
     ¦¦
O ==S== O
```

Lewis Structures of Ionic Compounds

Remember, ionic compounds are not composed of molecules. The formula of an ionic compound tells us the simplest ratio of cations to anions. When writing Lewis structures of ionic compounds, we must show the cations and anions separately (Section 6.8).

Lewis structures of monatomic ions were covered in Section 6.8.

Some ionic compounds include diatomic or polyatomic ions with atoms that are covalently bonded together. As with binary ionic compounds such as KI, the Lewis structures of the ions are written separately. An example is sodium hydroxide, NaOH. The cation is sodium ion, Na^+ (Group 1A). The anion is hydroxide ion, OH^-. Note that the -1 charge (add one electron) on the anion can be determined by recognizing that the sum of the ionic charges equals zero for an ionic compound:

NaOH

$+1 - 1 = 0$

In writing the Lewis structure for OH^-, follow the steps for covalent compounds.

Step 1 Calculate the number of valence electrons:

oxygen (Group 6A)	6 e^-
hydrogen	1 e^-
-1 charge	1 e^-
total =	8 e^-

Step 2 Predict 1 bond for H and 2 bonds for O (Group 6A).

Step 3 Join the atoms with a single bond:

O—H

Step 4 Because one bond for H agrees with what was predicted, do not add another bond. (Hydrogen can have only one bond, so oxygen cannot have two bonds here.)

Step 5 Add pairs of dots (octet rule). Remember to show the charge on the ion.

$[:\ddot{\underset{..}{O}}—H]^-$

Step 6 Check electrons. Because the formula has 8 electrons, it checks. The Lewis structure for NaOH is

$Na^+ [:\ddot{\underset{..}{O}}—H]^-$

Example 8.8

Write the Lewis structure for Na_2O. ■

Solution Remember, a metal cation generally has no valence electrons, and monatomic anions have an octet. The charges on the ions can be determined from the group number for the element.

Charges: Na^+ (Group 1A)
O^{2-} (Group 6A)

Lewis structure: $2Na^+[:\ddot{\underset{..}{O}}:]^{2-}$

Example 8.9

Write the Lewis structure for $NaNO_2$. ■

Solution The cation is Na^+, and the anion is NO_2^- (sum of ionic charges equals zero). To write the Lewis structure of the anion:

Calculate the number of valence electrons:

Step 1	nitrogen (Group 5A)	$5\ e^- \times 1$ atom $= 5\ e^-$
	oxygen (Group 6A)	$6\ e^- \times 2$ atoms $= 12\ e^-$
	-1 charge (add electron)	$= 1\ e^-$
		total $= 18\ e^-$

Step 2 Predict: 3 bonds for N (Group 5A)
2 bonds for O (Group 6A)

Step 3 Arrange the atoms for symmetry and connect with single bonds:

O—N—O

Step 4 Add another bond to agree with the number of bonds predicted for nitrogen (Step 2). Note that both oxygens cannot have two bonds without giving nitrogen more bonds than predicted:

O—N═O

Step 5 Add pairs of dots (octet rule). Include the charge on the ion:

$[:\ddot{\underset{..}{O}}—\ddot{N}═\ddot{O}:]^-$

Step 6 Check valence electrons. The formula shows 18 e^-. It checks.
The Lewis structure for sodium nitrite is

$$Na^+[:\ddot{\underset{\cdot\cdot}{O}}-\ddot{N}=\ddot{O}:]^-$$

Note that NO_2^-, as with SO_3, can be represented by resonance:

$$[:\ddot{\underset{\cdot\cdot}{O}}-\ddot{N}=\ddot{O}: \leftrightarrow :\underset{\cdot\cdot}{O}=\ddot{N}-\ddot{\underset{\cdot\cdot}{O}}:]^-$$

Practice Problem 8.4 Write Lewis structures for the following compounds:
a. NaSH
b. KNO_3

Exceptions to the Octet Rule

Some substances do not follow the octet rule. For example, NO and NO_2, compounds implicated in photochemical smog, have odd numbers of valence electrons:

$$\cdot\ddot{N}=\ddot{O}: \quad :\ddot{\underset{\cdot\cdot}{O}}-\dot{N}=\ddot{O}:$$

It is impossible to draw Lewis formulas of these compounds in which all atoms have an octet (an even number) of electrons. However, NO_2 molecules combine to form a dimer, N_2O_4, in which all atoms have eight valence electrons.

Dimer: Two parts (di = *"two"*; mer *from* meros = *"parts"*).

$$2NO_2(g) \rightarrow N_2O_4(g)$$

(Brown) (Colorless)

Lewis formulas of other compounds, such as PCl_5 and SF_6 (Figure 8.9), have more than eight electrons around the central atom. Moreover, some compounds have an atom with as few as four valence electrons (BeF_2), and others have an atom with just six valence electrons (BF_3):

$$:\ddot{\underset{\cdot\cdot}{F}}-Be-\ddot{\underset{\cdot\cdot}{F}}: \qquad :\ddot{\underset{\cdot\cdot}{F}}-B(-\ddot{F}:)_2$$

Oxygen (O_2) is a unique example. We can draw a Lewis formula in which both atoms have an octet of electrons, but experimental evidence suggests it is incorrect. According to the octet rule, there is a double bond.

$$:\ddot{O}=\ddot{O}:$$

However, oxygen is **paramagnetic,** meaning it is attracted by a magnetic field. Studies have shown that paramagnetic substances have unpaired electrons. The following Lewis formula has unpaired electrons and accounts for the paramagnetism of O_2:

$$\cdot\ddot{\underset{\cdot\cdot}{O}}-\ddot{\underset{\cdot\cdot}{O}}\cdot$$

But this formula conflicts with other evidence indicating that the bond is much like a double bond. In short, a Lewis formula simply cannot be drawn that accounts for all the properties of O_2.

8.5 Electronegativity

Learning Objectives

- Identify the most electronegative element.
- Using the periodic table, tell which atom of a pair is more electronegative.

Linus Pauling (1901–)

We have noted that a covalent bond is an attractive force between two atoms of a molecule that is the result of their sharing a pair of electrons. When the atoms are identical, as is true for H_2 and Cl_2, the atoms share electrons equally. However, when the atoms that are bonded together are different, the electrons are usually unequally shared. One atom may attract the bonding electrons more strongly than the other.

Unequal sharing of electrons is explained by the concept of **electronegativity** (electron greediness), a measure of the capacity of an atom to attract its bonding electrons. An element with high electronegativity attracts bonding electrons more strongly than one with lower electronegativity. This concept was first introduced in 1932 by Linus Pauling (1901–) of the California Institute of Technology. Note (Table 8.2) that fluorine is the most electronegative element and is assigned a value of 4.0 on a relative scale. Notice also that metals have relatively low values compared with nonmetals. Because the noble gases do not usually form bonds, only the electronegativities of krypton and xenon are shown in the table.* As we look at the periodic table, electronegativity increases from left to right across a period and from bottom to top in a group.

Example 8.11

Which is more electronegative:

a. O or F?
b. O or Cl?
c. P or Cl?
d. H or Cl? ■

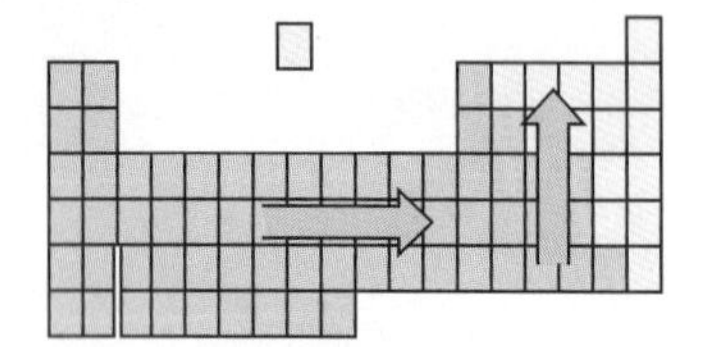

Arrows show direction of increasing electronegativity.

*A few noble gas compounds (e.g., XeF_2 and XeF_4) are known. However, the first such compound was not synthesized until 1962, 30 years after Linus Pauling introduced the concept of electronegativity.

TABLE 8.2 Electronegativities of the Elements

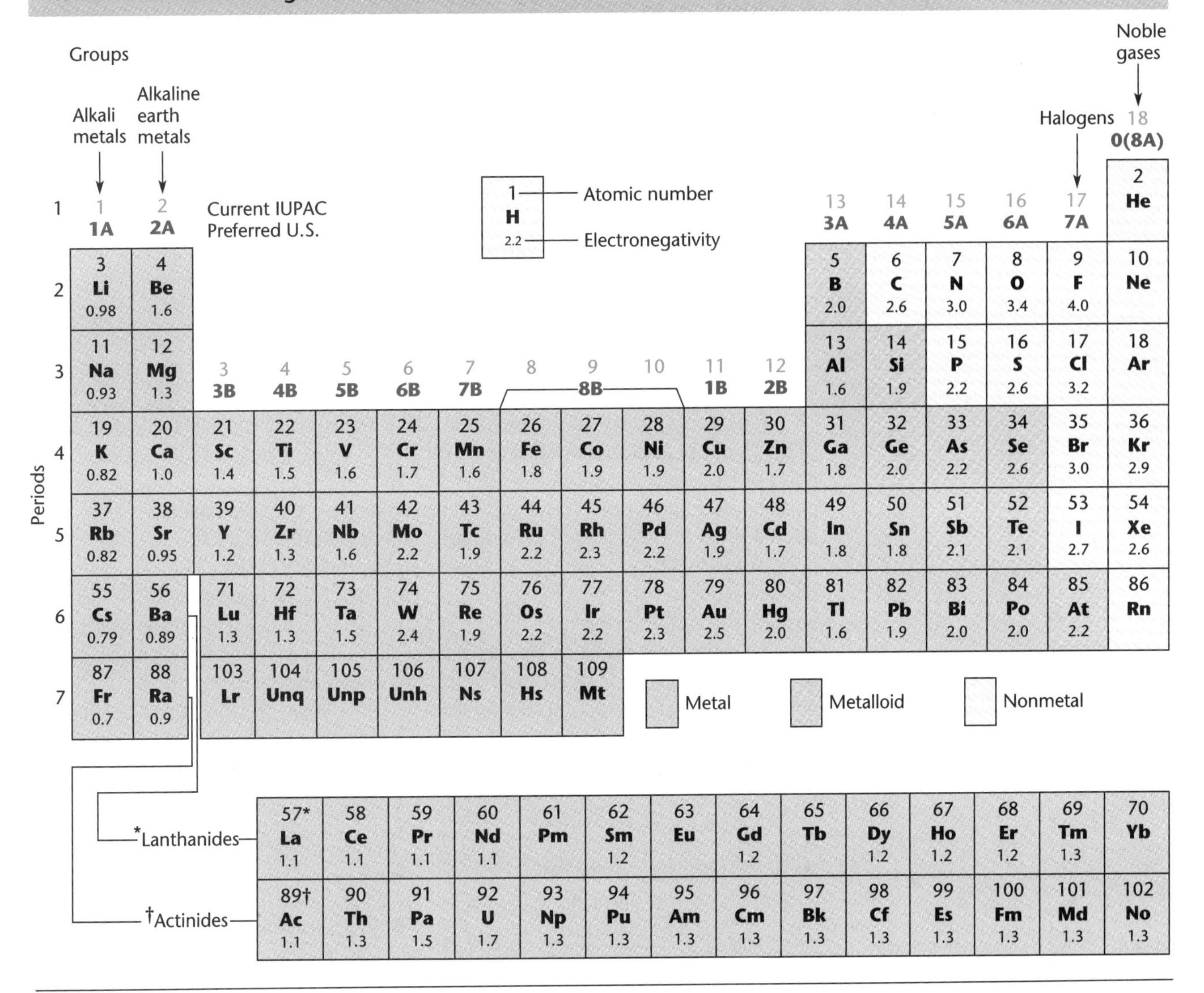

Key: 1 (Atomic number) H 2.2 (Electronegativity). Groups: Current IUPAC / Preferred U.S. Alkali metals = Group 1 (1A); Alkaline earth metals = Group 2 (2A); Halogens = Group 17 (7A); Noble gases = Group 18 (0(8A)).

Period	1 1A	2 2A	3 3B	4 4B	5 5B	6 6B	7 7B	8 8B	9 8B	10 8B	11 1B	12 2B	13 3A	14 4A	15 5A	16 6A	17 7A	18 0(8A)
1																		2 **He**
2	3 **Li** 0.98	4 **Be** 1.6											5 **B** 2.0	6 **C** 2.6	7 **N** 3.0	8 **O** 3.4	9 **F** 4.0	10 **Ne**
3	11 **Na** 0.93	12 **Mg** 1.3											13 **Al** 1.6	14 **Si** 1.9	15 **P** 2.2	16 **S** 2.6	17 **Cl** 3.2	18 **Ar**
4	19 **K** 0.82	20 **Ca** 1.0	21 **Sc** 1.4	22 **Ti** 1.5	23 **V** 1.6	24 **Cr** 1.7	25 **Mn** 1.6	26 **Fe** 1.8	27 **Co** 1.9	28 **Ni** 1.9	29 **Cu** 2.0	30 **Zn** 1.7	31 **Ga** 1.8	32 **Ge** 2.0	33 **As** 2.2	34 **Se** 2.6	35 **Br** 3.0	36 **Kr** 2.9
5	37 **Rb** 0.82	38 **Sr** 0.95	39 **Y** 1.2	40 **Zr** 1.3	41 **Nb** 1.6	42 **Mo** 2.2	43 **Tc** 1.9	44 **Ru** 2.2	45 **Rh** 2.3	46 **Pd** 2.2	47 **Ag** 1.9	48 **Cd** 1.7	49 **In** 1.8	50 **Sn** 1.8	51 **Sb** 2.1	52 **Te** 2.1	53 **I** 2.7	54 **Xe** 2.6
6	55 **Cs** 0.79	56 **Ba** 0.89	71 **Lu** 1.3	72 **Hf** 1.3	73 **Ta** 1.5	74 **W** 2.4	75 **Re** 1.9	76 **Os** 2.2	77 **Ir** 2.2	78 **Pt** 2.3	79 **Au** 2.5	80 **Hg** 2.0	81 **Tl** 1.6	82 **Pb** 1.9	83 **Bi** 2.0	84 **Po** 2.0	85 **At** 2.2	86 **Rn**
7	87 **Fr** 0.7	88 **Ra** 0.9	103 **Lr**	104 **Unq**	105 **Unp**	106 **Unh**	107 **Ns**	108 **Hs**	109 **Mt**									

*Lanthanides	57* **La** 1.1	58 **Ce** 1.1	59 **Pr** 1.1	60 **Nd** 1.1	61 **Pm**	62 **Sm** 1.2	63 **Eu**	64 **Gd** 1.2	65 **Tb**	66 **Dy** 1.2	67 **Ho** 1.2	68 **Er** 1.2	69 **Tm** 1.3	70 **Yb**
†Actinides	89† **Ac** 1.1	90 **Th** 1.3	91 **Pa** 1.5	92 **U** 1.7	93 **Np** 1.3	94 **Pu** 1.3	95 **Am** 1.3	96 **Cm** 1.3	97 **Bk** 1.3	98 **Cf** 1.3	99 **Es** 1.3	100 **Fm** 1.3	101 **Md** 1.3	102 **No** 1.3

Metal · Metalloid · Nonmetal

Solutions Electronegativity increases from left to right and from bottom to top in the periodic table. Choice (b) is difficult because oxygen is above chlorine but to its left, and you must look in Table 8.2 for the answer.

a. F
b. O
c. Cl
d. Cl

Practice Problem 8.5 Which is more electronegative:

a. Cl or I?
b. S or C?
c. N or As?
d. Se or Br?

Refer to Table 8.2.

8.6 Polar Covalent Bonds

Learning Objectives

- Specify bonds as nonpolar covalent, polar covalent, or ionic.
- Using appropriate notation, label polar covalent bonds in the Lewis formula of a compound.

Atoms having the same electronegativity share bonding electrons equally, resulting in a **nonpolar covalent bond.** We have already considered examples such as H_2, O_2, N_2, and Cl_2. However, when atoms having different electronegativities are covalently bonded, the more electronegative atom attracts the bonding electrons more strongly. The result is a **polar covalent bond,** a bond in which there is unequal sharing of electrons.

The words "polar" and "pole" are related and suggest opposites. We may refer to the opposite ends of a magnet as its north and south poles. We also speak of the poles of the earth, meaning the "opposite ends" of the earth. In a similar way, a polar bond is a bond that has ends that are opposite—they are oppositely charged. Such a bond is also said to have a **dipole,** which literally means "two poles."

A hydrogen chloride (HCl) molecule has a polar covalent bond. Because of the difference in their electronegativities (chlorine, 3.2 vs. hydrogen, 2.2), chlorine attracts the bonding electrons more strongly than hydrogen. This electron greediness of a chlorine atom causes it to have a small negative charge leaving the hydrogen atom with a small positive charge. Note, however, that these are only partial charges and are much smaller than -1 or $+1$. We can illustrate the polarity of a bond in two ways:

$\delta+$ and $\delta-$ mean partial charges of a dipole.

1. Use the symbols $\delta-$ and $\delta+$ at the ends of the bond in the formula:

$$\overset{\delta+}{H}—\overset{\delta-}{Cl}$$

2. Draw an arrow pointing toward the more electronegative element:

An arrow ($H \mapsto Cl$) indicates a dipole.

$$H \mapsto Cl$$

Delta, δ, is a Greek letter that is used here to mean "partial." Any bond between atoms that have different electronegativities is a polar covalent bond. For example, the bonds in HF, H_2O, NO, and CCl_4 are all polar.

The degree of polarity of a bond depends upon the difference in electronegativities of the atoms involved. For example, HF is more polar than HCl, and HCl is more polar than HBr. This order of polarity is explained by the order in electronegativity for the halogens: F $>$ Cl $>$ Br. Moreover, a C—O bond is more polar than a C—N bond because oxygen is more electronegative than nitrogen.

There is generally a large difference in the electronegativities of a metal and a nonmetal, and the bonding is ionic. For example, the difference in electronegativities of sodium (0.9) and chlorine (3.2) is 2.3, and NaCl is ionic. The relationship of bond type and the difference in electronegativities of bonded atoms is shown in Figure 8.8.

Bond type:	Nonpolar covalent	Polar covalent	Ionic
	Increasing polarity →		
Difference in electronegativities*:	Small (0.0–0.4)	Moderate (0.4–1.9)	Large (>1.9)
Examples: (electronegativity)	Cl—Cl (3.2, 3.2)	H—Cl (2.2, 3.2)	NaCl (0.9, 3.2)

*Ranges are approximate. Some examples fall outside these ranges.

FIGURE 8.8 Bond type as related to differences in electronegativity. Ranges are approximate, and some examples fall outside of the range.

Many molecules (H_2O, CCl_4, etc.) have two or more dipoles. How these dipoles are arranged in space is a major factor influencing the physical properties of the compound. To predict or explain the physical properties of these compounds, shapes of molecules (Section 8.7) must be considered. Molecules such as these will be discussed in more detail in Section 8.8.

Many fuels, pollutants, and biologically active compounds are organic (carbon) compounds. Most organic compounds have C—H bonds. The difference in the electronegativity of carbon (2.6) and hydrogen (2.2) is very small (0.4). This means that a C—H bond is practically nonpolar. Consequently, when evaluating the dipoles of an organic molecule, C—H bonds are usually ignored. On the other hand, polar bonds such as C—O, C—N, O—H, and N—H strongly influence physical properties and serve as reactive sites in chemical reactions.

A C—H bond is practically nonpolar.

Example 8.12

Identify each of the following bonds as nonpolar covalent, polar covalent, or ionic. Label the polar bonds to show the dipole of each.

a. KCl
b. HF
c. OH
d. I_2 ■

Solutions

a. ionic
b. polar covalent H ⟷ F
c. polar covalent, O ⟻ H
d. nonpolar covalent

Practice Problem 8.6 Which of the following are polar covalent bonds? Label the polar bonds to show the dipole of each. Refer to Table 8.2.

a. C—O
b. N—H
c. N_2

8.7 Shapes of Molecules

Learning Objective

- Predict the bond angles and the shape of a simple molecule using VSEPR theory.

Although we represent molecules by writing formulas on paper in two dimensions, we must remember that molecules are real objects. Molecules are three dimensional, and they have shapes. Knowing the shapes of molecules helps us to understand the chemical and physical properties of substances.

How can we know the shape of a molecule that cannot be seen? Scientists have answered this question using x-ray techniques to determine the shapes of

FIGURE 8.9 Geometry of simple molecules. In the Lewis formulas, A is the central atom.

Number of electron pairs	Lewis formula	Bond angle	Shape	Example
2	X—A—X	180°	linear	CO_2
3	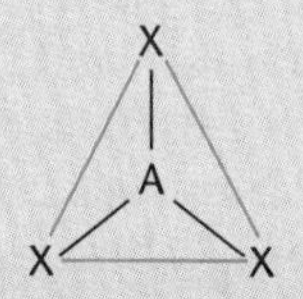	120°	trigonal planar	SO_3
4	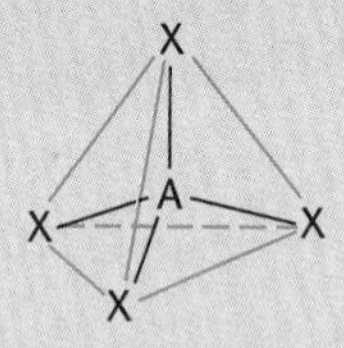	109°	tetrahedral	CH_4
5	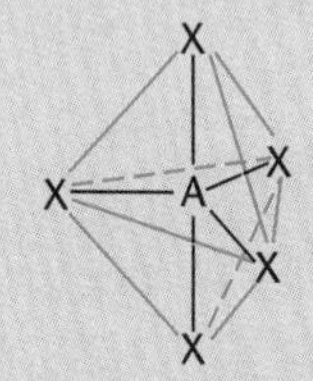	120° 90°	trigonal bipyramidal	PCl_5
6	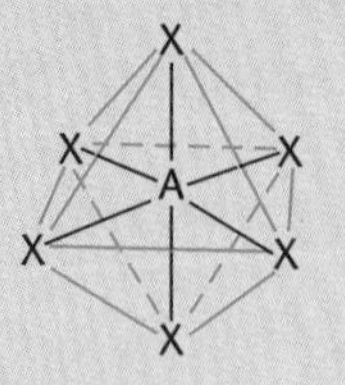	90°	octahedral	SF_6

many molecules in their crystalline states. However, because we could not possibly memorize all the known examples, we need a way of predicting shapes.

A useful theory for predicting shapes of molecules is called the valence shell electron pair repulsion (VSEPR) theory. Very simply, this theory recognizes that like charges repel each other. In other words, electron pairs repel each other, and the orbitals where we find them lie as far apart as possible in space. With a little knowledge of geometry, we can predict the shapes of molecules.

To predict the shape of a molecule, draw its Lewis formula and use the information in Figure 8.9. Count pairs (or sets) of electrons surrounding the atom in the center of the molecule, including both bonding and nonbonding electrons. From the number of sets of electrons, predict the bond angle and the shape of the molecule.

The bond angles shown in Figure 8.9 were predicted using geometry and models of molecules. The angles were measured after the lines representing the bonds were positioned as far apart as possible in space. There is usually good agreement between angles predicted using VSEPR theory and those determined experimentally using x-ray diffraction, an instrumental method of analysis that uses x-rays to determine the positions of atoms and ions in a crystal.

Nonmetals, except hydrogen, generally have a stable electron structure of eight valence electrons (four pairs) when they are bonded in molecules. This explains why many molecules are tetrahedral in shape. For example, the four bonding pairs of electrons for methane, CH_4 (natural gas), and carbon tetrachloride, CCl_4 (a carcinogenic solvent), are at the corners of a tetrahedron, a four-sided figure. Four bonded atoms at the corners of a tetrahedron are farther apart than at the corners of a square.

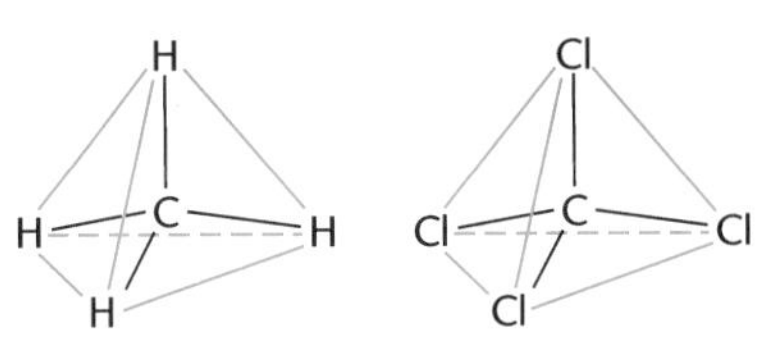

CH_4 and CCl_4 are tetrahedral molecules.

In predicting shapes, the electrons of a double or triple bond are counted as one set of electrons. For example, in carbon dioxide, CO_2, the carbon atom has two double bonds. Here, we count two sets of electrons and predict a bond angle of 180°. Experimental data confirm that carbon dioxide is a linear molecule.

CO_2 is a linear molecule

O=C=O

180°

The VSEPR theory is also useful in predicting the shape of a molecule that has a central atom with nonbonding pairs of electrons. For example, the oxygen of a water molecule has two bonds and two nonbonding pairs of electrons. Nonbonding pairs of electrons are counted even though no atoms are attached. With four pairs of electrons surrounding oxygen, you would predict a bond angle of 109°. The observed angle of 105° is very close to the predicted angle. With only three atoms in the molecule, the shape is described as bent rather than tetrahedral.*

H—Ö:
H

A water molecule is bent.

Example 8.13

Predict the bond angles and shapes for the following molecules:

a. BrCN
b. SO_3
c. SO_2 ■

*Note that the arrangement of valence *electrons* is tetrahedral, but the molecular structure (determined by positions of atoms) is described as bent.

Solutions

a. $:\ddot{\underset{..}{Br}}—C≡N:$ Two sets of e⁻ around C; 180°, linear

b. $:\ddot{O}:$ over $:\underset{..}{\ddot{O}}—S=\ddot{O}\!:$ Three sets of e⁻ around S; 120°, trigonal planar

c. $:\underset{..}{\ddot{O}}—\underset{..}{S}=\ddot{O}\!:$ Three sets of e⁻ (including nonbonding e⁻) around S; 120°, bent

---

Practice Problem 8.7 What are the approximate bond angles for the following molecules? Describe the shapes of the molecules.

a. $CHCl_3$
b. C_2H_2
c. C_2H_3Cl

8.8 Polar Molecules

Learning Objective

- Predict whether molecules are polar or nonpolar.

CO_2 is nonpolar because the dipoles cancel.

$\longleftarrow\!\!+ \quad +\!\!\longrightarrow$
O=C=O

Some molecules, although they have polar bonds, are nonpolar. In **nonpolar molecules,** the bonds are arranged around the central atom so that the dipoles cancel. Carbon dioxide is an example. Using the VSEPR theory, we have already predicted that a CO_2 molecule is linear (Section 8.7). Because the arrows (dipoles) are pointing in opposite directions, the dipoles cancel, resulting in a nonpolar molecule.

In **polar molecules,** the dipoles do not cancel. To decide whether or not a molecule is polar, you need to be able to look at the Lewis formula of the molecule. Our strategy for this kind of problem is summarized in Figure 8.10.

Carbon tetrachloride, CCl_4, is a molecule that has four equivalent polar bonds. Because of the shape of the molecule, the dipoles cancel. Another example is sulfur trioxide, SO_3, which has three polar bonds—all equivalent due to resonance (Section 8.4). Both of these molecules are nonpolar. From these and similar examples we can state the following rule:

When all bonds in a molecule of regular shape (linear, trigonal, tetrahedral, etc.) have the same dipole, the dipoles cancel, resulting in a nonpolar molecule.

Another example is chloroform, $CHCl_3$, a toxic solvent used as an anesthetic during the Civil War. C—Cl bonds are polar, but a C—H bond is practically nonpolar. The dipoles for the C—Cl and C—H bonds do not cancel, and the result is a polar molecule. Remember, dipoles cancel only when all bonds have the same dipole.

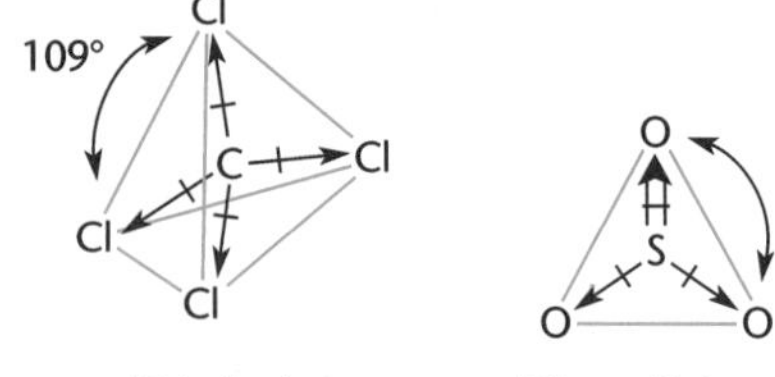

CCl_4 and SO_3 are nonpolar molecules because their dipoles cancel.

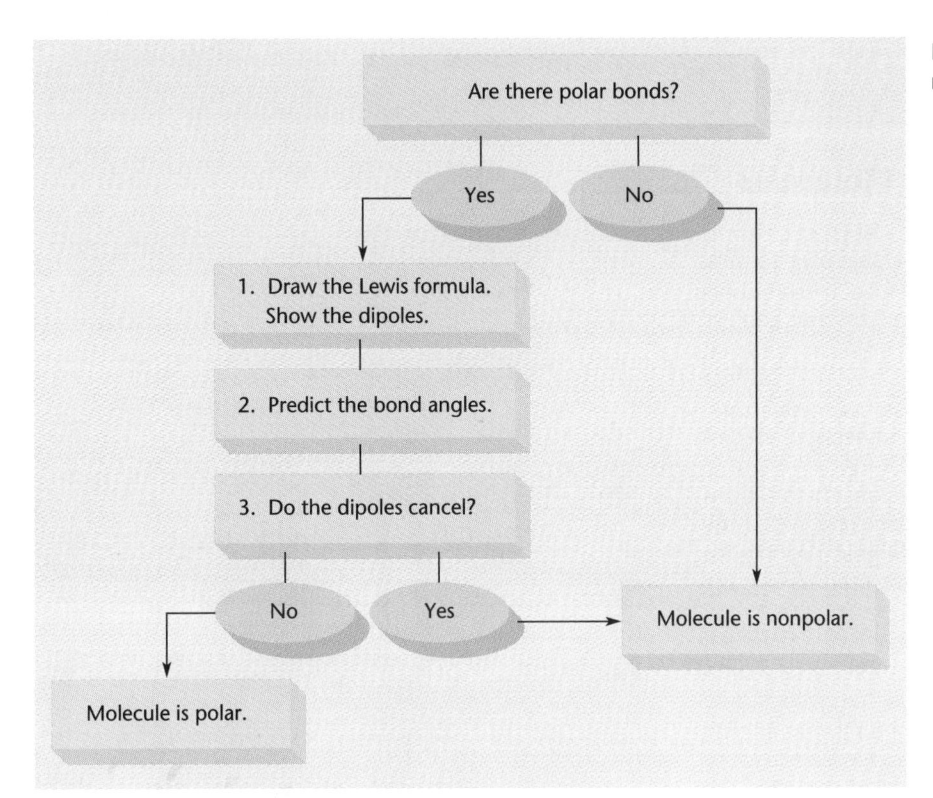

FIGURE 8.10 Identifying polar molecules.

Ammonia, NH_3, is a common fertilizer and an ingredient in household cleaning solutions. N—H bonds are polar, and the arrows do not cancel because of the irregular shape of the molecule. There are four sets of electrons including the nonbonding pair, and we can predict the bond angles to be approximately 109°. The shape is described as pyramidal because the four atoms occupy the corners of a pyramid having triangular faces. Both the presence of polar bonds and the shape of the molecule are factors contributing to the polarity of an ammonia molecule.

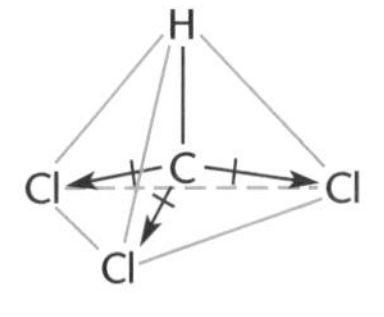

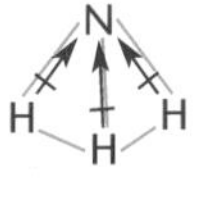

$CHCl_3$ and NH_3 are polar molecules because their dipoles do not cancel.

Knowing whether or not a substance is polar helps us to understand its physical and chemical properties. Table 8.3 summarizes some of the physical properties

TABLE 8.3 Typical Properties of Nonpolar Covalent, Polar Covalent, and Ionic Compounds

	Type of Compound		
Property	*Nonpolar Covalent*	*Polar Covalent*	*Ionic*
Melting point	low	higher	very high
Boiling point	low	higher	very high
Vapor pressure[a]	high	low	negligible
Water solubility	insoluble	soluble[b]	soluble[c]

[a]See Section 13.3.
[b]Compounds that have high molar mass may not be soluble.
[c]Many ionic compounds are insoluble for other reasons.

CHEMICAL WORLD

Microwaves and Polar Molecules

You're late! You wrap a couple of rashers of bacon in a paper towel, toss them in the microwave oven, and push the buttons. A few minutes later the bell rings. You take out the lukewarm package and, as usual, burn your fingers on the hot bacon as you put it on bread. Wait a minute! You've done this a hundred times and never thought about it, but now you're taking a science class and such phenomena are beginning to catch your attention.

How can the towel be only warm and the bacon hot? An ordinary oven would be a smoking mess. How do microwaves cook the bacon without heating the towel? Here's a hint. Maybe you can figure out the answer, or at least part of it.

Your laboratory instructor (not you without supervision and plenty of ventilation) fills a flask with water, another flask with octane, a third with carbon tetrachloride, and a fourth with isopropyl alcohol. The instructor places all four in a microwave oven in a laboratory exhaust hood and turns it on. After a while, the water and isopropyl alcohol are hot—but the octane and carbon tetrachloride are relatively very cool (and so is the air in the oven). What is the difference between these pairs of molecules? Think a moment with your new knowledge before you read on.

Right! Carbon tetrachloride and octane are nonpolar molecules, whereas water and isopropyl alcohol are polar molecules. The molecules in air are also nonpolar.

So now you know that microwave ovens heat polar substances, but how do they do this? Microwaves, like all forms of electromagnetic radiation (including x-rays and visible, infrared, and ultraviolet light), can be thought of as oscillating waves that alternately have positive and negative charges. These alternating charges of microwaves interact with the positive and negative charges on polar molecules and make the polar molecules move faster. Their faster motion—that is, their greater average kinetic energy—means (as you probably know) that the polar molecules are hotter.

Microwave radiation has other uses. Chemists are developing techniques that use microwaves to eliminate air pollution caused by toxic polar gases

Toxic substances in industrial smoke can be decomposed by microwave radiation.

such as sulfur dioxide and nitrogen dioxide, and by waste products of industries such as petroleum refining and electric power generation. Such molecules can be excited by microwaves to such a degree that they decompose into their nontoxic elements before they leave their smoke stack. For example, microwaves can decompose hydrogen sulfide into hydrogen and sulfur, both of which can be recovered and sold to offset pollution-control costs. Microwaves also can decompose nitrogen dioxide into nitrogen and oxygen, the main components of air.

Here's a question that you can answer with a small amount of thought and research, or an experiment at home. Is your microwaved bacon fried, boiled, or both?

of nonpolar covalent, polar covalent, and ionic compounds. The polarity of molecules affects intermolecular forces (Section 13.2), which in turn affect the physical properties of a substance. The relationship between physical properties and intermolecular forces is considered in more detail in Chapter 13.

Example 8.14

Which of the following molecules are polar and which are nonpolar?

a. CS_2
b. CH_2Cl_2
c. C_2Cl_4 ■

Solutions

a. S=C=S — *Nonpolar* molecule. C and S have the same electronegativity, and the bonds are nonpolar.

b.

```
     Cl
      ↖
H—C ↔ Cl
     /
    H
```

Polar molecule. This is a tetrahedral molecule, and the dipoles do not cancel.

c.

```
Cl       Cl
  ↖     ↗
   C=C
  ↙     ↘
Cl       Cl
```

Nonpolar molecule. This is a planar molecule (120° angles), and the dipoles cancel.

Practice Problem 8.8 Which of the following are polar molecules?

a. CH_3OH
b. C_2H_6
c. C_2H_3F

■ Chapter Review

Directions: From the list of choices at the end of the Chapter Review, choose the word or term that best fits in each blank. Use each choice only once. The answers appear at the end of the chapter.

Section 8.1 Chemical __________ (1) are attractive forces that maintain the definite composition of a compound.

Sections 8.2, 8.3 The attractive force between the oppositely charged ions of a compound is called an __________ (2) bond. Ionic bonding occurs when a __________ (3) is bonded with a nonmetal. Ionic bonding is __________ (4), in contrast to a covalent bond which is directed between two specific atoms. A __________ (5) bond is an attractive force between two atoms of a molecule that is characterized by their __________ (6) of a pair of electrons. Covalent bonds join together atoms that are __________ (7).

The energy needed to break a covalent bond is called its __________ __________ (8) energy. The strengths of different bonds vary widely. The formation of a covalent bond is an __________ (9) process. Bond breaking, however, is __________ (10).

Section 8.4 The most common type of structural formula is called a __________ (11) formula. This kind of formula uses a line to represent a covalent bond and dots to represent __________ (12) valence electrons. We can predict the number of bonds for an atom from its __________ (13) number in the __________ (14) table and the __________ (15) rule. When two or more Lewis formulas can be written for a substance that differ only in the position of __________ (16), there is __________ (17). Because it is __________ (18), oxygen (O_2) has __________ (19) electrons and is an exception to the octet rule.

Section 8.5 The capacity of an atom to attract its bonding electrons is called its __________ (20). __________ (21) have low values compared with nonmetals. The most electronegative element is __________ (22).

Section 8.6 A covalent bond joining atoms of different electronegativities is said to be __________ (23). This kind of bond is characterized by __________ (24) sharing of electrons. The polarity of a bond is shown in the formula using partial charges or by drawing an __________ (25) pointing toward the more electronegative atom.

Section 8.7 Bond __________ (26) and the __________ (27) of a molecule can be predicted using the __________ (28) theory. For example, when an atom in the center of a molecule has four single bonds, the shape is __________ (29) and the bond angle is approximately __________ (30).

Section 8.8 Molecules made up of only nonpolar bonds must be __________ (31). However, some molecules that have polar bonds are also nonpolar. To predict whether or not a molecule is polar, you must determine if the __________ (32) cancel.

Choices: angles, arrow, bond dissociation, bonds, covalent, dipoles, electronegativity, electrons, endothermic, exothermic, group, fluorine, ionic, Lewis, metal, metals, multidirectional, nonbonding, nonmetals, nonpolar, 109°, octet, paramagnetic, periodic, polar, resonance, shape, sharing, tetrahedral, unequal, unpaired, VSEPR

Study Objectives

After studying Chapter 8, you should be able to:

Section 8.2

1. Identify compounds that are ionic.
2. Describe the crystal structure for sodium chloride.

Section 8.3

3. Describe the bonding for a compound as ionic or covalent.
4. Describe the kind of energy changes that occur when bonds are formed and when bonds are broken.
5. Describe differences in bond length and in the strength of single, double, and triple bonds.

Section 8.4

6. Draw Lewis structures for simple molecules and ionic compounds.
7. Identify several substances that are exceptions to the octet rule.
8. Draw resonance structures to represent simple molecules and ions.

Section 8.5

9. Identify the most electronegative element.
10. Using the periodic table, tell which atom of a pair has greater electronegativity.

Section 8.6

11. Specify bonds as nonpolar covalent, polar covalent, or ionic.
12. Using appropriate notation, label polar covalent bonds in the Lewis formula of a compound.

Section 8.7

13. Predict the bond angles and the shape of a simple molecule using VSEPR theory.

Section 8.8

14. Predict whether molecules are polar or nonpolar.

Key Terms

Review the definition of each of the terms listed here by chapter section number. You may use the glossary if necessary.

8.1 chemical bond

8.2 ionic bond, electrolyte

8.3 covalent bond, bond dissociation energy, double bond, single bond, triple bond

8.4 Lewis formula, paramagnetic, resonance, resonance structures

8.5 electronegativity

8.6 dipole, nonpolar covalent bond, polar covalent bond

8.8 nonpolar molecule, polar molecule

Questions and Problems

The answers to questions and problems with blue numbers appear in Appendix C. Asterisks indicate the more challenging problems.

Ionic and Covalent Bonds (Sections 8.1, 8.2, 8.3)

1. What are the two main types of chemical bonds?
2. What is another term for a molecular compound?
3. Explain what is meant by an ionic bond; by a covalent bond. What feature of ionic bonding helps to distinguish it from a covalent bond?
4. In a crystal of sodium chloride, how many chloride ions are attracted to one sodium ion? How many sodium ions are attracted to one chloride ion?
5. What factors affect the packing arrangement in an ionic crystal?
6. What kind of bonding is found in a compound made up of a metal and a nonmetal?
7. What are some characteristic physical properties of ionic compounds?
8. What kind of bonding is found in molecules?
9. What kind of bonding joins two or more nonmetals?
10. Identify the following compounds as either ionic or covalent:
    a. NaF d. CaO
    b. $CaCl_2$ e. CuS
    c. CCl_4 f. CS_2
11. Identify the bonding as ionic or covalent:
    a. KBr d. PH_3
    b. N_2 e. $SOCl_2$
    c. BaS f. FeO
12. Identify which of the following compounds have ionic bonds, covalent bonds, or (possibly) both.
    a. $Ni(OH)_2$ c. NH_3
    b. CH_4 d. HNO_3
13. When a sample of chlorine is exposed to ultraviolet light, chlorine molecules dissociate into chlorine atoms:

    $Cl_2 \longrightarrow 2Cl$

    Is this reaction exothermic or endothermic?
14. Which is more stable? Explain.
    a. Cl_2 or 2Cl
    b. 2H or H_2
    c. HCl or H and Cl
15. Which is more stable? Explain.
    a. CCl_4 or C and 4Cl
    b. CCl_4 or Cl and CCl_3
16. What kinds of atomic orbitals overlap in the formation of the following bonds?
    a. Cl—Cl c. H—Cl
    b. H—H d. H—I

Lewis Formulas (Section 8.4)

17. What is the most commonly used type of structural formula?
18. Draw Lewis formulas for the following:
    a. HBr d. H_2S
    b. SCl_2 e. C_2H_2
    c. $POCl_3$ f. Na_2S
19. Represent each of the following using Lewis notation:
    a. NH_4Cl (ionic) d. KOH
    b. AsH_3 e. SeO_2
    c. NCl_3 f. $MgBr_2$
20. Epsom salts, hydrated magnesium sulfate ($MgSO_4$), is used for soaking infected toes and fingers. Use Lewis notation to represent $MgSO_4$.
21. Baking soda, $NaHCO_3$, is used in recipes to cause dough to rise. Use Lewis notation to represent $NaHCO_3$. (Note that hydrogen is bonded to oxygen in this compound.) Draw resonance structures for the anion.
22. Sodium carbonate (Na_2CO_3), or washing soda, is a detergent additive. Draw resonance structures for the anion.
23. Sulfur dioxide, SO_2, is a serious air pollutant resulting from the burning of fuels containing sulfur compounds. Draw resonance structures for SO_2.
24. Draw the Lewis formula for NO_2, a gas that gives smog its brown color. How many valence electrons surround the nitrogen atom in a NO_2 molecule?
25. Oxygen (O_2) is paramagnetic, but nitrogen (N_2) is not. What structural feature accounts for these observations?

Electronegativity, Polar Covalent Bonds (Sections 8.5, 8.6)

26. What is meant by electronegativity?
27. Which is the most electronegative element?
28. Which is more electronegative:
    a. H or S? c. Si or C?
    b. Se or O? d. H or P?

29. Identify the following bonds as polar or nonpolar. Refer to Table 8.2 as necessary.
   a. H—F c. H—O
   b. C—H d. N—Cl
30. Which of the following bonds are polar and which are nonpolar?
   a. N—O c. C=O
   b. C—C d. N≡N
31. A N—Cl bond is practically nonpolar even though two different atoms are joined together. Explain.
32. A C—Cl bond is polar, but a C—I bond is practically nonpolar. Explain.
33. Which of the following bonds in each pair is more polar?
   a. H—F or H—Cl d. N—O or C—O
   b. S—O or C—O e. P—Cl or C—Cl
   c. N—H or O—H

Shapes of Molecules (Section 8.7)

34. Predict the bond angles for the following molecules:
   a. CH_3Cl c. C_2H_4
   b. H_2S d. PCl_3
35. Identify the shape of each of the following as linear, trigonal, or tetrahedral:
   a. CH_2Cl_2 c. CS_2
   b. C_2H_2 d. CH_2O
36. Draw Lewis formulas for sulfur trioxide, SO_3, and sulfite ion, SO_3^{2-}. What are the shapes of these two substances? Why are the shapes different?
37. Hydrogen cyanide, HCN, is the lethal gas used to execute criminals in prison gas chambers. Draw its Lewis formula, show the bond angles, and describe its shape.
38. Limestone is calcium carbonate, $CaCO_3$, an ionic compound. Draw formulas for the cation and anion using Lewis notation. Show the bond angles for the anion, and describe its shape.

Polar Molecules (Section 8.8)

39. Carbon disulfide, CS_2, is a solvent that must be handled with great care because it is toxic and flammable. Is the molecule polar? Why or why not? Why is it not necessary to draw its Lewis formula to determine whether or not it is polar?
40. NF_3 is a polar compound. Draw its Lewis formula and explain.
41. Sulfur dioxide is polar, but carbon dioxide is not. Draw Lewis formulas for these compounds and explain.
42. H_2O is very polar, but H_2S (the odor of rotten eggs) is only slightly polar. Draw Lewis formulas for these compounds and explain.
43. Propane, C_3H_8, is a bottled gas used as a fuel for camp stoves, lanterns, and home heating in rural areas where natural gas is not available. Draw the Lewis formula for propane. Is it polar? Explain.
44. Chlorine is used to treat water in swimming pools and for domestic uses. When chlorine dissolves in water, some hypochlorous acid (HOCl) is formed. Draw the Lewis formula for this compound and describe its shape. Is it polar or nonpolar?

Additional Problems

*45. Three different Lewis formulas can be drawn for $C_2H_2Cl_2$. Draw these formulas and identify which is/are polar and which is/are nonpolar. Explain why these are not resonance structures. (Note that these compounds have the same empirical formula but different structural formulas and are called *isomers*.)
46. Draw a Lewis formula for SF_6 showing the shape of the molecule (Figure 8.9). How many valence electrons surround sulfur in this formula?
47. Draw Lewis formulas for the following. For each, two different arrangements of the atoms are possible.
   a. C_2H_6O
   b. C_3H_6 (Hint: One formula is a ring.)
48. Identify which of the following compounds have ionic bonds, covalent bonds, or (possibly) both?
   a. $POCl_3$ c. $CaSO_4$
   b. $Fe(NO_3)_3$ d. P_2S_5
49. Draw the Lewis formula for $SiCl_4$. What are the bond angles? Describe the shape of the molecule. Is a $SiCl_4$ molecule polar?
50. Draw Lewis formulas for $CHCl_3$ and CHF_3. What is the shape of these molecules? Which is more polar? Explain the difference in polarities.
51. Draw Lewis formulas for CCl_4 and CBr_4. Chlorine is more electronegative than bromine, causing a C—Cl bond to be more polar than a C—Br bond, yet both compounds are nonpolar. Explain.
52. Oxalic acid, $H_2C_2O_4$, is used to remove rust stains from plumbing fixtures. In the Lewis formula, each H is bonded to an O, and the C's are bonded to each other. Draw the Lewis formula.

Solutions to Practice Problems

PP 8.1

ionic (metal + nonmetal): a, d, e
covalent (only nonmetals): b, c

PP 8.2

Double bonds are stronger and shorter than single bonds.
a. C=O is stronger and shorter.
b. N=O is stronger and shorter.

PP 8.3

a. **Step 1** nitrogen (Group 5A) $5\ e^- \times 1$ atom $= 5\ e^-$
hydrogen (element 1) $1\ e^- \times 3$ atoms $= 3\ e^-$
total $= 8\ e^-$

Step 2 Predict: 3 bonds for nitrogen (Group 5A)
1 bond for each hydrogen (limited by the capacity of the first shell—2 e^-)

Step 3 The most symmetrical structure would have nitrogen as the central atom:

```
   H
   |
H—N—H
```

Step 4 Both hydrogen and nitrogen are shown with their predicted number of bonds, so no bonds should be added.

Step 5 Add 1 pair of dots to N (octet rule).

```
   H
   |
H—N—H
  ..
```

Step 6 Check valence electrons. Because the formula has 8 e^-, it checks.

b. **Step 1** hydrogen (element 1) $1\ e^- \times 1$ atom $= 1\ e^-$
carbon (Group 4A) $4\ e^- \times 1$ atom $= 4\ e^-$
nitrogen (Group 5A) $5\ e^- \times 1$ atom $= 5\ e^-$
total $= 10\ e^-$

Step 2 Predict: 3 bonds for nitrogen (Group 5A)
4 bonds for carbon (Group 4A)
1 bond for hydrogen (limited by the capacity of the first shell—2 e^-)

Step 3 Carbon, with more bonds, is the central atom:

H—C—N

Step 4 Add two bonds to C – N to give nitrogen 3 bonds and carbon 4 bonds (see Step 2):

H—C≡N

Step 5 Add 1 pair of dots to N (octet rule):

H—C≡N:

Step 6 Check valence electrons. Because the formula has 10 e^-, it checks.

PP 8.4

a. Represent the cation and anion separately. Remember, the cation has no valence electrons: Na^+. Follow the steps outlined previously in writing the Lewis formula of the anion. Because the cation has a +1 charge, the anion is −1.

Step 1 Calculate the number of valence electrons:
sulfur (Group 6A) $6\ e^-$
hydrogen $1\ e^-$
−1 charge (add 1 e^-) $1\ e^-$
total $= 8\ e^-$

Step 2 Predict 1 bond for H and 2 bonds for S (Group 6A).

Step 3 Join the atoms with a single bond:

S—H

Step 4 Because one bond for H agrees with what is predicted, do not add another bond. (Hydrogen can have only one bond.)

Step 5 Add 3 pairs of dots (octet rule):

:S̤̈—H

Step 6 Check electrons. Because the formula has 8 electrons, it checks. Remember to show the charge on the ion. The Lewis structure for NaSH is

Na^+ [:S̤̈—H]$^-$

b. The cation is K^+, and the anion is NO_3^- (sum of ionic charges equals zero). To write the Lewis structure of the anion:

Step 1 nitrogen (Group 5A) $5\ e^- \times 1$ atom $= 5\ e^-$
oxygen (Group 6A) $6\ e^- \times 3$ atoms $= 18\ e^-$
−1 charge (add 1 e^-) $= 1\ e^-$
total $= 24\ e^-$

Step 2 Predict: 3 bonds for N (Group 5A)
2 bonds for O (Group 6A)

Step 3 Arrange the atoms for symmetry and connect with single bonds:

```
O—N—O
  |
  O
```

Step 4 Because the number of bonds for N in the formula agrees with the number predicted, no additional bonds are added. Go to Step 5.

Step 5 Add pairs of dots to give all but the central atom (N in this case) an octet of electrons. Counting electrons shows 24 e^-, the correct number. To give nitrogen an octet, change a pair of dots to a bond:

```
:Ö—N—Ö:          :Ö—N=Ö:
 ¨  |  ¨   changes to   ¨  |
   :O:                   :O:
    ¨                     ¨
```

Include the charge on the ion:

```
[:Ö—N=Ö:]⁻
   ¨  |
     :O:
      ¨
```

Step 6 Check valence electrons. The formula shows $24e^-$. It checks. The Lewis structure for potassium nitrate is

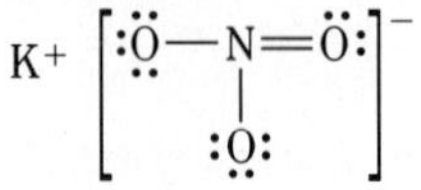

$$K^+ \left[:\ddot{O}-N=\ddot{O}: \atop \quad | \atop \quad :\underset{\cdot\cdot}{O}: \right]^-$$

Note that NO_3^- can be represented by resonance:

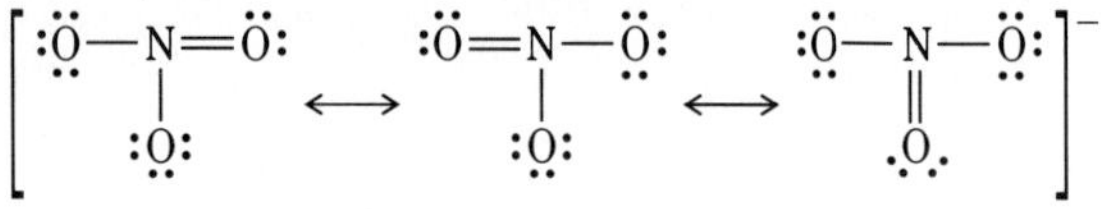

$$\left[:\ddot{O}-N=\ddot{O}: \;\longleftrightarrow\; :\ddot{O}=N-\ddot{O}: \;\longleftrightarrow\; :\ddot{O}-N-\ddot{O}: \right]^-$$

(in each structure N also bears a third O: single-bonded $:\underset{\cdot\cdot}{O}:$ in the first two, double-bonded O in the third)

PP 8.5

Remember, electronegativity increases from left to right and from bottom to top in the periodic table.

a. Cl
b. They have the same electronegativity
c. N
d. Br

PP 8.6

If the difference in electronegativities of the atoms is greater than 0.4 (Figure 8.8), the bond is polar. Refer to Table 8.2

a. polar: $C \mapsto O$
b. polar: $N \leftarrow\!\!+ H$
c. nonpolar

PP 8.7

Draw the Lewis formula and count the number of sets of electrons around the central atom.

a. $H-C(:\ddot{Cl}:)_3$ — H—C bonded to three :Cl: atoms (four sets of e^- around C; 109°, tetrahedral)

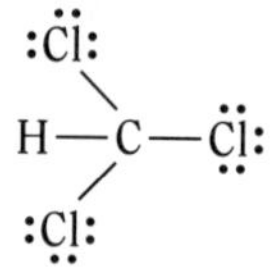

b. $H-C\equiv C-H$ (two sets of e^- around each C; 180°, linear)

c. $H_2C=CH(:\ddot{Cl}:)$ — H, H on one C; H and :Cl: on the other C (three sets of e^- around each C; 120°, planar)

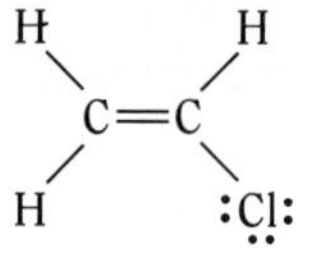

PP 8.8

a. $H_3C \mapsto \ddot{O}: \leftarrow\!\!+ H$ — C bonded to three H and to O; O bonded to H

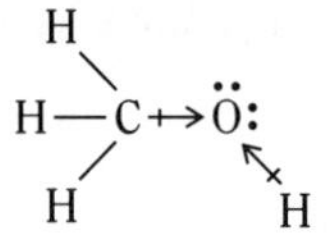

The bonds are tetrahedral around carbon and bent around oxygen. The dipoles do not cancel, leaving a *polar* molecule.

b. C—H bonds are practically nonpolar, and the C—H dipoles are very small. The bonds are tetrahedral, resulting in a *nonpolar* molecule:

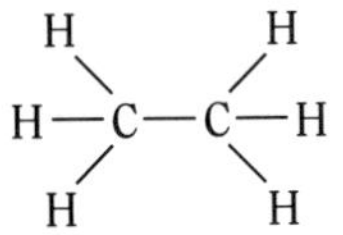

H_3C-CH_3 (H—C—C—H with two further H on each C)

c. $H_2C=CH(\mapsto :\ddot{F}:)$ — H, H on one C; H and :F: on the other C

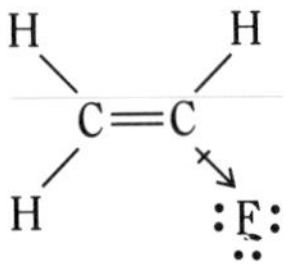

This is a planar molecule (120° angles), C—H bonds are practically nonpolar, C—F dipole causes the molecule to be *polar*.

Answers to Chapter Review

1. bonds
2. ionic
3. metal
4. multidirectional
5. covalent
6. sharing
7. nonmetals
8. bond dissociation
9. exothermic
10. endothermic
11. Lewis
12. nonbonding
13. group
14. periodic
15. octet
16. electrons
17. resonance
18. paramagnetic
19. unpaired
20. electronegativity
21. metals
22. fluorine
23. polar
24. unequal
25. arrow
26. angles
27. shape
28. VSEPR
29. tetrahedral
30. 109°
31. nonpolar
32. dipoles

9

Names and Formulas of Inorganic Compounds

Contents

A mushroom cloud of ash and volcanic gas billows into the atmosphere from Mt. Pinatubo in the Philippines on June 13, 1991. Cities and villages within 100 kilometers of the volcano were buried in knee-deep ash. Many people suffered severe respiratory irritation from breathing air contaminated with sulfur dioxide and a mist of sulfuric acid.

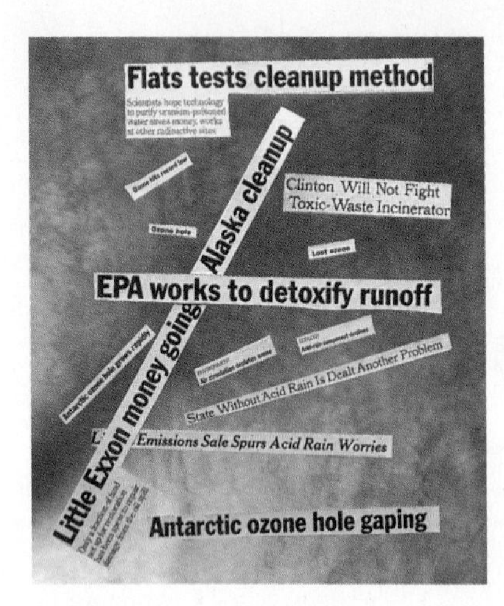

Chemical substances are in the news.

You Probably Know . . .

- Chemical terms appear in the news almost daily: acid rain, air pollutants such as carbon monoxide and nitrogen oxides, cholesterol in foods, nitrite preservatives in meats, alcohol in gasoline (gasohol), and so on. Knowing the chemical language will help us to understand the news.
- You have learned the common names of many ordinary chemical substances such as water, table salt, baking soda, and aspirin. A common name usually doesn't tell us much about the composition of a compound, but it sometimes gives a hint regarding how a compound is used.
- A chemical name, such as sodium chloride or carbon dioxide, usually tells us what elements make up the compound and sometimes in what ratio.

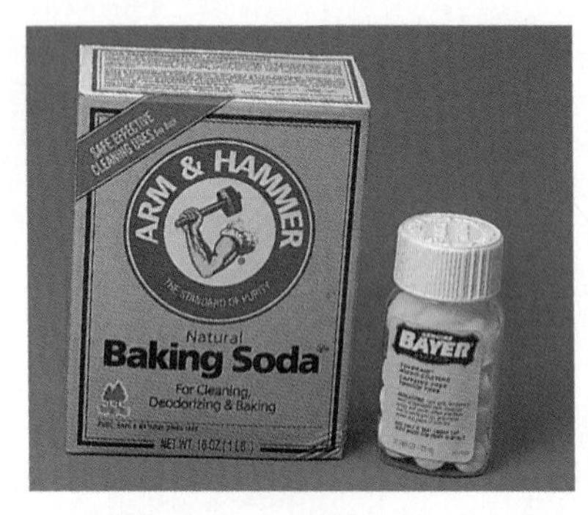

Baking soda and aspirin are common names.

Carbon dioxide
|
(one) carbon and two oxygen *atoms*

Why Should We Study Chemical Names and Formulas?

As modern technology has developed in this century, consumer products produced from chemicals have become commonplace. Plastics, fibers, rubber products, building materials, paints, drugs, cosmetics, fuels, and foods are all chemical substances, and most of these are the products of chemical technology. Furthermore, in the manufacture of these products, chemicals have been released into the environment. In our efforts to be informed consumers and to become knowledgeable about environmental issues, we are increasingly confronted with chemical terminology.

Chemistry, like any technical subject, has its own language. All languages have some kind of organization that includes sentence structure, rules for grammar, roots for words, prefixes, suffixes, and so on. When you know the rules of a language, you are able to decide what is correct and what is incorrect as you speak and write. But languages also include words and colloquial expressions that do not fit their rules. These words and expressions have to be memorized and practiced as you talk and listen to others. Because of the rules, the vocabulary, and the variations in a language, learning a language requires concentrated effort.

As you begin your study of the names and formulas of inorganic compounds,* think of it as similar to studying a foreign language. First, you must learn some rules. Then you need to apply these rules as you learn to read, write, and speak the names and formulas of compounds. You must also memorize some of the common names of substances that do not follow the rules. These are often based on a property or a use of the substance. Both

*Organic (carbon) compounds have their own system of nomenclature and are not included in this discussion.

systematic names (they follow the rules) and common names are used by chemists today.

Chemical Names

Names and formulas of compounds are the core of the language of chemistry. The rules for this language are set by the *International Union of Pure and Applied Chemistry* **(IUPAC).** The purpose of the IUPAC is to systematize chemical names. *Systematic names* tell us information about the composition of the compound. For example, when you hear the name sodium chloride, you know the compound is composed of sodium and chlorine.

In spite of efforts to systematize the names of compounds, common names of many substances continue to be widely used. This is particularly true for those substances that are used by people who have not studied the IUPAC system and would not recognize the systematic names. For example, water is the name we all use for H_2O. You probably learned this word when you first learned to talk. Your parents didn't teach you any IUPAC rules before you learned the word. You simply associated the word with the liquid that you drank and used for washing. In a similar way, you have learned other names such as aspirin, sugar, alcohol, and cholesterol. The common names of some widely used compounds are listed in Table 9.1.

Ordinary kitchen chemicals include *(left to right)* salt, NaCl; sugar, $C_{12}H_{22}O_{11}$; baking soda, $NaHCO_3$; and vinegar, an aqueous solution of acetic acid, $HC_2H_3O_2$.

TABLE 9.1 Common and Systematic Names for Common Compounds

Formula	*Common Name*	*Systematic Name*
NaCl	salt	sodium chloride
$C_{12}H_{22}O_{11}$	sugar	sucrose
C_2H_5OH	alcohol	ethanol
NaOH	lye	sodium hydroxide
$NaHCO_3$	baking soda	sodium hydrogen carbonate
$CaCO_3$	marble, limestone, chalk	calcium carbonate
CaO	lime	calcium oxide
NH_3	ammonia	ammonia[a]
$HC_2H_3O_2$	acetic acid (vinegar)	ethanoic acid
C_3H_6O	acetone	propanone
CH_2O	formaldehyde	methanal
HCl(aq)	muriatic acid	hydrochloric acid
$NaAl(SO_4)_2 \cdot 12H_2O$	alum	sodium aluminum sulfate dodecahydrate
$CaSO_4 \cdot 2H_2O$	gypsum	calcium sulfate dihydrate

[a]Ammonia is the officially accepted IUPAC name.

Alum, $NaAl(SO_4)_2 \cdot 12H_2O$, forms octahedral crystals. It is used in making dill pickles and in treating mouth sores.

9.2 Oxidation Numbers

Learning Objective

- Assign oxidation numbers for uncombined and combined elements.

Previously, when we considered electron transfer or redox reactions (Section 6.8), we saw that the oxidation state of an element can vary. The oxidation state of an element is expressed by a positive or negative number, or zero, called its **oxidation number.** Oxidation numbers can be used to explain or predict properties of an element. They also can be used to help us write and balance redox equations (Chapter 17). Oxidation numbers also tell us the combining capacity or **valence** of an element, information needed in writing formulas of compounds. Furthermore, oxidation numbers are used in the names of some compounds.

The following rules can be used in assigning oxidation numbers.

In formulas of ions, charges are written +, −, 2+, 2−, and so on, as superscripts. However, oxidation numbers are written +1, −2, −3, and so on.

Rule 1 The oxidation number of an uncombined element is zero whether it exists naturally as atoms, as diatomic molecules, or as larger molecules. (Examples: Na, Ca, Fe, N_2, O_2, P_4)

Rule 2 The oxidation number of a monatomic ion is the same as the charge on the ion. The charge on the ion of a representative element (A-group) can be determined from the periodic table. See Table 9.2. (Examples: Na^+, Ca^{2+}, S^{2-}, Cl^-)

Rule 3 The oxidation number for oxygen in a compound is usually −2.* (Examples: H_2O, CaO, Na_2O)

Rule 4 The oxidation number for hydrogen in a compound is +1 (e.g., H_2O, HCl) except in metal hydrides where it is −1 (e.g., NaH, AlH_3).

Rule 5 The oxidation number of a halogen (Group 7A) is −1 in binary compounds (two elements), both covalent and ionic, *when bonded to a less electronegative element.*

"Polyatomic" means made up of several atoms (Section 8.4).

Rule 6 For a compound, the sum of the oxidation numbers equals zero. For a polyatomic ion (Section 8.4), the sum of the oxidation numbers equals the charge on the ion.

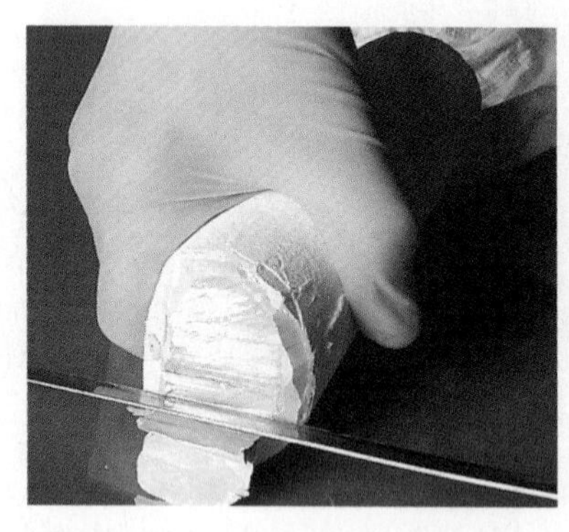

When sodium is exposed to air, it oxidizes to give a surface coating that contains Na^+ ions. All the alkali metals (Group 1A) form +1 ions.

Now let's look at some examples of each rule. Rule 1 applies to all uncombined elements whether they exist naturally as atoms, as diatomic molecules such as N_2, or as larger molecules such as P_4 or S_8. We sometimes write an oxidation number (zero for an element) above the symbol of the element:

Oxidation number:	0	0	0	0	0	0
Element:	K	Fe	N_2	H_2	P_4	S_8

The oxidation number of a monatomic ion (Rule 2) of an element in an A-group can be determined from the group number (Table 9.2). Thus, all alkali metal ions (Group 1A) have an oxidation number of +1, and alkaline earth metal ions (Group 2A) are +2. The oxidation number of a monatomic anion equals −(8

*An exception is the oxygen in a peroxide such as hydrogen peroxide, H_2O_2, which has an oxidation number of −1. Peroxides have an O—O bond, and all are unstable.

TABLE 9.2 Oxidation Numbers (Charges) of Monatomic Ions by Group Number

Group	Oxidation Number
1A	+1
2A	+2
3A	+3
4A	—[a]
5A	−3
6A	−2
7A	−1
8A	0

[a]Of the 4A elements, carbon and silicon form covalent bonds. Tin and lead form +2 cations, but in covalent compounds they are +4.

In rust (Fe_2O_3), iron has an oxidation number of +3.

− group number). For example, the oxidation number for a halide ion (Group 7A) is −1 (e.g., Br^-), and for sulfide ion (Group 6A), it is −2 (S^{2-}).

Many elements have different oxidation numbers in different compounds. This is true for a number of metals, especially those in the *d*-block in the periodic table (Table 9.3), and for many of the nonmetals when they are covalently bonded (Table 9.4).

Using Rules 2–5 with Rule 6 allows us to identify the oxidation number of an element bonded with oxygen, hydrogen, or another element of known oxidation

TABLE 9.3 Oxidation Numbers (Charges) of Monatomic Ions*

1A	2A											3A	4A	5A	6A	7A	
H^+																H^-	
Li^+	Be^{2+}													N^{3-}	O^{2-}	F^-	
Na^+	Mg^{2+}											Al^{3+}		P^{3-}	S^{2-}	Cl^-	
K^+	Ca^{2+}		Ti^{3+} Ti^{4+}		Cr^{2+} Cr^{3+}	Mn^{2+} Mn^{3+}	Fe^{2+} Fe^{3+}	Co^{2+} Co^{3+}	Ni^{2+}	Cu^+ Cu^{2+}	Zn^{2+}					Br^-	
Rb^+	Sr^{2+}									Ag^+	Cd^{2+}					I^-	
Cs^+	Ba^{2+}									Au^+ Au^{3+}	Hg^{2+} Hg^{2+}		Pb^{2+}				

*Some metals that form two cations are highlighted in green.

TABLE 9.4 Oxidation Numbers of Nonmetals and Metalloids

	3A	4A	5A	6A	7A	8A
1 **H** +1, −1						2 **He**
	5 **B** +3	6 **C** +4, +2, −2, −4	7 **N** +5, +4, +3, +2, +1, −1, −2, −3	8 **O** +2, −1, −2	9 **F** −1	10 **Ne**
		14 **Si** +4, −4	15 **P** +5, +3, −3	16 **S** +6, +4, +2, −2	17 **Cl** +7, +6, +5, +4, +3, +1, −1	18 **Ar**
		32 **Ge** +4, −4	33 **As** +5, +3, −3	34 **Se** +6, +4, −2	35 **Br** +7, +5, +1, −1	36 **Kr** +4, +2
			51 **Sb** +5, +3, −3	52 **Te** +6, +4, −2	53 **I** +7, +5, +1, −1	54 **Xe** +6, +4, +2
					85 **At** −1	86 **Rn**

number. To determine an unknown oxidation number of an element in a compound:

1. Write the known oxidation number for each atom (Rules 2–5).
2. Multiply each oxidation number by the number of atoms of that element shown in the formula.
3. Write an equation, following Rule 6. Remember, the sum of the oxidation numbers is zero for a compound.

Example 9.1

What is the oxidation number of sulfur in SO_2? ■

Solution Note that the oxidation number of oxygen is -2 (Rule 3) and is *per atom of oxygen.* Multiply the oxidation number by 2, the number of oxygen atoms in the formula:

$$SO_2$$

Step 1 $\quad -2$

Step 2 $\quad (-2)2$

Step 3 $\quad S + (-4) = 0$

$$S = +4$$

The oxidation number of sulfur in SO_2 is $+4$.

Example 9.2

What is the oxidation number of carbon in carbonic acid, H_2CO_3? ■

Solution Oxidation numbers: Hydrogen is $+1$; oxygen is -2.

	H_2CO_3	
Step 1	$+1$	-2
Step 2	$2(+1)$	$(-2)3$
Step 3	$+2 + C - 6 = 0$	
	$C = +4$	

The oxidation number of carbon is H_2CO_3 is $+4$.

This strategy works for polyatomic ions as well, as shown in the following example. However, remember that the sum of the oxidation numbers is equal to the charge on the ion.

Example 9.3

Determine the oxidation number of phosphorus in the phosphate ion, PO_4^{3-}. ■

Solution The oxidation number of oxygen is -2. Remember that the sum of the oxidation numbers is equal to the ionic charge (Rule 6).

	PO_4^{3-}
Step 1	-2
Step 2	$(-2)4$
Step 3	$P + (-8) = -3$
	$P = +5$

The oxidation number of phosphorus in PO_4^{3-} is $+5$.

Example 9.4

What is the oxidation number of sulfur in hydrogen sulfate ion, HSO_4^-? ■

Solution Oxidation numbers: Hydrogen is $+1$; oxygen is -2.

	HSO_4^-	
Step 1	$+1$	-2
Step 2	$+1$	$(-2)4$
Step 3	$+1 + S + (-8) = -1$	
	$S = +6$	

The oxidation number of sulfur in HSO_4^- is $+6$.

These examples illustrate that the oxidation number of a covalently bonded nonmetal can be different from the charge on its monatomic anion (Table 9.4). In HSO_4^-, the oxidation number of sulfur is +6 compared with −2 for the sulfide ion, S^{2-}.

Example 9.5

Identify the oxidation numbers of the following:

a. P in PCl_3
b. S in H_2S
c. K in KCl
d. Cl in $HClO_4$ ■

Solutions

a. PCl_3

Step 1 −1

Step 2 (−1)3

Step 3 P + (−3) = 0

P = +3

b. H_2S

Step 1 +1

Step 2 (+1)2 = 0

Step 3 (+2) + S = 0

S = −2

c. K is always +1 in its compounds (Table 9.2).

d. $HClO_4$

Step 1 +1 −2

Step 2 +1 (−2)4

Step 3 +1+ Cl −8 = 0

Cl = +7

Practice Problem 9.1 Identify the oxidation number of:
a. P in H_3PO_3
b. C in CH_4
c. Br in BrO_3^-
(Solutions to practice problems appear at the end of the chapter.)

9.3 Classification of Compounds

Classifying compounds helps us to be systematic in our approach to naming them. One of the simplest means of classification is according to the number of elements making up the compound. Using this approach, the vast majority of inorganic compounds would fall into two classes, binary or ternary. A **binary compound** is made up of two elements, such as NaCl and $CaCl_2$, although the formula unit may have more than two atoms. A **ternary compound,** such as $NaNO_3$, is composed of three elements. These classifications are very broad, so they have been subdivided into smaller, more useful classifications that group compounds according to certain properties. Outlines of binary and ternary compounds are shown in Figures 9.1 and 9.2.

Binary Compounds
Covalent (two nonmetals)
Ionic (metal–nonmetal)
Acids
Nonacids
Salts
Miscellaneous (oxides, peroxides)
Metal ion (fixed charge)
Metal ion (two possible charges)

FIGURE 9.1 Binary compounds.

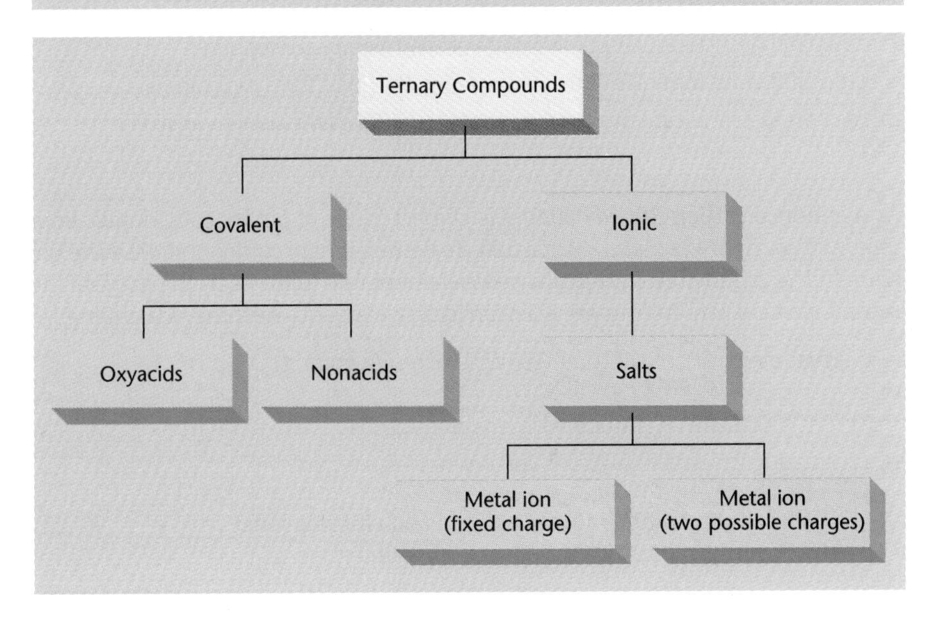

FIGURE 9.2 Ternary compounds.

General Rules for Naming Compounds

The systematic names and formulas of inorganic compounds follow two general rules:

1. In writing the name or formula of a compound, the cation or less electronegative element is written first followed by the anion or more electronegative element.*

*There are some exceptions to this rule. For example, in NH_3, nitrogen is more electronegative than hydrogen, yet N is written first.

TABLE 9.5 Suffixes for Names of Inorganic Compounds

Suffix	*Compound*
-ide	binary compounds (Section 9.5)
-ate, -ite	salts of oxyacids (Section 9.8)
-ic	binary acids (Section 9.6)
-ous/ic	oxyacids (Section 9.7)

2. The suffix of the name of the compound (*-ide, -ate,* etc.) identifies its classification.*

You must know these suffixes to be able to write formulas from the names of the compounds and vice versa. A summary of these endings is given in Table 9.5.

9.5 Binary Compounds

Learning Objectives

- Name and write formulas of ionic and covalent binary compounds.
- Name and write formulas of the cations of metals that form two cations.
- Use a prefix to indicate the number of atoms of an element in a molecule.

As mentioned earlier, the *-ide* ending is characteristic of binary compounds, both ionic and covalent (Sections 2.3 and 5.8). To name a binary compound, we write the name of the cation or less electronegative element first followed by the name of the second element (anion or more electronegative element) with the *-ide* ending:

Monatomic anions, -ide *ending (Section 2.3):*

Cl^-, chloride ion
S^{2-}, sulfide ion

(first element) + (second element with *-ide* ending)

or

(cation) + (anion with *-ide* ending)

For example:

NaCl: sodium chlor*ide* (ionic; sodium and chloride ions)
HBr: hydrogen brom*ide* (covalent compound)

Sodium chloride, ordinary table salt, is a typical salt.**

The names of metal cations (Table 9.6), except for metals that form two cations, are the same as the names of the metals. These include the cations formed from the 1A and 2A metals, Al^{3+}, and certain of the transition metal ions. However, the names of some polyatomic cations do not follow any particular pattern and must be memorized (e.g., NH_4^+, ammonium ion).

Remember that the name of a monatomic anion (Table 9.7) is formed by changing the ending of the name of the nonmetal to *-ide.* However, some *-ide* anions are diatomic, for example:

*With some compounds, a prefix also may be important in its classification. Some of these prefixes are mentioned later (Sections 9.7 and 9.8).

A **salt is an ionic compound that results when a metal ion or ammonium ion replaces a hydrogen of an acid (Section 9.8). Salts include most anions except OH^- and O^{2-}.

OH^-, hydroxide ion
CN^-, cyanide ion

These should be memorized.

TABLE 9.6 Some Common Cations

Charge +1		*Charge +2*		*Charge +3*	
Alkali metals		*Alkaline earths*		*Group 3A*	
Li^+	lithium	Be^{2+}	beryllium	Al^{3+}	aluminum
Na^+	sodium	Mg^{2+}	magnesium	Ga^{3+}	gallium
K^+	potassium	Ca^{2+}	calcium		
Rb^+	rubidium	Sr^{2+}	strontium		
Cs^+	cesium	Ba^{2+}	barium		
Transition metals		*Transition metals*		*Transition metals*	
Ag^+	silver	Cr^{2+}	chromium(II)	Cr^{3+}	chromium(III)
		Mn^{2+}	manganese(II)	Mn^{3+}	manganese(III)
		Fe^{2+}	iron(II)	Fe^{3+}	iron(III)
		Co^{2+}	cobalt(II)	Co^{3+}	cobalt(III)
		Ni^{2+}	nickel(II)		
Cu^+	copper(I)	Cu^{2+}	copper(II)		
		Zn^{2+}	zinc		
Au^+	gold(I)	Cd^{2+}	cadmium	Au^{3+}	gold(III)
Hg_2^{2+}	mercury(I)[a]	Hg^{2+}	mercury(II)		
		Group 4A[b]			
		Sn^{2+}	tin(II)		
		Pb^{2+}	lead(II)		
Polyatomic ions					
NH_4^+	ammonium				
H_3O^+	hydronium				

[a]Mercury(I) is a diatomic ion; its formula is Hg_2^{2+}, or $[Hg—Hg]^{2+}$. The overall ionic charge is +2, resulting in an oxidation number of +1 for each mercury atom.
[b]Tin and lead in the +4 state are covalently bonded. The +4 cations of tin and lead probably do not exist.

TABLE 9.7 Common Anions with *-ide* Ending

Charge −1		*Charge −2*		*Charge −3*	
Group 7A		*Group 6A*		*Group 5A*	
F^-	fluoride	O^{2-}	oxide	N^{3-}	nitride
Cl^-	chloride	S^{2-}	sulfide	P^{3-}	phosphide
Br^-	bromide				
I^-	iodide				
Others					
H^-	hydride	O_2^{2-}	peroxide		
OH^-	hydroxide				
CN^-	cyanide				

Example 9.6

Name the following compounds:

a. KI
b. MgS
c. LiF
d. $BaCl_2$ ■

Solutions Name the cation followed by the anion. Remember to use the *-ide* suffix for a binary compound.

a. potassium iodide
b. magnesium sulfide
c. lithium fluoride
d. barium chloride

Practice Problem 9.2 Name the following compounds:
a. $MgBr_2$
b. SrO
c. LiCl

Writing Formulas of Ionic Compounds

In ionic compounds, the sum of the charges of the cations and anions equals zero.

Regardless of the type of substance, the sum of the oxidation numbers of the atoms in a compound equals zero. It is also true that the sum of the charges of all the cations and anions making up an ionic compound must equal zero (Section 5.8). Using this idea, we can determine the ratio of cations and anions needed in the formula of the compound.

For example, consider the formula of calcium chloride, a salt made up of calcium ions (Ca^{2+}) and chloride ions (Cl^-). To have zero charge for the compound, the formula must show two Cl^- ions for each Ca^{2+} ion. Remember that the cation is written first in the formula and that a subscript of "1" is not shown.

$$\begin{array}{cc} Ca^{2+} & Cl^- \\ & Cl^- \end{array}$$

$$\mathbf{1}(+2) + \mathbf{2}(-1) = 0$$

Not written in the formula (arrow to 1); Subscript for Cl (arrow to 2)

Formula: $CaCl_2$

The steps in writing the formula of an ionic compound are:

1. Write the formula of the cation. Calcium (Group 2A) has a +2 charge: Ca^{2+} (Table 9.3).
2. Write the formula of the anion. Chloride (Group 7A) has a −1 charge: Cl^- (Table 9.3).
3. Write a charge equation, multiplying each ionic charge by the appropriate factor so that the sum equals zero. The factors are the subscripts for the cation and

anion in the formula. Remember that a subscript of "1" is not written in a formula.

In sodium oxide, the cation (Na^+) has a $+1$ charge and the anion (O^{2-}) has a -2 charge, resulting in a 2:1 ratio of Na^+ ions to O^{2-} ions in the formula:

$$\begin{array}{ll} Na^+ & O^{2-} \\ Na^+ & \end{array}$$

$$\mathbf{2}(+1) + \mathbf{1}(-2) = 0$$

Formula: Na_2O

The formula for magnesium nitride is a bit more complicated. To arrive at a net charge of zero for the compound, a ratio of three magnesium ions (Mg^{2+}) to two nitride ions (N^{3-}) is needed:

$$\begin{array}{ll} Mg^{2+} & N^{3-} \\ Mg^{2+} & N^{3-} \\ Mg^{2+} & \end{array}$$

$$\mathbf{3}(+2) + \mathbf{2}(-3) = 0$$

Formula: Mg_3N_2

In compounds involving polyatomic cations and anions, parentheses are used to indicate that a subscript applies to the ion rather than only to the preceding element. This is illustrated in the formula of calcium phosphate.

$$\begin{array}{ll} Ca^{2+} & PO_4^{3-} \\ Ca^{2+} & PO_4^{3-} \\ Ca^{2+} & \end{array}$$

$$\mathbf{3}(+2) + \mathbf{2}(-3) = 0$$

Formula: $Ca_3(PO_4)_2$

As you work the following examples, refer to the periodic table to determine charges on monatomic cations and anions of A-group elements. The formulas of many polyatomic anions are given in Table 9.10.

Example 9.7

Write the formula of aluminum chloride. ■

Solution Charges for monatomic ions of A-group elements can be determined from the group number (Table 9.2).

$$\begin{array}{ll} Al^{3+} & Cl^- \\ & Cl^- \\ & Cl^- \end{array}$$

$$\mathbf{1}(+3) + \mathbf{3}(-1) = 0$$

Formula: $AlCl_3$

a

b

c

Certain forms of alumina (Al_2O_3) with metal-ion impurities are valuable gems: (a) sapphire (Fe^{3+} and Ti^{4+} ions), (b) ruby (Cr^{3+} ion), (c) topaz (Fe^{3+}).

Example 9.8

Write the formula of aluminum oxide, also called alumina. ■

Solution Refer to the periodic table to determine ionic charges (also Tables 9.2 and 9.3).

Al^{3+}	O^{2-}
Al^{3+}	O^{2-}
	O^{2-}

$$\mathbf{2}(+3) + \mathbf{3}(-2) = 0$$

Formula: Al_2O_3

Practice Problem 9.3 Write the formulas for the following compounds:
a. potassium bromide
c. magnesium phosphate
b. strontium chloride
d. ammonium sulfide

■ Metals That Form Two Cations

Most of the transition metals form two cations, for example, Fe^{2+} and Fe^{3+}.

We have already noted that most metals found in the A-groups form only one cation. These include the metals in Groups 1A and 2A and aluminum in Group 3A (Tables 9.2 and 9.3).

However, most of the transition metals and some metals in the *p*-block form two cations having different charges (Table 9.6 and Table 9.8). Some of these are common metals such as iron and copper.

When writing the systematic name for a compound containing a metal that forms two cations, we place the charge (or oxidation number) of the cation in parentheses after the name of the metal (Table 9.6). Thus, the two copper ions can be named as

Cu^{+}: copper(I) ion
Cu^{2+}: copper(II) ion

Remember, the sum of the charges of the cations and anions equals zero in a compound. The charge on the cation can be determined by first identifying the charge on the anion and then solving the equation of ionic charges. For example, to

TABLE 9.8 Some Metals That Form Two Cations

Metal	*Cations*
copper	Cu^{+}, Cu^{2+}
mercury	Hg_2^{2+}, Hg^{2+}
iron	Fe^{2+}, Fe^{3+}
gold	Au^{+}, Au^{3+}

determine the charge on copper in CuCl, note that chloride (Cl^-) has a charge of -1, and the equation of ionic charges is

$$Cu - 1 = 0$$

The charge on copper in CuCl is $+1$, indicated by the Roman numeral I in the name of the compound.

CuCl: copper(I) chloride

Always put the Roman numeral in parentheses. There is no space between the name of the metal and the parenthesis.

Note the steps involved in writing the name:

When writing names, classify the compound and choose the suffix.

1. Classify the compound and determine the ending for the name. This is a binary compound, so use *-ide;* the name of the anion is chlor*ide.*
2. Determine the charges on the ions.
   a. Chloride ion (Group 7A), Cl^-
   b. Copper ion (write an equation of ionic charges)

   $$Cu - 1 = 0; \quad Cu = +1 \quad (Cu^+)$$
3. Write the names of the cation and anion and combine them to give the name of the compound:

   copper(I) chloride

Latin names for common metals:

Metal	Latin name
copper	*cuprum*
iron	*ferrum*
tin	*stannum*
lead	*plumbum*
gold	*aurum*
silver	*argentum*

The old system for naming the two cations of a metal uses Latin names and two endings: *-ous* for the smaller charge and *-ic* for the larger charge. For example, the Latin name for iron is *ferrum.* So the older names for iron(II) and iron(III) ions are *ferrous* and *ferric,* respectively.

$FeCl_2$: iron(II) chloride, ferr*ous* chloride
$FeCl_3$: iron(III) chloride, ferr*ic* chloride

Note, however, that the systematic names of the cations are preferred. The names of many of the common metal ions are shown in Table 9.6.

Latin names of metal cations:

	Suffix	
Metal	-ous (smaller charge)	-ic (larger charge)
copper	Cu^+, cupr*ous*	Cu^{2+}, cupr*ic*
iron	Fe^{2+}, ferr*ous*	Fe^{3+} ferr*ic*

Iron is the fourth most abundant element in the earth's crust. *Left:* An open pit mine. *Right:* Hematite, Fe_2O_3, an iron ore.

Example 9.9

Name the following compounds using systematic names.

a. $FeCl_3$

b. SnO

c. $HgBr_2$

d. Fe_2S_3 ■

Solutions These are binary compounds, so use the *-ide* suffix. Determine the charge on the metal ion from the charge on the anion. Combine the names of the cation and anion to give the name of the compound.

a. $FeCl_3$

$Fe + 3(-1) = 0$

$Fe = +3$

Name: iron(III) chloride

b. SnO

$Sn + (-2) = 0$

$Sn = +2$

Name: tin(II) oxide

c. $HgBr_2$

$Hg + 2(-1) = 0$

$Hg = +2$

Name: mercury(II) bromide

d. Fe_2S_3

$2(Fe) + 3(-2) = 0$

$2(Fe) = +6$

$Fe = +3$

Name: iron(III) sulfide

Example 9.10

Write the formulas for the following compounds:

a. gold(III) chloride

b. tin(II) sulfide

c. chromium(III) oxide ■

Solutions Write the formulas for the cation and the anion, write a charge equation to determine the subscripts, and write the formula.

a. Au^{3+} Cl^-
Cl^-
Cl^-

$(+3) + \mathbf{3}(-1) = 0$

$AuCl_3$

b. Sn^{2+} S^{2-}

$(+2) + (-2) = 0$

SnS

c. Cr^{3+} O^{2-}
Cr^{3+} O^{2-}
O^{2-}

$\mathbf{2}(+3) + \mathbf{3}(-2) = 0$

Cr_2O_3

Practice Problem 9.4 Give systematic names for the following compounds.

a. Hg_2Cl_2
b. Cu_2O
c. Fe_2O_3
d. $PbBr_2$

Practice Problem 9.5 Write formulas for the following compounds.

a. nickel(II) sulfide
b. chromium(III) bromide
c. tin(II) oxide

Binary Compounds of Two Nonmetals

Covalent binary compounds are made up of two nonmetals. Let's review what we discussed about these compounds in earlier chapters. The names for these compounds also use the characteristic *-ide* ending. This ending is used with the more electronegative of the two elements. Moreover, prefixes are used to indicate the numbers of each atom making up the molecule (Table 9.9). As with ionic binary compounds, we begin the process by classifying the compound and choosing the suffix:

1. Classify the compound and determine the suffix for the name. For binary compounds, use *-ide*.
2. Use prefixes to indicate the numbers of atoms in the molecule.
3. Write the name using the correct suffix and prefixes.

Use counting prefixes only with binary compounds of two nonmetals, not with ionic binary compounds.

TABLE 9.9 Counting Prefixes

Prefix	*Number of Atoms*
mono-	1
di-	2
tri-	3
tetra-	4
penta-	5
hexa-	6
hepta-	7
octa-	8
nona-	9
deca-	10

Carbon monoxide, a byproduct of incomplete combustion, is quite toxic. It is the cause of many accidental deaths and is a serious air pollutant.

The oxides of carbon are familiar examples:

CO_2: carbon *dioxide*
CO: carbon *monoxide*

Note that mono- is rarely used and is never used for the first element. Sometimes the vowel ending of the prefix is dropped when the stem of the name of the element begins with a vowel. No prefixes are used with hydrogen compounds such as HF, HCl, HBr, and H_2S.

Common names are used for some covalent binary compounds, and these must be memorized. Examples are water (H_2O) and ammonia (NH_3).

Naming Binary Compounds (suffix *-ide*)

Ionic (Metal + Nonmetal)		*Covalent (Two Nonmetals)*
1. Cation (fixed charge)	Cation (two possible charges)	Use counting prefixes
Name of metal	Name of metal + Roman numeral	
2. Anion, *-ide* suffix		

Example 9.11

Name the following compounds:

a. SO_2
b. P_4O_{10}
c. CS_2
d. N_2O_5 ■

Solutions The *-ide* suffix is used to name binary compounds, and because these compounds involve only nonmetals, you must use counting prefixes.

a. sulfur dioxide
b. tetraphosphorus decoxide
c. carbon disulfide
d. dinitrogen pentoxide

Practice Problem 9.6 Name the following compounds:

a. SCl_2
b. PCl_3
b. OF_2
d. $CaBr_2$

Binary Acids

9.6

Learning Objective

- Name and write formulas of binary acids.

An **acid** is a compound that ionizes in water to give hydrogen ions and anions. The formula of an acid usually has H written first. The following equations illustrate the ionization of aqueous HCl:

Ionize: To form ions.
Aqueous: Dissolved in water.

$$HCl \xrightarrow{\text{water}} H^+(aq) + Cl^-(aq)$$

or

$$HCl(aq) + H_2O(l) \rightarrow H_3O^+(aq) + Cl^-(aq)$$

Hydronium ion, H_3O^+, results from the association of H^+ with H_2O, so $H^+(aq)$ is an abbreviation for $H_3O^+(aq)$.

Both $H^+(aq)$ and $H_3O^+(aq)$ represent the hydronium ion.

Acids are reactive compounds and have many uses. They are found naturally in foods and have a sour taste. For example, citrus juices contain citric acid ($H_3C_6H_5O_7$) and vinegar contains acetic acid ($HC_2H_3O_2$). Acids are discussed in greater detail in Chapter 15.

A **binary acid** is an acid composed of two elements—hydrogen and an electronegative element. We can recognize the pattern for naming binary acids by considering the following example:

HCl(aq): *hydro*chlor*ic acid*
binary acid: *hydro—ic acid*

The name of a binary acid has three parts:

1. the prefix *hydro-*
2. the stem of the name of the other nonmetal + *-ic* suffix
3. the word *acid*

The prefix *hydro-* and the *-ic* suffix are characteristic of the names of binary acids.

Notice that HCl is the formula for both hydrodgen chloride and hydrochloric acid. However, binary hydrogen compounds such as HCl are named as binary acids *only when they are dissolved in water (aqueous).* When they are nonaqueous, these compounds are named as binary compounds with the *-ide* ending.

HCl(aq): *hydro*chlor*ic acid*
HCl(g): hydrogen chlor*ide*

Binary Acids

Group 6A	*Group 7A*
H_2S(aq): *hydro*sulfur*ic acid*	HF(aq): *hydro*fluor*ic acid*
H_2Se(aq): *hydro*selen*ic acid*	HCl(aq): *hydro*chlor*ic acid*
	HBr(aq): *hydro*brom*ic acid*
	HI(aq): *hydro*iod*ic acid*

An abnormally high concentration of hydrochloric acid in the stomach contributes to the formation of duodenal ulcers such as the one shown here.

FIGURE 9.3 Common oxyacids (*-ic* ending).

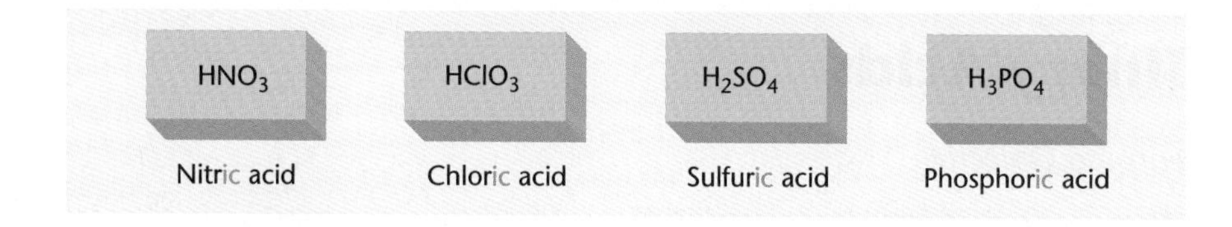

9.7 Oxyacids

Learning Objective

- Name and write formulas of oxyacids.

Sulfuric acid is the number one chemical produced in the United States: 89 billion pounds per year!

Oxyacids are ternary compounds made up of hydrogen (true for all acids), oxygen, and a third element. Oxyacids are some of the most important inorganic compounds. They have many uses, and some of them are important commercial chemicals. For example, the production of sulfuric acid (H_2SO_4) far exceeds that of any other chemical. Because of the importance of oxyacids and the salts related to them, you must know how to name these compounds.

The best way for you to begin to learn the language of oxyacids and their salts is to memorize the names of some of the more common ones. Four of these are shown in Figure 9.3. These acids are named by adding *-ic acid* to the stem of the third element. You may call these the *-ic* acids. Notice that these have the *-ic* ending like the binary acids such as hydrochloric acid. However, the prefix *hydro-* is not used with oxyacids. *Memorize these names and formulas.* They are the basis of a system for naming a large number of oxyacids and their salts.

The prefix hydro- *is used with binary acids,* not with oxyacids.

Other oxyacids related to the *-ic* acids use the *-ous* ending in their names. Each of the *-ous* acids has *one less oxygen* than the *-ic* acid made up of the same three elements. Consider the oxyacids made up of hydrogen, oxygen, and nitrogen:

HNO_3: nitr*ic* acid
HNO_2: nitr*ous* acid

Some nonmetals form three or four oxyacids. In each series of oxyacids, an *-ic* acid is the reference for the names and formulas of the others.

If an acid has one less oxygen than the *-ous* acid, then the prefix *hypo-* is used; the *-ous* ending is retained. If an acid has *one more oxygen* than the *-ic* acid, then the prefix *per-* is used; the *-ic* ending is retained. This system is illustrated in Figure 9.4 for four series of oxyacids. The reference acids (to be memorized) are highlighted inside the box.

H^+ is a nucleus with one proton and has no electrons. It is called either 'hydrogen ion' or a 'proton.'

Notice that the oxyacids containing chlorine and nitrogen have only one hydrogen. These are **monoprotic** acids, which have one acidic hydrogen (proton) in each formula unit. Sulfuric acid (H_2SO_4) is an example of a **diprotic** acid because it has two protons in each formula unit. Phosphoric acid (H_3PO_4) is a **triprotic** acid.*
Although the number of oxygens per formula unit varies, the number of hydrogens is the same in all the formulas in a series.

*Although H_3PO_3 and H_3PO_2 might appear to be triprotic, they are not. Only the hydrogens bonded to oxygen are acidic: H_3PO_3 is diprotic and H_3PO_2 is monoprotic. Their formulas might more properly be written as $H_2(HPO_3)$ and $H(H_2PO_2)$.

FIGURE 9.4 Naming oxyacids.

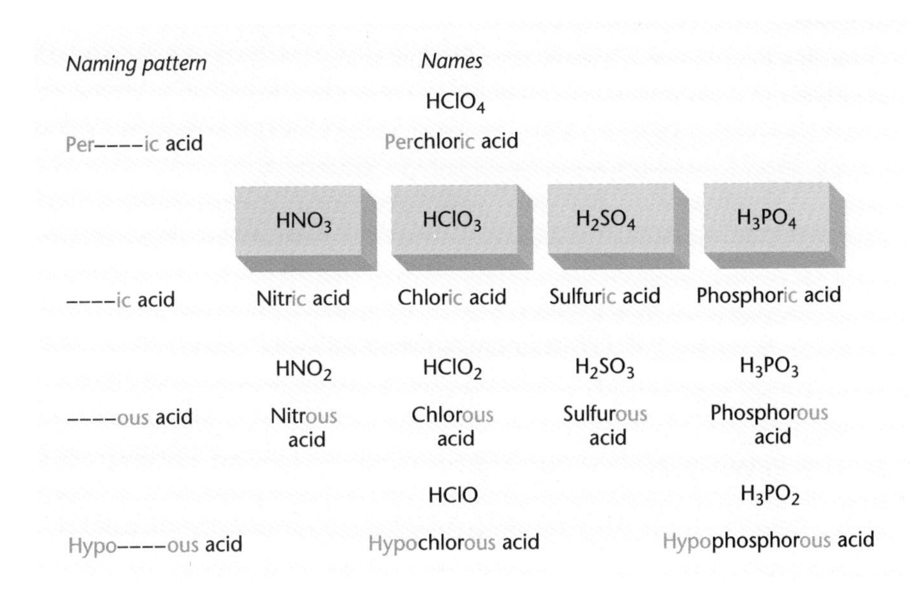

Many oxyacids not shown in Figure 9.4 can be related to one of these four series by noting the group in the periodic table in which the third element is found. For example, because Br and Cl are both in Group 7A, $HBrO_3$ is related to $HClO_3$. These compounds have similar names.

$HClO_3$: chloric acid
$HBrO_3$: bromic acid

Naming Oxyacids

Number of Oxygen Atoms Compared with -ic *Acid*	*Prefix/Suffix*	*Example*
one more	*per—ic*	$HClO_4$
same	*—ic*	$HClO_3$
one fewer	*—ous*	$HClO_2$
two fewer	*hypo—ous*	$HClO$

Example 9.12

Name the following acids.

a. H_3PO_4
b. HIO_3
c. HF(aq)
d. HClO ■

Solutions

a. phosphoric acid (memorize)
b. iodic acid (related to $HClO_3$, chloric acid)
c. hydrofluoric acid (binary acid)
d. hypochlorous acid (two fewer oxygens than chloric acid)

Practice Problem 9.7 Name the following acids:
a. HNO_2
b. HIO_4
c. H_2SO_3
d. HBr(aq)

Practice Problem 9.8 Write formulas for the following acids:
a. nitric acid
b. hydrochloric acid
c. phosphorous acid
d. hydrosulfuric acid

9.8 Salts of Oxyacids

Learning Objective

- Name and write formulas of salts of oxyacids.

The salt of an oxyacid is composed of a metal cation and an **oxyanion,** an anion formed when an oxyacid loses one or more hydrogen ions, or protons (H^+). Many salts of oxyacids are important compounds. In learning their names and formulas, we have to choose between memorizing the names and formulas of many oxyanions (Table 9.10) or learning a strategy to help us. Because the large number of oxyanions makes memorizing difficult, our approach will be to develop a strategy for naming oxyanions and writing their formulas.

Naming Salts of Oxyacids

Remember the endings for acid/oxyanion:

-ic/-ate
-ous/-ite

As with binary salts, naming salts of oxyacids involves naming the cation and the anion. The names of cations were discussed previously (Section 9.5). The name of the anion is determined by relating the suffix of the name of the acid to the suffix of the name of the oxyanion. When the name of the acid ends with *-ic,* the name of the oxyanion ends with *-ate.* For example:

$NaNO_3$ Cation: sodium
Anion: 1. related to HNO_3
2. nitr*ic* acid
3. *-ic/-ate:* nitr*ate*
Name: sodium nitrate

TABLE 9.10 Oxyanions

Group 5A		*Group 6A*		*Group 7A*	
				ClO_4^-	perchlorate
NO_3^-	nitrate	SO_4^{2-}	sulfate	ClO_3^-	chlorate
NO_2^-	nitrite	HSO_4^-	hydrogen sulfate	ClO_2^-	chlorite
PO_4^{3-}	phosphate	SO_3^{2-}	sulfite	ClO^-	hypochlorite
OHP_4^{2-}	hydrogen phosphate	HSO_3^-	hydrogen sulfite	BrO_4^-	perbromate
$H_2PO_4^-$	dihydrogen phosphate	SeO_4^{2-}	selenate	BrO_3^-	bromate
AsO_4^{3-}	arsenate			BrO^-	hypobromite
AsO_3^{3-}	arsenite			IO_4^-	periodate
Other oxyanions				IO_3^-	iodate
CO_3^{2-}	carbonate			IO^-	hypoiodite
HCO_3^-	hydrogen carbonate (bicarbonate)				
$C_2H_3O_2^-$	acetate				
MnO_4^-	permanganate				
CrO_4^{2-}	chromate				
$Cr_2O_7^{2-}$	dichromate				

Thus, the name of $NaNO_3$ is sodium nitr*ate.* The *-ic*/*-ate* relationship of the acid and anion (Table 9.11) is a key step in the strategy for naming salts of oxyacids.

Some oxyanions are related to *-ous* acids. For example, $NaNO_2$ is related to HNO_2. When the name of the oxyacid ends with *-ous,* the name of the related oxyanion ends with *-ite* (Table 9.11).

$NaNO_2$	Cation:	sodium
	Anion:	1. related to HNO_2
		2. nitr*ous* acid
		3. *-ous*/*-ite:* nitr*ite* ion
	Name:	sodium nitrite

The name of $NaNO_2$ is sodium nitrite. Recognizing the *-ous*/*-ite* relationship is important in naming many salts of oxyacids.

The strategy for naming salts of oxyacids is outlined in Figure 9.5. Practice this strategy as you name the compounds in the following exercises.

TABLE 9.11 Oxyacids and Oxyanions of Chlorine

Oxyacid		*Oxyanion*	
perchloric acid	$HClO_4$	ClO_4^-	perchlorate
chloric acid	$HClO_3$	ClO_3^-	chlorate
chlorous acid	$HClO_2$	ClO_2^-	chlorite
hypochlorous acid	$HClO$	ClO^-	hypochlorite

FIGURE 9.5 Strategy for naming salts of oxyacids, in this case, writing the name for K_2SO_3.

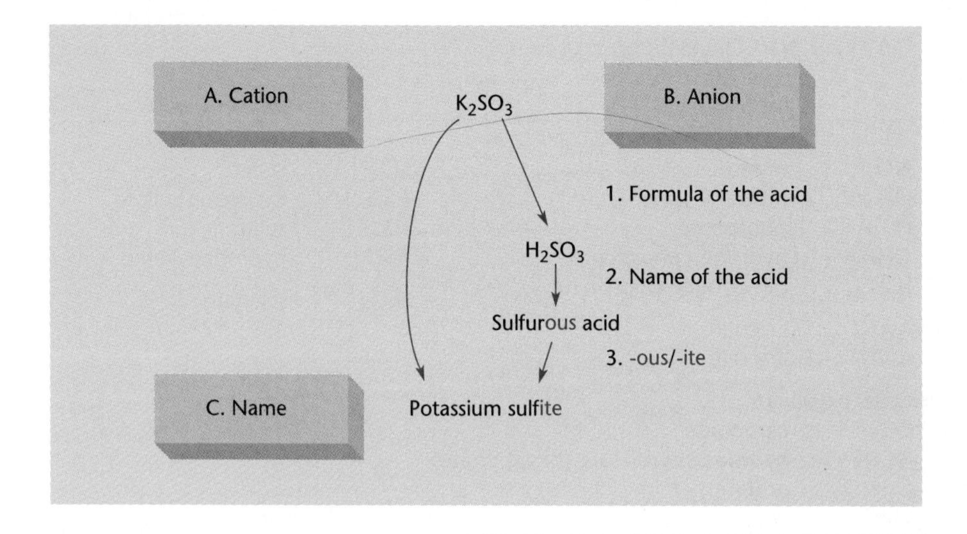

Example 9.13

Name the following compounds:

a. $KClO_3$
b. $CaSO_3$
c. Na_3PO_4
d. $Mg(NO_3)_2$ ■

Solutions

a. Cation: K^+ is potassium ion
Anion: 1. ClO_3^- is related to $HClO_3$
2. chlor*ic* acid
3. *-ic/-ate:* chlor*ate* ion
Name: potassium chlorate

b. Cation: Ca^{2+} is calcium ion
Anion: 1. SO_3^{2-} is related to H_2SO_3
2. sulfur*ous* acid
3. *-ous/-ite:* sulf*ite* ion
Name: calcium sulfite

c. Cation: Na^+ is sodium ion
Anion: 1. PO_4^{3-} is related to H_3PO_4
2. phosphor*ic* acid
3. *-ic/-ate:* phosph*ate* ion
Name: sodium phosphate

d. Cation: Mg^{2+} is magnesium ion
Anion: 1. NO_3^- is related to HNO_3
2. nitr*ic* acid
3. *-ic/-ate:* nitr*ate* ion
Name: magnesium nitrate

Practice Problem 9.9 Name the following compounds.

a. $KClO_2$
b. $Ba_3(PO_4)_2$
c. $AgClO_4$
d. $HgSO_4$

Acid Salts

A salt formed from a di- or triprotic acid that has an anion capable of giving up a proton is called an **acid salt.** Examples include $NaHSO_4$ and NaH_2PO_4. Typically, an acid salt is formed when a metal cation replaces a proton of a diprotic or triprotic acid. The systematic names of these salts use the words *hydrogen* or *dihydrogen* to indicate the presence of one or two hydrogens in the formula of the anion. This system is illustrated in Table 9.12.

Acid salts of diprotic acids also have common names. In the common names, the anion is named using *bi* in place of *hydrogen.* So, hydrogen sulfate ion is also called bisulfate ion, and hydrogen carbonate ion is named bicarbonate ion:

$NaHSO_4$: sodium *hydrogen* sulfate or sodium *bi*sulfate
$NaHCO_3$: sodium *hydrogen* carbonate or sodium *bi*carbonate

Sodium bicarbonate, or baking soda, is probably the most well known example of an acid salt.

Remember that the use of *bi-* is limited to the acid salts of diprotic acids. *Bi-* is not used in naming the acid salts of triprotic acids.

TABLE 9.12 Salts of Diprotic and Triprotic Oxyacids

Acid	*Salt*	*IUPAC Name*
H_2SO_4	$NaHSO_4$	sodium hydrogen sulfate
	Na_2SO_4	sodium sulfate
H_2CO_3	$NaHCO_3$	sodium hydrogen carbonate
	Na_2CO_3	sodium carbonate
H_3PO_4	NaH_2PO_4	sodium dihydrogen phosphate
	Na_2HPO_4	sodium hydrogen phosphate
	Na_3PO_4	sodium phosphate

Example 9.14

Give the systematic and common names (if appropriate) for the following compounds.

a. $KHSO_3$
b. Rb_2HPO_4
c. $Ca(H_2PO_4)_2$
d. $Sr(HCO_3)_2$ ■

Solutions

a. Cation: K^+ is potassium ion
Anion: 1. HSO_3^- is related to H_2SO_3
2. sulfur*ous* acid
3. *-ous/-ite:* hydrogen sulf*ite* ion
Name: potassium hydrogen sulfite or potassium bisulfite

b. Cation: Rb^+ is rubidium ion
Anion: 1. HPO_4^{2-} is related to H_3PO_4
2. phosphor*ic* acid
3. *-ic/-ate:* hydrogen phosph*ate* ion
Name: rubidium hydrogen phosphate

c. Cation: Ca^{2+} is calcium ion
Anion: 1. $H_2PO_4^-$ is related to H_3PO_4
2. phosphor*ic* acid
3. *-ic/-ate:* dihydrogen phosph*ate* ion
Name: calcium dihydrogen phosphate

d. Cation: Sr^{2+} is strontium ion
Anion: 1. HCO_3^- is related to H_2CO_3
2. carbon*ic* acid
3. *-ic/-ate:* hydrogen carbon*ate* ion
Name: strontium hydrogen carbonate or strontium bicarbonate

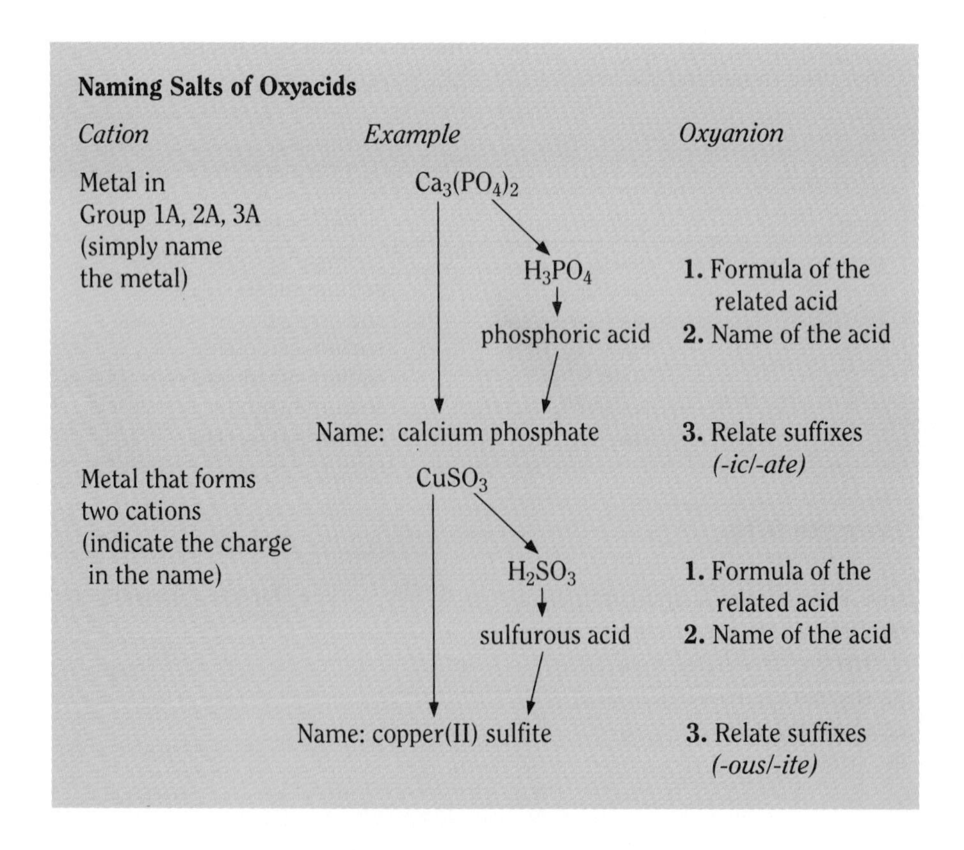

Naming Salts of Oxyacids
Cation
Example
Oxyanion
Metal in Group 1A, 2A, 3A (simply name the metal)
Ca3(PO4)2
H3PO4
phosphoric acid
1. Formula of the related acid
2. Name of the acid
Name: calcium phosphate
3. Relate suffixes (-ic/-ate)
Metal that forms two cations (indicate the charge in the name)
CuSO3
H2SO3
sulfurous acid
1. Formula of the related acid
2. Name of the acid
Name: copper(II) sulfite
3. Relate suffixes (-ous/-ite)

Practice Problem 9.10 Write systematic names for the following compounds:

a. NaH_2PO_4
b. $Ca(HSO_4)_2$
c. $Al(HCO_3)_3$

Writing Formulas of Salts of Oxyacids

In the previous section, we discussed how to name salts of oxyacids by relating the anion to its parent acid. Now we will reverse this process and consider how to write the formula of a salt of an oxyacid from its name.

To write the formula of a salt, we first need to write the formulas of the ions making up the compound. Formulas of cations and monatomic anions were discussed in Section 9.2.

Formulas of oxyanions (Table 9.10) are more complex than formulas of monatomic anions. In a good news–bad news story, that is the bad news. The good news is we have a strategy for writing most of these formulas, and we do not have to memorize them. Furthermore, because we have already studied oxyacids, the foundation for our strategy has already been laid. Here are the steps for writing the formula of an oxyanion:

1. From the ending of the name of the oxyanion, write (or think of) the ending of the related oxyacid. Remember: *-ic*/*-ate, -ous*/*-ite.* Write the name of the oxyacid.
2. Write the formula of the oxyacid. Remember the strategy given in Section 9.7.
3. By deleting the H's from the formula of the acid, write the formula for the oxyanion. Because H has an oxidation number of $+1$, *the charge on the anion is numerically equal to the number of H's deleted.*

For example, to write the formula of sodium nitrate (Figure 9.6):

Cation: Na^+ (Group 1A)
Anion:
1. name of the acid: relate suffixes (*-ic*/*-ate*); nitr*ate* is related to nitr*ic* acid.
2. formula of the acid: HNO_3 (You memorized this one.)
3. formula of oxyanion: deleting one H gives NO_3^-

Formula: $NaNO_3$ (the sum of the charges must be zero)

Practice this strategy as you work the following examples.

Example 9.15

Write formulas for the following oxyanions:

a. chlorate ion
b. sulfite ion
c. phosphate ion
d. nitrite ion

FIGURE 9.6 Writing formulas of salts of oxyacids

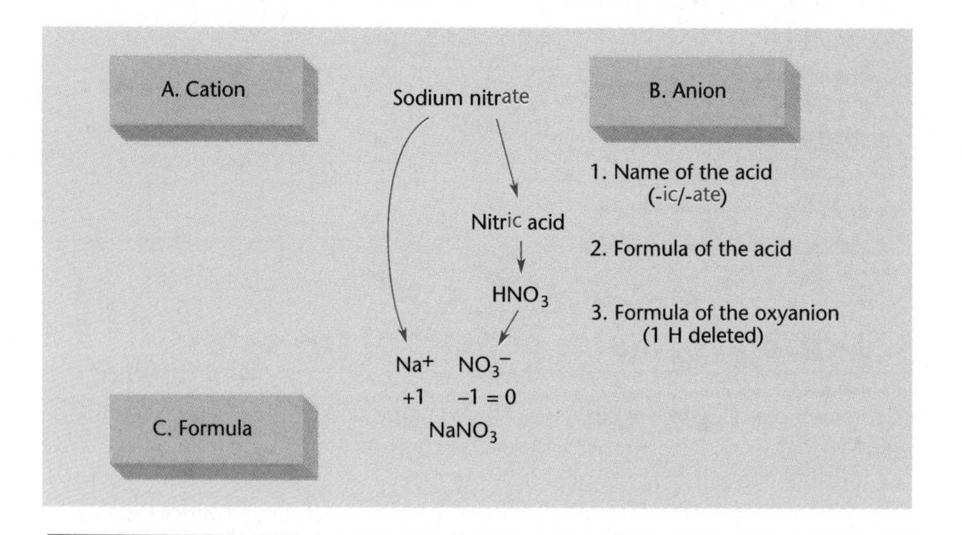

Solutions

a. 1. name of the acid (*-ic*/*-ate*): chloric acid
2. formula of the acid: $HClO_3$
3. formula of the oxyanion: ClO_3^-

b. 1. name of the acid (*-ous*/*-ite*): sulfurous acid
2. formula of the acid: H_2SO_3
3. formula of the oxyanion: SO_3^{2-}

c. 1. name of the acid (*-ic*/*-ate*): phosphoric acid
2. formula of the acid: H_3PO_4
3. formula of the oxyanion: PO_4^{3-}

d. 1. name of the acid (*-ous*/*-ite*): nitrous acid
2. formula of the acid: HNO_2
3. formula of the oxyanion: NO_2^-

Example 9.16

Write the formula for iron(III) sulfate. Use the strategy outlined in Figure 9.6. ■

Solution

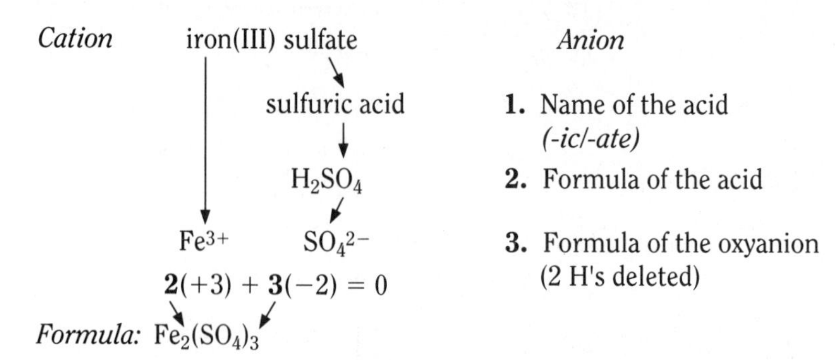

Remember, the sum of the ionic charges equals zero resulting in subscripts of "2" for Fe^{3+} and "3" for SO_4^{2-}. Remember also to use parentheses for sulfate ion.

You must memorize a few oxyanions that are not related to the oxyacids you have studied. Some of the more important ones are permanganate, MnO_4^-, chromate, CrO_4^{2-}, and dichromate, $Cr_2O_7^{2-}$ (Table 9.10). These are all good oxidizing agents and are frequently used in laboratory experiments. Acetate ($C_2H_3O_2^-$) is also quite common.

Memorize these oxyanions:

$C_2H_3O_2^-$	*acetate*
MnO_4^-	*permanganate*
CrO_4^{2-}	*chromate*
$Cr_2O_7^{2-}$	*dichromate*

Example 9.17

Write formulas for the following compounds.

a. potassium chlorate
b. magnesium chlorite
c. iron(II) sulfate ■

Solutions

a. Cation: K^+
Anion:
1. name of the acid (*-ic*/*-ate*): chloric acid
2. formula of the acid: $HClO_3$
3. formula of the oxyanion: ClO_3^-

K^+ ClO_3^-
$(+1) + (-1) = 0$

Formula: $KClO_3$

b. Cation: Mg^{2+}
Anion:
1. name of the acid (*-ous*/*-ite*): chlorous acid
2. formula of the acid: $HClO_2$
3. formula of the oxyanion: ClO_2^-

Mg^{2+} ClO_2^-
$(+2) + \mathbf{2}(-1) = 0$

Formula: $Mg(ClO_2)_2$

c. Cation: Fe^{2+}
Anion:
1. name of the acid (*-ic*/*-ate*): sulfuric acid
2. formula of the acid: H_2SO_4
3. formula of the oxyanion: SO_4^{2-}

Fe^{2+} SO_4^{2-}
$(+2) + (-2) = 0$

Formula: $FeSO_4$

Practice Problem 9.11 Write formulas for the following compounds:
a. tin(II) hydrogen phosphate
b. mercury(I) nitrate
c. sodium chromate

CHEMICAL WORLD

Sulfuric Acid

Sulfuric acid is the number one chemical produced in the United States, more than 40 million tons annually. It is commercially the most important chemical substance manufactured in the world because it is used, directly or indirectly, in the production of nearly all key chemical substances (Figure 1).

Sulfuric acid is a colorless, odorless liquid with a density of 1.84 g/mL and a high boiling point (317°C). It is a strong acid and therefore very corrosive. The concentrated acid is 96–98% H_2SO_4. Because it is practically anhydrous, sulfuric acid has advantages in certain applications over other strong acids, such as nitric acid and hydrochloric acid, that are available only as aqueous solutions.

About 70% of the sulfuric acid produced is used in making fertilizers from phosphate rock. This rock, which is available in abundance, is mainly calcium phosphate, $Ca_3(PO_4)_2$, and fluorapatite, $Ca_5(PO_4)_3F$. After the rock is crushed and ground, it is treated with sulfuric acid to produce phosphoric acid. The following equations illustrate the principal reactions in the process:

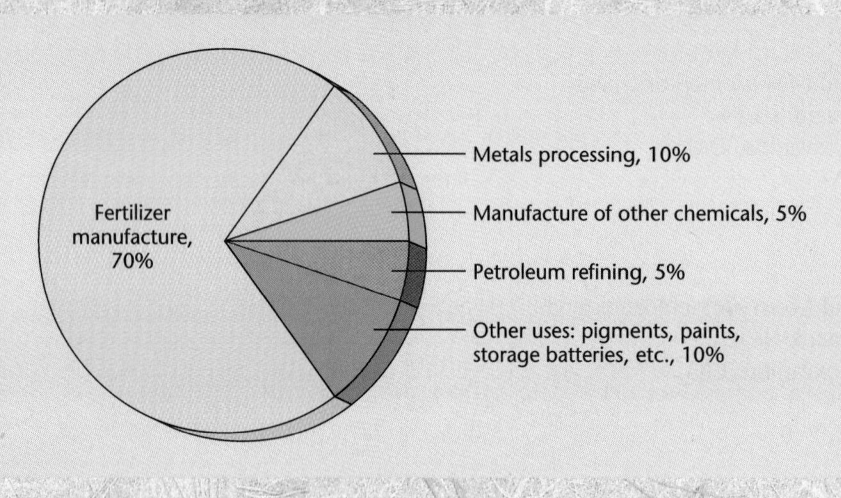

FIGURE 1 Uses of sulfuric acid, H_2SO_4.

Chapter Review

Directions: From the list of choices at the end of the Chapter Review, choose the word or term that best fits in each blank. Use each choice only once. The answers appear at the end of the chapter.

Section 9.1 Chemical nomenclature includes systematic or __________ (1) names and __________ (2) names. To learn systematic names, you must learn a set of __________ (3). Common names must be __________ (4).

Section 9.2 The oxidation state of an element is expressed by its __________ (5) number. The oxidation number of an uncombined element is __________ (6). When combined with other elements, oxygen generally has an oxidation number of __________ (7). The oxidation number of hydrogen is usually __________ (8) except in metal hydrides where it is __________ (9). The sum of the oxidation numbers for a compound is zero, but the sum for a polyatomic ion is equal to its __________ (10).

$$Ca_3(PO_4)_2(s) + 3H_2SO_4(aq) + 6H_2O(l) \longrightarrow 2H_3PO_4(aq) + 3CaSO_4 \cdot 2H_2O(s)$$

$$Ca_5(PO_4)_3F(s) + 5H_2SO_4(aq) + 10H_2O(l) \longrightarrow HF(aq) + 5CaSO_4 \cdot 2H_2O(s) + 3H_3PO_4(aq)$$

Gypsum ($CaSO_4 \cdot 2H_2O$), a material used to make plasterboard (sheetrock) for the construction industry, is a valuable byproduct of the process. The reaction of ammonia with phosphoric acid gives ammonium dihydrogen phosphate, $NH_4H_2PO_4$, a very efficient, soluble fertilizer that furnishes nitrogen and phosphorus, two nutrients essential for plant growth.

FIGURE 2 The electrolyte in the lead storage battery is aqueous sulfuric acid.

$$NH_3(g) + H_3PO_4(aq) \longrightarrow NH_4H_2PO_4(aq)$$

Sulfuric acid is also used in processing metals, in petroleum refining, in the manufacture of other chemicals, and in batteries. Sulfuric acid is known to most of us as battery acid because it is the electrolyte in the lead–acid storage battery used in cars, trucks, and other motor vehicles (Figure 2).

Sulfuric acid is manufactured by the contact process, starting with the combustion of sulfur to give sulfur dioxide.

Step 1 $S(s) + O_2(g) \longrightarrow SO_2(g)$

Step 2 $2SO_2(g) + O_2(g) \xrightarrow[400°C]{V_2O_5} 2SO_3(g)$

Step 3 $SO_3(g) + H_2SO_4(l) \longrightarrow H_2S_2O_7(l)$ pyrosulfuric acid

Step 4 $H_2S_2O_7(l) + H_2O(l) \longrightarrow 2H_2SO_4(l)$

Sulfur dioxide is an air pollutant formed during the combustion of coal that contains sulfur. It is slowly coverted to sulfuric acid in the atmosphere and contributes to the acid rain problem (Chapter 15). Clean-air laws now require factories and electricity-generating plants that burn sulfur-containing coal to install pollution control equipment to remove SO_2 from their stack gases. Sulfur dioxide from this source can be converted to sulfuric acid, which can be used by the generating plant or sold to offset the cost of pollution control equipment.

Section 9.3 Most inorganic compounds can be classified as either binary or __________ (11). They may be either covalent or __________ (12).

Section 9.4 Two general rules apply to the systematic names of all inorganic compounds. First, the __________ (13) or less electronegative element is written and named first. Second, the __________ (14) of the name is used to classify the compound.

Section 9.5 The ending of the name of a __________ (15) compound ends with *-ide* which is also the suffix for a __________ (16) anion. Some __________ (17) anions (e.g., OH^- and CN^-) are exceptions to this pattern and also use the *-ide* ending. The names of metal cations are simply the names of the metals except for metals that form __________ (18) cations.

When a compound contains a metal that forms two cations, the IUPAC name indicates the charge (or oxidation number) using a __________ __________ (19).

A __________ (20) results when a metal ion or ammonium ion replaces a __________ (21) of an acid. When writing the formula of a salt, remember that the __________ (22) of the ionic charges is zero.

For binary compounds of two nonmetals, __________ (23) are used to indicate the numbers of each atom making up the molecule. For example, __________ (24) means three atoms, __________ (25) means five atoms, and so on. These counting prefixes are not used in compounds involving metals.

Section 9.6 The pattern *hydro—ic acid* is for binary __________ (26). This pattern is followed only when these compounds are dissolved in water: such as HCl(aq), hydrochloric acid. Otherwise, the compound is named with the *-ide* suffix.

Section 9.7 Oxyacids are ternary compounds made up of hydrogen, __________ (27), and a third element. Either the *-ic* or the *-ous* ending is used in the name of an oxyacid, but the prefix __________ (28) is not used. You must memorize the names and formulas of *-ic* acids and then learn the rules you used to name the *-ous* acids. An *-ous* acid has one __________ (29) oxygen than an *-ic* acid. When the formula contains only one hydrogen, the acid is __________ (30). The terms diprotic and __________ (31) refer to acids that have two and three acidic hydrogens, respectively. The prefix *per-* indicates one __________ (32) oxygen than the *-ic* acid, and the prefix __________ (33) means one less oxygen than the *-ous* acid.

Section 9.8 Salts of oxyacids are classified by their suffixes, either *-ate* or *-ite*. These suffixes are related to the suffixes of the names of __________ (34). Remember to associate *-ic/* __________ (35) and *-ous/* __________ (36) when naming salts of oxyacids.

A salt that has an acidic hydrogen is called an __________ __________ (37). The IUPAC names of these salts use the words *hydrogen* or __________ (38) to indicate the presence of one or two hydrogens per formula unit. The name of an acid salt of a diprotic acid sometimes uses __________ (39) in place of *hydrogen*. So, hydrogen carbonate ion is also called __________ (40) ion.

Writing formulas of oxyanions is easier when you follow a strategy. Our strategy uses the *-ic/-ate* and *-ous/-ite* relationships for oxyacids and the oxyanions. The charge on an anion is determined from the number of __________ (41) deleted from the formula of the oxyacid.

Choices: acid salt, acids, -ate, bi-, bicarbonate, binary, cation, charge, common, diatomic, dihydrogen, H's, hydro-, hydrogen, hypo-, ionic, -ite, IUPAC, less, memorized, monatomic, monoprotic, more, oxidation, oxyacids, oxygen, penta-, prefixes, Roman numeral, rules, salt, suffix, sum, ternary, tri-, triprotic, two, zero, −1, −2, +1

Study Objectives

After studying Chapter 9, you should be able to:

Section 9.2

1. Assign oxidation numbers for uncombined or combined elements.

Section 9.5

2. Name and write formulas of ionic and covalent binary compounds.
3. Name and write formulas of the cations of metals that form two cations.
4. Use a prefix to indicate the number of atoms of an element in a molecule.

Section 9.6

5. Name and write formulas of binary acids.

Section 9.7

6. Name and write formulas of oxyacids.

Section 9.8

7. Name and write formulas of salts of oxyacids.

Key Terms

Review the definition of each of the terms listed here by chapter section number. You may use the glossary if necessary.

9.1 IUPAC

9.2 oxidation number, valence

9.3 binary compound, ternary compound

9.5 salt

9.6 acid, binary acid

9.7 diprotic, monoprotic, oxyacid, triprotic

9.8 oxyanion, acid salt

Questions and Problems

The answers to questions and problems with blue numbers appear in Appendix C.

Common Names (Section 9.1)

1. Give the common names for the following:
   a. NH_3 c. H_2O
   b. $NaHCO_3$ d. $NaOH$
2. Chalk is used in almost every classroom. What is the formula of this substance?
3. The formula for the stone used to build the Lincoln Memorial is $CaCO_3$. What is the common name for this compound?
4. The acid found in vinegar has the formula $HC_2H_3O_2$. What is the common name of this compound?

Oxidation Numbers (Section 9.2)

5. What is the oxidation number of nitrogen in each of the following:
   a. NH_3 4 c. NCl_3
   b. NO_2 d. HNO_2
6. Determine the oxidation number of each element in the following compounds:
   a. H_2SO_4 d. SO_3
   b. N_2O_4 e. CO_2
   c. $Ca_3(PO_4)_2$ f. $HBrO_3$
7. What is the oxidation number of chlorine in each of the following oxyacids?
   a. $HClO_4$ c. $HClO_2$
   b. $HClO_3$ d. $HClO$
8. What is the oxidation number of chromium in each of the following compounds?
   a. CrO_3 c. K_2CrO_4
   b. $CrCl_3$ d. $Na_2Cr_2O_7$
9. What is the oxidation number of oxygen in each of the following compounds?
   a. SO_2 c. Fe_2O_3
   b. H_2O_2 d. Na_3PO_4
10. What is the oxidation number of hydrogen in each of the following compounds?
    a. HCl c. CaH_2
    b. $NaHCO_3$ d. AlH_3
11. Oxidation numbers of elements in a compound generally can be calculated knowing the oxidation numbers of oxygen and hydrogen. Identify the oxidation number of the other element in each of the following compounds:
    a. H_2SO_3 c. Cl_2O_7
    b. PH_3 d. H_2Se
12. The oxidation number of oxygen in each of the following is -2. What is the oxidation number of the other element?
    a. NO_2^- c. PO_4^{3-}
    b. OH^- d. AsO_3^{3-}
13. What is the oxidation number of each element in the following ions:
    a. IO_4^- c. AsO_4^{3-}
    b. NH_4^+ d. MnO_4^-
14. Give the oxidation numbers of all elements in each of the following compounds:
    a. K_2SeO_4 c. $Cr(NO_3)_3$
    b. $Au(ClO_3)_3$ d. $Ni(BrO_3)_2$

Binary Compounds (Section 9.5)

15. Name the following compounds:
    a. KI d. SnS_2
    b. P_4O_{10} e. MgO
    c. FeO
16. Name the following compounds.
    a. $CuBr$ d. Cl_2O_7
    b. CO e. PCl_3
    c. NO_2
17. Name the following cations using systematic names:
    a. Fe^{3+} c. Sn^{4+}
    b. Hg_2^{2+} d. Cu^+
18. Give the systematic names for the following compounds:
    a. $HgCl_2$ c. Cu_2O
    b. SnO d. $FeCl_3$
19. Name the following compounds.
    a. $CuCl_2$ c. SnI_4
    b. Fe_3N_2 d. Cu_2S
20. Write the formulas of the following compounds:
    a. potassium sulfide c. calcium bromide
    b. iron(II) iodide d. mercury(I) oxide
21. Give the formulas of the following compounds:
    a. aluminum sulfide d. tin(II) chloride
    b. lead(IV) bromide e. magnesium nitride
    c. copper(II) cyanide
22. Hydrogen peroxide, a common antiseptic, is unstable and decomposes to give water and oxygen. What is the formula of hydrogen peroxide?
23. Name the following compounds:
    a. KCN c. $Cu(OH)_2$
    b. $Mg(OH)_2$ d. $Fe(OH)_3$
24. Name the following compounds:
    a. $Al(OH)_3$ c. $Ca(OH)_2$
    b. $Sn(OH)_2$ d. $RbOH$

25. Write the formulas of the following compounds:
    a. tin(IV) hydroxide
    b. ammonium bromide
    c. iron(II) hydroxide
26. Write the formulas of the following compounds:
    a. copper(II) cyanide
    b. potassium hydroxide
    c. iron(II) hydroxide

Binary Acids (Section 9.6)

27. Which of the following are binary acids? Note: (g) means "gas" and (aq) means "dissolved in water."
    a. $HCl(g)$
    b. $HBr(aq)$
    c. $H_2S(g)$
    d. $H_2S(aq)$
    e. NH_3
    f. NaH
    g. HNO_3
28. Name the following compounds:
    a. $HCl(aq)$
    b. $H_2S(aq)$
    c. $HBr(aq)$
    d. $HF(aq)$
29. Write formulas for the following:
    a. hydroiodic acid
    b. hydrobromic acid
    c. hydroselenic acid

Oxyacids (Section 9.7)

30. Name the following acids:
    a. H_3PO_4
    b. H_3PO_2
    c. HNO_2
    d. $HClO_2$
    e. HNO_3
    f. $HClO_3$
31. Name the following acids. Note that they are all related to oxyacids in Figure 9.4.
    a. H_3AsO_4
    b. $HBrO$
    c. H_2SeO_4
    d. H_2SeO_3
    e. HIO_3
    f. H_3AsO_3
32. Sulfur trioxide dissolves in water to give an oxyacid with an *-ic* ending. Write the formula for this acid.
33. Write the formulas of the following:
    a. nitric acid
    b. chlorous acid
    c. periodic acid
    d. sulfurous acid
34. Write the formulas of the following compounds:
    a. bromic acid
    b. arsenic acid
    c. hypochlorous acid
    d. bromous acid

Salts of Oxyacids (Section 9.8)

35. Name the following compounds:
    a. $NaClO$
    b. $CaSO_3$
    c. KNO_3
    d. $NaNO_2$
36. Name the following compounds:
    a. $Ba(ClO_3)_2$
    b. $MgSO_4$
    c. $AlPO_4$
37. Name the following salts:
    a. $Cu(NO_3)_2$
    b. $SnSO_4$
    c. $Cu_3(PO_4)_2$
38. Name the following compounds:
    a. $Hg_2(NO_3)_2$
    b. $LiClO_2$
    c. $Fe(NO_2)_2$
39. Sodium nitrite is a preservative used in lunchmeats, bacon, sausage, and ham. When ingested, sodium nitrite reacts with the acid in the stomach to form its related oxyacid. Give the name and formula of this acid.
40. Name the following salts:
    a. $NaHSO_4$
    b. $Cu(HSO_3)_2$
    c. $Hg(H_2PO_4)_2$
    d. K_2HPO_4
    e. $CaHPO_3$
41. Using common names for the anions, name the following salts:
    a. $KHSO_3$
    b. $Cu(HSO_4)_2$
    c. $NaHSeO_4$
42. Carbonic acid, H_2CO_3, is found in soft drinks. Carbonic acid also forms in your stomach when you swallow baking soda, $NaHCO_3$. Explain what happens in your stomach when the baking soda mixes with stomach acid (HCl).
43. Write formulas for the following oxyanions:
    a. nitrate
    b. perchlorate
    c. nitrite
    d. iodate
    e. sulfate
    f. chlorite
    g. hypochlorite
    h. bromate
44. Write formulas for and name the oxyanions derived from the following acids:
    a. perchloric acid
    b. nitrous acid
    c. phosphoric acid
    d. sulfuric acid
    e. sulfurous acid
    f. carbonic acid
45. Write formulas for the following oxyanions:
    a. permanganate
    b. chlorate
    c. dichromate
    d. chromate
    e. phosphate
46. Write formulas for the following salts:
    a. calcium chlorate
    b. magnesium bicarbonate
    c. copper(II) nitrate
    d. potassium sulfate
    e. sodium sulfite
    f. mercury(I) chlorite
47. Write formulas for the following salts.
    a. iron(III) sulfate
    b. copper(I) phosphate
    c. chromium(III) sulfate
    d. tin(II) chlorite
    e. sodium carbonate

48. Write formulas for the following compounds.
 a. potassium perchlorate
 b. sodium bisulfite
 c. potassium dihydrogen phosphate
 d. calcium hydrogen sulfate
 e. nickel(II) nitrate

49. The following compounds are all strong oxidizing agents. Write the formulas of these compounds.
 a. potassium permanganate c. sodium chromate
 b. potassium dichromate d. sodium chlorate

50. Ammonium perchlorate is a powerful oxidizing agent used in solid propellants for rockets. Write the formula of this compound.

51. Because of its high nitrogen content, ammonium nitrate is used as a fertilizer. Write the formula of ammonium nitrate.

52. Write the formula of ammonium dihydrogen phosphate, a fertilizer.

53. Liquid bleach is a solution of sodium hypochlorite. Write the formula of this compound.

Formula Writing Exercise

54. *Directions:* In each box, write the formula of the compound that would result from the combination of the cation in the left column with the anion at the top of the column. Write the names of the compounds on a separate sheet of paper. (Some of these compounds are unknown.)

Ions	Br^-	CO_3^{2-}	OH^-	PO_4^{3-}	HSO_4^-	SO_3^{2-}	MnO_4^-	$Cr_2O_7^{2-}$	ClO^-
NH_4^+									
K^+									
Ca^{2+}									
Cu^{2+}									
Sn^{2+}									
Al^{3+}									
Fe^{3+}									

Solutions to Practice Problems

PP 9.1

a. H_3PO_3

Step 1 $+1$ -2
Step 2 $3(+1)$ $(-2)3$
Step 3 $+3 + P + (-6) = 0$
$P = +3$

b. CH_4

Step 1 $+1$
Step 2 $(+1)4$
Step 3 $C + (+4) = 0$
$C = -4$

c. The sum of the oxidation numbers is equal to the charge on the ion.

BrO_3^-

Step 1 -2
Step 2 $(-2)3$
Step 3 $Br + (-6) = -1$
$Br = +5$

PP 9.2

Name the cation, followed by the anion. Remember to use the *-ide* suffix for a binary compound.
a. magnesium bromide
b. strontium oxide
c. lithium chloride

PP 9.3

a. KBr (from K^+ and Br^-)

b. Sr^{2+} Cl^- Cl^-

$(+2) + \mathbf{2}(-1) = 0$

Formula: $SrCl_2$

c. Mg^{2+} PO_4^{3-} Mg^{2+} PO_4^{3-} Mg^{2+}

$\mathbf{3}(+2) + \mathbf{2}(-3) = 0$

Formula: $Mg_3(PO_4)_2$

d. NH_4^+ S^{2-} NH_4^+

$\mathbf{2}(+1) + (-2) = 0$

Formula: $(NH_4)_2S$

PP 9.4

For a metal that forms two cations, determine the charge or oxidation number for the cation from the charge on the anion. Remember, the name of a monatomic anion ends with *-ide*.

a. Hg_2Cl_2
   $2(Hg) + 2(-1) = 0$
   $2(Hg) = +2$
   $Hg = +1$
   Name: mercury(I) chloride

b. Cu_2O
   $2(Cu) + (-2) = 0$
   $2(Cu) = +2$
   $Cu = +1$
   Name: copper(I) oxide

c. Fe_2O_3
   $2(Fe) + 3(-2) = 0$
   $2(Fe) = +6$
   $Fe = +3$
   Name: iron(III) oxide

d. $PbBr_2$
   $Pb + 2(-1) = 0$
   $Pb = +2$
   Name: lead(II) bromide

PP 9.5

Write the formulas for the cation and the anion, write a charge equation to determine the subscripts, and then write formula.

a. Ni^{2+} S^{2-}
   $(+2) + (-2) = 0$
   NiS

b. Cr^{3+} Br^-
   $(+3) + (-1) = 0$
   $CrBr_3$

c. Sn^{2+} O^{2-}
   $(+2) + (-2) = 0$
   SnO

PP 9.6

Counting prefixes are used in the names of binary compounds of two nonmetals.

a. sulfur dichloride
b. phosphorus trichloride
c. oxygen difluoride
d. calcium bromide (ionic, do not use counting prefixes)

PP 9.7

a. nitrous acid (one less oxygen than nitric acid)
b. periodic acid (related to $HClO_4$, perchloric acid)
c. sulfurous acid (one less oxygen than sulfuric acid)
d. hydrobromic acid (a binary acid)

PP 9.8

a. HNO_3 (memorized)
b. HCl(aq) (a binary acid)
c. H_3PO_3 (one less oxygen than phosphoric acid)
d. H_2S(aq) (a binary acid)

PP 9.9

a. Cation: K^+ is potassium ion
   Anion: 1. ClO_2^- is related to $HClO_2$
   2. chlor*ous* acid
   3. *-ous/-ite:* chlor*ite* ion
   Name: potassium chlorite

b. Cation: Ca^{2+} is calcium ion
   Anion: 1. PO_4^{3-} is related to H_3PO_4
   2. phosphor*ic* acid
   3. *-ic/-ate:* phosph*ate* ion
   Name: calcium phosphate

c. Cation: Ag^+ is silver ion
   Anion: 1. ClO_4^- is related to $HClO_4$
   2. perchlor*ic* acid
   3. *-ic/-ate:* perchlor*ate* ion
   Name: silver perchlorate

d. Cation: Because the anion is -2 (related to H_2SO_4), the cation is $+2$. Hg^{2+} is mercury(II) ion.
   Anion: 1. SO_4^{2-} is related to H_2SO_4
   2. sulfur*ic* acid
   3. *-ic/-ate:* sulf*ate* ion
   Name: mercury(II) sulfate

PP 9.10

a. Cation: sodium ion
   Anion: 1. $H_2PO_4^-$ is related to H_3PO_4
   2. phosphor*ic* acid
   3. *-ic/-ate:* dihydrogen phosph*ate* ion
   Name: sodium dihydrogen phosphate

b. Cation: calcium ion

Anion: 1. HSO_4^- is related to H_2SO_4

2. sulfur*ic* acid

3. *-ic*/*-ate:* hydrogen sulf*ate* ion

Name: calcium hydrogen sulfate

c. Cation: aluminum ion

Anion: 1. HCO_3^- is related to H_2CO_3

2. carbon*ic* acid

3. *-ic*/*-ate:* hydrogen carbon*ate* ion

Name: aluminum hydrogen carbonate

PP 9.11

a. Cation: Sn^{2+}

Anion: 1. name of the acid (*-ic*/*-ate*): phosphoric acid

2. formula of the acid: H_3PO_4

3. formula of the oxyanion: HPO_4^{2-}

Sn^{2+} HPO_4^{2-}

$(+2) + (-2) = 0$

Formula: $SnHPO_4$

b. Cation: Hg_2^{2+}

Anion: 1. name of the acid (*-ic*/*ate*): nitric acid

2. formula of the acid: HNO_3

3. formula of the oxyanion: NO_3^-

Hg_2^{2+} NO_3^-

$(+2) + \mathbf{2}(-1) = 0$

Formula: $Hg_2(NO_3)_2$

c. Cation: Na^+

Anion: CrO_4^{2-} (chromate ion should be memorized)

Na^+ CrO_4^{2-}

$\mathbf{2}(+1) + (-2) = 0$

Na_2CrO_4

Answers to Chapter Review

1. IUPAC
2. common
3. rules
4. memorized
5. oxidation
6. zero
7. −2
8. +1
9. −1
10. charge
11. ternary
12. ionic
13. cation
14. suffix
15. binary
16. monatomic
17. diatomic
18. two
19. Roman numeral
20. salt
21. hydrogen
22. sum
23. prefixes
24. tri-
25. penta-
26. acids
27. oxygen
28. hydro-
29. less
30. monoprotic
31. triprotic
32. more
33. hypo-
34. oxyacids
35. -ate
36. -ite
37. acid salt
38. dihydrogen
39. bi-
40. bicarbonate
41. H's

10

Chemical Equations

Contents

The reaction of copper with aqueous silver nitrate produces silver crystals, which cling to the copper.

Fire continues to captivate the attention of people.

You Probably Know . . .

- One of the first chemical reactions you were aware of was probably combustion. Fire has captivated the attention of young and old throughout the ages. As we continue to grapple with the issues of energy and the environment, combustion is one of the most important reactions to understand.
- Numerous chemical reactions, all interrelated, occur within the human body. Compounds in the food we eat are broken down in our bodies, releasing life-sustaining energy and producing body growth.
- Thousands of chemicals and consumer products are produced by industry. All of these are the result of chemical reactions carried out on a large scale in manufacturing facilities all over the world.
- As chemists carry out experiments in their laboratories, they describe chemical reactions using chemical equations.

A laboratory experiment.

Why Should We Study Chemical Reactions and Equations?

In the previous chapter, we studied the names and formulas of inorganic compounds. These are the basis of the chemical language—the words and formulas we use in talking and writing about chemical substances.

However, we have learned that chemistry is about change. Chemical changes are described by chemical equations. A chemical equation tells us what substances enter into a reaction and what new substances are formed. Furthermore, it tells us relative quantities of the substances involved. In other words, a chemical equation is a detailed qualitative and quantitative description of a reaction.

In this chapter, we study various classifications of reactions and how to write equations to describe them. A knowledge of chemical equations allows us to understand what chemists have learned about numerous chemical reactions that occur in our bodies, in the environment, and in the manufacture of countless consumer products.

10.1 Chemical Reactions and Equations

Chemical reactions are the heart of chemistry. Your study of chemistry to this point has been leading up to a study of chemical reactions. You have studied the structure of atoms and of compounds, both those that are molecular and those that are ionic. You have studied the periodic table and used it to write formulas of simple ions. You have also studied rules and strategies for naming various types of compounds and for writing their formulas. All this is necessary background for understanding chemical reactions.

Chemical reactions were introduced in Chapter 2. From that brief introduction you may remember that a chemical reaction always produces new substances. There is a change in composition of the substances entering into a reaction. This distinguishes a chemical change from a physical change.

In our earlier discussion of chemical reactions, we learned that we often begin our description of a reaction with a word equation. A word equation uses words in equation form to describe the reaction. For the reaction that occurs when charcoal (carbon) burns, the word equation is

carbon + oxygen ⟶ carbon dioxide

This reaction also occurs when coke, a form of carbon, is burned in a blast furnace during the conversion of iron ore to iron (Figure 10.1).

FIGURE 10.1 Coke is a form of carbon formed by heating coal. Very high temperatures are produced when coke is burned in a blast furnace.

Remember, a chemical equation is an abbreviated form of a word equation that is written using chemical symbols and formulas (Section 2.4). With an understanding of symbols of elements and formulas of compounds, you are now ready to write chemical equations for a variety of reactions. To write an equation to describe burning charcoal, you must know the symbol for carbon and the formulas of oxygen (a diatomic element) and carbon dioxide:

$$C + O_2 \longrightarrow CO_2$$

For this reaction, carbon and oxygen are the reactants, and carbon dioxide is the product. Remember, the *reactants* are the substances entering into the chemical change, and the *products* are the substances formed in the reaction. The arrow symbolizes change and is read "produce" or "yield."

reactants ⟶ products (Section 2.4)

Read an arrow in an equation as "produce(s)" or "yield(s)."

Before you practice writing and balancing equations, you need to become familiar with the abbreviations and symbols that are used in writing equations.

10.2 Notation Used in Equations

Learning Objective

- Explain the meaning of the symbols commonly used in chemical equations.

In addition to the chemical symbols and formulas that represent the substances involved in the reaction, other symbols are used in chemical equations to give additional information. For example, double arrows are used to indicate a *reversible reaction,* in which products react to form the original reactants. When a product is a gas, an arrow pointing up (↑) may be used. An arrow pointing down (↓) indicates that the product formed is a solid (called a precipitate). However, generally the physical state of a substance is shown by using the symbols (s) for solid, (l) for liquid, or (g) for gas. The symbol (aq) is used to show that the substance has been dissolved in water to give an aqueous solution. A delta (Δ) is written above or below the arrow to show that heat is being added to the reactants to cause them to react. These symbols are shown in Table 10.1.

$\rightleftarrows$ indicates a reversible reaction.

a

b

Calcium carbonate, $CaCO_3$, is found naturally as marble, limestone, and in seashells. (a) Marble gravestones have lasted for centuries. (b) Seashells.

TABLE 10.1 Symbols Used in Chemical Equations

Symbol	*Meaning*
+	Read "plus" or "and" (written between formulas).
$\rightarrow$	Arrow symbolizes change; read "yields" or "produces" (separates reactants and products).
$\rightleftarrows$	Reaction is reversible.
$\uparrow$	Product is evolved as a gas.
$\downarrow$	Product is a solid or a precipitate.
(s)	Substance is a solid.
(l)	Substance is a liquid.
(g)	Substance is a gas.
(aq)	Substance is in aqueous solution.
Δ	Heat is added (written above or below the arrow).

The following equation for the reaction of calcium carbonate with phosphoric acid illustrates the use of some of these symbols:

$$3CaCO_3(s) + 2H_3PO_4(aq) \rightarrow Ca_3(PO_4)_2(s) + 3H_2O(l) + 3CO_2(g)$$

This equation tells us that solid calcium carbonate reacts with aqueous phosphoric acid to produce solid calcium phosphate plus liquid water and gaseous carbon dioxide.

You must learn the symbols in Table 10.1 to be able to read and write chemical equations.

Writing and Balancing Equations

Learning Objectives

- Explain the law of conservation of matter.
- Outline a three-step strategy for writing balanced chemical equations.
- Balance chemical equations.

When writing chemical equations, you must consider the law of conservation of matter (Section 2.5). This law says that matter is neither gained (created) nor lost (destroyed) during a chemical reaction. In other words, the total mass of products from a chemical reaction is equal to the total mass of reactants. This law resulted from experiments involving careful measurements made by Antoine-Laurent Lavoisier (Section 1.2) two centuries ago. His work, more than that of any other individual of his time, led to chemistry becoming the precise science that it is today. The law of conservation of matter has been confirmed by chemists over the years in countless numbers of experiments.

A mass balance is required by the law of conservation of matter.

When charcoal is burned, the law of conservation of matter tells us that the mass of carbon dioxide produced is equal to the mass of carbon burned plus the mass of oxygen consumed. A mass balance of reactants and products is shown in

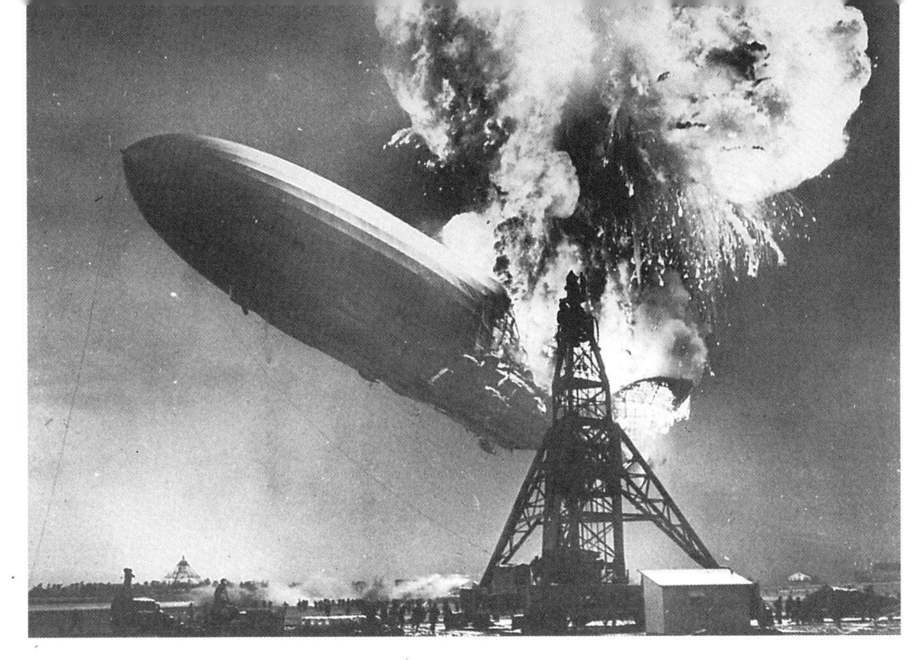

FIGURE 10.2 The German airship *Hindenberg,* the largest ever built, caught fire when it was landing at Lakehurst, New Jersey, on May 6, 1937. The use of highly flammable hydrogen gas in dirigibles was discontinued after this accident. Today, helium is used in place of hydrogen.

the equation by showing the same number of atoms of each element on both sides of the arrow. You can see that the equation for burning charcoal is balanced by simply writing the formulas for the reactants and products.

$$C(s) + O_2(g) \longrightarrow CO_2(g)$$

However, many equations for reactions are not balanced by simply writing the formulas for reactants and products. Let's consider the burning of hydrogen. In the 1930s, hydrogen was used in balloons and dirigibles to make them rise in the atmosphere. However, hydrogen is very flammable, and fatal accidents with hydrogen-filled dirigibles such as the German airship *Hindenberg* resulted in discontinuation of the use of hydrogen for this purpose (Figure 10.2).

Through careful observations, chemists have learned that water is the only compound produced when hydrogen burns. Thus, the word equation for burning hydrogen is

hydrogen + oxygen ⟶ water

From the word equation, write the skeletal (unbalanced) chemical equation. Note that hydrogen and oxygen are diatomic.

$$H_2 + O_2 \longrightarrow H_2O$$

As you inspect the skeletal equation, you can quickly see that it is not balanced because there are different numbers of oxygen atoms on the left and on the right. Now we need to consider how to balance an equation.

To balance an equation, you must show equal numbers of atoms on both sides of the equation by placing numbers in front of formulas. In this equation, because there are two oxygen atoms on the left side, you need two of them on the right side. Thus, you need two water molecules for each oxygen molecule reacting. Then the equation looks like this:

$$H_2 + O_2 \longrightarrow 2H_2O$$

By counting the hydrogen atoms you can see that hydrogen is no longer balanced because there are four hydrogen atoms on the right side but only two on the left side. Two more hydrogen atoms are needed on the left side. The balanced equation looks like this:

$$2H_2(g) + O_2(g) \longrightarrow 2H_2O(g)$$

Write the symbol for a gas (g) after the formulas. At the reaction temperature, water is also a gas.

This process is sometimes called "balancing by inspection." The numbers that you have written in front of the formulas to balance the equation are called balancing coefficients. They were needed to balance the equation but are not part of the formulas for the elements and compounds.

A balancing coefficient is not part of a formula.

Your success in writing chemical equations depends upon your understanding the steps involved and practice. You should carefully follow the three-step strategy shown in Figure 10.3. The following examples illustrate this three-step strategy and provide additional pointers that will help you to successfully balance equations.

1. Write (or think of) the word equation for the reaction.
2. Write the correct formulas for the reactants and products (skeletal or unbalanced equation).
3. Balance the equation using balancing coefficients.

FIGURE 10.3 The basic steps for writing and balancing equations

Example 10.1

Ammonia (NH_3) is a valuable fertilizer. It is manufactured (Haber process) by the reaction of nitrogen with hydrogen at high temperature. Write the balanced equation for the manufacture of ammonia. ■

Solution

1. Word equation:

nitrogen + hydrogen ⟶ ammonia

2. Correct formulas (skeletal equation):

$$N_2 + H_2 \longrightarrow NH_3$$

3. Balance the equation:

Step 1 Start with the most complicated formula, NH_3. (In general, balance H and O last.)

■ Balance N: Starting with 2N (in N_2) on the left, you need **$2NH_3$** on the right. (With more complex formulas, begin with the element having the most atoms.)

2N needed

$$N_2 + H_2 \longrightarrow \mathbf{2}NH_3$$

Step 2

■ Balance H: On the right, there are 6H in $2NH_3$ ($2 \times 3H = 6H$). To have 6H on the left side, you must write **$3H_2$**. Write the symbol for a gas (g) after the formulas.

$$N_2(g) + \mathbf{3}H_2(g) \longrightarrow \mathbf{2}NH_3(g)$$

$2 \times 3H = 6H$

6H needed on each side

Check: Reverse the steps. Check H; then check N. 

Example 10.2

Propane (C_3H_8) is a liquefied petroleum gas (LPG) used as a cooking and heating fuel in rural areas or wherever natural gas is not available. Write the balanced equation for the combustion of propane to give carbon dioxide and water. ■

Solution

1. Word equation:

propane + oxygen ⟶ carbon dioxide + water

2. Correct formulas (skeletal equation):

$$C_3H_8 + O_2 \longrightarrow CO_2 + H_2O$$

3. Balance the equation:

Step 1 Start with the most complicated formula, C_3H_8.

- Balance C: Starting with 3C on the left, you need **3**CO_2 on the right.

3C needed

$$C_3H_8 + O_2 \longrightarrow \mathbf{3}CO_2 + H_2O$$

Step 2 Balance H and O.

- Balance H: Starting with 8H on the left, **4**H_2O are needed on the right.

8H needed

$$C_3H_8 + O_2 \longrightarrow 3CO_2 + \mathbf{4}H_2O$$

$4 \times 2H = 8H$

- Balance O: On the right side, calculate 10 O-atoms; **5**O_2 are needed on the left side. Write the symbol for a gas (g) after the formulas.

$$C_3H_8(g) + \mathbf{5}O_2(g) \rightarrow 3CO_2(g) + \mathbf{4}H_2O(g)$$

$3 \times 2O + 4O = 10O$

10O needed

Check: Check O, H, and C, in that order. 

Baking soda ($NaHCO_3$) can be used to neutralize battery acid (H_2SO_4).

Example 10.3

Write a balanced chemical equation for the reaction of sodium bicarbonate (baking soda) with aqueous sulfuric acid (battery acid) to give sodium sulfate, carbon dioxide, and water. ■

Solution

1. Word equation:

sodium bicarbonate + sulfuric acid ⟶ sodium sulfate + carbon dioxide + water

2. Correct formulas (skeletal equation):

$$NaHCO_3 + H_2SO_4 \longrightarrow Na_2SO_4 + CO_2 + H_2O$$

3. Balance the equation:

Step 1 Start with $NaHCO_3$.

- Balance Na (remember, balance H and O last):

2Na needed

$$\mathbf{2}NaHCO_3 + H_2SO_4 \longrightarrow Na_2SO_4 + CO_2 + H_2O$$

- Balance C: 2C are on the left side; **2**C are needed on the right side.

$$\mathbf{2}NaHCO_3 + H_2SO_4 \longrightarrow Na_2SO_4 + \mathbf{2}CO_2 + H_2O$$

$2 \times 1C = 2C$

2C needed

- Balance SO_4 (actually SO_4^{2-}): Rather than balancing S and O separately, it is faster to balance the polyatomic ion as a unit when it appears on both sides. One SO_4 on the left side is balanced with one SO_4 on the right side.

Step 2 Balance H and O.

- Balance H:

$$2NaHCO_3 + H_2SO_4 \longrightarrow Na_2SO_4 + 2CO_2 + \mathbf{2}H_2O$$

$2 \times 1H + 2H = 4H$

4H needed

- Balance O: Remember, do not count O in SO_4. The 6O on the left side are balanced by 6O on the right side. Add symbols after the formulas to indicate the physical states of the substances.

$$2NaHCO_3(s) + H_2SO_4(aq) \longrightarrow Na_2SO_4(aq) + 2CO_2(g) + 2H_2O(l)$$

$2 \times 3O = 6O$

6O needed

Check: Check O, H, SO_4, C, and Na. 

To avoid mistakes commonly made by beginning students, remember to follow the three-step strategy. Do not try to balance the equation until you are sure the formulas are correct. Students often make mistakes when they try to invent formulas that will give them a balanced equation. Continue to practice this three-step strategy as you write equations for the following reactions.

Remember, write a skeletal equation before balancing.

Practice Problem 10.1 Write the balanced equations for the following reactions. (Solutions to practice problems appear at the end of this chapter.)

a. Potassium chlorate decomposes to give potassium chloride and oxygen.
b. Iron(III) oxide plus carbon monoxide yield iron and carbon dioxide.
c. Nitric acid plus calcium hydroxide give water and calcium nitrate.

Writing and Balancing Equations

1. Write the word equation.
2. Write the skeletal equation.
3. Balance the equation.

Step 1 Start with the most complicated formula. (Balance H and O last.)
- Start with the element having the most atoms.
- Balance a polyatomic ion as a unit when it appears on both sides of the equation.

Step 2 Balance H and O.

Check: Check the elements in reverse order. 

Classifying Chemical Reactions

10.4

Learning Objective

- List five classifications of common chemical reactions.

Information is easier to understand and remember when it is organized and classified. Over the years chemists have studied thousands of chemical reactions. Many of the more common reactions fit the classifications listed in Figure 10.4.

All but the third classification involve inorganic reactions. Some of the reactions included in these classifications fit into a broader classification of oxidation–reduction reactions (Section 6.8). More complex oxidation–reduction reactions are covered in Chapter 17.

1. Combination
2. Decomposition
3. Combustion of organic compounds
4. Single replacement
5. Double replacement

FIGURE 10.4 Classifications for common reactions.

10.5 Combination Reactions

Learning Objective

- Complete and balance equations for selected combination reactions.

A **combination reaction** is a reaction in which two or more substances (elements or compounds) combine to form a single compound.

$$A + B \rightarrow AB$$

The combination of two elements gives a binary compound.

The combining substances can be elements or compounds. The examples shown here illustrate the most common type of combination reaction, the combination of two elements to form a binary compound:

$$2Na(s) + Cl_2(g) \rightarrow 2NaCl(s)$$

$$2Mg(s) + O_2(g) \rightarrow 2MgO(s)$$

$$2H_2(g) + O_2(g) \rightarrow 2H_2O(g)$$

$$Mg(s) + S(s) \rightarrow MgS(s)$$

Writing formulas for binary compounds was covered in Section 9.5. Remember, the charge on many monatomic ions (A-groups) can be determined using the periodic table. Other more complicated examples will not concern us here.

Some combination reactions involve the reaction of two compounds. For example, some metal oxides and nonmetal oxides combine with water (Section 15.2):

$$MgO(s) + H_2O(l) \rightarrow Mg(OH)_2(s)$$

$$SO_2(g) + H_2O(l) \rightarrow H_2SO_3(aq)$$

Example 10.4

Write the balanced equation for the combination of magnesium with bromine. ■

Solution The product will be magnesium bromide.

Cation: Mg^{2+} (Group 2A)
Anion: Br^- (Group 7A)
Formula: $MgBr_2$

Watch out for the diatomic elements: H_2, N_2, O_2, F_2, Cl_2, Br_2, I_2.

Salts and most metals are solids.

Remember, bromine, a liquid, is one of the diatomic elements. You can determine the physical state for each of the substances by remembering that most metals and all salts are solids. When you do not know the physical state, you can find the information in a chemistry handbook.

$$Mg(s) + Br_2(l) \rightarrow MgBr_2(s)$$

All of the balancing coefficients are "1" (not written).

Example 10.5

Write a balanced equation for the combination of sulfur with oxygen. ■

Solution $S(s) + O_2(g) \longrightarrow SO_2(g)$

All of the balancing coefficients are 1.

Sulfur dioxide, a serious air pollutant, is formed when coal that contains sulfur is burned. The reaction involves a combination of sulfur with oxygen to give sulfur dioxide.

Practice Problem 10.2 Write balanced equations for:
a. the reaction of aluminum with sulfur
b. the reaction of calcium with chlorine
c. the reaction of magnesium with nitrogen

10.6 Decomposition Reactions

Learning Objective

- Complete and balance equations for selected decomposition reactions.

In a **decomposition reaction,** a single compound decomposes or breaks down to give two or more elements or compounds.

$$AB \longrightarrow A + B$$

A decomposition reaction may be thought of as the opposite of a combination reaction. Decomposition reactions are often carried out by strongly heating a compound, although other forms of energy such as ultraviolet radiation and electricity can be used. Oxidation–reduction is involved when elements are formed in decomposition reactions. The following are examples of compounds that decompose when heated:

Metal oxides Some metal oxides decompose to give the metal and oxygen. For example, heating mercury(II) oxide results in the formation of mercury and oxygen. It was by studying this reaction that Joseph Priestley (Section 1.2) discovered oxygen.

$$2HgO(s) \longrightarrow 2Hg(s) + O_2(g)$$

Salts of oxyacids Some salts of this type yield oxygen when heated. Cases are known where explosions have resulted, so caution should be exercised when using these compounds in experiments. In some cases, as with $NaNO_3$, not all of the oxygen is lost in the decomposition.

$$2KClO_3(s) \longrightarrow 2KCl(s) + 3O_2(g)$$

$$2NaNO_3(s) \longrightarrow 2NaNO_2(s) + O_2(g)$$

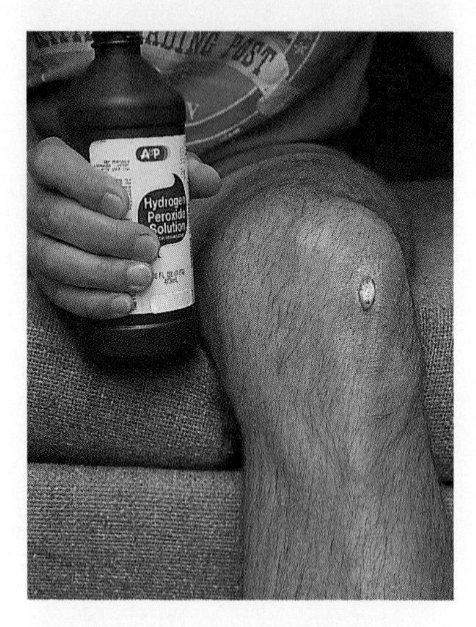

A dilute solution of hydrogen peroxide is a common antiseptic. When placed in an open wound, it decomposes to produce bubbles of oxygen gas.

Carbonates Some carbonates decompose when heated to produce carbon dioxide. For example, heating limestone, $CaCO_3$, gives lime, CaO, and carbon dioxide, a reaction that occurs in a blast furnace during the manufacture of steel. Note that oxidation–reduction is not involved in this reaction.

$$CaCO_3(s) \longrightarrow CaO(s) + CO_2(g)$$

Hydrogen peroxide and other peroxides The fizzing that you may have observed when you put hydrogen peroxide on a wound results from oxygen, a product of its decomposition. Hydrogen peroxide decomposes when it comes in contact with impurities or when heated.

$$2H_2O_2(aq) \longrightarrow 2H_2O(l) + O_2(g)$$

The storage life of hydrogen peroxide can be extended by keeping it under refrigeration. Because light also initiates its decomposition, it is usually packaged in dark bottles. However, even with these precautions, it gradually loses its strength.

Example 10.6

Write the chemical equation to describe the decomposition of water by electrolysis to give hydrogen and oxygen. Electrical energy breaks the strong O—H bonds (Section 2.3). ■

Solution The problem describes the reaction in words. Write a skeletal equation (correct formulas) and then balance the equation.

$$2H_2O(l) \xrightarrow{\text{electrolysis}} 2H_2(g) + O_2(g)$$

Example 10.7

Write the chemical equation for the decomposition of silver oxide (Ag_2O) to give silver and oxygen. ■

Solution $2Ag_2O(s) \longrightarrow 4Ag(s) + O_2(g)$

Practice Problem 10.3 Write an equation to describe the decomposition of strontium carbonate, $SrCO_3$.

10.7 Combustion of Organic Compounds

Learning Objective

- Complete and balance equations for the combustion of a hydrocarbon or a C, H, O compound.

Combustion is a chemical reaction that releases both heat and light. Combustion occurs when we burn wood, paper, and a variety of other fuels. In these reactions, oxygen (from the air) is an oxidizing agent, and the fuel is a reducing agent.

Oxidizing agent: An electron acceptor. Reducing agent: An electron donor (Section 6.8).

Most organic compounds burn. This means they are fuels. Most of our common fuels are either hydrocarbons, such as methane (in natural gas), propane, and gasoline, or compounds made up of carbon, hydrogen, and oxygen, such as alcohol and carbohydrates. **Hydrocarbons** are compounds that are made up only of hydrogen and carbon. **Carbohydrates**—composed of C, H, and O—include sugars (sucrose, glucose), starches (amylose), and cellulose. When these compounds are burned, a chemical reaction occurs with oxygen to produce water and carbon dioxide.

Carbohydrate: Carbo = *carbon,* hydrate = *water. From their empirical formula (CH_2O), many carbohydrates appear to have water combined with carbon. Their molecular formulas are more complicated.*

Methane (CH_4) and propane (C_3H_8) can be burned efficiently in modern furnaces and are common heating fuels:

$$CH_4(g) + 2O_2(g) \longrightarrow CO_2(g) + 2H_2O(g)$$

$$C_3H_8(g) + 5O_2(g) \longrightarrow 3CO_2(g) + 4H_2O(g)$$

Ethanol (C_2H_5OH), often simply called alcohol (grain alcohol), is added to gasoline to produce a fuel called gasohol (Figure 10.5). Because of the oxygen content of alcohol, a lower O_2-to-fuel ratio than for gasoline alone is needed for combustion. Using gasohol and other oxygenated fuels in motor vehicles reduces carbon monoxide emissions resulting from incomplete combustion:

$$C_2H_5OH(l) + 3O_2(g) \longrightarrow 2CO_2(g) + 3H_2O(g)$$

FIGURE 10.5 Gasohol is a motor fuel sold mostly in corn-producing states and in cities to combat air pollution. The alcohol used in this fuel is manufactured from corn.

Example 10.8

Write a balanced equation for the complete combustion of butane (C_4H_{10}), a fuel used in campstoves, lanterns, and cigarette lighters. ■

Solution

1. Word equation:

butane + oxygen ⟶ carbon dioxide + water

2. Correct formulas:

$$C_4H_{10} + O_2 \longrightarrow CO_2 + H_2O$$

3. Balanced equation:

Step 1
Start with C_4H_{10}.
- Balance C and H:

4C are needed

$$C_4H_{10} + O_2 \longrightarrow \mathbf{4}CO_2 + \mathbf{5}H_2O$$

10H are needed

Step 2

- Balance O:

$$C_4H_{10} + \frac{13}{2} O_2 \longrightarrow 4CO_2 + 5H_2O$$

$\frac{13}{2} \times 20 = 130$

$4 \times 20 + 50 = 130$

130 are needed

To clear the equation of the fraction, multiply all coefficients by 2. Write the symbols for a gas (g) after the formulas.

$$2C_4H_{10}(g) + 13O_2(g) \longrightarrow 8CO_2(g) + 10H_2O(g)$$

Practice Problem 10.4 Write a balanced equation for the combustion of acetone, C_3H_6O.

10.8 Single Replacement Reactions

Learning Objective

- Using an activity series, complete and balance single replacement reactions.

In a **single replacement reaction** an element reacting with a compound replaces one of the elements making up the compound:

$$A + BC \longrightarrow B + AC$$

Here, element A replaces element B in compound BC, resulting in the formation of element B and compound AC.

Most examples of single replacement reactions involve a metal reacting with a salt or acid in aqueous solution. The following are examples of this type of reaction:

$$Zn(s) + Cu(NO_3)_2(aq) \longrightarrow Cu(s) + Zn(NO_3)_2(aq)$$

$$Ni(s) + 2AgNO_3(aq) \longrightarrow 2Ag(s) + Ni(NO_3)_2(aq)$$

$$Mg(s) + 2HCl(aq) \longrightarrow H_2(g) + MgCl_2(aq)$$

Note that the reverse reactions do not occur. For example:

$$Cu(s) + Zn(NO_3)_2(aq) \longrightarrow \text{no reaction}$$

This illustrates that copper is less reactive than zinc.

Some of the most active metals are able to replace hydrogen from water. The reaction of sodium with water is an example:

$$2Na(s) + 2H_2O(l) \longrightarrow H_2(g) + 2NaOH(aq)$$

TABLE 10.2 Activity Series of Selected Metals

Most active (best reducing agents)	Li
Will liberate H_2 from cold water, steam, or acids	K
	Ba
	Ca
	Na
Will liberate H_2 from steam or acids	Mg
	Al
	Zn
	Fe
Will liberate H_2 only from acids	Ni
	Sn
	Pb
	H
Will *not* liberate H_2 from acids	Cu
	Hg
	Ag
	Pt
Least active (poorest reducing agents)	Au

Sodium hydroxide, a base, is also formed. A **base** is a metal hydroxide or other ionic compound in which the anion is hydroxide ion (OH^-). This reaction can be violent, and the hydrogen that forms is explosive when mixed with air. Serious laboratory accidents have resulted from careless handling of sodium.

Sodium reacts violently with water.

Metals have been ranked according to their ability to displace other metals or hydrogen from aqueous solutions of salts or acids. The result is an *activity series of metals* (Table 10.2).

Metals high in the series liberate hydrogen upon reaction with cold water. Other metals lower in the series do not react with cold water but liberate hydrogen from acids. An example is the reaction of zinc with hydrochloric acid:

$$Zn(s) + 2HCl(aq) \longrightarrow ZnCl_2(aq) + H_2(g)$$

On the other hand, a metal below hydrogen in the series will *not* displace hydrogen from an aqueous solution of an acid. Moreover, when a metal is added to a salt solution, it will displace a metal from the solution if the metal making up the salt is lower in the series. Thus, copper wire reacts with aqueous silver nitrate to give elemental silver (Figure 10.6):

$$Cu(s) + 2AgNO_3(aq) \longrightarrow Cu(NO_3)_2(aq) + 2Ag(s)$$

FIGURE 10.6 Copper wire reacts with a solution of silver nitrate to produce silver crystals. The blue color of the solution is due to the presence of copper(II) ions.

$$Cu(s) + 2AgNO_3(aq) \longrightarrow 2Ag(s) + Cu(NO_3)_2(aq)$$

In this example, copper is higher in the activity series and is more active than silver which is lower in the series. Copper is oxidized and is the reducing agent. Silver is reduced and is the oxidizing agent.

Remember that metals are reducing agents (electron donors). The best reducing agents are high in the activity series. In single replacement reactions, water (H^+) or metal cations in solution act as oxidizing agents (electron acceptors) and are reduced. Thus, single replacement reactions are oxidation–reduction reactions.

Another type of single replacement reaction involves the reaction of a halogen with a solution containing halide ions. For example, chlorine reacts with a solution of sodium bromide to give bromine:

$$Cl_2(aq) + 2NaBr(aq) \longrightarrow Br_2(aq) + 2NaCl(aq)$$

Reactivity of halogens: $F_2 > Cl_2 > Br_2 > I_2$.

The oxidizing agent is Cl_2, and bromide ion (Br^-) is undergoing oxidation. The reactivity of the halogens follows the order F_2, Cl_2, Br_2, I_2, where F_2 is the most reactive. The reaction will occur when the reacting halogen is more reactive than the halogen formed in the reaction but not when the reverse is true. For example:

$$Br_2(aq) + NaCl(aq) \longrightarrow \text{no reaction}$$

Fluorine is not used in these reactions because it reacts explosively with water.

Example 10.9

Write an equation to describe the reaction of zinc with silver nitrate solution. ■

Solution Zinc is above silver in the activity series, so it replaces silver in solution. Note the charge on Zn^{2+} (Table 9.6).

$$Zn(s) + 2AgNO_3(aq) \longrightarrow 2Ag(s) + Zn(NO_3)_2(aq)$$

When an aqueous solution of Cl_2 and NaBr is shaken with an organic solvent, the red color of Br_2 appears in the organic layer.

Example 10.10

Write an equation for the reaction of aluminum with aqueous sulfuric acid. ■

Solution Aluminum is above hydrogen in the activity series, so it replaces hydrogen in the solution to give H_2. Al (Group 3A) is oxidized to Al^{3+}.

$$2Al(s) + 3H_2SO_4(aq) \longrightarrow 3H_2(g) + Al_2(SO_4)_3(aq)$$

Practice Problem 10.5 Write equations for the following reactions. Write "NR" if no reaction occurs.

a. $K(s) + H_2O(l) \longrightarrow$

b. $Cu(s) + HCl(aq) \longrightarrow$

c. $I_2(aq) + NaCl(aq) \longrightarrow$

d. $Al(s) + Cu(NO_3)_2(aq) \longrightarrow$

10.9 Double Replacement Reactions

Learning Objectives

- Complete and balance double replacement reactions including precipitation reactions, acid–base reactions, and reactions in which a gas is formed.
- Identify these reactions: combination, decomposition, combustion, single replacement, and double replacement.

CHEMICAL WORLD

Happier Troops Through Chemistry

Front-line rations. American GIs have never had warm feelings toward cold rations. During World War II K-rations were the most popular, mainly because they contained a chocolate powder meant for use when hot water was available, a rare occasion. But the GIs made a paste of the chocolate with a little water and spread it on the dry round crackers: instant chocolate cookies!

No one thought much about the relationship between food and morale until the war was winding down. After all, war is supposed to be hell, isn't it? Anyway, soldiers haven't been able to build fires at the front since about 1850 when wars, at least civilized wars, stopped for (hot) tea in the afternoon. The problem since then is that artillery and mortars have become accurate, and smoke or flame draws enemy fire. So cold rations were the only way to go.

But through chemistry, there's now a way to get smokeless heat—and hot meals. Less than an ounce of magnesium (24 g) mixed with ordinary water generates enough heat to boil a liter of water:

$$Mg(s) + 2H_2O(l) \longrightarrow Mg(OH)_2(s) + H_2(g) + 355 \text{ kJ}$$

No flame, no smoke, no toxic chemicals. For the first time, during Operation Desert Storm, GIs were issued Flameless Ration Heaters (FRH), which, with a small amount of magnesium, heat U.S. Army Standard Food Packs to about 100°F in 12 minutes, with enough residual heat to keep the pack warm for about an hour. This is very handy for soldiers, who often have their meals interrupted. The FRH produces no poisonous fumes, and so it can be used in closed quarters or even in a GI's pocket.

Pure magnesium won't work in the FRH because, like aluminum, magnesium reacts with air to form an airtight film of metallic oxide (MgO) that clings to the surface of the metal and protects it from further oxidation (good for other uses). To avoid this problem, other chemicals are added to the FRH to dissolve the oxide.

An American soldier using a Flameless Ration Heater (FRH) during Operation Desert Storm.

Psychologists working with the armed forces believe these warm meals give a needed morale boost. Well, no doubt. But the FRH is nothing more than a simple device that makes use of an exothermic reaction. The reaction has been known for a long time; if someone had thought of it, the doughboys in World War I could have had warm rations, too. One wonders what other simple chemical applications may exist that haven't been thought of.

In a **double replacement reaction,** the cation of each reactant combines with the anion of the other to give two new compounds. The general form of the reaction is

$$AB + CD \longrightarrow AD + CB$$

The two cations are sometimes described as trading partners (anions) in the reaction. The driving force for these reactions is the formation of a **precipitate**

TABLE 10.3 Solubility of Salts in Water at 20°C

Soluble in Water	*Exceptions*
Cations	
1A metal cations (Na^+, K^+, etc.), NH_4^+	Some Li^+ salts
Anions	
1. nitrates (NO_3^-), chlorates (ClO_3^-), and acetates ($C_2H_3O_2^-$)	$AgC_2H_3O_2$
2. chlorides (Cl^-), bromides (Br^-), iodides (I^-)	Ag^+, Hg_2^{2+}, Pb^{2+}
3. sulfates (SO_4^{2-})	Ba^{2+}, Sr^{2+}, Pb^{2+} (Ca^{2+}, Ag^+ are slightly soluble)
Acids	
All the common acids	

Insoluble in Water	*Exceptions*
carbonates (CO_3^{2-}), phosphates (PO_4^{3-}), sulfides (S^{2-}), oxides (O^{2-})	1A cations, NH_4^+
hydroxides[a] (OH^-)	1A cations (Ca^{2+}, Sr^{2+}, Ba^{2+} are slightly soluble)

[a]Note that hydroxides are bases (Section 15.2), not salts.

(insoluble product), a gas, or other molecular substance such as H_2O. Double replacement reactions do not involve oxidation and reduction, so there are no changes in ionic charges or oxidation numbers.

The following types of reactions are included in this classification:

1. *Precipitation reactions.* One of the products of this type of reaction is an insoluble solid (precipitate).

$$CaCl_2(aq) + 2AgNO_3(aq) \longrightarrow 2AgCl(s) + Ca(NO_3)_2(aq)$$

$$Pb(NO_3)_2(aq) + K_2CO_3(aq) \longrightarrow PbCO_3(s) + 2KNO_3(aq)$$

$$Na_2S(aq) + CuCl_2(aq) \longrightarrow CuS(s) + 2NaCl(aq)$$

You can predict whether or not a precipitate will form when two solutions are mixed by consulting a table of solubilities. The solubility behavior of common salts is given in Table 10.3. When reading the table, note that salts are often classified according to the ions they have in common. For example, NH_4Cl and NH_4Br are ammonium salts; KCl, NaCl, and $CaCl_2$ are chlorides.

2. *Acid–base (neutralization) reactions.* **Neutralization** refers to the combination of H^+ (from an acid) and OH^- (from a base) to form H_2O. Neutralization is always exothermic, and the heat released can often be detected by simply touching the reaction container. Acids and bases are discussed further in Chapter 15.

$$HCl(aq) + NaOH(aq) \longrightarrow H_2O(l) + NaCl(aq)$$

$$H_2SO_4(aq) + 2KOH(aq) \longrightarrow 2H_2O(l) + K_2SO_4(aq)$$

3. *Formation of a gas.* One of the products of this type of reaction is a gas. In some reactions, a gas is formed indirectly by the decomposition of one of the products.

Carbonates and bicarbonates liberate carbon dioxide upon reaction with acids. This is the most common double replacement reaction in which a gas is evolved. Carbon dioxide is formed indirectly from the decomposition of carbonic acid:

$$NaHCO_3(aq) + HCl(aq) \longrightarrow NaCl(aq) + H_2CO_3(aq)$$

Then:

$$H_2CO_3(aq) \longrightarrow H_2O(l) + CO_2(g)$$

Adding these equations gives the following equation:

$$NaHCO_3(aq) + HCl(aq) \longrightarrow NaCl(aq) + H_2O(l) + CO_2(g)$$

This reaction occurs when someone drinks a solution of baking soda for an upset stomach. Some of the acid (HCl) in the stomach is neutralized by reaction with the baking soda. A similar reaction occurs when baking soda in sourdough pancake batter reacts with an acid in the batter to produce carbon dioxide, which causes the batter to rise (Figure 10.7).

FIGURE 10.7 Baking soda ($NaHCO_3$) reacts with lactic acid ($HC_3H_5O_3$) in sourdough batter to form carbon dioxide, which causes the batter to rise.

$$NaHCO_3(aq) + HC_3H_5O_3(aq) \longrightarrow NaC_3H_5O_3(aq) + H_2O(l) + CO_2(g)$$

The chemical reactions discussed in this chapter are summarized at the end of this section. Use the summary as a reference when you work the following examples.

Example 10.11

Complete and balance the following precipitation reactions. Refer to Table 10.3 to determine which product is insoluble. If both possible products are soluble, write "no reaction."

a. $Na_3PO_4(aq) + CaCl_2(aq) \longrightarrow$
b. $Pb(NO_3)_2(aq) + K_2S(aq) \longrightarrow$
c. $CuCl_2(aq) + NaOH(aq) \longrightarrow$ ■

Solutions

a. $2Na_3PO_4(aq) + 3CaCl_2(aq) \longrightarrow Ca_3(PO_4)_2(s) + 6NaCl(aq)$
b. $Pb(NO_3)_2(aq) + K_2S(aq) \longrightarrow PbS(s) + 2KNO_3(aq)$
c. $CuCl_2(aq) + 2NaOH(aq) \longrightarrow Cu(OH)_2(s) + 2NaCl(aq)$

Example 10.12

Complete and balance the following equations. All are acid-base reactions.

a. $H_2SO_4(aq) + Ca(OH)_2(aq) \longrightarrow$
b. $HNO_3(aq) + Ba(OH)_2(aq) \longrightarrow$
c. $HClO_3(aq) + KOH(aq) \longrightarrow$ ■

Solutions

a. $H_2SO_4(aq) + Ca(OH)_2(aq) \longrightarrow 2H_2O(l) + CaSO_4(s)$
b. $2HNO_3(aq) + Ba(OH)_2(aq) \longrightarrow 2H_2O(l) + Ba(NO_3)_2(aq)$
c. $HClO_3(aq) + KOH(aq) \longrightarrow H_2O(l) + KClO_3(aq)$

Example 10.13

In each of the following examples a gas is formed when the solutions are mixed. Write balanced equations for the reactions.

a. $Na_2CO_3(aq) + HNO_3(aq) \longrightarrow$
b. $K_2S(aq) + HBr(aq) \longrightarrow$ ■

Solutions

H_2S is emitted from rotten eggs. Stench!

a. $Na_2CO_3(aq) + 2HNO_3(aq) \longrightarrow 2NaNO_3(aq) + H_2O(l) + CO_2(g)$
b. $K_2S(aq) + 2HBr(aq) \longrightarrow 2KBr(aq) + H_2S(g)$

Practice Problem 10.6 Complete and balance the following equations.
a. $HCl(aq) + Ca(OH)_2(aq) \longrightarrow$
b. $CuCl_2(aq) + Na_2CO_3(aq) \longrightarrow$
c. $K_2CO_3(aq) + HBr(aq) \longrightarrow$

Common Chemical Reactions

Combination Reactions (A + B ⟶ AB)

1. element A + element B ⟶ binary compound

$$2Mg(s) + O_2(g) \longrightarrow 2MgO(s)$$

2. Compound A + compound B ⟶ ternary compound

$$CaO(s) + H_2O(l) \longrightarrow Ca(OH)_2(s)$$

$$SO_2(g) + H_2O(l) \longrightarrow H_2SO_3(aq)$$

Decomposition Reactions (AB ⟶ A + B)

1. metal oxide ⟶ metal + oxygen

$$2HgO(s) \longrightarrow 2Hg(l) + O_2(g)$$

2. salt of oxyacid ⟶ oxygen + binary commpound

$$2KClO_3(s) \longrightarrow 3O_2(g) + 2KCl(s)$$

3. carbonate ⟶ carbon dioxide + metal oxide

$$CaCO_3(s) \longrightarrow CO_2(g) + CaO(s)$$

4. hydrogen peroxide ⟶ water + oxygen

$$2H_2O_2(aq) \longrightarrow 2H_2O(l) + O_2(g)$$

5. water ⟶ hydrogen + oxygen

$$2H_2O(l) \xrightarrow{\text{electrolysis}} O_2(g) + 2H_2(g)$$

Combustion of Organic Compounds
(fuel + oxygen $\longrightarrow$ carbon dioxide + water)

1. hydrocarbon + oxygen $\longrightarrow$ carbon dioxide + water

$$CH_4(g) + 2O_2(g) \longrightarrow CO_2(g) + 2H_2O(l)$$

2. C, H, O compounds + oxygen $\longrightarrow$ carbon dioxide + water

$$C_2H_5OH(l) + 3O_2(g) \longrightarrow 2CO_2(g) + 3H_2O(l)$$

Single Replacement Reactions (A + BC $\longrightarrow$ B + AC)

1. $metal_1(s) + salt_2(aq) \longrightarrow metal_2(s) + salt_1(aq)$

$$Mg(s) + CuCl_2(aq) \longrightarrow Cu(s) + MgCl_2(aq)$$

2. metal(s) + acid(aq) $\longrightarrow$ hydrogen(g) + salt(aq)

$$Mg(s) + 2HCl(aq) \longrightarrow H_2(g) + MgCl_2(aq)$$

3. active metal(s) + water(l) $\longrightarrow$ hydrogen(g) + base(aq)

$$2K(s) + 2H_2O(l) \longrightarrow H_2(g) + 2KOH(aq)$$

4. $halogen_1(aq)$ + halide $salt_2(aq) \longrightarrow halogen_2(aq)$ + halide $salt_1(aq)$

$$Cl_2(aq) + 2NaBr(aq) \longrightarrow Br_2(aq) + 2NaCl(aq)$$

Double Replacement Reactions (AB + CD $\longrightarrow$ AD + CB)

1. two compounds(aq) $\longrightarrow$ precipitate + compound(aq or s)

$$NaCl(aq) + AgNO_3(aq) \longrightarrow AgCl(s) + NaNO_3(aq)$$

2. acid(aq) + base(aq) $\longrightarrow$ water + salt(aq or s)

$$HCl(aq) + NaOH(aq) \longrightarrow H_2O(l) + NaCl(aq)$$

3. two compounds(aq) $\longrightarrow$ gas + salt(aq or s)

$$Na_2S(aq) + 2HCl(aq) \longrightarrow H_2S(g) + 2NaCl(aq)$$

4. acid(aq) + carbonate(aq or s) $\longrightarrow$ carbon dioxide(g) + water + salt(aq or s)

$$2HCl(aq) + CaCO_3(s) \longrightarrow CO_2(g) + H_2O(l) + CaCl_2(aq)$$

Net Ionic Equations

10.10

Learning Objective

- Write balanced net ionic equations for reactions occurring in aqueous solution.

Earlier, we noted that when ionic compounds are dissolved in water, the ions dissociate and move about independently of one another (Section 8.2). Dissociation of a dissolved compound can be illustrated by writing an equation. For example:

$$NaCl(s) \xrightarrow{H_2O} Na^+(aq) + Cl^-(aq)$$

$$NaNO_3(s) \xrightarrow{H_2O} Na^+(aq) + NO_3^-(aq)$$

$$K_2SO_3(s) \xrightarrow{H_2O} 2K^+(aq) + SO_3^{2-}(aq)$$

$$CaBr_2(s) \xrightarrow{H_2O} Ca^{2+}(aq) + 2Br^-(aq)$$

$$(NH_4)_2SO_4(s) \xrightarrow{H_2O} 2NH_4^+(aq) + SO_4^{2-}(aq)$$

Notice that some subscripts in formulas indicate the numbers of cations and anions in the formula unit. These subscripts become the balancing coefficients for the ions in the equation. Do not confuse these subscripts (e.g., in $(NH_4)_2SO_4$, the 2 outside the parentheses) with those that are part of the formula unit of a polyatomic ion (the 4 inside the parentheses). When a compound dissolves in water, the composition of a polyatomic ion remains unchanged:

$$K_2SO_3(s) \longrightarrow \begin{matrix} K^+(aq) \\ K^+(aq) \end{matrix} + SO_3^{2-}(aq)$$

$$(NH_4)_2SO_4(s) \longrightarrow \begin{matrix} NH_4^+(aq) \\ NH_4^+(aq) \end{matrix} + SO_4^{2-}(aq)$$

Many chemical reactions occur in aqueous solution. For example, mixing solutions of sodium chloride and silver nitrate results in the precipitation of silver chloride. The *total equation* is

$$NaCl(aq) + AgNO_3(aq) \longrightarrow AgCl(s) + NaNO_3(aq)$$

Showing the ions present in solution gives an **ionic equation:**

$$Na^+(aq) + Cl^-(aq) + Ag^+(aq) + NO_3^-(aq) \longrightarrow AgCl(s) + Na^+(aq) + NO_3^-(aq)$$

Because silver chloride is undissolved, it is not dissociated and is therefore written as AgCl(s). Furthermore, because $Na^+(aq)$ and $NO_3^-(aq)$ appear on both sides of the equation, they may be canceled from the equation:

$$\cancel{Na^+(aq)} + Cl^-(aq) + Ag^+(aq) + \cancel{NO_3^-(aq)} \longrightarrow AgCl(s) + \cancel{Na^+(aq)} + \cancel{NO_3^-(aq)}$$

Spectator ions: Identical *formulas on both sides of the equation.*

By canceling these ions, we are showing that they are not part of the reaction. Ions in solution that do not take part in the reaction are called **spectator ions.** Only ions that have *identical* formulas on both sides of the equation qualify as spectator ions. The equation that results when spectator ions are canceled from an ionic equation is called a **net ionic equation:**

$$Cl^-(aq) + Ag^+(aq) \longrightarrow AgCl(s)$$

A net ionic equation helps to focus our attention on the ions that are actually involved in the reaction. Silver chloride precipitates when a solution containing chloride ion is mixed with a solution of silver ion regardless of the identity of the other ions in solution. For example, mixing solutions of calcium chloride and silver chlorate also gives a precipitate of silver chloride.

Total equation:

$$CaCl_2(aq) + 2AgClO_3(aq) \longrightarrow 2AgCl(s) + Ca(ClO_3)_2(aq)$$

Total ionic equation:

$$\cancel{Ca^{2+}(aq)} + 2Cl^-(aq) + 2Ag^+(aq) + \cancel{2ClO_3^-(aq)} \longrightarrow 2AgCl(s) + \cancel{Ca^{2+}(aq)} + \cancel{2ClO_3^-(aq)}$$

Net ionic equation:

$$2Cl^-(aq) + 2Ag^+(aq) \longrightarrow 2AgCl(s)$$

This equation may be reduced to:

$$Cl^-(aq) + Ag^+(aq) \longrightarrow AgCl(s)$$

Note that net ionic equations can be written for any reaction involving ionic substances in aqueous solution. Other examples include acid–base reactions, gas-forming reactions, and single replacement reactions.

Example 10.14

Write the net ionic equation for the reaction of potassium hydroxide and hydrochloric acid in aqueous solution. ■

Solution

Total equation:

$$KOH(aq) + HCl(aq) \longrightarrow H_2O(l) + KCl(aq)$$

Total ionic equation:

$$\cancel{K^+(aq)} + OH^-(aq) + H^+(aq) + \cancel{Cl^-(aq)} \longrightarrow H_2O(l) + \cancel{K^+(aq)} + \cancel{Cl^-(aq)}$$

Net ionic equation:

$$OH^-(aq) + H^+(aq) \longrightarrow H_2O(l)$$

Note that water is molecular and is not dissociated.

Example 10.15

Write the net ionic equation for the reaction of magnesium with hydrochloric acid. ■

Solution This is a single replacement reaction. Refer to the Activity Series (Table 10.2).

Total equation:

$$Mg(s) + 2HCl(aq) \longrightarrow H_2(g) + MgCl_2(aq)$$

Total ionic equation:

$$Mg(s) + 2H^+(aq) + \cancel{2Cl^-(aq)} \longrightarrow H_2(g) + Mg^{2+}(aq) + \cancel{2Cl^-(aq)}$$

Net ionic equation:

$$Mg(s) + 2H^+(aq) \longrightarrow H_2(g) + Mg^{2+}(aq)$$

Note that this is a redox reaction; Mg loses 2 e^- and is oxidized, and each H^+ gains 1 e^- and is reduced. Magnesium is the reducing agent (electron donor), and hydrochloric acid (or H^+) is the oxidizing agent (electron acceptor).

Note that Mg(s) is not an ion (no charge), nor is H_2 (a molecule, no charge).

Practice Problem 10.7 Write net ionic equations for the following reactions in aqueous solution (see Practice Problem 10.6):

a. $HCl(aq) + Ca(OH)_2(aq) \longrightarrow$

b. $CuCl_2(aq) + Na_2CO_3(aq) \longrightarrow$

c. $K_2CO_3(aq) + HBr(aq) \longrightarrow$

Chapter Review

Directions: From the list of choices at the end of the Chapter Review, choose the word or term that best fits in each blank. Use each choice only once. The answers appear at the end of the chapter.

Section 10.1 In a chemical reaction, __________ (1) substances are formed. This distinguishes a chemical change from a __________ (2) change. We begin our description of a reaction with a __________ (3) equation. A __________ (4) equation is an abbreviated form of a word equation using chemical symbols and formulas.

Section 10.2 In addition to the chemical symbols and formulas used to represent elements and compounds, other symbols are used in equations. For example, the symbols (s), (l), and (g) mean __________ (5), liquid, and __________ (6). A substance in aqueous (water) solution is indicated by __________ (7). Knowledge of these symbols is necessary to be able to read and write equations.

Section 10.3 When writing chemical equations, you must consider the law of conservation of __________ (8). This law tells us that the total mass of reactants equals the total mass of __________ (9) in a reaction. Thus, to be correct, an equation must be __________ (10). This is done using balancing __________ (11). A balanced equation shows equal numbers of __________ (12) of each element on both sides of the equation. In writing equations, follow a three-step strategy that includes a word equation, writing correct __________ (13) for reactants and products (skeletal equation), and balancing the equation.

Section 10.4 Many chemical reactions fit into five classifications: __________ (14), decomposition, __________ (15) of organic compounds, __________ (16) replacement, and double replacement.

Section 10.5 A reaction in which two __________ (17) or two compounds combine to form a single compound is called a combination reaction. The most common type is the combination of two elements to form a binary compound. For many examples the formula of the product can be written by referring to the periodic table to determine the __________ (18) on monatomic ions.

Section 10.6 In a __________ (19) reaction, a single compound breaks down to give two or more elements or compounds. Examples of compounds that decompose when heated include some metal oxides, salts of __________ (20), __________ (21), and hydrogen peroxide. __________ (22) decomposes by electrolysis to give hydrogen and oxygen.

Section 10.7 Combustion is a chemical reaction that releases both __________ (23) and light. Most of our common fuels are either __________ (24) or compounds made up of C, H, and O. Hydrocarbons are compounds made up of only hydrogen and __________ (25). Complete combustion of these organic fuels produces water and __________ __________ (26).

Section 10.8 In a single __________ (27) reaction one element displaces another from a compound in aqueous solution. This type of reaction can be predicted by referring to an ________ ________ (28) of metals. A metal below ________ (29) in the series will *not* react with an aqueous acid to liberate hydrogen. In another type of single replacement reaction, a halogen reacts with a solution of a ________ (30) ion. A reaction occurs only when the reacting halogen is more reactive than the halogen formed. The most reactive halogen, ________ (31), reacts explosively with water.

Section 10.9 In a double replacement reaction, the ________ (32) of each reactant combines with the anion of the other to give two new compounds. Included in this classification are ________ (33) reactions, acid–________ (34) reactions, and reactions in which a gas is formed. The products of an acid-base reaction are water and a ________ (35). When mixed with an ________ (36), a carbonate reacts to produce carbon dioxide.

Section 10.10 Ionic compounds dissociate when dissolved in water. A reaction that occurs when two solutions of ionic compounds are mixed can be represented by writing a ________ (37) ionic equation. In writing a net ionic equation, ________ (38) ions are canceled.

Choices: acid, activity series, (aq), atoms, balanced, base, carbon, carbon dioxide, carbonates, cation, charges, chemical, coefficients, combination, combustion, decomposition, elements, fluorine, formulas, gas, halide, heat, hydrocarbons, hydrogen, matter, net, new, oxyacids, physical, precipitation, products, replacement, salt, single, solid, spectator, water, word

Study Objectives

After studying Chapter 10, you should be able to:

Section 10.2

1. Explain the meaning of the symbols commonly used in chemical equations.

Section 10.3

2. Explain the law of conservation of matter.
3. Outline a three-step strategy for writing balanced chemical equations.
4. Balance chemical equations.

Section 10.4

5. List five classifications of common chemical reactions.

Section 10.5, 10.6

6. Complete and balance equations for selected combination and decomposition reactions.

Section 10.7

7. Complete and balance equations for the combustion of a hydrocarbon or a C, H, O compound.

Section 10.8

8. Using an activity series, complete and balance single replacement reactions.

Section 10.9

9. Complete and balance double replacement reactions including precipitation reactions, acid–base reactions, and reactions in which a gas is formed.
10. Identify these reactions: combination, decomposition, combustion, single replacement, and double replacement.

Section 10.10

11. Write balanced net ionic equations for reactions occurring in aqueous solution.

Key Terms

Review the definition of each of the terms listed here by chapter section number. You may use the glossary if necessary.

10.5 combination reaction

10.6 decomposition reaction

10.7 carbohydrate, combustion, hydrocarbon

10.8 base, single replacement reaction

10.9 double replacement reaction, neutralization, precipitate

10.10 ionic equation, net ionic equation, spectator ion

Questions and Problems

Answers to questions and problems with blue numbers appear in Appendix C.

Writing and Balancing Equations (Sections 10.2, 10.3)

1. Explain the meaning of each of the following symbols used in writing chemical equations:
   a. $\longrightarrow$
   b. $\rightleftarrows$
   c. $\uparrow$
   d. $\downarrow$
   e. (aq)
   f. (l)
   g. (s)
   h. Δ
   i. (g)
2. Balance the following equations:
   a. $Cu + O_2 \longrightarrow CuO$
   b. $Mg + HBr \longrightarrow H_2 + MgBr_2$
   c. $HI + Mg(OH)_2 \longrightarrow H_2O + MgI_2$
   d. $Na_2O + H_2O \longrightarrow NaOH$
   e. $CO_2 + H_2O \longrightarrow H_2CO_3$
3. Balance the following equations:
   a. $C_5H_{12} + O_2 \longrightarrow CO_2 + H_2O$
   b. $Ca(OH)_2 + HCl \longrightarrow CaCl_2 + H_2O$
   c. $CaCl_2 + AgNO_3 \longrightarrow AgCl + Ca(NO_3)_2$
   d. $Na_2CO_3 + HBr \longrightarrow NaBr + H_2O + CO_2$
   e. $Al + O_2 \longrightarrow Al_2O_3$
4. Balance the following equations:
   a. $CH_4O + O_2 \longrightarrow CO_2 + H_2O$
   b. $Al + Cu(NO_3)_2 \longrightarrow Cu + Al(NO_3)_3$
   c. $Na + Cl_2 \longrightarrow NaCl$
   d. $CaCO_3 \longrightarrow CaO + CO_2$
   e. $BaCl_2 + Na_2S \longrightarrow BaS + NaCl$
5. Balance the following equations:
   a. $NH_3 + O_2 \longrightarrow NO + H_2O$
   b. $Ni + HCl \longrightarrow NiCl_2 + H_2$
   c. $Ca(OH)_2 + H_3PO_4 \longrightarrow H_2O + Ca_3(PO_4)_2$
   d. $NaI + Cl_2 \longrightarrow NaCl + I_2$
6. Balance the following equations:
   a. $CaS + HCl \longrightarrow H_2S + CaCl_2$
   b. $NaCN + H_2SO_4 \longrightarrow HCN + Na_2SO_4$
7. Write a word equation for each reaction described:
   a. When aqueous solutions of iron(II) acetate and potassium sulfide are mixed, a black solid precipitates.
   b. When dinitrogen tetroxide gas is heated, it decomposes to give nitrogen dioxide, a brown gas.
   c. A white precipitate forms when a solution of mercury(I) nitrate is mixed with a solution of sodium chloride.
8. Write a word equation for each of the following reactions:
   a. Calcium oxide (lime) reacts with water to form calcium hydroxide (slaked lime).
   b. Magnesium reacts with steam to produce magnesium hydroxide and hydrogen gas.
   c. Stomach acid (hydrochloric acid) is neutralized by milk of magnesia (magnesium hydroxide) to give water and magnesium chloride.
9. Write balanced chemical equations for the reactions described in problem 7.
10. Write balanced chemical equations for the reactions described in problem 8.
11. Write word equations for each of the following reactions:
    a. Battery acid (sulfuric acid) reacts with iron to yield iron(II) sulfate and hydrogen gas.
    b. Electrolysis of melted sodium chloride yields sodium metal and chlorine gas.
    c. Copper smelting involves a reaction of copper(I) sulfide with oxygen to produce copper and sulfur dioxide.
12. Write balanced chemical equations for the reactions described in problem 11.
13. Liquid bleach (an aqueous solution of sodium hypochlorite and sodium chloride), reacts with vinegar (a solution of acetic acid) to produce chlorine gas, water, and sodium acetate. Write a balanced equation for this reaction.
14. Write balanced chemical equations for the following reactions:
    a. hexane (C_6H_{14}) + oxygen $\longrightarrow$ carbon dioxide + water
    b. copper + oxygen $\longrightarrow$ copper(II) oxide
    c. mercury(I) nitrate + sodium chloride $\longrightarrow$ mercury(I) chloride + sodium nitrate (aqueous)
15. Write balanced chemical equations for the following reactions in aqueous solution:
    a. potassium hydroxide + phosphoric acid $\longrightarrow$ water + potassium phosphate
    b. magnesium + nickel(II) chloride $\longrightarrow$ magnesium chloride + nickel
    c. magnesium carbonate + hydrochloric acid $\longrightarrow$ water + magnesium chloride + carbon dioxide
16. Glycerin (glycerol), $C_3H_8O_3$, obtained from the hydrolysis of fats, reacts with oxygen to give carbon dioxide and water. Write a balanced equation for this reaction.
17. Write balanced chemical equations for the following reactions in aqueous solution:
    a. potassium + water $\longrightarrow$ potassium hydroxide + hydrogen
    b. aluminum + sulfuric acid $\longrightarrow$ hydrogen + aluminum sulfate
    c. magnesium carbonate $\longrightarrow$ magnesium oxide + carbon dioxide
    d. sodium sulfate + barium chloride $\longrightarrow$ sodium chloride + barium sulfate

Combination Reactions (Section 10.5)

18. Complete and balance each of the following combination reactions:
    a. sulfur + calcium $\longrightarrow$
    b. magnesium + bromine $\longrightarrow$
    c. carbon + sulfur $\longrightarrow$
    d. aluminum + oxygen $\longrightarrow$
19. Write balanced chemical equations for the following reactions:
    a. In an automobile engine, nitrogen reacts with oxygen to form nitrogen oxide, NO.
    b. Sulfur trioxide reacts with water to produce sulfuric acid.
    c. The combination of phosphorus and chlorine produces phosphorus trichloride.

Decomposition Reactions (Section 10.6)

20. Write balanced equations for the decomposition of the following compounds when they are heated:
    a. sodium chlorate
    b. barium carbonate
21. Write balanced equations for the following reactions:
    a. Sodium peroxide decomposes to give sodium oxide and oxygen.
    b. Carbonic acid, H_2CO_3, in carbonated beverages, decomposes to give carbon dioxide and water.
22. Table sugar, $C_{12}H_{22}O_{11}$, decomposes when heated to give carbon and water. Write a chemical equation for the reaction.

Combustion of Organic Compounds (Section 10.7)

23. Write a balanced chemical equation for each reaction described:
    a. The combustion of glucose, $C_6H_{12}O_6(s)$
    b. Octane, $C_8H_{18}(l)$, a fuel, burns.
24. Write balanced equations for the following combustion reactions:
    a. Acetylene (C_2H_2) is burned in a welder's torch.
    b. Methyl alcohol (CH_3OH) is burned.
25. Write a balanced equation for the complete combustion of toluene (C_7H_8), a high octane compound in gasoline.

Single Replacement Reactions (Section 10.8)

26. Write chemical equations for the following reactions occurring in aqueous solution. If no reaction occurs, write "no reaction."
    a. silver + lead(II) nitrate $\longrightarrow$
    b. aluminum + hydrobromic acid $\longrightarrow$
    c. copper + hydrochloric acid $\longrightarrow$
    d. nickel + copper(II) nitrate $\longrightarrow$
27. Write chemical equations for the following reactions in aqueous solution. If no reaction occurs, write "no reaction."
    a. zinc + sulfuric acid $\longrightarrow$
    b. calcium + water $\longrightarrow$
    c. aluminum + water $\longrightarrow$
    d. magnesium + silver nitrate $\longrightarrow$
28. Write chemical equations for the following reactions in aqueous solution. If no reaction occurs, write "no reaction."
    a. chlorine + potassium bromide $\longrightarrow$
    b. bromine + sodium iodide $\longrightarrow$
    c. bromine + potassium chloride $\longrightarrow$

Double Replacement Reactions (Section 10.9)

29. Determine which of the following will react in aqueous solution. Refer to a table of solubilities as necessary. Write chemical equations for the reactions. If no reaction occurs, write "no reaction."
    a. silver nitrate + hydrochloric acid $\longrightarrow$
    b. copper(II) chloride + sodium sulfide $\longrightarrow$
    c. iron(III) nitrate + sodium chloride $\longrightarrow$
    d. lead(II) nitrate + potassium chromate $\longrightarrow$
30. Write a balanced equation for the reaction of acetic acid, $HC_2H_3O_2(aq)$, the acid in vinegar, with baking soda, $NaHCO_3$.
31. Write balanced equations for the following precipitation reactions. Write "no reaction" for those that do not react.
    a. sodium phosphate + calcium chloride $\longrightarrow$
    b. potassium sulfate + barium nitrate $\longrightarrow$
    c. mercury(II) nitrate + sodium chloride $\longrightarrow$
    d. aluminum nitrate + sodium sulfide $\longrightarrow$

Net Ionic Equations (Section 10.10)

32. Write net ionic equations for the reactions in problem 26.
33. Write net ionic equations for the reactions in problem 29.
34. Write net ionic equations for the reactions in problem 31.
35. Write the net ionic equation for the reaction of sodium with water.
36. When chlorine is dissolved in aqueous sodium hydroxide, a reaction occurs to give sodium hypochlorite (bleach), sodium chloride, and water. Write the net ionic equation for this reaction.
37. Ammonium chloride, as well as other ammonium salts, reacts with aqueous sodium hydroxide with the evolution of ammonia gas. Write a net ionic equation for this reaction.
38. Acid rain, containing sulfuric acid, reacts with the marble ($CaCO_3$) of buildings, statues, and gravestones, resulting in irreparable damage. Write a net ionic equation for this reaction.

Additional Problems

39. Complete and balance the following equations:
    a. $Ca + O_2 \longrightarrow$
    b. $HClO_3 + Ba(OH)_2 \longrightarrow$
    c. $HgO \longrightarrow O_2 + ?$
    d. $Ca + H_2O \longrightarrow H_2 + ?$

40. Classify each of the reactions in problem 2 as combination, decomposition, combustion, single replacement, or double replacement.
41. Classify each of the reactions in problem 3 as combination, decomposition, combustion, single replacement, or double replacement.
42. Identify the reactions in problem 3 that involve oxidation–reduction.
43. Complete and balance the following equations. All except b. are in aqueous solution.
 a. $Mg + HCl \longrightarrow$
 b. $C_6H_6 + O_2 \longrightarrow$
 c. $Cd(NO_3)_2 + H_2S \longrightarrow$
 d. $FeCl_3 + Na_3PO_4 \longrightarrow$
44. Identify the reactions in problem 43 that involve oxidation–reduction. You may refer to Section 6.8.
45. For each of the reactions noted in problem 44, identify the element oxidized (ox) and the element reduced (red).
46. For each of the reactions in problem 44, identify the oxidizing agent (ox-agent) and the reducing agent (red-agent).
47. Complete and balance the following equations:
 a. $Al(NO_3)_3(aq) + NaOH(aq) \longrightarrow$
 b. $Ni(s) + HBr(aq) \longrightarrow$
 c. $C_4H_{10}O(l) + O_2(g) \longrightarrow$
48. Write balanced equations for the following reactions:
 a. $CrCl_3(aq) + KOH(aq) \longrightarrow$
 b. $Cu(NO_3)_2(aq) + Al(s) \longrightarrow$
49. While working in a laboratory, a chemist accidently dropped a gold ring in a beaker of hydrochloric acid. What do you think happened to the ring? Could the ring be recovered, and if so, how?

Solutions to Practice Problems

PP 10.1

a. 1. Word equation:

potassium chlorate $\longrightarrow$ potassium chloride + oxygen

2. Correct formulas (skeletal equation):

$KClO_3 \longrightarrow KCl + O_2$

3. Balance the equation:

Step 1 Start with the most complicated formula, $KClO_3$.

- Both K and Cl are already balanced.
- Balance O: Starting with 3O on the left and 2O on the right, a total of 6O are needed.

$\mathbf{2}KClO_3 \longrightarrow KCl + \mathbf{3}O_2$

$\mathbf{2} \times 3O = 6O$ $\quad$ $\mathbf{3} \times 2O = 6O$

6O are needed

Step 2 At this point, K and Cl are no longer balanced.

- Balance K and Cl. Add the state symbols. Remember, salts are solids.

2K and 2Cl are needed

$\mathbf{2}KClO_3(s) \longrightarrow \mathbf{2}KCl(s) + \mathbf{3}O_2(g)$

Check: Check K, Cl, and O in that order.

b. 1. Word equation:

iron(III) oxide + carbon monoxide $\longrightarrow$ iron + carbon dioxide

2. Correct formulas (skeletal equation):

$Fe_2O_3 + CO \longrightarrow Fe + CO_2$

3. Balance the equation:

Step 1 Start with Fe_2O_3.

- Balance Fe:

2Fe are needed

$Fe_2O_3 + CO \longrightarrow \mathbf{2}Fe + CO_2$

- Balance C: C is balanced.

Step 2

- Balance O: Increasing O on the right also increases C. 3CO are needed to balance both O and C. Add the state symbols.

$Fe_2O_3(s) + 3CO(g) \longrightarrow 2Fe(s) + 3CO_2(g)$

Check: Check Fe, C, and O.

c. 1. Word equation:

nitric acid + calcium hydroxide $\longrightarrow$ water + calcium nitrate

2. Correct formulas (skeletal equation):

$HNO_3 + Ca(OH)_2 \longrightarrow H_2O + Ca(NO_3)_2$

3. Balance the equation:

Step 1 Start with HNO_3.

- Balance NO_3 (actually NO_3^- ion):

2NO_3 are needed

$$\mathbf{2}HNO_3 + Ca(OH)_2 \longrightarrow H_2O + Ca(NO_3)_2$$

- Balance Ca: Ca is balanced.

Step 2 Balance H and O:

- Balance H:

$$\mathbf{2}HNO_3(aq) + Ca(OH)_2(aq) \longrightarrow \mathbf{2}H_2O(l) + Ca(NO_3)_2(aq)$$

2H + 2H = 4H

4H are needed

- Balance O: O is balanced. Add the state symbols.

Check: Check NO_3, Ca, H, and O.

PP 10.2

a. The product is a binary compound, aluminum sulfide.
 Cation: Al^{3+}
 Anion: S^{2-}
 Formula: Al_2S_3
 Balanced equation:

2Al are needed

$$\mathbf{2}Al(s) + \mathbf{3}S(s) \longrightarrow Al_2S_3(s)$$

3S are needed

b. The product is a binary compound, calcium chloride.
 Cation: Ca^{2+}
 Anion: Cl^-
 Formula: $CaCl_2$
 Balanced equation:

$$Ca(s) + Cl_2(g) \longrightarrow CaCl_2(s)$$

c. The product is a binary compound, magnesium nitride.
 Cation: Mg^{2+}
 Anion: N^{3-}
 Formula: Mg_3N_2
 Balanced equation:

3Mg are needed

$$\mathbf{3}Mg(s) + N_2(g) \longrightarrow Mg_3N_2(s)$$

PP 10.3

Remember, CO_2 is formed.

$$SrCO_3(s) \longrightarrow SrO(s) + CO_2(g)$$

PP 10.4

1. Word equation:

 acetone + oxygen ⟶ carbon dioxide + water

2. Correct formulas:

 $$C_3H_6O + O_2 \longrightarrow CO_2 + H_2O$$

3. Balanced equation:

 Step 1 Start with C_3H_6O.
 - Balance C and H:

 3C are needed

 $$C_3H_6O + O_2 \longrightarrow \mathbf{3}CO_2 + \mathbf{3}H_2O$$

 6H are needed

 Step 2
 - Balance O. Add the state symbols.

 $$C_3H_6O(l) + \mathbf{4}O_2(g) \longrightarrow 3CO_2(g) + 3H_2O(g)$$

 1O + 8O = 9O 3 × 2O + 3O = 9O

 9O are needed

PP 10.5

Refer to Table 10.2 to determine the relative reactivity of the metals and hydrogen. Remember the relative reactivities of the halogens: $F_2 > Cl_2 > Br_2 > I_2$.

a. Potassium will displace hydrogen from water.

$$2K(s) + 2H_2O(l) \longrightarrow H_2(g) + 2KOH(aq)$$

b. Copper is below hydrogen in the activity series, so no reaction occurs.

$$Cu(s) + HCl(aq) \longrightarrow NR$$

c. Iodine is the least reactive halogen.

$$I_2(aq) + NaCl(aq) \longrightarrow NR$$

d. Aluminum is higher in the activity series than copper.

$$2Al(s) + 3Cu(NO_3)_2(aq) \longrightarrow 3Cu(s) + 2Al(NO_3)_3(aq)$$

PP 10.6

Classify the reaction and write correct formulas for the products. Then balance the equation.

a. Acid–base reaction; the products are water and a salt.

$$2HCl(aq) + Ca(OH)_2(aq) \longrightarrow 2H_2O(l) + CaCl_2(aq)$$

b. Precipitation reaction; the precipitate is copper(II) carbonate.

$$CuCl_2(aq) + Na_2CO_3(aq) \longrightarrow CuCO_3(s) + 2NaCl(aq)$$

c. Carbon dioxide (a gas) is formed.

$$K_2CO_3(aq) + 2HBr(aq) \longrightarrow CO_2(g) + H_2O(l) + 2KBr(aq)$$

PP 10.7

The total equations are shown in the solution to PP 10.5.

a. *Total ionic equation:*

$$2H^+(aq) + \cancel{2Cl^-(aq)} + \cancel{Ca^{2+}(aq)} + 2OH^-(aq) \longrightarrow 2H_2O(l) + \cancel{Ca^{2+}(aq)} + \cancel{2Cl^-(aq)}$$

Net ionic equation:

$$H^+(aq) + OH^-(aq) \longrightarrow H_2O(l)$$

b. *Total ionic equation:*

$$Cu^{2+}(aq) + \cancel{2Cl^-(aq)} + \cancel{2Na^+(aq)} + CO_3^{2-}(aq) \longrightarrow CuCO_3(s) + \cancel{2Na^+(aq)} + \cancel{2Cl^-(aq)}$$

Net ionic equation:

$$Cu^{2+}(aq) + CO_3^{2-}(aq) \longrightarrow CuCO_3(s)$$

c. *Total ionic equation:*

$$\cancel{2K^+(aq)} + CO_3^{2-}(aq) + 2H^+(aq) + \cancel{2Br^-(aq)} \longrightarrow CO_2(g) + H_2O(l) + \cancel{2K^+(aq)} + \cancel{2Br^-(aq)}$$

Net ionic equation:

$$CO_3^{2-}(aq) + 2H^+(aq) \longrightarrow CO_2(g) + H_2O(l)$$

Answers to Chapter Review

1. new
2. physical
3. word
4. chemical
5. solid
6. gas
7. (aq)
8. matter
9. products
10. balanced
11. coefficients
12. atoms
13. formulas
14. combination
15. combustion
16. single
17. elements
18. charges
19. decomposition
20. oxyacids
21. carbonates
22. water
23. heat
24. hydrocarbons
25. carbon
26. carbon dioxide
27. replacement
28. activity series
29. hydrogen
30. halide
31. fluorine
32. cation
33. precipitation
34. base
35. salt
36. acid
37. net
38. spectator

11

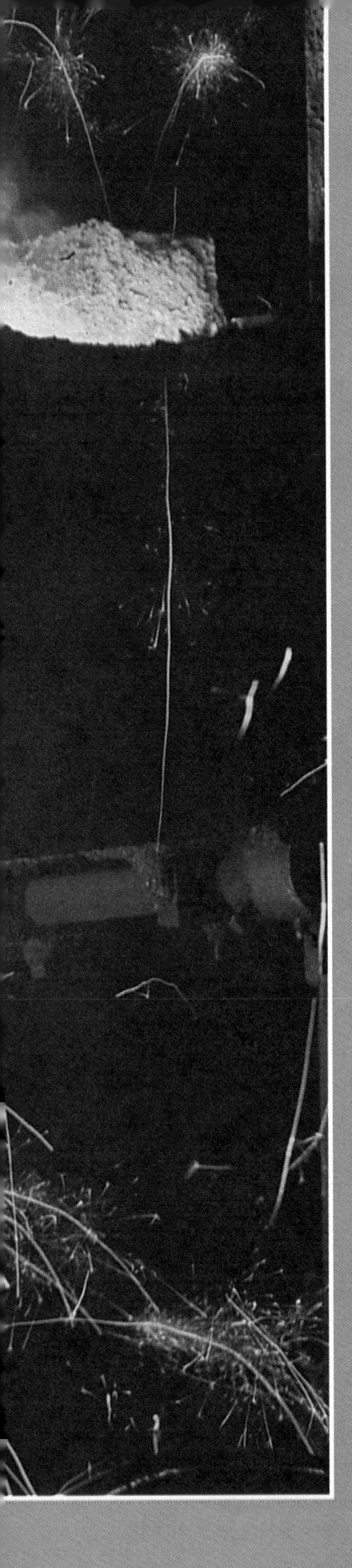

Calculations Involving Chemical Equations

Contents

Iron is produced from iron ore in a blast furnace.

The number of cupcakes obtained from a recipe depends upon the quantities of ingredients.

A match produces only a little smoke.

You Probably Know . . .

- The number of cupcakes obtained from a recipe depends upon the quantities of flour, sugar, and other ingredients.
- A small fire produces only a little smoke, but a large fire can give a lot of smoke. The quantities of reactants used in a reaction determine the quantities of products formed.

Why Should We Study Calculations Involving Chemical Equations?

Chemists and cooks have something in common. Both use raw materials to prepare new substances. Like cooks, chemists are concerned with the quantity as well as the kind and quality of their products. How much hydrogen and nitrogen (reactants) are needed to prepare 1.00 kg of ammonia (product)? What quantity of iron can be produced from 125 g of iron(III) oxide? Chemists must be able to determine the quantity of reactants required to prepare a needed quantity of a product as well as the quantity of product that can be prepared from a given quantity of reactant. Calculations such as these are the focus of this chapter.

11.1 Interpreting Chemical Equations

Learning Objective

- Write mole ratios from a balanced chemical equation.

A balanced chemical equation tells us both qualitative and quantitative information about a reaction. The formulas for the reactants and products show us what substances enter into the reaction and what new substances are formed. This is qualitative information. The balancing coefficients provide quantitative information that allows us to calculate amounts of reactants to be used and amounts of products we expect from a reaction.

To illustrate the significance of balancing coefficients, consider the combination of hydrogen and oxygen to produce water, a highly exothermic reaction that can occur explosively.

$$2H_2(g) + O_2(g) \rightarrow 2H_2O(g)$$

$$2 \text{ molecules } H_2 + 1 \text{ molecule } O_2 \rightarrow 2 \text{ molecules } H_2O$$

$$2 \text{ moles } H_2 + 1 \text{ mole } O_2 \rightarrow 2 \text{ moles } H_2O$$

$$4 \text{ g } H_2 + 32 \text{ g } O_2 \rightarrow 36 \text{ g } H_2O$$

A balanced equation tells us relative numbers of *molecules* of reactants and products, and relative numbers of *moles* of reactants and products. Thus, this equation tells us 1 molecule of oxygen is needed for 2 molecules of hydrogen to react. We also can see that 2 molecules of water will form for each molecule of oxygen that reacts. Moreover, the sum of the masses of hydrogen and oxygen (reactants) equals the mass of water (product) formed, in agreement with the law of conservation of mass. A description of the relative quantities of substances taking part in a reaction is called the *stoichiometry* of the reaction. The stoichiometry of a reaction is best summarized by its balanced chemical equation.

"Stoichiometry" is pronounced stoi' kē ahm' ĕ trē.

It is useful to express the stoichiometry of a reaction in terms of mole ratios of reactants and products. Mole ratios express relative numbers of moles of reactants and products. For example, the equation for the reaction of hydrogen with oxygen tells us that 1 mole of oxygen is needed for 2 moles of hydrogen to react, or:

$$\frac{1 \text{ mol } O_2}{2 \text{ mol } H_2} \quad \text{and} \quad \frac{2 \text{ mol } H_2}{1 \text{ mol } O_2}$$

Note that mole ratios, like conversion factors, always come in pairs. One ratio is simply the inverse of the other. We can also see that 2 moles of water will form for each mole of oxygen that reacts. The mole ratios are

$$\frac{2 \text{ mol } H_2O}{1 \text{ mol } O_2} \quad \text{and} \quad \frac{1 \text{ mol } O_2}{2 \text{ mol } H_2O}$$

A similar pair of mole ratios can also be written expressing the relationship between H_2 and H_2O.

Mole ratios will be used to calculate the number of moles of reactant needed or product formed in a reaction.

Mole Calculations

11.2

Learning Objective

- Calculate the number of moles of a reactant or product from the number of moles given for another reactant or product.

Calculating the number of moles of one substance from the number of moles of another taking part in a reaction is a relatively simple matter using the dimensional analysis strategy. For example, when 2.8 mol H_2 are combined with O_2, we can calculate the number of moles of water formed. First, outlining the problem:

Find: number of moles H_2O = ?

Given: 2.8 mol H_2

Known: $\frac{2 \text{ mol } H_2O}{2 \text{ mol } H_2}$ (from balanced equation in Section 11.1)

Solution Starting with the unknown (a key element in our problem-solving strategy), the calculation looks like this:

$$\text{number of moles } H_2O = \left(\frac{2 \text{ mol } H_2O}{2 \cancel{\text{mol } H_2}}\right)(2.8 \cancel{\text{mol } H_2}) = 2.8 \text{ mol } H_2O$$

We can also calculate the number of moles of O_2 needed to react with 2.8 mol

H_2. As before, the balanced equation for the reaction gives us the required mole ratio:

$$\text{number of moles } O_2 = \left(\frac{1 \text{ mol } O_2}{2 \text{ mol } H_2}\right)(2.8 \text{ mol } H_2) = 1.4 \text{ mol } O_2$$

Unknown moles = mole ratio × given moles.

To summarize, the unknown number of moles of a substance is found by multiplying the appropriate mole ratio (from the balanced chemical equation) by the number of moles given for another substance in the reaction. In setting up the calculation, remember our strategy (Section 4.4):

1. Start with the unknown (with units) followed by an equal sign.
2. Choose the first term of the equation by matching its units with the units of the unknown.
3. Select the next term of the equation to cancel unwanted units in the preceding term, leaving the units of the answer.

In working the following problems, remember to outline the problem. Then begin the solution process with the unknown.

Example 11.1

Ammonia can be prepared by the reaction of hydrogen with nitrogen at high temperature according to the following equation:

$$3H_2(g) + N_2(g) \longrightarrow 2NH_3(g)$$

How many moles of ammonia can be prepared in the reaction of 1.5 mol of hydrogen with excess nitrogen? ■

Find: number of moles NH_3 = ?

Given: 1.5 mol H_2

Known: From the balanced equation we can write a mole ratio relating the given and the unknown:

$$\frac{2 \text{ mol } NH_3}{3 \text{ mol } H_2}$$

Solution: $\text{number of moles } NH_3 = \left(\frac{2 \text{ mol } NH_3}{3 \text{ mol } H_2}\right)(1.5 \text{ mol } H_2) = 1.0 \text{ mol } NH_3$

Example 11.2

How many moles of silver will form when 1.2 mol of copper react with aqueous silver nitrate? Note that $Cu(NO_3)_2(aq)$ is also formed in the reaction. ■

Find: number of moles Ag = ?

Given: 1.2 mol Cu

Known: Write the balanced chemical equation for the reaction:

$$\underset{1.2 \text{ mol}}{Cu(s)} + 2AgNO_3(aq) \longrightarrow \underset{? \text{ mol}}{2Ag(s)} + Cu(NO_3)_2(aq)$$

Mole ratio: $\frac{2 \text{ mol Ag}}{1 \text{ mol Cu}}$

Solution: number of moles Ag $= \left(\frac{2 \text{ mol Ag}}{1 \cancel{\text{mol Cu}}}\right)(1.2 \cancel{\text{mol Cu}}) = 2.4 \text{ mol Ag}$

Practice Problem 11.1 Butane (C_4H_{10}) is a fuel used in campstoves and lanterns. How many moles of carbon dioxide will form when 3.0 mol of butane burn? You may refer to Section 10.7 when writing the chemical equation for the reaction. (Solutions to practice problems appear at the end of the chapter.)

Mass Calculations

11.3

Learning Objective

- Calculate the mass of a reactant or product from either the mass or the number of moles given for another reactant or product.

When you understand mole relationships for a reaction, you are ready to calculate the mass of a reactant needed for a reaction or the mass of product formed. These problems usually involve a three-step calculation:

1. Calculate the number of moles of the given substance from its mass using its molar mass, *MM*.
2. Calculate the number of moles of the unknown substance from the number of moles calculated in step 1 using the appropriate mole ratio from the balanced chemical equation.
3. Calculate the mass of the unknown substance from the number of moles of that substance using its molar mass.

Mole calculations (step 1) were introduced in Section 7.4. You may want to review those calculations before continuing with this chapter. Remember, mole ratios are determined from the balanced chemical equation. You must also know the *molar masses, MM,* of the unknown and known substances. Recall, also, that the units for molar mass are g/mol.

Review mole calculations in Section 7.4.

At this point, we can summarize the overall plan for calculating the mass of substance B from the mass of substance A:

$$\text{mass}_A \xrightarrow[(1)]{MM_A} \text{moles}_A \xrightarrow[(2)]{\frac{\text{mol B}}{\text{mol A}}} \text{moles}_B \xrightarrow[(3)]{MM_B} \text{mass}_B$$

Good problem solvers divide complex problems into smaller, simpler problems (Section 4.1). Example 11.3 illustrates how this strategy can be used successfully. The solution involves three steps as described in the overall plan. Each of these steps can be set up using dimensional analysis.

The Civic Bank building in Los Angeles. The metallic portions of the building exterior are anodized aluminum, a type of aluminum coated with aluminum oxide (Al_2O_3).

Example 11.3

Aluminum combines with oxygen to form aluminum oxide. Find the mass of aluminum oxide formed from the oxidation of 35.1 g of aluminum. ■

Find: mass (g) Al_2O_3 = ?

Given: 35.1 g Al

Word equation: aluminum + oxygen → aluminum oxide

Known: Chemical equation: $4Al(s) + 3O_2(g) \rightarrow 2Al_2O_3(s)$

35.1 g ? g

Mole ratio: $\dfrac{2\text{ mol }Al_2O_3}{4\text{ mol Al}}$

$MM\text{-Al} = \dfrac{27.0\text{ g Al}}{1\text{ mol}}$ (from the periodic table)

$MM\text{-}Al_2O_3 = \dfrac{102\text{ g }Al_2O_3}{1\text{ mol}}$ (calculated from *AM*'s)

Solution: This is a three-step calculation similar to the one outlined in Section 7.10. Dimensional analysis is used in writing each equation. The first equation is

$$\text{mass (g) }Al_2O_3 = \left(\frac{102\text{ g }Al_2O_3}{1\text{ mol}}\right)(?\text{ mol }Al_2O_3) \quad (1)$$

Because mol Al_2O_3 is unknown, set up the second equation to calculate the number of moles of Al_2O_3 using dimensional analysis:

$$\text{number of moles }Al_2O_3 = \left(\frac{2\text{ mol }Al_2O_3}{4\text{ mol Al}}\right)(?\text{ mol Al}) \quad (2)$$

Because mol Al is unknown, set up the third equation to find the number of moles Al:

$$\text{number of moles Al} = \left(\frac{1\text{ mol}}{27.0\text{ }\cancel{\text{g Al}}}\right)(35.1\text{ }\cancel{\text{g Al}}) = 1.30\text{ mol Al} \quad (3)$$

Substitute 1.30 mol Al in equation (2) and calculate the number of moles Al_2O_3:

$$\text{number of moles }Al_2O_3 = \left(\frac{2\text{ mol }Al_2O_3}{4\text{ }\cancel{\text{mol Al}}}\right)(1.30\text{ }\cancel{\text{mol Al}}) \quad (2)$$

$$= 0.650\text{ mol }Al_2O_3$$

Substitute 0.650 mol Al_2O_3 in equation (1) and calculate the mass of Al_2O_3:

$$\text{mass (g) }Al_2O_3 = \left(\frac{102\text{ g}}{1\text{ }\cancel{\text{mol }Al_2O_3}}\right)(0.650\text{ }\cancel{\text{mol }Al_2O_3}) \quad (1)$$

$$= 66.3\text{ g}$$

These three calculations can be combined into one:

$$\text{mass (g) }Al_2O_3 = \left(\frac{102\text{ g }Al_2O_3}{1\text{ }\cancel{\text{mol }Al_2O_3}}\right)\left(\frac{2\text{ }\cancel{\text{mol }Al_2O_3}}{4\text{ }\cancel{\text{mol Al}}}\right)\left(\frac{1\text{ }\cancel{\text{mol Al}}}{27.0\text{ }\cancel{\text{g Al}}}\right)(35.1\text{ }\cancel{\text{g Al}})$$

$$= 66.3\text{ g }Al_2O_3$$

Check: *Units:* The problem asked for *mass; grams* are correct units.
Significant figures: The mass of Al was given to 3 significant figures; the answer should have 3 significant figures.
Is the answer reasonable? Yes. Aluminum is combining with oxygen, and the mass of Al_2O_3 should be greater than the mass of Al.

To work this type of problem, you need to bring together a number of skills and strategies. You may want to review the sections of the text in which these were discussed.

1. Outline the problem (Section 4.2).
2. Write formulas for substances involved in a reaction (Chapter 9).
3. Write the balanced equation for the reaction (Section 10.3).
4. Write the mole ratios (Section 11.1).
5. Set up equations to calculate unknowns (solution strategies, Sections 4.4 and 4.5).
6. Calculate the number of moles of a substance from its mass (Section 7.4).
7. Calculate the number of moles of an unknown substance from the number of moles of a given substance involved in a reaction (Section 11.2).

Multiple-step problems can be complex. Remember, with a complex problem, outlining and solution strategies are very important! Your success in solving problems depends upon consistent application of these strategies.

Remember:

1. *Outline the problem.*
2. *Use solution strategies.*
3. *Begin with the unknown.*

Although outlining a problem is important, the specific format of an outline is not critical. It is sometimes helpful to be creative in outlining various types of problems. The format of the type shown in Figure 11.1 will be useful in solving multiple-step problems involving chemical equations. To solve the problem, simply

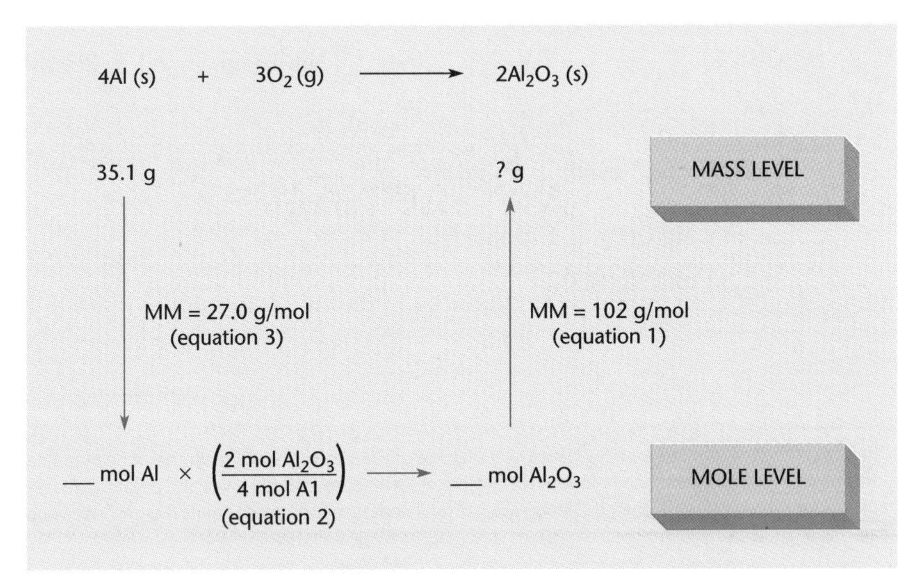

FIGURE 11.1 Outline for problems involving chemical equations (see Example 11.3).

follow the arrows in the outline and fill in the blanks with the answers to each step. The equations for the three steps are shown in Example 11.3.

To construct an outline like that shown in Figure 11.1, follow these steps:

1. Write the balanced chemical equation for the reaction.
2. Begin the outline with the unknown. Identify the unknown from the statement of the problem. If the unknown is the mass of one of the substances in the reaction, write "? g" under the formula of that substance in the equation.
3. If a mass of another substance is given, write its mass under its formula.
4. The mass of the unknown will be calculated from (a) its molar mass (*MM*) and (b) the number of moles of that substance. (Example 11.3, equation 1). Under "? g," write "______ mol." Draw a vertical arrow pointing to "? g." Write the *MM* next to the arrow.
5. The number of moles of the unknown will be calculated using a mole ratio and the number of moles of the given substance (Example 11.3, equation 2). Write "______ mol" beneath the formula of the unknown at the mole level. Draw a horizontal arrow pointing to moles of the unknown. Over the arrow write "×" and the mole ratio relating the given and unknown substances.
6. The number of moles of the given substance will be calculated using (a) its *MM* and (b) its mass (Example 11.3, equation 3). Draw a vertical arrow pointing from the mass given to "______ mol." Write the *MM* of the substance next to the arrow.

The outline in Figure 11.1 is a convenient format for multiple-step problems.* It clearly identifies the unknown and includes all the given and known information needed to solve the problem.

Example 11.4

What mass of magnesium hydroxide (milk of magnesia) would be needed to neutralize 1.70 mol of excess stomach acid (HCl)? ■

First, write the equation (acid–base reaction, Section 10.9) and outline the problem.

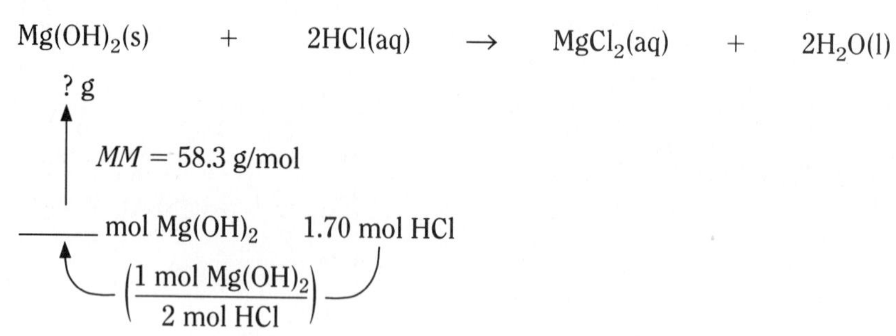

Milk of magnesia contains magnesium hydroxide.

*Before you began your study of chemistry, you may have been acquainted with another kind of mole, a critter that lives underground. Looking at the outline, if we imagine that the given and unknown masses are written "above ground," then the moles are written "underground" in the "mole burrow." We could call this a "mole burrow outline."

Solution Because the number of moles of HCl (rather than its mass) was given, the calculation involves only two steps:

$$\text{number of moles Mg(OH)}_2 = \left(\frac{1 \text{ mol Mg(OH)}_2}{2 \cancel{\text{mol HCl}}}\right)(1.70 \cancel{\text{mol HCl}})$$

$$= 0.850 \text{ mol Mg(OH)}_2 \qquad (1)$$

$$\text{mass (g) Mg(OH)}_2 = \left(\frac{58.3 \text{ g Mg(OH)}_2}{1 \cancel{\text{mol}}}\right)(0.850 \cancel{\text{mol}})$$

$$= 49.6 \text{ g Mg(OH)}_2 \qquad (2)$$

Check: *Units:* Grams are mass units.
Significant figures: 3 significant figures are needed.
Is the answer reasonable? Yes, 1.70 mol HCl requires less than 1 mol (58.3 g) $Mg(OH)_2$.

Practice Problem 11.2 Iron can be prepared from iron(III) oxide in a blast furnace. Calculate the mass of iron that could be obtained by the reaction of 2.50 kg of iron(III) oxide with carbon monoxide. The balanced equation for the reaction is

$$Fe_2O_3(s) + 3CO(g) \longrightarrow 2Fe(s) + 3CO_2(g)$$

Practice Problem 11.3 When limestone ($CaCO_3$) is heated strongly, carbon dioxide is driven off leaving lime (CaO), a substance used to make cement and plaster. The process is described by the following equation:

$$CaCO_3(s) \xrightarrow{\Delta} CO_2(g) + CaO(s)$$

How much lime could be prepared starting with 3.75 kg of limestone?

Lime (CaO) is used to make cement and plaster.

Reaction Yields

11.4

Learning Objectives

- Calculate the theoretical yield of a reaction product.
- Calculate the percent yield of a reaction product.

In Example 11.3, we found that the oxidation of 35.1 g of aluminum will produce 66.3 g of aluminum oxide. This calculated mass is called the theoretical yield of Al_2O_3 for the reaction. The **theoretical yield** is the quantity of product that forms when the reactant completely reacts. The theoretical yield is always a *calculated*

CHEMICAL WORLD

A Disaster Curtailed

Six A.M. Sunday morning on the outskirts of Denver, Colorado. A tank car is punctured by the coupling of another car and 20 000 gallons of nitric acid spill onto the railroad bed. The situation is so dangerous that 5000 people in the surrounding 500 blocks are summarily moved out of their homes. What high-tech equipment, what complexities of chemistry are called for in such a desperate situation? Of course, money is no object. The Denver firefighters' answer is soda ash (Na_2CO_3), in hydrated form known as just plain old washing soda. Just spread it on the spill:

$$Na_2CO_3(s) + 2HNO_3(aq) \longrightarrow H_2O(l) + CO_2(g) + 2NaNO_3(aq)$$

and wash what's left down the drain.

That equation will solve any problem you may have with acid spills in the lab—just throw some soda ash or baking soda on it. But 120 *tons* of nitric acid? You might use a much stronger base, such as sodium hydroxide (lye). Less of that would neutralize more acid. But if you use too much, or if some of the stuff doesn't contact the acid and isn't neutralized, you'll have as great a problem with that caustic base as you have with the acid. No, soda ash is better. It works well and it's safe.

But how much do you need? Well, the equation tells you how many moles of soda ash you need, so you can figure that out. Go ahead, figure it out. Concentrated nitric acid (density 1.42 g/mL) is 72% HNO_3.

The next problem is where to get that much—raid all the grocery stores in town? No, find a glass works, or a paper mill, or a soap factory. They all use a lot of soda ash and should have large supplies. Find all three. You need a lot in a hurry.

Another problem: How do we efficiently get the soda ash in contact with the acid? Dump trucks won't work. Quick thinking made it easy in Denver. The fire department borrowed snow blowers from the airport and got the job done in short order. What would you do in Arizona, Alabama, or southern California?

Soda ash (Na_2CO_3) is spread over spilled nitric acid (HNO_3) using snow blowers.

Did you figure out how much Na_2CO_3 you need for this emergency? Assuming an unlikely 100% contact between soda ash and the acid, you'll need 73 tons of sodium carbonate. The stuff is relatively harmless,* so maybe you'd better pick up a few extra tons.

*However, sodium carbonate can cause severe irritation to the eyes.

quantity. On the other hand, when an experiment is performed, the **actual yield** is the *measured* quantity of product obtained.

The actual yield of a reaction is usually less than the theoretical yield. This is due to a number of factors including incomplete reaction and loss of product in transferring from container to container. For example, any product spilled is not included in the mass measured with a balance. The **percent yield** is the ratio of the actual yield to the theoretical yield expressed as a percent. The following equation defines percent yield:

$$\text{percent yield} = \frac{\text{actual yield}}{\text{theoretical yield}} \times 100$$

If the actual yield of Al_2O_3 from the reaction described in Example 11.3 had been 55.6 g, the percent yield would be

$$\text{percent yield} = \frac{55.6\ \text{g}}{66.3\ \text{g}} \times 100 = 83.9\%$$

Example 11.5

(a) What is the theoretical yield of magnesium chloride formed in the reaction described in Example 11.4? **(b)** If the actual yield is 76.3 g, what is the percent yield? ■

Solution

a. First, write the balanced chemical equation and outline the problem:

$$Mg(OH)_2(s) + 2HCl(aq) \longrightarrow MgCl_2(aq) + 2H_2O(l)$$

MASS LEVEL ? g

$MM = 95.3$ g/mol

MOLE LEVEL $1.70\ \text{mol} \times \left(\frac{1\ \text{mol}\ Mg(OH)_2}{2\ \text{mol HCl}}\right) \rightarrow$ ______ mol

This outline shows the two steps in the calculation of the theoretical yield of $MgCl_2$:

$$\text{number of moles } MgCl_2 = \left(\frac{1\ \text{mol}\ MgCl_2}{2\ \cancel{\text{mol HCl}}}\right)(1.70\ \cancel{\text{mol HCl}})$$
$$= 0.850\ \text{mol}\ MgCl_2 \quad (1)$$

$$\text{mass (g)}\ MgCl_2 = \left(\frac{95.3\ \text{g}}{1\ \cancel{\text{mol}\ MgCl_2}}\right)(0.850\ \cancel{\text{mol}\ MgCl_2})$$
$$= 81.0\ \text{g (theoretical yield)} \quad (2)$$

Check: Is the answer reasonable? Yes, the theoretical yield should be *less* than the mass of 1 mol (95.3 g) of $MgCl_2$.

b. $$\text{percent yield} = \frac{76.3\ \text{g}}{81.0\ \text{g}} \times 100 = 94.2\%$$

Example 11.6

The final step in the production of sodium carbonate (soda ash) by the Solvay process is described by the following equation:

$$2NaHCO_3(s) \xrightarrow{\Delta} CO_2(g) + H_2O(g) + Na_2CO_3(s)$$

How much $NaHCO_3$ would be needed to prepare 1.50 kg of Na_2CO_3 if the process gives an 88.6% yield? ■

Begin by outlining the problem under the chemical equation.

$$2NaHCO_3(s) \xrightarrow{\Delta} CO_2(g) + H_2O(g) + Na_2CO_3(s)$$

? g (NaHCO₃) ↑ $MM = 84.0$ g/mol ↑ ____ mol ← $\left(\frac{2 \text{ mol } NaHCO_3}{1 \text{ mol } Na_2CO_3}\right)$ ← ____ mol ↑ $MM = 106$ g/mol ↑ ____ g (Na_2CO_3)

Note that 1.50 kg of Na_2CO_3 is the *actual yield* (act., measured not calculated). This looks like the usual three-equation problem, but the theoretical (theor.) yield of Na_2CO_3 needed to begin the calculation is not given in the problem. Therefore, it must be calculated. This problem can be outlined as follows:

Find: mass (g, theor.) Na_2CO_3 = ?

Given: 88.6% yield, which means $\frac{88.6 \text{ g } Na_2CO_3\text{, act.}}{100 \text{ g } Na_2CO_3\text{, theor.}}$

Known: actual yield = 1.50 kg (1500 g) Na_2CO_3

Solution: Using dimensional analysis gives

$$\text{mass (g, theor.) } Na_2CO_3 = \left(\frac{100 \text{ g } Na_2CO_3\text{, theor.}}{88.6 \cancel{\text{ g } Na_2CO_3\text{, act.}}}\right)(1500 \cancel{\text{ g } Na_2CO_3\text{, act.}})$$

$$= 1690 \text{ g } Na_2CO_3 \text{ (theor.)} \qquad \text{(3 significant figures)}$$

Check: Is this answer reasonable? Yes, the theoretical yield (1690 g) would be *greater* than the actual yield (1.50 kg).

Now, write "1690 g" under Na_2CO_3 in the outline. Then follow the arrows through the three steps to calculate the mass (g) $NaHCO_3$ needed.

$$2NaHCO_3(s) \longrightarrow CO_2(g) + H_2O(g) + Na_2CO_3(s)$$

? g (NaHCO₃) ↑ $MM = 84.0$ g/mol ↑ ____ mol ← $\left(\frac{2 \text{ mol } NaHCO_3}{1 \text{ mol } Na_2CO_3}\right)$ ← ____ mol ↑ $MM = 106$ g/mol ↑ 1690 g (Na_2CO_3)

$$\text{number of moles } Na_2CO_3 = \left(\frac{1 \text{ mol } Na_2CO_3}{106 \cancel{\text{ g}}}\right)(1690 \cancel{\text{ g}}) = 15.9 \text{ mol} \qquad (1)$$

$$\text{number of moles } NaHCO_3 = \left(\frac{2 \text{ mol } NaHCO_3}{1 \cancel{\text{ mol } Na_2CO_3}}\right)(15.9 \cancel{\text{ mol } Na_2CO_3}) \qquad (2)$$

$$= 31.8 \text{ mol } NaHCO_3$$

$$\text{mass (g) } NaHCO_3 = \left(\frac{84.0 \text{ g}}{1 \cancel{\text{ mol } NaHCO_3}}\right)(31.8 \cancel{\text{ mol } NaHCO_3})$$

$$= 2760 \text{ g} \quad \text{or} \quad 2.67 \text{ kg } NaHCO_3 \qquad (3)$$

Practice Problem 11.4 Chromium, a metal used in the manufacture of stainless steel, can be produced from chromium(III) oxide as described by the following equation:

$$Cr_2O_3(s) + 2Al(s) \longrightarrow 2Cr(s) + Al_2O_3(s)$$

Calculate **(a)** the theoretical yield and **(b)** the percent yield of chromium if 825 g of chromium is produced from 1250 g of Cr_2O_3.

A chrome-plated faucet.

Limiting Reactant Problems

11.5

Learning Objectives

- Determine which is the limiting reactant in a reaction.
- From the quantity of a limiting reactant, calculate the theoretical yield of product.

Sodium reacts with water to produce hydrogen and sodium hydroxide (Section 10.8):

$$2Na(s) + 2H_2O(l) \longrightarrow H_2(g) + 2NaOH(aq)$$

Imagine that someone tossed a small piece of sodium into a large body of water such as Lake Michigan. Which reactant, sodium or water, would determine the theoretical yield of hydrogen from the reaction? To answer this question, we must consider the relative amounts of the reactants. It is apparent that more water is available in Lake Michigan than is needed for the reaction. In other words, there is an excess of water. Thus, the quantity of sodium will limit how much hydrogen will form, and sodium is called the limiting reactant. The **limiting reactant** is the reactant that is *not* in excess. The quantity of limiting reactant used in the reaction is used to calculate the theoretical yield of product.

For many reactions we can quickly determine which is the limiting reactant from the description of the reaction. For example, when propane (C_3H_8) is completely burned, we assume that oxygen (in the air) is available in excess of that needed for the reaction.

$$C_3H_8(g) + 5O_2(g) \longrightarrow 3CO_2(g) + 4H_2O(g)$$

We can decide almost without thinking about it that propane is the limiting reactant. To calculate the theoretical yield of carbon dioxide formed, we use the quantity of propane in the calculation, not a quantity of oxygen.

However, for many reactions performed in the laboratory or used in the manufacture of chemicals, the limiting reactant must be determined by calculation. You can recognize that you must calculate the limiting reactant when the amounts of *two reactants* are given in the problem. Remember, to calculate the theoretical yield for a reaction, it is necessary to know which reactant is the limiting reactant.

Consider the following combination reaction used to make calcium chloride, a substance widely used to melt ice and snow from sidewalks:

$$Ca(s) + Cl_2(g) \longrightarrow CaCl_2(s)$$

What is the theoretical yield of calcium chloride when 52.1 g of calcium is allowed to react with 78.1 g of chlorine?

Calcium chloride is used to melt ice and snow from sidewalks.

Because masses of *both* reactants are given, this is a limiting reactant problem. There are two parts to the problem:

1. Determine the limiting reactant by calculation.
2. Calculate the theoretical yield of the product ($CaCl_2$).

Part 1 The limiting reactant is determined from the *number of moles* of reactants. The first step is to outline the problem (Figure 11.2). Next, calculate the number of moles of each reactant from the masses given in the problem:

$$\text{number of moles Ca} = \left(\frac{1 \text{ mol Ca}}{40.1 \text{ g Ca}}\right)(52.1 \text{ g Ca}) = 1.30 \text{ mol Ca (moles given)}$$

$$\text{number of moles } Cl_2 = \left(\frac{1 \text{ mol } Cl_2}{70.9 \text{ g } Cl_2}\right)(78.1 \text{ g } Cl_2) = 1.10 \text{ mol } Cl_2 \text{ (moles given)}$$

Now, assume one of the reactants to be the limiting reactant (we will choose Ca) and find the number of moles of the other reactant needed (Cl_2 in this example):

$$(1.30 \text{ mol Ca})\left(\frac{1 \text{ mol } Cl_2}{1 \text{ mol Ca}}\right) = 1.30 \text{ mol } Cl_2 \text{ needed}$$

Compare the number of moles Cl_2 given with the number of moles Cl_2 needed. We can see that insufficient Cl_2 is available. Therefore, Ca is not the limiting reactant. We must conclude that Cl_2 is the limiting reactant.

Part 2 Because Cl_2 is the limiting reactant, we will use the number of moles of Cl_2 given to calculate the theoretical yield of $CaCl_2$.

$$\begin{array}{cccc} Ca(s) & + \; Cl_2(g) & \rightarrow & CaCl_2(s) \\ 52.1 \text{ g} & 78.1 \text{ g} & & 122 \text{ g} \\ & \downarrow \; MM = 70.9 \text{ g/mol} & & \uparrow \; MM = 111 \text{ g/mol} \\ & 1.10 \text{ mol } Cl_2 & \xrightarrow{\times\left(\frac{1 \text{ mol } CaCl_2}{1 \text{ mol } Cl_2}\right)} & 1.10 \text{ mol } CaCl_2 \end{array}$$

Thus, the theoretical yield is 122 g $CaCl_2$.

Check:

1. Consider an alternative hypothesis: Ca is the limiting reactant.
2. Calculate the number of moles of $CaCl_2$ that could form if Ca is the limiting reactant:

$$\text{number of moles } CaCl_2 = \left(\frac{1 \text{ mol } CaCl_2}{1 \text{ mol Ca}}\right)(1.30 \text{ mol Ca}) = 1.30 \text{ mol } CaCl_2$$

3. The reactant that would give *less* product is the limiting reactant. In this example, Cl_2 would give 1.10 mol $CaCl_2$; so Cl_2 is the limiting reactant.

1. *Question:* Which reactant is the limiting reactant?
2. *Given/known (outline):*

Ca(s)	+	Cl_2(g)	→	$CaCl_2$(s)
52.1 g		78.1 g		? g
↓ MM = 40.1 g/mol		↓ MM = 70.9 g/mol		MM = 111 g/mol
1.30 mol		1.10 mol (moles given)		

3. *State the hypothesis:* Ca is the limiting reactant.
4. *Test the hypothesis:*

$$\text{mole ratio} = \frac{1 \text{ mol } Cl_2}{1 \text{ mol Ca}}$$

Therefore, to react 1.30 mol of Ca, 1.30 mol of Cl_2 is needed. But only 1.10 mol of Cl_2 is given.

5. *Conclusion:* No, Ca is not the limiting reactant. Therefore, Cl_2 is the limiting reactant.

FIGURE 11.2 Steps of a limiting reactant calculation.

Now, let's find out if this reaction agrees with the law of conservation of mass. Note that since Cl_2 is the limiting reactant, Ca is in excess. In other words, not all of the Ca will react. Because the equation shows a 1:1 mole ratio for the reactants, 1.10 mol of Ca will react, and 1.30 mol − 1.10 mol = 0.20 mol of Ca is left unreacted. The mass of Ca remaining is

$$\text{mass (g) Ca} = \left(\frac{40.1 \text{ g Ca}}{1 \cancel{\text{mol Ca}}}\right)(0.20 \cancel{\text{mol Ca}}) = 8.0 \text{ g Ca}$$

The mass of product (122 g) plus the mass of Ca unreacted (8.0 g) equals the total mass of reactants used (52.1 g + 78.1 g = 130.2 g), to 3 significant figures. This is in agreement with the law of conservation of mass.

The steps in determining a limiting reactant are outlined in Figure 11.2.

Determination of the limiting reactant follows the scientific method outlined in Chapter 1. The steps of the scientific method and their application to this problem are:

1. State the problem: Which is the limiting reactant?
2. Identify what is known about the problem: The information given and known was outlined.
3. State a hypothesis: Ca is the limiting reactant.*
4. Test the hypothesis: Calculate the number of moles of the other reactant (Cl_2) needed and compare with the number of moles given. Note that insufficient Cl_2 was given.
5. State a conclusion: The hypothesis was incorrect; Ca was not the limiting reactant. Therefore, Cl_2 is the limiting reactant.

*With two reactants, two hypotheses are possible. Our hypothesis could have been: Cl_2 is the limiting reactant. We can test this hypothesis by calculating the number of moles of Ca needed and comparing with the number of moles of Ca given.

Acetylene is used as a fuel for the welder's torch.

Example 11.7

Calcium carbide (CaC_2) reacts with water to produce calcium hydroxide ($Ca(OH)_2$) and acetylene (C_2H_2), the gas used as a fuel for the welder's torch.

a. Which is the limiting reactant when 125 g of CaC_2 is mixed with 125 g of H_2O?
b. What is the theoretical yield of acetylene? ■

Solution

a. Determine which is the limiting reactant. Write the balanced equation and outline the problem.

Outline:

$CaC_2(s)$	+	$2H_2O(l)$	$\rightarrow$	$Ca(OH)_2(s)$	+	$C_2H_2(g)$
125 g		125 g				? g
↓ $MM = 64.1$ g/mol		↓ 18.0 g/mol				26.0 g/mol
1.95 mol CaC_2		6.94 mol H_2O (moles given)				

Hypothesis: CaC_2 is the limiting reactant.

Test:

$$(1.95 \cancel{\text{mol CaC}_2})\left(\frac{2 \text{ mol H}_2\text{O}}{1 \cancel{\text{mol CaC}_2}}\right) = 3.90 \text{ mol H}_2\text{O needed if CaC}_2 \text{ is the limiting reactant.}$$

The number of moles of water given is more than the number of moles needed.

Conclusion: CaC_2 is the limiting reactant. (It is the reactant that is not in excess.)

b. Calculate the theoretical yield.

$CaC_2(s)$	+	$2H_2O(l)$	$\longrightarrow$	$Ca(OH)_2(s)$	+	$C_2H_2(g)$
125 g						50.7 g
↓ $MM = 64.1$ g/mol						↑ 26.0 g/mol
1.95 mol	×	$\left(\frac{1 \text{ mol Ca(OH)}_2}{1 \text{ mol CaC}_2}\right)$	$\longrightarrow$			1.95 mol

Thus, the theoretical yield is 50.7 g C_2H_2.

Check: If H_2O were the limiting reactant, then

$$\text{number of moles C}_2\text{H}_2 = \left(\frac{1 \text{ mol C}_2\text{H}_2}{2 \cancel{\text{mol H}_2\text{O}}}\right)(6.94 \cancel{\text{mol H}_2\text{O}})$$

$$= 3.47 \text{ mol C}_2\text{H}_2$$

Therefore, CaC_2 must be the limiting reactant because it gives less C_2H_2.

Practice Problem 11.5 Calcium dihydrogen phosphate, an ingredient in some fertilizers, is made by reacting calcium phosphate with phosphoric acid. Calculate the theoretical yield of the product, starting with 835 g of calcium phosphate and 1350 g of phosphoric acid. What mass of excess reactant remains when the reaction is complete? The equation is

$$Ca_3(PO_4)_2(s) + 4H_3PO_4(aq) \longrightarrow 3Ca(H_2PO_4)_2(s)$$

Chapter Review

Directions: From the list of choices at the end of the Chapter Review, choose the word or term that best fits in each blank. Use each choice only once. The answers appear at the end of the chapter.

Section 11.1 The balancing __________ (1) in a chemical equation can be used to write mole __________ (2). These are used to calculate amounts of reactants to be used and amounts of __________ (3) we expect from a reaction. In calculations, a mole ratio can be used like a __________ (4) factor. Mole ratios, like conversion factors, always come in __________ (5).

Section 11.2 When a calculation is based upon a chemical reaction, we must write a __________ (6) chemical equation. When setting up mole calculations, as in most other calculations, we usually begin with the __________ (7). The unknown number of moles can be calculated by multiplying a mole ratio by the __________ (8) number of moles.

Section 11.3 Calculation of the mass of a reactant or product involved in a reaction usually involves __________ (9) steps:

1. Calculate the number of moles of a substance from its __________ (10).
2. Calculate the number of __________ (11) of a second substance from the number of moles of the first substance.
3. Calculate the mass of the __________ (12) substance from the number of moles of that substance.

In following these steps, we may use one of the strategies of good problem solvers: we may divide a __________ (13) problem into smaller, simpler problems. Remember, with a complex problem it is important to __________ (14) the problem, use good solution __________ (15), and begin with the unknown.

Section 11.4 The quantity of product that will form when a reactant completely reacts is called the __________ __________ (16) of the product. The theoretical yield must be __________ (17), whereas the actual yield is the __________ (18) quantity of product. The __________ (19) yield equals the actual yield divided by the theoretical yield times 100.

Section 11.5 When the quantities of two reactants are given, the reactant that is not in excess is called the __________ (20) reactant. The quantity of this reactant is used to calculate the __________ (21) yield of the product. The limiting reactant is determined by considering the numbers of moles of both __________ (22). After outlining the problem, we state a __________ (23) and __________ (24) the hypothesis to decide which is the limiting reactant.

Choices: balanced, calculated, coefficients, complex, conversion, given, hypothesis, limiting, mass, measured, moles, outline, pairs, percent, products, ratios, reactants, second, strategies, test, theoretical, theoretical yield, three, unknown

Study Objectives

After studying Chapter 11, you should be able to:

Section 11.1

1. Write mole ratios from a balanced chemical equation.

Section 11.2

2. Calculate the number of moles of a reactant or product from the number of moles given for another reactant or product.

Section 11.3

3. Calculate the mass of a reactant or product from either the mass or the number of moles given for another reactant or product.

Section 11.4

4. Calculate the theoretical yield of a reaction product.
5. Calculate the percent yield of a reaction product.

Section 11.5

6. Determine which is the limiting reactant in a reaction.
7. From the quantity of a limiting reactant, calculate the theoretical yield of product.

Key Terms

Review the definition of each of the terms listed here by chapter section number. You may use the glossary if necessary.

11.4 actual yield, percent yield, theoretical yield

11.5 limiting reactant

Questions and Problems

Answers to questions and problems with blue numbers appear in Appendix C.

Mole Calculations (Section 11.2)

1. How many moles of hydrogen are needed to react with 3 mol of oxygen to form water?

$$2H_2(g) + O_2(g) \longrightarrow 2H_2O(g)$$

2. How many moles of water will form when 5 mol of hydrogen react with excess oxygen?
3. Hydrogen peroxide decomposes to give water and oxygen. How many moles of oxygen will form when 6 mol of hydrogen peroxide decompose?
4. How many moles of carbon dioxide will form when 15 mol of toluene (C_7H_8), a compound in gasoline, burn in air?

$$C_7H_8(g) + 9O_2(g) \longrightarrow 7CO_2(g) + 4H_2O(g)$$

5. How many moles of baking soda ($NaHCO_3$) are needed to neutralize 1.3 mol of battery acid (H_2SO_4)?

$$H_2SO_4(aq) + 2NaHCO_3(s) \longrightarrow Na_2SO_4(aq) + 2CO_2(g) + 2H_2O(l)$$

6. Vinegar, containing acetic acid ($HC_2H_3O_2$), reacts with baking soda to produce carbon dioxide, water, and sodium acetate. How many moles of acetic acid will be neutralized by 2.75 mol of baking soda?
7. Phosphorus (P_4) burns to give tetraphosphorus decoxide (P_4O_{10}).
   a. How many moles of oxygen are needed to burn 0.655 mol of phosphorus?
   b. How many moles of tetraphosphorus decoxide will form when 1.50 mol of phosphorus burn?
8. Limestone ($CaCO_3$) reacts with phosphoric acid to yield carbon dioxide, water, and calcium phosphate.
   a. How many moles of phosphoric acid will react with 1.25 mol of limestone?
   b. How many moles of carbon dioxide will form from the reaction of 1.5 mol of limestone?
9. Sodium sulfite is used in fish aquariums to remove chlorine from the water. The equation for the reaction is

$$Cl_2(aq) + Na_2SO_3(aq) + H_2O(l) \longrightarrow 2HCl(aq) + Na_2SO_4(aq)$$

   a. How many moles of sodium sulfite are needed to remove 0.250 mol of chlorine from the water in an aquarium?
   b. Find the number of moles of hydrochloric acid that will form when 0.250 mol of chlorine react.
10. How many moles of stomach acid, HCl(aq), will be neutralized by two antacid tables each containing 0.0050 mol of aluminum hydroxide?

$$3HCl(aq) + Al(OH)_3(s) \longrightarrow AlCl_3(aq) + 3H_2O(l)$$

11. The fermentation of glucose ($C_6H_{12}O_6$) produces ethyl alcohol (C_2H_5OH), called grain alcohol, and carbon dioxide.

$$C_6H_{12}O_6(s) \longrightarrow 2C_2H_5OH(l) + 2CO_2(g)$$

   a. How many moles of ethyl alcohol are produced from the fermentation of 6.35 mol of glucose?
   b. Find the number of moles of carbon dioxide that would form from 6.35 mol of glucose.
12. In photosynthesis, carbon dioxide and water react to yield oxygen and glucose ($C_6H_{12}O_6$), the building block of cellulose.

$$6CO_2(g) + 6H_2O(l) \longrightarrow 6O_2(g) + C_6H_{12}O_6(s)$$

   a. When 2.85 mol of carbon dioxide enter into photosynthesis, how many moles of glucose will form?
   b. How many moles of oxygen will form when 2.85 mol of carbon dioxide react?

Mass Calculations (Section 11.3)

13. Methanol (CH_3OH), an experimental fuel for motor vehicles, burns to give carbon dioxide and water. What mass of

carbon dioxide would be formed from the combustion of 675 g of methanol?

14. When a marshmallow catches fire, sugar ($C_{12}H_{22}O_{11}$) burns to give carbon dioxide and water. What mass of carbon dioxide would be produced from the complete combustion of a marshmallow having a mass of 6.2 g? Assume the marshmallow is 100% sugar.

15. Calculate the mass of baking soda ($NaHCO_3$) that would react with sulfuric acid to produce 7.50 mol of CO_2.

$$H_2SO_4(aq) + 2NaHCO_3(s) \longrightarrow Na_2SO_4(aq) + 2CO_2(g) + 2H_2O(l)$$

16. Find the mass of calcium phosphate that could be produced from the reaction of 175 g of calcium hydroxide with excess phosphoric acid.

$$3Ca(OH)_2(aq) + 2H_3PO_4(aq) \longrightarrow Ca_3(PO_4)_2(s) + 6H_2O(l)$$

17. What mass of silver chloride will be produced from the reaction of 0.478 g of silver nitrate with calcium chloride? The equation (unbalanced) for the reaction is

$$CaCl_2(aq) + AgNO_3(aq) \longrightarrow AgCl(s) + Ca(NO_3)_2(aq)$$

18. What mass of sodium hydroxide is needed to completely neutralize 0.0345 mol of phosphoric acid? The balanced equation is

$$H_3PO_4(aq) + 3NaOH(aq) \longrightarrow Na_3PO_4(aq) + 3H_2O(l)$$

19. Magnesium burns with a very bright flame to give magnesium oxide. Find the mass of magnesium oxide formed when a 0.13-g piece of magnesium is burned.

20. Sodium bisulfite will remove the purple color of iodine from a solution according to the following reaction:

$$NaHSO_3(aq) + I_2(aq) + H_2O(l) \longrightarrow 2HI(aq) + NaHSO_4(aq)$$

What mass of sodium bisulfite is needed to react with 0.379 g of iodine?

21. Copper(II) oxide reacts with hydrogen to give copper and water according to the following equation:

$$CuO(s) + H_2(g) \longrightarrow Cu(s) + H_2O(g)$$

Calculate the mass of CuO needed to produce 17.5 g of copper.

22. Copper(I) sulfide is a compound found in copper ores. Copper is produced by "roasting" the ore as described by the following equation:

$$Cu_2S(s) + O_2(g) \xrightarrow{\text{heat}} 2Cu(s) + SO_2(g)$$

How much copper can be formed by "roasting" 10.5 kg of copper(I) sulfide?

23. Heating limestone (calcium carbonate) at high temperature produces lime (CaO) and carbon dioxide. What mass of lime, in kg, can be produced from 15.0 kg of limestone?

24. Boron (B) can be produced in a reaction of the oxide with magnesium according to the following equation:

$$B_2O_3(s) + 3Mg(s) \longrightarrow 2B(s) + 3MgO(s)$$

What mass of boron can be produced from 125 g of B_2O_3?

25. Phosphorus (P_4) reacts with chlorine to give phosphorus trichloride. What mass of phosphorus is needed to make 5.67 kg of phosphorus trichloride?

26. Potassium chlorate decomposes when heated to give potassium chloride and oxygen. What mass of oxygen is formed from the decomposition of 15.7 g of $KClO_3$?

27. The electrolysis of melted sodium chloride yields sodium and chlorine. Calculate the masses of sodium and chlorine produced from the electrolysis of 225 g of NaCl.

28. Ammonia (NH_3) is manufactured by the reaction of nitrogen and hydrogen at high temperature (Haber process). Calculate the mass of nitrogen, in kg, needed to produce 3.0 kg of ammonia.

29. For the reaction

$$4FeS_2(s) + 11O_2(g) \longrightarrow 2Fe_2O_3(s) + 8SO_2(g)$$

What mass of iron(III) oxide will be produced from 65.6 g of FeS_2?

30. When heated, silver oxide decomposes as described by the following equation:

$$2Ag_2O(s) \longrightarrow 4Ag(s) + O_2(g)$$

What mass of silver oxide is needed to produce 2.68 g of silver?

Theoretical and Percent Yield (Section 11.4)

31. Calculate the theoretical yield of silver from the decomposition of 4.50 g silver oxide. The equation is given in problem 30.

32. What is the theoretical yield of alumina (Al_2O_3) in the reaction of 15.0 g aluminum with excess oxygen?

33. Carbon tetrachloride, a toxic organic solvent, is manufactured by the reaction of methane with chlorine as described by the following equation:

$$CH_4(g) + 4Cl_2(g) \longrightarrow CCl_4(l) + 4HCl(g)$$

What is the theoretical yield of carbon tetrachloride from the reaction of 50.4 g of methane with excess chlorine?

34. In the synthesis described in problem 33, if the actual yield of CCl_4 was 415 g, what was the percent yield?

35. Mercury(II) oxide can be decomposed by heating to give mercury and oxygen. Heating 1.45 g of HgO resulted in an actual yield of 1.24 g of mercury. Calculate the theoretical and percent yields of mercury.

36. Trinitrotoluene (TNT) is manufactured by the reaction of toluene with nitric acid as described by the following equation:

$$C_7H_8(l) + 3HNO_3(aq) \longrightarrow C_7H_5(NO_2)_3(s) + 3H_2O(l)$$

a. Calculate the theoretical yield of TNT when 69.0 g of toluene and an excess of nitric acid are used in a synthesis.
b. What is the percent yield if the synthesis results in the formation of 135 g of TNT?

37. Aluminum reacts with $TiCl_4$ to produce aluminum chloride ($AlCl_3$) and titanium.
a. What is the theoretical yield of titanium when 2.00 mol of $TiCl_4$ react with excess aluminum?
b. If 71.0 g of titanium are recovered from the reaction, what is the percent yield?

38. Lard is rich in the triglyceride, stearin ($C_{57}H_{110}O_6$). Reaction with sodium hydroxide (lye) converts stearin to sodium stearate ($NaC_{18}H_{35}O_2$), a soap, and glycerol as described by the following equation:

$$C_{57}H_{110}O_6(s) + 3NaOH(aq) \longrightarrow 3NaC_{18}H_{35}O_2(s) + C_3H_5(OH)_3(l)$$

Soap making was common practice in the homes of past generations, and the product of the reaction is sometimes referred to as "Grandma's lye soap." When Grandma started with 1.00 lb (454 g) of lard (assume 100% stearin) and a slight excess of lye, what was the theoretical yield of soap (sodium stearate)?

Limiting Reactant (Section 11.5)

39. Which is the limiting reactant in the reaction of 7.50 g of oxygen with 10.0 g of methane (CH_4)? What is the theoretical yield of CO_2 in the reaction?

$$CH_4(g) + 2O_2(g) \longrightarrow CO_2(g) + 2H_2O(g)$$

40. Which is the limiting reactant in the reaction of 15.5 g of oxygen with 10.0 g of butane (C_4H_{10})? What is the theoretical yield of water?

$$2C_4H_{10}(g) + 13O_2(g) \longrightarrow 8CO_2(g) + 10H_2O(g)$$

41. Ammonium phosphate, a fertilizer, can be made by the reaction of ammonia with phosphoric acid:

$$3NH_3(g) + H_3PO_4(aq) \longrightarrow (NH_4)_3PO_4(s)$$

a. Which is the limiting reactant in a reaction of 51.0 g of ammonia with 196 g of phosphoric acid?
b. What is the theoretical yield of ammonium phosphate?

42. Ethylene glycol, automobile antifreeze, is manufactured by the reaction of ethylene oxide (C_2H_4O) with water:

$$C_2H_4O(g) + H_2O(l) \longrightarrow C_2H_4(OH)_2(l)$$

a. What is the limiting reactant when 50.0 g of ethylene oxide react with 28.0 g of water?
b. What is the theoretical yield of ethylene glycol from the reaction?

43. Isooctane, a component of gasoline having an octane rating of 100, can be manufactured by the reaction of isobutylene (C_4H_8) with isobutane (C_4H_{10}):

$$C_4H_8(g) + C_4H_{10}(g) \longrightarrow C_8H_{18}(l)$$

What is the theoretical yield of isooctane starting with 2.0 kg of each reactant?

44. A solution containing 0.815 g of silver nitrate ($AgNO_3$) was mixed with a solution containing 0.450 g of potassium chloride (KCl) resulting in the following reaction:

$$AgNO_3(aq) + KCl(aq) \longrightarrow KNO_3(aq) + AgCl(s)$$

a. Which was the limiting reactant?
b. What is the theoretical yield of silver chloride?
c. How much of the excess reactant remained unreacted?

45. Freon-12 (CCl_2F_2), used as a refrigerant in air conditioners, refrigerators, and freezers, is suspected of depleting the ozone layer in the earth's atmosphere. This compound is prepared by the reaction:

$$3CCl_4(l) + 2SbF_3(s) \longrightarrow 3CCl_2F_2(g) + 2SbCl_3(s)$$

a. What is the limiting reactant in a synthesis involving 50.0 g of each reactant?
b. What is the theoretical yield of Freon-12 for this synthesis?
c. What mass of excess reactant remained when the reaction was complete?

Additional Problems

46. Sulfur dioxide, an air pollutant, reacts with sodium hydroxide solution as described by the following equation:

$$SO_2(g) + 2NaOH(aq) \longrightarrow Na_2SO_3(aq) + H_2O(l)$$

What mass of sodium hydroxide is needed to react with the sulfur dioxide in 100. m^3 of air containing a concentration of 5.4 mg/m^3 of SO_2?

47. A petroleum company produced a gasoline containing 1.10 mL of tetraethyl lead (TEL), $(C_2H_5)_4Pb$, per liter of gasoline. The density of TEL is 1.66 g/mL. How much ethyl chloride, C_2H_5Cl, is needed to make sufficient TEL for 1.0 L of gasoline according to the following equation?

$$4C_2H_5Cl(l) + 4NaPb(s) \longrightarrow (C_2H_5)_4Pb(l) + 4NaCl(s) + 3Pb(s)$$

48. The tarnish on silver objects is a coating of silver sulfide (Ag_2S). A particular silver tarnish remover is a mixture that contains powdered aluminum. The equation for the reaction of aluminum with the tarnish is:

$$3Ag_2S(s) + 2Al(s) \longrightarrow 6Ag(s) + Al_2S_3(s)$$

What mass of aluminum would be needed to remove 0.125 g of Ag_2S from a silver candlestick?

49. Ethanol (C_2H_5OH) is added to gasoline to make gasohol, an oxygenated fuel. Ethanol can be produced by the fermentation of glucose ($C_6H_{12}O_6$) as described by the following equation:

$$C_6H_{12}O_6(s) \longrightarrow 2C_2H_5OH(l) + 2CO_2(g)$$

a. Starting with 2.00 kg of glucose, what is the theoretical yield of ethanol?

b. If the actual yield of ethanol from 2.00 kg of glucose is 625 g, what is the percent yield?

50. Calculate the theoretical yield of lead(II) oxide (PbO) that can be produced from 3.50 mol of PbS according to the following equation:

$$2PbS(s) + 3O_2(g) \longrightarrow 2PbO(s) + 2SO_2(g)$$

51. Methanol (CH_3OH) is a synthetic fuel that is used to power buses and other fleet vehicles. It can be synthesized from carbon monoxide according to the following equation:

$$CO(g) + 2H_2(g) \longrightarrow CH_3OH(l)$$

a. Starting with 56.0 g of CO and 6.50 g of H_2, which is the limiting reactant?
b. Calculate the theoretical yield of methanol in this synthesis.

52. Acid rain, which contains sulfuric acid (H_2SO_4), can erode statues and buildings made of marble ($CaCO_3$) as described by the following equation:

$$CaCO_3(s) + H_2SO_4(aq) \longrightarrow CaSO_4(s) + H_2O(l) + CO_2(g)$$

In a laboratory study of this reaction, 120. g of sulfuric acid was mixed with 175 g of $CaCO_3$.
a. What was the limiting reactant in the experiment?
b. What mass of the $CaCO_3$ reacted in the experiment?
c. Calculate the mass of $CaSO_4$ that theoretically could form.

53. Hydrogen sulfide (H_2S) is a highly toxic gas that is sometimes found in natural gas wells. Oxidation of hydrogen sulfide with nitric acid is described by the following equation:

$$H_2S(g) + 8HNO_3(aq) \longrightarrow H_2SO_4(aq) + 8NO_2(g) + 4H_2O(l)$$

a. How many moles of HNO_3 are needed to oxidize 100. g of hydrogen sulfide?
b. What is the theoretical yield of sulfuric acid (H_2SO_4) in a reaction of 100. g of H_2S?

54. Iron reacts with hydrochloric acid as described by the following equation:

$$Fe(s) + 2HCl(aq) \longrightarrow FeCl_2(aq) + H_2(g)$$

a. What mass of hydrogen could be formed when 75.0 g of iron is mixed with 2.50 mol of hydrochloric acid?
b. What is the theoretical yield of iron(II) chloride?
c. Calculate the mass of excess reactant remaining when the reaction is complete.
d. Verify the law of conservation of mass by showing that the mass of products equals the mass of reactants entering into the reaction.

Solutions to Practice Problems

PP 11.1

Find: number of moles CO_2 = ?

Given: 3.0 mol C_4H_{10}

Known: Combustion of a hydrocarbon (Section 10.7) produces carbon dioxide and water:

$$2C_4H_{10}(g) + 13O_2(g) \longrightarrow 8CO_2(g) + 10H_2O(g)$$

Mole ratio: $\dfrac{8 \text{ mol } CO_2}{2 \text{ mol } C_4H_{10}}$

Solution:

$$\text{number of moles } CO_2 = \left(\frac{8 \text{ mol } CO_2}{2 \text{ mol } C_4H_{10}}\right)(3.0 \text{ mol } C_4H_{10}) = 12 \text{ mol } CO_2$$

PP 11.2

Outline the problem under the balanced equation for the reaction. Note that 2.50 kg is equivalent to 2.50 × 10³ g. Molar masses can be calculated from atomic masses.

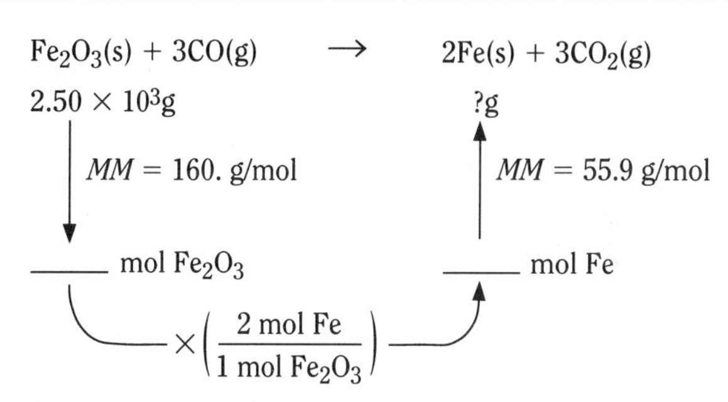

This problem involves three equations:

$$\text{number of moles } Fe_2O_3 = \left(\frac{1 \text{ mol}}{160. \text{ g } Fe_2O_3}\right)(2.50 \times 10^3 \text{ g } Fe_2O_3) = 15.6 \text{ mol } Fe_2O_3 \quad (1)$$

$$\text{number of moles Fe} = \left(\frac{2 \text{ mol Fe}}{1 \text{ mol } Fe_2O_3}\right)(15.6 \text{ mol } Fe_2O_3) = 31.2 \text{ mol Fe} \quad (2)$$

$$\text{mass (g) Fe} = \left(\frac{55.9 \text{ g Fe}}{1 \text{ mol Fe}}\right)(31.2 \text{ mol Fe}) = 1740 \text{ g Fe} = 1.74 \text{ kg Fe} \quad \text{(3 significant figures)} \quad (3)$$

PP 11.3

Begin by outlining the problem under the equation. Note that 3.75 kg is equivalent to 3750 g. Molar masses can be calculated from the atomic masses of the elements in the compound.

$$CaCO_3(s) \xrightarrow{\Delta} CO_2(g) + CaO(s)$$

3750 g → (MM = 100. g/mol) → ___ mol $\times \left(\frac{1 \text{ mol CaO}}{1 \text{ mol } CaCO_3}\right)$ → ___ mol → (MM = 56.1 g/mol) → ?g

$$\text{number of moles } CaCO_3 = \left(\frac{1 \text{ mol } CaCO_3}{100. \text{ g } CaCO_3}\right)(3750 \text{ g } CaCO_3)$$
$$= 37.5 \text{ mol} \quad (1)$$

$$\text{number of moles CaO} = \left(\frac{1 \text{ mol CaO}}{1 \text{ mol } CaCO_3}\right)(37.5 \text{ mol } CaCO_3)$$
$$= 37.5 \text{ mol} \quad (2)$$

$$\text{mass (g) CaO} = \left(\frac{56.1 \text{ g CaO}}{1 \text{ mol CaO}}\right)(37.5 \text{ mol CaO})$$
$$= 2104 \text{ g}$$
$$= 2.10 \times 10^3 \text{ g or } 2.10 \text{ kg} \quad \text{(3 significant figures)} \quad (3)$$

PP 11.4

a. To calculate the theoretical yield, first outline the problem, and calculate molar masses from atomic masses:

$$Cr_2O_3(s) + 2Al(s) \longrightarrow 2Cr(s) + Al_2O_3(s)$$

1250 g → (MM = 152 g/mol) → ___ mol $\times \left(\frac{2 \text{ mol Cr}}{1 \text{ mol } Cr_2O_3}\right)$ → ___ mol → (MM = 52.0 g/mol) → ?g

The calculation involves three equations:

$$\text{number of moles } Cr_2O_3 = \left(\frac{1 \text{ mol } Cr_2O_3}{152 \text{ g } Cr_2O_3}\right)(1250 \text{ g } Cr_2O_3)$$
$$= 8.22 \text{ mol} \quad (1)$$

$$\text{number of moles Cr} = \left(\frac{2 \text{ mol Cr}}{1 \text{ mol } Cr_2O_3}\right)(8.22 \text{ mol } Cr_2O_3)$$
$$= 16.4 \text{ mol} \quad (2)$$

$$\text{mass (g) Cr} = \left(\frac{52.0 \text{ g Cr}}{1 \text{ mol Cr}}\right)(16.4 \text{ mol Cr})$$
$$= 853 \text{ g Cr (theor. yield)} \quad (3)$$

This three-step calculation can also be set up using a string of factors:

$$\text{mass (g) Cr} = \left(\frac{52.0 \text{ g}}{1 \text{ mol Cr}}\right)\left(\frac{2 \text{ mol Cr}}{1 \text{ mol } Cr_2O_3}\right)\left(\frac{1 \text{ mol } Cr_2O_3}{152 \text{ g } Cr_2O_3}\right)(1250 \text{ g } Cr_2O_3)$$
$$= 855 \text{ g (theor. yield)}$$

The answer from the stepwise calculation differs slightly because of rounding of numbers in the intermediate steps.

b. $$\%\text{ yield} = \frac{\text{act. yield}}{\text{theor. yield}} \times 100 = \frac{825 \text{ g}}{855 \text{ g}} \times 100 = 96.5\% \quad \text{(3 significant figures)}$$

PP 11.5

1. Determine which is the limiting reactant.

Outline:

$$Ca_3(PO_4)_2(s) + 4H_3PO_4(aq) \longrightarrow 3Ca(H_2PO_4)_2(s)$$

835 g → (MM = 310. g/mol) → 2.69 mol; 1350 g → (98.0 g/mol) → 13.8 mol (moles given); ?g (234 g/mol)

Hypothesis: $Ca_3(PO_4)_2$ is the limiting reactant.

Test:

$$(2.69 \text{ mol } Ca_3(PO_4)_2)\left(\frac{4 \text{ mol } H_3PO_4}{1 \text{ mol } Ca_3(PO_4)_2}\right) = 10.8 \text{ mol } H_3PO_4 \text{ needed}$$

The calculation shows that 10.8 mol of H_3PO_4 are needed to react all the $Ca_3(PO_4)_2$. Because 13.8 mol of H_3PO_4 are given, there is an excess of 3.0 mol of H_3PO_4.

Conclusion: $Ca_3(PO_4)_2$ is the limiting reactant.

2. Calculate the theoretical yield.

$$Ca_3(PO_4)_2(s) + 4H_3PO_4(aq) \longrightarrow 3Ca(H_2PO_4)_2(s)$$

835 g → (MM = 310. g/mol) → 2.69 mol $\times \left(\frac{3 \text{ mol } Ca(H_2PO_4)_2}{1 \text{ mol } Ca_3(PO_4)_2}\right)$ → 8.07 mol → (234 g/mol) → 1890 g

The answer is 1890 g $Ca(H_2PO_4)_2$ to 3 significant figures.

3. Find the mass of excess reactant (H_3PO_4) remaining when the reaction is complete.

$$\text{mass (g) } H_3PO_4 = \left(\frac{98.0 \text{ g } H_3PO_4}{1 \text{ mol } H_3PO_4}\right)(3.0 \text{ mol } H_3PO_4)$$
$$= 290 \text{ g} \quad \text{(2 significant figures)}$$

Answers to Chapter Review

1. coefficients
2. ratios
3. products
4. conversion
5. pairs
6. balanced
7. unknown
8. given
9. three
10. mass
11. moles
12. second
13. complex
14. outline
15. strategies
16. theoretical yield
17. calculated
18. measured
19. percent
20. limiting
21. theoretical
22. reactants
23. hypothesis
24. test

12

The Gaseous State

Contents

A hot-air balloon rises in the atmosphere because the density of the warm air inside the balloon is less than the density of the cooler surrounding air.

Some people with respiratory illnesses must breathe oxygen.

You Probably Know . . .

- The earth's atmosphere is a mixture of gases made up mostly of nitrogen (N_2) and oxygen (O_2), together with argon (Ar), carbon dioxide (CO_2), ozone (O_3), and other trace gases (see Table 12.1).
- Human and animal life are dependent upon oxygen. Some people with respiratory illnesses must breathe oxygen instead of air.
- Carbon dioxide (CO_2) is exhaled by people and animals and is used by plants in photosynthesis, a process that utilizes sunlight to produce sugars, starches, and cellulose (carbohydrates).
- Ozone (O_3), a gas in the upper atmosphere, absorbs harmful ultraviolet radiation, thereby providing protection to life on the earth's surface.
- Air and other gases can be compressed. Compressed air is used to inflate tires and athletic balls. Pneumatic tools, such as the wrenches used to mount wheels on cars in a tire shop, are powered by compressed air.
- Some fuels are gases. Natural gas (mostly methane, CH_4) is used to heat homes and cook food. Propane (C_3H_8) is a common fuel in some rural areas where natural gas is not available, for gas barbeques, and for camping stoves and heaters.

A hand pump can be used to compress air and inflate a bicycle tire.

Why Should We Study Gases?

Gases are a major part of our environment. We live in an atmosphere of gases that is essential for the support and protection of life (Table 12.1). Although you may be well acquainted with the role of oxygen and carbon dioxide in supporting human, animal, and plant life, other atmospheric gases are also important.

Nitrogen is one of these gases. Although nitrogen as N_2 is not absorbed by plants, bacteria in the root nodules of some plants such as clover and beans convert atmospheric nitrogen into compounds that can be absorbed. This conversion is called *nitrogen fixation.* It also occurs during thunderstorms when N_2 combines with oxygen and is converted into nitrogen oxides. These are washed from the atmosphere by rainfall, thereby providing natural fertilization for plants. Nitrogen oxides also form by the reaction of nitrogen and oxygen in motor vehicle engines and contribute to the air pollution of major cities.

TABLE 12.1 The Composition of Dry Air at Sea Level

	Percentage Composition	
Component	*Volume*	*Mass*
nitrogen, N_2	78.08	75.51
oxygen, O_2	20.95	23.14
argon, Ar	0.93	1.29
carbon dioxide, CO_2	0.03	0.05
trace gases	0.01	0.01

Ozone, O_3, is an important trace gas in the atmosphere. In sufficient concentrations it is a severe respiratory irritant and is an air pollutant that should be controlled. However, ozone in the upper atmosphere provides protection to life on the earth's surface by absorbing harmful ultraviolet radiation.

In addition to their importance in our environment and our daily lives, gases have a strong impact on the economies of industrialized nations. Of the five chemicals produced in greatest quantity in the United States,* four are gases. After sulfuric acid (#1), a liquid, they are nitrogen (#2), which is extracted from air (Figure 12.1), oxygen (#3), also obtained from air and used in great quantities in steel production, ethylene (#4), which is used to make plastics, fibers, and antifreeze, and ammonia (#5), prepared from nitrogen and used in fertilizers.

FIGURE 12.1 A liquid air plant. Nitrogen and oxygen, two of the top five chemicals produced in the United States, are extracted from air.

Thus, understanding the behavior of gases is critical to understanding our environment and how chemicals impact our daily lives. We begin our study of gases by examining their physical properties. We then consider a useful theory that effectively explains gas behavior both qualitatively and quantitatively. Finally, we examine some chemical reactions involving gases.

Properties of Gases

12.1

Learning Objective

- Summarize the general properties of gases.

In an earlier discussion (Section 2.1), we recognized that gases are the most mobile of the three states of matter. As they exist in the atmosphere or when spilled from a container, gases disperse rapidly. Gases are fluids. They flow through pipes and ventilation ducts. In general, attractions between gaseous atoms or molecules are negligible and have no effect on their mobility. As a result, gases are not rigid like solids, and they have no definite shape. A gas fills the entire space of a container and therefore assumes the container's shape. Thus, the volume of a gas is the same as the volume of its container.

In addition to being the most mobile of the three states of matter, gases have the lowest densities. This is because molecules of a gas are much farther apart than those of a liquid or solid. The wide spacing of its molecules allows a gas to be compressed by applying pressure (Figure 12.2). Furthermore, because of the space between gas molecules, gases mix (diffuse) more readily than do liquids. **Diffusion** is the process of spontaneous mixing of gases (and liquids) that occurs because of the random movement of the molecules. You can smell the pleasant fragrances of a delicious dinner or a perfume at a distance because the fragrant gaseous molecules move freely between the widely spaced molecules of the air to your nose.

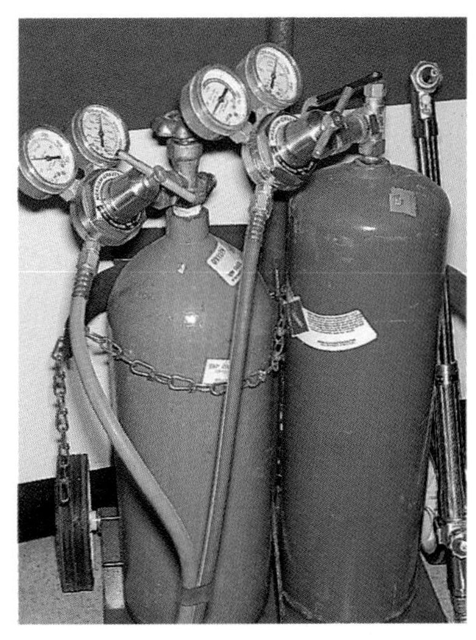

FIGURE 12.2 Gases can be compressed and stored in heavy steel cylinders at high pressures. The regulator attached to the top of each cylinder controls the pressure of the gas flowing from the cylinder. One gauge indicates the pressure of the gas in the cylinder, and the other gives the pressure of the gas flowing through the regulator.

The general properties of gases are summarized as follows:

1. *Gases fill their containers completely.* Changing the volume of a container filled with a gas (such as a balloon or a gas cylinder) changes the volume of the gas.

*1993 data compiled by the American Chemical Society and published in *Chemical and Engineering News,* April 12, 1993, p. 11.

2. *Gases flow readily.* Gases such as air or natural gas flow through ventilation ducts and pipes.
3. *Gases have low densities.* The density of air is about 1/1000 the density of many liquids.
4. *Gases are compressible.* Compressed air is used to inflate tires and athletic balls.
5. *Gases mix spontaneously.* Motor vehicle exhaust and smoke from chimneys diffuse rapidly in air.

12.2 The Kinetic-Molecular Theory of Gases

Learning Objectives

- State the main features of the kinetic-molecular theory.
- Compare the rates of effusion or diffusion of two gases from their densities or molar masses.
- State the conditions of temperature and pressure at which real gases deviate from ideal behavior.

The properties of gases are explained by the **kinetic-molecular theory,** a term meaning a theory of "moving molecules." This theory effectively describes gas behavior in both qualitative and quantitative terms. A gas that behaves exactly according to the kinetic-molecular theory is called an **ideal gas** (Figure 12.3). Under most conditions, gases are observed to behave in agreement with this theory. The main features of the kinetic-molecular theory are:

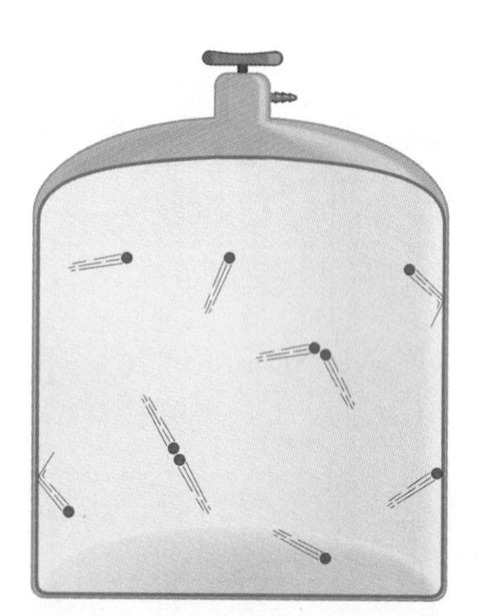

FIGURE 12.3 The molecules of an ideal gas are widely separated and are moving randomly in straight lines. The molecules collide with the walls of the container and with each other with no loss in total energy. The velocity of the molecules depends upon the absolute temperature (K). In this drawing, the sizes of the molecules are greatly exaggerated.

1. *Gases are composed of atoms or molecules moving rapidly and randomly in straight lines.* The molecules frequently collide with each other and the walls of their container.
2. *The atoms or molecules of a gas are widely separated.* Most of the volume of a gas sample is empty space.
3. *The attractive forces between the atoms or molecules of a gas are negligible.*
4. *There is no loss of energy when atoms or molecules of a gas collide with each other or the walls of their container.* Although individual molecules may gain or lose energy when they collide, the total kinetic energy of the molecules remains the same at a given temperature.
5. *The average kinetic energy of the atoms or molecules of a gas depends upon the Kelvin temperature of the gas.* When the temperature increases, the molecules move more rapidly, and their average kinetic energy increases.

Practically all gas behavior can be explained by the kinetic-molecular theory. Consider again the general properties summarized in Section 12.1. Gases expand to fill their containers completely because gas molecules are widely spaced and have negligible attraction for each other. For the same reason, gases flow readily. The low densities of gases are also explained in terms of the large spaces between molecules, which gives rise to a low mass per unit volume of gas. Gases diffuse rapidly because of the large spaces between molecules and their rapid, random motion.

Gas effusion is a property closely related to gas diffusion. **Effusion** is the passage of a gas through a tiny hole or porous membrane from a region of high pressure into a region of low pressure. The velocity (v) of a molecule is related to its kinetic energy (KE) by the equation:

$$KE = \tfrac{1}{2}mv^2$$

where m is the mass of the molecule. Because two different gases at the same temperature have the same average kinetic energy, the molecules of smaller mass have greater average velocity. Thus, the rates of effusion and diffusion are greater for the gas having smaller molar mass. For example, if a balloon were filled with equal volumes of hydrogen (MM = 2.0 g/mol) and nitrogen (MM = 28.0 g/mol), hydrogen would be lost through a small leak faster than nitrogen.

Thomas Graham (1805–1869) observed that the rate of effusion depends upon the density of a gas. *Graham's law of effusion* states that the rates of effusion of two gases at a given temperature and pressure are inversely proportional to the square roots of their densities (D) or molar masses (MM).

$$\frac{\text{rate of gas}_A}{\text{rate of gas}_B} = \sqrt{\frac{D_B}{D_A}} = \sqrt{\frac{MM_B}{MM_A}}$$

Applying Graham's law to the mixture of hydrogen and nitrogen in a balloon gives

$$\frac{\text{rate of } H_2}{\text{rate of } N_2} = \sqrt{\frac{MM\text{-}N_2}{MM\text{-}H_2}} = \sqrt{\frac{28.0 \text{ g/mol}}{2.0 \text{ g/mol}}} = 3.7$$

Hydrogen leaks from the balloon 3.7 times faster than nitrogen. Note that relative rates of diffusion can also be calculated using Graham's law.

At most temperatures and pressures, real gases behave in good agreement with the ideal model described by the kinetic-molecular theory. However, gases deviate from ideal behavior at very low temperatures and very high pressures. At these conditions gas molecules are closer together, and the size of the molecules and attractions between them contribute to nonideal behavior.

Example 12.1

Calculate the relative rates of effusion of helium (He) and methane (CH_4). ■

Solution

$$\frac{\text{rate of He}}{\text{rate of } CH_4} = \sqrt{\frac{MM\text{-}CH_4}{MM\text{-He}}} = \sqrt{\frac{16.0}{4.00}} = 2.00$$

The rate of effusion of helium is 2.00 times that of methane.

Practice Problem 12.1 Which gas diffuses faster in air:

a. Cl_2 or Ar?

b. He or H_2?

(Solutions to practice problems appear at the end of the chapter.)

Practice Problem 12.2 A balloon originally filled with equal volumes of helium and nitrogen developed a leak.
a. Which gas will leak faster from the balloon?
b. Calculate the relative rates of effusion of the gases.

12.3 Gas Measurements

Learning Objectives

- Briefly describe a barometer and a manometer and what is measured with each.
- Convert pressures from torr to atmospheres and vice versa.

We can learn about the properties of gases from the behavior of balloons.

We can learn something about properties of gases from the behavior of balloons. Anyone who has blown up a balloon has observed that the balloon gets bigger when more air is blown into it. As the balloon expands, the pressure of the confined air in the balloom makes it more difficult to blow more air into the balloon. After the balloon is filled and as the warm air inside cools, the balloon shrinks in size. In other words, our experiences with balloons have shown us that the quantity, volume, pressure, and temperature of a gas are interrelated.

Laboratory experiments with gases involve measurements of some or all of these properties. The quantity of a gas, expressed as mass or number of moles, can be determined directly by weighing or by calculation from its volume. The volume (V) of a gas is determined from the volume of its container, generally expressed in liters or milliliters (Section 3.5). Although we usually measure the temperature of a gas in °C, we always use the Kelvin temperature (T) in calculations. We will consider the effect of temperature changes on a gas in Sections 12.5 and 12.6. The pressure (P) of a gas is also easily measured as described in the following paragraphs.

Pressure

The pressure of a gas in a balloon is the result of molecules or atoms colliding with the surface of the balloon. **Pressure** is defined as the force exerted per unit area of surface:

$$\text{pressure } (P) = \frac{\text{force}}{\text{area}}$$

Units of pressure are obtained by dividing a unit of force by a unit of area. When force is expressed in pounds (lb) and area in square inches (in.^{2}), we have the common English unit of pressure, lb/in.^{2}, often abbreviated "psi." The SI unit of pressure is the pascal (Pa), which is defined as one newton (SI unit of force, N) per square meter (area), N/m^2.

In our calculations involving gases, we will use units of millimeters of mercury (mmHg, equivalent to torr) and atmospheres (atm). These commonly used pressure units may seem unusual at first because they do not appear to express force per unit area. To understand these units, consider barometric pressure.

Barometric pressure is the pressure of the earth's atmosphere, or **atmospheric pressure,** as measured with a barometer. A **barometer** (Figure 12.4) is a pressure-measuring instrument designed specifically to measure atmospheric pressure. It was invented by Evangelista Torricelli (1608–1647), an Italian physicist for whom the torr was named. The mercury stays up in the tube because the force of the atmosphere upon the surface of the mercury in the dish balances the force of gravity on the mercury in the tube. The height of the column of mercury in the tube is proportional to the atmospheric pressure. When the height of the mercury column is 628 mm, we say the barometric pressure, or atmospheric pressure, is 628 mmHg, or 628 torr. Changes in barometric pressure are used to predict weather changes.

Barometric pressure = atmospheric pressure.

1 atm = 760 mm Hg
= 760 torr
= 14.69 psi
= 1.013×10^5 Pa

One atmosphere (atm) is a pressure of 760 mmHg (exactly), or 760 torr.* This is equivalent to the barometric pressure measured at sea level on an average day. Atmospheres are particularly useful in expressing very high pressures. One atmosphere is equivalent to 14.69 lb/in.^{2} (14.69 psi).

Measuring the pressure of a gas in a balloon or other similar container would not be possible with a barometer. Such measurements are made with a **manometer,** an instrument used to measure the pressure of a confined gas (Figure 12.5). The difference, *d,* between the mercury levels in the two sides of the J-tube is a measure of the pressure of the confined gas relative to atmospheric pressure. Note that to determine the absolute pressure of the gas, *d* is either added to or subtracted from the barometric pressure.

A manometer measures the pressure of a confined gas.

A water manometer, rather than a mercury manometer, can be used to measure pressures that differ only slightly from barometric pressure. Remember that the density of water is 1.00 g/mL compared with 13.6 g/mL for mercury. To measure a difference in pressure of 20 torr, the difference in liquid levels in a mercury

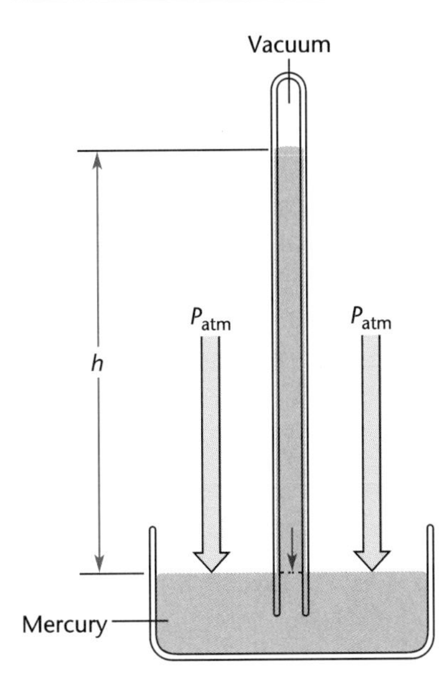

FIGURE 12.4 A mercury barometer. A barometer can be made by filling a glass tube with mercury and inverting the tube in a dish of mercury. Mercury is supported in the tube by the atmospheric pressure, P_{atm}. The space in the tube above the mercury is a vacuum. The height, *h,* of the column of mercury is proportional to the atmospheric pressure.

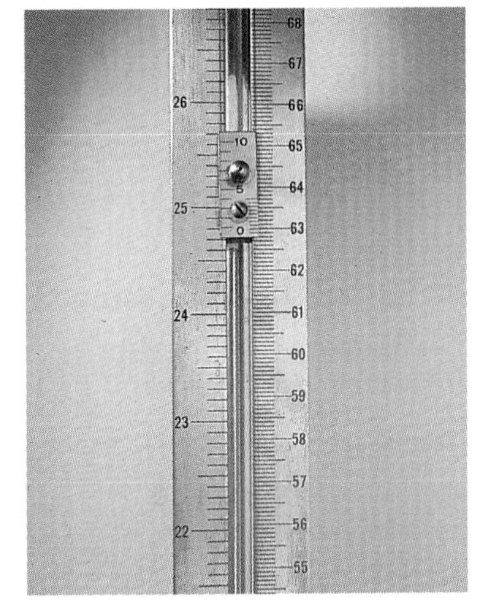

The mercury level in a laboratory barometer.

*The U.S. weather bureau usually expresses atmospheric pressures in inches of mercury. 760 mm Hg = 29.92 in. Hg

FIGURE 12.5 A manometer measures the difference between the pressure of the gas and atmospheric (or barometric) pressure. (a) P_{gas} is greater than the barometric pressure. (b) P_{gas} is less than the barometric pressure.

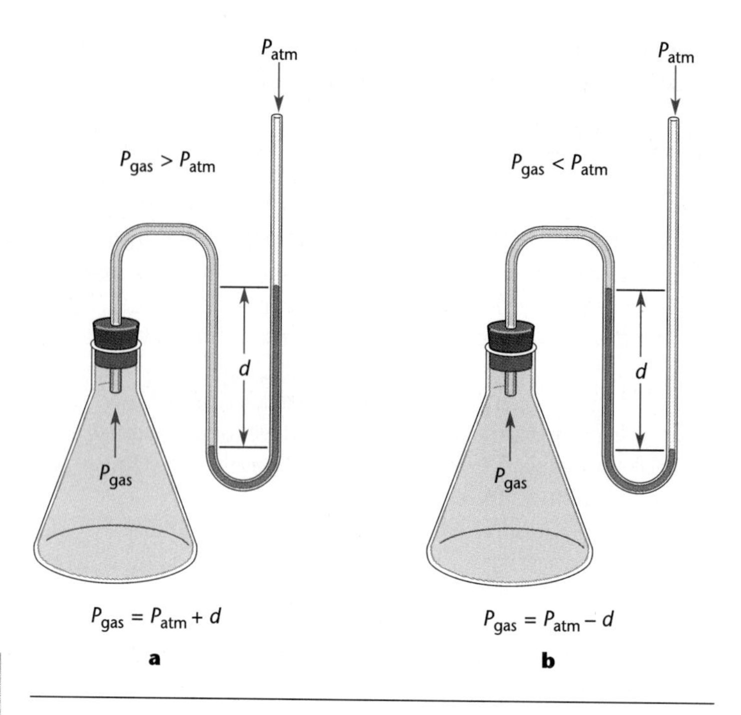

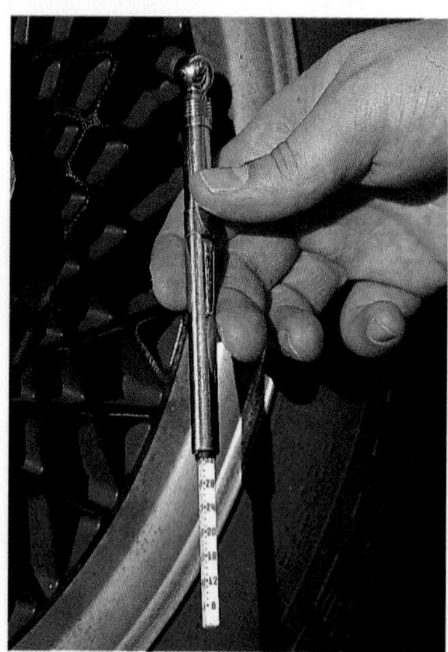

A tire gauge measures the pressure *above* atmospheric pressure.

manometer would be only 2 cm (about 1 in.), whereas the difference in liquid levels in a water manometer would be about 27 cm (about 11 in.). Thus, to measure small differences in pressure, the water manometer is preferred because it gives a larger response.

Despite the cost and toxicity of mercury, it continues to be used in manometers and barometers instead of water because of the difference in the densities of the two liquids. A water barometer would be impractical because of its height. To have the same mass as a column of mercury, a column of water would be 13.6 times higher. For example, to measure a pressure of 1.00 atm, a water barometer would have to be 10.3 m high (about 34 ft)!

Mechanical gauges are used outside laboratories to measure the pressures of gases. A tire gauge is the most familiar mechanical pressure gauge. The pressure measured by a tire gauge is the pressure *above* atmospheric pressure. Thus, a gauge pressure of 32 psi for an inflated tire at sea level is actually an absolute pressure of 47 psi (32 psi shown on the gauge plus about 15 psi atmospheric pressure). The pressure measured with a tire gauge is sometimes called "gauge pressure."

Example 12.2

Express 632 torr in atmospheres. ■

Find: $P\,(\text{atm}) = ?$

Given: 632 torr

Known: $760 \text{ torr} = 1 \text{ atm, or } \dfrac{760 \text{ torr}}{1 \text{ atm}}$

Solution: $P\,(\text{atm}) = \left(\dfrac{1 \text{ atm}}{760 \not{\text{torr}}}\right)(632 \not{\text{torr}}) = 0.832 \text{ atm}$

Boyle's Law: The Relationship of Pressure and Volume

12.4

Learning Objectives

- State Boyle's law.
- Given the initial pressure and volume of a gas, calculate the final volume when the pressure is changed or the final pressure when the volume is changed.

Robert Boyle (1627–1691) used a manometer to study the pressure–volume relationship of a gas.

Experiments conducted by Robert Boyle (1627–1691) established the relationship between the pressure and the volume of a gas. Boyle found that when a constant temperature is maintained for a fixed quantity of a gas, the pressure and volume of a gas are inversely proportional:

$$V \propto \frac{1}{P} \qquad \text{or} \qquad PV = \text{constant}$$

This relationship of the pressure and volume of a gas is often called **Boyle's law.** Another way of expressing Boyle's law is to say that the product of the pressure and the volume of a gas is constant. This is shown in the following equation:

$$P_1V_1 = \text{constant} = P_2V_2$$

or simply

$$P_1V_1 = P_2V_2$$

where P_1 and V_1 are the original pressure and volume of a gas and P_2 and V_2 are the final values. To illustrate the use of this equation, consider a 4.0-L sample of a gas at a pressure of 1.0 atm (Figure 12.6). When the volume of the gas is reduced to 2.0 L, what is the final pressure at constant temperature? For this problem we have:

Boyle's law: $V\uparrow$, $P\downarrow$ or $V\downarrow$, $P\uparrow$.

$$P_1 = 1.0 \text{ atm} \qquad P_2 = ?$$

$$V_1 = 4.0 \text{ L} \qquad V_2 = 2.0 \text{ L}$$

Dimensional analysis can be used to solve this kind of problem. The final pressure of the gas can be calculated by multiplying the original pressure times a ratio of the volumes.

$$\text{final pressure} = \text{original pressure} \times \text{volume ratio}$$

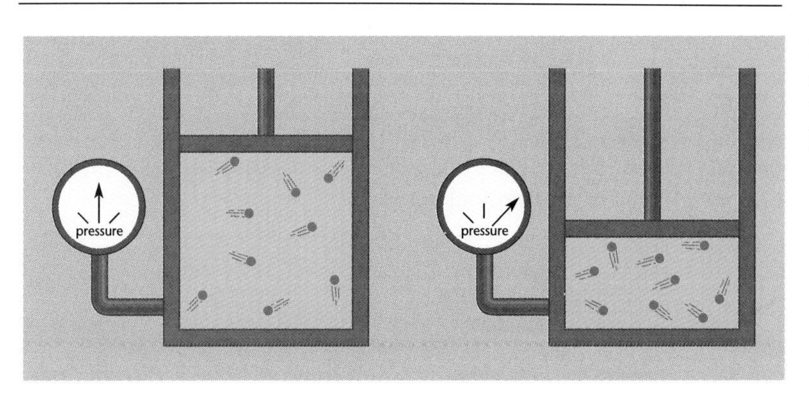

FIGURE 12.6 For a sample of gas at constant temperature, reducing the volume by one-half causes the pressure to double.

Boyle's law tells us that when the volume of the gas is reduced, the pressure increases ($V\downarrow$, $P\uparrow$). Thus, the original pressure must be multiplied by a volume ratio that would give a final pressure *greater* than the original pressure. The larger volume must be in the numerator:

$$P_2 = (1.0 \text{ atm}) \left(\frac{4.0 \cancel{\text{L}}}{2.0 \cancel{\text{L}}}\right) = 2.0 \text{ atm}$$

Because the Boyle's law equation defines the relationship between the quantities involved in this problem, algebra can be used as an alternative strategy. To solve the Boyle's law equation for P_2, we must divide both sides of the equation by V_2:

$$\frac{P_1V_1}{V_2} = \frac{P_2\cancel{V_2}}{\cancel{V_2}}$$

or

$$P_2 = \frac{P_1V_1}{V_2}$$

Substituting gives:

$$P_2 = \frac{(1.0 \text{ atm})\ (4.0 \cancel{\text{L}})}{(2.0 \cancel{\text{L}})} = 2.0 \text{ atm}$$

To check the problem, make sure the volume units cancel to give atm (a pressure unit). Also check significant figures. Is the answer reasonable? Yes, decreasing the volume results in a greater final pressure.

Whether calculating a final pressure (P_2) or a final volume (V_2), you can use either dimensional analysis or algebra.

Example 12.3

The volume of a sample of a gas was 1.65 L at a pressure of 745 torr. If the temperature is maintained constant, at what volume would the pressure be 688 torr? ■

Find: $V_2(\text{L}) = ?$

Given: $V_1 = 1.65 \text{ L}$ $\quad P_1 = 745 \text{ torr}$

$P_2 = 688 \text{ torr}$

Known: $P_1V_1 = P_2V_2$ or $P\downarrow, V\uparrow$

Solution

Dimensional analysis Recognize that as the pressure of a gas at constant temperature decreases, its volume increases ($P\downarrow$, $V\uparrow$). Multiply the initial volume by a pressure ratio that will give a *greater* final volume:

final volume = initial volume × pressure ratio

$$V_2(\text{L}) = (1.65 \text{ L}) \left(\frac{745 \cancel{\text{torr}}}{688 \cancel{\text{torr}}}\right) = 1.79 \text{ L}$$

Algebra Solve Boyle's law for the unknown, V_2:

$$V_2 = \frac{P_1V_1}{P_2} = \frac{(745\ \cancel{\text{torr}})\ (1.65\ \text{L})}{(688\ \cancel{\text{torr}})} = 1.79\ \text{L}$$

Check: Volume units (L) are needed; the answer should have 3 significant figures. Is the answer reasonable? Yes, decreasing the pressure causes an increase in the volume of a gas.

Practice Problem 12.3 A sample of nitrogen occupied a volume of 3.20 L at a pressure of 675 torr. At what pressure would its volume be 4.00 L when the temperature is held constant?

Charles' Law: The Relationship of Volume and Temperature

12.5

Learning Objectives

- State Charles' law.
- Given the initial temperature and volume of a gas, calculate the final volume when the temperature is changed or the final temperature when the volume is changed.

When a sample of gas is heated at constant pressure, its volume increases. This is a statement of **Charles' law** and can be expressed as:

Charles' law:
$V\uparrow, T\uparrow$ *or* $V\downarrow, T\downarrow$

$$V \propto T \qquad \text{or} \qquad V = \text{constant} \times T \qquad \text{or} \qquad \frac{V}{T} = \text{constant}$$

The relationship of the volume and temperature of a gas was first observed by the French physicist Jacques Charles (1746–1823), for whom this law is named (Figure 12.7). In the equation expressing this relationship, T stands for the absolute temperature (K).

Graphing the volumes of a gas versus the temperatures results in a straight line. When this line is extended by extrapolation to the point at which $V = 0$, then at this same point $T = 0$ K (-273.15°C). Figure 12.8 shows the lines for helium at three different pressures. All intersect at the same point, where $V = 0$ and $T = 0$ K. This is how absolute zero was first determined. Because volume cannot be negative, -273.15°C must be the lowest possible temperature. Regardless of the gas investigated, extrapolation of the line to $V = 0$ always gave a temperature of -273.15°C. William Thomson (1824–1907), a Scottish physicist who is better known by his title, Lord Kelvin, set the zero of his temperature scale at this point.

In a laboratory, temperatures are measured in degrees Celsius (°C). However, Charles discovered that the volume of a gas is proportional to its Kelvin temperature. Remember, the kelvin (K) is the SI unit of temperature. A Kelvin temperature is calculated (Section 4.5) from a Celsius temperature using the equation:

Use Kelvin temperatures when working gas problems.

$$T_K = T_C + 273$$

Kelvin temperatures must be used when solving gas problems.

According to the kinetic-molecular theory, raising the temperature increases the average kinetic energy of a gas. T↑, KE↑

To understand the effect of temperature on the volume of a gas, we must remember that raising the temperature increases the average kinetic energy ($KE = \frac{1}{2}mv^2$) of a gas. Because the mass (m) of a gas molecule doesn't vary, increasing the temperature of a gas must increase the velocity (v) of the gas molecules. This increases the frequency of collisions with the walls of the container, thereby increasing the pressure. On the other hand, to maintain a constant pressure when the temperature is increased, the volume must also increase. An increase in volume increases the area over which the collisions occur, thereby compensating for the higher frequency of collisions at the smaller volume.

FIGURE 12.7 Jacques Charles (1746–1823) built the first hydrogen-filled balloon, which he named the *Charliere.* It was flown by both Charles and Joseph-Louis Gay-Lussac (1778–1850), who set an altitude record of 23 018 ft in 1804. It was largely their interest in ballooning that led these men to study the effect of temperature on the volume, pressure, and density of a gas.

Charles' law can also be written in terms of initial conditions (V_1 and T_1) and final conditions (V_2 and T_2) for a gas at constant pressure:

$$\frac{V_1}{T_1} = \text{constant} = \frac{V_2}{T_2}$$

or simply

$$\frac{V_1}{T_1} = \frac{V_2}{T_2}$$

As with Boyle's law calculations, either dimensional analysis or algebra can be used in solving problems involving volume and temperature changes.

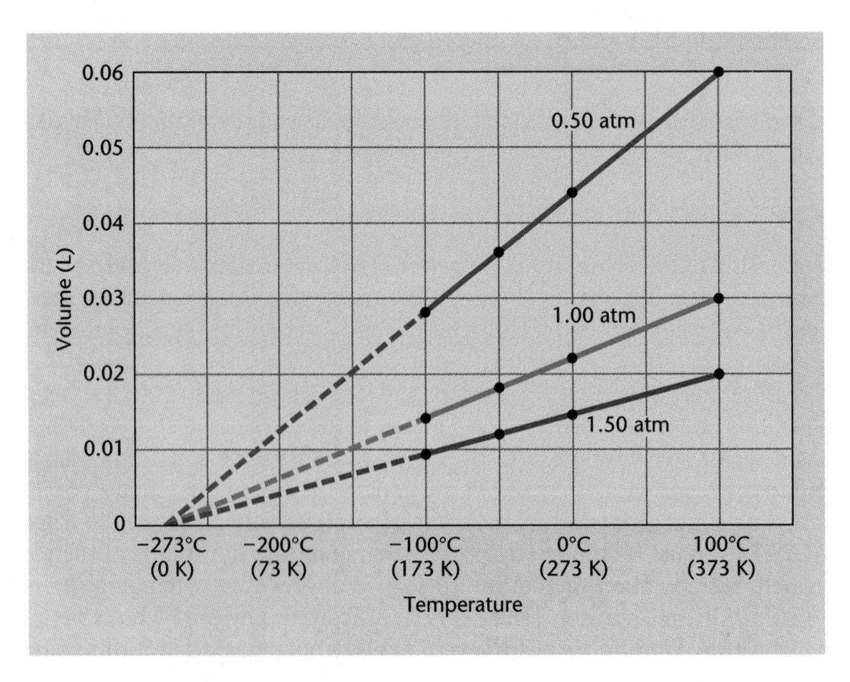

FIGURE 12.8 A graph of volume versus temperature for helium at three pressures. The straight lines show that the volume of a gas is directly proportional to its temperature. At $V = 0$, the line for each pressure intersects the temperature axis at −273.15°C.

Example 12.4

Calculate the volume of a sealed balloon at 45°C that had a volume of 1.00 L at 18°C. Assume the pressure of the air in the balloon remains constant. ■

Find: $V_2(\text{L}) = ?$

Given: $V_1 = 1.00\ \text{L}$ $\quad T_1 = 18 + 273 = 291\ \text{K}$

$T_2 = 45 + 273 = 318\ \text{K}$

Known: $\dfrac{V_1}{T_1} = \dfrac{V_2}{T_2}$ or $T\uparrow, V\uparrow$

Solution

Dimensional analysis At constant pressure, as the temperature increases, the volume increases ($T\uparrow, V\uparrow$). Therefore, multiply the initial volume by a temperature ratio that will give a *greater* final volume.

$$V_2\ (\text{L}) = (1.00\ \text{L})\left(\frac{318\ \cancel{\text{K}}}{291\ \cancel{\text{K}}}\right) = 1.09\ \text{L}$$

Algebra Solve for the unknown, V_2, and substitute:

$$V_2 = \frac{T_2V_1}{T_1} = \frac{(318\ \cancel{\text{K}})\ (1.00\ \text{L})}{(291\ \cancel{\text{K}})} = 1.09\ \text{L}$$

Example 12.5

The volume of a sample of gas in an experimental test cylinder was measured to be 465 mL at 29°C. If the volume of the gas is reduced to 225 mL, to what temperature must the sample be cooled to maintain constant pressure? ■

Find: $T_2\ (\text{K}) = ?$

Given: $V_1 = 465\ \text{mL}$ $\quad T_1 = 29 + 273 = 302\ \text{K}$

$V_2 = 225\ \text{mL}$

Known: $\dfrac{V_1}{T_1} = \dfrac{V_2}{T_2}$ or $V\downarrow, T\downarrow$

Solution

Dimensional analysis At constant pressure, as the volume decreases, the temperature decreases ($V\downarrow, T\downarrow$). Therefore, multiply the initial temperature by a volume ratio that will give a *lower* final temperature.

$$T_2\ (\text{K}) = (302\ \text{K})\left(\frac{225\ \cancel{\text{mL}}}{465\ \cancel{\text{mL}}}\right) = 146\ \text{K}$$

Algebra Solve for the unknown, T_2, and substitute:

$$T_2 = \frac{V_2T_1}{V_1} = \frac{(225\ \cancel{\text{mL}})\ (302\ \text{K})}{(465\ \cancel{\text{mL}})} = 146\ \text{K}$$

CHEMICAL WORLD

Misleading Advertisements and Charles' Law

One of the important rules to remember in applying Charles' law to gas problems is that the temperature must always be given in kelvins. Because converting Celsius to Kelvin temperatures is an extra step in the solution process, you may be tempted to save time by using the Celsius temperatures. Don't do it! To learn why, consider a ploy of the advertising industry.

Figure 1 shows the percentage of customers satisfied with two consumer products, Brand A and Brand B. A quick, uncritical scan of this bar graph gives the impression that customers are about twice as likely to be satisfied by Brand A as by Brand B. However, notice that the vertical scale begins at the 94% level.

The importance of this scale becomes evident when you compare Figure 1 with Figure 2. Both figures convey the same information, but the impressions they give are very different. Figure 2 presents the ratings with the absolute minimum value of 0% at the bottom of the scale. Now we can clearly see that the difference between the two brands is small. Twice as many customers do not prefer Brand A over Brand B.

The Kelvin temperature scale, like the percentage scale in Figure 2, uses an absolute zero point, namely 0 K. There are no negative temperatures on the Kelvin scale. Thus, when a Kelvin temperature is compared with another Kelvin temperature, the ratio between the two values reflects the actual change in temperature. A temperature of 100 K is twice as high as a temperature of 50 K.

The zero point (0°C) on the Celsius scale, however, is 273 K. So increasing the temperature from 1°C to 2°C is not really doubling the temperature. It is only a change from 274 Celsius degrees above absolute zero (0 K) to 275 Celsius degrees above absolute zero. The volume of the gas, rather than doubling, increases only slightly. See Figure 3.

Charles' law states that if the temperature of a gas is doubled, the volume of the gas also doubles. What a mistake you'd make if you were to apply this law to an experiment using the Celsius scale!

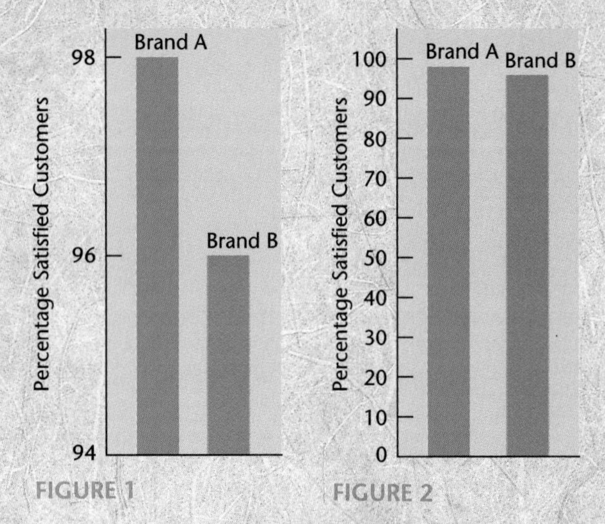

FIGURE 1

FIGURE 2

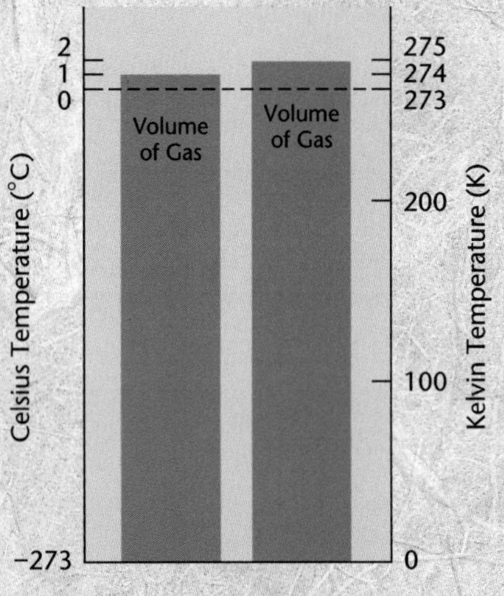

FIGURE 3 Volume of gas at different temperatures.

Practice Problem 12.4 The volume of a sample of helium is 8.50 L at a temperature of 25°C. What volume would it occupy when its temperature is raised to 65°C at constant pressure?

Gay-Lussac's Law: The Relationship of Pressure and Temperature

12.6

Learning Objectives

- State Gay-Lussac's law.
- Given the initial temperature and pressure of a gas, calculate the final pressure when the temperature is changed or the final temperature when the pressure is changed.

You may have noticed that the air pressure in your car's tires is higher after driving on a hot day. Studies by Jacques Charles and Joseph Gay-Lussac showed that the pressure of a gas at constant volume is directly proportional to the absolute temperature of the gas. This statement, known as **Gay-Lussac's law,** may be abbreviated as follows:

$$P \propto T \qquad \text{or} \qquad P = \text{constant} \times T \qquad \text{or} \qquad \frac{P}{T} = \text{constant}$$

Gay-Lussac's law: $P\uparrow$, $T\uparrow$, or $P\downarrow$, $T\downarrow$. T must be in kelvins.

Remember, *the temperature (T) must be expressed in kelvins.* As the temperature of a gas at constant volume is increased its pressure increases ($T\uparrow$, $P\uparrow$). The final pressure can be calculated by multiplying the initial pressure by a temperature ratio that will give a *greater* final pressure.

final pressure = initial pressure × temperature ratio

As with Boyle's law and Charles' law, the relationship of pressure and temperature can be expressed in terms of the initial and final states of a gas:

$$\frac{P_1}{T_1} = \text{constant} = \frac{P_2}{T_2}$$

or simply

$$\frac{P_1}{T_1} = \frac{P_2}{T_2}$$

Problems that involve the pressure–temperature relationship of a gas may be solved using either dimensional analysis or algebra.

Joseph-Louis Gay-Lussac (1778–1850) studied the effect of temperature on the pressure of a gas.

Example 12.6

The pressure of propane in a gas cylinder was 525 psi at a temperature of 22°C. Because of a fire in the building where the cylinder was stored, the temperature of the propane cylinder rose to 135°C. What was the final pressure of the propane in the cylinder? ■

Find: final pressure, P_2 (psi) = ?
Given: $P_1 = 525$ psi $T_1 = 22 + 273 = 295$ K
$T_2 = 135 + 273 = 408$ K

Known: $\frac{P_1}{T_1} = \frac{P_2}{T_2}$ or $T\uparrow, P\uparrow$

Solution

Dimensional analysis Recognize that as the temperature of a gas at constant volume increases, its pressure will increase ($T\uparrow, P\uparrow$). Therefore, multiply the initial pressure times a temperature ratio that will give a *greater* final pressure.

$$P_2 \text{ (psi)} = (525 \text{ psi}) \left(\frac{408 \text{ K}}{295 \text{ K}}\right) = 726 \text{ psi}$$

Algebra Solve the Gay-Lussac's law equation for the final pressure (P_2) and substitute:

$$P_2 = \frac{P_1 T_2}{T_1} = \frac{(525 \text{ psi})(408 \text{ K})}{(295 \text{ K})} = 726 \text{ psi}$$

Example 12.7

A gas in a steel cylinder at constant volume was at 10.°C and a pressure of 8.50 atm. If the gas was heated, what would be the final temperature when the pressure had increased to 10.5 atm? ■

Find: final temperature, T_2(K) = ?
Given: $P_1 = 8.50$ atm $T_1 = 10. + 273 = 283$ K
$P_2 = 10.5$ atm

Known: $\frac{P_1}{T_1} = \frac{P_2}{T_2}$ or $P\uparrow, T\uparrow$

Solution

Dimensional analysis $P\uparrow, T\uparrow$. Therefore, multiply the initial temperature by a pressure ratio that will give a *higher* final temperature.

$$T_2 = (283 \text{ K}) \left(\frac{10.5 \text{ atm}}{8.50 \text{ atm}}\right) = 350. \text{ K}$$

Algebra Solve the Gay-Lussac's law equation for the unknown, T_2, and substitute:

$$T_2 = \frac{T_1 P_2}{P_1} = \frac{(283 \text{ K})(10.5 \text{ atm})}{(8.50 \text{ atm})} = 350. \text{ K}$$

Practice Problem 12.5 The pressure of neon in a 10.0-L steel cylinder was 1250 atm at 20°C. Find the pressure of the neon after the temperature had risen to 35°C.

The Combined Gas Law

12.7

Learning Objective

- Given the initial pressure, volume, and temperature of a gas and two of the final measurements, calculate the third final measurement.

Our experience with balloons has shown us that the volume of a gas depends upon its temperature and pressure. Earlier in this chapter, this was stated by Boyle's and Charles' laws. Combination of these laws results in the **combined gas law:** The volume of a gas is directly proportional to its temperature and inversely proportional to its pressure.

$$V \propto \frac{T}{P} \quad \text{or} \quad \frac{PV}{T} = \text{constant}$$

Once again, using the subscripts 1 and 2 for the initial and final states of a gas, we have

$$\frac{P_1V_1}{T_1} = \text{constant} = \frac{P_2V_2}{T_2}$$

or simply

$$\frac{P_1V_1}{T_1} = \frac{P_2V_2}{T_2}$$

Included in the combined gas law equation is the pressure–temperature relationship expressed by Gay-Lussac's law. When five of the variables are known, either dimensional analysis or algebra can be used to solve for the unknown. The examples that follow illustrate the dimensional analysis approach. For example, to find the final volume of a gas after its pressure and temperature have been changed, multiply the initial volume times a pressure ratio and a temperature ratio:

$$\text{final volume} = \text{initial volume} \times \text{pressure ratio} \times \text{temperature ratio}$$

In a similar manner, you can calculate a final pressure or a final temperature of a gas:

$$\text{final pressure} = \text{initial pressure} \times \text{volume ratio} \times \text{temperature ratio}$$

$$\text{final temperature} = \text{initial temperature} \times \text{volume ratio} \times \text{pressure ratio}$$

The ratios can be determined using the reasoning practiced in earlier problems.

Example 12.8

A balloon at 23°C was inflated to a volume of 765 mL at a pressure of 743 torr. It was then submerged in water at 11°C to a depth that resulted in a pressure of 1265 torr for the air confined in the balloon. What was the final volume of the balloon? ■

Find: V_2 (mL) = ?
Given: $V_1 = 765$ mL
$P_1 = 743$ torr $P_2 = 1265$ torr
$T_1 = 23 + 273 = 296$ K $T_2 = 11 + 273 = 284$ K
Known: $\frac{P_1V_1}{T_1} = \frac{P_2V_2}{T_2}$ or $P\uparrow, V\downarrow$ and $T\downarrow, V\downarrow$

Solution

Dimensional analysis Recognize that as the pressure of a gas increases its volume decreases ($P\uparrow, V\downarrow$). Multiply the initial volume times a pressure ratio that will give a *smaller* final volume. Then, because decreasing the temperature of a gas results in a decrease in its volume ($T\downarrow, V\downarrow$), multiply the initial volume times a temperature ratio that will give a *smaller* final volume. The two ratios may be included in one equation:

$$V_2\text{ (mL)} = (765\text{ mL})\left(\frac{743\text{ torr}}{1265\text{ torr}}\right)\left(\frac{284\text{ K}}{296\text{ K}}\right) = 431\text{ mL}$$

Algebra Solve the combined gas law equation for the unknown, V_2:

$$V_2 = \frac{P_1V_1T_2}{T_1P_2} = \frac{(743\text{ torr})(765\text{ mL})(284\text{ K})}{(296\text{ K})(1265\text{ torr})} = 431\text{ mL}$$

Standard Temperature and Pressure

The volume of a gas varies depending upon its temperature and pressure. Therefore, when expressing the quantity of a gas in volume units, we must state the temperature and pressure at which the volume was measured. As a matter of convenience when comparing volumes of gases, reference conditions have been chosen for gases, called **standard conditions,** or **standard temperature and pressure (STP).** Standard temperature is 273 K (0°C) and standard pressure is 1 atm (760 torr). To compare quantities of gases, it is common practice to convert their volumes to STP.

Example 12.9

The gas in a 1.00-L cylinder was at 45° C and a pressure of 1.25 atm. What would be the volume of the gas at STP? (Note that two sets of conditions are given here since STP is a temperature and pressure.) ■

Find: V_2 (L) = ?
Given: $V_1 = 1.00$ L
$P_1 = 1.25$ atm $P_2 = 1.00$ atm
$T_1 = 45 + 273 = 318$ K $T_2 = 273$ K
Known: $\frac{P_1V_1}{T_1} = \frac{P_2V_2}{T_2}$ or $P\downarrow, V\uparrow$ and $T\downarrow, V\downarrow$

Solution

Dimensional analysis $T\downarrow, V\downarrow$. Therefore, multiply the initial volume times a temperature ratio that will give a *smaller* final volume (Charles' law). $P\downarrow, V\uparrow$, so multiply also by a pressure ratio that will give a *larger* final volume (Boyle's law).

$$V_2\ (\text{L}) = (1.00\ \text{L})\left(\frac{273\ \cancel{\text{K}}}{318\ \cancel{\text{K}}}\right)\left(\frac{1.25\ \cancel{\text{atm}}}{1.00\ \cancel{\text{atm}}}\right) = 1.07\ \text{L}$$

Algebra Solve the combined gas law equation for the unknown, V_2, and substitute:

$$V_2 = \frac{P_1V_1T_2}{T_1P_2} = \frac{(1.25\ \cancel{\text{atm}})\ (1.00\ \text{L})\ (273\ \cancel{\text{K}})}{(318\ \cancel{\text{K}})\ (1.00\ \cancel{\text{atm}})} = 1.07\ \text{L}$$

Practice Problem 12.6 A sample of propane occupied a volume of 6.50 L at STP. Calculate its volume at 22°C and 715 torr.

12.8 Avogadro's Hypothesis and Molar Volume of Gases

Learning Objectives

- State Avogadro's hypothesis.
- Calculate the number of moles of a gas at STP from its volume and vice versa.
- State the law of combining volumes.
- Using a balanced chemical equation, perform volume–volume calculations involving gaseous reactants and products at the same temperature and pressure.

You have observed that the more air you blow into a balloon the bigger it gets. In other words, the volume of the air in the balloon increases as the quantity of air increases. When we express the quantity of a gas in terms of the number of moles (n) of the gas, then we can say that the volume of a gas is proportional to the number of moles of gas at constant temperature and pressure:

$$V \propto n \qquad \text{or} \qquad V = \text{constant} \times n$$

Avogadro's hypothesis:
$n\uparrow, V\uparrow$ *or* $n\downarrow, V\downarrow$

Our observations with balloons are in agreement with an idea first stated almost two centuries ago (1811) by Amedeo Avogadro (Section 7.1). **Avogadro's hypothesis** was that equal volumes of different gases at the same temperature and pressure contain the same number of molecules. This hypothesis supports the kinetic-molecular theory that, you remember, describes an ideal gas. Because the molecules of an ideal gas are widely spaced and have negligible attraction for each other, the volume should depend upon the number of molecules in a sample, regardless of what kind of molecules they are. Thus, 1.0 L of hydrogen (H_2) and 1.0 L of carbon dioxide (CO_2) at the same temperature and pressure contain the same number of molecules and respond to changes in pressure and temperature in the same way.

Molar Volume of Gases

Molar volume = 22.4 L/mol at STP.

The kinetic-molecular theory tells us that all gas molecules, regardless of kind or size, are widely spaced and have negligible attraction for one another. Because one mole of a gas, regardless of what kind, contains 6.02×10^{23} molecules, one mole of any gas will occupy the same volume at a given temperature and pressure. At standard conditions, the volume of one mole of a gas has been determined experimentally to be 22.4 L. This volume is called the **molar volume** of an ideal gas at STP.

The molar volume of an ideal gas is a ratio that allows us to calculate the number of moles of a gas from its volume and vice versa. For example, we can determine the number of moles in 2.5 L of nitrogen at STP:

$$\text{number of moles } N_2 = \left(\frac{1 \text{ mol}}{22.4 \cancel{L}}\right)(2.5 \cancel{L}) = 0.11 \text{ mol } N_2$$

Note that the molar volume of a gas is a different value at conditions other than STP. For example, at 25°C (298 K) and 1 atm, using Charles' law ($T\uparrow$, $V\uparrow$), we would have

$$V = (22.4 \text{ L})\left(\frac{298 \cancel{K}}{273 \cancel{K}}\right) = 24.5 \text{ L}$$

Thus, the molar volume of a gas at 25°C and 1 atm is 24.5 L/mol.

Example 12.10

What is the volume of 0.549 g H_2 at STP?

Find: V-H_2 (L) = ?

Given: 0.549 g H_2, STP conditions

Known: 22.4 L/mol, $MM(H_2)$ = 2.02 g/mol

Solution

First, write an equation for the unknown:

$$V\text{-}H_2 \text{ (L)} = \left(\frac{22.4 \text{ L}}{1 \text{ mol } H_2}\right)(? \text{ mol } H_2) \qquad (1)$$

Because the number of moles of hydrogen was not given, it must be calculated:

$$\text{number of moles } H_2 = \left(\frac{1 \text{ mol}}{2.02 \cancel{g\,H_2}}\right)(0.549 \cancel{g\,H_2}) = 0.272 \text{ mol} \qquad (2)$$

Then, substituting into equation (1), we get

$$V\text{-}H_2 \text{ (L)} = \left(\frac{22.4 \text{ L}}{1 \cancel{\text{mol } H_2}}\right)(0.272 \cancel{\text{mol } H_2}) = 6.09 \text{ L}$$

Practice Problem 12.7 What mass of nitrogen (N_2) occupies a volume of 11.2 L at STP?

Law of Combining Volumes

Avogadro's hypothesis offered an explanation for observations made two years earlier by Gay-Lussac that led to the **law of combining volumes** of gases: The volumes of gases reacting with each other, when measured at the same temperature and pressure, are in ratios of small, whole numbers. Consider, for example, the reaction that occurs when methane (CH_4) burns:

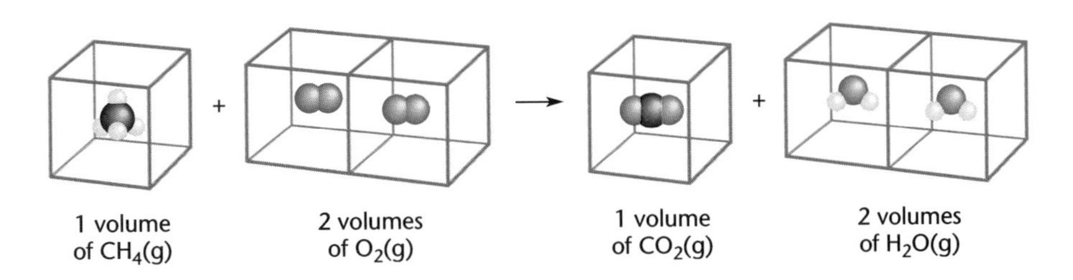

Balancing coefficients give volume ratios of gases ($n \propto V$).

The volumes of gaseous reactants and products are in whole-number ratios that correspond to the balancing coefficients in the chemical equation. Avogadro reasoned that volume ratios of reacting gases are the same as their mole ratios because equal volumes of different gases contain the same number of molecules (or moles of molecules).

The law of combining volumes allows us to determine the volume of one gas needed to react with a given volume of another *when the volumes are all measured at the same temperature and pressure.* For example, what volume of oxygen is needed to burn 2.3 L of methane (measured at the same temperature and pressure)? Because the balancing coefficients tell us volume ratios of gases, we have

$$V\text{-}O_2\ (L) = \left(\frac{2\ L\ O_2}{1\ \cancel{L\ CH_4}}\right)(2.3\ \cancel{L\ CH_4}) = 4.6\ L\ O_2$$

Example 12.11

What volume of ammonia (NH_3) can be produced from the reaction of 15.0 L hydrogen with excess nitrogen? The volumes of all three gases are measured at the same temperature and pressure. ■

Solution First, write the balanced equation for the reaction. Next, outline the problem under the equation as you would for a mole problem. For this kind of problem, the volume ratio replaces the mole ratio:

$$3H_2(g) \quad + \quad N_2(g) \quad \longrightarrow 2NH_3(g)$$

$$15.0\ \cancel{L\ H_2} \times \left(\frac{2\ L\ NH_3}{3\ \cancel{L\ H_2}}\right) \longrightarrow ?\ L\ NH_3$$

Thus,

$$V\text{-}NH_3 = 10.0\ L\ NH_3$$

Practice Problem 12.8 Hydrogen and chlorine combine to form hydrogen chloride: $H_2(g) + Cl_2(g) \rightarrow 2HCl(g)$. What volume of hydrogen chloride could be formed from the reaction of 7.50 L of chlorine with excess hydrogen if all volumes are measured at the same temperature and pressure?

12.9 The Ideal Gas Equation

Learning Objectives

- Write the ideal gas equation.
- Use the ideal gas equation to find the pressure, volume, temperature, or quantity of a gas from data given.
- Calculate the molar mass and the density of a gas from experimental data.

We noted earlier in the chapter that our experiences with balloons have shown us that the quantity, volume, pressure, and temperature of a gas are interrelated. The relationship among these four variables for an ideal gas is expressed by the **ideal gas equation:**

$$\frac{PV}{T} = nR \qquad \text{or} \qquad PV = nRT$$

In this equation, n equals the number of moles of the gas, and R is a constant called the gas constant. R can be evaluated by solving the equation (algebra strategy) and substituting the values for one mole of an ideal gas at STP, $V = 22.4$ L, $n = 1$ mol, $P = 1$ atm, $T = 273$ K:

Gas constant, $R = 0.0821 \frac{L \cdot atm}{mol \cdot K}$.

$$R = \frac{PV}{nT} = \frac{(1\ \text{atm})(22.4\ \text{L})}{(1\ \text{mol})(273\ \text{K})} = 0.0821\ \frac{\text{L} \cdot \text{atm}}{\text{mol} \cdot \text{K}}$$

When using this value for R, you must remember to express P, V, and T using the units of atm, L, and K. When other units are used for P and V, then the numerical value of R is different. For example, when P is in torr and V is in mL, then

$$R = 6.24 \times 10^4\ \frac{\text{mL} \cdot \text{torr}}{\text{mol} \cdot \text{K}}$$

Use of other units results in still other numerical values for R. In solving problems in this book, we will use only

$$R = 0.0821\ \frac{\text{L} \cdot \text{atm}}{\text{mol} \cdot \text{K}}$$

Example 12.12

What is the volume of 0.500 mol of carbon dioxide at 20°C and a pressure of 725 torr? ■

Find: V (L) = ?

Given: $n = 0.500$ mol $\quad T = 20 + 273 = 293$ K

$P = 725$ torr

Known: $PV = nRT$

$$P = \frac{725 \text{ torr}}{760 \text{ torr/atm}} = 0.954 \text{ atm}$$

Solution Because the ideal gas equation relates n, P, V, and T, we will use algebra to solve the equation for the unknown, V:

$$V = \frac{nRT}{P} = \frac{(0.500 \text{ mol})\left(0.0821 \frac{\text{L} \cdot \text{atm}}{\text{mol} \cdot \text{K}}\right)(293 \text{ K})}{(0.954 \text{ atm})} = 12.6 \text{ L}$$

Check: Make sure the units cancel to give the volume in L. 

Example 12.13

Calculate the temperature of an 11.0-g sample of propane (C_3H_8) contained in a 5.00-L cylinder under a pressure of 1.25 atm. ■

Find: T (K) = ?

Given: $V = 5.00$ L $\quad P = 1.25$ atm

11.0 g C_3H_8

Known: $PV = nRT$

MM-(C_3H_8) = 44.1 g/mol C_3H_8 (from atomic masses)

$$n \text{ (mol)} = \frac{11.0 \text{ g } C_3H_8}{44.1 \text{ g/mol } C_3H_8} = 0.249 \text{ mol } C_3H_8$$

Solution Using algebra, solve the ideal gas equation for the unknown, T. Then substitute:

$$T = \frac{PV}{nR} = \frac{(1.25 \text{ atm})(5.00 \text{ L})}{(0.249 \text{ mol})\left(0.0821 \frac{\text{L} \cdot \text{atm}}{\text{mol} \cdot \text{K}}\right)} = 306 \text{ K}$$

Check: See that the units cancel to give K. 

Practice Problem 12.9 Find the pressure inside a 5.00-L steel cylinder that contains 125 g of helium at 23°C.

Molar Mass of a Gas

Earlier, we illustrated how the molecular formula of a compound can be determined from its empirical formula and its molar mass (*MM*). Now, we are prepared to understand how the molar mass of a compound can be determined from gas measurements. We can identify the measurements needed for the calculation by starting with the units of molar mass:

$$MM = \frac{\text{g}}{\text{mol}}$$

The mass of a gas can be measured directly. The number of moles of a gas can be calculated from its volume using the ideal gas equation.

Example 12.14

A 0.598-g sample of a gas occupied a volume of 0.489 L at 752 torr and 27°C. Calculate the molar mass of the gas. ■

Find: MM (g/mol) = ?

Given: 0.598 g $V = 0.489\text{ L}$ $T = 27 + 273 = 300.\text{ K}$

$$P = \frac{752\ \cancel{\text{torr}}}{760\ \cancel{\text{torr}}/\text{atm}} = 0.990\text{ atm}$$

Known: $PV = nRT$ $R = 0.0821\ \frac{\text{L}\cdot\text{atm}}{\text{mol}\cdot\text{K}}$

Solution The units of molar mass tell us that we must divide mass (g) by the number of moles of the gas.

$$MM = \frac{\text{g}}{\text{mol}}$$

From the ideal gas equation:

$$n = \frac{PV}{RT} = \frac{(0.990\ \cancel{\text{atm}})\ (0.489\ \cancel{\text{L}})}{\left(0.0821\frac{\cancel{\text{L}}\cdot\cancel{\text{atm}}}{\text{mol}\cdot\cancel{\text{K}}}\right)(300.\ \cancel{\text{K}})} = 0.0197\text{ mol}$$

$$MM = \frac{0.598\text{ g}}{0.0197\text{ mol}} = 30.4\text{ g/mol}$$

Practice Problem 12.10 Calculate the molar mass of a gas if 0.356 g of the gas occupies a volume of 219 mL at 25°C and 1.00 atm.

Density of a Gas

Imagine a cool morning when the air is very still. Several colorful balloons are rising into the clear, blue sky. The intermittent "whoosh" of the gas burners can be heard as they heat the air inside the giant nylon bags. The crews in the gondolas wave to friends as they rise into the air and drift slowly away.

Why do the balloons rise in the air? The answer lies with the gas burners. The hot air inside an ascending balloon is "lighter" than the cool air of the atmosphere outside the balloon. In other words, the density of the hot air is less than the density of the surrounding cooler air. The difference in densities causes a balloon to rise in the same way that smoke from a fire rises into the sky.

The ideal gas equation can be used to calculate the density of a gas from gas measurements. Starting with the defining equation for density, algebra can be used to combine this equation with the ideal gas equation:

A gas burner heats air inside a balloon as it rises in a clear, blue sky. The density of hot air is less than the density of cooler surrounding air, so the balloon rises.

$$D = \frac{m}{V} \quad \text{and} \quad V = \frac{m}{D}$$

Substituting into the ideal gas equation gives

$$PV = nRT$$

$$\frac{Pm}{D} = nRT$$

Solving for D, we get

$$D = \frac{Pm}{nRT}$$

Because

$$MM = \frac{\text{g}}{\text{mol}} = \frac{m}{n}$$

then

$$D = MM\,\frac{P}{RT}$$

This equation shows that the density of a gas is directly related to its molar mass and inversely proportional to its absolute temperature. Note that the units of density for gases are generally g/L.

CHEMICAL WORLD

Space Fire

Lucky you! You've been selected as the first first-year chemistry student to be launched into space. The experiment is to see if you can survive out there with the knowledge you've acquired so far. Of course, you'll be with experienced astronauts, but they're instructed to interfere in your activities only if you're about to blow up the spacecraft, launch yourself, or starve.

Dinner the first day out. To celebrate your launch, you're equipped with a steak (or a chunk of tough tofu if you're a vegetarian) and a special barbecue in which you're to build a real fire to cook your steak. No flameless ration heater here! Sucking their food tubes, the old timers watch you with amusement.

You arrange your paper and kindling as you know how to do, and strike a match. The match goes out. You strike another and it goes out. Faulty matches? NASA has flung you out here a hundred-odd miles from earth, and they can't get decent matches? You light another one and put it quickly to the paper, which flares for a second, as it should—after all you're in a near-100% oxygen atmosphere—and then it goes out. Check the paper. It's not wet. Try again. Same result. What's wrong? You look at the astronauts, but they're absorbed in studying various dials and parts of the bulkhead.

All right, that's what you're here for. Figure it out!

Is it something about the spacecraft? It has automatic fire retardant? But this one is generally no different from the Apollo craft through which a sudden fire raced and killed three astronauts. Of course, the flammables are gone now, but there is plenty of oxygen, and this paper and wood are certainly flammable.

Wait. The Apollo had been on the ground for a test. This spacecraft is in space. So what's the difference between a grounded spacecraft and an orbiting one? Go ahead. Make a hypothesis about it before you read on. Why won't the paper and kindling burn? Why do the matches go out so fast?

The difference in the two craft is the lack of gravity. Now think a bit more before you go on.

Gas density! On earth we say warm air rises, though what's actually happening is that cold air is sinking, because it's heavier than warm air, which forces the warm air upward. The effect is that fresh cold air rushing in at the base of a fire brings oxygen to the flames and pushes the warm products of combustion upward. This can't happen without gravity, so the warm air just sits there. The fire, what there is of it, makes CO_2, which doesn't support combustion. In fact, it's used in fire extinguishers. So the fire dies for lack of oxygen.

You look around and find the fan that came with the barbecue; then you start your fire, and fan it constantly while you burn your steak.

An artist's view of a future space station.

Example 12.15

What is the density of helium at 22°C and a pressure of 0.650 atm?

Find: $D\ (\mathrm{g/L}) = ?$

Given: $T = 22 + 273 = 295\ \mathrm{K} \qquad P = 0.650\ \mathrm{atm}$

Known: $D = MM\,\dfrac{P}{RT}$ (from ideal gas equation)

$MM(\mathrm{He}) = 4.00\ \mathrm{g/mol}$

$R = 0.0821\ \dfrac{\mathrm{L \cdot atm}}{\mathrm{mol \cdot K}}$

Solution:
$$D = \frac{(4.00\ \mathrm{g/\cancel{mol}})\ (0.650\ \cancel{\mathrm{atm}})}{(0.0821\ \mathrm{L \cdot \cancel{atm}/\cancel{mol} \cdot \cancel{K}})(295\ \cancel{\mathrm{K}})} = 0.107\ \mathrm{g/L}$$

Example 12.16

Which gas has a greater density at 25°C and 1.00 atm?

a. N_2 or He
b. CH_4 or C_3H_8

Solutions The density of a gas at a given temperature and pressure depends upon its molar mass.

a. N_2
b. C_3H_8

Practice Problem 12.11 Calculate the density of methane (CH_4) at 22°C and 715 torr.

Practice Problem 12.12 Which gas is more dense at a given temperature and pressure:
a. F_2 or Cl_2?
b. O_2 or Ar?

Dalton's Law of Partial Pressures

12.10

Learning Objectives

- State Dalton's law of partial pressures.
- Given the total pressure of a gas mixture and all but one of the partial pressures, calculate the unknown partial pressure.

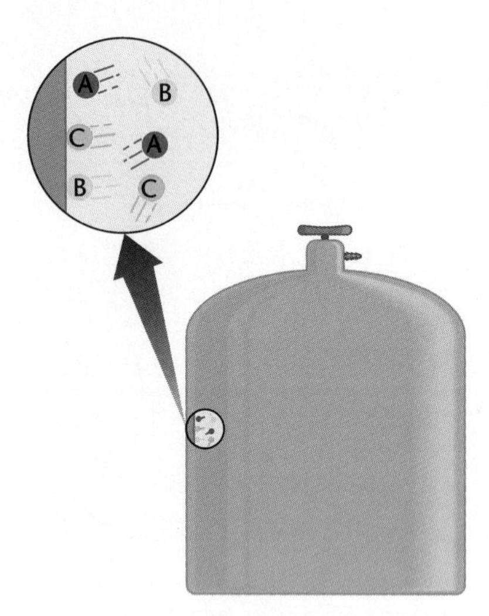

The total pressure equals the sum of the partial pressures:

$P_T = P_A + P_B + P_C$

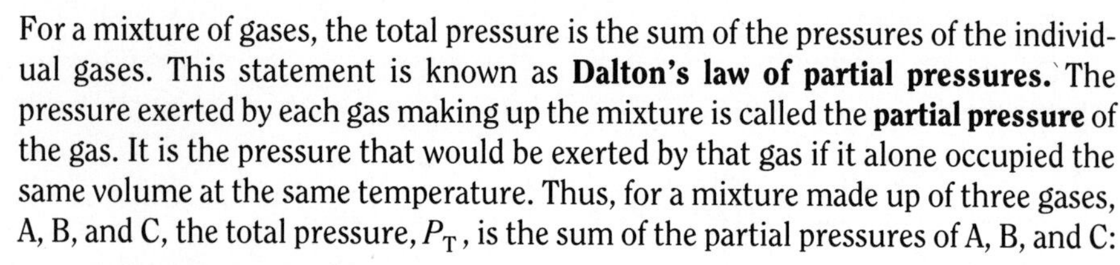

For a mixture of gases, the total pressure is the sum of the pressures of the individual gases. This statement is known as **Dalton's law of partial pressures.** The pressure exerted by each gas making up the mixture is called the **partial pressure** of the gas. It is the pressure that would be exerted by that gas if it alone occupied the same volume at the same temperature. Thus, for a mixture made up of three gases, A, B, and C, the total pressure, P_T, is the sum of the partial pressures of A, B, and C:

$$P_T = P_A + P_B + P_C$$

For example, if a sample of air is made up of nitrogen at a partial pressure of 582 torr, oxygen at a partial pressure of 156 torr, and argon at a partial pressure of 7.1 torr, the total pressure is

$$P_T = 582 \text{ torr} + 156 \text{ torr} + 7.1 \text{ torr} = 745 \text{ torr}$$

In laboratory experiments, insoluble gases are frequently collected by displacement of water from a bottle or gas buret. An apparatus for collecting hydrogen is shown in Figure 12.9. Because the hydrogen collected in the buret was bubbled through water, it is mixed with water vapor. Thus, P_T, which is equal to the barometric pressure, is the sum of the partial pressures of hydrogen and water vapor:

$$P_T = P_{H_2} + P_{H_2O}$$

Vapor pressures of water at various temperatures are given in Table 12.2.

Example 12.17

A 48.5-mL sample of hydrogen was collected over water at 23°C and a barometric pressure of 685 torr. Find the volume of the hydrogen at STP. The vapor pressure of water at 23°C is 21.2 torr. ■

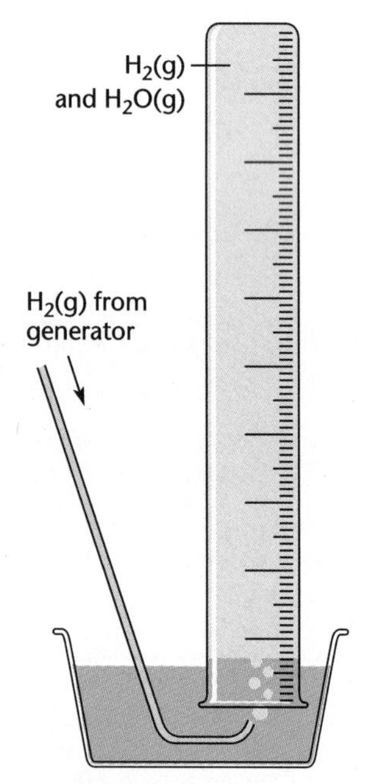

FIGURE 12.9 Hydrogen collected by displacement of water.

TABLE 12.2 Vapor Pressure of Water at Various Temperatures

Temperature (°C)	*Vapor Pressure (torr)*
5	6.5
10	9.2
15	12.8
20	17.5
25	23.8
30	31.8
35	42.2
40	55.3
50	92.5
60	149.4
70	233.7
80	355.1
90	525.8
100	760.0

Find: V_2 (mL) = ? at STP

Given: $V_1 = 48.5$ mL

$T_1 = 23 + 273 = 296$ K $\qquad T_2 = 273$ K

$P_{bar} = 685$ torr $\qquad P_2 = 760$ torr

$P_{H_2O} = 21.2$ torr

Known: $\dfrac{P_1V_1}{T_1} = \dfrac{P_2V_2}{T_2} \qquad P_T = P_{H_2} + P_{H_2O} \qquad P\uparrow, V\downarrow$ and $T\downarrow, V\downarrow$

Solution The partial pressure of hydrogen at 23°C can be calculated using Dalton's law:

$$P_{H_2} = P_T - P_{H_2O} = 685 \text{ torr} - 21.2 \text{ torr} = 664 \text{ torr}$$

This is the initial pressure for hydrogen, or P_1. Then:

$$V_2 = V_1 \times \text{pressure ratio} \times \text{temperature ratio}$$

$P\uparrow, V\downarrow$: Therefore, multiply the initial volume times a pressure ratio that gives a *smaller* final volume. $T\downarrow, V\downarrow$: Therefore multiply the initial volume times a temperature ratio that gives a *smaller* final volume.

$$V_2 \text{ (mL)} = (48.5 \text{ mL}) \left(\frac{664 \text{ torr}}{760 \text{ torr}}\right)\left(\frac{273 \text{ K}}{296 \text{ K}}\right) = 39.1 \text{ mL}$$

Practice Problem 12.13 Oxygen can be prepared in the laboratory by the decomposition of potassium chlorate ($KClO_3$):

$$2KClO_3(s) \longrightarrow 2KCl(s) + 3O_2(g)$$

If 225 mL of O_2 was collected over water at 22°C and a barometric pressure of 634 torr, what is the volume of dry O_2 at STP? The vapor pressure of water at 22°C is 19.8 torr.

Calculations from Chemical Equations Involving Gases

12.11

Learning Objective

- Perform mole-volume and mass-volume calculations using balanced chemical equations involving gaseous reactants and/or products.

In Chapter 11 we used a three-step outline (mole burrow) for solving problems involving chemical equations. These steps involved calculation of the:

1. number of moles of a given substance from its mass
2. number of moles of the unknown from the number of moles of the given substance
3. mass of the unknown from the number of moles of the unknown

Depending on the problem, the calculation may have involved one, two, or all three of these steps.

In the problems worked previously, the number of moles of a given substance was calculated from its mass and its molar mass. However, in this chapter we have shown that the number of moles of a gaseous substance can be determined from its volume, or vice versa, using either the ideal gas equation or the molar volume of an ideal gas:

$$PV = nRT \qquad \text{or} \qquad 22.4 \text{ L/mol at STP}$$

Thus, when the reactants and products are gases, volumes may be used instead of masses in mole calculations.

Example 12.18

What mass of sodium will react with water to produce 15.0 L of hydrogen measured at STP? The chemical equation is

$$2\text{Na(s)} + 2\text{H}_2\text{O(l)} \longrightarrow 2\text{NaOH(aq)} + \text{H}_2\text{(g)}$$ ■

Solution Outline the problem under the balanced chemical equation:

$$2\text{Na(s)} + 2\text{H}_2\text{O(l)} \longrightarrow 2\text{NaOH(aq)} + \text{H}_2\text{(g)}$$

? g ↑ $MM = 23.0$ g/mol ← ____ mol Na ← $\left(\frac{2 \text{ mol Na}}{1 \text{ mol H}_2}\right)$ ← ____ mol H_2 ← 22.4 L/mol ← 15.0 L

Follow the steps of the outline to set up equations to calculate the mass of sodium needed. Use the molar volume to calculate the number of moles of H_2:

$$\text{number of moles H}_2 = \left(\frac{1 \text{ mol}}{22.4 \cancel{\text{L}}}\right)(15.0 \cancel{\text{L}}) = 0.670 \text{ mol H}_2 \qquad (1)$$

$$\text{number of moles Na} = \left(\frac{2 \text{ mol Na}}{1 \cancel{\text{mol H}_2}}\right)(0.670 \cancel{\text{mol H}_2}) = 1.34 \text{ mol Na} \qquad (2)$$

$$\text{mass (g) Na} = \left(\frac{23.0 \text{ g}}{1 \cancel{\text{mol Na}}}\right)(1.34 \cancel{\text{mol Na}}) = 30.8 \text{ g} \qquad (3)$$

After the calculations have been completed, the outline looks like this:

$$2\text{Na(s)} + 2\text{H}_2\text{O(l)} \longrightarrow 2\text{NaOH(aq)} + \text{H}_2\text{(g)}$$

30.8 g ↑ $MM = 23.0$ g/mol ← 1.34 mol ← $\left(\frac{2 \text{ mol Na}}{1 \text{ mol H}_2}\right)$ ← 0.670 mol ← 22.4 L/mol ← 15.0 L

Practice Problem 12.14 What mass of ammonium sulfate, $(NH_4)_2SO_4$, forms when 75.0 L of ammonia, measured at 20°C and 1.00 atm, reacts with sulfuric acid?

Gas Laws

Graham's law: $\frac{r_1}{r_2} = \sqrt{\frac{MM_2}{MM_1}} = \sqrt{\frac{D_2}{D_1}}$

Boyle's law: $P_1V_1 = P_2V_2$

Charles' law: $\frac{V_1}{T_1} = \frac{V_2}{T_2}$

Gay-Lussac's law: $\frac{P_1}{T_1} = \frac{P_2}{T_2}$

Combined gas law: $\frac{P_1V_1}{T_1} = \frac{P_2V_2}{T_2}$

Ideal gas equation: $PV = nRT$

Dalton's law: $P_T = P_A + P_B + P_C + \cdots$

Chapter Review

Directions: From the list of choices at the end of the Chapter Review, choose the word or term that best fits in each blank. Use each choice only once. The answers appear at the end of the chapter.

The earth's atmosphere is made up mostly of oxygen and ___________ (1), and other gases in smaller amounts. Oxygen is essential for human and animal life. The nitrogen in the atmosphere can be converted to nitrogen compounds that are plant nutrients. Gases have a major economic impact on industrialized nations. Of the top five chemicals produced in the United States, ___________ (2) are gases. These include nitrogen, ___________ (3), ethylene, and ___________ (4).

Section 12.1 Of the three states of matter, gases are the most ___________ (5) and have the lowest ___________ (6). Gases flow, mix readily, and fill their containers completely. Thus, the ___________ (7) of a gas is the same as the volume of its container.

Section 12.2 The properties of gases can be explained by the ___________ ___________ (8) theory. According to this theory, gas molecules are moving ___________ (9) in straight lines, are widely ___________ (10), and are not ___________ (11) to each other. The average kinetic energy of the molecules depends upon the ___________ (12) temperature. A gas that behaves according to the kinetic-molecular theory is called an ___________ (13) gas. The behavior of real gases is essentially ideal except at low temperatures and ___________ (14) pressures.

The relative rates of effusion or diffusion of two gases are ___________ (15) proportional to the square roots of their densities or their molar masses.

Section 12.3 The quantity, volume, pressure, and temperature of a gas are interrelated. The temperature depends upon the average ___________ (16) energy of a gas, and raising the temperature increases the ___________ (17) of the molecules. The coldest possible temperature is called ___________ ___________ (18). ___________ (19) is defined as force per unit area. The common English unit is lb/in.^{2}, or ___________ (20). Other common units are mmHg, or ___________ (21), and atmospheres. These units are used to express either atmospheric pressure or the pressure of a confined gas. One __________ (22) is a pressure of 760 torr. A __________ (23) is an instrument used to measure atmospheric pressure. Because of its high density, __________ (24) is the liquid used in barometers. In a laboratory, a __________ (25) is used to measure the pressure of a confined gas.

Section 12.4 __________ (26) law states that the volume of a gas is inversely proportional to its pressure. Thus, increasing the volume of a quantity of gas at constant temperature results in a __________ (27) in its pressure.

Section 12.5 Charles' law states that the volume of a gas is __________ (28) proportional to the absolute temperature. Thus, __________ (29) the temperature of a quantity of gas at constant pressure results in an increase in its volume.

Section 12.6 Gay-Lussac's law states that the pressure of a gas at constant volume is directly proportional to the __________ __________ (30) of the gas. This means that increasing the temperature of a gas confined in a cylinder of fixed volume would result in a(n) __________ (31) in the pressure of the gas.

Section 12.7 Combination of Boyle's, Charles', and __________ (32) laws resulted in the combined gas law. This law expresses the relationship of the pressure, volume, and temperature of a gas. Comparison of volumes of gases is often made at standard temperature and pressure, or __________ (33), where $T =$ __________ (34) and $P =$ __________ (35).

Section 12.8 Equal volumes of gases at the same temperature and pressure contain the same number of __________ (36). This is a statement of __________ (37) hypothesis. From this hypothesis we recognize that the volume of a gas is also proportional to the number of __________ (38) of the gas. At standard conditions, the volume of one mole of an ideal gas, called its __________ (39) volume, is __________ (40). The molar volume of an ideal gas is a __________ (41) factor that we can use to find volume from moles and vice versa.

The volumes of gases reacting with each other, when measured at the same temperature and pressure, are in __________ (42) of small, whole numbers. This law of __________ (43) volumes allows us to determine the volume of one gas needed to react with a given volume of another.

Section 12.9 The ideal gas equation interrelates the volume, pressure, temperature and quantity (moles) of a gas. The units of the __________ __________ (44), R, determine what units must be used for volume and pressure when using this equation in calculations.

The ideal gas equation can be used to calculate the molar mass and the density of a gas. To calculate its molar mass, the pressure, volume, and temperature of a certain __________ (45) of a gas must be measured. The density of a gas can be calculated from its __________ __________ (46), pressure, and temperature. As the temperature of a gas increases, its density __________ (47). At a given pressure and temperature, a gas having a high molar mass has a __________ (48) density than a gas having a low molar mass.

Section 12.10 The total pressure of a mixture of gases is the sum of the __________ (49) pressures of the gases making up the mixture. This is known as __________ (50) law of partial pressures.

Section 12.11 In calculations involving gaseous reactants or __________ (51) of a chemical reaction, the number of moles of a gas may be calculated from its volume or vice versa using either the molar volume at STP or the ideal gas equation. Thus, the number of moles of a gas can be calculated either from its mass (using its molar mass) or its volume.

Choices: absolute temperature, absolute zero, ammonia, atmosphere, attracted, Avogadro's, barometer, Boyle's, combining, conversion, Dalton's, decrease, decreases, density, directly, four, gas constant, Gay-Lussac's, high, higher, ideal, increase, increasing, inversely, Kelvin, kinetic, kinetic-molecular, manometer, mass, mercury, mobile, molar, molar mass, molecules, moles, nitrogen, 1 atm, oxygen, partial, pressure, products, psi, rapidly, ratios, separated, STP, torr, 273 K, 22.4 L, velocity, volume

Study Objectives

After studying Chapter 12, you should be able to:

Introduction

1. Give the approximate percentage of nitrogen, oxygen, and argon in dry air.

Section 12.1

2. Summarize the general properties of gases.

Section 12.2

3. State the main features of the kinetic-molecular theory.
4. Compare the rates of effusion or diffusion of two gases from their densities or molecular weights.
5. State the conditions of temperature and pressure at which real gases deviate from ideal behavior.

Section 12.3

6. Briefly describe a barometer and a manometer and what is measured with each.
7. Convert pressures from torr to atmospheres and vice versa.

Section 12.4

8. State Boyle's law.
9. Given the initial pressure and volume of a gas, calculate the final volume when the pressure is changed or the final pressure when the volume is changed.

Section 12.5

10. State Charles' law.
11. Given the initial temperature and volume of a gas, calculate the final volume when the temperature is changed or the final temperature when the volume is changed.

Section 12.6

12. State Gay-Lussac's law.
13. Given the initial temperature and pressure of a gas, calculate the final pressure when the temperature is changed or the final temperature when the pressure is changed.

Section 12.7

14. State the combined gas law.
15. Given an initial pressure, volume, and temperature of a gas and two of the final measurements, calculate the third final measurement.

Section 12.8

16. State Avogadro's hypothesis.
17. Calculate the number of moles of a gas at STP from its molar volume and vice versa.
18. State the law of combining volumes.
19. Using a balanced chemical equation, perform volume–volume calculations involving gaseous reactants and products at the same temperature and pressure.

Section 12.9

20. Write the ideal gas equation.
21. Use the ideal gas equation to find the pressure, volume, temperature, or quantity of a gas from data given.
22. Calculate the molar mass and the density of a gas from experimental data.

Section 12.10

23. State Dalton's law of partial pressures.
24. Given all but one of the partial pressures and the total pressure of a gas mixture, calculate the unknown partial pressure.

Section 12.11

25. Perform mole–volume and mass–volume calculations using balanced chemical equations involving gaseous reactants and/or products.

Key Terms

Review the definition of each of the terms listed here by section number. You may use the glossary if necessary.

12.1 diffusion

12.2 effusion, ideal gas, kinetic-molecular theory

12.3 atmospheric pressure, barometer, barometric pressure, manometer, pressure

12.4 Boyle's law

12.5 Charles' law

12.6 Gay-Lussac's law

12.7 combined gas law, standard conditions, standard temperature and pressure (STP)

12.8 Avogadro's hypothesis, law of combining volumes, molar volume

12.9 ideal gas equation

12.10 Dalton's law of partial pressures, partial pressure

Questions and Problems

Answers to questions and problems with blue numbers appear in Appendix C. Asterisks indicate the more challenging problems.

1. What is the approximate composition of dry air?
2. Briefly explain the role of ozone in the upper atmosphere.
3. Describe nitrogen fixation that occurs during thunderstorms.

Properties of Gases (Section 12.1)

4. Gases are fluids. Explain how we use this property of gases to advantage in handling air and other gases such as propane (a fuel), oxygen (for medical and other purposes), and chlorine (used in water treatment).
5. The molecules of gases are much more widely spaced compared with the molecules of a liquid such as water, resulting in much lower densities for gases. Explain how the difference in density of air and water influences the velocity of a bullet shot into a lake.
6. Explain why it is easier to compress a gas than a liquid or a solid.
7. Many animals have a keen sense of smell. What general property of gases does the sense of smell depend upon? When an animal is swimming under water, why is its sense of smell practically ineffective?
8. Of the top five chemicals produced in the United States—H_2SO_4, N_2, O_2, C_2H_4, and NH_3—which are gases with densities less than the density of air? Explain your choices.
9. A volume of a liquid can be measured by pouring it into a graduated cylinder. Explain why a volume of gas cannot be measured in the same way.

Kinetic-Molecular Theory (Section 12.2)

10. Summarize the main features of the kinetic-molecular theory.
11. Samples of Ar, O_2, CH_4, CO_2, and N_2 were at 25°C. Arrange the gases in order of increasing molecular velocity. Explain.
12. Under what conditions of temperature and pressure do real gases deviate significantly from ideal behavior? Why?
13. Gasoline was poured into a pan and left uncovered in a corner of a closed garage. In the opposite corner of the garage, about 30 ft from the pan of gasoline, was a space heater with a gas pilot light. In a short time, an explosion occurred, and the garage was engulfed in flames. Explain, in terms of the kinetic-molecular theory, how the gasoline ignited even though the pan was 30 ft from the pilot light.
14. Find the relative rates of diffusion of Cl_2 and H_2 gases.
15. Air is a mixture of gases, including nitrogen, oxygen, argon, and carbon dioxide. Arrange these gases in order of increasing velocity at a given temperature and pressure.

Gas Measurements (Section 12.3)

16. Carbon tetrachloride is a toxic organic liquid having a density of 1.6 g/mL. Discuss possible problems with its use in a barometer or manometer instead of mercury.
17. Define pressure. Explain what causes the pressure of a gas.
18. A manometer was connected to a balloon as illustrated in Figure 12.5 (a). The distance between the mercury levels was 165 mm. The barometer reading was 632 torr. What was the pressure of the gas in the balloon?
19. An open-end manometer was connected to a rigid metal gas cylinder. The mercury level in the side connected to the gas cylinder was 132 mm higher than the mercury level in the open side of the manometer. The barometric pressure was 706 torr. What was the pressure of the gas?
20. Convert 693 mmHg to:
    a. torr
    b. centimeters of Hg
    c. atm
    d. psi
21. A pressure gauge for a cylinder of oxygen reads 2245 psi. What is the gauge pressure in atmospheres? in torr?
22. A new tire was mounted on a wheel, inflated to 32 psi, and tested to determine that there were no leaks. It was then shipped from Denver (barometric pressure = 622 torr) to Los Angeles (barometric pressure = 755 torr). Even though the tire was not leaking, and the temperature was unchanged, the gauge pressure had dropped to 29 psi. Explain.
23. At a certain altitude, a weather balloon recorded a barometric pressure of 113 torr. What was the barometric pressure in atm?

Boyle's Law (Section 12.4)

24. A sample of gas occupied a volume of 135 mL at a pressure of 633 torr. What was the volume of the gas when its pressure increased to 695 torr?
25. A cylinder containing nitrogen gas was fitted with a movable piston. The pressure in the cylinder was 2.35 atm when the volume was 1.25 L. What would be the new pressure after the piston was moved to give a new volume of 3.15 L?
26. A 265-mL sample of air had a pressure of 725 mmHg. What would be the new volume of the air when the pressure was increased to 2.00 atm?
27. The volume of a sample of argon was 45.6 mL at 282 K and 0.945 atm. If its pressure were allowed to change to 0.675 atm with no change in temperature, what was the final volume of the gas?
28. The initial pressure, volume, and temperature of a gas were 452 torr, 50.8 mL, and 292 K, respectively. What was the final pressure of the gas when the final volume and temperature were 62.3 mL and 292 K?

Charles' Law (Section 12.5)

29. A 6.35-L sample of a gas at 272 K was heated to 325 K at constant pressure. What was the final volume of the gas?
30. A 3.55-L sample of a gas at 298 K was compressed to a volume of 1.68 L at constant pressure. What was the final temperature of the gas?
31. A sample of helium had a volume of 125 mL at 26°C. What was the volume of the helium after it was cooled to −186°C at constant pressure?
32. The temperature of a sample of gas was 23°C. At what temperature would the gas have expanded to three times its original volume at constant pressure?

Gay-Lussac's Law (Section 12.6)

33. Nitrogen in a 10.0-L rigid steel cylinder at a pressure of 10.2 atm was heated from 21°C to 103°C. What was the final pressure of the nitrogen?
34. A steel cylinder was filled with oxygen in Phoenix, where the temperature was 94°F, to a pressure of 2450 psi. It was then shipped to Denver. What was the pressure of the oxygen after the cylinder had been stored in a warehouse for a week at a temperature of 34°F?
35. The pressure of a 25.0-L cylinder of propane sitting on an open delivery truck was 22.3 atm. If the temperature outside was 2.0°C, what would be the pressure when the cylinder was brought inside and allowed to warm to 27°C?

Combined Gas Law (Section 12.7)

36. A sample of a gas occupied a volume of 265 mL at 49°C and 1.00 atm. What would be the volume of the gas at 0°C and 1.85 atm?
37. Volumes of gases are often compared at standard temperature and pressure (STP). Explain what is meant by STP.
38. Which of the following are standard conditions?
    a. 0 K, 760 torr
    b. 0°C, 670 torr
    c. 670 torr, 273°C
    d. 273 K, 760 mmHg
39. A 1.00-L rigid cylinder was filled with nitrogen at 25°C to give a pressure of 1.50 atm. What would be the volume of the nitrogen at STP?
40. The volume of a gas was 10.0 L at STP. What would be the volume of the gas at 298 K and a pressure of 825 torr?
41. The gas in a 0.538-L engine cylinder had a pressure of 695 torr at 96°C. What was the final pressure of the gas after it was compressed to a volume of 0.125 L as the temperature rose to 345°C?
42. A sample of air occupied a volume of 25.0 L at 298 K and 632 torr. What would be the volume of the air at 15°C and 1.00 atm?
43. The volume of a quantity of a gas was 10.0 L measured at 3.50×10^{-3} torr and 135 K. What volume would the gas occupy at 10.0 torr and 298 K?
44. A 725-L cylinder was filled with ammonia at 55°C and a pressure of 165 atm at a manufacturing plant. What would be the volume of ammonia at 25°C and 730. torr?

Avogadro's Hypothesis, Molar Volume, Combining Volumes (Section 12.8)

45. If 0.325 mol of a gas occupied a volume of 6.50 L, how many moles of the gas occupy a volume of 28.5 L at the same temperature and pressure?
46. A 1.00-L balloon contained 0.179 g of helium. How many grams of helium are needed to fill a 5.00-L balloon at the same temperature and pressure?
47. If a 1.00-L balloon is filled with 28.0 g of nitrogen, what size balloon will be filled with 4.00 g of hydrogen at the same temperature and pressure?
48. How many moles of helium are in a 10.0-L sample at STP?
49. What is the volume of 0.230 mol of methane at STP?
50. Find the volume of 1.00 g of Ar at STP.
51. What is the volume of 18.0 g of oxygen at STP?
52. What is the mass of 45.6 mL of hydrogen measured at STP?
53. How many moles of helium are needed to fill a balloon having a volume of 3.75×10^3 m^3 at STP? (1 m^3 = 1000 L)
54. What volume of oxygen is needed to burn 15.0 L hydrogen, both measured at the same temperature and pressure?

    $$2H_2(g) + O_2(g) \longrightarrow 2H_2O(g)$$

55. Propane (C_3H_8) burns to give carbon dioxide and water. What volume of propane can be completely burned with 22 L of oxygen? Assume both gases are measured at the same temperature and pressure.
56. Hydrogen combines with chlorine to produce hydrogen chloride, a gas. What volume of hydrogen is needed to produce 75 L of hydrogen chloride when both are measured at 25°C and 1.00 atm?
57. Carbon disulfide, CS_2, is a highly flammable solvent. What volume of sulfur dioxide will be formed when a quantity of carbon disulfide burns to give 6.5 L of CO_2 measured at the same conditions of temperature and pressure?
58. Oxidation of glucose, $C_6H_{12}O_6$, is described by the following equation:

    $$C_6H_{12}O_6(s) + 6O_2(g) \longrightarrow 6CO_2(g) + 6H_2O(l)$$

    What volume of carbon dioxide will be formed when 13.5 L of oxygen react with glucose? Both gases are measured at the same temperature and pressure.

Ideal Gas Equation (Section 12.9)

59. A 0.128-mol sample of neon in a 1.00-L container was at a temperature of 23°C. What was the pressure of the neon?
60. A sample of hydrogen was collected in a bottle having a volume of 245 mL at a pressure of 686 torr and a temperature of 19°C. How many moles of hydrogen were collected?
61. How many moles of CO_2 are in a 3.42-L container if the pressure is 2.65 atm and the temperature is 298 K?

62. The pressure exerted by 2.80 g N_2 in a 10.0-L gas collection bag was 725 torr. Calculate the temperature of the gas.
63. What mass of methane (CH_4) would fill a 2.2-L balloon at 28°C and a pressure of 755 torr?
64. Find the mass of SO_2 in a 1.00-L bottle at a pressure of 2.60 atm and a temperature of 301 K.
65. What volume would be occupied by 3.20 g of ammonia (NH_3) at 733 torr and a temperature of 293 K?
66. Find the volume of a balloon filled with 12.3 g of helium at 688 torr and a temperature of 11°C.
67. The volume of 0.234 g of a gas at STP was 119 mL. Calculate the molar mass of the gas.
68. A 1.85-g sample of a gas was collected in a bottle having a volume of 961 mL at a pressure of 743 torr and a temperature of 291 K. What was the molar mass of the gas?
69. The molar mass of methane (CH_4) is 16.0 g/mol. What volume would 3.88 g of methane occupy at 719 torr and 23°C?
70. A 2.68-g sample of a gas occupied a volume of 1.50 L at STP. Calculate the molar mass of the gas.
71. Arrange the following gases in order of increasing density at a given temperature and pressure: O_2, CH_4, HCl, HI, SO_2.
72. What is the density of carbon dioxide at 12°C and 625 torr?
73. The density of a gas is 0.759 g/L at STP. What is the molar mass of the gas?
74. What is the density of methane (CH_4) at 686 torr and 292 K?
75. The density of a gas is 1.25 g/L at STP. What is its density at 1.00 atm and 295 K?

Dalton's Law of Partial Pressures (Section 12.10)

76. At a barometric pressure of 645 torr, the partial pressure of nitrogen and oxygen were 503 torr and 135 torr, respectively. Assuming the remainder was argon, what was the partial pressure of the argon?
77. In a laboratory experiment, a sample of oxygen was collected over water in a gas buret at 25°C and a barometric pressure of 643 torr. The volume of oxygen in the buret was measured to be 41.6 mL. The vapor pressure of water at 25°C is 23.8 torr.
    a. What was the partial pressure of oxygen in the buret?
    b. Assuming ideal behavior, how many moles of oxygen were collected in the buret?
78. An atmosphere of nitrogen and oxygen was to be used in an experiment with laboratory mice. What partial pressure of nitrogen would be needed to provide a partial pressure of oxygen equal to 356 torr at a total pressure of 715 torr?
79. A mixture of 0.15 mol helium and 0.25 mol neon was placed in a container at 28°C and at a total pressure of 1.00 atm. What was the volume of the container?

Calculations from Chemical Equations (Section 12.11)

80. Marble, $CaCO_3$, dissolves in hydrochloric acid as described by the following equation:

    $CaCO_3(s) + 2HCl(aq) \longrightarrow CaCl_2(aq) + CO_2(g) + H_2O(l)$

    What volume of carbon dioxide measured at STP will be formed by dissolving 1.75 g of marble in hydrochloric acid?
81. The decomposition of a quantity of hydrogen peroxide produced 245 mL of oxygen measured at 24°C and 683 torr.

    $2H_2O_2(aq) \longrightarrow 2H_2O(l) + O_2(g)$

    How many moles of hydrogen peroxide decomposed?
82. 0.240 g of sodium hydride (NaH) reacted with water to produce sodium hydroxide and hydrogen. What volume of hydrogen was produced if the reaction was carried out at STP?
83. Mercury(II) oxide decomposes upon heating to give mercury and oxygen. Calculate the mass of mercury(II) oxide needed to yield 10.0 L of oxygen at STP.

*84. Acetylene (C_2H_2) reacts with hydrogen chloride to give vinyl chloride, a compound used to make polyvinyl chloride (PVC):

    $C_2H_2(g) + HCl(g) \longrightarrow C_2H_3Cl(g)$

    When 3.0 mol acetylene and 1.2 mol HCl were mixed in a 25-L reaction vessel at 25°C, vinyl chloride was formed. What was the pressure when the reaction was complete?

85. In school laboratories, ammonia and hydrogen chloride that are released from reagent bottles will react in the air to produce ammonium chloride, a white solid that coats windows and other surfaces:

    $NH_3(g) + HCl(g) \longrightarrow NH_4Cl(s)$

    What mass of ammonium chloride would be formed from the reaction of 5.0 L ammonia and 5.0 L hydrogen chloride measured at STP?
86. Sulfur trioxide, SO_3, reacts with water to produce sulfuric acid.

    $SO_3(g) + H_2O(l) \longrightarrow H_2SO_4(aq)$

    What volume of sulfur trioxide measured at 293 K and 0.870 atm is needed to produce 2.43 mol sulfuric acid?
87. One of the steps in the preparation of copper from a copper sulfide ore involves roasting the ore to oxidize the copper sulfide:

    $2Cu_2S(s) + 3O_2(g) \longrightarrow 2Cu_2O(s) + 2SO_2(g)$

    What mass of copper(I) sulfide can be roasted using 3.00 m^3 of oxygen measured at STP? (1 m^3 = 1000 L)
88. Baking soda ($NaHCO_3$) reacts with the acid (represented by HA) in a sourdough recipe to produce the carbon dioxide that causes the dough to rise. What volume of carbon dioxide will form at 26°C and 633 torr when 4.82 g of baking soda reacts in a sourdough mix?

    $NaHCO_3(aq) + HA(aq) \longrightarrow NaA(aq) + H_2O(l) + CO_2(g)$

89. Carbon dioxide is absorbed by a solution of barium hydroxide to form barium carbonate according to the following equation:

$Ba(OH)_2(aq) + CO_2(g) \longrightarrow BaCO_3(s) + H_2O(l)$

What mass of barium carbonate will result from the reaction of 325 mL carbon dioxide measured at 21°C and 1.00 atm with a solution of barium hydroxide?

90. In an industrial process, methanol is produced by the reaction of hydrogen with carbon monoxide:

$2H_2(g) + CO(g) \longrightarrow CH_3OH(l)$

What volumes of hydrogen and carbon monoxide measured at STP are needed to produce 1.00 kg of methanol?

91. Liquid sodium chloride can be decomposed by electrolysis to produce sodium metal and chlorine gas:

$2NaCl(l) \xrightarrow{\text{electric current}} 2Na(l) + Cl_2(g)$

What volume of chlorine measured at 125°C and 1.00 atm will be produced by the electrolysis of 425 g of NaCl?

Additional Problems

92. An 8.00-L expandable balloon was filled with helium at STP and sealed. At what temperature would the balloon expand to 10.0 L if the pressure remained constant?

93. Calculate the molar volume of an ideal gas at 24°C and a pressure of 632 torr.

94. A 1.25-L container was filled with a mixture of 2.80 g of nitrogen and 2.00 g of argon at 24°C. What was the total pressure of the mixture?

95. What mass of hydrogen is needed to fill a balloon having a volume of 1.35 L at 24°C and a pressure of 632 torr?

96. What are the densities of methane (CH_4) and propane (C_3H_8) at 298 K and 0.965 atm? Which gas tends to rise in air and which gas tends to sink in air? Explain your answer.

97. The explosion of nitroglycerin, described by the following equation, results in the formation of many gaseous molecules:

$4C_3H_5N_3O_9(l) \longrightarrow 6N_2(g) + O_2(g) + 12CO_2(g) + 10H_2O(g)$

What total volume of gases from the explosion contains 1.0 L of CO_2?

*98. An explosion of nitroglycerin (described in problem 97) resulted in the formation of 29.0 L of gaseous products measured at STP. What mass of nitroglycerin exploded?

*99. Oxygen was generated in a laboratory experiment by the decomposition of potassium chlorate and collected over water in a gas buret. The volume of oxygen was measured to be 46.8 mL at 15°C and a barometric pressure of 748 torr. The vapor pressure of water at this temperature is 12.8 torr. What was the mass of potassium chlorate decomposed in the experiment?

$2KClO_3(s) \longrightarrow 2KCl(s) + 3O_2(g)$

*100. In another experiment, 1.23 g $KClO_3$ was decomposed (see problem 99) to produce 385 mL oxygen collected over water at a barometric pressure of 756 torr and a temperature of 15°C. The vapor pressure of water at this temperature is 12.8 torr. From these data, calculate the molar volume of oxygen at STP.

*101. A sample of 1.00 L hydrogen was collected at 25°C and 1.00 atm. How many molecules of hydrogen were in the sample?

102. A 20.1-g sample of a gas occupied a volume of 12.7 L at 313 K and a pressure of 698 torr. What is the molar mass of the gas?

103. Calculate the volume of 36.3 g of CO_2 at 33°C and a pressure of 1.25 atm.

*104. How many oxygen (O_2) molecules are in a 1.00-L balloon filled with air at 24°C and a pressure of 0.924 atm? Note that the volume percent of oxygen in air is 20.95%.

Solutions to Practice Problems

PP 12.1

The gas with the smaller molar mass diffuses faster.

a. Ar b. H_2

PP 12.2

a. The molar mass of helium is smaller, and it leaks faster.

b. $$\frac{\text{rate of He}}{\text{rate of } N_2} = \sqrt{\frac{MM\text{-}N_2}{MM\text{-He}}} = \sqrt{\frac{28.0}{4.00}} = 2.65$$

Helium leaks 2.65 times faster than nitogen.

PP 12.3

Find: P_2 (torr) = ?

Given: $V = 3.20$ L $P_1 = 675$ torr

$V_2 = 4.00$ L

Known: $P_1V_1 = P_2V_2$ or $V\uparrow, P\downarrow$

Solution

Dimensional analysis Recognize that as the volume of a gas at constant temperature increases, its pressure decreases ($V\uparrow, P\downarrow$). Multiply the initial pressure times a volume ratio that will give a *smaller* final pressure:

final pressure = initial pressure × volume ratio

$$P_2 \text{ (torr)} = (675 \text{ torr})\left(\frac{3.20 \cancel{L}}{4.00 \cancel{L}}\right) = 540. \text{ torr}$$

Algebra Solve Boyle's law for the unknown, P_2:

$$P_2 = \frac{P_1V_1}{V_2} = \frac{(675 \text{ torr})(3.20 \cancel{L})}{(4.00 \cancel{L})} = 540. \text{ torr}$$

PP 12.4

Find: V_2 (L) = ?

Given: $T_1 = 25 + 273 = 298$ K $\quad V_1 = 8.50$ L

$T_2 = 65 + 273 = 338$ K

Known: $\frac{V_1}{T_1} = \frac{V_2}{T_2}$ or $T\uparrow, V\uparrow$

Solution

Dimensional analysis Recognize that when the temperature of a gas at constant pressure is increased, the volume increases ($T\uparrow, V\uparrow$). Therefore, multiply the initial volume times a temperature ratio that will give a *greater* final volume:

$$V_2 = (8.50 \text{ L})\left(\frac{338 \cancel{K}}{298 \cancel{K}}\right) = 9.64 \text{ L}$$

Algebra Solve the Charles' law equation for the unknown, V_2:

$$V_2 = \frac{T_2V_1}{T_1} = \frac{(338 \cancel{K})(8.50 \text{ L})}{(298 \cancel{K})} = 9.64 \text{ L}$$

PP 12.5

Find: final pressure, P_2 (atm) = ?

Given: $P_1 = 1250$ atm $\quad T_1 = 20 + 273 = 293$ K

$T_2 = 35 + 273 = 308$ K

Known: $\frac{P_1}{T_1} = \frac{P_2}{T_2}$ or $T\uparrow, P\uparrow$

Solution

Dimensional analysis Recognize that as the temperature of a gas at constant volume increases, its pressure increases ($T\uparrow$, $P\uparrow$). Therefore, multiply the initial pressure times a temperature ratio that will give a *greater* final pressure:

$$P_2 \text{ (atm)} = (1250 \text{ atm})\left(\frac{308 \cancel{K}}{293 \cancel{K}}\right) = 1310 \text{ atm or } 1.31 \times 10^3 \text{ atm}$$

Algebra Solve the Gay-Lussac's law equation for the final pressure (P_2) and substitute:

$$P_2 = \frac{P_1T_2}{T_1} = \frac{(1250 \text{ atm})(308 \cancel{K})}{(293 \cancel{K})} = 1310 \text{ atm or } 1.31 \times 10^3 \text{ atm}$$

PP 12.6

Find: V_2 (L) = ?

Given: $V_1 = 6.50$ L

$P_1 = 760$ torr $\quad P_2 = 715$ torr

$T_1 = 273$ K $\quad T_2 = 22 + 273 = 295$ K

Known: $\frac{P_1V_1}{T_1} = \frac{P_2V_2}{T_2}$ $\quad P\downarrow, V\uparrow$ and $T\uparrow, V\uparrow$

Solution

Dimensional analysis $T\uparrow$, $V\uparrow$; therefore, multiply the initial volume times a temperature ratio that will give a *larger* final volume (Charles' law.) $P\downarrow$, $V\uparrow$, so multiply the initial volume times a pressure ratio that will give a *larger* final volume (Boyle's law).

$$V_2 \text{ (L)} = (6.50 \text{ L})\left(\frac{295 \cancel{K}}{273 \cancel{K}}\right)\left(\frac{760 \cancel{\text{torr}}}{715 \cancel{\text{torr}}}\right) = 7.47 \text{ L}$$

Algebra Solve the combined gas law equation for the unknown, V_2:

$$V_2 = \frac{P_1V_1T_2}{T_1P_2} = \frac{(760 \cancel{\text{torr}})(6.50 \text{ L})(295 \cancel{K})}{(273 \cancel{K})(715 \cancel{\text{torr}})} = 7.47 \text{ L}$$

PP 12.7

Find: mass N_2 (g) = ?

Given: 11.2 L at STP

Known: 22.4 L/mol at STP

MM-N_2 = 28.0 g/mol (from atomic masses)

Solution:

$$\text{mass } N_2 \text{ (g)} = \left(\frac{28.0 \text{ g}}{1 \text{ mol } N_2}\right)(? \text{ mol } N_2) \qquad (1)$$

The number of moles of N_2 must be calculated:

$$\text{number of moles } N_2 = \left(\frac{1 \text{ mol}}{22.4 \cancel{L}}\right)(11.2 \cancel{L}) = 0.500 \text{ mol} \qquad (2)$$

Substitution into the first equation gives the answer:

$$\text{mass } N_2\text{(g)} = \frac{28.0 \text{ g}}{(1 \cancel{\text{mol } N_2})}(0.500 \cancel{\text{mol } N_2}) = 14.0 \text{ g} \qquad (1)$$

PP 12.8

The coefficients in the balanced equation give us the volume ratio needed:

$$V\text{(L)-HCl} = \left(\frac{2 \text{ L HCl}}{1 \cancel{\text{L } Cl_2}}\right)(7.50 \cancel{\text{L } Cl_2}) = 15.0 \text{ L HCl}$$

PP 12.9

Find: P (atm) = ?

Given: $V = 5.00$ L, $T = 23 + 273 = 296$ K

125 g He

Known: $PV = nRT$, $R = 0.0821 \frac{\text{L} \cdot \text{atm}}{\text{mol} \cdot \text{K}}$

MM-He = 4.00 g/mol

Solution: Rearrange the equation to solve for P.

$$P = \frac{nRT}{V} \qquad (1)$$

Because n, the number of moles of He, is not given, it must be calculated:

$$\text{number of moles He} = \left(\frac{1\text{ mol}}{4.00\text{ g}}\right)(125\text{ g}) = 31.3\text{ mol} \quad (2)$$

Substitution into the former equation gives the pressure of the helium:

$$P = \frac{(31.3\text{ mol})\left(0.0821\frac{\text{L}\cdot\text{atm}}{\text{mol}\cdot\text{K}}\right)(296\text{K})}{(5.00\text{ L})} = 152\text{ atm} \quad (1)$$

PP 12.10

Find: MM (g/mol) = ?

Given: 0.356 g $V = 219$ mL (0.219 L), $T = 273 + 25 = 298$ K $P = 1.00$ atm

Known: $PV = nRT$ $R = 0.0821 \frac{\text{L}\cdot\text{atm}}{\text{mol}\cdot\text{K}}$

Solution: $MM = \frac{\text{g}}{\text{mol}}$ (1)

Mass is given, so find the number of moles, n:

$$n = \frac{PV}{RT} = \frac{(1.00\text{ atm})(0.219\text{ L})}{\left(0.0821\frac{\text{L}\cdot\text{atm}}{\text{mol}\cdot\text{K}}\right)(298\text{ K})} = 0.00895\text{ mol} \quad (2)$$

$$MM = \frac{(0.356\text{ g})}{(0.00895\text{ mol})} = 39.8\text{ g/mol} \quad (3)$$

PP 12.11

Find: D (g/L) = ?

Given: $T = 22 + 273 = 295$ K, $P = 715$ torr (0.941 atm)

Known: $MM(CH_4) = 16.0$ g/mol

Solution: $$D = MM\frac{P}{RT} = \frac{(16.0\text{ g/mol})(0.941\text{ atm})}{\left(0.0821\frac{\text{L}\cdot\text{atm}}{\text{mol}\cdot\text{K}}\right)(295\text{ K})} = 0.622\text{ g/L}$$

PP 12.12

The gas of greater molar mass has the greater density.

a. Cl_2 is more dense than F_2.
b. Ar is more dense than O_2.

PP 12.13

Find: V_2 (mL) = ? at STP

Given: $V_1 = 225$ mL
$T_1 = 22 + 273 = 295$ K $T_2 = 273$ K
$P_{bar} = 634$ torr $P_2 = 760$ torr
$P_{H_2O} = 19.8$ torr

Known: $\frac{P_1V_1}{T_1} = \frac{P_2V_2}{T_2}$; $T\downarrow$, $V\downarrow$, and $P\uparrow$, $V\downarrow$
$P_T = P_{O_2} + P_{H_2O}$

Solution: The partial pressure of oxygen at 22°C can be calculated using Dalton's law:

$$P_{O_2} = P_T - P_{H_2O} = 634\text{ torr} - 19.8\text{ torr} = 614\text{ torr} \quad (1)$$

This is the initial pressure for oxygen, or P_1. Then:

$$V_2 = V_1 \times \text{pressure ratio} \times \text{temperature ratio} \quad (2)$$

$P\uparrow$, $V\downarrow$; therefore, multiply the initial volume by a pressure ratio that gives a *smaller* final volume. $T\downarrow$, $V\downarrow$; therefore multiply by a temperature ratio that gives a *smaller* final volume.

$$V_2\text{ (mL)} = (225\text{ mL})\left(\frac{614\text{ torr}}{760\text{ torr}}\right)\left(\frac{273\text{ K}}{295\text{ K}}\right) = 168\text{ mL} \quad (2)$$

Note that 760 torr is an exact number, and the answer should have 3 significant figures.

PP 12.14

Outline the problem under the balanced chemical equation:

$$2NH_3(g) + H_2SO_4(aq) \rightarrow (NH_4)_2SO_4(aq)$$

75.0 L → ($PV = nRT$) → ____ mol × $\left(\frac{1\text{ mol }(NH_4)_2SO_4}{2\text{ mol }NH_3}\right)$ → ____ mol → ($MM = 132$ g/mol) → ? g

Using algebra, solve the ideal gas equation for n, number of moles NH_3. Note that $T = 20 + 273 = 293$ K.

$$n = \frac{PV}{RT} = \frac{(1.00\text{ atm})(75.0\text{ L})}{(0.0821\text{ L}\cdot\text{atm/mol}\cdot\text{K})(293\text{ K})} = 3.12\text{ mol }NH_3 \quad (1)$$

Then, follow the steps of the outline:

$$\text{number of moles }(NH_4)_2SO_4 = \left(\frac{1\text{ mol }(NH_4)_2SO_4}{2\text{ mol }NH_3}\right)(3.12\text{ mol }NH_3) = 1.56\text{ mol }(NH_4)_2SO_4 \quad (2)$$

$$\text{mass (g) }(NH_4)_2SO_4 = \left(\frac{132\text{ g}}{\text{mol }(NH_4)_2SO_4}\right)(1.56\text{ mol }(NH_4)_2SO_4) = 206\text{ g }(NH_4)_2SO_4 \quad (3)$$

After completing the calculations, the outline looks like this:

$$2NH_3(g) + H_2SO_4(aq) \rightarrow (NH_4)_2SO_4(aq)$$

75.0 L → ($n = \frac{PV}{RT}$) → 3.12 mol × $\left(\frac{1\text{ mol }(NH_4)_2SO_4}{2\text{ mol }NH_3}\right)$ → 1.56 mol → ($MM = 132$ g/mol) → 206 g

Answers to Chapter Review

1. nitrogen
2. four
3. oxygen
4. ammonia
5. mobile
6. density
7. volume
8. kinetic-molecular
9. rapidly
10. separated
11. attracted
12. absolute
13. ideal
14. high
15. inversely
16. kinetic
17. velocity
18. absolute zero
19. pressure
20. psi
21. torr
22. atmosphere
23. barometer
24. mercury
25. manometer
26. Boyle's
27. decrease
28. directly
29. increasing
30. absolute temperature
31. increase
32. Gay-Lussac's
33. STP
34. 273 K
35. 1 atm
36. molecules
37. Avogadro's
38. moles
39. molar
40. 22.4 L
41. conversion
42. ratios
43. combining
44. gas constant
45. mass
46. molar mass
47. decreases
48. higher
49. partial
50. Dalton's
51. products

13

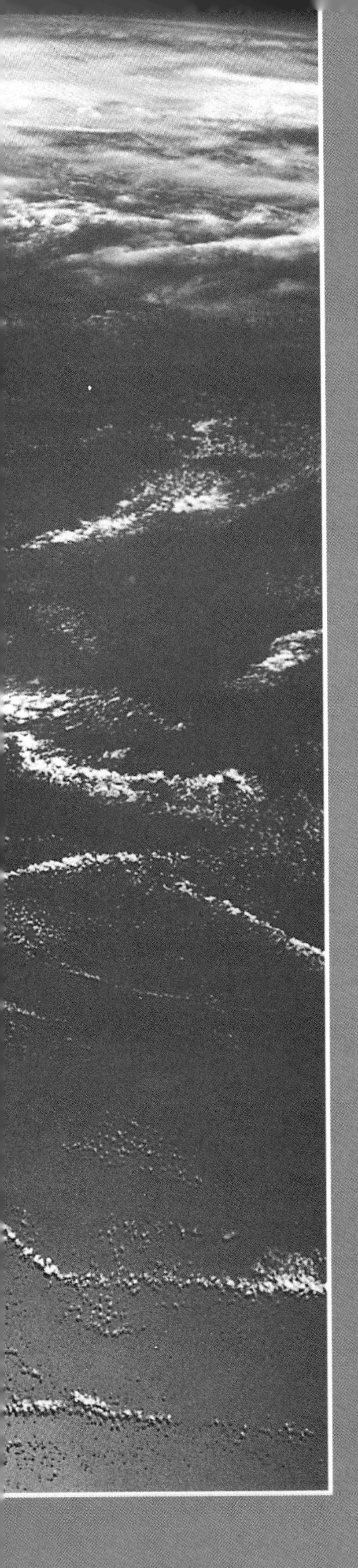

Liquids, Solids, and Changes of State

Contents

The Pacific Ocean surrounds the chain of Hawaiian Islands. More than 70% of the surface of the earth is covered with water.

Many fruits and vegetables have a high water content.

Clothes dry faster at higher temperatures.

Sodium chloride forms cubic crystals.

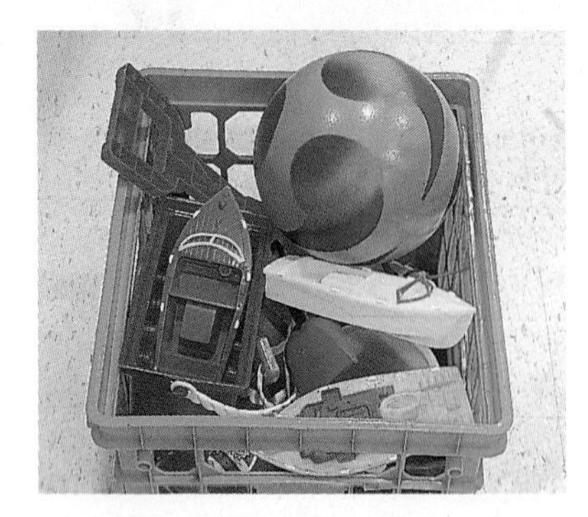

Plastics are not crystalline.

You Probably Know . . .

- Water is the most abundant compound on earth. More than 70% of the surface of the earth is covered by water. The oceans contain most of the earth's water (about 97%) with the rest in glaciers, lakes, rivers and small streams, and groundwater.
- The body of an adult person is mostly (about 70%) water. Most of the mass of plants is water as well. For example, many fruits and vegetables are composed of approximately 90% water.
- Although water exists in all three physical states, most of it on earth is liquid water. Only about 2% of the water covering the earth is found in glaciers and polar ice. A very small percentage exists as water vapor in the earth's atmosphere.
- Some liquids, such as gasoline, are volatile and evaporate quickly at ordinary temperatures. Others, such as water, evaporate slowly at room temperature but more rapidly at higher temperatures. For example, to dry our clothes quickly, we heat them in a clothes dryer. Liquids such as vegetable oil and motor oil do not evaporate much even when heated.
- Some solids, such as sodium chloride, form crystals having regular geometric shapes. They are generally hard and brittle. Many other solids are not crystalline. Examples include plastics (polyethylene, polystyrene) and fibers (polyesters, nylon).
- Ionic compounds melt only at extremely high temperatures. Examples include sodium chloride (table salt) and calcium carbonate (for instance, marble). By comparison, molecular solids—such as ice, citric acid, and cholesterol—melt at relatively low temperatures.

Why Should We Study Liquids and Solids?

We see many liquids and solids in the world around us. We use many liquid substances daily, including gasoline, motor oil, and antifreeze in our cars and vegetable oil, vinegar, and various beverages in the kitchen and at our dining table. However, of all the known liquids, water is the most common and most important. We use water for bathing, cleaning, heating and cooling, drinking, and cooking. Water quality in our environment affects all of us. Water is crucial in sustaining all kinds of life since most of the chemical reactions of living things occur in water.

Most of the elements are solids at ordinary temperatures. The structural materials of our bodies, buildings, vehicles, and natural formations of the earth are solids. We take solid raw materials from the earth and the forests and refine them to produce steel, lumber, glass, and other construction materials.

Knowledge of the properties of liquids and solids is necessary to understand changes we observe in substances. To interpret chemical and physical changes in the laboratory or elsewhere we must understand the relationship of the structure of substances and their properties in the solid, liquid, and gaseous states. Our focus in this chapter is the properties of liquids and solids.

Properties of Liquids

13.1

Learning Objective

- Describe the properties of liquids that distinguish them from solids and gases.

In the previous chapter we found that the molecules of a gas are moving rapidly, are far apart, and exert little attraction for one another. Consequently, gases have no specific shape and completely fill their containers. They flow readily, have low densities, diffuse rapidly, and are compressible.

Our experience tells us that solids are very different from gases. Solids are rigid and have a definite shape. They have densities much greater than the densities of gases and are practically incompressible. These properties suggest that the particles of a solid are relatively immobile and are closely packed.

The properties of a liquid are somewhere between those of gases and solids. A liquid has no definite shape, and it does not always completely fill its container. A liquid flows readily but is practically incompressible. The density of a liquid substance is much greater than its density in the gaseous state, but usually is only slightly less than the density of the substance in the solid state. Liquids diffuse but do so much more slowly than gases. The properties of liquids suggest that their molecules are moving more slowly than gas molecules, are attracted to one another, and are relatively close together.

Water is unusual in that the density of ice is less than the density of liquid water.

Properties of Solids, Liquids, and Gases

Solid	*Liquid*	*Gas*
Definite shape	Takes the shape of the part of the container it fills	Takes the shape of its container
Definite volume; does not always completely fill its container	Definite volume; does not always completely fill its container	No definite volume; completely fills its container
Density much higher than of a gas	Density similar to that of a solid	Very low density
Incompressible	Incompressible	Compressible
Practically no diffusion	Diffuses more slowly than a gas	Diffuses readily
Does not flow	Generally flows readily	Flows very readily

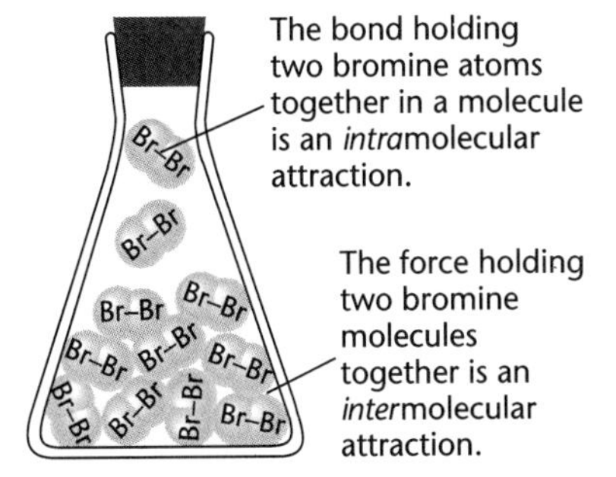

The bond holding two bromine atoms together in a molecule is an *intra*molecular attraction.
The force holding two bromine molecules together is an *inter*molecular attraction.
Br–Br

Intramolecular and intermolecular attractive forces.

Before we discuss specific properties of liquids in detail, we need to understand how the structures of substances affect their properties. Previously we considered chemical bonds, both ionic attractions and *intra*molecular forces that hold the atoms of molecules together. In the following section we will consider the nature of *inter*molecular forces and how they affect the properties of liquids and solids.

13.2 Intermolecular Forces

Learning Objectives

- Describe three types of intermolecular forces.
- Identify the types of intermolecular forces for a given compound.
- Compare the relative strengths of intermolecular forces for various compounds.

The properties of gases are very different from those of solids and liquids because gas molecules are extremely mobile and widely spaced. As a gas is cooled, its molecules slow down, the distances between the molecules decrease, the attractions between molecules become more important, and the gas eventually condenses to a liquid. As we continue to study solids and liquids, we will see that the strength of the attractive forces among the molecules determines their physical properties.

Intermolecular forces:
dipole–dipole attractions
hydrogen bonding
dispersion forces

The physical properties of a molecular substance are determined by three types of intermolecular forces: dipole–dipole attractions, hydrogen bonding, and dispersion forces.

Dipole–Dipole Attractions

Dipole–dipole attractions are electrostatic attractive forces between polar molecules (Section 8.8). Remember, a molecule is polar if (1) it has polar covalent bonds (consider electronegativities), and if (2) the polar bonds do not cancel one another (consider molecular geometry). Figure 13.1 illustrates dipole–dipole attractions

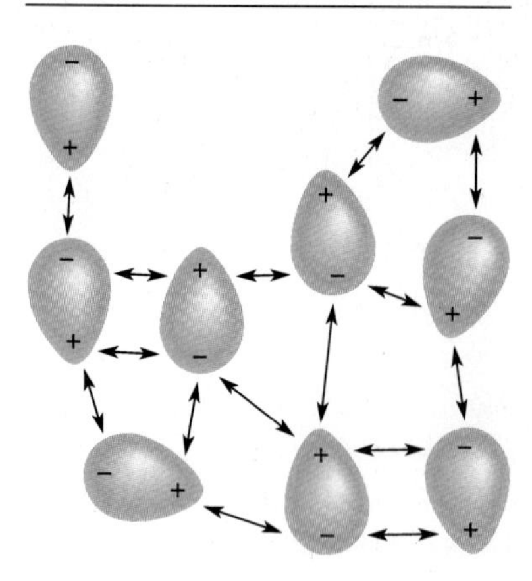

FIGURE 13.1 Dipole–dipole attractive forces in a liquid. The positive and negative charges of a polar molecule are attracted to opposite charges of neighboring polar molecules.

between the molecules of a polar liquid. Ordinary dipole–dipole attractive forces are rather weak (about 4 kJ/mol) compared with the strength of covalent bonds (about 300–400 kJ/mol). Consequently, dipole–dipole forces are effective only when molecules are close together, as is true in a liquid or solid. These forces have almost no effect on gaseous molecules at ordinary temperatures and pressures.

Dipole–dipole forces are weak.

Dipole–dipole attractions are evident in liquids such as $CHCl_3$, CH_2Cl_2, CH_3OCH_3, PCl_3, and ICl. Because ordinary dipole–dipole intermolecular forces are weak, these substances are volatile, and some must be kept at low temperatures to maintain their liquid state.

Volatile: Readily vaporized at low temperature.

Hydrogen Bonding

The strongest dipole–dipole attractive force, occurring between molecules having H—F, H—O, and H—N bonds, is called **hydrogen bonding.** Hydrogen bonding is much stronger than other dipole–dipole attractions (about 20–30 kJ/mol) but is much weaker than covalent bonding. Note that hydrogen bonding is an intermolecular attractive force, not a covalent bond.

Hydrogen bonding is the strongest dipole–dipole attraction.

Attractive forces between hydrogen fluoride (HF) molecules are examples of hydrogen bonding. An HF molecule is polar because the electronegative fluorine atom strongly attracts its bonding pair of electrons. The fluorine atom is the negative end of the dipole and the small hydrogen atom the positive end. The small, partially positive hydrogen atom is attracted by the negative pole of another molecule, in this case a fluorine atom. Hydrogen bonding is sometimes indicated by a dashed line between the formulas:

$$\overset{\delta+}{H}—\overset{\delta-}{F} \cdots\cdots \overset{\delta+}{H}—\overset{\delta-}{F}$$

Remember, hydrogen bonding occurs only with molecules having H—F, H—O, and H—N bonds. The very large dipole required for hydrogen bonding results only when hydrogen, a very small atom of low electronegativity, is covalently bonded to F, O, or N, which are small atoms of high electronegativity.

Hydrogen bonding between water molecules is illustrated in Figure 13.2. Because water molecules are strongly attracted to one another, water is sometimes referred to as an associated liquid. Because of hydrogen bonding, water is much less volatile than other compounds of similar molar mass. As we continue to study water, we will see that it has a number of unusual properties resulting from hydrogen bonding.

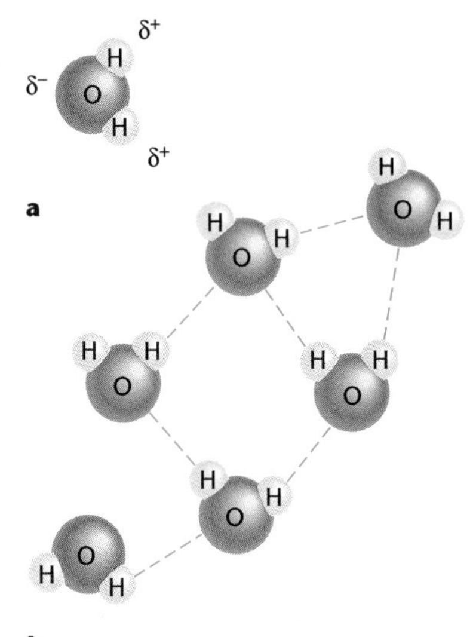

FIGURE 13.2 Hydrogen bonding in water occurs because an electronegative oxygen atom of one molecule attracts partially positive hydrogen atoms of neighboring molecules. (a) A polar water molecule. (b) A network of hydrogen bonds results in the association of water molecules with one another in the liquid.

Hydrogen bonding can also exist between molecules of different substances. For example, in aqueous solutions of ammonia, NH_3 molecules are hydrogen bonded to H_2O molecules and vice versa. You may have used an ammonia solution in laboratory work or as a cleaning solution (household ammonia) in your home.

```
                                H
                                |
H—N—H - - - - O—H - - - - N—H
  |           |             |
  H           H             H
```

Aqueous ammonia, $NH_3(aq)$, contains low concentrations of NH_4^+ and OH^- ions. Although this solution is sometimes called ammonium hydroxide, there are no NH_4OH molecules.

Covalent compounds that cannot form hydrogen bonds with water are relatively insoluble. For example, hydrocarbons such as methane (CH_4), in natural gas, and octane (C_8H_{18}), in gasoline, have no oxygen or nitrogen to which the H—O of water can hydrogen bond. As a result, hydrocarbons are repelled by water (**hydrophobic**) and do not dissolve to any great extent. However, other compounds such as ethyl alcohol (C_2H_5OH, grain alcohol) and acetone (C_3H_6O) readily dissolve in water because of hydrogen bonding with water molecules. Substances that are attracted to water are **hydrophilic.**

Hydrophobic: Repelled by water.

Hydrophilic: Attracted to water.

Dispersion Forces

Attractions between nonpolar molecules are called **dispersion forces,** or London forces. The strength of these forces depends on the sizes of the molecules involved and increases with increasing molar mass. For example, dispersion forces are weaker in pentane (C_5H_{12}) and other compounds in gasoline having low molar mass and are stronger in octane (C_8H_{18}) and other compounds having high molar mass. As a result, pentane is more volatile than octane. Dispersion forces also exist between molecules of polar substances, and the total intermolecular forces are the sum of the dispersion forces and the dipole–dipole attractions.

Hydrogen bonding in aqueous solutions of (a) ethyl alcohol and (b) acetone.

Dispersion forces can be understood by considering the continuous movement of electrons within a molecule. Even though there is no permanent dipole in a nonpolar molecule such as pentane, the movement of its electrons can result in a momentary dipole (Figure 13.3). The partial charges of this dipole can then cause a temporary distortion of the electron cloud of a neighboring molecule. The molecule that is momentarily polarized by another is said to have an *induced dipole,* an idea first proposed by Fritz London in 1937. The induced dipoles result in dipole–dipole attractions between molecules. Because large molecules have many electrons, induced dipoles form frequently, resulting in larger dispersion forces. These forces can be significant between large molecules but are very weak between small molecules.

Dispersion forces between small molecules are very weak.

The momentary polarization of an otherwise nonpolar molecule should not be confused with the permanent dipole of a polar molecule such as HF. In a nonpolar molecule, the average position of its electrons never results in a permanent dipole, even though momentary polarization occurs. On the other hand, HF has a permanent dipole because the average position of its electrons is closer to the electronegative fluorine atom. In addition to its permanent dipole, it also has a

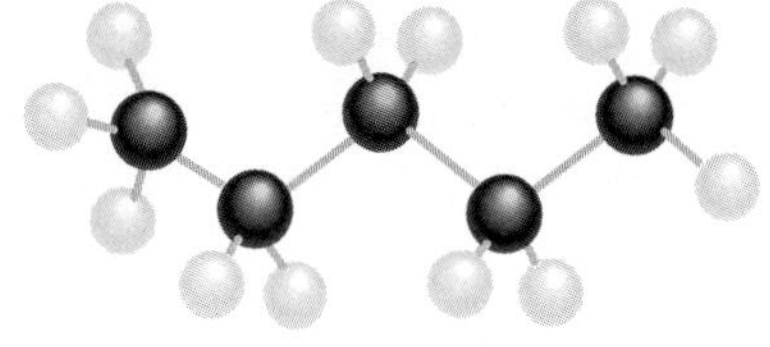

$CH_3CH_2CH_2CH_2CH_3$

Pentane

$CH_3CH_2CH_2CH_2CH_2CH_2CH_2CH_3$

Octane

FIGURE 13.3 Dispersion forces. Smaller pentane molecules have fewer induced dipoles, resulting in weaker intermolecular forces. As a result, pentane is more volatile than octane.

Intermolecular Forces

Type of Compound	*Example*	*Type of Force*	*Note*	*Relative Strength*
nonpolar	C_5H_{12} CO_2	dispersion	depends upon molar mass	very weak for small molecules
polar	$CHCl_3$	dipole–dipole (& dispersion)	depends upon polarity	weak
polar with H—F, H—O, H—N bonds	HF, H_2O, CH_3OH	H-bonding (& dispersion)	only with HF and compounds with H—O and H—N bonds	strong

small, temporary dipole that is created by its fluctuating electrons. We must remember that dispersion forces are always a part of the total intermolecular forces.

An understanding of intermolecular forces is critical in understanding the properties of liquids and solids. As you are evaluating intermolecular forces, remember that they are always weaker forces than either covalent or ionic bonds. Interionic attractions between the cations and anions of a compound are very strong compared with intermolecular forces. For example, sodium chloride is a nonvolatile solid because of the strong multidirectional attractions between neighboring Na^+ and Cl^- ions.

The following principles can be used to evaluate the relative importance of the three types of intermolecular forces:

1. For nonpolar compounds, dispersion forces are the only intermolecular forces. Compare molar masses. Dispersion forces between smaller molecules are weaker than between larger molecules.
2. For polar compounds, those having no H—F, H—O, or H—N bonds, the intermolecular forces are a combination of dispersion forces and dipole–dipole attractions. When molar masses are similar, compare polarities (Section 8.8). Dipole–dipole attractions are greater between molecules having large dipoles.
3. For other polar compounds, such as H—F and compounds that have H—O and H—N bonds, the intermolecular forces are a combination of dispersion forces and hydrogen bonds. For small molecules, hydrogen bonds are much stronger than dispersion forces and dipole–dipole attractions.

Example 13.1

What type of intermolecular force is involved with hexane (C_6H_{14}) and octane (C_8H_{18})? Which of these compounds has stronger intermolecular forces? ■

Solution Dispersion forces are the only intermolecular forces because these are nonpolar compounds. Octane has stronger dispersion forces because of its higher molar mass.

Example 13.2

What types of intermolecular forces are present in ethyl alcohol (C_2H_5OH) and in propyl alcohol (C_3H_7OH)? Which of these compounds has stronger intermolecular forces? ■

Solution Because of the OH groups, both compounds exhibit hydrogen bonding. Dispersion forces are involved with all compounds. Propyl alcohol has stronger dispersion forces because of its higher molar mass.

Practice Problem 13.1 The intermolecular forces between ethyl alcohol (C_2H_5OH) molecules are stronger than between ethyl chloride (C_2H_5Cl) molecules, even though the molar mass of ethyl chloride (64.5 g/mol) is greater than the molar mass of ethyl alcohol (46.0 g/mol). Describe the intermolecular forces involved and explain. (Solutions to practice problems appear at the end of this chapter.)

13.3 Vaporization of Liquids

Learning Objectives

- Compare the vapor pressure of a liquid at various temperatures.
- Compare the vapor pressures, heats of vaporization, and boiling points of two substances.
- From knowledge of their boiling points, compare the vapor pressures and the volatility of two liquids.

Vapor: A substance in the gaseous state above its liquid.

Consider a liquid in a jar, as shown in Figure 13.4. Its molecules are moving randomly, and some of the more energetic surface molecules overcome intermolecular attractions with other molecules and escape into the space above the liquid. A substance in the gaseous state above its liquid is called a **vapor,** and the formation of a vapor from a liquid is called **vaporization.** As the temperature is raised, greater numbers of liquid molecules have sufficient energy to escape to the vapor phase, thereby increasing vaporization of the liquid.

$$\textit{liquid} \underset{\text{condensation}}{\overset{\text{vaporization}}{\rightleftharpoons}} \textit{vapor}$$

When a liquid is in a covered container, vapor cannot escape and disperse. Vapor (gas) molecules move randomly, and in so doing, some of them collide with the surface of the liquid, slow down, and become part of the liquid. This process of a vapor becoming a liquid is called **condensation,** which literally means becoming more dense.

When the jar is open, the rate of condensation is lower than the rate of vaporization because many vapor molecules escape to the surroundings. When the jar is covered, the number of vapor molecules in the jar increases, and the rate of

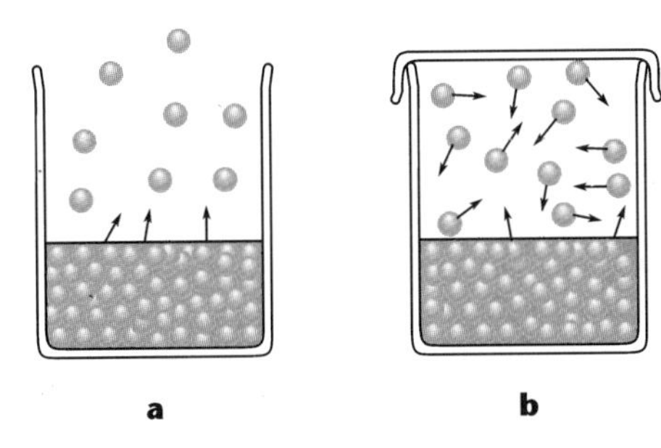

FIGURE 13.4 (a) Vaporization of a liquid from an open jar continues until the liquid is gone because the vapor is allowed to disperse. (b) In a covered jar, both vaporization and condensation occur. Initially, molecules are escaping the liquid faster than vapor molecules are returning to the liquid state. As the number of vapor molecules increases, the rate of their return to the liquid increases until it equals the rate of vaporization. This is a state of dynamic equilibrium for the liquid and vapor.

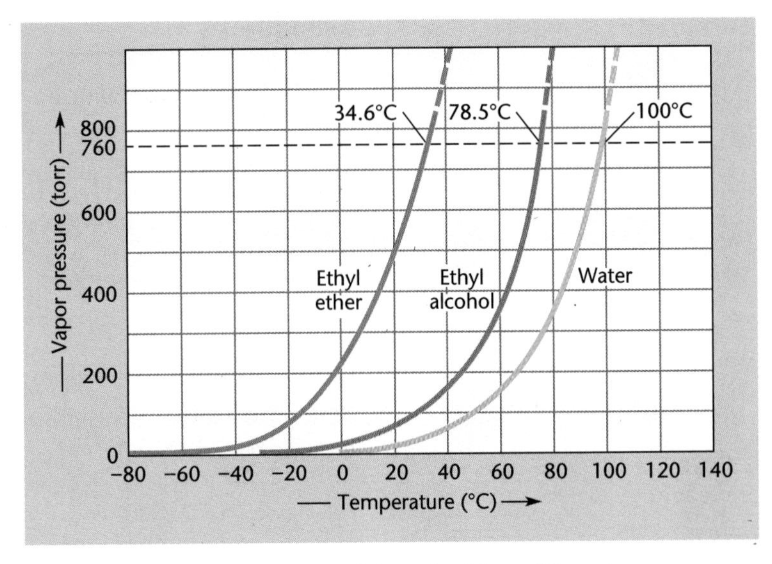

FIGURE 13.5 Vapor pressures of three liquids at different temperatures.

condensation increases until eventually it is equal to the rate of vaporization. When the two rates are equal, the number of vapor molecules remains unchanged with time. This is a state of **dynamic equilibrium** for the liquid and vapor, a state in which the opposing changes are occurring at equal rates. At equilibrium, the number of vapor molecules remains constant.

Some solids vaporize directly without first melting. This process is called **sublimation** and is quite common for nonpolar solids. Iodine (I_2) and *para*-dichlorobenzene ($C_6H_4Cl_2$, moth crystals) are two examples of solids that readily sublime.

Solid iodine vaporizes (sublimes) readily, then condenses on a cool glass surface.

$$\text{solid} \underset{\text{condensation (deposition)}}{\overset{\text{sublimation}}{\rightleftarrows}} \text{vapor}$$

Vapor Pressure

The pressure exerted by vapor molecules is called the **vapor pressure** of a liquid. When a liquid and vapor are in a state of equilibrium, the pressure is called the equilibrium vapor pressure of the liquid. The equilibrium vapor pressure of a liquid is a measure of its volatility. It is a definite value at a specific temperature and increases with increasing temperature. Figure 13.5 shows how the equilibrium vapor pressures of several liquids depends upon temperature. Vaporization is much faster at higher temperatures, and a small increase in temperature can result in a very large increase in the vapor pressure of a liquid. This is why clothes dry faster at higher temperatures.

Vapor pressure is greater at higher temperatures.

At a given temperature, the vapor pressure of a liquid depends upon the strength of its intermolecular forces. Evaluation of dipole–dipole attractions, hydrogen bonding, and dispersion forces enables us to understand and predict differences in vapor pressures of liquids.

Example 13.3

Which has a higher vapor pressure at a given temperature, pentane (C_5H_{12}) or octane (C_8H_{18})? ■

Solution Pentane (C_5H_{12}). These are nonpolar compounds, so dispersion forces are the only type of intermolecular force. Pentane has a lower molar mass than octane and therefore has weaker dispersion forces and a higher vapor pressure.

Example 13.4

Which has a lower vapor pressure at a given temperature, ethyl alcohol (C_2H_5OH) or acetone ($CH_3\underset{\underset{O}{\|}}{C}CH_3$) ? ■

Solution Ethyl alcohol. Hydrogen bonds exist between ethyl alcohol molecules but not between molecules of acetone. The molar mass of acetone is only slightly greater, so dispersion forces are practically the same for the two compounds.

Practice Problem 13.2 Which has a higher vapor pressure at a given temperature, methyl alcohol (CH_3OH) or isopropyl alcohol (C_3H_7OH, rubbing alcohol)? Both compounds have OH groups, typical of alcohols (Section 19.6).

■ Heat of Vaporization

When a liquid absorbs heat from its surroundings, energetic molecules escape from the surface, and the liquid vaporizes. The energy needed to vaporize one mole of liquid is called its **molar heat of vaporization** (ΔH_{vap}). It is also common to express the heat of vaporization per gram of substance vaporized. The heat of vaporization is high for a liquid with hydrogen bonding such as water (Table 13.1) and is low for a liquid such as ethyl ether that has weak intermolecular forces (dipole–dipole attractions and dispersion forces).

Note that vaporization of water and other liquids is endothermic. The escape of high-energy molecules from the liquid causes cooling of the remaining liquid. An example is the evaporative cooling that occurs on the surface of our skin as we perspire, a process that is critical in maintaining a stable body temperature. Evaporative cooling is used in dry climates to cool homes and other buildings.

Condensation of a vapor, the opposite of vaporization of a liquid, is exothermic. When water vapor comes in contact with a cold glass, it loses energy and condenses. When humid air cools, water vapor condenses into fog. Both the glass and the air are warmed as they absorb the energy lost in the condensation of water. The energy lost when one mole of water vapor condenses is numerically equal to the molar heat of vaporization of water.

TABLE 13.1 Physical Properties of Some Liquids

Substance	*Intermolecular Forces*	*Vapor Pressure, torr (25°C)*	*Boiling Point, °C (1 atm)*	*Heat of Vaporization, kJ/mol*
ethyl ether, $C_2H_5OC_2H_5$	weakest	460	35	26
carbon tetrachloride, CCl_4	↓	97	77	30
benzene, C_6H_6	↓	94	80	29
ethyl alcohol, C_2H_5OH	↓	54	78	39
water, H_2O	strongest	24	100	41

Boiling Point

When cold water is heated, its vapor pressure gradually increases as its temperature rises. When its vapor pressure equals the atmospheric pressure, vapor bubbles form in the liquid, rise to the surface, and the water boils. The **boiling point (b.p.)** is the temperature at which the vapor pressure of a liquid is equal to atmospheric pressure. The boiling temperature of a liquid at one atmosphere pressure is called its **normal boiling point.** The normal boiling point of a liquid is one of its characteristic physical properties (Table 13.1).

In Denver, the mile-high city, the atmospheric pressure is lower than at sea level, and water boils at approximately 95°C, well below its normal boiling point. Other liquids also boil below their normal boiling points at locations where the atmospheric pressure is less than one atmosphere. The boiling points of water at several elevations are shown in Table 13.2. At high elevations, foods cooked in boiling water must be cooked longer because of the lower cooking temperature.

A liquid that has a lower boiling point is more volatile at any temperature than one that has a higher boiling point.

A liquid that has a low boiling point has a high vapor pressure and is more volatile at a given temperature than a liquid with a higher boiling point. Refer to the vapor pressure–temperature graph in Figure 13.5. At any temperature, ethyl ether has the highest vapor pressure. At any pressure, ethyl ether has the lowest boiling point of the three liquids. At any temperature, water has the lowest vapor pressure and, at any pressure, the highest boiling point of the three liquids. We can compare the volatilities of liquids from their boiling points. If a liquid (such as motor oil) has a high boiling point compared with another liquid (such as gasoline), it has a lower vapor pressure at any temperature and is not as volatile.

TABLE 13.2 Boiling Point of Water at Different Elevations

Location	*Elevation, ft*	*Boiling Point, °C*	*Atmospheric Pressure, torr*
Mt. McKinley, Alaska	20 320	79	342
Mt. Elbert, Colorado	14 431	85	433
Leadville, Colorado	10 150	89	507
Denver, Colorado	5 280	95	634
Los Angeles, California	20	100	760

A vapor pressure–temperature graph such as that shown in Figure 13.5 can be used to determine boiling points at various pressures. From the graph, the normal boiling point of ethyl alcohol (at 760 torr) is 78.5°C, and the boiling point at 400 torr is about 63°C.

Example 13.5

Using Figure 13.5, estimate the boiling point of water at 550 torr. ■

Solution 91°C

The boiling points of two compounds can be compared by evaluating their intermolecular forces. For example, the dispersion forces for pentane (C_5H_{12}, *MM* = 72.1 g/mol, b.p. = 36°C) are weaker than for octane (C_8H_{18}, *MM* = 114 g/mol, b.p. = 126°C), resulting in a lower boiling point for pentane. In comparing propane (C_3H_8, *MM* = 44.1 g/mol, b.p. = −42°C) with ethyl alcohol (C_2H_5OH, *MM* = 46.0 g/mol, b.p. = 78°C), we find that dispersion forces are about the same because their molar masses are almost the same. However, hydrogen bonding causes ethyl alcohol to boil at a much higher temperature.

Example 13.6

Which has a higher boiling point, CCl_4 or CBr_4? ■

Solution CBr_4. Both compounds are nonpolar (Section 8.8). Because CBr_4 has a higher molar mass, dispersion forces are stronger.

Example 13.7

Which has a higher boiling point, ethyl alcohol (C_2H_5OH, *MM* = 46.0) or methyl ether (CH_3OCH_3, *MM* = 46.0)? ■

Solution Ethyl alcohol. Because their molar masses are identical, dispersion forces are equal. However, ethyl alcohol exhibits hydrogen bonding.

Practice Problem 13.3 The molar mass of octane (C_8H_{18}) is 114 g/mol, whereas the molar mass of isopropyl chloride (C_3H_7Cl) is 78.5 g/mol. Which compound has the higher boiling point? Explain.

Surface Tension and Viscosity

13.4

Learning Objectives

- Compare the surface tensions and viscosities of two liquids based upon intermolecular forces.
- Explain why the surface tensions and viscosities of two liquids are different.

In the previous section, we considered how intermolecular forces affect the vapor pressure and boiling point of a liquid. Intermolecular forces also affect surface tension and viscosity—physical properties of liquids that do not involve changes in state. We need to understand these properties to be able to interpret our observations of water and other liquids.

Surface Tension

A water strider skims across the surface of the still water of a pond. Its body seems to be supported on a transparent elastic coating on the surface of the water. What keeps this insect from sinking? It is kept afloat by the surface tension of the water.

Surface tension is the tendency of a liquid surface to resist stretching and behave like an elastic skin. This tendency is measured by the amount of energy required to stretch or expand the surface by a certain amount. The surface tension of water is about three times that of other common liquids, and this enables some insects to walk on water. Even though a sewing needle's density is about eight times that of water, it can float on water because of surface tension.

The intermolecular forces in water are strong because of hydrogen bonding. A molecule beneath the surface is attracted in all directions by neighboring molecules (Figure 13.6). However, a surface molecule experiences a net attractive force away from the surface toward the center of the liquid. The effect of this force is to keep the surface area to a minimum. The weight of a floating water strider acts to stretch or expand the surface, but this is resisted by attraction of surface molecules by other molecules in the liquid.

Surface tension, like most other physical properties of liquids, depends upon intermolecular forces. Most common liquids, such as acetone (CH_3COCH_3) and gasoline (a mixture of hydrocarbons), have surface tensions less than that of water because of weaker intermolecular forces. On the other hand, the surface tension of mercury is about six times that of water because of the strong forces of metallic bonding (Section 13.5).

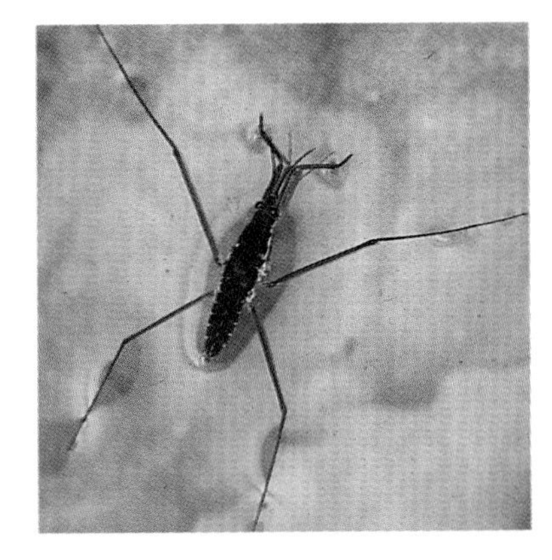

Surface tension enables a water strider to walk on water.

The surface tension of water allows a sewing needle to float even though its density is much greater than the density of water.

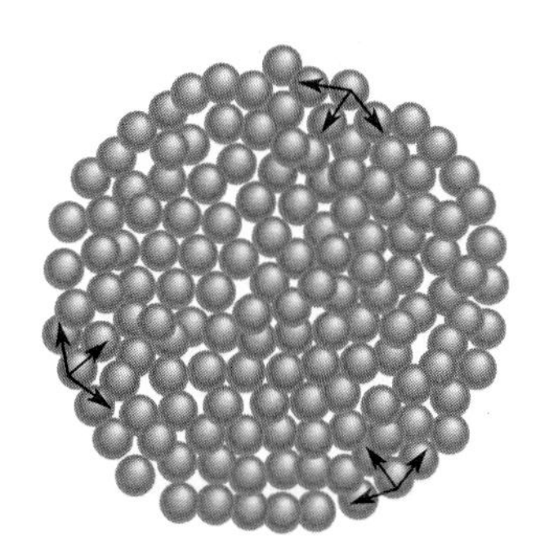

FIGURE 13.6 Surface tension is the result of a net attractive force on surface molecules away from the surface of the liquid. Stretching or expanding of the surface is resisted by this inward force, resulting in an elastic skin on the surface.

Hydrogen bonding in water and in methyl alcohol.

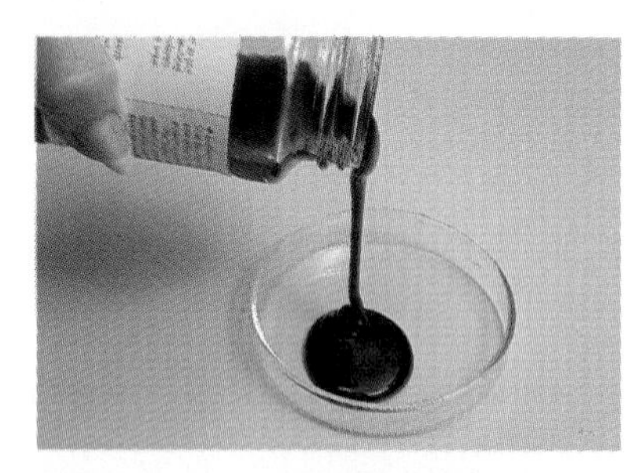

Molasses is a viscous liquid at room temperature.

Example 13.8

Explain why the surface tension of methyl alcohol (CH_3OH) is much less than that of water. ■

Solution Methyl alcohol has only one H participating in hydrogen bonding, whereas water has two H's. With less hydrogen bonding, the intermolecular forces are weaker in methyl alcohol.

■ Viscosity

The **viscosity** of a liquid is its resistance to flow. A liquid that flows slowly and does not spread easily when spilled is said to be **viscous.** Molasses is an example of a viscous liquid.

For most liquids, viscosity decreases with increasing temperature. For example, molasses and pancake syrup become less viscous when they are heated and more viscous when they are cooled. You may have heard the saying "as slow as molasses in January." The viscosity of water at 100°C is about one-sixth its viscosity near 0°C. Another example is motor oil. A 30-weight motor oil flows freely at the operating temperature of an engine but is very viscous at below-zero temperatures on a cold winter morning. A multiviscosity oil flows freely at both low and high temperatures, thereby providing good lubrication when an engine is either cold or hot. This kind of oil is preferred by many car owners because it allows the engine to be started easily even at low temperatures.

The high viscosity of some liquids is explained by strong intermolecular forces that restrict the mobility of molecules. As might be expected, hydrogen bonding plays a significant role. For example, the viscosity of water is much greater than the viscosity of pentane (C_5H_{12}) which has no hydrogen bonding. Glycerol (or glycerin), a lubricant used in laboratories and found in suppositories available in pharmacies, is very viscous because of extensive intermolecular hydrogen bonding (Figure 13.7).

FIGURE 13.7 The viscosity of glycerol,

OH
|
$HOCH_2CHCH_2OH$, is high because of extensive intermolecular hydrogen bonding.

Hydrogen bonding also explains the high viscosity of cold pancake syrup, which contains a high concentration of sucrose, $C_{12}H_{22}O_{11}$. A sucrose molecule has eight OH groups, resulting in a network of intermolecular hydrogen bonding with water and other sucrose molecules.

Why, then, do many motor oils and lubricating greases have high viscosities even though they are composed of nonpolar molecules? Once again, the answer lies in an understanding of the intermolecular forces involved. These substances are mixtures of hydrocarbons (C, H compounds), and hydrogen bonding plays no role whatsoever. The question has a two-part answer. First, these substances are mixtures of large molecules, ranging from $C_{16}H_{34}$ to $C_{22}H_{46}$, and dispersion forces are strong. Second, the molecules have long chains of carbon atoms that are tangled together much like a bowl full of cooked spaghetti (Figure 13.8). The molecules can flow only slowly because they are so entangled.

FIGURE 13.8 Hydrocarbon molecules of a motor oil are tangled together and can flow only slowly.

Example 13.9

Which is more viscous at any given temperature, methyl alcohol (CH_3OH) or ethylene glycol ($HOCH_2CH_2OH$)? Explain. ■

Solution Ethylene glycol is more viscous because it has two OH groups per molecule that can be involved in hydrogen bonding, whereas methyl alcohol has only one OH group per molecule.

13.5 The Solid State

Learning Objectives

- Compare the properties of crystalline and amorphous solids.
- Describe the four types of crystalline solids in terms of the particles involved and the types of attractive forces in the crystal.
- Give examples of each type of crystalline solid and contrast their properties.

FIGURE 13.9 Table salt, sodium chloride, under magnification.

We noted earlier that the properties of solids are very different from those of liquids and gases. The particles of a solid are the least mobile of the three physical states, and solids do not flow as gases and liquids do. Solids are characterized as being rigid. Solid objects have definite shapes. Solids are incompressible. These properties are characteristic of concrete, wood, steel, glass, and many other solid substances that are a part of our daily lives.

Many substances exist as **crystalline solids,** those that have regular, geometric patterns to their structures. When we look at sodium chloride under magnification, we see cubic crystals of various sizes (Figure 13.9). A partial crystal structure of sodium chloride is shown in Figure 13.10.

An **amorphous solid** has no identifiable repeating geometric pattern. Rubber, plastics, and glass are familiar examples of amorphous solids. Many amorphous solids have properties in between the properties of crystalline solids and those of

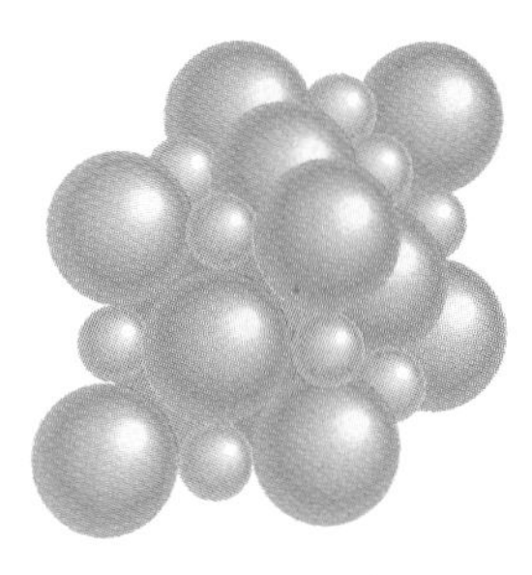

FIGURE 13.10 The crystal structure of sodium chloride has an alternating arrangement of sodium ions (gray) and chloride ions (green).

TABLE 13.3 Melting Points of Selected Substances

Substance	*Melting Point, °C*
ethyl alcohol (C_2H_5OH)	−117
carbon tetrachloride (CCl_4)	−23
water	0
benzene (C_6H_6)	6
naphthalene ($C_{10}H_8$)	80
sodium	98
aspirin	135
citric acid	153
aluminum	658
copper	1083
iron	1535

a

b

c

FIGURE 13.11 Ionic crystals. (a) Calcite ($CaCO_3$), (b) fluorite (CaF_2), (c) galena (PbS).

liquids. For example, some are flexible and others are elastic, whereas crystalline solids are generally hard and brittle. This suggests that the molecules of an amorphous solid are more mobile than those of a crystalline solid.

Melting Point

The temperature at which a solid melts, and at which the solid and liquid are in equilibrium, is called its **melting point.** Pure, crystalline solids melt sharply (Table 13.3), a feature that distinguishes them from amorphous solids. For example, sodium chloride melts sharply at 801°C in contrast to glass which gradually softens when heated and slowly becomes liquid over a broad temperature range.

The melting point of an element or compound is one of its characteristic physical properties. For example, one of the characteristics of water that we all learned at an early age is its melting point. Note that the melting point of a solid and the freezing point of a liquid are the same temperature.

Types of Crystalline Solids

Crystalline solids are classified by whether their particles are atoms, molecules, or ions, and by the type of attractive forces in the crystal. There are four types of crystalline solids: ionic crystals, molecular crystals, network solids, and metallic crystals.

Ionic Crystals

In **ionic crystals,** cations and anions are held together by strong electrostatic forces (ionic bonding). Examples include NaCl, $CaCO_3$ (calcite, limestone, marble), and PbS (galena). Ionic crystals have high melting points. They are hard and rigid, but shatter easily. Some are water soluble and some are not. Their melts and aqueous solutions conduct electricity by the movement of ions, and this characteristic is a common test for an ionic compound. A substance that conducts electricity when melted or dissolved is called an electrolyte (Section 8.2). Several ionic crystals are shown in Figure 13.11.

Molecular Crystals

Molecules of **molecular crystals** are rigidly held in the crystal by intermolecular forces: dipole–dipole attractions, hydrogen bonding, and dispersion forces. Remember that intermolecular forces are weak compared with ionic and covalent bonds. Molecular crystals are characteristically low melting compared with ionic crystals. If hydrogen bonding with water is possible, crystalline substances having small molecules are soluble in water. Their aqueous solutions and melts are nonconducting due to the absence of ions. Ice, sugar, and naphthalene (moth crystals) are familiar examples (Figure 13.12).

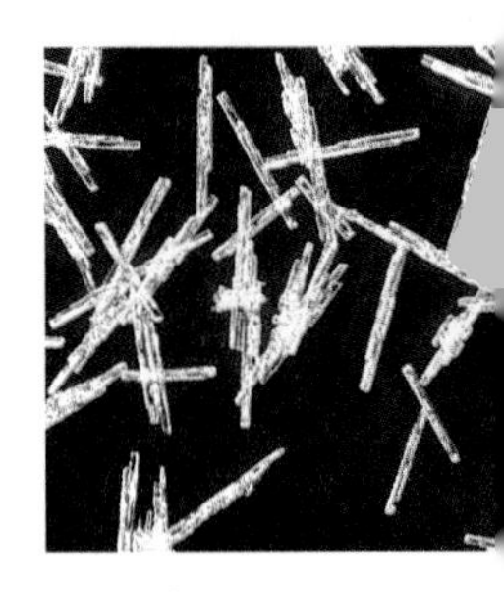

FIGURE 13.12 Sugar (left) and ice (right) are examples of molecular crystals.

Network Solids

The atoms of **network solids** are covalently bonded in an extended network. For example, in diamond, each carbon atom is joined to four other carbon atoms by single bonds to give a continuous structure (Figure 13.13). In graphite, another

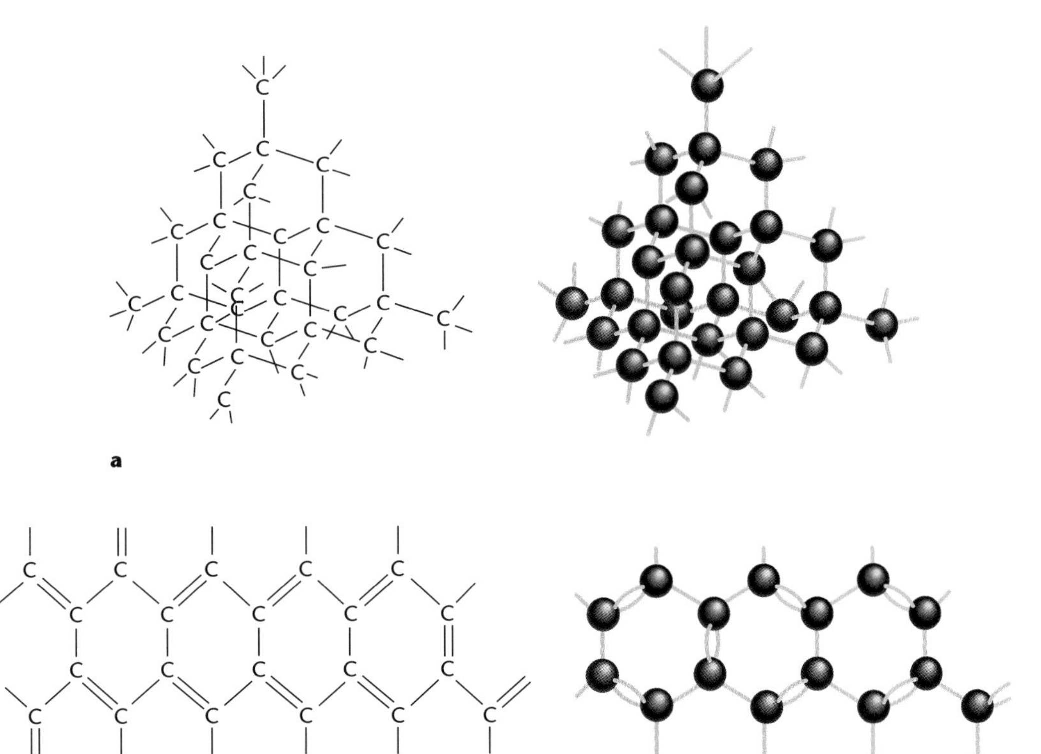

FIGURE 13.13 Network solids. (a) Diamond. Each carbon atom is covalently bonded to four other carbon atoms in a continuous structure. There are no molecules. (b) Graphite. Each carbon atom is covalently bonded to three other carbon atoms in an extended planar structure.

FIGURE 13.14 Quartz crystals. Each silicon atom (black) is covalently bonded to four oxygen atoms and each oxygen atom (red) is bonded to two silicon atoms.

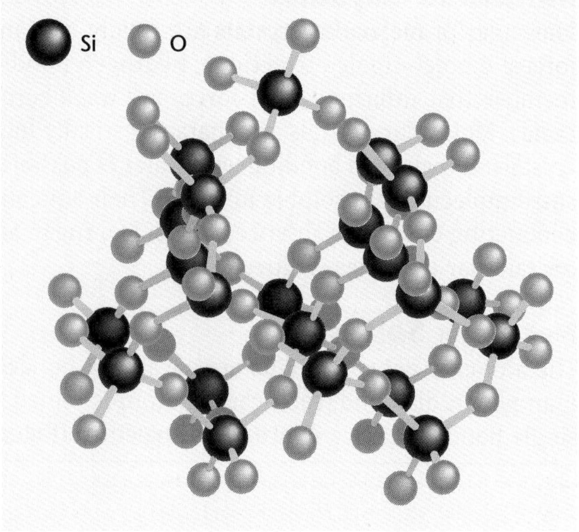

form of carbon, each carbon atom is joined to three neighboring carbons giving sheets of continuous structure. The structure of quartz (SiO_2) is similar to that of diamond. Each silicon atom is covalently bonded to four oxygen atoms (Figure 13.14).

Because of their extended networks of covalent bonds, network solids can be heated to very high temperatures without melting. For example, quartz remains a solid well above 1000°C but finally melts at about 1600°C with the breaking of many Si—O bonds. Upon cooling, an amorphous, glassy material forms rather than the orderly structure of crystalline quartz. Furthermore, network solids are typically insoluble in water and poor conductors of electricity, either as solids or when melted.

Metallic Crystals

The bonding in metals, called **metallic bonding,** is neither ionic nor covalent. A simple model pictures a **metallic crystal** as consisting of metal ions at regular, fixed positions surrounded by mobile valence electrons. In this "electron-sea model,"

TABLE 13.4 Types of Crystalline Solids

Type	*Examples*	*Particles*	*Attractive Forces*	*Properties*
ionic	NaCl, CaF_2	cations, anions	ionic bonds	high melting, often water soluble, hard, brittle, electrolytes
molecular	water, sugar	molecules	intermolecular forces	low melting, nonelectrolytes
network	diamond, quartz	atoms	covalent bonds	high melting, insoluble, poor conductors
metallic	iron, copper	metal ions, electrons	metallic bonds	good electrical conductors

electrons are not confined to individual bonds or metal atoms (Figure 13.15). The freely flowing electrons account for the electrical conductivity of metals even in the solid state.

General properties of the four types of crystalline solids are summarized in Table 13.4.

Example 13.10

For each of the following, identify the type of solid and the type of attractive forces in the crystal:

a. graphite, elemental carbon
b. iodine, I_2
c. silver, Ag
d. hematite, Fe_2O_3 ■

Solutions

a. network solid, covalent bonds between carbon atoms
b. molecular crystal, dispersion forces
c. metallic crystal, metallic bonding
d. ionic crystal, ionic bonding

Practice Problem 13.4 Identify each of the following solids as an ionic, molecular, network, or metallic solid. What is the principal type of attractive force in each solid?

a. benzoic acid, $C_6H_5CO_2H$
b. barium oxide, BaO
c. copper, Cu

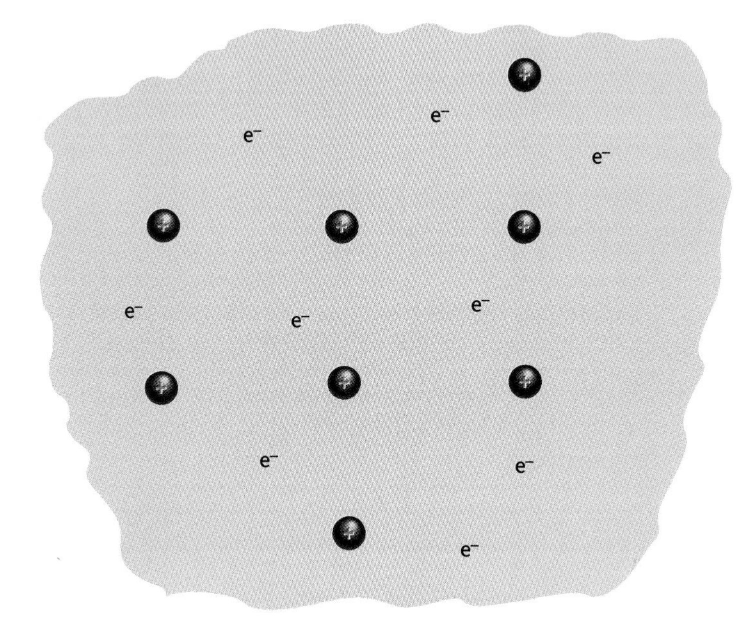

FIGURE 13.15 Electron-sea model of a metallic crystal. Metal ions (black) in a regular geometric pattern are surrounded by a sea of freely flowing electrons.

CHEMICAL WORLD

Biofeedback Cards

Ah, the wonders of science! You're worried about school and you're sure that stress isn't good for you. Of course, physicians aren't in agreement about what stress is, what causes it, or even whether it's really bad for you—but now you can find out instantly whether you're experiencing it. Step right up and buy this plastic card at the drugstore, put your thumb on a special spot, and if the spot turns blue, you're OK. If it turns black, you'd better calm down. Forget the chemistry test! Go swimming.

A biofeedback card.

Something new is discovered almost every day now, and many people immediately start racking their brains to think of uses for it. Many other people rack their brains over how to make a buck with it. The state of scientific knowledge is such that it's not hard to sell the latest miracle in some form.

You know now about the three states of matter: that most matter can exist in all three states and some matter, such as iodine, sublimes from solid to gas without bothering with the liquid state. Now chemists have discovered a material, called *liquid crystal,* that hovers between being a solid crystal and a liquid.

Well, you know what crystals are and what a liquid is. Now let's try to understand liquid crystals. Picture a polar substance made up of long thin molecules that tend to lie next to each other and form layers like pencils in a box. As you tilt a pencil box, the pencils continue to lie next to each other up to a point, but if you tilt far enough they roll over each other.

Do you have the picture? A liquid crystal is made up of long, thin molecules that are weakly attracted to each other by their polarity. If they're undisturbed and closely confined, like pencils in a pencil box, they lie close together in layers. If you raise the temperature, they vibrate faster, like all molecules, and this overcomes some of their attractions. They begin to flow over each other, liquid-like, and the layers of molecules get farther apart.

Now think of oil on water. When you put a couple drops of oil in a dish of water, it spreads over the whole surface of the water in a layer about one molecule thick. This layer displays rainbow colors. Why?

Incoming light reflects from the layer of oil and, only one molecule beneath it, from the surface of the water. Perhaps you know that light moves in waves and that its color depends on the length of the waves. White light contains all the visible wavelengths of light, and colored light contains only one or a few of them. Reflecting from the two surfaces (the oil and water), the wavelengths run into each other. If the wavelengths of the two reflected beams are the same, they reinforce each other and reflect that color. If they're different, they cancel out, and that color doesn't reach your eyes.

The layers of molecules in liquid crystals are similar to the surfaces of oil and water. Incoming light reflects from the close-lying layers and the reflections interfere with each other, positively or negatively. Unlike the oil, however, liquid-crystal molecules are arranged so that they reflect only certain colors, depending on the distance between them. This distance depends on temperature (though, because of polarity, electric current affects it too).

What has this to do with stress? Very little. The excuse for the existence of stress cards is that stress can cause the blood vessels in your hand to constrict, lowering the temperature of your thumbs (and so will a hundred other conditions). However, exercise raises your temperature. What if you're stressed and active? Maybe you'd better study for your chemistry test after all. Passing it will warm your thumbs—and your cockles, too.

Heating Water and Changes of State 13.6

Learning Objectives

- Calculate the energy needed to melt a quantity of a substance from its heat of fusion.
- Calculate the energy needed to vaporize a quantity of a substance from its heat of vaporization.
- Calculate the total energy change for heating ice or other solid at a given temperature to steam or other vapor at a given temperature.

If you take a block of ice from a freezer at −15°C, place it in a covered pan on a stove, and turn on the burner, five changes occur:

Step 1 The temperature of the ice rises to its melting point (0°C).
Step 2 The ice melts.
Step 3 The temperature of the water rises to its boiling point (100°C at 1.00 atm).
Step 4 The water boils (changes to steam).
Step 5 The temperature of the steam rises.

All of these changes are endothermic, that is, the ice (Steps 1 and 2), the water (Steps 3 and 4), and the steam (Step 5) are absorbing heat from the stove.

These changes are reversible. Simply allow the steam to cool (Step 5) until it condenses to water (Step 4), cool the water to its freezing point (Step 3), continue to cool the water in a freezer until it's completely frozen (Step 2), and cool the resulting ice to the temperature of the freezer (Step 1). All of these reverse steps are exothermic, that is, heat is removed from the steam, the water, and the ice.

The changes just described can be summarized in a heating curve for water (Figure 13.16). This graph shows the temperature changes that occur as heat is

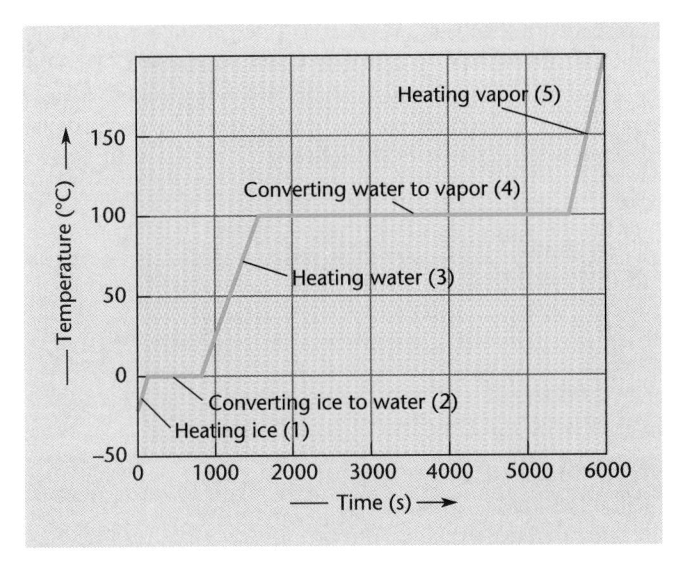

FIGURE 13.16 A heating curve for water. The sample is heated slowly. Steps 1, 3, and 5 involve temperature changes. No temperature change occurs during changes of state—Steps 2 and 4—even though heat is being absorbed by the sample.

absorbed (reading from left to right) and as heat is lost (reading from right to left) by the water. Note that during changes of state—melting and boiling—the temperature remains constant.

Fusion means melting.

The heat absorbed in melting one mole of a solid is its **molar heat of fusion,** ΔH_{fus}. The molar heat of fusion of water is 6.03 kJ/mol. The heat of fusion of a substance may also be expressed as energy/g, which is 335 J/g for water. Using the heat of fusion, we can calculate the quantity of heat (q) needed to melt a certain quantity of a substance.

Example 13.11

Find the quantity of heat needed to melt 50.0 g of ice at 0°C. ■

Begin by outlining the problem.

Find: $q(\text{J}) = ?$

Given: 50.0 g ice at 0°C

Known: $\Delta H_{fus} = 335\ \text{J/g}$

Solution: The quantity of heat needed is

$$q(\text{J}) = \left(335\frac{\text{J}}{\cancel{\text{g}}}\right)(50.0\ \cancel{\text{g}}) = 16\ 800\ \text{J} \quad \text{or} \quad 16.8\ \text{kJ}$$

Specific heat: The quantity of heat involved in changing the temperature of one gram of a substance one Celsius degree (Section 3.7).

The heat absorbed in vaporizing a quantity of water is calculated using its heat of vaporization (Section 13.3), 40.7 kJ/mol or 2.26 kJ/g. Thus, to vaporize 50.0 g of water requires

$$q(\text{kJ}) = \left(\frac{2.26\ \text{kJ}}{\cancel{\text{g}}}\right)(50.0\ \cancel{\text{g}}) = 113\ \text{kJ}$$

The total amount of heat absorbed, starting with ice at -15°C and ending with steam at 112°C, can be calculated by finding the sum of the heat absorbed in the five steps. The heat absorbed in Steps 1, 3, and 5 can be calculated from the specific heats of ice, water, and steam: $SH = 2.06\ \text{J/g}\cdot\text{C°}$ (s), $4.18\ \text{J/g}\cdot\text{C°}$ (l), and $2.03\ \text{J/g}\cdot\text{C°}$ (g). The heat absorbed in Steps 2 and 4 can be calculated from the heats of fusion and vaporization of water. Calculation of the heat absorbed by 36.0 g of ice starting at -15°C and ending with vapor at 112°C follows:

Find: $q_{tot}\ (\text{kJ}) = ?$

Given: 36.0 g H_2O (2.00 mol)

$T_i = -15\text{°C},\ T_f = 112\text{°C}$

Note: $\Delta T_{ice} = 15\ \text{C°},\ \Delta T_{water} = 100\ \text{C°}$

$\Delta T_{steam} = 12\ \text{C°}$

Known: $SH(\text{ice}) = 2.06\ \text{J/g}\cdot\text{C°}$

$SH(\text{water}) = 4.18\ \text{J/g}\cdot\text{C°}$

$SH(\text{steam}) = 2.03\ \text{J/g}\cdot\text{C°}$

$\Delta H_{fus} = 6.03\ \text{kJ/mol}$

$\Delta H_{vap} = 40.7\ \text{kJ/mol}$

Solution: For the five steps of the process we have

$$q_{tot}\text{ (kJ)} = q_1 + q_2 + q_3 + q_4 + q_5$$

For Step 1:

$$q_1\text{ (J)} = \left(\frac{2.06\text{ J}}{\cancel{g}\cdot\cancel{C°}}\right)(36.0\,\cancel{g})(15\,\cancel{C°}) = 1100\text{ J} = 1.1\text{ kJ}$$

For Step 2:

$$q_2\text{ (kJ)} = \left(\frac{6.03\text{ kJ}}{\cancel{mol}}\right)(2.00\,\cancel{mol}) = 12.1\text{ kJ}$$

For Step 3:

$$q_3\text{ (J)} = \left(\frac{4.18\text{ J}}{\cancel{g}\cdot\cancel{C°}}\right)(36.0\,\cancel{g})(100\,\cancel{C°}) = 1.50 \times 10^4\text{ J} = 15.0\text{ kJ}$$

For Step 4:

$$q_4\text{ (kJ)} = \left(\frac{40.7\text{ kJ}}{\cancel{mol}}\right)(2.00\,\cancel{mol}) = 81.4\text{ kJ}$$

For Step 5:

$$q_5\text{ (J)} = \left(\frac{2.03\text{ J}}{\cancel{g}\cdot\cancel{C°}}\right)(36.0\,\cancel{g})(12\,\cancel{C°}) = 88\text{ J} = 0.88\text{ kJ}$$

For the total:

$$q_{tot}\text{ (kJ)} = 1.1\text{ kJ} + 12.1\text{ kJ} + 15.0\text{ kJ} + 81.4\text{ kJ} + 0.88\text{ kJ} = 110.5\text{ kJ}$$

Example 13.12

When steam contacts exposed skin, serious burns can occur because of the heat released in condensation of steam and the high temperature of the resulting water. Calculate the heat released when 2.00 g of steam are condensed at 100°C. ■

Find: q(kJ) = ?

Given: 2.00 g water

Known: $\Delta H_{vap} = 2.26$ kJ/g

Solution: $q\text{(kJ)} = \left(\frac{2.26\text{ kJ}}{\cancel{g}}\right)(2.00\,\cancel{g}) = 4.52\text{ kJ}$

Practice Problem 13.5 How much heat would be absorbed in vaporizing 75.0 g of ethyl alcohol, C_2H_5OH? $\Delta H_{vap} = 38.6$ kJ/mol.

Practice Problem 13.6 Find the amount of heat absorbed:

a. in melting 63.5 g of ice at 0°C. $\Delta H_{fus} = 6.03$ kJ/mol.

b. when the temperature of 63.5 g of water rises from 0°C to 38°C. $SH = 4.18$ J/g·C°

c. starting with 63.5 g of ice at 0°C and ending with water at 38°C.

13.7 Water—A Remarkable Liquid

Learning Objective

- Identify the properties of water that distinguish it from other common liquids.

Water is not only the most abundant substance on earth, it is one of the most remarkable. It is the only common substance that we regularly encounter in all three physical states. Its unique properties (Table 13.5) have a significant impact on all living systems and on the environment. Water is also of great economic importance because of the large amount we use for domestic purposes, for irrigation, and in manufacturing processes.

An adult human body is composed of 60–70% water, and the body of an infant can be as high as 75% water. Most of the body's water acts as a solvent (Chapter 14) for salts, molecules, and gases in intracellular fluid, various organs, blood, urine, and so on. Everyday, you lose between 1.5 and 3.0 L of water from your body. Water is lost in the urine, by expiration from the lungs in breathing, and in perspiration. It is important to replace the water lost by the body to avoid dehydration.

A stable body temperature is possible because of the unusually high heat of vaporization of water, 40.7 kJ/mol. If you lose approximately 300 mL (300 g, about 17 mol) of water in perspiration in a day, your body's heat loss due to perspiration is approximately 700 kJ. This is enough energy to raise the temperature of a block of gold, equal in mass to your body mass, from 20°C to 100°C. Furthermore, perspiration increases dramatically during vigorous exercise in order to keep your body from overheating.

TABLE 13.5 Properties of Water

Property	*Value for Water*
melting point	0°C (1 atm)
boiling point	100°C (1 atm)
vapor pressure	17.5 torr at 20°C
molar heat of vaporization	40.7 kJ/mol at 100°C
molar heat of fusion	6.03 kJ/mol at 0°C
specific heat (liquid)	4.18 J/g·C° at 15°C
density	1.000 g/mL at 4°C (liquid) 0.917 g/mL at 0°C (solid)

The high specific heat of water means that it can absorb large amounts of heat without large changes in temperature. This property, combined with water's high heat of vaporization, is important not only in the temperature regulation of the human body. It is also important in the environment. High humidity and large bodies of water have a moderating influence on climate. Energy from the sun is absorbed by the water during the heat of the day, thereby preventing a large increase in land temperature (Figure 13.17). Heat stored by the water is released gradually at night when the air is cool, preventing an extreme drop in the nighttime temperature. Temperature variations in coastal areas are generally much smaller than in inland areas. Just the opposite is true in arid regions where temperature differences between day and night are large because there is little water to absorb heat during the day.

Water expands when it freezes, and the density of ice is less than that of liquid water (see Table 13.5). As a result, ice forms on the surface of lakes and rivers and insulates the water below. If ice were a typical solid and were more dense than liquid water, ice would sink. As a result, many lakes would freeze solid. A cold winter could have a disastrous effect on marine life.

Hydrogen bonding, responsible for the high surface tension of water, is also responsible for the ability of water to wet surfaces. This wetting ability (or **adhesion,** an attractive force between a liquid and a surface) is the basis for capillary action that carries water and nutrients to the leaves of plants. Adhesive forces between water and glass also cause a **meniscus,** a curved surface of water (Figure 13.18).

Because of its remarkable properties, water is used in many industrial processes. The largest single use is as a coolant. Because of its high specific heat and high heat of vaporization, and its great abundance and low cost, water is used to cool processing equipment, engines, and machinery. To conserve water and to reduce thermal pollution of natural waters, much of the cooling water used by large

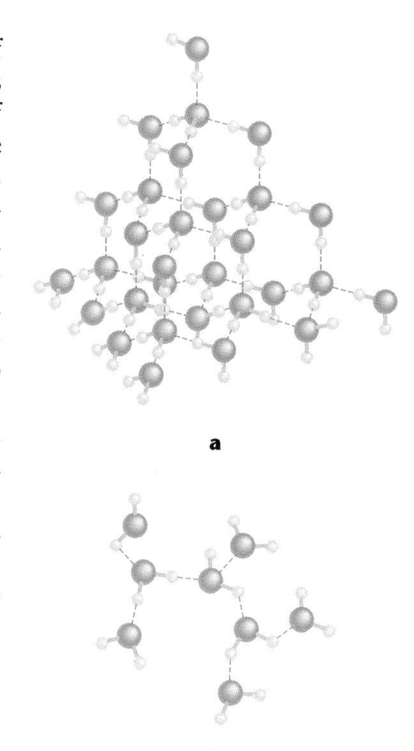

The hydrogen bonding in ice (a) results in an expanded structure and a lower density compared with liquid water (b). The same degree of hydrogen bonding is not possible in the liquid because of the constant movement of the molecules.

FIGURE 13.17 Energy from the sun is absorbed by a large body of water, both in raising the water temperature, and in vaporization, thereby preventing a large increase in land temperature.

FIGURE 13.18 When the adhesive forces between water and the glass tube are greater than the attractive forces (cohesive forces) within the liquid, the surface of the water forms a meniscus. Similar adhesive forces carry water and nutrients by capillary action to the leaves of plants.

FIGURE 13.19 Industrial cooling water is recycled through cooling towers such as this one. Some of the water is vaporized to cool the rest of the water, which is recycled.

industries is recycled. For example, the steel industry uses about 50 000 gallons of water per ton of steel produced. Of this quantity, 60% is recycled, 25% is vaporized in cooling towers (Figure 13.19), and the remainder is released as wastewater.

13.8 Hydrates

Learning Objectives

- Name and write formulas for hydrates.
- Write equations for hydration and dehydration reactions.
- Describe the energy changes involved in hydration and dehydration.
- Calculate molar masses of hydrates.

Because of its polarity, water has the ability to associate strongly with ions and with dissolved molecules through hydrogen bonding. For example, when copper(II) sulfate, a white solid, is dissolved in water, a deep-blue solution results. The color of this solution is due to hydrated copper(II) ions, $Cu(H_2O)_4^{2+}$. **Hydration** is the attraction or bonding of water molecules to an ion or molecule. When water is partially evaporated from a solution of copper(II) sulfate, beautiful blue crystals of copper(II) sulfate pentahydrate, $CuSO_4 \cdot 5H_2O$, separate from the solution (Figure 13.20). A **hydrate** is a substance containing water molecules in its crystal structure. Water in a hydrate is called **water of hydration** or **water of crystallization.**

The formula of a hydrate is written with a dot separating the formula of the salt and the number of water molecules per formula unit, as shown for copper(II) sulfate pentahydrate. Calcium chloride dihydrate is $CaCl_2 \cdot 2H_2O$. A hydrate, like all compounds, has a definite composition as indicated by its formula. When calculat-

a

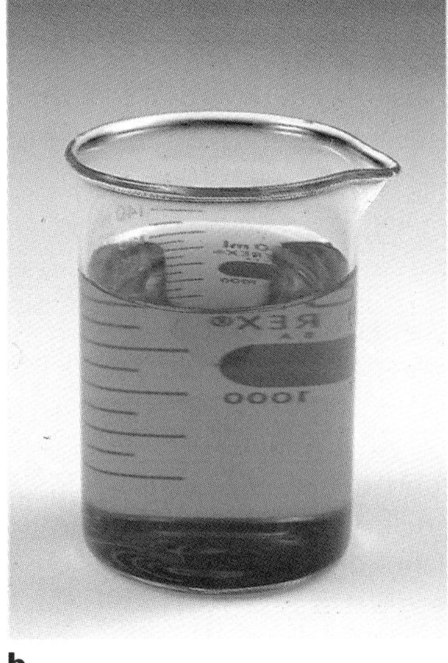

b

c

FIGURE 13.20 When copper(II) sulfate (a) is dissolved in water, a deep-blue solution (b) forms. This solution contains $Cu(H_2O)_4^{2+}$ ions. Evaporation and crystallization results in the formation of beautiful crystals of $CuSO_4 \cdot 5H_2O$, as in (c).

ing the molar mass, you must include the indicated number of moles of water. For $CaCl_2 \cdot 2H_2O$:

$$MM = 40.1 \text{ g/mol Ca} + 2(35.5 \text{ g/mol Cl}) + 2(18.0 \text{ g/mol } H_2O)$$
$$= 147 \text{ g/mol } CaCl_2 \cdot 2H_2O$$

Note that a hydrate is named with the name of the salt (Chapter 9), followed by a counting prefix and the word "hydrate." The counting prefix indicates the number of molecules of water per formula unit (Table 13.6). The number of water molecules per formula unit varies depending upon the cation and anion involved.

Hydration is generally reversible, and heating results in dehydration to form anhydrous copper(II) sulfate:

Anhydrous: Without water.

$$\underset{\text{blue crystal}}{CuSO_4 \cdot 5H_2O(s)} \xrightarrow{\Delta} \underset{\text{white powder}}{CuSO_4(s)} + 5H_2O(g)$$

Because dehydration is endothermic, hydration (the reverse process) is exothermic.

TABLE 13.6 Some Common Hydrates

Formula	*Common Name*	*Systematic Name*
$CuSO_4 \cdot 5H_2O$	blue vitriol	copper(II) sulfate pentahydrate
$CaCl_2 \cdot 2H_2O$		calcium chloride dihydrate
$Na_2B_4O_7 \cdot 10H_2O$	borax	sodium tetraborate decahydrate
$MgSO_4 \cdot 7H_2O$	Epsom salt	magnesium sulfate heptahydrate
$CaSO_4 \cdot \frac{1}{2}H_2O$, or $(CaSO_4)_2 \cdot H_2O$	plaster of Paris	calcium sulfate hemihydrate
$Na_2SO_4 \cdot 10H_2O$	washing soda	sodium sulfate decahydrate

Another example of a hydrate is plaster of Paris, $CaSO_4 \cdot \frac{1}{2}H_2O$, used in making molds, figurines, and casts for immobilizing broken bones. When mixed with the proper quantity of water, it sets to a hard, rigid solid. The following equation describes the hydration, an exothermic process:

$$2CaSO_4 \cdot \tfrac{1}{2}H_2O(s) + 3H_2O(l) \rightarrow 2CaSO_4 \cdot 2H_2O(s)$$

Some compounds attract water so strongly that they are **hygroscopic,** meaning that they absorb moisture from the air. One example is silica gel, a form of SiO_2 that has a large surface area. After shoes are assembled during manufacturing, a small packet of silica gel is usually placed in each box to keep the shoes dry and to prevent the formation of mildew during storage. Another hygroscopic compound is calcium chloride, which is used in laboratories as a drying agent for organic liquids and to maintain a dry atmosphere for laboratory experiments. Calcium chloride is so hygroscopic that it is **deliquescent,** which means it absorbs enough water to form a solution.

Example 13.13

Write formulas for the following hydrates:

a. strontium chloride hexahydrate

b. chromium(III) nitrate nonahydrate ■

Solutions

a. $SrCl_2 \cdot 6H_2O$

b. $Cr(NO_3)_3 \cdot 9H_2O$

Example 13.14

Write a balanced equation for the hydration of calcium chloride to form its dihydrate. ■

Solution $CaCl_2(s) + 2H_2O(l) \rightarrow CaCl_2 \cdot 2H_2O(s)$

Practice Problem 13.7 Cobalt(II) chloride hexahydrate (pink) and cobalt(II) chloride tetrahydrate (blue) are used in weather forecasting bulbs. When the humidity is high, the blue tetrahydrate absorbs moisture to form the pink hexahydrate. When the humidity is low, the reaction reverses to give the blue tetrahydrate. Write an equation to illustrate this reaction.

■ Chapter Review

Directions: From the list of choices at the end of the Chapter Review, choose the word or term that best fits in each blank. Use each choice only once. The answers appear at the end of the chapter.

__________ (1) is the most abundant compound on earth, covering about 75% of the earth's surface. It is the only common substance found naturally in all three physical states.

Section 13.1 The properties of __________ (2) are between those of solids and __________ (3). A liquid has no definite __________ (4) but does not always fill its container. A liquid __________ (5) readily and is practically incompressible. The density of a liquid substance is much __________ (6) than the density of the gaseous substance. Compared with molecules of a gas, liquid molecules move __________ (7) and are closer together.

Section 13.2 Three types of intermolecular forces are important: dipole–dipole attractions, hydrogen bonding, and __________ __________ (8). Dipole–dipole attractions are weak electrostatic attractive forces between __________ (9) molecules. Polar liquids of low molar mass are quite volatile because these forces are weak. __________ (10) bonding occurs between HF molecules and those having H—O and __________ (11) bonds. Water has unusual properties because of strong hydrogen bonding. For example, it is __________ (12) volatile than other liquids of similar molar mass. Hydrogen bonding affects the solubility of substances in water. For example, ammonia is very soluble in water because it is __________ (13) or strongly attracted to water. Hydrocarbons are __________ (14), or repelled by water, and are insoluble. Attractions between __________ (15) molecules are called dispersion forces. These forces can be significant between __________ (16) molecules but are very weak between small molecules.

Section 13.3 A substance in the gaseous state above its liquid is called a __________ (17). The formation of a vapor from a liquid is called __________ (18), which is the opposite of __________ (19).

The pressure exerted by vapor molecules is called the __________ __________ (20) of a liquid. When the rates of vaporization and condensation are equal, the pressure is called the __________ (21) vapor pressure. The vapor pressure of a liquid increases as the __________ (22) increases. It also depends upon the strength of the intermolecular forces. For example, the vapor pressure of water is relatively __________ (23) due to strong hydrogen bonding. The vapor pressure of pentane is high due to __________ (24) dispersion forces.

The energy needed to vaporize one mole of a liquid is called its __________ (25) heat of vaporization. Water has a high heat of vaporization due to strong hydrogen bonding. Vaporization is __________ (26) and condensation is exothermic.

The __________ __________ (27) of a liquid is the temperature at which its vapor pressure is equal to atmospheric pressure. At one atmosphere, this temperature is called the __________ (28) boiling point of the liquid. A liquid that has a low vapor pressure has a higher boiling point than one that has a high vapor pressure at the same temperature. Because of hydrogen bonding, the boiling point of ethyl alcohol (C_2H_5OH) is __________ (29) than that of pentane (C_5H_{12}).

Section 13.4 The tendency of a liquid surface to resist stretching and behave like a liquid skin is called its __________ __________ (30). The surface tension of water is high because of strong hydrogen __________ (31).

The __________ (32) of a liquid is its resistance to flow. Liquids are more __________ (33) at low temperatures. Strong __________ (34) attractive forces such as hydrogen bonding restrict the mobility of the molecules of a liquid, resulting in high viscosities.

Section 13.5 A __________ (35) solid, such as sodium chloride, has a regular geometric pattern to its structure. In contrast, rubber, plastics, and glass are examples of __________ (36) solids.

The four types of crystalline solids are __________ (37) crystals, molecular crystals, __________ (38) solids, and __________ (39) crystals. Ionic crystals have high melting points, and their melts and aqueous solutions are __________ (40). Molecular crystals have lower melting points, and their melts and aqueous solutions are __________ (41). Network solids have atoms __________ (42) bonded in an extended network. Examples include two forms of carbon, __________ (43) and graphite, and __________ (44) (SiO_2). In metallic crystals, atoms are held in regular positions by metallic bonding involving __________ (45) that flow freely through the crystal.

Section 13.6 A __________ (46) curve describes in five steps what happens when a solid such as ice is heated to produce vapor such as steam. The temperature at which a solid melts, when the solid and liquid are in equilibrium, is called its __________ __________ (47). The energy needed to melt one mole of solid is called its molar heat of __________ (48).

Section 13.7 Water has unusual properties compared with most common liquids. Its high __________ (49) heat and high heat of vaporization are important in cooling our bodies, machinery, and industrial equipment. Lakes and other large bodies of water have a __________ (50) effect on climate. The density of __________ (51) is less than the density of liquid water. The __________ (52) ability of water is the basis for capillary action that carries nutrients to the leaves of plants.

Section 13.8 The bonding of water molecules to an ion or molecule is called __________ (53). A substance containing water molecules in its crystal structure is called a __________ (54). Water in a hydrate is called water of hydration. Copper(II) sulfate __________ (55), $CuSO_4 \cdot 5H_2O$, is an example of a hydrate. Dehydration is endothermic, and hydration is __________ (56).

Compounds that attract water so strongly that they absorb water from the air are said to be __________ (57). Silica gel and calcium chloride are examples. A hygroscopic compound that forms a solution with absorbed water is __________ (58).

Choices: amorphous, boiling point, bonding, condensation, covalently, crystalline, deliquescent, diamond, dispersion forces, electrolytes, electrons, endothermic, equilibrium, exothermic, flows, fusion, gases, greater, H—N, heating, higher, hydrate, hydration, hydrogen, hydrophilic, hydrophobic, hygroscopic, ice, intermolecular, ionic, large, less, liquids, low, melting point, metallic, moderating, molar, network, nonconducting, nonpolar, normal, pentahydrate, polar, quartz, shape, slower, specific, surface tension, temperature, vapor, vapor pressure, vaporization, viscosity, viscous, water, weak, wetting

Study Objectives

After studying Chapter 13, you should be able to:

Section 13.1

1. Describe the properties of liquids that distinguish them from solids and gases.

Section 13.2

2. Describe three types of intermolecular forces.
3. Identify the types of intermolecular forces for a given compound.
4. Compare the relative strengths of intermolecular forces for various compounds.

Section 13.3

5. Compare the vapor pressure of a liquid at various temperatures.
6. Compare the vapor pressures, heats of vaporization, and boiling points of two substances.
7. From knowledge of their boiling points, compare the vapor pressures and the volatility of two liquids.

Section 13.4

8. Compare the surface tensions and viscosities of two liquids, based upon intermolecular forces.
9. Explain why the surface tensions and viscosities of two liquids are different.

Section 13.5

10. Compare crystalline and amorphous solids.
11. Describe the four types of crystalline solids in terms of the particles involved and the types of attractive forces in the crystal.
12. Give examples of each type of crystalline solid and contrast their properties.

Section 13.6

13. Calculate the energy needed to melt a quantity of a substance from its heat of fusion.
14. Calculate the energy needed to vaporize a quantity of a substance from its heat of vaporization.
15. Calculate the total energy change for heating ice or other solid at a given temperature to steam or other vapor at a given temperature.

Section 13.7

16. Identify the properties of water that distinguish it from other common liquids.

Section 13.8

17. Name and write formulas for hydrates.
18. Write equations for hydration and dehydration reactions.
19. Describe the energy changes involved in hydration and dehydration.
20. Calculate molar masses of hydrates.

Key Terms

Review the definition of each of the terms listed here by chapter section number. You may use the glossary if necessary.

13.2 dipole–dipole attractions, dispersion forces, hydrogen bonding, hydrophilic, hydrophobic

13.3 boiling point, condensation, dynamic equilibrium, molar heat of vaporization, normal boiling point, sublimation, vapor, vapor pressure, vaporization

13.4 surface tension, viscosity, viscous

13.5 amorphous solid, crystalline solid, ionic crystal, melting point, metallic bonding, metallic crystal, molecular crystal, network solid

13.6 molar heat of fusion

13.7 adhesion, meniscus

13.8 anhydrous, deliquescent, hydrate, hydration, hygroscopic, water of hydration, water of crystallization

Questions and Problems

Answers to questions and problems with blue numbers appear in Appendix C. Asterisks indicate the more challenging problems.

Properties of Liquids (Section 13.1)

1. Describe the properties of a typical liquid.
2. Why is a liquid practically incompressible?
3. Why does a liquid flow readily?
4. Explain why the density of a liquid is much greater than the density of a gas.
5. Why does a liquid diffuse more slowly than a gas?
6. Explain why a liquid does not completely fill its container as does a gas.

Intermolecular Forces (Section 13.2)

7. Identify the types of intermolecular forces.
8. Using Lewis formulas (Section 8.4), illustrate the dipole–dipole attractions among HCl molecules.
9. Describe the types of intermolecular forces in a sample of HBr.
10. Compare the dipole–dipole attractions in HCl and HBr.
11. Explain why dipole–dipole attractions are practically ineffective with gases.
12. Which is the strongest type of intermolecular force possible between small molecules?
13. Explain why polar liquids, except those O—H and N—H bonds, are generally volatile.
14. Which of the following have dipole–dipole attractions? *Hint:* Draw Lewis formulas (Section 8.4) and determine which are polar molecules.
    a. CH_2Cl_2
    b. C_2Cl_4
    c. ClBr
    d. C_3H_8
15. What molecular features are necessary for hydrogen bonding?
16. Why is water less volatile at a given temperature than other compounds of similar molar mass?
17. Using Lewis formulas, illustrate hydrogen bonding between water molecules, and between ammonia (NH_3) molecules.
18. Using Lewis formulas, illustrate hydrogen bonding between methyl alcohol (CH_3OH) molecules.
19. Which of the following compounds are hydrophilic? Explain.
    a. ethyl alcohol, C_2H_5OH
    b. methyl amine, CH_3NH_2
    c. ammonia, NH_3
    d. methyl ether, CH_3OCH_3
20. Which of the following compounds are hydrophobic? Explain.
    a. butane, C_4H_{10}
    b. hydrogen peroxide, H_2O_2
    c. hexachloroethane, C_2Cl_6
21. Explain how nonpolar molecules such as C_5H_{12} are attracted to one another. What are these intermolecular forces called?
22. Explain what is meant by an induced dipole.
23. Dispersion forces between small molecules are very weak. Give an example of a compound that has strong dispersion forces.
24. Choose the compound in each pair that has stronger intermolecular forces:
    a. CH_4, CO_2
    b. H_2O, H_2S
    c. HF, HCl
    d. $HOCH_2CH_2OH$, C_2H_5OH

Vaporization of Liquids (Section 13.3)

25. Explain why the rate of vaporization of a liquid increases with increasing temperature.
26. Why is vaporization of a liquid in an open container faster than condensation?

27. Explain what is meant by the dynamic equilibrium of a liquid and its vapor. What is meant by the equilibrium vapor pressure of a liquid?
28. Why is the vapor pressure of a liquid greater at higher temperatures than at lower temperatures?
29. What is meant by sublimation? Give examples of substances that sublime.
30. Naphthalene ($C_{10}H_8$), a compound used as moth crystals, sublimes readily. Why is this property important in the use of naphthalene as moth crystals?
31. Which liquid in each pair has a higher vapor pressure at a given temperature?
    a. benzene (C_6H_6) or toluene (C_7H_8)
    b. ethyl ether ($C_2H_5OC_2H_5$) or butyl alcohol (C_4H_9OH)
    c. chloroform ($CHCl_3$) or carbon tetrachloride (CCl_4)
    d. toluene (C_7H_8) or chlorobenzene (C_6H_5Cl)
32. For each pair of compounds in problem 31, predict which has the higher molar heat of vaporization at a given temperature.
33. For each pair of compounds in problem 31, predict which has the higher normal boiling point.
34. The molar heat of vaporization of water is 40.7 kJ/mol.
    a. Calculate its molar heat of vaporization in kcal/mol.
    b. What is the heat of vaporization of water in cal/g?
35. How much energy would be required to vaporize 1.00 L of water at its normal boiling point? Assume the density of water is 0.998 g/mL.
36. How much ethyl ether could be vaporized at its boiling point by the energy needed to vaporize 1.00 L of water? Refer to Table 13.1 for the necessary physical properties.
37. Sodium chloride and other ionic compounds generally have very low vapor pressures at ordinary temperatures and very high normal boiling points. Explain, considering the type of attractive forces involved.
38. Why are interionic attractive forces much greater than dipole–dipole attractions?
39. Predict which compound, NaCl or CH_3Cl, has the higher normal boiling point. Explain.
40. Explain why HF has a higher normal boiling point than HCl, HBr, or HI.
41. Explain why HI has a higher normal boiling point than either HBr or HCl.
42. Why are longer cooking times needed at high altitudes for foods cooked in boiling water?

The Solid State (Section 13.5)

43. What is meant by a crystalline solid? Give several examples.
44. What is meant by an amorphous solid? Give several examples.
45. What physical property distinguishes a crystalline solid from an amorphous solid.
46. Identify the following solids as crystalline or amorphous:
    a. polystyrene (a plastic)
    b. a glass marble
    c. a diamond
    d. chewing gum
47. Name the four types of crystalline solids.
48. What particles make up the structure of an ionic crystal such as KI? Describe the attractive forces in the crystal.
49. What type of crystal is formed by sucrose, $C_{12}H_{22}O_{11}$? The structure of sucrose has eight OH groups. Describe the intermolecular forces in a sugar (sucrose) crystal.
50. Describe the particles and attractive forces in a network solid such as diamond.
51. Why do network solids have very high melting points compared with molecular crystals?
52. What is an electrolyte? Under what conditions does sodium chloride conduct an electric current?
53. Why is an aqueous sugar solution a nonconductor of electricity?
54. Briefly explain metallic bonding.
55. Identify the following solids as ionic crystals, molecular crystals, or network solids.
    a. ice
    b. fluorite, CaF_2
    c. quartz, SiO_2
    d. iodine, I_2
    e. dry ice, CO_2
    f. marble, $CaCO_3$

Heating Water and Changes of State (Section 13.6)

56. Calculate the energy needed to vaporize 0.250 L of water (density = 1.00 g/mL). Refer to Table 13.5 for needed data.
57. How much energy is needed to melt a 1.00-kg block of ice at 0°C?
58. Find the quantity of energy absorbed by a 30.0-g ice cube initially at −15°C when it warms to 0°C and melts.
59. What is the total energy needed to warm 245 g of water from 20°C to its normal boiling point and then to convert it to steam at 100°C?
60. How much energy will be lost in the condensation of 1.00 kg of steam followed by cooling of the resulting water to 50.0°C?
61. What mass of ice could be melted at 0°C by the energy lost in condensation of 50.0 g of steam and cooling the resulting water to 0°C?
62. The heat of vaporization of ethyl alcohol is 38.6 kJ/mol.
    a. How much energy would be needed to vaporize 25.0 g of ethyl alcohol, C_2H_5OH?
    b. If the energy needed to vaporize the ethyl alcohol was absorbed from 0.500 kg of water initially at 95°C, what would be the final temperature of the water?

63. The melting point of benzene is 5.5°C and its boiling point is 80.0°C. Draw a heating curve for benzene starting at 0°C and ending at 90°C. Label the temperature scale and show the melting point and boiling point on the graph.

Additional Exercises

64. Explain why the surface tension of ethyl alcohol, C_2H_5OH, is significantly less than that of water.
65. Recognizing that heating a liquid increases the average velocity of its molecules, why does the surface tension of water decrease when it is heated?
66. Why does the viscosity of a liquid decrease when it is heated?
67. Ordinary vegetable oils contain molecules having two or three 18-carbon chains joined together (Section 19.9). Explain why vegetable oils are more viscous than water even though many of them have no hydrogen bonding.
68. Explain why glycerol, $C_3H_5(OH)_3$, is more viscous than ethylene glycol, $HOCH_2CH_2OH$.
69. What properties of water are particularly important in its use as a coolant in industrial processes?
70. What property of water is particularly important to the cooling mechanism of our bodies?
71. If the density of ice were greater than the density of liquid water, what would be the effect on marine life?
72. Explain how a large lake can have a moderating effect on the climate of the surrounding area.
73. Explain what is meant by hydration.
74. What is meant by water of crystallization?
75. What is meant by a hydrate? Give an example.
76. Name the following compounds:
    a. $CoCl_2 \cdot 6H_2O$
    b. $SnCl_2 \cdot 2H_2O$
77. Write formulas for the following compounds:
    a. sodium carbonate decahydrate
    b. magnesium sulfate heptahydrate
    c. barium hydroxide octahydrate
78. What is a hygroscopic compound? Describe some uses for a hygroscopic compound.
79. What is a deliquescent compound? Give an example.
80. Calculate the molar mass of each of the compounds in problems 76 and 77.
81. Suggest a reason why the boiling point of hydrogen peroxide (H_2O_2, 152°C) is higher than the boiling point of water.
82. Why does water have good adhesion to glass (made from SiO_2) but not to a dishpan made from polyethylene ($\sim CH_2CH_2 \sim$), a hydrocarbon (contains only hydrogen and carbon)?
83. Explain why water in a narrow glass tube has a meniscus but pentane (C_5H_{12}) does not.
84. Calculate the heat absorbed in doing the following:
    a. melting 475 g of ice. (The heat of fusion of ice is 6.03 kJ/mol.)
    b. boiling 750. g of water. (The heat of vaporization of water is 40.7 kJ/mol.)
85. The equilibrium vapor pressure of ethyl ether ($C_2H_5OC_2H_5$) is 460 torr at 25°C, whereas the vapor pressure of ethyl alcohol (C_2H_5OH) is 54 torr at this temperature. Explain the difference in terms of the intermolecular forces involved.
86. Which substance in problem 85 has the higher normal boiling point? Explain in terms of the definition of boiling point.
*87. What is the final temperature of a 0.100-kg sample of water initially at 90°C after it has absorbed 175 kJ of energy?
88. Calculate the quantity of heat involved in condensing 350. g of steam at 100°C and cooling the resulting liquid water to 28°C. Is the process endothermic or exothermic? Refer to Table 13.5 for the necessary physical constants for water.
*89. A fire department investigated the cause of a house fire on a subzero winter day. Upon questioning the residents, the investigator was told that the family car would not start that morning. Someone had drained motor oil from the car and heated the oil in a pan on the kitchen stove just before the fire started. The driver reasoned that the car would start if warm oil was added to the car's engine. The investigator could not ignite a small volume of motor oil with a match. Explain how the fire might have started.

Solutions to Practice Problems

PP 13.1

The intermolecular forces present in ethyl alcohol are a combination of dispersion forces and hydrogen bonding, whereas dispersion forces and dipole–dipole attractions are involved in ethyl chloride. The intermolecular forces are stronger in ethyl alcohol because of hydrogen bonding. Dipole–dipole attractions are generally weak, as are dispersion forces in small molecules.

PP 13.2

Methyl alcohol has a higher vapor pressure at any temperature because of its weaker intermolecular forces. Hydrogen bonding occurs in both compounds, but dispersion forces are weaker in methyl alcohol because of its smaller molar mass.

PP 13.3

The normal boiling point of octane (125°C) is higher than the normal boiling point of isopropyl chloride (36°C). Although isopropyl chloride is polar, dipole–dipole attractions are weak, and dispersion forces are stronger in octane because of its higher molar mass.

PP 13.4

a. molecular crystal, hydrogen bonding, and dispersion forces
b. ionic crystal, ionic bonding
c. metallic crystal, metallic bonding

PP 13.5

Find: $q(\text{kJ}) = ?$
Given: $\Delta H_{vap} = 38.6$ kJ/mol, 75.0 g C_2H_5OH
Known: MM-$C_2H_5OH = 46.0$ g/mol (from AM's)
Solution:

$$q(\text{kJ}) = \left(\frac{38.6 \text{ kJ}}{\text{mol}}\right)(? \text{ mol}) \quad (1)$$

$$\text{number of moles} = \left(\frac{1 \text{ mol}}{46.0 \text{ g}}\right)(75.0 \text{ g}) = 1.63 \text{ mol} \quad (2)$$

Substituting in equation (1) gives

$$q(\text{kJ}) = \left(\frac{38.6 \text{ kJ}}{\text{mol}}\right)(1.63 \text{ mol}) = 62.9 \text{ kJ}$$

PP 13.6

a. *Find:* $q(\text{kJ}) = ?$
Given: 63.5 g water
Known: $\Delta H_{fus} = 6.03$ kJ/mol, MM-$H_2O = 18.0$ g/mol
Solution:

$$q(\text{kJ}) = \left(\frac{6.03 \text{ kJ}}{\text{mol}}\right)(? \text{ mol}) \quad (1)$$

$$\text{number of moles} = \left(\frac{1 \text{ mol}}{18.0 \text{ g}}\right)(63.5 \text{ g}) = 3.53 \text{ mol} \quad (2)$$

Substituting in equation (1) gives

$$q(\text{kJ}) = \left(\frac{6.03 \text{ kJ}}{\text{mol}}\right)(3.53 \text{ mol}) = 21.3 \text{ kJ}$$

b. *Find:* $q(\text{J}) = ?$
Given: 63.5 g water, 0°C to 38°C, $\Delta T = 38$ C°
Known: $SH = 4.18$ J/g·C°
Solution:

$$q(\text{J}) = \left(\frac{4.18 \text{ J}}{\text{g} \cdot \text{C}^\circ}\right)(63.5 \text{ g})(38 \text{ C}^\circ) = 1.0 \times 10^4 \text{ J} = 10. \text{ kJ}$$

c. The total heat absorbed is the sum of the heats absorbed in the two steps:

$$q(\text{total}) = 21.3 \text{ kJ} + 10. \text{ kJ} = 31 \text{ kJ}$$

PP 13.7

$$\underset{\text{blue}}{CoCl_2 \cdot 4H_2O(s)} + 2H_2O(g) \rightleftarrows \underset{\text{pink}}{CoCl_2 \cdot 6H_2O(s)}$$

Answers to Chapter Review

1. water
2. liquids
3. gases
4. shape
5. flows
6. greater
7. slower
8. dispersion forces
9. polar
10. hydrogen
11. H–N
12. less
13. hydrophilic
14. hydrophobic
15. nonpolar
16. large
17. vapor
18. vaporization
19. condensation
20. vapor pressure
21. equilibrium
22. temperature
23. low
24. weak
25. molar
26. endothermic
27. boiling point
28. normal
29. higher
30. surface tension
31. bonding
32. viscosity
33. viscous
34. intermolecular
35. crystalline
36. amorphous
37. ionic
38. network
39. metallic
40. electrolytes
41. nonconducting
42. covalently
43. diamond
44. quartz
45. electrons
46. heating
47. melting point
48. fusion
49. specific
50. moderating
51. ice
52. wetting
53. hydration
54. hydrate
55. pentahydrate
56. exothermic
57. hygroscopic
58. deliquescent

14
40 ml
±5%
30
20
10

Solutions

Contents

Mixing aqueous solutions of potassium chromate (K_2CrO_4) and lead(II) nitrate ($Pb(NO_3)_2$) results in a precipitate of yellow lead(II) chromate ($PbCrO_4$).

A distance runner is breathing hard.

You Probably Know . . .

- Solutions are homogeneous mixtures. Air is a gaseous solution essential to our life. We inhale air to obtain needed oxygen. We exhale air to expel carbon dioxide, a waste product of metabolism.
- Water is also essential to life. Almost all natural waters—whether in lakes, streams, the oceans, or falling rain—are impure. Natural waters are solutions. For example, ocean water contains sodium chloride and other dissolved salts in such concentrations that it is unfit to drink.
- Beverages such as apple juice, wine, and soft drinks are aqueous solutions. If apple juice is spilled and partially evaporates, the residue may be sticky due to the dissolved sugar. An aqueous solution has some of the properties of water (a liquid) and the dissolved substances such as sugar.
- Aqueous solutions are essential in sustaining life in living systems. For example, blood carries dissolved oxygen and nutrients to the cells of our bodies and carries away dissolved waste products.
- Solutions are frequently used in laboratory experiments. For example, mixing aqueous solutions of potassium chromate (K_2CrO_4) and lead(II) nitrate ($Pb(NO_3)_2$) results in the formation of a precipitate of lead(II) chromate ($PbCrO_4$).

Natural waters are solutions.

Why Should We Study Solutions?

The material world is composed mostly of mixtures—some heterogeneous, some homogeneous. A **solution** is a homogeneous mixture of one or more substances (as atoms, molecules, or ions) uniformly dispersed (dissolved) in another substance. Solutions are the most useful type of mixture because their components do not separate on standing. Solutions are very common, and it would be difficult to overstate their importance to all of us. Air, the oceans, lakes and streams, gasoline and other petroleum products—all are examples of solutions.

A solution is a homogeneous mixture.

Aqueous solutions, in which substances are dissolved in water, are the primary focus of this chapter. As discussed earlier, many chemical reactions occur in aqueous solution. In our bodies, nutrients and oxygen are transported in the blood, an aqueous solution, to the cells, and solutions carry away waste products. This is also true in plants. Most of the chemical reactions in human, animal, and plant life occur in aqueous solution.

Other apparently homogeneous mixtures called colloids are also important to living things. The dispersed particles of a colloid are larger than those of a solution but small enough that they do not separate readily as do coarser particles of other heterogeneous mixtures. Examples of colloids such as milk, mayonnaise, and smoke are discussed later in this chapter.

14.1 Terminology of Solutions

Learning Objective

- Know and be able to use the key terms of solution terminology.

The **solvent,** usually the major component, is the substance that determines the physical state of the solution. The **solute** is the substance that is dispersed in the solvent and is usually the minor component of the solution. Although our focus here is aqueous solutions, the terminology also applies to other solutions such as those involving organic solvents (alcohol, acetone, paint thinner, gasoline, and so forth).

Solvent: Substance that determines the physical state of a solution, usually the major component.

Solute: Substance dispersed in the solvent, usually the minor component.

The **concentration** of a solution tells us the amount of solute in a given amount of solution (or solvent). The quantitative expression of concentration is discussed in Section 14.4. A **dilute** solution is one that contains a small amount of solute compared with the amount of solution. A **concentrated** solution contains a large amount of solute. These are relative terms, and it is not possible to specify numerical values that would be generally applicable. For example, two 1-L solutions—containing 1 g and 3 g of sugar, respectively—could both be considered dilute solutions even though one is more concentrated than the other. A 1-L solution that contains 250 g of sugar might be considered concentrated by comparison.

There is a limit to how much solid solute a given amount of solution can contain at a particular temperature. For example, as you add sugar to a glass of

TABLE 14.1 Water Solubility of Selected Substances

Substance	*Solubility at 20°C (g solute/100 g water)*
Gases (at 1 atm)	
oxygen (O_2)	0.0043
nitrogen (N_2)	0.0019
carbon dioxide (CO_2)	0.145
ammonia (NH_3)	51.8
Solids	
sodium chloride (NaCl)	36.0
sodium nitrate ($NaNO_3$)	88.0
baking soda ($NaHCO_3$)	9.6
sucrose ($C_{12}H_{22}O_{11}$)	200
Liquids	
acetone (CH_3COCH_3)	infinite
ethanol (C_2H_5OH)	infinite
1-butanol (C_4H_9OH)	7.9

A saturated sugar solution.

The solubility of a substance is its concentration in a saturated solution.

water, it dissolves quickly at first, slows as you add more, and eventually stops dissolving; the excess sugar simply settles to the bottom of the glass. The resulting solution is **saturated,** meaning that it contains the most solute normally possible at that temperature. For a saturated solution, the undissolved and dissolved solute are in a state of dynamic equilibrium (Section 13.3), meaning that molecules of undissolved solute are entering the solution at the same rate that molecules of dissolved solute are crystallizing.

$$\text{solute(s)} + \text{water} \underset{\text{crystallizing}}{\overset{\text{dissolving}}{\rightleftharpoons}} \text{solute(aq)}$$

The concentration of a saturated solution is the **solubility** of the solute at a given temperature (for example, 200 g sugar/100 g of water at 20°C). The solubilities of selected substances in water are given in Table 14.1. General solubility rules are found in Section 10.9.

A solution is **unsaturated** when it contains less than the equilibrium amount of solute. For example, a sugar solution of any concentration that is below the solubility of sugar in water is unsaturated.

14.2 Characteristics of Solutions

Learning Objective

- Describe the general properties of solutions.

A solution can be classified according to its physical state as a solid, liquid, or gaseous solution (Table 14.2). The physical state of a solution is the same as that of its solvent. Air is the gaseous solution most familiar to us. Well-known metals such

TABLE 14.2 Common Solutions

Physical State	*Example*	*Components*[a]
gas	air	nitrogen, oxygen, argon, trace gases
solid	brass	copper, zinc
solid	steel	iron, carbon, manganese (chromium in stainless)
liquid	sea water	water, sodium chloride, other salts
liquid	vinegar	water, acetic acid
liquid	rubbing alcohol	isopropyl alcohol, water

[a]The major component is listed first.

as brass and steel are examples of solid solutions. We noted previously that our drinking water is really a mixture or liquid solution, not a single substance.

In general, the physical properties or characteristics of a liquid solution closely resemble those of its solvent. For example, a sugar solution prepared by dissolving a teaspoon of sugar in a glass of water is a liquid that has the appearance of water. Its melting point and boiling point are similar but not identical to those of water. A sugar solution and water have similar viscosities and surface tensions as well. The most noticeable difference is the sweet taste of the solution, a property of the dissolved sugar.

As with liquid solutions, the physical properties of a gaseous solution such as air resemble those of the major component. However, solid solutions can have physical properties very different from those of the pure components. For example, the melting point of a solution of two crystalline solids is usually lower than the melting points of the components. The behavior of solid solutions of two or more metals, called alloys, is very complex and above the level of this text.

The overall chemical properties of a solution are a result of the chemical properties of each component. In addition, we must consider possible chemical reactions among the solution components. A number of these chemical reactions were covered in Chapter 10.

Factors That Affect Solubility

14.3

Learning Objectives

- Identify the factors that affect the solubility of a solute in water or other solvent.
- Explain how to prepare a supersaturated solution.
- Identify the factors that influence the rate at which a solid solute dissolves.

When sugar is dissolved in water, the solid sugar seems to disappear as sugar molecules disperse in the water. However, when sand is stirred with water, it remains undissolved. What factors cause one substance to be soluble and another to be insoluble? To answer this question, we take a closer look at the formation of a solution.

■ Intermolecular Forces

The solubility of a molecular solid in water depends upon its ability to hydrogen bond with water.

The solubility of a molecular substance in water depends upon its ability to hydrogen bond with water. In the crystalline state, sugar molecules are strongly attracted to each other by hydrogen bonding (Figure 14.1). When sugar dissolves in water, the hydrogen bonding among molecules in the crystal is replaced by hydrogen bonding with water molecules. A molecular solid does not dissolve in water unless hydrogen bonding with water replaces the intermolecular forces in the crystal.

Disssolving sugar in water illustrates the importance of intermolecular forces on the solubility of a molecular solid. In fact, intermolecular and other solvent–solute attractions play a major role in the solubilities of all types of solids—and also in the solubilities of liquid and gaseous solutes in liquid solvents such as water. The discussion that follows considers the types of solvent–solute attractions that occur with various solutes.

Hydration of ions is indicated by (aq).

Solvent–solute attractions are important when an ionic solid dissolves in water. For example, when sodium chloride dissolves in water, the ionic attractions in the crystal are replaced by strong ion–dipole attractions (hydration) with water (Figure 14.2). The solubility of an ionic solid depends in large part upon the strength of these ion–dipole attractions compared with the strength of the ionic attractions in the crystal. Ionic solids are insoluble in nonpolar solvents because ion–dipole attractions are not possible.

Hydrated ions are dissociated and dispersed in solution.

Remember, as an ionic solid dissolves in water, cations and anions dissociate. Cations and anions become associated with water molecules, or hydrated, and the hydrated ions disperse among water molecules in the solution. An equation shows the dissociation of cations and anions as an ionic solute dissolves.

$$\mathrm{NaCl(s)} \xrightarrow{\mathrm{H_2O(l)}} \mathrm{Na^+(aq)} + \mathrm{Cl^-(aq)}$$

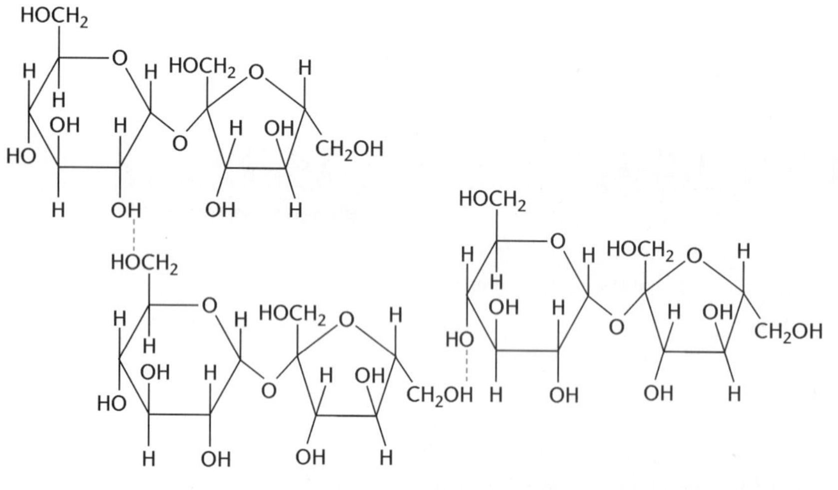

a

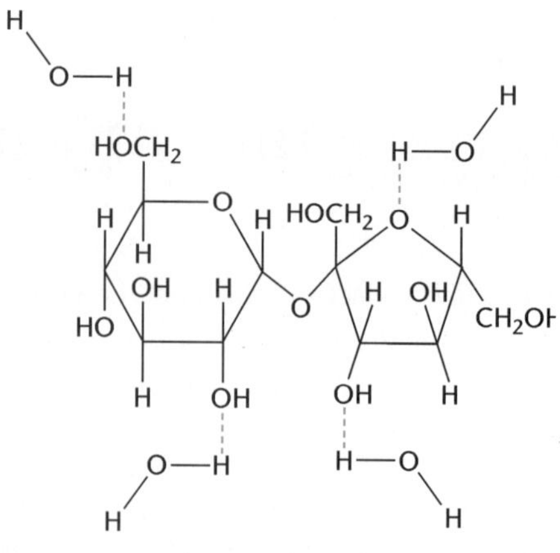

b

FIGURE 14.1 (a) Hydrogen bonding holds sugar molecules in place in a crystal. (b) When sugar dissolves in water, hydrogen bonding in the crystal is replaced by hydrogen bonding with water molecules.

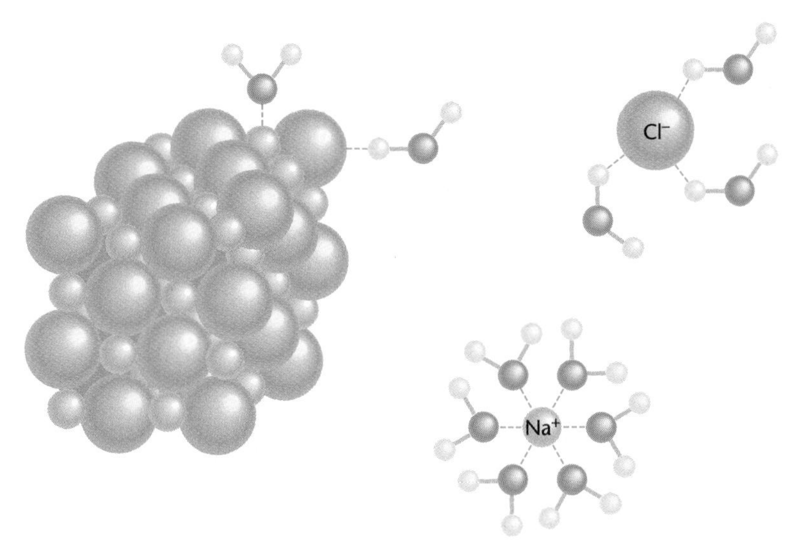

FIGURE 14.2 When sodium chloride dissolves, ionic attractions in the crystal are replaced by ion-dipole attractions with water.

Because of the mobility of the ions, aqueous solutions of ionic solutes conduct electric current (Section 8.2).

Many ionic substances, such as AgCl and $BaSO_4$, are practically insoluble in water. For these substances, the ionic attractions in the solid are too great to be overcome by ion–dipole attractions in solution, and the solids fail to dissolve to any appreciable extent.

Network solids are insoluble in water.

In general, network solids such as silica (sand) are insoluble in water and other solvents. Even though hydrogen bonding with water is a very strong intermolecular force, it is insufficient to cause the covalent bonds of a network solid to break.

Gases that hydrogen bond with water are very soluble.

A gas such as ammonia that can hydrogen bond with water is very soluble (see Table 14.2). You should note that there are no significant intermolecular forces between the molecules of an ideal gas as there are in solids. However, there is a natural tendency for the molecules of a gas to remain in their disordered state. Because of this, only gases that are readily hydrated dissolve in water to any appreciable extent. The water solubility of nonpolar gases such as oxygen and nitrogen is very low.

Ionization: Formation of ions by breaking a covalent bond.

Some gases ionize when they dissolve in water. The resulting ion–dipole attractions play an important role in the solubility of these gases just as they do with ionic solids. The ionization of hydrogen chloride in water is illustrated by the following equation:

$$HCl(g) \xrightarrow{H_2O(l)} H^+(aq) + Cl^-(aq)$$

Note the difference between the ionization of hydrogen chloride, a molecular compound, and the dissociation of sodium chloride, which is ionic in the solid state.

Miscible liquids dissolve in all proportions.

Hydrophilic: Attracted to water. Hydrophobic: Repelled by water. (Section 13.2)

Many liquids are soluble in water. For example, ethyl alcohol (C_2H_5OH) and acetone (CH_3COCH_3) are **miscible** with water, meaning that they dissolve in all proportions. As with molecular solids, water solubility of liquids depends upon hydrogen bonding of water with solute molecules. Butyl alcohol (C_4H_9OH) is **immiscible** with water, meaning it is only slightly soluble. Although the OH group is hydrophilic due to hydrogen bonding, the difference in solubility is due to the chain of four carbons, which is hydrophobic. The water solubility of alcohols and

Petroleum leaking from a ship forms an oil slick on the surface of the ocean.

Cooling a hot saturated solution results in a supersaturated solution, which often crystallizes. Shown here is the crystallization of sodium acetate ($NaC_2H_3O_2$).

other organic compounds decreases as the number of carbons in the chain increases.

As was noted in the previous chapter, hydrocarbons such as octane (C_8H_{18}) are nonpolar and are hydrophobic. Consequently, they are insoluble in water. Because the densities of these compounds are less than the density of water, they float on the surface of water. For example, when petroleum leaks from a damaged oil tanker, an oil slick forms on the surface of the ocean.

These examples illustrate the effect of intermolecular forces on the solubility of a substance in water or other solvent. Remember, when a substance dissolves, the solvent–solute attractions replace the intermolecular forces or ionic attractions in the undissolved solute. A substance that does not associate with water through hydrogen bonding or ion–dipole attractions is hydrophobic and practically insoluble.

Temperature

The solubility of a solid solute is generally greater at higher temperatures.

The solubility of most substances is affected by temperature. In general, the solubility of a solid solute is greater at higher temperatures (Figure 14.3), but there are exceptions. For example, the solubility of NaCl changes very little when the temperature is changed. Furthermore, the solubilities of some solids (for example, Na_2SO_4) are actually lower at higher temperatures.

Careful cooling of a hot saturated solution can result in a solution having a concentration greater than that of a saturated solution at lower temperatures. Such a solution is **supersaturated.** A supersaturated solution is unstable and often crystallizes. Crystallization continues until the solution once again is saturated, a stable condition. Because some of the solute crystallized, the remaining solution is less concentrated than the hot saturated solution.

Gases are less soluble at higher temperatures.

In contrast to the behavior of solid solutes, gases are always less soluble at higher temperatures. Familiar examples are root beer and other soft drinks, which

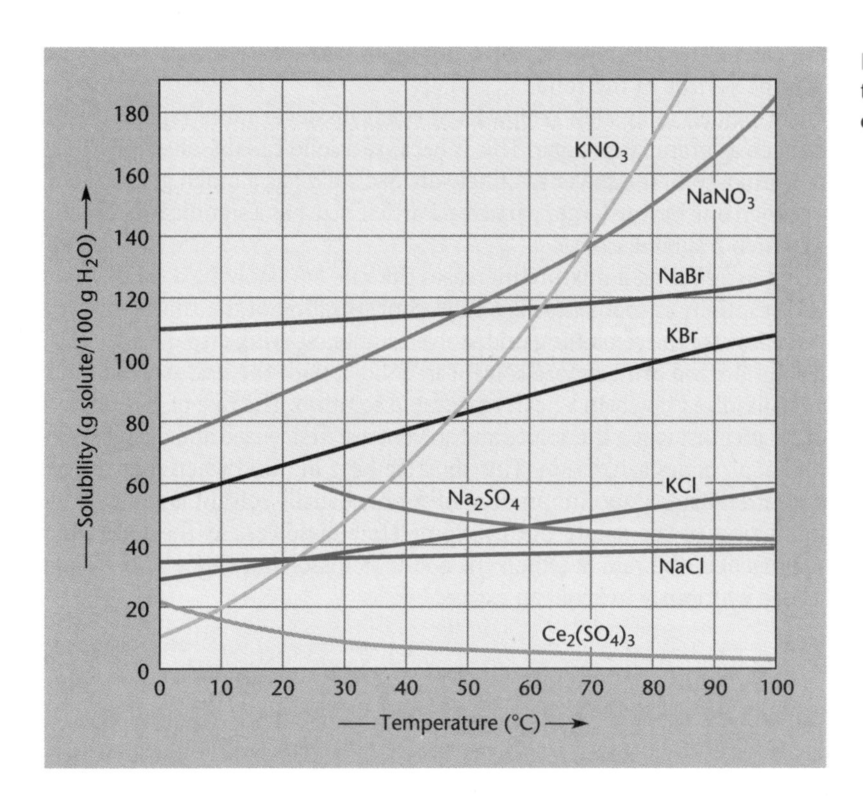

FIGURE 14.3 Effect of temperature on the solubility of compounds in water.

rapidly lose their fizz (dissolved carbon dioxide) when they get warm. Keeping soft drinks cold slows the loss of carbon dioxide.

Partial Pressure of a Gaseous Solute

The partial pressure of a gas significantly affects its solubility in a liquid solvent such as water. For example, when you open a bottle of a carbonated beverage, you can see gas bubbles rising to the surface. The release of pressure when you remove the cap allows some of the dissolved CO_2 to escape from the solution. During bottling, the beverage is "carbonated" by introducing a partial pressure of CO_2 that is slightly greater than atmospheric pressure.

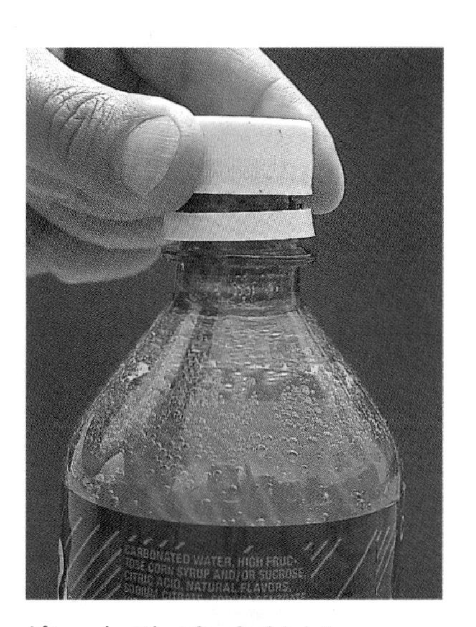

After a bottle of soft drink is opened, bubbles of CO_2 rise to the surface.

Dissolving Solids: How Fast?

The rate at which a solid dissolves in a solvent depends upon temperature, particle size of the solid, stirring or agitation of the mixture, and how much of the solid has already dissolved. Note, however, that *how fast* a solid dissolves is not the same as its solubility. Although it is true that very soluble solids generally dissolve rapidly, particle size and stirring do not affect the quantity of solid that will dissolve at a certain temperature.

Solids generally dissolve faster at higher temperatures due to a kinetic effect. At higher temperatures, solvent molecules are moving faster and collide more frequently with the surfaces of the undissolved solute. Hydrated solute molecules or ions are carried rapidly away into the solution, and more solvent molecules collide with the surface of the solid.

A finely divided solid such as powdered sugar dissolves more rapidly than a coarse one such as granulated sugar. This is because a solid can dissolve only when its surface is exposed to the solvent. A finely divided solid has a much greater total surface area than one having larger particles. Particle size has a significant effect on the rate at which a solid dissolves.

Stirring or agitating a mixture increases the rate of dissolving a solid. When solute dissolves, there is momentarily a high concentration of the solution in the immediate vicinity of the remaining undissolved solute. Stirring disperses the concentrated solution and brings more solvent in contact with the undissolved solid.

A solid dissolves slowly in a nearly saturated solution. The rate of dissolving a solid solute is greatest when the solute and solvent are first mixed and decreases as the solution approaches saturation. This must be kept in mind when preparing a saturated solution using a specific amount of solute. Usually solvent is added gradually with stirring to the solute until it completely dissolves. As the saturation point is approached, the rate at which the solute is dissolving slows, and solvent must be added cautiously to avoid an excess.

14.4 Concentration of Solutions

Learning Objectives

- Calculate mass percent, parts per million, and molarity from solution data.
- Calculate the mass or number of moles of solute from data for a solution.
- Calculate the mass of solvent or volume of solution from the mass or number of moles of solute and the concentration of a solution.
- Solve dilution problems for either an unknown volume or an unknown concentration of a solution.

Earlier in this discussion of solutions you learned that concentration expresses the amount of solute in a given amount of solution (or solvent). In other words, a concentration is expressed as a *ratio* of a quantity of solute to a quantity of solution. Because concentrations are ratios, they can be used like conversion factors to solve problems about quantities of solute, solvent, or solution. The most frequently used concentration expressions are defined in this section.

Mass Percent

A common way of expressing solution concentration is by **mass percent** (mass %), the grams of solute per 100 g of solution:

$$\text{mass }\% = \frac{\text{mass of solute}}{\text{mass of solution}} \times 100$$

$$= \frac{\text{mass of solute}}{\text{mass of solute} + \text{mass of solvent}} \times 100$$

For example, if 2.0 g of sucrose was dissolved in 48 g of water, the concentration in mass percent would be:

$$\text{mass \%} = \frac{2.0 \text{ g}}{2.0 \text{ g} + 48 \text{ g}} \times 100 = 4.0\% \text{ sucrose}$$

This solution contains 4.0 g of sucrose per 100 g of solution. In other words, mass percent expresses the grams of solute per 100 g of solution, or parts per hundred.

Mass % means g of solute per 100 g solution.

$$4.0\% \text{ sucrose means } \frac{4.0 \text{ g sucrose}}{100 \text{ g solution}}$$

In this form, the concentration is like a conversion factor, which can be used in problem solving.

Example 14.1

Calculate the mass % of sodium chloride in a solution containing 5.65 g of NaCl in 78.0 g of water. ■

Solution

$$\text{mass \%} = \frac{5.65 \text{ g}}{5.65 \text{ g} + 78.0 \text{ g}} \times 100 = 6.75\% \text{ NaCl}$$

Example 14.2

A 15.0-g sample of sodium bicarbonate, $NaHCO_3$ (baking soda), was stirred with 100.0 g of water at room temperature until no further solid would dissolve. The undissolved solute was removed by filtration and dried, and its mass was measured to be 4.4 g.

a. Calculate the solubility.
b. Calculate the mass % of sodium bicarbonate in water at room temperature. ■

Solutions

a. The solubility of $NaHCO_3$ is its concentration (g $NaHCO_3$/100 g of water) in a saturated solution. First, calculate the mass of $NaHCO_3$ that dissolved in 100.0 g of water:

$$\text{mass (g) dissolved } NaHCO_3 = 15.0 \text{ g} - 4.4 \text{ g} = 10.6 \text{ g}$$

or

$$10.6 \text{ g } NaHCO_3/100 \text{ g of water}$$

b. $\text{mass \%} = \frac{10.6 \cancel{\text{g}}}{110.6 \cancel{\text{g}}} \times 100 = 9.58\% \ NaHCO_3$

Note that the mass of the solution (denominator) is the sum of the masses of the solvent and the solute.

$$\text{mass of solution} = 100.0 \text{ g} + 10.6 \text{ g} = 110.6 \text{ g}$$

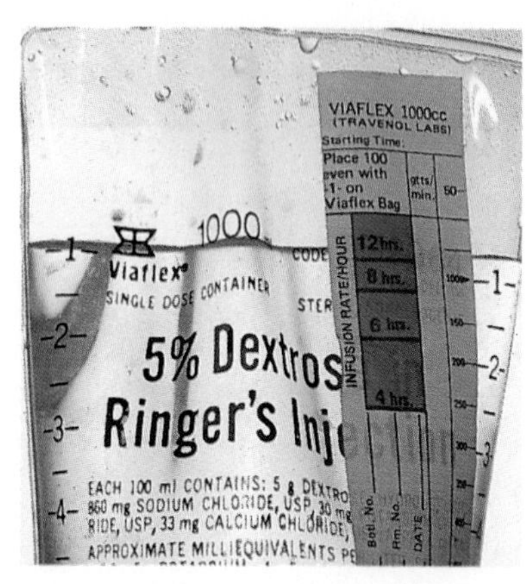

Dextrose is a nutrient that can be fed to a patient intravenously.

Example 14.3

What mass of potassium iodide is needed to prepare 75 g of a 5.0% solution? ▪

Find: mass (g) KI = ?

Given: 5.0% KI, or $\dfrac{5.0 \text{ g KI}}{100 \text{ g solution}}$, 75 g solution

Solution: $\text{mass (g) KI} = \left(\dfrac{5.0 \text{ g KI}}{100 \cancel{\text{ g solution}}}\right)(75 \cancel{\text{ g solution}})$

$= 3.75 \text{ g KI}$

which is rounded to 3.8 g KI.

Practice Problem 14.1 An intravenous feeding solution contains 5.0% dextrose (glucose). What mass of the feeding solution is needed to provide a patient with 23 g of dextrose? (Solutions to practice problems appear at the end of the chapter.)

▪ Parts per Million

To express the concentrations of very dilute solutions it is convenient to use **parts per million,** ppm, defined as grams of solute per one million (10^6) grams of solution:

$$\text{ppm} = \frac{\text{mass of solute}}{\text{mass of solution}} \times 10^6$$

Similar ratios include parts per billion, ppb (g solute/10^9 g solution), and parts per trillion, ppt (g solute/10^{12} g solution). The use of these units is becoming more common because of the ever-increasing capability of modern instruments to detect trace amounts of substances. Pesticide residues in food, air and water pollutants, and drugs in body fluids are just a few of the substances that are measured at trace levels. In calculations involving both mass percent and parts per million, be careful to use mass of solution, not mass of solvent.

Example 14.4

The concentration (conc) of a pesticide in water was found to be 1.3×10^{-4} g/L solution. What was the pesticide concentration in ppm? Assume the density of solution is 1.0 g/mL (because there is so little of the pesticide dissolved in it). ▪

Solution Because the density is 1.0 g/mL, the mass of one liter (1000 mL, an exact number) of the solution is 1.0×10^3 g.

$$\text{conc (ppm)} = \frac{\text{g pesticide}}{\text{g solution}} \times 10^6 = \left(\frac{1.3 \times 10^{-4} \cancel{\text{g}}}{1.0 \times 10^3 \cancel{\text{g}}}\right) \times 10^6$$

$$= 0.13 \text{ ppm}$$

Practice Problem 14.2 A sample of gasoline was found to contain 0.106 g of benzene (a carcinogen) per liter. Calculate the parts per million of benzene in the gasoline. The density of gasoline is 0.748 g/mL.

Molarity

It is usually more convenient to measure the quantity of a solution by volume rather than by mass. Moreover, when we carry out chemical reactions in solution, we often need to know the number of moles of solute in a solution. For these reasons, the most often used solution concentration is **molarity,** M, defined as the number of moles of solute per liter of solution:

$$\text{molarity, M} = \frac{\text{moles of solute}}{\text{liter of solution}}$$

For example, a solution of 0.0150 M $KMnO_4$ contains 0.0150 mol $KMnO_4$ dissolved in exactly 1 L of solution (Figure 14.4). Note that the volume of a solution is not necessarily the same as the volume of water used to prepare it. A volume change often occurs when substances are dissolved in water.

Molarity can be used as a conversion factor to find the number of moles of solute from the volume of a solution or vice versa. For example, if you need to know the number of moles of NaOH in 125 mL of a 2.00 M solution, you can use dimensional analysis to set up the calculation:

$$\text{number of moles NaOH} = \left(\frac{2.00 \text{ mol NaOH}}{\cancel{\text{L soln}}}\right)(0.125 \cancel{\text{ L soln}}) = 0.250 \text{ mol}$$

Conversion of 125 mL to 0.125 L was necessary to cancel volume units.

Example 14.5

What is the molarity of a solution that contains 0.20 mol of KOH in 150 mL of solution? ■

Solution

$$\text{M} = \frac{\text{mol KOH}}{\text{L soln}} = \frac{0.20 \text{ mol KOH}}{0.15 \text{ L soln}} = 1.3 \text{ M KOH}$$

Example 14.6

Calculate the molarity of a solution containing 3.50 g of NaCl in 225 mL of solution. ■

Begin by outlining the problem.

Find: $\text{M}\left(\frac{\text{mol NaCl}}{\text{L soln}}\right) = ?$

Given: 3.50 g NaCl, 225 mL (0.225 L) of solution

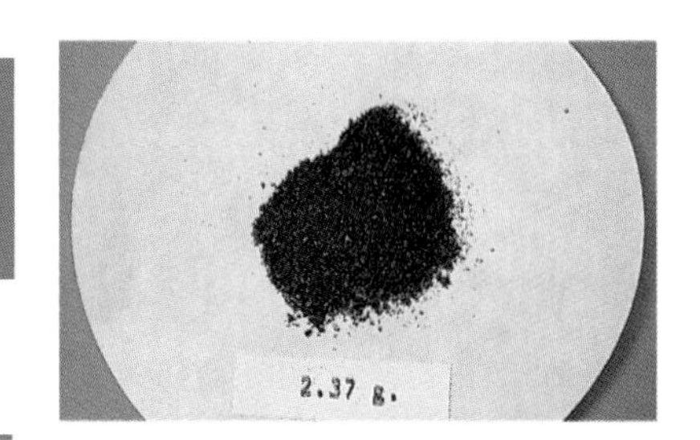

a

b

c

FIGURE 14.4 Preparation of 0.0150 M $KMnO_4$ solution in a 1-L volumetric flask. (a) A measured mass of $KMnO_4$. (b) The $KMnO_4$ is dissolved in distilled water. (c) Water is added to the mark on the flask.

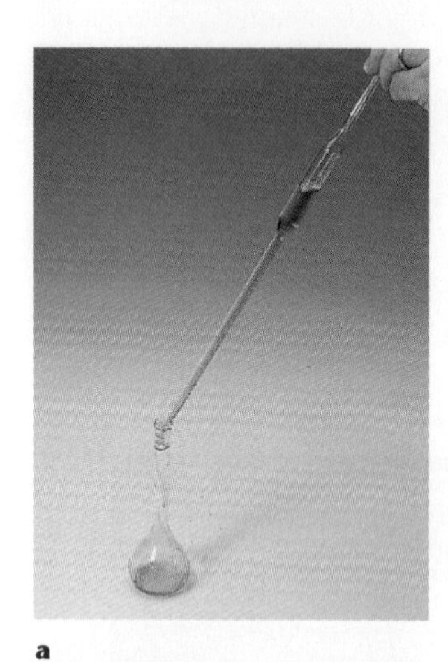

a

b

FIGURE 14.5 Preparation of a potassium dichromate ($K_2Cr_2O_7$) solution by dilution of a stock solution. (a) A specific volume (10.0 mL) of a concentrated solution is measured using a pipet and transferred into a 50-mL volumetric flask. (b) Water is added to the mark on the flask with mixing.

Known: MM-NaCl = 58.5 g/mol NaCl (from atomic masses)

Solution: The volume of the solution is given, but the number of moles of NaCl must be calculated:

$$\text{number of moles NaCl} = \left(\frac{1 \text{ mol}}{58.5 \cancel{\text{g NaCl}}}\right)(3.50 \cancel{\text{g NaCl}})$$

$$= 0.0598 \text{ mol}$$

Then, to calculate molarity:

$$\text{M} = \left(\frac{\text{mol NaCl}}{\text{L soln}}\right) = \frac{0.0598 \text{ mol NaCl}}{0.225 \text{ L soln}} = 0.266 \text{ M}$$

Example 14.7

What mass of Na_2SO_4 is needed to prepare 175 mL of 2.50 M solution? ■

After outlining the problem, set up the calculation starting with the unknown.

Find: mass (g) Na_2SO_4 = ?

Given: 175 mL (0.175 L) of solution, 2.50 mol Na_2SO_4/L solution

Known: MM-Na_2SO_4 = 142 g/mol (from atomic masses)

Solution: Using dimensional analysis gives

$$\text{mass (g) Na}_2\text{SO}_4 = \left(\frac{142 \text{ g}}{\cancel{\text{mol Na}_2\text{SO}_4}}\right)\left(\frac{2.50 \cancel{\text{mol Na}_2\text{SO}_4}}{\cancel{\text{L soln}}}\right)(0.175 \cancel{\text{L soln}}) = 62.1 \text{ g}$$

Practice Problem 14.3 What is the molarity of a potassium iodide solution containing 45.8 g KI in 500. mL of solution?

Practice Problem 14.4 What is the mass of solute in 225 mL of 0.150 M $Hg(NO_3)_2$ solution?

■ Dilution of Solutions

Diluting solutions is a common laboratory procedure. Laboratory stockrooms usually buy or prepare a supply of concentrated solutions called stock solutions. Solutions for laboratory experiments are then prepared as needed by diluting the stock solutions with water (Figure 14.5). When a certain volume of a solution is diluted with water, the volume of the solution increases, but the amount of solute does not change. In other words:

$$\underset{\text{(before dilution)}}{\text{number of moles solute}} = \underset{\text{(after dilution)}}{\text{number of moles solute}}$$

Understanding this simple relationship is crucial to the calculations required for preparing solutions by dilution.

Example 14.8

What is the concentration of a solution prepared by diluting 20.0 mL of 12 M hydrochloric acid to a final volume of 250. mL? ■

Use M_c and V_c for the molarity and volume of the concentrated solution and M_d and V_d for the molarity and volume of the dilute solution.

Find: $M_d \left(\frac{\text{mol HCl}}{\text{L soln}}\right) = ?$

Given: $M_c = 12\text{ M or } \frac{12\text{ mol HCl}}{\text{L soln}}$

$V_c = 20.0\text{ mL}, V_d = 250.\text{ mL}$

Known: The number of moles of HCl is not changed in dilution.

Solution: The volume of the dilute solution is given. Use dimensional analysis to find the number of moles of HCl:

$$\text{number of moles HCl} = \left(\frac{12\text{ mol HCl}}{\cancel{\text{L soln}}}\right)(20.0 \times 10^{-3}\,\cancel{\text{L soln}})$$
$$= 0.24\text{ mol}$$

Then calculate the molarity of the dilute solution:

$$M_d = \frac{0.24\text{ mol HCl}}{0.250\text{ L soln}} = 0.96\text{ M}$$

Example 14.9

What volume of 18 M sulfuric acid is needed to prepare 500. mL of 3.0 M solution? ■

Find: $V_c\text{ (mL)} = ?$

Given: $M_c = 18\text{ M or }\left(\frac{18\text{ mol }H_2SO_4}{\text{L soln}}\right),\ M_d = 3.0\text{ M or }\left(\frac{3.0\text{ mol }H_2SO_4}{\text{L soln}}\right)$

$V_d = 500.\text{ mL (0.500 L)}$

Known: The number of moles of H_2SO_4 remains unchanged.

Solution: Using dimensional analysis gives

$$V_c\text{ (L)} = \left(\frac{1\text{ L soln}}{18\text{ mol }H_2SO_4}\right)(?\text{ mol }H_2SO_4) \qquad (1)$$

$$\text{number of moles }H_2SO_4 = \left(\frac{3.0\text{ mol }H_2SO_4}{\cancel{\text{L soln}}}\right)(0.500\,\cancel{\text{L soln}})$$
$$= 1.5\text{ mol }H_2SO_4 \qquad (2)$$

$$V_c\text{ (L)} = \left(\frac{1\text{ L soln}}{18\,\cancel{\text{mol }H_2SO_4}}\right)(1.5\,\cancel{\text{mol }H_2SO_4}) = 0.083\text{ L, or 83 mL} \qquad (1)$$

Practice Problem 14.5 What is the final concentration after diluting 10.0 mL of 8.50 M H_2SO_4 to a final volume of 50.0 mL?

Solution Concentrations

Concentration	*Definition*
mass percent, mass %	$\frac{\text{g solute}}{\text{g solution}} \times 100$
parts per million, ppm	$\frac{\text{g solute}}{\text{g solution}} \times 10^6$
molarity, M	$\frac{\text{mol solute}}{\text{L solution}}$

14.5 Calculations Involving Reactions in Solution

Learning Objective

- Use concentrations of solutions to solve problems involving reactions occurring in aqueous solution.

As emphasized in Chapter 11, a balanced chemical equation expresses mole ratios of reactants and products. Previously, in working problems, the number of moles of a reactant was usually calculated from its mass. However, when a reactant is in a solution of known molarity, the number of moles can be calculated from the volume of the solution, using the molarity as a conversion factor. For example, to find the number of moles of NaOH in 35.0 mL of 6.0 M solution:

$$\text{number of moles NaOH} = \left(\frac{6.0 \text{ mol NaOH}}{\cancel{\text{L soln}}}\right)(0.0350 \cancel{\text{ L soln}}) = 0.21 \text{ mol}$$

Note that the volume was converted to liters to allow cancellation of the volume units.

Example 14.10

How many moles of H_2SO_4 are in 20.0 mL of 4.00 M solution? ■

Solution

$$\text{number of moles } H_2SO_4 = \left(\frac{4.00 \text{ mol } H_2SO_4}{\cancel{\text{L soln}}}\right)(0.0200 \cancel{\text{ L soln}}) = 0.0800 \text{ mol}$$

Example 14.11

Calculate the mass of $BaSO_4$ that forms upon mixing 15.0 mL of 0.100 M $Ba(OH)_2$ solution with an excess of 3.00 M H_2SO_4 solution. ■

Solution First, outline the problem under the balanced equation:

$$Ba(OH)_2(aq) + H_2SO_4(aq) \longrightarrow BaSO_4(s) + 2H_2O(l)$$

15.0 mL ↓ 0.100 mol $Ba(OH)_2$/L soln → ____ mol $\times \left(\frac{1 \text{ mol } BaSO_4}{1 \text{ mol } Ba(OH)_2}\right)$ → ____ mol ↑ MM = 233 g/mol → ? g

Because excess H_2SO_4 was used, it is not the limiting reactant and does not enter into the calculation. The calculation involves three steps. Following the outline, first calculate the number of moles of $Ba(OH)_2$, then the number of moles of $BaSO_4$, and finally the mass of $BaSO_4$:

$$\text{number of moles } Ba(OH)_2 = \left(\frac{0.100 \text{ mol } Ba(OH)_2}{\cancel{\text{L soln}}}\right)(0.0150 \cancel{\text{ L soln}})$$

$$= 0.001\ 50 \text{ mol } Ba(OH)_2 \quad (1)$$

$$\text{number of moles } BaSO_4 = \left(\frac{1 \text{ mol } BaSO_4}{1 \cancel{\text{ mol } Ba(OH)_2}}\right)(0.001\ 50 \cancel{\text{ mol } Ba(OH)_2})$$

$$= 0.001\ 50 \text{ mol } BaSO_4 \quad (2)$$

$$\text{mass (g) } BaSO_4 = \left(\frac{233 \text{ g}}{1 \cancel{\text{ mol } BaSO_4}}\right)(0.001\ 50 \cancel{\text{ mol } BaSO_4})$$

$$= 0.350 \text{ g} \quad (3)$$

Example 14.12

What volume of 6.0 M HCl is needed to react with and dissolve 75 g of calcium carbonate? ■

Solution Begin by writing a balanced equation for the reaction and outlining the problem:

$$CaCO_3(s) + 2HCl(aq) \longrightarrow CaCl_2(aq) + CO_2(g) + H_2O(l)$$

75 g ↓ MM = 100. g/mol → ____ mol $\times \left(\frac{2 \text{ mol HCl}}{1 \text{ mol } CaCO_3}\right)$ → ____ mol ↑ 6.0 mol HCl/L soln → ? mL

Calcium carbonate reacts with hydrochloric acid, forming carbon dioxide, which bubbles from the solution.

$$\text{number of moles } CaCO_3 = \left(\frac{1 \text{ mol}}{100. \cancel{\text{g } CaCO_3}}\right)(75 \cancel{\text{g } CaCO_3}) = 0.75 \text{ mol} \quad (1)$$

$$\text{number of moles HCl} = \left(\frac{2 \text{ mol HCl}}{1 \cancel{\text{mol } CaCO_3}}\right)(0.75 \cancel{\text{mol } CaCO_3}) = 1.5 \text{ mol} \quad (2)$$

$$V(\text{L}) \text{ HCl} = \left(\frac{1 \text{ L soln}}{6.0 \cancel{\text{mol HCl}}}\right)(1.5 \cancel{\text{mol HCl}}) = 0.25 \text{ L, or } 250 \text{ mL} \quad (3)$$

Practice Problem 14.6 What mass of silver chloride (AgCl) forms when you mix 25.0 mL of 0.112 M $AgNO_3$ solution with excess 0.100 M HCl?

14.6 Colligative Properties of Solutions

Learning Objectives

- Compare the vapor pressures, boiling points, freezing points, and osmotic pressures of solutions of various concentrations.
- Calculate the boiling point and freezing point of a solution from its concentration (molality).

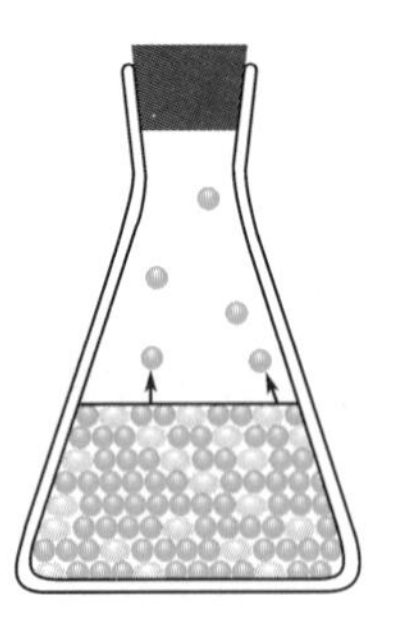

Solution

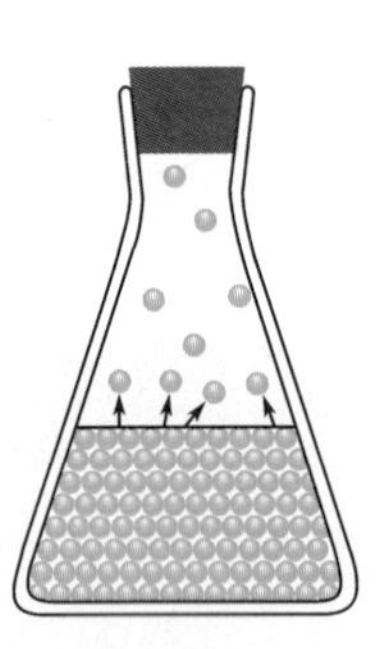

Water

FIGURE 14.6 Vapor pressure lowering for an aqueous sugar solution. Because sugar molecules (white) occupy positions on the surface, fewer water molecules (blue) can escape to the vapor phase. As a result, the vapor pressure of the solution is lower than that of water.

Certain physical properties of solutions depend upon the concentration of solute particles but are independent of the type of solute. These properties are called **colligative properties.** Vapor pressure lowering, boiling point elevation, freezing point depression, and osmotic pressure are physical properties that are the same for solutions of all nonvolatile solutes. Only the concentration of a solution is important, not the identity of the solute.

Vapor Pressure of a Solution

When a nonvolatile molecular solute such as sugar is dissolved in water, the vapor pressure above the solution is a result only of vaporization of the water. Because sugar molecules occupy some of the positions on the surface of the solution, fewer water molecules escape from the solution than from the pure solvent (Figure 14.6). Consequently, the vapor pressure of the solution is *lower* than the vapor pressure of pure water (Figure 14.7).

Boiling Point of a Solution

Because an aqueous solution of a nonvolatile molecular solute has a lower vapor pressure than water, it boils at a *higher* temperature (see Figure 14.7). The difference in boiling point, ΔT_b, can be calculated from the concentration, m, of the solution and a constant, K_b:

$$\Delta T_b = mK_b$$

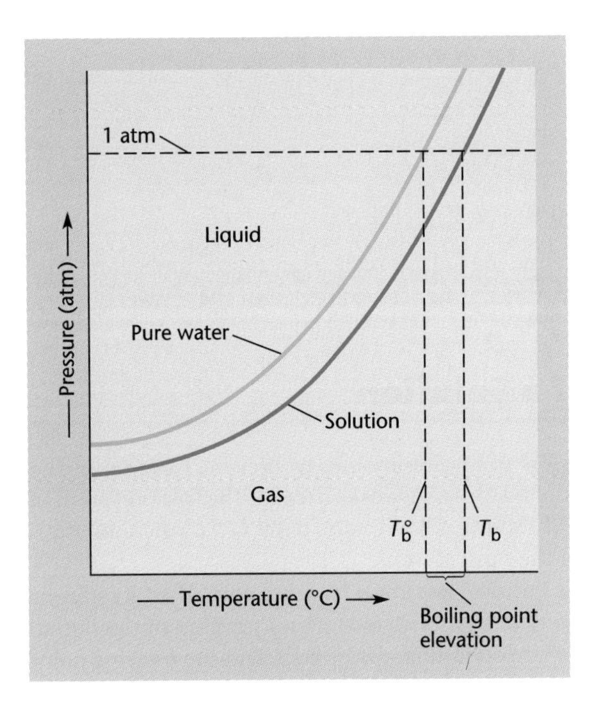

FIGURE 14.7 The vapor pressure of an aqueous sugar solution is lower than the vapor pressure of water. As a result, the solution has a higher boiling point.

where m represents the **molality** of the solution, the number of moles of solute per kilogram of solvent. The constant, K_b, called the molal boiling point elevation constant, is different for various solvents. For water, K_b is 0.51 C°/m. The boiling point of the solution (T_b) is determined by adding ΔT_b to 100°C, the normal boiling point of water.

Molality, $m = \dfrac{\text{mol solute}}{\text{kg solvent}}$.

For an aqueous solution,

$T_b = \Delta T_b + 100°C$

Example 14.13

Calculate the normal boiling point of a solution of 75 g of ethylene glycol (EG, $C_2H_6O_2$) in 250 g of water.

Find: T_b (solution) = ? °C

Given: 75 g EG, 250 g water

Known: MM-EG = 62.0 g EG/mol (from atomic masses)

$$K_b = 0.51\text{C°}/m,\ m = \frac{\text{mol EG}}{\text{kg H}_2\text{O}}$$

T_b (solution) = 100°C + ΔT_b

Solution: To find T_b (solution), ΔT_b must be calculated. Dividing the problem into smaller parts, we have:

$$\Delta T_b = mK_b \quad (1)$$

$$m = \frac{?\ \text{mol EG}}{0.250\ \text{kg H}_2\text{O}} \quad (2)$$

$$\text{number of moles EG} = \left(\frac{1\ \text{mol}}{62.0\ \cancel{\text{g EG}}}\right)(75\ \cancel{\text{g EG}}) = 1.2\ \text{mol} \quad (3)$$

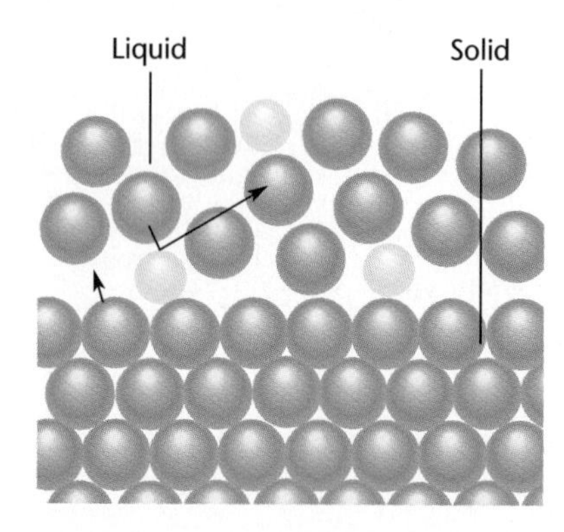

FIGURE 14.8 A solution freezing. Solvent molecules freeze at a reduced rate because of the interference of solute particles. Because the rate of melting of the solid solvent is unaffected by solute, melting occurs faster than freezing and the solid melts. Melting and freezing occur at equal rates again when the temperature is lowered.

$$m = \frac{1.2 \text{ mol EG}}{0.250 \text{ kg } H_2O} = 4.8\ m \qquad (2)$$

$$\Delta T_b = (4.8\ \cancel{m})\left(\frac{0.51\ C°}{\cancel{m}}\right) = 2.5\ C° \qquad (1)$$

Then, T_b (solution) $= 100°C + 2.5° = 102.5°C$

Note that 100°C, used to define the Celsius scale, is an exact number and does not limit the number of significant figures expressed in the answer.

Freezing Point of a Solution

Antifreeze is added to the radiator of an automobile to prevent freezing of the coolant during the winter season. An antifreeze works because the freezing point of water is lowered by the addition of a solute. The freezing point (T_f) of any solution is *lower* than the freezing point of the pure solvent.

We can understand why a solution has a lower freezing point than its solvent by considering the rates of melting of the solid solvent and freezing of the liquid. These rates are equal (an equilibrium condition, Section 13.3) at the freezing point of the pure solvent. However, for a solution, freezing of the solvent is slowed because solute particles occupy places in the solution otherwise occupied by solvent molecules in the pure solvent (Figure 14.8). Because melting of the solid solvent is unaffected by the solute in solution, melting occurs faster than freezing, and the solid melts. Equilibrium can be restored by lowering the temperature. Thus, the freezing point of the solution is lower than the freezing point of pure solvent.

At equilibrium, the rates of forward and reverse processes are equal.

The change in freezing point, ΔT_f, can be calculated using the equation:

$$\Delta T_f = mK_f$$

where K_f is the molal freezing point depression constant for the solvent. For water, K_f is 1.86 C°/*m*. Table 14.3 gives freezing points, boiling points, and values for K_f and K_b for several solvents.

For an aqueous solution,

$T_f = 0°C - \Delta T_f$

Solutions of Ionic Solutes

Assuming complete dissociation, 0.10 molal sodium chloride solution has a total ion concentration of 0.20 molal.

TABLE 14.3 Freezing Point and Boiling Point Data for Selected Solvents

Solvent	*Freezing Point, °C*	K_f, C°/*m*	*Normal Boiling Point, °C*	K_b, C°/*m*
acetone	−95	2.40	56	1.71
benzene	5.5	4.90	80	2.53
camphor	179	39.7	205	5.61
cyclohexane	6.5	20.1	81	2.79
ethanol	−117	1.99	78	1.22
water	0	1.86	100	0.51

$$NaCl(s) \xrightarrow{H_2O} Na^+(aq) \quad + \quad Cl^-(aq)$$

$$0.10\ m\ NaCl \quad 0.10\ m\ Na^+ + 0.10\ m\ Cl^-$$

$$\text{or } 0.20\ m \text{ total ion concentration}$$

Compared with the solution of a molecular solute at the same concentration, a solution of an ionic solute such as sodium chloride has a lower freezing point and a higher boiling point. Thus, the change in freezing point, ΔT_f, of 0.10 m NaCl is about twice the ΔT_f of a 0.10 m sugar solution. Furthermore, the ΔT_f and ΔT_b of 0.10 m $CaCl_2$ are about three times those of a 0.10 m sugar solution.

$CaCl_2$ is commonly used to melt ice on sidewalks and driveways.

Example 14.14

Estimate the freezing point of the ethylene glycol solution described in Example 14.13. ■

Solution To determine the freezing point of the solution, ΔT_f must be calculated.

$$T_f \text{ (solution)} = 0°C - \Delta T_f \qquad (1)$$

$$\Delta T_f = mK_f = (4.8\ \cancel{m}) \left(\frac{1.86\ C°}{\cancel{m}}\right) = 8.9\ C° \qquad (2)$$

$$T_f \text{ (solution)} = 0° - 8.9° = -8.9°C \qquad (1)$$

Practice Problem 14.7 Calculate the freezing point of a solution of 1.56 g of naphthalene ($C_{10}H_8$) in 75.0 g of cyclohexane. Refer to Table 14.3 for needed data.

■ Osmosis and Osmotic Pressure

Osmosis is the net movement of a solvent through a semipermeable membrane from a dilute solution to a more concentrated solution. A semipermeable membrane is a membrane that allows solvent molecules to pass through it, but not solute molecules. Many plant and animal tissues are semipermeable membranes.

Osmosis can be demonstrated by using a hollowed-out carrot filled with molasses and placed in a glass of water (Figure 14.9). Water molecules pass through the carrot in both directions, but faster from the water into the molasses. The molasses is diluted and the liquid level rises in the tube in the center of the carrot. The pressure exerted by the liquid in the tube causes an increase in the rate the water molecules pass from the molasses solution into the water. Eventually, the two rates become equal, the system reaches equilibrium, and no further change occurs in the liquid level in the tube.

The pressure required to stop osmosis is called the **osmotic pressure.** Osmotic pressure depends upon the concentration of the solution involved, but like other colligative properties, it is unrelated to the type of solute. In other words, using a solution more concentrated than molasses would cause liquid to rise higher in the tube, resulting in higher osmotic pressure.

FIGURE 14.9 Osmosis can be demonstrated using a hollowed-out carrot filled with molasses and fitted with a glass tube. When the carrot is placed in a glass of water, water passes through the carrot into the molasses, causing the liquid to rise in the tube. The difference between the water level in the glass and the liquid level in the tube is equal to the osmotic pressure.

14.7 Colloids

Learning Objectives

- Describe and give examples of various types of colloids.
- Describe the Tyndall effect.

A solution is a homogeneous mixture of molecule- or ion-size particles dispersed in a solvent. The properties of a solution are distinctly different from those of a heterogeneous mixture. For example, solute particles do not settle on standing, in contrast to those of a heterogeneous suspension such as finely divided sand in water. Furthermore, sand can be separated from water by filtration (Section 2.6), unlike the solute and solvent of a solution.

FIGURE 14.10 *Left:* Like a solution, a colloid passes through an ordinary filter. However, a colloid scatters light (the Tyndall effect), a property that distinguishes it from a solution. *Right:* The light from the setting sun produces a Tyndall effect.

CHEMICAL WORLD

Second-Hand Colloids

One aspect of a colloid is that the particles comprising it have surface—a lot of surface. Try a thought experiment. Imagine a 1-centimeter cube. The area of each surface is 1 cm^2, which, multiplied by a cube's six faces, gives a total area of 6 cm^2. Cut the cube in half. That makes two new faces: total area of the cut cube is 8 cm^2. Now (to make the math easy) slice the pieces parallel to the 1-cm^2 sides. With every slice you add two more square centimeters of surface. Now forget the math and chop the pieces about a trillion times (colloid particles are *small*), and you'll end up with a surface area of about 1.5 acres. From a 1-cm^2 cube!

Now, the important thing about surfaces is that molecules tend to stick to them. Molecules that stick to the surfaces of a colloid in air (an aerosol) remain suspended in the air, attached to the colloidal particles.

Remember radon, the radioactive gas? We discussed it in the Chemical World essay in Chapter 1. What we didn't mention is that radon is actually fairly harmless by itself. Like most other gases, you inhale it one moment and exhale it the next. However, the trouble with radon is that it decays into radioactive particles such as polonium. Radioactive particles behave differently from gases. If radioactive particles get into your lungs, they stay there for a while and emit harmful radiation.

Still, the radioactive particles won't necessarily get into your lungs because most of them will stick to the surfaces of the walls, floor, and furniture in your house. Unless, of course, your house contains an aerosol. Then they stick to the colloidal surfaces and stay in the air. And you breathe the aerosol.

Radon levels in the air in your home can be measured with a radon detector such as the one shown here.

Now, guess what colloid is found most often in the air of houses and buildings. That's right—tobacco smoke. You've probably heard people doubt whether second-hand smoking can be harmful. Well, five hours after a cigarette has been burned in a room, the concentration of polonium particles in the air has increased by 25%, and this increase remains constant for about 9 hours before it begins to trail off. When a pack-a-day smoker smokes those 20 cigarettes in 24 hours, the concentration of polonium and other radioactive particles in a room increases by 300%.

Direct cigarette smoking is still the leading cause of lung cancer. Radon comes in second. It sounds as if nonsmokers really have something to complain about.

A **colloid** is a homogeneous mixture containing suspended dispersed particles that are aggregates of molecules, ions, or atoms. Although generally larger in size than solute particles, the aggregates are small enough that their random motions can keep them suspended. In many respects, colloids resemble solutions, and the suspended particles are not visible by ordinary means. Like solutions, liquid colloids cannot be separated by ordinary filtration. However, liquid colloids scatter light, and this allows them to be distinguished easily from liquid solutions. A beam of light shining into a colloid is easily visible from a side view (Figure 14.10). Light scattering by a colloid is known as the **Tyndall effect.**

Colloids exist in all three physical states (Table 14.4). Some colloids have special names such as **sol,** a liquid colloid in which a solid is dispersed in a liquid.

TABLE 14.4 Colloids

Dispersing Medium	*Dispersed Substance*	*Type of Colloid*	*Examples*
gas	solid	aerosol	smoke
gas	liquid	aerosol	fog, clouds, aerosol sprays
liquid	solid	sol	paint, milk of magnesia, gelatin
liquid	liquid	emulsion	milk, mayonnaise
liquid	gas	foam	soap suds, shaving cream
solid	solid	solid sol	glass, gems
solid	liquid	solid emulsion	butter, cheese
solid	gas	solid foam	marshmallow, foamed plastics (styrofoam)

Paints are examples of sols. A **gel,** such as the familiar fruit-flavored gelatin served in salads and desserts, is similar to a sol except that it has a semirigid structure. Milk and mayonnaise are colloids in which one liquid is dispersed in another and are examples of **emulsions.** An **aerosol** has colloidal particles dispersed in a gas. Examples include smoke, in which solid particles are dispersed in combustion gases, and fog, in which tiny water droplets are suspended in air. Smog is a mixture of air with water droplets, solid particles, and polluting gases. Many common household products are aerosol sprays.

Chapter Review

Directions: From the list of choices at the end of the Chapter Review, choose the word or term that best fits in each blank. Use each choice only once. The answers appear at the end of the chapter.

A __________ (1) is a homogeneous mixture of one or more substances uniformly dispersed in another substance.

Section 14.1 A solvent is the __________ (2) component of a solution, whereas the __________ (3) is the minor component. The __________ (4) of a solution is an expression of the amount of solute in a given amount of solvent or solution. A solution that contains a large amount of solute is said to be __________ (5), whereas a solution that has a small amount of solute compared with the amount of solution is __________ (6). A __________ (7) solution contains the most solute possible at a given temperature. In a saturated solution, dissolved and undissolved solute are in a state of dynamic __________ (8). The concentration of a saturated solution at any given temperature is the __________ (9) of the solute. A solution having a concentration less than the solubility of the solute is said to be __________ (10).

Section 14.2 Solutions can be solids, liquids, or gases. In general, the physical properties of a solution closely resemble those of the __________ (11).

Section 14.3 The solubility of a solute in water depends upon intermolecular forces and __________ (12). The solubility of a molecular substance depends upon __________ (13) bonding with water. When an ionic compound such as NaCl dissolves in water, the ionic attractions in the crystal are replaced by __________ (14) attractions in solution. Sodium chloride is insoluble in solvents in which ion–dipole attractions are not possible (that is, __________ (15) solvents). Gases dissolve in water if hydrogen bonding is possible or if __________ (16) of gas molecules occurs. In the latter case, ion–dipole attractions play an important role just as they do with ionic solutes.

Some liquids, such as ethyl alcohol and acetone, dissolve in all proportions and are said to be ______ (17) with water. As the hydrophobic part of the molecules increases in size, solubility decreases. Thus, although some liquids (such as butyl alcohol) are slightly soluble in water, they are __________ (18).

The solubility of most substances is affected by temperature. In general, the solubility of a solid solute __________ (19) with increasing temperature. However, gases are always less soluble at __________ (20) temperatures. Furthermore, the solubility of a gas is directly proportional to its __________ (21) pressure above a solution.

The rate at which a solid dissolves depends upon the temperature, its __________ (22) size, and __________ (23).

Section 14.4 A solution concentration is always expressed as a __________ (24) of a quantity of solute to a quantity of solvent or solution. Parts per hundred is generally expressed as __________ __________ (25). The concentration of a dilute solution may be expressed in parts per million, abbreviated __________ (26). The concentration most frequently used in laboratory work is __________ (27) which has units of __________ (28).

Section 14.5 Many chemical reactions occur in solution, and solution concentrations can be used as __________ (29) factors in problem solving. For example, molarity, or mol solute/L soln, can be used to find the number of moles of solute (a reactant) from the __________ (30) of a solution. Then, the quantity of product can be calculated from the quantity of reactant.

Section 14.6 A solution property that depends upon the concentration of solute particles but is independent of the nature of the solute is called a __________ (31) property. Compared with the properties of the solvent, the vapor pressure of the solution is __________ (32), its boiling point is higher, and its __________ (33) point is lower. Calculations of changes in boiling points and freezing points of solutions involve the concentration expressed as __________ (34), the number of moles of solute per kilogram of solvent.

The net movement of solvent such as water through a semipermeable membrane from a dilute solution into a more concentrated solution is called __________ (35). The pressure required to stop osmosis is called the __________ (36) pressure.

Section 14.7 A homogeneous mixture that contains particles that are aggregates of molecules, ions, or atoms dispersed in a solid, liquid, or gas is called a __________ (37). Although the dispersed particles are larger than those in a solution, they are small enough that they do not settle on standing and cannot be separated by ordinary __________ (38). A colloid can be distinguished from a solution by the __________ (39) effect. Different types of colloids include a solid dispersed in a liquid, called a __________ (40), and a solid or liquid dispersed in a gas, called an __________ (41).

Choices: aerosol, colligative, colloid, concentrated, concentration, conversion, dilute, equilibrium, filtration, freezing, higher, hydrogen, immiscible, increases, ion–dipole, ionization, lower, major, mass percent, miscible, molality, molarity, mol solute/L soln, nonpolar, osmosis, osmotic, partial, particle, ppm, ratio, saturated, sol, solubility, solute, solution, solvent, stirring, temperature, Tyndall, unsaturated, volume

Study Objectives

After studying Chapter 14, you should be able to:

Section 14.2

1. Describe the general properties of solutions.

Section 14.3

2. Identify the factors that affect the solubility of a solute in water or other solvent.
3. Explain how to prepare a supersatured solution.
4. Identify the factors that influence the rate at which a solid solute dissolves.

Section 14.4

5. Calculate mass percent, parts per million, and molarity from solution data.
6. Calculate the mass or number of moles of solute from data for a solution.
7. Calculate the mass of solvent or volume of solution from the mass or number of moles of solute and the concentration of a solution.
8. Solve dilution problems for either an unknown volume or an unknown concentration of a solution.

Section 14.5

9. Use concentrations of solutions to solve problems involving reactions occurring in aqueous solution.

Section 14.6

10. Compare the vapor pressure, boiling point, freezing point, and osmotic pressure of solutions of various concentrations.
11. Calculate the boiling point and the freezing point of a solution from its concentration (molality).

Section 14.7

12. Describe and give examples of various types of colloids.
13. Describe the Tyndall effect.

Key Terms

Review the definition of each of the terms listed here by chapter section number. You may use the glossary if necessary.

14.1 concentrated, concentration, dilute, saturated, solute, solubility, solvent, unsaturated

14.3 immiscible, miscible, supersaturated

14.4 mass percent, molarity, parts per million

14.6 colligative property, molality, osmosis, osmotic pressure

14.7 aerosol, colloid, emulsion, gel, sol, Tyndall effect

Questions and Problems

Answers to questions and problems with blue numbers appear in Appendix C. Asterisks indicate the more challenging problems.

Solution Terminology (Section 14.1)

1. What is the most common solvent for liquid solutions?
2. What kind of solute particles are present in a solution of sugar in water?
3. What kind of solute particles are present in a solution of sodium bicarbonate in water?
4. A recipe called for one teaspoon of sugar dissolved in a cup of water. Another recipe, for pancake syrup, called for one cup of sugar mixed with one cup of water. Which solution was dilute? Which one was concentrated?
5. Explain how to prepare a saturated NaCl solution. How could this solution be made unsaturated?
6. Explain or describe the condition of dynamic equilibrium for a saturated NaCl solution.

Solubility (Section 14.3)

7. What factors influence the solubility of a solid or liquid in water?
8. Explain what is meant by two miscible liquids. Give examples.
9. Water and vegetable oil are immiscible. Explain what this means.
10. Ethylene glycol, $HOCH_2CH_2OH$, is miscible with water, but hexane, C_6H_{14}, is immiscible. Explain the difference in solubility of these solutes.
11. Potassium iodide, KI, is readily soluble in water. Explain what happens when KI dissolves and why it is soluble.
12. Explain why NaCl is insoluble in hexane, C_6H_{14}.
13. Why is ammonia gas, NH_3, very soluble in water but nitrogen gas, N_2, is practically insoluble?
14. According to the kinetic-molecular theory, there are no attractive forces between the molecules of an ideal gas. Why, then, is oxygen gas only slightly soluble in water?
15. Hydrogen chloride, HCl(g), is very soluble in water even though there is no hydrogen bonding with water. Explain.
16. What kind of attractive forces are important in a solution of HBr in water?
17. A mixture of 40.0 g of NaCl and 100.0 g of water was stirred until no more salt would dissolve. After the undissolved solid was separated by filtration and dried, its mass was found to be 3.61 g. What was the solubility of the NaCl expressed in g NaCl/100.0 g of water, at the temperature of the experiment?
18. After thoroughly stirring 56.0 g of ammonium nitrate with 50.0 g of water, 8.60 g of solid was recovered by filtration. What was the solubility of the ammonium nitrate expressed in g NH_4NO_3/100.0 g of water, at the temperature of the experiment?

19. Describe the effect of increasing the temperature on the solubility of most solids in water.
20. Describe the effect of increasing the temperature on the solubility of gases in water.
21. What is a supersaturated solution?
22. Explain how to prepare a supersaturated solution.
23. What usually happens when a supersaturated solution is allowed to stand for a period of time?
24. When a can of soft drink is opened, a release of pressure can be heard. What is the gas that is released? Why is the can pressurized before it is sealed?
25. Why do goldfish die when placed in a fishbowl containing cool water that was previously boiled?
26. What factors increase the rate at which a solid solute dissolves in water?
27. Why does a finely divided solid dissolve more rapidly than one composed of coarse particles?
28. Explain why increasing the temperature increases the rate at which a solid dissolves in water.

Concentrations of Solutions (Section 14.4)

Mass Percent

29. What mass of dextrose is dissolved in 100. g of a 5.0% aqueous solution?
30. Calculate the mass percent of KI in each of the following solutions:
    a. 1.0 g KI dissolved in 35 g water
    b. 2.0 g KI dissolved in 48 g water
    c. 3.00 g KI dissolved in 105 g water
    d. 0.155 g KI dissolved in 11.0 g water
31. Find the mass of NaCl in each of the following solutions:
    a. 75 g of 5.0% NaCl solution
    b. 125 g of 1.50% NaCl solution
    c. 55 g of 0.25% NaCl solution
    d. 675 g of 12.0% NaCl solution
32. Glucose (dextrose) is a nutrient used in intravenous feeding solutions. Calculate the mass of glucose in 1.50 L of a 5.0% solution. The density of the solution is 1.06 g/mL.
33. Calculate the mass of water and the mass of NaOH needed to prepare 100. g of a 5.0% solution.
34. What mass of water and mass of NaCl are needed to prepare 600. g of a 10.0% solution?
35. Calculate the mass of $KMnO_4$ and mass of water needed to prepare 25.0 g of 2.0% solution.
36. What is the mass of water in 20.0 mL of 96% H_2SO_4 solution that has a density of 1.82 g/mL?
37. What is the mass of HCl in 20.0 mL of 36.0% HCl solution that has a density of 1.18 g/mL?
38. Find the mass of HNO_3 in 5.0 mL of 32% nitric acid solution that has a density of 1.19 g/mL.
39. Vinegar is a 5.0% solution of acetic acid ($HC_2H_3O_2$). What mass of acetic acid is in 5.0 mL (1 tsp) of vinegar? The density of vinegar is 1.006 g/mL.

Parts per Million

40. The concentration of arsenic in a water sample was 0.050 ppm. Find the mass of arsenic in 2.0 L of water. Assume the density of solution is 1.0 g/mL.
41. Nitrates from fertilizers contaminate many lakes and streams in agricultural areas. Calculate the mass of nitrate ion (NO_3^-) in 1.0 L of water in which the concentration is 25 ppm. The density of the solution is 1.0 g/mL.
42. Find the mass of lead in 10.0 L of water in which the concentration of lead is 0.030 ppm. The density of the solution is 1.0 g/mL.
43. What is the concentration in ppm of a solution of 2.0 mg of fluoride ion in 0.50 L of water? Density = 1.0 g/mL.
44. Calculate the concentration (ppm) of cyanide ion (CN^-) in 1.0 L of water containing 0.30 mg of the ion. The density of the solution is 1.0 g/mL.
45. Chromium contamination of water resulted from the manufacture of chrome bumpers for cars. What is the concentration (ppm) of chromium in water containing 0.060 mg of chromium per liter? The density of the solution is 1.0 g/mL.
46. A U.S. Public Health Service standard for dissolved organic matter in water is 0.20 ppm. A sample of water was found to contain 3.0×10^{-4} g of dissolved organics per liter of water. Does this concentration exceed the limit set by the standard? The density is 1.0 g/mL.

Molarity

47. Find the molarity of a solution that contains 4.50 mol of sulfuric acid in 325 mL of solution.
48. What is the molarity of a solution that contains 6.2 mol of phosphoric acid in 175 mL of solution?
49. What is the molarity of a solution of hydrochloric acid that contains 36.0 g of HCl in 125 mL of solution?
50. Calculate the molarity of a nitric acid solution containing 26.0 g HNO_3 in 100. mL of solution.
51. Find the molarity of a solution of NaOH that contains 15.0 g of solute in 75.0 mL of solution.
52. Calculate the molarity of each of the following solutions:
    a. 1.3 g of KBr in 250. mL of solution
    b. 2.65 g of NaOH in 65.0 mL of solution
    c. 1.5 g of $NaHCO_3$ in 40.0 mL of solution
    d. 1.20 kg of $NaNO_3$ in 5.5 L of solution
53. Find the molarity of each of the following solutions:
    a. 2.35 g of KCl in 35.0 mL of solution
    b. 10.6 g of $Ca(NO_3)_2$ in 125 mL of solution
    c. 3.50 g of $BaCl_2$ in 250. mL of solution
    d. 4.75 g of KOH in 25.0 mL of solution

54. Find the number of moles of solute in each of the following solutions:
   a. 10.0 mL of 6.0 M HCl solution
   b. 20.0 mL of 3.0 M H_2SO_4 solution
   c. 45.0 mL of 1.25 M $NaHCO_3$ solution
   d. 125 mL of 0.100 M $AgNO_3$ solution
55. What mass of solute is needed to prepare 100.0 mL of each of the following solutions?
   a. 3.0 M KI solution
   b. 1.50 M Na_2SO_4 solution
   c. 1.1 M K_3PO_4 solution
   d. 1.25 M $BaCl_2$ solution
56. What volume of each of the following solutions contains 10.0 g of solute?
   a. 1.50 M $CaCl_2$ solution
   b. 0.65 M KBr solution
   c. 1.80 M $NaC_2H_3O_2$ solution
   d. 0.15 M $Fe(NO_3)_3$ solution
57. What is the molarity of a solution prepared by dissolving 75.0 g of glucose, $C_6H_{12}O_6$, in water to make 1.20 L of solution?
58. How many moles of solute are found in each of the following solutions?
   a. 15.0 mL of 17.0 M $HC_2H_3O_2$ solution
   b. 16.58 mL of 0.678 M NaOH solution
   c. 22.35 mL of 0.132 M HCl solution
   d. 14.59 mL of 0.244 M $H_2C_2O_4$ solution
59. Find the number of moles of sucrose in 65.0 mL of 0.750 M solution.
60. What is the mass of solute in each of the following solutions?
   a. 25.8 mL of 0.0875 M $CaCl_2$ solution
   b. 13.6 mL of 1.22 M HCl solution
   c. 20.0 mL of 3.0 M NaOH solution
   d. 10.0 mL of 1.20 M $Fe(NO_3)_3$ solution
61. What mass of solute is needed to prepare each of the following solutions?
   a. 100. mL of 1.20 M $Ca(NO_3)_2$ solution
   b. 250. mL of 2.00 M NaOH solution
   c. 500. mL of 0.250 M NaCl solution
   d. 1.50 L of 3.00 M Na_2SO_3 solution
62. What volume of each of the following solutions will contain 0.250 mol of solute?
   a. 1.20 M $Ca(NO_3)_2$ solution
   b. 2.00 M NaOH solution
   c. 0.250 M NaCl solution
   d. 3.00 M Na_2SO_3 solution
63. What volume of 0.0165 M HCl solution is needed to provide 0.00235 mol of HCl?
64. What volume of 0.867 M acetic acid solution will provide 0.0125 mol of acetic acid ($HC_2H_3O_2$)?

Dilution of Solutions

65. What is the molarity of a solution prepared by diluting 10.0 mL of 6.00 M sulfuric acid solution to a final volume of 25.0 mL?
66. What is the molarity of the dilute solution prepared by diluting
   a. 10.0 mL of 6.00 M hydrochloric acid to 50.0 mL.
   b. 2.00 mL of 3.00 M NaOH solution to 25.0 mL.
   c. 1.00 mL of 6.00 M acetic acid solution to 25.0 mL.
   d. 5.00 mL of 0.875 M potassium iodide solution to 25.0 mL.
67. What is the concentration of a solution prepared by diluting 10.0 mL of 15 M ammonia solution to 250. mL?
68. Find the concentration of a solution prepared by diluting 5.00 mL of 3.00 M phosphoric acid solution to 25.0 mL.
69. What volume of 0.125 M $AgNO_3$ solution is needed to prepare 25.0 mL of 0.0100 M solution?
70. What volume of 6.00 M sulfuric acid solution is needed to prepare 100.0 mL of 2.00 M solution?
71. What volume of 12 M HCl solution is needed to prepare 1.00 L of 3.00 M solution?

Reactions in Solution (Section 14.5)

72. What is the theoretical yield of barium sulfate ($BaSO_4$) formed in the reaction of 22.3 mL of 0.0500 M barium hydroxide with excess 0.122 M sulfuric acid solution?
73. Calculate the theoretical yield of calcium sulfate formed in the reaction of 18.6 mL of 6.75 M sulfuric acid with excess CaO (lime). The equation for this reaction is:

$$CaO(s) + H_2SO_4(aq) \longrightarrow H_2O(l) + CaSO_4(s)$$

74. What mass of oxalic acid dihydrate, $H_2C_2O_4 \cdot 2H_2O$ (a diprotic acid), is needed to neutralize 27.3 mL of 0.102 M NaOH solution?
75. What mass of $CaCl_2$ is needed to react with 10.0 mL of 0.112 M $AgNO_3$ solution? The equation (unbalanced) for the reaction is:

$$CaCl_2(aq) + AgNO_3(aq) \longrightarrow AgCl(s) + Ca(NO_3)_2(aq)$$

76. What mass of calcium oxalate (CaC_2O_4) can be prepared in a reaction of 25.0 mL of 0.244 M $H_2C_2O_4$ solution with excess $CaCl_2$?
77. What volume of 0.765 M NaOH solution is needed to neutralize 0.250 mol of sulfuric acid solution?
78. What volume of 0.125 M HCl solution is needed to neutralize 15.6 mL of 0.143 M KOH solution?
79. Citric acid ($H_3C_6H_5O_7$), found in lemon juice, is a triprotic acid that can be used to remove rust stains (Fe_2O_3) from plumbing fixtures. What volume of 0.122 M citric acid solu-

tion will dissolve 1.23 g of rust? The equation for the reaction is

$$Fe_2O_3(s) + 4H_3C_6H_5O_7(aq) \longrightarrow 6H^+ + 2Fe(C_6H_5O_7)_2{}^{3-}(aq) + 3H_2O(l)$$

80. What mass of magnesium hydroxide ($Mg(OH)_2$, milk of magnesia) is needed to neutralize 25.0 mL of 0.111 M HCl solution (stomach acid)?
81. What volume of 0.122 M citric acid (problem 79) is needed to completely neutralize 14.2 mL of 0.135 M NaOH solution?
82. Calculate the mass of NaOH needed to neutralize 16.2 mL of 0.322 M sulfuric acid solution.
83. What mass of $NaHCO_3$ (baking soda) is needed to neutralize 25.0 mL of 0.234 M HCl solution? The equation for the reaction is

$$NaHCO_3(s) + HCl(aq) \longrightarrow NaCl(aq) + CO_2(g) + H_2O(l)$$

Colligative Properties of Solutions (Section 14.6)

84. Calculate the freezing point of a solution of 175 g of ethylene glycol ($C_2H_6O_2$) dissolved in 0.500 kg of water. See Table 14.3.
85. What is the boiling point of the solution described in problem 84? See Table 14.3.
86. Estimate the freezing point of a solution containing 225 g of sucrose ($C_{12}H_{22}O_{11}$) in 750. g of water. See Table 14.3.
87. What is the boiling point of the solution described in problem 86? See Table 14.3.
88. What is the molality of an aqueous solution of ethylene glycol ($C_2H_6O_2$) having a freezing point of $-8.6°C$?
89. What is the molality of an aqueous solution of ethylene glycol ($C_2H_6O_2$) having a boiling point of 103.2°C?
90. Estimate the boiling point of a solution of 2.50 g of urea (CH_4N_2O) in 50.0 g of ethanol. See Table 14.3.
91. What is the freezing point of the solution described in problem 90? See Table 14.3.
92. Consider solutions separated from water by a semipermeable membrane. Which solution would give the higher osmotic pressure:
    a. 10.0 g urea (CH_4N_2O) dissolved in 100. g water, or
    b. 10.0 g sucrose ($C_{12}H_{22}O_{11}$) dissolved in 100. g water?
93. Consider solutions separated from water by a semipermeable membrane. Which solution would give the higher osmotic pressure:
    a. 10.0 g urea (CH_4N_2O) dissolved in 100. g water, or
    b. 50.0 g sucrose ($C_{12}H_{22}O_{11}$) dissolved in 100. g water?

Colloids (Section 14.7)

94. Explain how a colloid differs from a solution. What properties of colloids and solutions are similar?
95. What is the Tyndall effect?
96. Explain or describe a smoke. What is another name for a smoke?
97. Mayonnaise and homogenized milk are examples of colloids. What type of colloid are they?
98. Spray cans of lubricants are available in auto parts stores. What type of colloid are these products?

Additional Problems

*99. Calculate the molarity of acetic acid ($HC_2H_3O_2$, found in vinegar) in a 5.00% solution. Density = 1.006 g/mL.

*100. Calculate the molarity of a 25.0% solution of sulfuric acid solution. Density = 1.18 g/mL.

*101. What is the molarity of a 6.00% solution of $NH_3(aq)$? Density = 0.960 g/mL.

*102. A 6.0-M solution of NaOH has a density of 1.22 g/mL. What is the mass % of NaOH in the solution?

*103. A 6.0-M solution of HCl has a density of 1.10 g/mL. What is the mass % of HCl in the solution?

*104. A 32% solution of nitric acid was 6.0 M HNO_3. What was the density of the solution?

*105. A 34% solution of acetic acid was 6.0 M $HC_2H_3O_2$. What was the density of the solution?

*106. Calculate the mass % of Na^+ in 0.050 M Na_2SO_4. The density of the solution is 0.998 g/mL.

*107. A 22.13-mL sample of 0.250 M NaOH solution was mixed with 18.65 mL of 0.400 M H_2SO_4 solution. How many moles of water formed in the reaction?

*108. What mass of water would be formed upon mixing 10.00 mL of 0.675 M NaOH with 12.00 mL of 0.500 M HCl?

Solutions to Practice Problems

PP 14.1

Find: mass (g) solution = ?

Given: 5.0% dextrose or $\dfrac{5.0 \text{ g dextrose}}{100 \text{ g solution}}$, 23 g dextrose

Solution:

$$\text{mass (g) solution} = \left(\frac{100 \text{ g solution}}{5.0 \text{ g dextrose}}\right)(23 \text{ g dextrose}) = 4.6 \times 10^2 \text{ g solution}$$

PP 14.2

$$\text{ppm} = \left(\frac{0.106 \text{ g benzene}}{748 \text{ g gasoline}}\right) \times 10^6 = 1.42 \times 10^3 \text{ ppm}$$

PP 14.3

Find: $M \left(\frac{\text{mol KI}}{\text{L soln}}\right) = ?$

Given: 45.8 g KI, 500. mL (0.500 L) of solution

Known: *MM*-KI = 166 g/mol KI (from atomic masses)

Solution: The volume of the solution is given, but the number of moles of KI must be calculated:

$$\text{number of moles KI} = \left(\frac{1 \text{ mol}}{166 \cancel{\text{g KI}}}\right)(45.8 \cancel{\text{g KI}}) = 0.276 \text{ mol KI} \quad (1)$$

Then, to calculate molarity:

$$M\left(\frac{\text{mol KI}}{\text{L soln}}\right) = \frac{0.276 \text{ mol KI}}{0.500 \text{ L soln}} = 0.552 \text{ M KI} \quad (2)$$

PP 14.4

Find: mass (g) $Hg(NO_3)_2$ = ?

Given: 225 mL (0.225 L) of solution, 0.150 mol $Hg(NO_3)_2$/L soln

Known: *MM*-$Hg(NO_3)_2$ = 325 g/mol (from atomic masses)

Solution: Using dimensional analysis gives

$$\text{mass (g) } Hg(NO_3)_2 = \left(\frac{325 \text{ g}}{\cancel{\text{mol } Hg(NO_3)_2}}\right)\left(\frac{0.150 \cancel{\text{mol } Hg(NO_3)_2}}{\cancel{\text{L soln}}}\right)(0.225 \cancel{\text{L soln}}) = 11.0 \text{ g}$$

PP 14.5

Find: $M_d \left(\frac{\text{mol } H_2SO_4}{\text{L soln}}\right) = ?$

Given: 8.50 M or 8.50 mol H_2SO_4/L soln, V_c = 10.0 mL, V_d = 50.0 mL

Known: The number of moles of H_2SO_4 does not change when the solution is diluted.

Solution: The volume of the dilute solution is given. The number of moles of H_2SO_4 must be calculated.

$$\text{number of moles } H_2SO_4 = \left(\frac{8.50 \text{ mol } H_2SO_4}{\cancel{\text{L soln}}}\right)(0.0100 \cancel{\text{L soln}}) = 0.0850 \text{ mol } H_2SO_4 \quad (1)$$

$$M_d\left(\frac{\text{mol } H_2SO_4}{\text{L soln}}\right) = \left(\frac{0.0850 \text{ mol } H_2SO_4}{0.0500 \text{ L soln}}\right) = 1.70 \text{ M} \quad (2)$$

PP 14.6

Begin by outlining the problem under the balanced equation:

$AgNO_3(aq) + HCl(aq) \longrightarrow AgCl(s) + HNO_3(aq)$

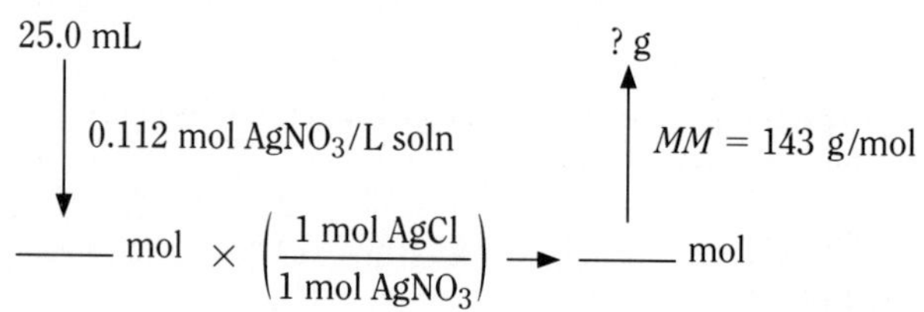

Because excess HCl was used, it is not the limiting reactant and does not enter into the calculation. The calculation involves three steps. Following the outline, first calculate the number of moles of $AgNO_3$, then the number of moles of AgCl, and finally the mass of AgCl:

$$\text{number of moles } AgNO_3 = \left(\frac{0.112 \text{ mol } AgNO_3}{\cancel{\text{L soln}}}\right)(0.0250 \cancel{\text{L soln}}) = 0.00280 \text{ mol} \quad (1)$$

$$\text{number of moles AgCl} = \left(\frac{1 \text{ mol AgCl}}{1 \cancel{\text{mol } AgNO_3}}\right)(0.002\ 80 \cancel{\text{mol } AgNO_3}) = 0.002\ 80 \text{ mol AgCl} \quad (2)$$

$$\text{mass (g) AgCl} = \left(\frac{143 \text{ g}}{\cancel{\text{mol AgCl}}}\right)(0.002\ 80 \cancel{\text{mol AgCl}}) = 0.400 \text{ g} \quad (3)$$

PP 14.7

Find: T_f (solution) = ? °C

Given: 1.56 g $C_{10}H_8$, 75.0 g cyclohexane

Known: K_f = 20.1 C°/*m*, T_f (cyclohexane) = 6.5°C (Table 14.3)

MM-$C_{10}H_8$ = 128 g/mol (from atomic masses)

Solution:

$$T_f \text{ (solution)} = 6.5°C - \Delta T_f \quad (1)$$

$$\Delta T_f = K_f m \quad (2)$$

The molality was not given and must be calculated.

$$m = \frac{\text{mol } C_{10}H_8}{\text{kg cyclohexane}} \quad (3)$$

$$\text{number of moles } C_{10}H_8 = \left(\frac{1 \text{ mol}}{128 \cancel{\text{g}}}\right)(1.56 \cancel{\text{g}}) = 0.0122 \text{ mol } C_{10}H_8 \quad (4)$$

Substitute in equations (3), (2), and (1) in succession:

$$m = \frac{0.0122 \text{ mol } C_{10}H_8}{0.075 \text{ kg cyclohexane}} = 0.163\ m \quad (3)$$

$$\Delta T_f = K_f m = \left(\frac{20.1\ C^\circ}{\cancel{m}}\right)(0.163\cancel{m}) = 3.28\ C^\circ \qquad (2)$$

$$T_f\ (\text{solution}) = 6.5^\circ - 3.28^\circ = 3.2^\circ C \qquad (1)$$

Note how starting with the unknown in equation (1) allows you to develop a plan for dividing the problem into several parts.

Answers to Chapter Review

1. solution
2. major
3. solute
4. concentration
5. concentrated
6. dilute
7. saturated
8. equilibrium
9. solubility
10. unsaturated
11. solvent
12. temperature
13. hydrogen
14. ion–dipole
15. nonpolar
16. ionization
17. miscible
18. immiscible
19. increases
20. higher
21. partial
22. particle
23. stirring
24. ratio
25. mass percent
26. ppm
27. molarity
28. mol solute/L soln
29. conversion
30. volume
31. colligative
32. lower
33. freezing
34. molality
35. osmosis
36. osmotic
37. colloid
38. filtration
39. Tyndall
40. sol
41. aerosol

15

Acids and Bases

Contents

Controlling the pH of the water in an aquarium is essential in maintaining the health of tropical fish.

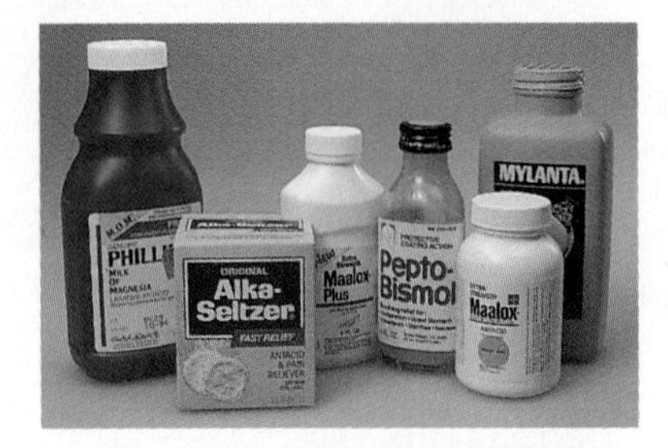

Some common stomach antacids.

A number of household cleaners contain acids or bases.

You Probably Know . . .

- Acids are found in fruits and vegetables and other food. The sour taste of lemons, vinegar, and sourdough is due to the acids in each.
- Gastric juice in the stomach contains hydrochloric acid, which is largely responsible for its sour taste.
- Common stomach antacids contain bases such as magnesium hydroxide (milk of magnesia) and aluminum hydroxide.
- A number of household cleaners contain acids or bases. Examples include household ammonia (NH_3), a base, tile cleaners (H_3PO_4), and drain cleaners (NaOH, lye).
- A car battery contains sulfuric acid (H_2SO_4).
- Acid rain, containing sulfuric and nitric acids, is a serious environmental problem.

Why Should We Study Acids and Bases?

Acids and bases are some of the most common and most useful chemical substances. They are found in food, household products, the environment, and

TABLE 15.1 Examples of Acids and Bases

food	lemons, oranges: citric acid, $H_3C_6H_5O_7$; ascorbic acid, $H_2C_6H_6O_6$
	vinegar: acetic acid, $HC_2H_3O_2$
	baking soda: sodium hydrogen carbonate, $NaHCO_3$ (acid or base)
	soft drinks: carbonic acid, H_2CO_3
household	ammonia: NH_3 (base)
	drain cleaner: sodium hydroxide, NaOH (base)
	tile cleaner: phosphoric acid, H_3PO_4
	washing soda: sodium carbonate, Na_2CO_3 (base)
automobile	battery acid: sulfuric acid, H_2SO_4
agriculture	ammonia fertilizer: NH_3 (base)
	lime to treat the soil: calcium oxide, CaO (base)
environment	acid rain: sulfuric acid, H_2SO_4; nitric acid, HNO_3
human body	gastric juice: hydrochloric acid, HCl
	blood and other body fluids (acidity is carefully controlled for proper bodily function)
medicine	antacids: magnesium hydroxide, $Mg(OH)_2$ (milk of magnesia); aluminum hydroxide, $Al(OH)_3$; acetylsalicylic acid, $HC_9H_7O_4$ (aspirin)
industry	among the top ten chemicals produced:
	acids: H_2SO_4, H_3PO_4
	bases: NH_3, NaOH, CaO

our bodies. We can see from the examples in Table 15.1 that acids and bases have a significant impact on our lives. Acids and bases also are important in many of the chemical reactions in our bodies. Furthermore, acids and bases have significant roles in many laboratory reactions and industrial processes. Five of the top ten chemicals produced in the United States are acids or bases. Some of these will be mentioned as we discuss the properties of acids and bases.

Properties of Acids and Bases

Learning Objectives

- Summarize the properties of acids.
- Summarize the properties of bases.

From common foods such as lemons and vinegar, we have learned to recognize the sour taste of acids. Citric acid is found in all citrus fruits, and acetic acid is responsible for the taste of vinegar. Gastric juice (in the stomach) contains hydrochloric acid. Moreover, acids cause color changes in certain organic compounds called acid–base indicators. An example is litmus, which turns red when moistened with an acid (Figure 15.1). Acids react with carbonates and bicarbonates to liberate carbon dixoide, a reaction important in baking (Figure 15.2). Acids also react with bases to form water and salts. This reaction, called neutralization, is the most important chemical reaction of acids.

Bases, sometimes called alkalis, have a bitter taste. Strong bases such as sodium hydroxide (NaOH, lye) are very corrosive and have a slippery feel when spilled on the skin. Bases also cause color changes with acid–base indicators. Examples are red litmus, which turns blue in basic solution, and phenolphthalein,

Chemicals for unclogging household drains are solutions of NaOH (lye). These solutions feel slippery when spilled on the skin and are very corrosive.

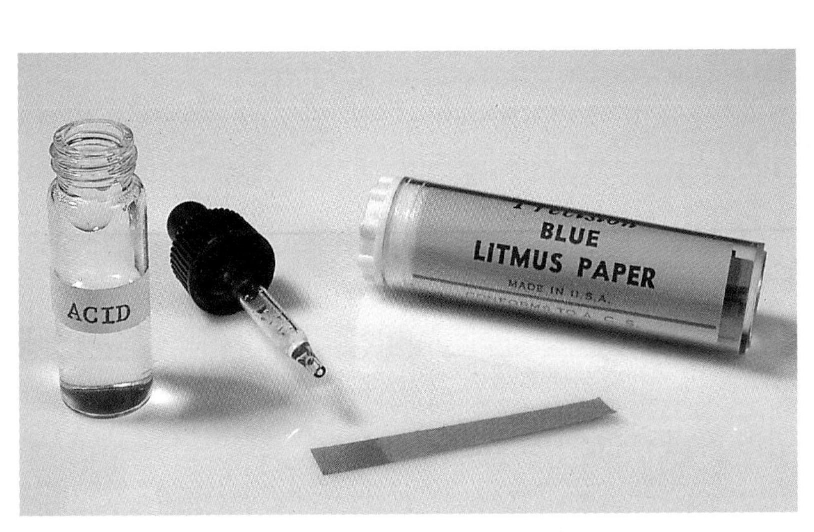

FIGURE 15.1 Blue litmus paper turns red when moistened with an acid.

FIGURE 15.2 Baking soda ($NaHCO_3$) reacts with hydrochloric acid to produce carbon dioxide, shown here bubbling from the solution.

FIGURE 15.3 (a) A solution of a base causes red litmus paper to turn blue. (b) Sodium hydroxide solution causes a solution of phenolphthalein, an acid–base indicator, to turn pink.

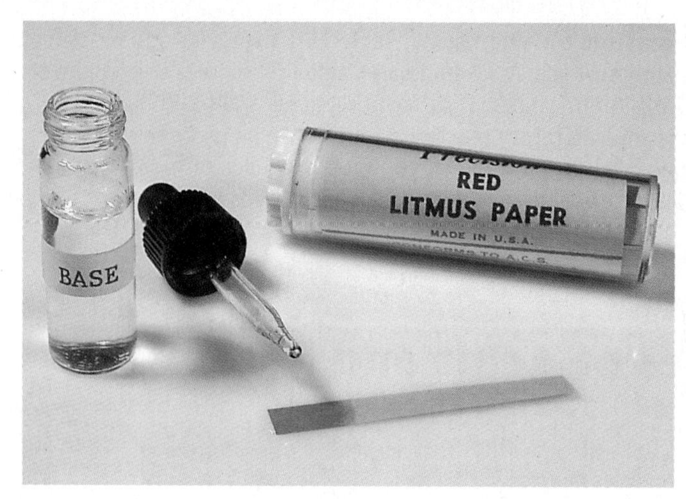

a

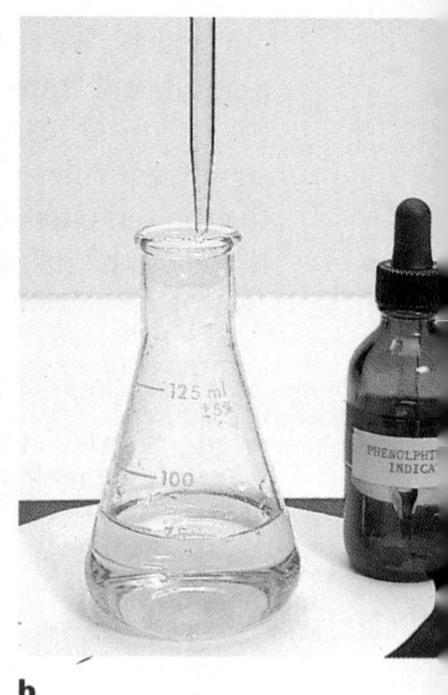

b

which turns pink in basic solution (Figure 15.3). In a reaction known for centuries, a strong base such as NaOH reacts with animal fats to produce soaps (Section 20.3). Bases also neutralize acids to form salts and water. This is their most characteristic chemical reaction. The properties of acids and bases are summarized in Table 15.2.

15.2 The Arrhenius Concept of Acids and Bases

Learning Objectives

- Recognize and give examples of Arrhenius acids and bases.
- Describe an acid–base titration.
- Calculate an unknown concentration of an acid or base using titration data.
- Recognize and give examples of acid and base anhydrides.
- Write balanced equations for the reactions of nonmetal oxides and metal oxides with water.

TABLE 15.2 Properties of Acids and Bases

Acids	*Bases*
taste sour	taste bitter
are corrosive	feel slippery, are corrosive
neutralize bases	neutralize acids
furnish $H^+(aq)$	furnish $OH^-(aq)$
turn litmus red	turn litmus blue
liberate CO_2 from carbonates	react with fats to form soaps

What do different acids have in common that gives them similar properties? Why do different bases behave so alike? To answer these questions, a widely used concept of acids and bases was proposed in 1884 by Svante Arrhenius (1859–1927), a Swedish chemist. According to the **Arrhenius** concept, an **acid** is defined as a substance that forms hydrogen ions (H^+) when dissolved in water. For example, when hydrogen chloride gas dissolves in water, it ionizes to form the solution we call hydrochloric acid (Section 14.3):

TABLE 15.3 Some Arrhenius Acids and Bases

Acids	*Bases*
HCl	$NaOH$
HBr	KOH
HNO_3	$Ca(OH)_2$
$HClO_3$	$Ba(OH)_2$
H_2SO_4	$Fe(OH)_3$
H_3PO_4	$Al(OH)_3$
HPO_4^{2-}	$Ni(OH)_2$
HCO_3^-	$Zn(OH)_2$

$$HCl(g) \xrightarrow{H_2O(l)} H^+(aq) + Cl^-(aq)$$

However, because hydrogen ions in aqueous solutions are strongly associated with water through ion–dipole attractions (Section 14.3), this reaction is more properly written as:

$$HCl(g) + H_2O(l) \longrightarrow H_3O^+(aq) + Cl^-(aq)$$

where $H_3O^+(aq)$, or $H(H_2O)^+(aq)$, is called a hydronium ion. The symbols $H^+(aq)$ and $H_3O^+(aq)$ represent the same ion and are often used interchangeably when writing equations.

An **Arrhenius base** is a substance that dissolves in water to liberate hydroxide ions (OH^-). One of the most familiar examples is sodium hydroxide.

The Arrhenius concept: An acid forms $H_3O^+(aq)$. A base liberates $OH^-(aq)$.

$$NaOH(s) \xrightarrow{H_2O(l)} Na^+(aq) + OH^-(aq)$$

Sodium hydroxide, like other ionic compounds, dissociates when it dissolves. A solution of a base is often said to be alkaline. A number of Arrhenius acids and bases are listed in Table 15.3.

We should note a distinction between the terms ionize and dissociate. When HCl (a covalent compound) dissolves, it ionizes *(that is, it forms ions). When NaOH (an ionic compound) dissolves, it* dissociates. *Because undissolved NaOH is ionic, it is incorrect to say it ionizes (forms ions) when it dissolves (Section 14.3).*

Remember that acids react with bases to form water and salts (Section 10.9). An example is the reaction of hydrochloric acid with sodium hydroxide:

$$HCl(aq) + NaOH(aq) \longrightarrow H_2O(l) + NaCl(aq)$$

Water is formed by the reaction of hydronium ion (H_3O^+) with hydroxide ion (OH^-), a neutralization reaction:

$$H_3O^+(aq) + OH^-(aq) \longrightarrow 2H_2O(l)$$

Another example is the reaction of milk of magnesia, $Mg(OH)_2$ (an antacid), with hydrochloric acid in the stomach:

$$Mg(OH)_2(s) + 2HCl(aq) \longrightarrow 2H_2O(l) + MgCl_2(aq)$$

Neutralization of a diprotic acid such as H_2SO_4 occurs in two steps. The first is

A diprotic acid has two ionizable hydrogens (protons) per formula unit (Section 9.9).

$$H_2SO_4(aq) + NaOH(aq) \longrightarrow H_2O(l) + NaHSO_4(aq)$$

Sodium hydrogen sulfate, $NaHSO_4$, is an acid salt (Section 9.8) and is capable of reacting further with a base such as NaOH. So the second step is

$$NaHSO_4(aq) + NaOH(aq) \longrightarrow H_2O(l) + Na_2SO_4(aq)$$

When two moles of NaOH are added to one mole of H_2SO_4, the two steps occur in rapid succession, and both protons react to give sodium sulfate, Na_2SO_4:

$$H_2SO_4(aq) + 2NaOH(aq) \longrightarrow 2H_2O(l) + Na_2SO_4(aq)$$

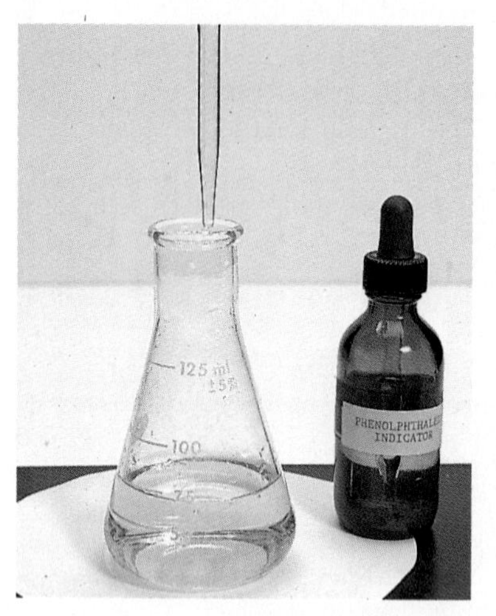

FIGURE 15.4 An acid–base titration. Sodium hydroxide solution is added from a buret to a solution of an acid containing a small amount of phenolphthalein indicator. When the acid has been neutralized, the indicator turns from colorless to pink, indicating that the equivalence point has been reached.

Acid–Base Titrations

A **titration** is an experimental procedure that involves the measurement of the volume of one solution that is required to react completely with another. Although titrations may involve different kinds of reactions, our discussion will be limited to acid–base titrations. Typically, a base solution of known concentration is added from a buret to a measured volume of an acid solution of unknown concentration in a reaction flask containing a small amount of an acid–base indicator. An **acid–base indicator** is a compound that changes color when the solution reaches a certain level of acidity. The titration is continued until the **endpoint** is reached, the point at which the indicator changes color (Figure 15.4). An acid–base indicator is selected that changes color at the **equivalence point,** the point in the titration when just enough base has been added to completely neutralize the acid. Phenolphthalein, used as a dilute solution in ethanol, is one of the most common acid–base indicators.

An acid–base titration is useful for determining an unknown concentration of an acid solution from a known concentration of a base solution, and vice versa. Molarities and mole ratios can be used in these calculations as shown in the following example.

Example 15.1

Titration of 20.0 mL of H_2SO_4 solution of unknown concentration required 22.3 mL of 0.633 M NaOH for complete neutralization. What is the concentration of the H_2SO_4 solution? ■

Solution The balanced equation for the reaction is

$$2NaOH(aq) + H_2SO_4(aq) \longrightarrow 2H_2O(l) + Na_2SO_4(aq)$$

The volume of the H_2SO_4 solution is known, but the number of moles of H_2SO_4 must be calculated. From the equation, we recognize the following mole ratio:

$$\frac{1 \text{ mol } H_2SO_4}{2 \text{ mol NaOH}}$$

Expressing the concentration as molarity, we have:

$$M\left(\frac{\text{mol } H_2SO_4}{\text{L soln}}\right) = ? \qquad (1)$$

$$\text{number of moles } H_2SO_4 = \left(\frac{1 \text{ mol } H_2SO_4}{2 \text{ mol NaOH}}\right)(? \text{ mol NaOH}) \qquad (2)$$

where

$$\text{number of moles NaOH} = \left(\frac{0.633 \text{ mol NaOH}}{\cancel{\text{L soln}}}\right)(0.0223\ \cancel{\text{L soln}}) \qquad (3)$$

$$= 0.0141 \text{ mol NaOH}$$

Substituting in equation (2) gives

$$\text{number of moles } H_2SO_4 = \left(\frac{1 \text{ mol } H_2SO_4}{2 \text{ }\cancel{\text{mol NaOH}}}\right)(0.0141 \text{ }\cancel{\text{mol NaOH}}) \quad (2)$$

$$= 0.00705 \text{ mol } H_2SO_4$$

Then, substituting in equation (1) gives

$$M \left(\frac{\text{mol } H_2SO_4}{\text{L soln}}\right) = \left(\frac{0.00705 \text{ mol } H_2SO_4}{0.0200 \text{ L soln}}\right) = 0.353 \text{ M } H_2SO_4 \quad (1)$$

Practice Problem 15.1 A 10.0-mL sample of 0.667 M oxalic acid ($H_2C_2O_4$) was titrated with a potassium hydroxide (KOH) solution of unknown concentration using phenolphthalein as an indicator. The solution turned pink indicating complete neutralization after the addition of 12.7 mL of the base. What is the concentration of the KOH solution? (Solutions to practice problems appear at the end of the chapter.)

Acid and Base Anhydrides

Why are carbonated beverages slightly acidic? The answer lies in the effect of the dissolved carbon dioxide in these beverages. Carbon dioxide acts as an acid anhydride when it is dissolved in water.

An **anhydride** is a compound derived from another by the removal of water. An **acid anhydride** combines with water to form an acid. Nonmetal oxides such as SO_2 and CO_2 are acid anhydrides. The following equations illustrate the combination of nonmetal oxides with water to give oxyacids:

Acid anhydride: A nonmetal oxide.

$$SO_2(g) + H_2O(l) \longrightarrow H_2SO_3(aq)$$

$$CO_2(g) + H_2O(l) \longrightarrow H_2CO_3(aq)$$

It is the carbonic acid (H_2CO_3) formed in the latter reaction that causes carbonated beverages to be acidic.

In some cases, it may not be obvious which oxyacid is formed when a nonmetal oxide reacts with water. For example, when P_4O_{10} reacts with water, an oxyacid containing phosphorus is formed. But which one? In Chapter 9 we mentioned three: H_3PO_4, H_3PO_3, and H_3PO_2. The correct answer is identified by noting that the oxidation number of P (+5) is the same in the oxide and the oxyacid. To calculate the oxidation number of P in an oxyacid, we must remember the oxidation numbers of H (+1) and O (−2) and that the sum of the oxidation numbers is zero for a compound.

$$H_3PO_4$$

$$3(+1) + P + 4(-2) = 0$$

for hydrogen — unknown — for oxygen

$$+3 + P \quad -8 = 0$$

$$P = +5$$

Carbonated beverages are slightly acidic because of dissolved CO_2.

An oxidation number describes the state of oxidation of an element (Section 9.2).

Because the oxidation numbers of P in H_3PO_4 and in P_4O_{10} match, H_3PO_4 is the reaction product. The oxidation number of P is +3 in H_3PO_3 and +1 in H_3PO_2. The balanced equation for the combination of P_4O_{10} with H_2O is:

$$P_4O_{10}(s) + 6H_2O(l) \longrightarrow 4H_3PO_4(l)$$

Base anhydride: A metal oxide.

A **base anhydride** is a metal oxide that combines with water to form a base. Examples are calcium oxide (lime), CaO, and magnesium oxide, MgO. These combination reactions, as with the acid anhydrides, do not involve oxidation–reduction, and the charge on the metal ion remains unchanged. The following equations illustrate the formation of bases from metal oxides:

$$CaO(s) + H_2O(l) \longrightarrow Ca(OH)_2(aq)$$

$$Na_2O(s) + H_2O(l) \longrightarrow 2NaOH(s)$$

Lime, CaO, can be added to neutralize or "sweeten" acidic (sour) soil.

Other less soluble metal oxides do not react readily with water but will react with strong acids. For example, iron(III) oxide reacts with hydrochloric acid:

$$Fe_2O_3(s) + 6HCl(aq) \longrightarrow 2FeCl_3(aq) + 3H_2O(l)$$

This reaction is used to remove the oxide coating (rust) from steel parts to provide a clean surface prior to painting. A similar reaction occurs when hydrochloric acid (also called muriatic acid) is used to clean cement (containing metal oxides) from new brickwork.

Example 15.2

Write formulas for the acids and bases formed by the combination of water with the following anhydrides:

a. SO_3
b. Cl_2O
c. P_4O_6
d. K_2O ■

Solutions Note that oxyacids are formed when nonmetal oxides react with water, and there is no oxidation–reduction.

a. H_2SO_4
b. HClO
c. H_3PO_3
d. KOH

Example 15.3

Complete and balance the following equations:

a. $BaO(s) + H_2O(l) \longrightarrow$
b. $As_2O_3(s) + H_2O(l) \longrightarrow$
c. $Al_2O_3(s) + H_2SO_4(aq) \longrightarrow$ ■

Solutions

a. $BaO(s) + H_2O(l) \longrightarrow Ba(OH)_2(aq)$
b. $As_2O_3(s) + 3H_2O(l) \longrightarrow 2H_3AsO_3(aq)$
c. $Al_2O_3(s) + 3H_2SO_4(aq) \longrightarrow 3H_2O(l) + Al_2(SO_4)_3(aq)$

Practice Problem 15.2 What compound results when copper(II) oxide, CuO, reacts with water?

Practice Problem 15.3 Write a balanced equation for the reaction of dichlorine heptoxide, Cl_2O_7, with water.

CHEMICAL WORLD

Acid Snow

Acid snow? That must be frozen acid rain. What's the difference?

First, consider acid rain. You now know that nonmetallic oxides form acids when combined with water. Nitrogen and oxygen in the air combine in the high operating temperatures of automobile engines and power plants to form nitrogen oxide, and the combustion of gasoline and other fuels produces carbon dioxide. Burning fuels such as certain kinds of coal that contain sulfur produces various sulfur oxides. Vent these compounds into the moist atmosphere and we get nitric acid, carbonic acid, and sulfuric acid. For example,

$$CO_2(g) + H_2O(l) \longrightarrow \underset{\text{(carbonic acid)}}{H_2CO_3(aq)}$$

Lovely stuff to have floating around in the air. When it rains, it isn't raining rain drops, you know, it's raining acid drops—into lakes and onto trees and crops. Acid rain kills fish and other water creatures or makes them unable to breed; it defoliates trees, damages crops, slowly dissolves marble and limestone buildings, and etches metals: tens of billions of dollars in damage.

Gargoyles on Notre Dame cathedral in Paris have been damaged by acid rain.

Acid snow? It's the same while it's falling. But snow accumulates for several months in some regions. Then, in the spring when plants are greening and animals are breeding—all at their most vulnerable time—the snow melts and releases an entire winter's accumulation of acid in a relatively short time. The damage locally can be even greater than that of acid rain.

15.3 Brønsted–Lowry Acids and Bases

Learning Objectives

- Recognize and give examples of Brønsted–Lowry acids and bases. Give the formula of the conjugate base of a given acid and the conjugate acid of a given base.
- Write balanced chemical equations for Brønsted–Lowry acid–base reactions.

The Arrhenius concept includes many acids but only one kind of base—hydroxide ion. In time, chemists recognized that this was a major limitation in explaining the behavior of bases. In 1923, a more general concept of acids and bases was proposed by the Danish chemist Johannes Brønsted (1897–1947) and the English chemist Thomas Lowry (1847–1936). The Brønsted–Lowry concept views acid–base reactions as simply the transfer of protons (H^+) from an acid to a base. Accordingly, a **Brønsted–Lowry acid** is a proton donor, and a **Brønsted–Lowry base** is a proton acceptor. For example, when HCl is dissolved in water to prepare hydrochloric acid, HCl donates a proton and is an acid, and H_2O accepts a proton and is a base.

Brønsted–Lowry concept:
An acid is a proton (H^+) donor.
A base is a proton (H^+) acceptor.

$$HCl(aq) + H_2O(l) \rightleftarrows H_3O^+(aq) + Cl^-(aq)$$

Furthermore, as we consider the reverse reaction, H_3O^+ donates a proton and is an acid, and Cl^- accepts a proton and is a base.

Because Brønsted–Lowry acid–base reactions are generally reversible, each reaction involves two acids and two bases, or two acid-base pairs. When HCl (an acid) donates a proton, Cl^- (a base) forms, and HCl/Cl^- is called a conjugate acid–base pair. A **conjugate acid–base pair** is made up of an acid and the base that results when the acid loses *one* proton. The second conjugate acid–base pair in this reaction is H_3O^+/H_2O. When H_3O^+ (an acid) donates a proton, H_2O (a base) results. Some conjugate acid–base pairs are shown in Table 15.4.

How can we recognize a Brønsted–Lowry base? Remember that a proton (H^+) has no electrons to share in forming a covalent bond with a base. Therefore, *a base must have a pair of nonbonding electrons* to be able to bond with a proton.

A Brønsted–Lowry base must have a pair of nonbonding electrons.

TABLE 15.4 Some Conjugate Acid–Base Pairs

Acid	*Base*
HCl	Cl^-
HI	I^-
H_2O	OH^-
H_3O^+	H_2O
NH_4^+	NH_3
H_2SO_4	HSO_4^-
HSO_4^-	SO_4^{2-}
HNO_3	NO_3^-
H_3PO_4	$H_2PO_4^-$
$H_2PO_4^-$	HPO_4^{2-}

Any molecule containing oxygen or nitrogen (such as H_2O and NH_3) can be a base, and all anions are capable of being proton acceptors. The proton transfer involving HCl and H_2O can be shown more clearly using Lewis formulas:

$$\text{H—}\ddot{\text{O}}\text{:(—H)} + \text{H—}\ddot{\underset{\cdot\cdot}{\text{Cl}}}\text{:} \rightleftarrows \left[\text{H—}\ddot{\text{O}}\text{(—H)—H}\right]^+ + \left[\text{:}\ddot{\underset{\cdot\cdot}{\text{Cl}}}\text{:}\right]^-$$

base acid acid base

In aqueous ammonia, water is an acid and ammonia is a base:

$$\text{H—}\ddot{\text{O}}\text{:(—H)} + \text{:N(H)(H)—H} \rightleftarrows \left[\text{H—}\ddot{\underset{\cdot\cdot}{\text{O}}}\text{:}\right]^- + \left[\text{H—N(H)(H)—H}\right]^+$$

acid base base acid

A substance, such as water, that can act as either an acid or a base is called an **amphoteric** substance. Note that an amphoteric substance must have one or more H's in its formula and a pair of nonbonding electrons. The conjugate base of a diprotic acid, such as HSO_4^-, is an example of an amphoteric anion. Equations for the stepwise ionization of sulfuric acid in water show HSO_4^- acting as both a base and an acid:

Amphoteric: Capable of acting as either an acid or a base.

$$H_2SO_4(aq) + H_2O(l) \rightleftarrows H_3O^+(aq) + HSO_4^-(aq)$$

acid base acid base

$$HSO_4^-(aq) + H_2O(l) \rightleftarrows H_3O^+(aq) + SO_4^{2-}(aq)$$

acid base acid base

Example 15.4

Write the formula of the conjugate base of each of the following acids:

a. HBr
b. NH_3
c. $H_2PO_4^-$
d. H_2S ■

Solutions The conjugate base of an acid results when the acid loses a proton:

a. Br^-
b. NH_2^-
c. HPO_4^{2-}
d. HS^-

Example 15.5

Write the formula of the conjugate acid of each of the following bases:

a. F^-
b. O^{2-}
c. NH_3
d. CH_3OH ■

Solutions Adding a proton (H^+) to a base gives its conjugate acid:

a. HF
b. OH^-
c. NH_4^+
d. $CH_3OH_2^+$

Example 15.6

Which of the substances in Example 15.5 are amphoteric? ■

Solution Both NH_3 and CH_3OH are bases—that is, they can gain a proton—as shown in Example 15.5. Furthermore, both can lose a proton to give NH_2^- and CH_3O^-, respectively.

Practice Problem 15.4 Write the formula for the conjugate base of each of the following acids:
a. HNO_2
b. CH_4
c. H_2SO_3

Practice Problem 15.5 HCO_3^- is an amphoteric ion. Write formulas for its conjugate acid and its conjugate base.

15.4 Strengths of Acids and Bases

Learning Objectives

- Recognize common strong acids such as HCl, HNO_3, H_2SO_4, and H_3O^+.
- Recognize common weak acids such as $HC_2H_3O_2$, H_2CO_3, and HCN.
- Rank the strengths of a series of bases using a ranking of the strengths of their conjugate acids, and vice versa.
- Determine whether the solution of a given salt is neutral, acidic, or basic.

A strong acid is almost completely ionized.
A weak acid is only slightly ionized.

In the previous section, HCl(aq) was described as a strong acid. The strength of an acid refers to its ability to donate protons. The strengths of two acids can be compared by allowing the acids to react with the same base, usually water. The concentration of H_3O^+(aq) in each solution is a measure of the strength of the acid. For example, in 0.10 M HCl, the concentration of H_3O^+(aq) is 0.10 M. However, in 0.10 M $HC_2H_3O_2$, the concentration of H_3O^+(aq) is only 0.0013 M. In other words, in aqueous solution, a **strong acid** is an acid that ionizes completely in dilute solutions, or almost so, whereas a weak acid is one that is only slightly ionized:

Strong acid:
$$HCl(aq) + H_2O(l) \rightarrow H_3O^+(aq) + Cl^-(aq)$$

Weak acid:
$$HC_2H_3O_2(aq) + H_2O(l) \rightleftarrows H_3O^+(aq) + C_2H_3O_2^-(aq)$$

In dilute solutions of $HC_2H_3O_2$, only a small percentage of the molecules ionize, resulting in a low concentration of $H_3O^+(aq)$. Notice that double arrows may be used to indicate that the ionization of an acid is reversible. Arrows of different lengths show that a reaction is favored in one direction over the other, thereby indicating that an acid is strong or weak. For a strong acid, it is common practice to use only a single arrow.

Note the inverse relationship between the strength of an acid and the strength of its conjugate base. A strong acid, HA, has a relatively weak conjugate base, A^-. In other words, an acid that readily donates protons (strong acid) has a conjugate base that is a poor proton acceptor (weak base). For example, because acetic acid ($HC_2H_3O_2$) is a weaker acid than hydrochloric acid, acetate ion ($C_2H_3O_2^-$) is a stronger base than chloride ion. Water is a very weak acid, and OH^- is a **strong base,** a good proton acceptor. When comparing the strengths of two acids and their conjugate bases, note that *the weaker acid has the stronger conjugate base.* Table 15.5 lists a number of acids and bases according to their strength.

Strengths of acids and their conjugate bases are inversely related:

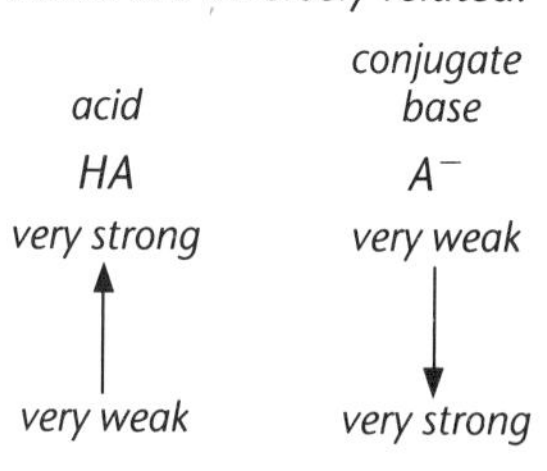

Although salts result from acid–base or neutralization reactions, not all salts dissolve in water to give neutral solutions. For example, a solution of sodium acetate, $NaC_2H_3O_2$, is slightly basic because of the reaction of acetate ion with water to form acetic acid, a weak acid, and hydroxide ion:

$$C_2H_3O_2^-(aq) + H_2O(l) \leftrightarrows HC_2H_3O_2(aq) + OH^-(aq)$$

This reaction is called **hydrolysis,** a general term for a reaction in which the water molecule is split.

Other examples of salts that give basic solutions are sodium phosphate (Na_3PO_4), a cleansing agent, and sodium hydrogen carbonate ($NaHCO_3$), baking

TABLE 15.5 Relative Strengths of Some Acids and Bases

Acid	*Conjugate Acid–Base Pairs*	*Base*
perchloric acid	$HClO_4/ClO_4^-$	perchlorate ion
nitric acid	HNO_3/NO_3^-	nitrate ion
hydrochloric acid	HCl/Cl^-	chloride ion
sulfuric acid	H_2SO_4/HSO_4^-	hydrogen sulfate ion
hydronium ion	H_3O^+/H_2O	water
oxalic acid	$H_2C_2O_4/HC_2O_4^-$	hydrogen oxalate ion
sulfurous acid	H_2SO_3/HSO_3^-	hydrogen sulfite ion
hydrogen sulfate ion	HSO_4^-/SO_4^{2-}	sulfate ion
phosphoric acid	$H_3PO_4/H_2PO_4^-$	dihydrogen phosphate ion
hydrofluoric acid	HF/F^-	fluoride ion
nitrous acid	HNO_2/NO_2^-	nitrite ion
acetic acid	$HC_2H_3O_2/C_2H_3O_2^-$	acetate ion
carbonic acid	H_2CO_3/HCO_3^-	hydrogen carbonate ion
hydrosulfuric acid	H_2S/HS^-	hydrogen sulfide ion
hydrocyanic acid	HCN/CN^-	cyanide ion
ammonium ion	NH_4^+/NH_3	ammonia
bicarbonate ion	HCO_3^-/CO_3^{2-}	carbonate ion
water	H_2O/OH^-	hydroxide ion
ammonia	NH_3/NH_2^-	amide ion
hydroxide ion	OH^-/O^{2-}	oxide ion

soda. As in the previous example, these anions accept a proton from water producing hydroxide ion:

$$PO_4^{3-}(aq) + H_2O(l) \rightleftarrows HPO_4^{2-}(aq) + OH^-(aq)$$

$$HCO_3^-(aq) + H_2O(l) \leftrightarrows H_2CO_3(aq) + OH^-(aq)$$

Note that other salts, such as sodium chloride (NaCl), dissolve in water to give neutral solutions. Furthermore, the solutions of some salts, such as ammonium chloride (NH_4Cl), are slightly acidic. How can we know whether the solution of a salt will be neutral, basic, or acidic? The following guidelines will help in answering this question:

1. A salt of a weak acid and a *strong base* (such as $NaC_2H_3O_2$) dissolves in water to give a *basic* solution. Note that Na^+ is the cation of NaOH, a strong base, and $C_2H_3O_2^-$ is the conjugate base of $HC_2H_3O_2$, a weak acid. Hydrolysis of acetate ion ($C_2H_3O_2^-$) produces hydroxide ion (OH^-).
2. A salt of a *strong acid* and a *strong base* (such as NaCl) dissolves in water to give a neutral solution. Note that Na^+ is the cation of NaOH, a strong base, and Cl^- is the conjugate base of HCl, a strong acid. Neither ion undergoes hydrolysis.
3. A salt of a *strong acid* and a weak base (such as NH_4Cl) dissolves in water to give an *acidic* solution. Note that NH_4^+ is the conjugate acid of NH_3, a weak base, and Cl^- is the conjugate base of HCl, a strong acid.

To understand why $NH_4Cl(aq)$ is an acidic solution, note that NH_4^+ is an acid, and hydronium ions are formed in aqueous solution:

$$NH_4^+(aq) + H_2O(l) \rightleftarrows NH_3(aq) + H_3O^+(aq)$$

Because Cl^- is a weak base, it does not hydrolyze:

$$Cl^-(aq) + H_2O(l) \longrightarrow \text{no reaction}$$

Thus, an aqueous solution of NH_4Cl is slightly acidic.

Table 15.6 summarizes the guidelines for determining whether a salt solution is acidic, basic, or neutral.

Example 15.7

Using Table 15.5, list the following acids in order of decreasing strength (strongest first): H_2S, H_2O, HCl, HF. ■

Solution HCl, HF, H_2S, H_2O

TABLE 15.6 Acidic, Basic, and Neutral Salts

Type of Salt			
Cation of:	*Anion of:*	*Solution*	*Example*
strong base	weak acid	basic	$NaC_2H_3O_2$
strong base	strong acid	neutral	NaCl
weak base	strong acid	acidic	NH_4Cl

Example 15.8

Using Table 15.5, list the following bases in order of decreasing strength (strongest first): F^-, Cl^-, CN^-, OH^-. ■

Solution OH^-, CN^-, F^-, H_2O, Cl^-

Example 15.9

Which of the following salts dissolves in water to give an acidic solution? A basic solution? A neutral solution? Explain your answers.

a. KBr
b. Na_2CO_3
c. NH_4NO_3 ■

Solutions

a. KBr forms a neutral solution. It is the salt of a strong base (KOH) and a strong acid (HBr).

b. Na_2CO_3 forms a basic solution. It is the salt of a strong base (NaOH) and a weak acid (HCO_3^-). The anion hydrolyzes to give hydroxide ions:

$$CO_3^{2-}(aq) + H_2O(l) \rightleftarrows HCO_3^-(aq) + OH^-(aq)$$

c. NH_4NO_3 forms an acidic solution. It is the salt of a weak base (NH_3) and a strong acid (HNO_3). The ammonium ion reacts with water to give hydronium ions:

$$NH_4^+(aq) + H_2O(l) \rightleftarrows NH_3(aq) + H_3O^+(aq)$$

Practice Problem 15.6 The following acids are listed in order of decreasing acidity, nitric acid being the strongest: HNO_3, $HC_2H_3O_2$, OH^- Arrange their conjugate bases in order of decreasing basicity.

Practice Problem 15.7 Is an aqueous solution of sodium amide, $NaNH_2$, acidic, basic, or neutral? Write an equation for the hydrolysis reaction.

Ionization of Water

15.5

Learning Objectives

- Write an equation for the ionization of water.
- State the molar concentrations of hydronium ion, $[H_3O^+]$, and hydroxide ion, $[OH^-]$, in neutral solutions.

Water is amphoteric—it can either donate protons or accept protons. Therefore, it is listed in Table 15.5 as both an acid and a base. The ionization of water illustrates its amphoteric behavior:

$$\underset{\text{acid}}{H_2O(l)} + \underset{\text{base}}{H_2O(l)} \longrightarrow \underset{\text{acid}}{H_3O^+(aq)} + \underset{\text{base}}{OH^-(aq)}$$

It is important to note that hydronium ion (H_3O^+) and hydroxide ion (OH^-) are formed in equal amounts. In pure water and in neutral solutions, neither ion is in excess. The ionization of water results in a solution that is 1.0×10^{-7} M H_3O^+ and 1.0×10^{-7} M OH^- at 25°C. These concentrations may be expressed as

$[H_3O^+]$ means mol H_3O^+/L soln.

$$[H_3O^+] = [OH^-] = 1.0 \times 10^{-7} \text{ mol/L soln}$$

A chemical formula in brackets means the molar concentration of that substance. Thus, $[H_3O^+]$ means mol H_3O^+/L soln.

15.6 The pH Scale

Learning Objectives

- Recognize a given pH as acidic, neutral, or basic.
- Calculate pH and pOH from appropriate concentrations, and vice versa.

The pH scale is used to express the acidity of dilute aqueous solutions. The scale commonly used ranges from 0 through 14, although values outside this range can be measured (Table 15.7). A pH of 7 indicates a neutral solution, one that is neither acidic nor basic. Acidic solutions have pH values less than 7, and the pH of a basic solution is greater than 7. For example, the pH of vinegar (5.0% acetic acid) is 2.8. Lemon juice has a pH of 2.3 due to its citric acid content. Household ammonia (NH_3), a weak base, has a pH of 11.0 (Table 15.8).

TABLE 15.7 The pH Scale

$[H_3O^+]$, mol/L soln	*pH*
1×10^{-15}	15
1×10^{-14}	14
1×10^{-13}	13
1×10^{-12}	12 basic
1×10^{-11}	11
1×10^{-10}	10
1×10^{-9}	9
1×10^{-8}	8
1×10^{-7}	7 neutral
1×10^{-6}	6
1×10^{-5}	5
1×10^{-4}	4
1×10^{-3}	3
1×10^{-2}	2 acidic
1×10^{-1}	1
1×10^{0}	0
1×10^{1} (or 10)	−1

TABLE 15.8 pH Values of Some Common Solutions[a]

Solution	*pH*
1 M NaOH	14.0
household ammonia	11.0
blood	7.4
saliva	6.5–7.5
milk	6.3–6.6
coffee	4.0–5.0
tomato juice	4.2
orange juice	3.7
vinegar	2.8
lemon juice	2.3
0.1 M HCl	1.0
gastric juice	1.0

[a]The pH values of juices and other natural solutions are approximate.

Until now, we have expressed acidity in terms of the concentration of H_3O^+ in a solution. Using pH to express the acidity of dilute solutions avoids the use of exponential notation and units of mol H_3O^+/L soln. The following equation is used to calculate pH from the molar concentration of H_3O^+:

$$pH = -\log[H_3O^+]$$

where log means the *common logarithm* or the logarithm to the base 10. Thus, **pH** is defined as the negative logarithm of the molar concentration of hydronium ion. The **logarithm** (log) of a number is the power to which 10 must be raised to give that number. For example:

A log is an exponent of 10.

$$\log(100) = \log(1 \times 10^2) = 2.0$$
$$\log(1000) = \log(1 \times 10^3) = 3.0$$
$$\log(10\ 000) = \log(1 \times 10^4) = 4.0$$
$$\log(0.01) = \log(1 \times 10^{-2}) = -2.0$$
$$\log(0.001) = \log(1 \times 10^{-3}) = -3.0$$

The logarithm of a number can be determined using a scientific calculator that has a log key. For example, $\log(63) = 1.80$. With most calculators, the calculation involves two steps:

1. Enter the number.
2. Press the *log* key.

Note that 63 is between 10 and 100, and log(63) is between $\log(10^1)$ and $\log(10^2)$—that is, between 1 and 2. Knowledge of logarithms that are integers is helpful in determining if the answers you obtain using a calculator are reasonable.

To determine the pH of a solution from the molar concentration of H_3O^+, simply substitute into the equation for pH. For example, to calculate the pH of 0.0010 M H_3O^+, use:

The "p" in pH stands for "power," as in exponential power. The "H" (symbol for hydrogen) is a capital letter, but "p" is a lower-case letter.

$$pH = -\log[H_3O^+] = -\log(1.0 \times 10^{-3}) = 3.00$$

After calculating the logarithm, multiply by -1 to obtain the pH. Using a calculator, the steps are:

1. Enter the $[H_3O^+]$.
2. Press the log key to find the logarithm.
3. Press the change-of-sign key to multiply by -1.

The determine the correct number of significant figures for pH, use the following rule: *the number of decimal places for a logarithm of a number must agree with the number of significant figures shown in the number.* In other words, in a logarithm, only the digits to the right of the decimal point are counted as significant figures. For example:

Number of decimal places (pH or pOH) = number of significant figures (concentration).

$$[H_3O^+] = 1.0 \times 10^{-3}\ M \qquad \text{(2 significant figures)}$$

and

$$pH = 3.00 \qquad \text{(2 decimal places)}$$

Keep in mind that, because the pH scale is logarithmic, *a change of one pH unit is a tenfold change in the concentration of hydronium ion.* Thus, when the pH of a solution is changed from 3.0 to 1.0 (a change of 2.0 pH units), the hydronium ion concentration is increased 10^2 or 100 times.

Another "p" value is **pOH,** defined as the negative logarithm of the OH^- concentration:

$$pOH = -\log[OH^-]$$

The relationship of pOH and pH can be illustrated by noting, for neutral solutions that

$$[H_3O^+] = [OH^-] = 1 \times 10^{-7}\ M$$

Thus, in neutral solution, $pH = -\log(1 \times 10^{-7}) = 7.0$, and $pOH = 7.0$. Furthermore, adding pH and pOH gives

pH + pOH = 14.0 at 25°C

$$pH + pOH = 7.0 + 7.0 = 14.0$$

It has been determined that the sum of pH + pOH = 14.0 in *all* aqueous solutions at 25°C, not just neutral solutions. Sometimes it is convenient to use this equation to calculate the pH of a basic solution when the hydroxide ion concentration is known.

When the pH or pOH is known, the corresponding concentration is found by taking the antilogarithm of the negative of the pH or pOH:

$$[H_3O^+] = \text{antilog } -pH = 10^{-pH}$$

and

$$[OH^-] = \text{antilog } -pOH = 10^{-pOH}$$

When a pH is an integer value, simply substitute to find $[H_3O^+]$. For example, for a solution having a pH of 3.0:

$$[H_3O^+] = 10^{-pH} = 1 \times 10^{-3}\ M \qquad \text{(1 significant figure)}$$

To calculate a concentration from pH or pOH using a calculator, the steps are:

1. Enter the pH or pOH.
2. Press the change-of-sign key.
3. Press the 10^x key. (Some calculators require a different key to take the antilog of a number. Consult the user's manual for your calculator.)

For example, to find $[H_3O^+]$ in a solution with a pH of 3.50:

1. Enter 3.50.
2. Press the change-of-sign key: -3.50
3. Press the 10^x key: $[H_3O^+] = 3.2 \times 10^{-4}$ (2 significant figures).

The following examples further illustrate these calculations.

Example 15.10

What is the pH of 0.010 M HCl? ■

Solution Remember that HCl is a strong acid and is completely ionized in dilute solutions.

$$HCl(aq) + H_2O(l) \longrightarrow H_3O^+(aq) + Cl^-(aq)$$

For each mole of HCl that ionizes, one mole of H_3O^+ results. Thus, $[H_3O^+] = 0.010\ M = 1.0 \times 10^{-2}\ M$.

$$pH = -\log(1.0 \times 10^{-2}) = -1(-2.00) = 2.00$$

Because the hydronium ion concentration has 2 significant figures, the pH has 2 decimal places.

Example 15.11

What is the pH of 0.010 M NaOH solution? ▪

Solution Remember that NaOH is very soluble in water and dissociates completely. Thus, $[OH^-] = 0.010\ M = 1.0 \times 10^{-2}\ M$. pOH = 2.00, and pH = 14.00 − 2.00 = 12.00 (2 significant figures).

Example 15.12

What is the $[H_3O^+]$ in a solution for which the pH is 4.2? What is the $[OH^-]$? ▪

Solution Use the expression $[H_3O^+] = 10^{-pH}$ and substitute. Using your calculator, carry out these operations:

1. Enter the pH (4.2).
2. Press the change-of-sign key.
3. Press the 10^x key.

You then have:

$$[H_3O^+] = 6 \times 10^{-5}\ M, \text{ or } 0.000\ 06\ M \qquad \text{(1 significant figure)}$$

To calcualte $[OH^-]$, first find pOH:

$$pH + pOH = 14.0$$
$$pOH = 14.0 - pH = 14 - 4.2 = 9.8$$

Now use your calculator as before, this time entering the pOH. You will then have:

$$[OH^-] = 2 \times 10^{-10}\ M \qquad \text{(1 significant figure)}$$

Practice Problem 15.8 What is the pH of $1.0 \times 10^{-11}\ M\ H_3O^+$ solution?

Practice Problem 15.9 Calculate the molarity of hydronium ion in a solution having a pH of 8.7.

The measurement of pH is very important in laboratories, in chemical processing, and in the environment. Acid–base indicators such as phenolphthalein

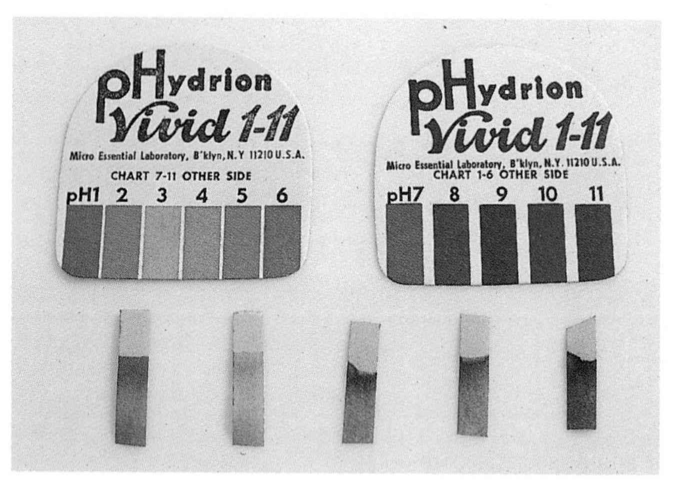

FIGURE 15.5 pH paper for checking the approximate pH of solutions. The center strip is a test of pure water. A strongly acidic solution gave a pH of 1 (far left), and a weakly acidic solution showed a pH of 4 (second from left). The light blue strip (second from right) showing a pH of 8 indicates a weakly basic solution, and a strongly basic solution (pH > 11) is indicated by the dark blue strip on the far right.

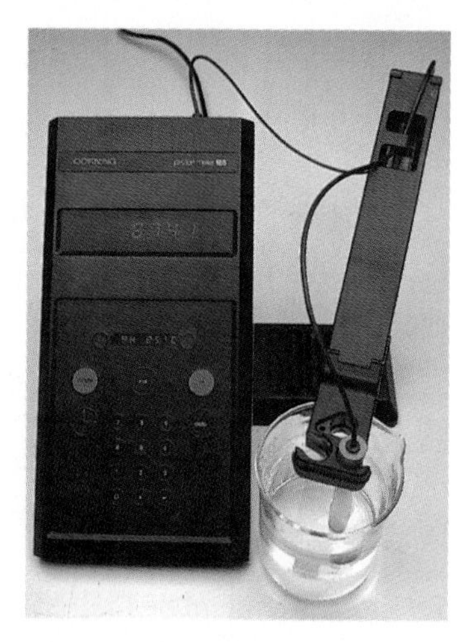

FIGURE 15.6 A modern pH meter.

(Section 15.2) and methyl orange are widely used in laboratories to indicate changes in pH during acid–base titrations. Indicator papers, such as litmus paper and pH paper (Figure 15.5), are commonly used to check the approximate pH of solutions.

A more accurate measurement of pH can be obtained using an instrument called a pH meter (Figure 15.6). This is a voltmeter calibrated to measure pH. Modern pH meters are easy to use and express pH to 0.01 unit. Portable pH meters are used to check the pH of lakes and streams in environmental or water quality surveys.

15.7 Buffer Solutions

Learning Objective

- Explain how a buffer solution resists a change in pH when a strong acid or a strong base is added.

A buffer is a material that reduces the shock of an impact. A chemical buffer *is a substance that minimizes the "shock" of adding a strong acid or base to a solution.*

Most biochemical reactions in the human body and in other living organisms occur in aqueous solution and are most efficient within a narrow pH range. For example, the pH of blood is about 7.4, and variation of more than 0.2 unit from this value would make someone seriously ill. Our bodies make use of buffer solutions to control the pH of blood and other body fluids. A **buffer solution** is a solution that resists changes in pH even when a strong acid or a strong base is added. A wide variety of buffer solutions have been prepared and studied in the laboratory.

The most common type of buffer solution can be prepared from a weak acid and its conjugate base. For example, an acetic acid buffer is prepared by dissolving acetic acid ($HC_2H_3O_2$) and sodium acetate ($NaC_2H_3O_2$) in water to give a solution containing a large supply of both acetic acid and its conjugage base, acetate ion.

$$HC_2H_3O_2(aq) + H_2O(l) \rightleftarrows C_2H_3O_2^-(aq) + H_3O^+(aq)$$
(large supply) (large supply)

How does a buffer solution resist a change in pH when a strong acid enters the solution? When a strong acid such as HCl is added to the buffer solution, acetate ion reacts with hydronium ions (the reverse reaction) to form acetic acid. Acetate ion, as the conjugage base of a weak acid, is basic and has an affinity for protons. The reaction of $C_2H_3O_2^-$ with hydronium ions prevents a large increase in the concentration of H_3O^+, thereby preventing a large change in pH.

How does a buffer solution resist a change in pH when a strong base (OH^-) is added? Because OH^-, a strong base, has an affinity for a proton, it readily removes H^+ from $HC_2H_3O_2$. As a result, when hydroxide ions are added to the buffer solution, they immediately react with acetic acid. The reaction prevents a large increase in the concentration of OH^-, and there is little change in pH.

$$OH^-(aq) + HC_2H_3O_2(aq) \longrightarrow H_2O(l) + C_2H_3O_2^-(aq)$$

The choice of weak acid to use in a buffer solution is determined by the pH to be maintained. For example, an acetic acid buffer can be used to maintain a pH of 4–5, depending upon the ratio of acetic acid and acetate ion. An acid weaker than acetic acid, such as bicarbonate ion (HCO_3^-), could be used to maintain a higher pH. In other words, selection of the buffer is based upon the strength of the weak acid. The calculations necessary to select an appropriate buffer are beyond the coverage of this book.

Buffer capacity, or the quantity of a strong acid or strong base that can be added to a buffer solution without causing a change in pH, is limited by the quantities of weak acid and its conjugate base in the solution.

Characteristics of a Buffer Solution

1. A buffer solution contains a large supply of a weak acid and its conjugate base: for example, $HC_2H_3O_2/C_2H_3O_2^-$.
2. A buffer solution resists changes in pH when either a strong acid or a strong base is added.
3. The base in the buffer solution reacts with any H_3O^+ from addition of a strong acid, thereby preventing an increase in acidity.

added $H_3O^+(aq)$

$$HC_2H_3O_2(aq) + H_2O(l) \longleftarrow$$

buffer: $HC_2H_3O_2(aq) + H_2O(l) \rightleftarrows H_3O^+(aq) + C_2H_3O_2^-(aq)$
(large supply) (large supply)

$$\longrightarrow H_2O(l) + C_2H_3O_2^-(aq)$$

added OH^-

4. The acid (for example, acetic acid) in the buffer solution reacts with any OH^- from the addition of a strong base, thereby preventing a decrease in acidity.

■ Chapter Review

Directions: From the list of choices at the end of this Chapter Review, choose the word or term that best fits in each blank. Use each choice only once. Answers appear at the end of the chapter.

Section 15.1 Acids have a __________ (1) taste. They also cause color changes with acid–base __________ (2) such as litmus, which is __________ (3) in acidic solution. Acids react with carbonates to liberate __________ (4). Acids also __________ (5) bases.

A base, typically a metal __________ (6), has a __________ (7) taste. In basic solution, litmus turns __________ (8) and phenolphthalein turns __________ (9). Bases react with fats to produce __________ (10), and bases neutralize __________ (11).

Section 15.2 According to the Arrhenius concept, an acid is a substance that forms __________ (12) in aqueous solution, whereas a __________ (13) liberates OH^-. Sodium hydroxide, also called __________ (14), is an example of a strong base.

An experimental procedure that involves the measurement of the volume of one solution required to react completely with another is called a __________ (15). The change in color of the acid–base indicator signals the __________ (16). An acid–base indicator is used that changes color when the stoichiometric amounts of acid and base have been added to the reaction flask (the __________ (17) point).

A compound derived from another by the removal of water is called an __________ (18). Examples of acid anhydrides are __________ (19) oxides such as SO_2 and CO_2. They react with water to form oxyacids. A base, or metal hydroxide, forms when water reacts with a base anhydride (__________ (20) oxide). Metal oxides also react with acids to form a salt and __________ (21).

Section 15.3 According to the Brønsted–Lowry concept, an acid is a proton __________ (22), and a base is a __________ (23) acceptor. When an acid loses a proton, what is left is called its __________ (24) base. Thus, an acid and its conjugate base differ by only one proton. For example, the conjugate base of H_2O is __________ (25). Every acid–base reaction (Brønsted–Lowry) involves __________ (26) conjugate acid–base pairs. A base must have a pair of __________ (27) to be able to bond with a proton. A substance that can act as either an acid or a base is called an __________ (28) substance. The conjugate base of a __________ (29) acid is an example of an amphoteric substance.

Section 15.4 A __________ (30) acid is one that __________ (31) completely, whereas a __________ (32) acid is only slightly ionized. A strong acid has a weak conjugate base, and a weak acid has a strong conjugate base. For example, because HCl is a strong acid, __________ (33) is a weak base. Because acetic acid ($HC_2H_3O_2$) is a weaker acid than hydrochloric acid (HCl), __________ (34) ion is a stronger base than chloride ion.

Section 15.5 In neutral solutions, the concentrations of both hydronium ion and hydroxide ion are __________ (35). These ions result from the __________ (36) of water, a reaction in which water acts as both an acid and a base.

Section 15.6 The range of the commonly used __________ (37) scale is 0–14. A solution with a pH of 7 is __________ (38). An __________ (39) solution has a pH below 7, and a solution with a $pH > 7$ is __________ (40). pH values can be calculated using the equation pH = __________ (41). Because the pH scale is logarithmic, a change of one pH unit is a __________ (42) change in the hydronium ion concentration. pOH is defined as __________ (43). The sum, pH + pOH = __________ (44). Acid–base indicators and indicator __________ (45) are used to indicate the approximate pH of a solution. Very accurate pH measurements can be made using a pH __________ (46).

Section 15.7 A __________ (47) solution is a solution that resists a __________ (48) in pH even when a strong acid or base is added. An __________ __________ (49)–acetate buffer is an example. The choice of weak acid to use in a buffer solution is based upon its __________ (50). Buffer __________ (51) is limited by the quantities of weak acid and its conjugate base in the solution.

Choices: acetate, acetic acid, acidic, acids, amphoteric, anhydride, base, basic, bitter, blue, buffer, capacity, change, Cl^-, CO_2, conjugate, diprotic, donor, electrons, endpoint, equivalence, 14, H_3O^+, hydroxide, indicators, ionization, ionizes, $-\log[H_3O^+]$, $-\log[OH^-]$, lye, metal, meter, neutral, neutralize, nonmetal, OH^-, 1×10^{-7} M, papers, pH, pink, proton, red, soaps, sour, strength, strong, tenfold, titration, two, water, weak

Study Objectives

After studying Chapter 15, you should be able to:

Section 15.1

1. Summarize the properties of acids.
2. Summarize the properties of bases.

Section 15.2

3. Recognize and give examples of Arrhenius acids and bases.
4. Describe an acid–base titration.
5. Calculate an unknown concentration of an acid or base using titration data.
6. Recognize and give examples of acid and base anhydrides.
7. Write balanced equations for the reactions of nonmetal oxides and metal oxides with water.

Section 15.3

8. Recognize and give examples of Brønsted–Lowry acids and bases. Give the formula of the conjugate base of a given acid and the conjugate acid of a given base.
9. Write balanced chemical equations for Brønsted–Lowry acid–base reactions.

Section 15.4

10. Recognize common strong acids such as HCl, HNO_3, H_2SO_4, and H_3O^+.
11. Recognize common weak acids such as $HC_2H_3O_2$, H_2CO_3, and HCN.
12. Rank the strengths of a series of bases using a ranking of the strengths of their conjugate acids, and vice versa.
13. Determine whether the solution of a given salt is neutral, acidic, or basic.

Section 15.5

14. Write an equation for the ionization of water.
15. State the molar concentrations of hydronium ion, $[H_3O^+]$, and hydroxide ion, $[OH^-]$, in neutral solutions.

Section 15.6

16. Recognize a given pH as acidic, neutral, or basic.
17. Calculate pH and pOH from appropriate concentrations, and vice versa.

Section 15.7

18. Explain how a buffer solution resists a change in pH when a strong acid or a strong base is added.

Key Terms

Review the definition of each of the terms listed here. For reference, they are listed by chapter section number. You may use the glossary as necessary.

15.2 acid anhydride, acid (Arrhenius), base (Arrhenius), acid–base indicator, anhydride, base anhydride, endpoint, equivalence point, titration

15.3 acid and base (Brønsted–Lowry), amphoteric, conjugate acid–base pair

15.4 hydrolysis, strong acid, strong base

15.6 logarithm, pH, pOH

15.7 buffer solution

Questions and Problems

Answers to questions and problems that have blue numbers appear in Appendix C. Asterisks indicate the more challenging problems.

Properties of Acids (Section 15.1)

1. What taste do we attribute to acids?
2. What kind of substance gives a lemon its sour taste?
3. How would you describe the taste of vinegar, which contains acetic acid?
4. What kind of substance causes gastric juice to be sour?
5. What taste do we attribute to bases?
6. What two characteristics of a strong base might you notice if a solution was spilled on your skin?

7. What kinds of compounds react with acids to liberate carbon dioxide?
8. What kind of compound reacts with an acid to give a salt and water?
9. What substances react with NaOH to form soaps?

Arrhenius and Brønsted–Lowry Acids and Bases (Sections 15.2, 15.3)

10. From their definitions, what distinguishes an Arrhenius acid from a Brønsted–Lowry acid?
11. Which of the following are Arrhenius bases?
    a. KOH
    b. SO_2
    c. $Ba(OH)_2$
    d. Cl^-
12. Which of the following are Arrhenius acids?
    a. HCl
    b. $AlCl_3$
    c. CH_4
    d. HSO_4^-
13. Which of the following are Brønsted–Lowry bases?
    a. OH^-
    b. H_2O
    c. Cl^-
    d. NH_3
14. Write the formula of the conjugate acid of ethyl alcohol, CH_3CH_2OH (grain alcohol).
15. Write the formula of the conjugate base of each of the following:
    a. HBr
    b. H_2SO_4
    c. CH_3NH_2
    d. $HClO_3$
16. Give the formula of the conjugate acid of each of the following:
    a. SO_4^{2-}
    b. H_2O
    c. Cl^-
    d. NH_3
17. Write a balanced equation for the reaction of each of the following bases with water:
    a. H^-
    b. NH_2^-
    c. O^{2-}
    d. NH_3
    e. CO_3^{2-}
    f. CN^-
18. Write a balanced equation for the reaction of each of the following acids with water:
    a. HBr
    b. H_2SO_4
    c. $HC_2H_3O_2$
19. Which of the following substances is/are amphoteric? Write the formulas of the conjugate acid and base of each.
    a. OH^-
    b. C_2H_6
    c. HS^-
    d. $H_2PO_4^-$

Acid–Base Titrations

20. A 20.0-mL sample of 0.136 M HCl was titrated with NaOH solution using phenolphthalein indicator. The endpoint was observed when 12.3 mL of base had been added to the reaction flask. Calculate the molarity of the NaOH solution.
21. A 20.0-mL sample of HCl solution required 18.9 mL of 0.133 M NaOH solution to reach the endpoint of a titration. What was the molarity of the HCl solution?
22. What is the molarity of a H_2SO_4 solution if 10.0 mL required 13.4 mL of 0.233 M KOH to reach the endpoint of a titration?
23. A 0.235-g sample of oxalic acid dihydrate ($H_2C_2O_4 \cdot 2H_2O$) was dissolved in water and titrated with 23.3 mL of NaOH solution to the endpoint. What was the molarity of the NaOH solution?
24. Using phenolphthalein as the indicator, a 10.0-mL sample of 0.144 M sulfuric acid solution was titrated with 13.6 mL of NaOH solution to the endpoint. What was the molarity of the NaOH solution?

Acid and Base Anhydrides

25. Explain what is meant by an acid anhydride. Give an example involving an element in Group 6A in the periodic table.
26. What is meant by a base anhydride? Give an example involving an element in Group 2A in the periodic table.
27. Identify the acid formed when SO_3 reacts with water.
28. What is the acid anhydride of H_3AsO_3? Write a balanced equation for the formation of this acid from its anhydride.
29. Write an equation for the reaction of Cl_2O with water to give an oxyacid. What is the oxidation number of chlorine in these compounds?
30. Which of the following are base anhydrides?
    a. Na_2O
    b. BaO
    c. CO_2
    d. CuO
31. For each of the base anhydrides in problem 30, what is the base formed upon reaction with water?
32. Write a balanced equation for the reaction of each of the following anhydrides with water.
    a. CaO
    b. SO_2
    c. CO_2
    d. K_2O
33. Write a balanced equation for the reaction of ZnO with hydrochloric acid.
34. Acid rain results from the reaction of air pollutants with water in the atmosphere. What kinds of substances are these gases that react with water to give an acidic solution? What are two pollutants that contribute significantly to acid rain?
35. Some metal oxides are only slightly soluble in water but very soluble in dilute acids. Write an equation to describe the reaction of sulfuric acid with Al_2O_3, an oxide that forms on the surface of polished aluminum.

Strengths of Acids and Bases (Section 15.4)

36. Explain what is meant by a strong acid. Write an equation for the ionization of nitric acid in water.

37. What is meant by a weak acid? Write an equation for the ionization of acetic acid in water. What notation may be used in the equation to indicate that acetic acid is a weak acid?
38. Nitric acid is a stronger acid than acetic acid. Which is the stronger base, nitrate ion or acetate ion? Explain.
39. Which is the stronger base, Cl^-, or F^-? Refer to Table 15.5.
40. Rank the following acids (refer to Table 15.5), listing the strongest first: HF, HCl, $HClO_4$, H_2O, NH_4^+.
41. Refer to Table 15.5 and rank the following bases, listing the strongest first: F^-, Cl^-, ClO_4^-, OH^-, NH_3.

Ionization of Water, pH (Sections 15.5, 15.6)

42. Water behaves as both an acid and a base. Write an equation for the ionization of water and justify this statement.
43. What is the hydronium ion concentration in a neutral solution? What is the hydroxide ion concentration?
44. Give the definition of pH and of pOH.
45. What is the pH of a neutral solution? The pOH?
46. The concentration of H_3O^+ in lemon juice is about 5×10^{-3} M. What is the pH of lemon juice?
47. The concentration of HCl in gastric juice is about 0.1 M. What is the pH of gastric juice?
48. Household ammonia (NH_3) has a pH of 11.0. What is the hydronium ion concentration of the solution? What is the pOH? What is the hydroxide ion concentration?
49. The pH of orange juice is about 3.7 compared with a pH of about 2.3 for lemon juice. Suggest a reason why orange juice is not as sour as lemon juice.
50. Calculate the pH and pOH of the following solutions:
    a. 0.01 M HBr
    b. 0.001 M NaOH
    c. 0.0001 M HCl
    d. 1×10^{-5} M KOH
51. Find the pOH and the hydroxide ion concentration for each of the following solutions:
    a. pH = 5.0
    b. pH = 8.0
    c. pH = 3.0
    d. pH = 12.0
52. By what factor does hydronium ion concentration change when the pH of a solution changes from a pH of 2 to a pH of 5?
53. What instrument is used to make accurate pH measurements?

Buffer Solutions (Section 15.7)

54. Many chemical reactions in the human body are efficient only within a narrow pH range. What is responsible for maintaining the pH in body fluids?
55. What is the most common type of buffer solution?
56. Describe the composition of a formic acid (HCO_2H) buffer.
57. What is the composition of a buffer in which $H_2PO_4^-$ is the acid?
58. Describe the composition of a buffer in which HCO_3^- is the acid.
59. Explain how an acetic acid buffer resists a change in pH when a strong acid is added to a buffer solution. Write an equation for the reaction.
60. How does an acetic acid buffer resist a change in pH when a strong base (OH^-) is added? Write an equation for the reaction.
61. Explain how a HCO_3^-/CO_3^{2-} buffer resists a change in pH when a strong acid is added. Write an equation for the reaction.
62. Write an equation for the reaction that occurs when a strong base (OH^-) is added to a HCO_3^-/CO_3^{2-} buffer. Explain how the buffer solution prevents a change in pH.

Additional Problems

63. What is the molarity of a formic acid (HCO_2H) solution if 20.0 mL of the acid solution required 23.4 mL of 0.133 M KOH to reach the endpoint of a titration?
64. A solution of sodium bicarbonate was tested with pH paper and found to be slightly basic. Write an equation to describe the reaction of bicarbonate ion with water and explain the observation. What is the conjugate acid of bicarbonate ion? (Note that this acid is unstable. See Section 10.9.)
*65. Sulfuric acid, H_2SO_4, loses its two protons in a stepwise fashion. Write equations for the stepwise ionization of sulfuric acid in water. Suggest a reason why HSO_4^- is a weaker acid (loses its proton less readily) than H_2SO_4. (*Hint:* A proton is a positive ion.)
66. What is the pH of a 0.0010 M KOH solution?
67. Calculate the pH of a 0.0015 M hydrobromic acid (HBr) solution.
68. In the following list of acids, the numbers indicate relative acid strength (sulfuric acid is the strongest acid). Write formulas for their conjugate bases and rank them according to base strength (#1 for the strongest base).

    H_2SO_4 (1), H_2O (3), C_2H_5OH (4), C_6H_5OH (2)
69. Identify each of the following as a base anhydride or an acid anhydride. Write the formula of the substance that results when the anhydride reacts with water.
    a. ZnO
    b. Al_2O_3
    c. As_2O_5
    d. SeO_2
70. A solution of acetic acid had a pH of 3.60. Calculate the hydronium ion concentration in the solution.
71. The pH of a solution of calcium hydroxide is 11.6. What is the hydroxide ion concentration?
72. Calculate the volume of 0.156 M KOH solution needed to just neutralize a solution containing 0.265 g of a diprotic acid (H_2A), that has a molar mass of 126 g/mol.

Solutions to Practice Problems

PP 15.1

Find: $M \left(\frac{\text{mol KOH}}{\text{L soln}}\right) = ?$

Given: $\frac{0.667 \text{ mol } H_2C_2O_4}{\text{L soln}}$, 10.0 mL (0.0100 L) $H_2C_2O_4$ soln

12.7 mL (0.0127 L) KOH soln

Known: $2KOH(aq) + H_2C_2O_4(aq) \longrightarrow 2H_2O(l) + K_2C_2O_4(aq)$

From the equation we have:

$\frac{1 \text{ mol } H_2C_2O_4}{2 \text{ mol KOH}}$

Solution:

$$M \left(\frac{\text{mol KOH}}{\text{L soln}}\right) = ? \qquad (1)$$

The volume of the KOH solution is given. To calculate its molarity, first find the number of moles of KOH.

$$\text{number of moles KOH} = \left(\frac{2 \text{ mol KOH}}{1 \text{ mol } H_2C_2O_4}\right)(? \text{ mol } H_2C_2O_4) \qquad (2)$$

$$\text{number of moles } H_2C_2O_4 = \left(\frac{0.667 \text{ mol } H_2C_2O_4}{\cancel{\text{L soln}}}\right)(0.0100 \cancel{\text{ L soln}}) \qquad (3)$$

$$= 0.006\,67 \text{ mol } H_2C_2O_4$$

Substituting in equation (2) gives

$$\text{number of moles KOH} = \left(\frac{2 \text{ mol KOH}}{1 \cancel{\text{ mol } H_2C_2O_4}}\right)(0.006\,67 \cancel{\text{ mol } H_2C_2O_4}) \qquad (2)$$

$$= 0.0133 \text{ mol KOH}$$

Now, calculate the molarity of KOH (equation 1):

$$M \left(\frac{\text{mol KOH}}{\text{L soln}}\right) = \frac{(0.0133 \text{ mol KOH})}{(0.0127 \text{ L soln})} = 1.05 \text{ M KOH} \qquad (1)$$

PP 15.2

Copper(II) hydroxide, $Cu(OH)_2$. Note that the oxidation number of copper is the same in both compounds.

PP 15.3

$Cl_2O_7(g) + H_2O(l) \longrightarrow 2HClO_4(aq)$

PP 15.4

A conjugate base differs from its acid by a single proton (H^+).
a. NO_2^- b. CH_3^- c. HSO_3^-

PP 15.5

Conjugate acid: H_2CO_3; conjugate base: CO_3^{2-}

PP 15.6

$O^{2-} > C_2H_3O_2^- > NO_3^-$

PP 15.7

The solution will be basic.

$NH_2^-(aq) + H_2O(l) \longrightarrow NH_3(aq) + OH^-(aq)$

PP 15.8

$pH = -\log(1.0 \times 10^{-11}) = 11.00$ (2 significant figures)

PP 15.9

$[H_3O^+] = 10^{-pH} = 10^{-8.7}$

Using your calculator:

1. Enter the pH (8.7).
2. Press the change-of-sign key.
3. Press the 10^x key.

Thus, $[H_3O^+] = 2 \times 10^{-9}$ M (1 significant figure)

Answers to Chapter Review

1. sour
2. indicators
3. red
4. CO_2
5. neutralize
6. hydroxide
7. bitter
8. blue
9. pink
10. soaps
11. acids
12. H_3O^+
13. base
14. lye
15. titration
16. endpoint
17. equivalence
18. anhydride
19. nonmetal
20. metal
21. water
22. donor
23. proton
24. conjugate
25. OH^-
26. two
27. electrons
28. amphoteric

29. diprotic
30. strong
31. ionizes
32. weak
33. Cl^-
34. acetate
35. 1×10^{-7} M
36. ionization
37. pH
38. neutral
39. acidic
40. basic
41. $-\log[H_3O^+]$
42. tenfold
43. $-\log[OH^-]$
44. 14
45. papers
46. meter
47. buffer
48. change
49. acetic acid
50. strength
51. capacity

16

Reaction Rates and Chemical Equilibrium

Contents

A forest fire burns rapidly in a strong wind.

You Probably Know . . .

- A forest fire rages in a strong wind. The fire burns faster in a strong wind because air (oxygen) is continuously supplied to the burning trees.
- When wood burns, it combines with oxygen in the air. Carbon dioxide and water are formed, together with smoke resulting from unburned particles dispersed in air. However, dry wood can be in contact with air for months or years without burning.
- A mixture of natural gas and air fails to react at ordinary temperatures, but a spark will cause the explosion of the mixture. Similar circumstances caused the explosions of hydrogen-filled dirigibles in the 1930s, resulting in the deaths of many people.
- Equilibrium is a balanced condition for a reversible process. Chapter 13 discussed equilibrium between the liquid and solid states of a substance. Many chemical reactions are also in a state of equilibrium.

Why Should We Study Reaction Rates and Chemical Equilibrium?

Why are certain mixtures sometimes stable and sometimes explosive? What factors influence how fast chemical reactions occur? Why in some reactions are products formed in very high yields, whereas in other reactions the product yields are very low? In general, the answers to these questions depend upon our understanding of reaction rates and chemical equilibrium.

16.1 Reaction Rates and Collision Theory of Reactions

Learning Objectives

- Briefly describe the collision theory of reactions.
- Explain the criteria for an effective collision.
- Explain why large, complex molecules tend to react more slowly than small molecules.

A reaction rate expresses how fast products are formed.

Reaction kinetics is the study of reaction rates.

Imagine blowing soap bubbles using hydrogen gas. When the soap bubbles break, hydrogen mixes with oxygen in the air. According to the kinetic molecular theory (Section 12.2), hydrogen and oxygen molecules are moving rapidly and are colliding with each other. Why doesn't the mixture burn? When a burning candle tied to the end of a long pole ignites a bubble, the gas mixture explodes in a ball of fire (Figure 16.1). Why did the mixture explode? How does a tiny amount of energy from a candle cause an explosion? The principles of **reaction kinetics,** or the study of reaction rates, provide answers to these questions. The rate of a reaction expresses how fast products are formed from reactants.

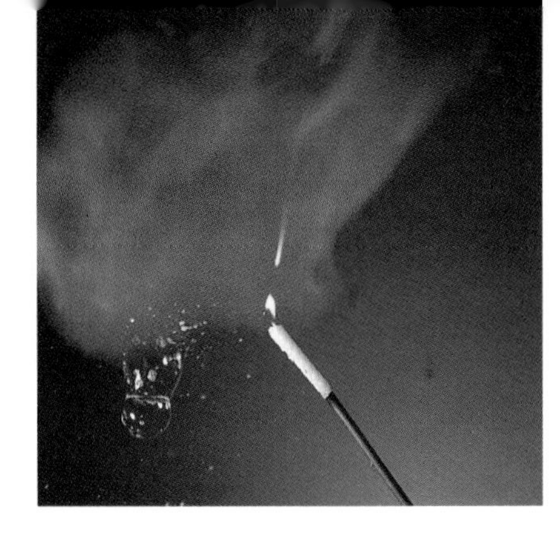

FIGURE 16.1 *Left:* Hydrogen gas is used to blow soap bubbles. *Right:* As the bubbles float upward, they are ignited using a candle on a long pole. An orange flame results from the reaction of hydrogen with oxygen in the air.

The **collision theory** of reactions assumes that reactant particles must collide for a chemical reaction to occur. However, not all collisions are effective. An **effective collision** causes bonds of the reactants to break, allowing new bonds of the products to form. Following an **ineffective collision,** the colliding molecules bounce apart unchanged; no bonds are broken and no products are formed.

An effective collision depends upon the

1. *kinetic energy, and*
2. *orientation of the colliding molecules.*

Whether or not a collision is effective is determined primarily by two factors: (1) the kinetic energy of the colliding molecules and (2) their orientation when they collide. Energy is needed to break the bonds of the reactants. When the colliding molecules have sufficient energy to react, they must also be oriented properly to allow new bonds to form to give the products.

To understand these factors, consider the reaction of the reactants A_2 and B_2 to give AB:

$$A_2 + B_2 \longrightarrow 2AB$$

Effective and ineffective collisions are illustrated in Figure 16.2. The effective collision shown in (a) has both sufficient kinetic energy to break bonds of the reactants

Reactants → High-energy complex → Products

a In an effective collision of A_2 and B_2, the colliding molecules have both sufficient kinetic energy to break A—A and B—B bonds and the proper orientation for the formation of A—B bonds.

Reactants → Collision → Reactants

b Even when oriented properly for the formation of A—B bonds, collision of A_2 and B_2 is ineffective because they have insufficient kinetic energy to break A—A and B—B bonds.

A — A B — B → A — A * B — B → A — A B — B

Reactants Collision Reactants

c Even though colliding molecules have sufficient energy to break bonds, the collision is ineffective because they do not have the proper orientation to form A—B bonds.

FIGURE 16.2 Effective and ineffective collisions of A_2 and B_2.

FIGURE 16.3 In the reaction of hydrogen (H_2) with the C=C double bond in a large, complex molecule (a), the reactants are less likely to be oriented properly for bonding to occur between specific atoms in the products. Consequently, large molecules tend to react more slowly than smaller molecules (b).

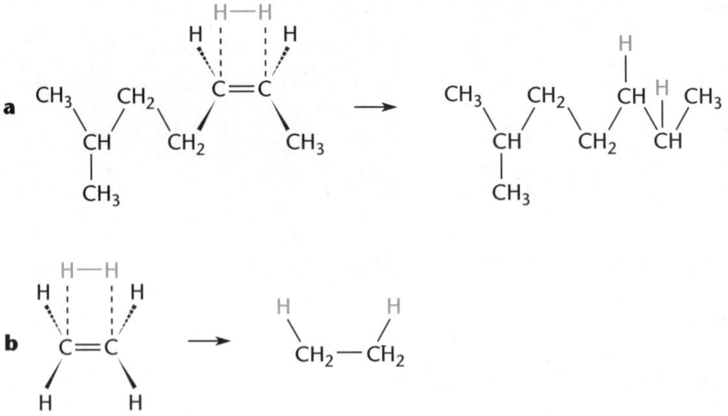

and the proper orientation of the reacting molecules (A_2 and B_2) to form the bonds of the product molecules (A—B). The collision shown in (b) is ineffective because, although the colliding molecules are oriented properly, they lack sufficient kinetic energy to break the bonds of the reactants, A_2 and B_2. The collision shown in (c) is ineffective because the colliding molecules lack the proper orientation to form A—B bonds. As they move randomly about, large, complex molecules are less likely to be oriented properly when they collide (Figure 16.3). Consequently, they tend to react more slowly than smaller, simpler molecules.

At this point in our discussion, we can see that how fast a reaction occurs depends upon the number of effective collisions occurring in a certain period of time. Moreover, the number of effective collisions is determined by both an energy factor and an orientation factor. The energy factor is discussed further in Section 16.2.

16.2 Energy Changes During a Reaction

Learning Objectives

- Draw potential energy diagrams for exothermic and endothermic reactions showing the reactants, products, transition state, E_a, and ΔH.
- Explain why a reaction that has a large E_a tends to be slower than a reaction that has a small E_a.

Transition state: A high-energy complex in which reactant bonds are breaking and product bonds are forming.

As we have noted, colliding molecules must have sufficient kinetic energy to break bonds for the collision to be effective. When A_2 and B_2 collide, the velocity of the molecules decreases, and the decrease in kinetic energy results in an increase in potential energy (stored energy). Momentarily, the substances exist in a high-energy state, called a **transition state,** T.S. (Figure 16.4). This high-energy complex was illustrated in Figure 16.2 (a). In the transition state, bonds of the reactants are breaking and bonds of the products are forming. When the bond breaking and forming process is completed to form AB, the potential energy decreases and the kinetic energy increases once again.

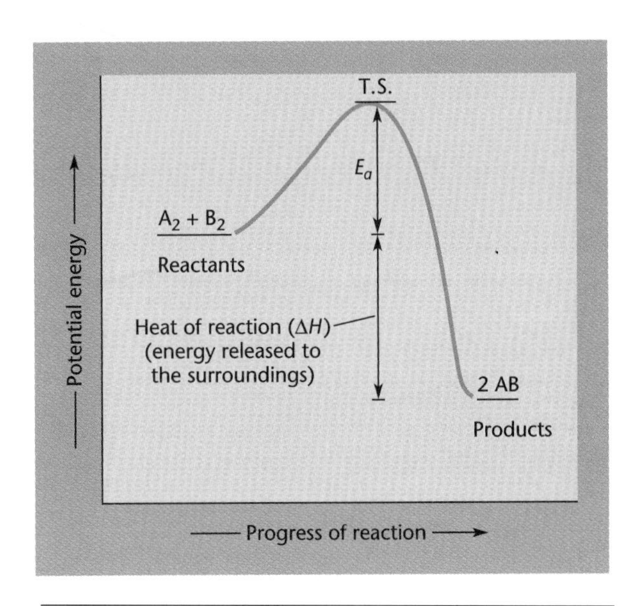

FIGURE 16.4 Energy changes for an exothermic reaction of A_2 with B_2. When A_2 and B_2 collide, bond breaking begins and A_2 and B_2 move to a high-energy state called a transition state (T.S.). When the new bonds of the product form, AB is at a lower energy state, as shown.

We can compare this description of a changing energy pathway for a chemical reaction to a road trip over a mountain pass. Suppose we have left a mountain town (elevation 9000 ft) and are traveling toward Denver (elevation about 5000 ft). This may appear to be an easy downhill ride, but the road goes over a mountain pass that is 11 000 ft high. To reach Denver, we must first ascend from 9000 ft to the top of the pass. If we start climbing this pass and run out of gas (that is, we do not have enough energy to reach the top), we will not reach Denver. However, once we reach the summit, we can coast downhill to Denver.

The "hump" of the diagram in Figure 16.4 represents an energy barrier between reactants and products. The difference between the potential energy of the reactants and that of the transition state is the **activation energy,** E_a, of the reaction. It is the minimum kinetic energy needed for colliding molecules to react. When the kinetic energy of colliding molecules is equal to or greater than E_a, the collision is effective and products form. When the kinetic energy of colliding molecules is less than E_a, the collision is ineffective and the molecules bounce apart unchanged.

Activation energy: The minimum kinetic energy needed for colliding molecules to react.

The activation energy of a reaction is related to the energy needed to begin breaking bonds of the reactants. Consequently, the activation energy is characteristic of a particular reaction and cannot be changed by changing reaction conditions such as temperature.

A reaction that has a small activation energy is faster at a given temperature than a reaction that has a larger activation energy. In a reaction with a smaller activation energy, a larger number of molecules have kinetic energies that exceed the activation energy of the reaction, and more collisions are effective than in a reaction with a larger activation energy (Figure 16.5). Reactions between oppositely charged ions in solution have small activation energies and are generally very fast. For example, the neutralization reaction

If E_a is small, the reaction is fast. If E_a is large, the reaction is slow.

$$H^+(aq) + OH^-(aq) \longrightarrow H_2O(l)$$

FIGURE 16.5 Energy distribution curve for colliding molecules. The shaded area under the curve represents the number of collisions in that energy range. The number of collisions having energies greater than E_a is smaller for a reaction with a larger E_a. Consequently, the reaction with a larger E_a is slower.

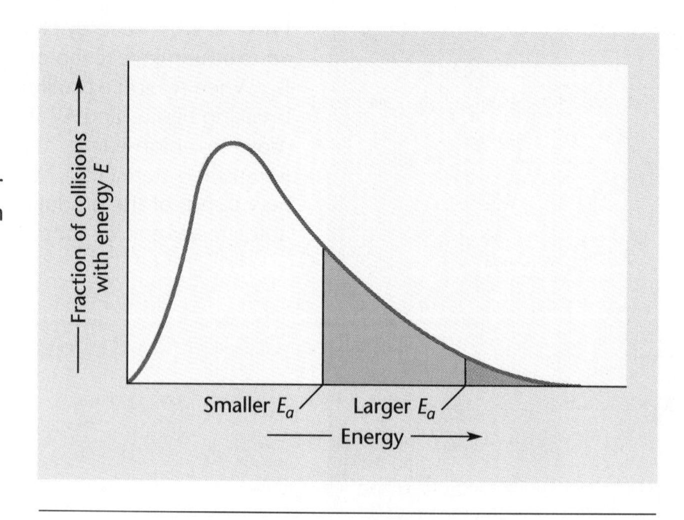

is one of the fastest reactions for which a rate has been measured. On the other hand, reactions between molecules involving breaking and forming of covalent bonds are generally slower, in large part because of greater activation energies.

The difference between the potential energy states of the reactants and products is related to the **heat of reaction,** ΔH. When the potential energy of the products is less than the potential energy of the reactants, as illustrated in Figure 16.4, the reaction is exothermic, and energy is released to the surroundings. Although activation energy is needed to start a reaction, the energy released by an exothermic reaction is often sufficient to keep it going at a rapid rate. For example, although a piece of paper must be heated to ignite it, the heat released as it burns keeps it burning rapidly.

An energy diagram for an endothermic reaction of C_2 with D_2 is shown in Figure 16.6.

$$C_2 + D_2 \longrightarrow 2CD$$

FIGURE 16.6 Energy changes for an endothermic reaction of C_2 with D_2. When C_2 and D_2 molecules collide, bond breaking begins and they move to the transition state, T.S. New C—D bonds form to give the product, CD. The energy state of the product (CD) is higher than that of the reactants (C_2 and D_2), indicating that energy has been absorbed from the surroundings.

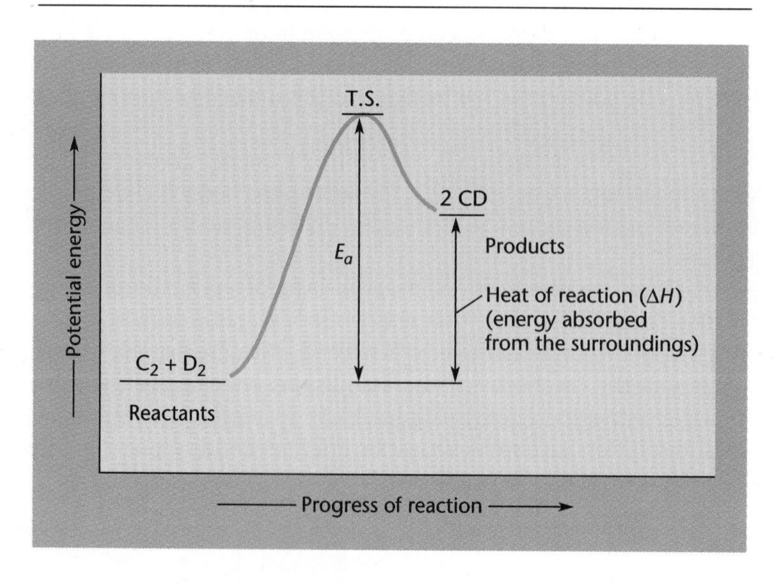

The potential energy of the products is greater than that of the reactants, and so energy is being absorbed from the surroundings. The energy of the transition state is always greater than that of the products, and the activation energy of a highly endothermic reaction is large. As a result, highly endothermic reactions are generally slower than exothermic reactions.

Variables That Affect Reaction Rate

16.3

Learning Objectives

- Describe three reaction variables that affect the rate of a reaction.
- Explain why raising the temperature increases the rate of a reaction.
- Explain how adding a catalyst increases the rate of a reaction.
- Explain why increasing the concentrations of reactants generally increases the rate of a reaction.

In our discussion to this point, two factors that affect the rate of a reaction have emerged. Both of these, the energy of activation (E_a) and the orientation factor, are characteristics of a particular reaction. They are not affected by changing reaction conditions such as temperature and the concentrations of the reactants.

E_a and the orientation factor are unaffected by changing the temperature or concentrations of reactants.

The effects on the reaction rate of temperature, a catalyst, and concentrations of reactants are considered in this section.

Temperature

Chemical reactions are always faster at higher temperatures. Forest fires burn faster during hot summer days than in cooler weather. Foods cook faster at higher temperatures, and they spoil only slowly when stored at lower temperatures.

Reactions are faster at higher temperatures.

The effect of temperature on reaction rate can be explained by considering Figure 16.7. When the temperature of a sample is raised from T_1 to T_2, the number

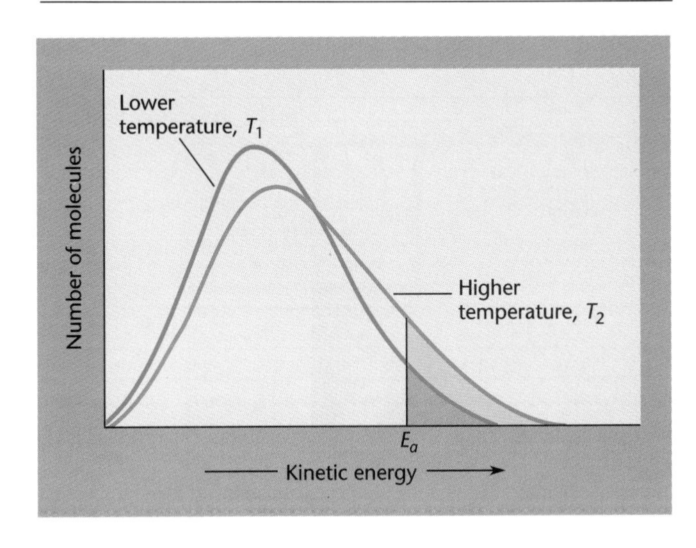

FIGURE 16.7 Distribution of kinetic energies of molecules at two temperatures. Only molecules with kinetic energies greater than the activation energy, E_a, for the reaction will react. At the higher temperature, T_2, a larger number of molecules (represented by the shaded area under the red line) have energies greater than E_a than at the lower temperature, T_1 (shaded area under the blue line). Thus, at T_2, more molecules react in a given time, and the reaction at T_2 is faster than at T_1.

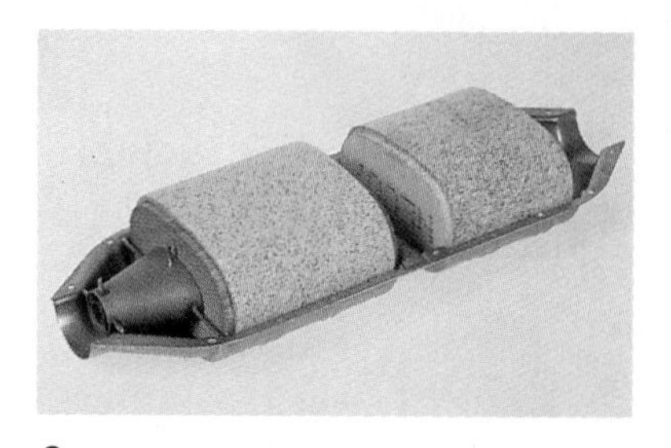

a

b

FIGURE 16.8 (a) An automotive catalytic converter is designed to reduce air pollution resulting from unburned hydrocarbons, carbon monoxide, and nitrogen oxides in the engine exhaust. (b) Hydrocarbons react to form CO_2 and H_2O, CO is converted to CO_2, and nitrogen oxides (NO_x) are changed back to N_2 and O_2.

Foods cook faster at higher temperatures.

of molecules having kinetic energies exceeding E_a increases. More collisions are effective, and so the reaction rate increases. For many reactions, increasing the temperature by 10°C causes the rate to increase two to three times. That is why foods cook faster at higher temperatures. Furthermore, many foods spoil rapidly when allowed to remain unrefrigerated on a warm summer day.

With an understanding of the effect of temperature on reaction rate, a chemist can select a temperature necessary for a reaction to be successful. For example, there is no detectable reaction of nitrogen with hydrogen at temperatures of 100–200°C. However, the reaction is quite fast at 500°C, and ammonia is prepared commercially at high temperature:

$$N_2(g) + 3H_2(g) \xrightarrow{500°C} 2NH_3(g)$$

A Catalyst

Catalyst: A substance that increases a reaction rate without being consumed.

A **catalyst** is a substance that increases the rate of a reaction without being consumed. Catalysts are quite common, and many chemical reactions would be impractical without them. Modern cars use catalytic converters to reduce exhaust emissions and to minimize air pollution (Figure 16.8). Our bodies use catalysts called **enzymes,** without which we could not live. Chemists use catalysts in labora-

a

b

FIGURE 16.9 (a) At room temperature, pure hydrogen peroxide undergoes no detectable decomposition. (b) When a lump of manganese(IV) oxide, a catalyst for the reaction, is immersed in the solution, decomposition occurs rapidly, as indicated by the evolution of O_2 gas.

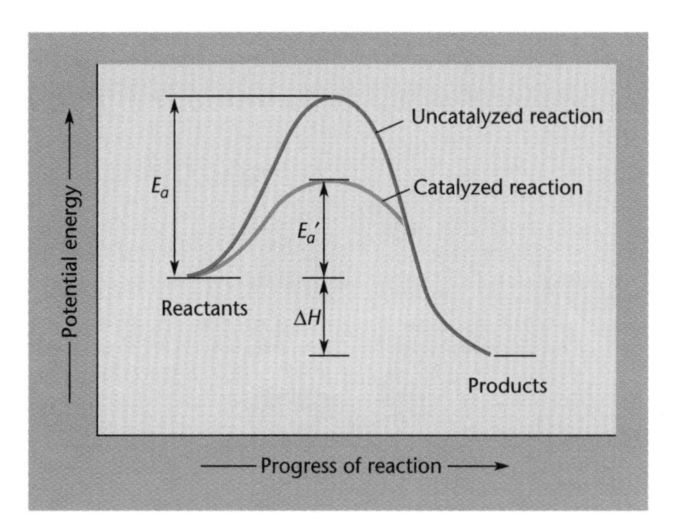

FIGURE 16.10 A catalyst allows a reaction to take place along a lower energy pathway. The activation energy, E_a', for a catalyzed reaction is smaller than E_a for an uncatalyzed reaction. Consequently, more molecules can react in a given time, and the reaction is faster.

tory reactions such as the decomposition of hydrogen peroxide (Figure 16.9), and chemical engineers use them in chemical processes for the manufacture of industrial chemicals. Catalysts have been known since the time of the alchemists, but only in recent years through chemical research have we begun to understand how they work.

A catalyst provides a different pathway for a reaction (Figure 16.10). The new reaction pathway leads to the same products but has a smaller activation energy. Because E_a is smaller, more molecules have kinetic energies greater than E_a, and the reaction is faster. Although a catalyst may react with reactants, it is reformed as the products are formed and is therefore not consumed in a reaction.

Adding a catalyst to a reaction is similar to boring a tunnel through a mountain (Figure 16.11). Both roads lead to the same resort, but the road through the tunnel is a lower energy pathway compared with the road over the mountain. Driving time through the tunnel is shorter than along the road over the mountain, resulting in a faster trip to the resort. In a similar way, a catalyzed reaction along a lower energy pathway is faster.

Because a catalyst is not consumed in a reaction, often only a small amount is needed. Furthermore, we do not need to continually add a catalyst to keep a reaction going as we do with the reactants. For example, to keep a car engine running,

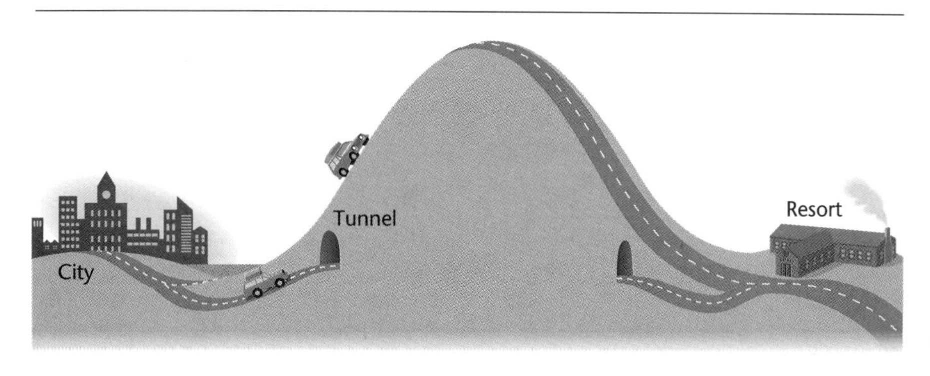

FIGURE 16.11 A catalyzed reaction is similar to traveling on the road through the tunnel to the resort. The trip through the tunnel is faster than driving over the mountain. Likewise, a catalyzed reaction along a lower energy pathway is faster than an uncatalyzed reaction along a higher energy pathway.

CHEMICAL WORLD

$H_2 + O_2 \rightarrow$!!!

Water is born of a violent chemical reaction. Most of us have seen the pictures of the flaming demise of the Hindenberg zeppelin and the explosion that destroyed the space shuttle Challenger, but few people are aware of the smaller hydrogen explosions that have caused more injuries than those two tragedies combined: car battery explosions.

While you run your car to school, your alternator charges your battery but also causes the electrolysis of water in the electrolyte to produce an explosive mixture of hydrogen and oxygen gases:

$$2H_2O(l) \longrightarrow 2H_2(g) + O_2(g)$$

No problem. Hydrogen gas can't blow up by itself. A flame or spark is needed to raise the kinetic energy of hydrogen above its activation energy, E_a. It's fine, as long as you don't check your battery's fluid level with a match—or jump-start your car incorrectly.

The instructions are there in your owner's manual, but you probably didn't read it. Anyway, can there be much hydrogen in a battery? No. But the trouble is that even a small explosion throws sulfuric acid and battery case fragments about. This can be very hard on eyes, which you will need to read the instructions about how to jump-start your car correctly. About 6000 eye injuries every year in the United States are caused by explosions of lead–acid batteries.[1]

The most important point to remember is *not* to make the final jumper-cable connection *between the two batteries.* Make the final connection between bare metal engine surfaces of the two cars. Make it a few feet away from the batteries. Then if there's a spark, it will jump from the cable to the engine and not to the battery.[2]

To avoid a hydrogen explosion, follow instructions carefully when jump-starting a car.

[1]Source: The National Society to Prevent Blindness (NSPB).
[2]Check your owner's manual. Some cars require a different procedure.

gasoline and air (the reactants) must continually flow into the engine. However, a catalyst in a car's catalytic converter will continue to work throughout the entire life of the car without being consumed. Although a catalyst may be expensive, its cost is generally a one-time cost, unlike the cost of gasoline to operate a car. Over the life of a car, the cost of fuel is far greater than the cost of a catalyst.

Concentrations of Reactants

Increasing the concentration of reactants increases the rate of a reaction.

Increasing the concentrations of reactants generally increases a reaction rate. The rate of a reaction depends upon the frequency of effective collisions of the reac-

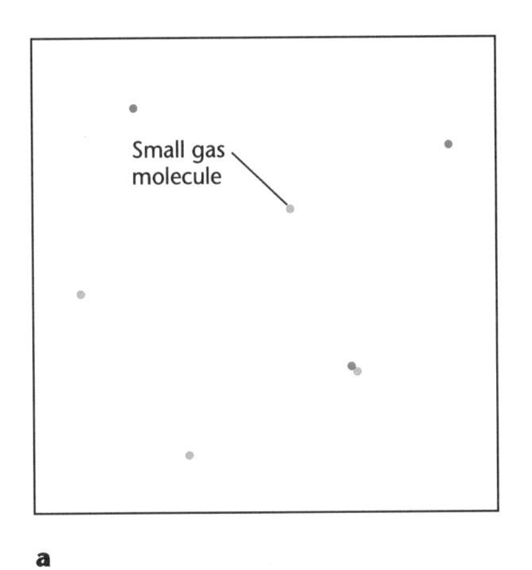

a

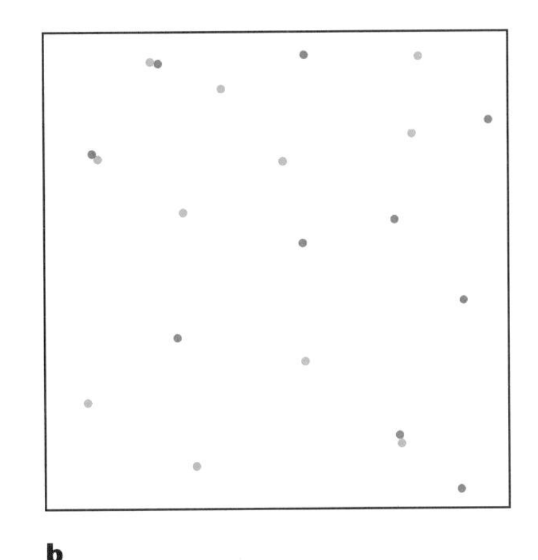

b

FIGURE 16.12 (a) Molecules at low concentration collide only occasionally. (b) At high concentration, however, molecules are crowded and collide more frequently. The number of effective collisions is also greater, resulting in a faster reaction.

tants. At higher concentrations, molecules are more crowded (Figure 16.12) and they collide more frequently. Increasing the total number of collisions also increases the number of effective collisions, resulting in a faster reaction.

The effect of concentration on reaction rate is apparent if we compare the reaction rate of a substance burned in an atmosphere of 100% oxygen with its reaction in air, about 21% oxygen. For example, hot steel wool (Fe) oxidizes only slowly in air but bursts into flame in 100% oxygen (Figure 16.13). The higher concentration of oxygen increases the collision frequency of oxygen molecules with the hot steel wool, resulting in a rapid reaction. Because of the rapid combustion of materials in oxygen-rich atmospheres, cigarette smoking and flames are prohibited in a hospital room where a patient is breathing oxygen.

a

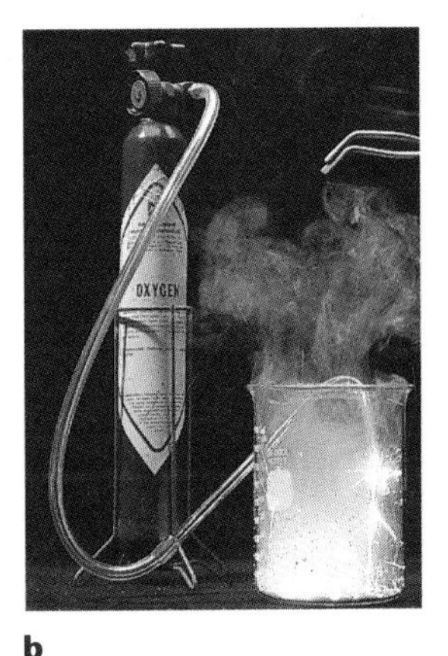

b

FIGURE 16.13 Hot steel wool in air (a) oxidizes rather slowly, but in an atmosphere of 100% oxygen (b) it bursts into flame.

Factors That Affect Reaction Rate

Factor	*Condition*	*Effect on Reaction Rate*
Reaction Characteristics		
activation energy	large	slow
	small	fast
molecular orientation	simple molecules	fast
	complex molecules	slow
Reaction Variables		
temperature	increase	increase
	decrease	decrease
catalyst	—	increase
concentration of reactants	increase	increase
	decrease	decrease

16.4 Reversible Reactions and Chemical Equilibrium

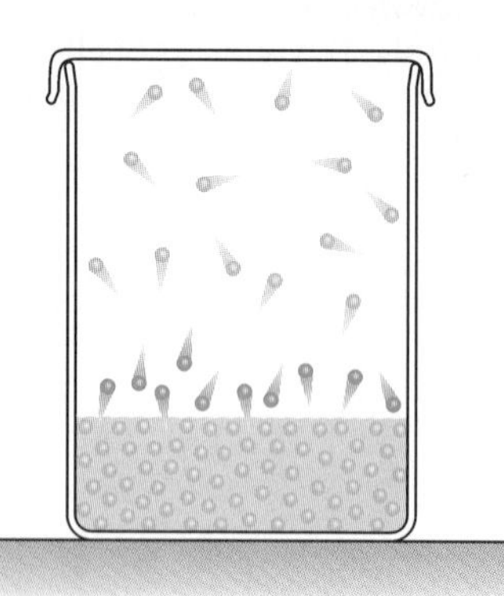

Liquid water and its vapor in dynamic equilibrium (Section 13.3). As long as the jar is tightly covered, the quantities of liquid water and its vapor remain constant.

The brown color of photochemical smog is largely due to NO_2.

Learning Objective

- Describe three important characteristics of chemical equilibrium.

Equilibrium is a balanced state or condition. To illustrate, consider the evaporation of water. If allowed to stand in an open jar, water will eventually completely evaporate. However, when water is stored in a tightly covered jar, vaporization of water is balanced by condensation of its vapor, and so the quantity of liquid water remains constant indefinitely. The liquid water and its vapor are in a state of dynamic equilibrium, a balanced condition in which the rates of opposing processes are equal.

Many chemical reactions are reversible. An example is the reaction of nitrogen dioxide (NO_2) to form dinitrogen tetroxide (N_2O_4):

$$\underset{\text{brown}}{2NO_2(g)} \rightleftarrows \underset{\text{colorless}}{N_2O_4(g)}$$

The progress of the reaction can be followed visually because NO_2 is brown and N_2O_4 is colorless. When NO_2 is placed in a closed glass vial (Figure 16.14), the forward reaction is faster because the NO_2 concentration is high. Initially, the rate of the reverse reaction is zero because there is no N_2O_4. The color begins to fade as NO_2 (brown) reacts and as N_2O_4 (colorless) forms. The rate of the forward reaction decreases as the concentration of NO_2 decreases, and the rate of the reverse reaction increases as the concentration of N_2O_4 increases. In time, the rates of the opposing reactions become equal. The color then remains unchanged, indicating that the concentration of NO_2 is no longer changing. At this point, the reaction is at equilibrium. **Chemical equilibrium** is the state in which the rates of the forward and reverse reactions are equal.

This discussion points out three important characteristics of chemical equilibrium:

1. The rates of the opposing reactions are equal.
2. Concentrations remain constant at equilibrium even though the opposing reactions are continuing.

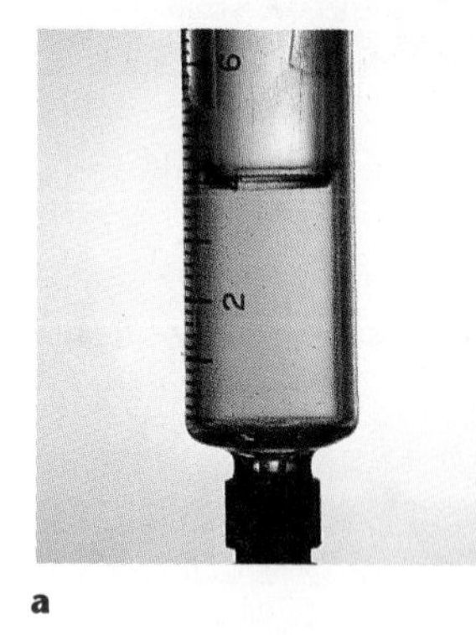

a

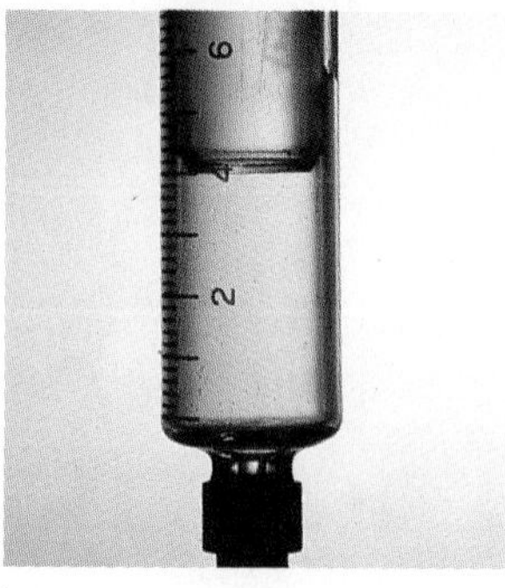

b

FIGURE 16.14 (a) A vial containing NO_2 gas. (b) A vial containing a mixture of NO_2 and N_2O_4. As NO_2 reacts to form colorless N_2O_4 gas, the brown color of the mixture fades. When equilibrium has been reached, the color remains unchanged, indicating the concentrations are no longer changing.

3. Attainment of equilibrium is not instantaneous and often requires an extended period of time.

All of these characteristics are important. The first establishes the requirement for chemical equilibrium. The third characteristic reminds us that not all reversible reactions are at equilibrium. We must remember that some reactions take considerable time to reach equilibrium. We will discuss the second characteristic involving concentrations of reactants and products in Section 16.5.

Characteristics of Chemical Equilibrium

- Opposing reactions are occurring at equal rates.
- Equilibrium concentrations of reactants and products are constant.
- Attainment of equilibrium is not immediate and often requires an extended period of time.

Equilibrium Constants

16.5

Learning Objectives

- Write equilibrium constant (K_{eq}) expressions for reactions at equilibrium.
- Calculate the value of K_{eq} from equilibrium concentrations.
- Predict the position of equilibrium from the value of K_{eq}.

When a reversible reaction is at equilibrium, the concentrations of the reactants and products are constant at a given temperature. An equilibrium constant expression can be written that relates the concentrations of the products to the concentrations of the reactants. Consider the general reaction:

$$\underbrace{aA + bB}_{\text{reactants}} \rightleftarrows \underbrace{cC + dD}_{\text{products}}$$

An expression for the **equilibrium constant,** K_{eq}, is written as follows:

$$K_{eq} = \frac{[C]^c[D]^d}{[A]^a[B]^b}$$

The quantitites in brackets are molar concentrations of products and reactants at equilibrium. In writing the expression for K_{eq}, each concentration is raised to a power equal to its balancing coefficient in the equation. The concentrations of the products are written in the numerator, and those of the reactants appear in the denominator. Because all concentrations are constant at equilibrium, K_{eq} is a constant at a given temperature.

Because equilibrium reactions are reversible, the chemical equation and equilibrium constant expression could be written:

$$cC + dD \rightleftarrows aA + bB$$

$$K_{eq}' = \frac{[A]^a[B]^b}{[C]^c[D]^d} = \frac{1}{K_{eq}}$$

Note that the equilibrium constant expression and the value of K_{eq} correspond to the chemical equation as written.

The expression for K_{eq} is known as the *law of chemical equilibrium* or the law of mass action. It was developed by two Norwegian chemists, Cato Maximilian Guldberg and Peter Waage, on the basis of observations of many chemical reactions. Since the expression for K_{eq} was first proposed in 1864, it has been verified experimentally with numerous chemical reactions.*

To illustrate how to write an expression for K_{eq}, consider the synthesis of ammonia from nitrogen and hydrogen:

$$3H_2(g) + N_2(g) \rightleftarrows 2NH_3(g)$$

When the reaction has reached equilibrium, the concentrations of H_2, N_2, and NH_3 all remain constant at a given temperature. An expression for K_{eq} can be written from the chemical equation for the reaction:

$$K_{eq} = \frac{[NH_3]^2}{[H_2]^3\,[N_2]} \quad \begin{matrix} \leftarrow \text{product} \\ \leftarrow \text{reactants} \end{matrix}$$

Remember, brackets are used with a formula, e.g., $[NH_3]$, to indicate the molarity of the substance in the equilibrium mixture. Each concentration is raised to a power equal to its balancing coefficient in the chemical equation. Concentrations of products appear in the numerator, and concentrations of reactants are written in the denominator of the expression.

Example 16.1

Write the expression for the equilibrium constant for the following reaction:

$$H_2(g) + Br_2(g) \rightleftarrows 2HBr(g)$$ ■

Solution

$$K_{eq} = \frac{[HBr]^2}{[H_2]\,[Br_2]}$$

The concentration of HBr is written in the numerator with an exponent of 2 in agreement with its balancing coefficient in the chemical equation. A balancing coefficient of 1 is not written in the chemical equation, nor are exponents of 1 written for $[H_2]$ and $[Br_2]$ in the equation for K_{eq}.

■ Calculation of an Equilibrium Constant

The value of an equilibrium constant for a reaction must be determined from experimental data. The value of K_{eq} can be calculated by substituting equilibrium concentrations of the reactants and products into the K_{eq} expression for the reaction. For example, in an experiment with the following reaction at 25°C,

$$2NO_2(g) \rightleftarrows N_2O_4(g)$$

*The expression for K_{eq} can also be theoretically derived.

the equilibrium concentration of NO_2 was 0.009 66 M, and the concentration of N_2O_4 was 0.0202 M. From these data, K_{eq} can be calculated as follows:

$$K_{eq} = \frac{[N_2O_4]}{[NO_2]^2} = \frac{(0.0202)}{(0.009\ 66)^2} = 216$$

Note that units generally are not used for equilibrium constants. The value of K_{eq} is constant at a given temperature but varies when the temperature is changed. The effect of temperature on the value of K_{eq} is discussed in Section 16.6.

Example 16.2

Calculate the equilibrium constant for the following reaction using the equilibrium concentrations given for the reactants and products.

$$2NOCl(g) \rightleftarrows 2NO(g) + Cl_2(g)$$

$[NOCl] = 1.0$ M, $[NO] = 0.044$ M, $[Cl_2] = 0.022$ M ■

Solution

1. Write the K_{eq} expression:

$$K_{eq} = \frac{[NO]^2\,[Cl_2]}{[NOCl]^2}$$

2. Substitute equilibrium concentrations into the equation and calculate the answer:

$$K_{eq} = \frac{(0.044)^2\,(0.022)}{(1.0)^2} = 4.3 \times 10^{-5}$$

Practice Problem 16.1 Nitrogen and hydrogen react to form ammonia as described by the following equation.

$$3H_2(g) + N_2(g) \rightleftarrows 2NH_3(g)$$

In one experiment, the equilibrium concentrations were:

$[NH_3] = 0.150$ M, $[H_2] = 0.575$ M, $[N_2] = 0.425$ M

Calculate K_{eq} at the temperature of the experiment. (Solutions to practice problems appear at the end of the chapter.)

■ Position of Equilibrium

The value of the equilibrium constant tells us something about the position of equilibrium, or whether the reaction favors products or reactants (Figure 16.15). If the equilibrium constant is a very large number, the position of equilibrium lies to the right. In other words, the concentrations of the products are high compared

FIGURE 16.15 The effect of K_{eq} on the position of equilibrium.

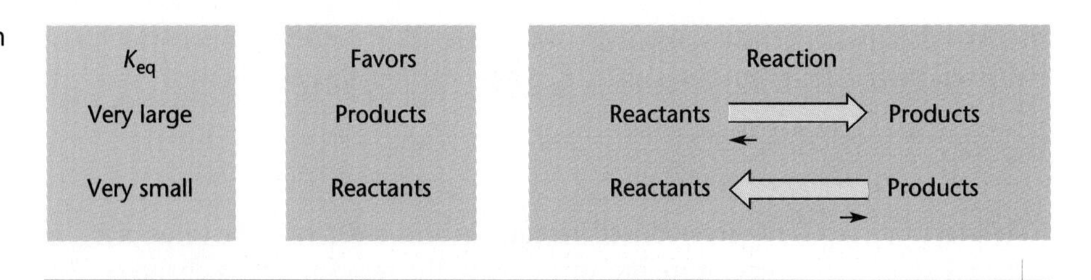

If K_{eq} is very large, the reaction favors products.

with the concentrations of the reactants, and formation of products is favored. For example, for the reaction:

$$A \rightleftarrows B$$

$$K_{eq} = \frac{[B]}{[A]}$$

If $K_{eq} = 10^5$, then $[B] >> [A]$.

If K_{eq} is very small, the reaction favors reactants.

On the other hand, if the equilibrium constant is a very small number, the position of equilibrium lies to the left, and very little of the reactants is converted to the products. For example, for the reaction:

$$B \rightleftarrows C$$

$$K_{eq} = \frac{[C]}{[B]}$$

If $K_{eq} = 10^{-5}$, then $[B] >> [C]$.

If the equilibrium constant is neither very large nor very small, that is, between about 100 and 0.01, neither products nor reactants are strongly favored.

Example 16.3

For the formation of HCl, described by the following equation, K_{eq} is 1.3×10^{34} at a certain temperature.

$$H_2(g) + Cl_2(g) \rightleftarrows 2HCl(g)$$

Write the K_{eq} expression. Is the formation of HCl favored? Explain. ■

Solution

$$K_{eq} = \frac{[HCl]^2}{[H_2]\,[Cl_2]} = 1.3 \times 10^{34} \qquad [HCl]^2 >> [H_2]\,[Cl_2]$$

Because K_{eq} is a very large value, the formation of HCl is favored.

Practice Problem 16.2 At room temperature K_{eq} is 1×10^{-30} for the formation of NO from N_2 and O_2. Write a balanced equation for the reaction and the K_{eq} expression. Is the formation of NO favored at room temperature?

Le Chatelier's Principle

16.6

Learning Objectives

- Use Le Chatelier's principle to predict how a change in concentration, a change in pressure or volume, and a change in temperature will affect the position of equilibrium.
- Describe how an increase or decrease in temperature affects the K_{eq} for a reaction.
- Describe the effect of a catalyst on equilibrium.

A chemical reaction at equilibrium will remain at equilibrium if undisturbed. However, imposing a change in reaction conditions or in the concentration of a reactant or a product can upset the balanced state of the opposing reactions, and the reaction will adjust to a new state of equilibrium. We can predict the effect of a change in a reaction variable by applying **Le Chatelier's principle:** If a change is imposed on a reaction at equilibrium, the reaction will adjust to counteract the effect of the change and establish a new equilibrium state. We will examine how this principle can be used to predict the effects of changes in concentration, pressure, volume, and temperature on a reaction at equilibrium.

Henri Le Chatelier (1850–1936) proposed a principle to predict the effects of changes imposed on a system at equilibrium.

Effect of Changes in Concentration on Reaction Equilibrium

Nitrogen oxide (NO) is formed when a mixture of nitrogen (N_2) and oxygen (O_2) is heated at high temperatures.

$$N_2(g) + O_2(g) \rightleftarrows 2NO(g)$$

After the reaction has reached equilibrium, adding O_2 to the mixture disturbs the balance of the forward and reverse reactions. Some of the added O_2 reacts with N_2 to form more NO. Using Le Chatelier's principle, we can predict that the position of equilibrium will shift to the right to counteract the increased O_2 concentration. As a result, the concentration of O_2 decreases, thereby counteracting the effect of the imposed change. The concentration of NO is higher at the new position of equilibrium.

A similar effect is observed if N_2 is added to the equilibrium mixture. Le Chatelier's principle predicts that the reaction will shift to the right. As a result, both the N_2 and the O_2 concentrations decrease after the addition of N_2, and the concentration of NO increases.

Consider again the preparation of ammonia from nitrogen and hydrogen:

$$3H_2(g) + N_2(g) \rightleftarrows 2NH_3(g)$$

Adding either H_2 or N_2 to the equilibrium mixture causes a shift in the position of equilibrium to the right, favoring the formation of NH_3. A shift to the right is also observed when NH_3 is removed from the equilibrium mixture. For example, NH_3 can be removed from the mixture by dissolving it in water, in which it is very soluble. Adding water has almost no effect on the concentrations of H_2 and N_2 because these molecules are nonpolar and are practically insoluble in water.

Le Chatelier's principle: A reaction will shift away *from a substance added and* toward *a substance removed from an equilibrium mixture.*

Consideration of these examples leads us to another way of stating Le Chatelier's principle: *When a reactant or product is added to an equilibrium mixture, the reaction will shift* away *from the substance added. When a reactant or product is removed from an equilibrium mixture, the reaction will shift* toward *the substance removed.*

In the previous example, when either H_2 or N_2 was added to the equilibrium mixture, the reaction shifted *away* from the substance added. When NH_3 was removed from the equilibrium mixture, the reaction shifted *toward* NH_3, the substance removed.

Although adding or removing a substance causes a shift in equilibrium, remember that the value of K_{eq} does not change. This can be understood by examining the K_{eq} expression:

$$K_{eq} = \frac{[NH_3]^2}{[H_2]^3\,[N_2]}$$

Adding N_2 to the equilibrium mixture causes a shift in equilibrium to the right. For example, after adding 4.72 mol N_2 to an equilibrium mixture in a 1.00-L container, the original and new equilibrium concentrations (determined experimentally) are as follows:

Original Equilibrium Position	*New Equilibrium Position*
$[N_2] = 0.300$ M	$[N_2] = 4.92$ M
$[H_2] = 0.400$ M	$[H_2] = 0.250$ M
$[NH_3] = 0.100$ M	$[NH_3] = 0.200$ M

Original Equilibrium

$$K_{eq} = \frac{(0.100)^2}{(0.400)^3\,(0.300)} = 0.521$$

New Equilibrium

$$K_{eq} = \frac{(0.200)^2}{(0.250)^3\,(4.92)} = 0.521$$

The values of K_{eq} calculated from the original and the new equilibrium concentrations are the same. The value of K_{eq} remained unchanged even though the concentrations of the reactants and the product are different at the new position of equilibrium.

Example 16.4

Predict the direction the reaction will shift to compensate for each change indicated in the equilibrium mixture.

$PCl_5(g) \rightleftarrows PCl_3(g) + Cl_2(g)$

a. Cl_2 is added.
b. Cl_2 is removed.
c. PCl_5 is added. ■

Solutions Le Chatelier's principle predicts the reaction will shift away from a substance added and toward a substance removed from the equilibrium mixture.

a. A shift to the left. More PCl_5 is formed from PCl_3 and Cl_2.
b. A shift to the right. More PCl_3 and Cl_2 are formed from PCl_5.
c. A shift to the right.

Practice Problem 16.3 Consider the following reaction at equilibrium at 200°C:

$CH_4(g) + H_2O(g) \rightleftarrows CO(g) + 3H_2(g)$

Predict how each of the following changes will influence the concentration of H_2:

a. Addition of H_2O to the system.
b. Removal of CO from the system.

Effect of Changing Pressure and Volume

According to the ideal gas equation (Section 12.9), the pressure of a gas is directly proportional to the number of moles of a gas at constant volume and temperature. Therefore, increasing the partial pressure of a gaseous reactant or product in an equilibrium mixture increases the number of moles of that substance and its concentration. When the concentration of a substance is increased, Le Chatelier's principle predicts that the reaction will shift away from that substance.

For example, consider the following reaction at equilibrium:

$$2NOCl(g) \rightleftarrows 2NO(g) + Cl_2(g)$$

Increasing the partial pressure of chlorine results in a shift to the left away from Cl_2. At the new position of equilibrium, the concentration of NO (and its partial pressure) are lower and the concentration of NOCl (and its partial pressure) are higher than the original equilibrium values.

On the other hand, decreasing the partial pressure of a gas in an equilibrium mixture results in a shift of the reaction toward that substance. This is exactly what happens when a bottle of a soft drink is opened.

$$H_2CO_3(aq) \rightleftarrows H_2O(l) + CO_2(g)$$

When the cap is removed, CO_2 escapes, its pressure drops, and the reaction shifts to the right toward CO_2. As CO_2 is lost from the solution, the fizz of the carbonated water is lost from the soft drink.

The partial pressure of CO_2 is reduced when a bottle of soft drink is opened. As a result, carbonic acid decomposes, CO_2 escapes from the solution, and the fizz of the carbonated water is lost.

Example 16.5

Predict the effect of each change on the concentration of HI in the following reaction at equilibrium:

$H_2(g) + I_2(g) \rightleftarrows 2HI(g)$

a. Increasing the partial pressure of H_2.
b. Increasing the partial pressure of I_2.
c. Decreasing the partial pressure of H_2.

Solutions

a. [HI] increases.
b. [HI] increases.
c. [HI] decreases.

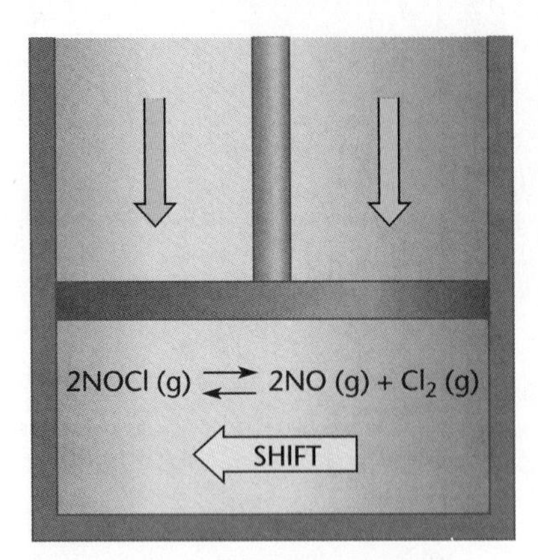

FIGURE 16.16 An equilibrium mixture of NOCl, NO, and Cl_2. When the volume is decreased, the reaction shifts toward the side having fewer molecules, in this case, toward the reactants.

$2NOCl(g) \rightleftarrows 2NO(g) + Cl_2(g)$

2 molecules 3 molecules

When the volume of a gaseous equilibrium mixture is decreased, the reaction shifts to the side having fewer molecules.

Practice Problem 16.4 Predict the effect of each change on the concentration of $H_2O(g)$ in the following reaction at equilibrium:

$$2H_2(g) + O_2(g) \rightleftarrows 2H_2O(g)$$

a. Increasing the partial pressure of O_2.
b. Decreasing the partial pressure of H_2.

Having considered the effect of changing the partial pressure of *one* of the gases in a mixture, let's now examine the effect of changing the total pressure of the system.

When the volume occupied by a gas is decreased at constant temperature, its pressure increases. Imagine that the following equilibrium mixture is in a cylinder with a movable piston:

$$2NOCl(g) \rightleftarrows 2NO(g) + Cl_2(g)$$

If we decrease the volume by pushing on the piston (Figure 16.16), the pressure momentarily increases. However, Le Chatelier's principle predicts that the reaction will compensate by shifting in a direction to give a new position of equilibrium having a lower pressure. Because the pressure of a gas is directly proportional to the number of molecules of a gas, *the reaction shifts in the direction that reduces the numbers of molecules.* In this case, the reaction shifts to the left toward the reactants because the balanced equation shows fewer gaseous reactant molecules than gaseous product molecules.

Example 16.6

For each of the following reactions, predict whether the reaction will shift left or right when the volume of the reaction container is decreased.

a. $3H_2(g) + N_2(g) \rightleftarrows 2NH_3(g)$
b. $2SO_3(g) \rightleftarrows 2SO_2(g) + O_2(g)$
c. $H_2(g) + I_2(g) \rightleftarrows 2HI(g)$ ■

Solutions Decreasing the volume increases the pressure of the equilibrium mixture. Le Chatelier's principle predicts the reaction will shift toward the side having fewer molecules.

a. Reaction shifts to the right.
b. Reaction shifts to the left.
c. There is no change in the position of equilibrium because there are equal numbers of molecules on both sides of the equation.

■ Effect of Changing Temperature

To raise the temperature of an equilibrium mixture, it must be heated. Le Chatelier's principle predicts that a reaction will shift in a direction that will compensate for the heat absorbed when the temperature is raised. Once again, consider the preparation of ammonia from hydrogen and nitrogen, an exothermic reaction:

$$3H_2(g) + N_2(g) \rightleftarrows 2NH_3(g) + 92 \text{ kJ}$$

When the temperature is raised (heat is added), a reaction at equilibrium shifts away from the energy term.

When the temperature is raised, this reaction shifts to the left to compensate for the heat absorbed by the equilibrium mixture. When the temperature is lowered, the reaction shifts to the right to compensate for the heat lost to the surroundings. In other words, *when the temperature is raised, the reaction shifts in the direction in which heat is absorbed (endothermic). When the temperature is lowered, the reaction shifts in the direction in which heat is lost (exothermic).*

Remember, the value of K_{eq} is not changed by adding more of a reactant or product to an equilibrium mixture, by changing the pressure of a gaseous substance, or by changing the volume of a gaseous equilibrium mixture. Imposing one of these changes causes a shift in the position of equilibrium but does not change K_{eq}.

However, the effect of temperature is fundamentally different because the value of K_{eq} changes when the temperature is changed. For an exothermic reaction, the value of K_{eq} is smaller at a higher temperature than at lower temperatures. For an endothermic reaction, just the reverse is true. Because the value of K_{eq} changes with temperature, the equilibrium concentrations of reactants and products also change.

Effect of Temperature on K_{eq}

Reaction Type	*Temperature Change*	*Change in* K_{eq}
exothermic	increase	decrease
	decrease	increase
endothermic	increase	increase
	decrease	decrease

Example 16.7

Predict the effect of an increase in temperature on (a) the equilibrium constant and (b) the position of equilibrium for the following *exothermic* reaction:

$$PCl_5(g) \rightleftarrows PCl_3(g) + Cl_2(g) + \text{heat energy}$$ ■

Solutions

a. Because this is an exothermic reaction, raising the temperature decreases K_{eq}.

b. For an exothermic reaction, the energy may be treated as if it were a product. Energy is added when the temperature is raised, and the reaction shifts to the left, resulting in a higher concentration of PCl_5 and lower concentrations of PCl_3 and Cl_2.

Practice Problem 16.5 What is the effect of raising the temperature on (a) the value of the equilibrium constant and (b) the position of equilibrium for the following *endothermic* reaction:

$$\text{heat energy} + CuSO_4 \cdot 5H_2O(s) \rightleftarrows CuSO_4(s) + 5H_2O(g)$$

Effect of a Catalyst on the Position of Equilibrium

A reaction at equilibrium is occurring at equal rates in opposite directions. A catalyst increases the rates of the opposing reactions equally and has no effect on the position of equilibrium.

If no shift in the position of equilibrium occurs when a catalyst is added to an equilibrium mixture, what is its effect on equilibrium? To answer this question, consider again the preparation of ammonia:

$$3H_2(g) + N_2(g) \rightleftarrows 2NH_3(g)$$

At 100°C, K_{eq} is very large and the formation of NH_3 is highly favored in an equilibrium mixture. However, the reaction is quite slow at this temperature. Because of the length of time required to reach equilibrium, the reaction is not a practical synthesis of ammonia at these conditions.

A catalyst shortens the time required to reach equilibrium.

In other words, the problem with the preparation of ammonia is one of kinetics, not equilibrium. To make the synthesis practical, chemists must increase the rate of the reaction. Adding a catalyst, iron powder in this case, increases the rates of the opposing reactions equally and shortens the time to reach equilibrium. Although a catalyst has no effect on the position of equilibrium or the value of K_{eq}, it may speed the attainment of equilibrium considerably.

16.7 Ionization Constants

Learning Objectives

- Write the K_a expression for a weak acid and the K_b expression for a weak base.
- Correlate the strength of an acid or base with the value of K_a or K_b.
- Calculate the value of K_a or K_b using the pH of a solution of an acid or base of known concentration.
- Calculate the percent ionization of a weak acid or weak base from concentration data.
- Determine the pH of a solution from its concentration and K_a or K_b.

Acid Ionization Constants

According to the Arrhenius concept, an acid is a substance that forms H^+ (or hydronium ions) when dissolved in water. For example, when HCl is dissolved in water, ionization occurs to form hydrogen ions:

$$HCl(aq) \longrightarrow H^+(aq) + Cl^-(aq)$$

Because HCl ionizes completely, it is considered to be a strong acid (Section 15.4). Although the reaction is reversible, in dilute solutions practically 100% of the HCl is converted to H^+ and Cl^-.

By comparison, although acetic acid is very soluble in water, it is a weak acid and ionizes only slightly:

$$HC_2H_3O_2(aq) \rightleftarrows H^+(aq) + C_2H_3O_2^-(aq)$$

Because proton transfers are both fast and reversible, equilibrium is reached very quickly when an acid is dissolved in water.

The equilibrium constant for this reaction is called an **acid ionization constant,** K_a. The expression for K_a has the same form as other K_{eq} expressions:

$$K_a = \frac{[H^+]\,[C_2H_3O_2^-]}{[HC_2H_3O_2]} = 1.8 \times 10^{-5} \qquad \text{at } 25°C$$

A small K_a indicates the position of equilibrium favors the reactant (acetic acid), which is what is expected for a weak acid. For example, 0.10 M acetic acid is only 1.3% ionized; 98.7% of dissolved acetic acid remains nonionized. Therefore, the equilibrium concentration of molecular acetic acid is approximately the same as before ionization. The ionization constants for a number of weak acids and their percent ionization in dilute solution are given in Table 16.1.

Equilibrium problems involving solutions of weak acids are simplified when we realize that the equilibrium concentration of the molecular acid is practically the same as its initial concentration. The data in Table 16.1 show that when K_a is 10^{-4} or smaller, more than 97% of the acid remains nonionized. In other words, when K_a is very small, almost all of the acid remains nonionized.

TABLE 16.1 Acid Ionization Constants and Percent Ionization of 0.10 M Solutions at 25°C

Acid	*Formula*	K_a	*Percent Ionization*
acetic	$HC_2H_3O_2$	1.8×10^{-5}	1.3
barbituric	$HC_4H_3N_2O_3$	9.8×10^{-5}	3.1
benzoic	$HC_7H_5O_2$	6.5×10^{-5}	2.6
carbolic (phenol)	HC_6H_5O	1.3×10^{-10}	0.003 6
formic	$HCHO_2$	1.9×10^{-4}	4.4
lactic	$HC_3H_5O_3$	1.4×10^{-4}	3.7
nicotinic	$HC_6H_4NO_2$	1.4×10^{-5}	1.2
nitrous	HNO_2	4.5×10^{-4}	6.7

Example 16.8

A 0.15 M solution of a weak acid, HA, had a pH of 3.0. What was the percent ionization of the acid? ■

Find: % ionization = ?

Given: 0.15 M HA, pH = 3.0

Known: $HA(aq) \rightleftarrows H^+(aq) + A^-(aq)$

Solution When pH = 3.0, $[H^+] = 1 \times 10^{-3}$ M, or 0.001 M. The chemical equation for the ionization of HA shows $[H^+] = [A^-]$, and only 0.001 mol/L of the 0.15 mol/L HA is ionized. Now we can calculate the percent ionization:

$$\% \text{ ionization} = \frac{(0.001)}{(0.15)} \times 100 = 0.7\%$$

Example 16.9

Calculate K_a for hydrocyanic acid, HCN, if the pH is 5.1 for a 0.12 M solution. ■

Solution $HCN(aq) \rightleftarrows H^+(aq) + CN^-(aq)$

When pH = 5.1, $[H^+] = 8 \times 10^{-6}$ M = $[A^-]$.

$[HCN] = 0.12\text{ M} - 0.000\ 008\text{ M} = 0.12\text{ M}$

Substituting into the K_a expression gives

$$K_a = \frac{[H^+][CN^-]}{[HCN]} = \frac{(8 \times 10^{-6})(8 \times 10^{-6})}{(0.12)} = 5 \times 10^{-10}$$

Example 16.10

Calculate the pH of a 0.10 M solution of nicotinic acid ($HC_6H_4NO_2$). $K_a = 1.4 \times 10^{-5}$ at 25°C. ■

Find: pH = ?

Given: 0.10 M solution $HC_6H_4NO_2$, $K_a = 1.4 \times 10^{-5}$

Known: $pH = -\log[H^+]$

$HC_6H_4NO_2(aq) \rightleftarrows H^+(aq) + C_6H_4NO_2^-(aq)$

Solution $pH = -\log[H^+]$ (1)

Let $[H^+] = [C_6H_4NO_2^-] = x$. Because K_a is smaller than 10^{-4}, the percent ionization of the acid is very small, and the equilibrium concentration of the acid $[HC_6H_4NO_2] \cong 0.10$ M. Then:

$$K_a = 1.4 \times 10^{-5} = \frac{[H^+][C_6H_4NO_2^-]}{[HC_6H_4NO_2]} = \frac{(x)(x)}{(0.10)} \quad (2)$$

$$x^2 = (0.10)(1.4 \times 10^{-5}) = 1.4 \times 10^{-6}$$

$$x = 1.2 \times 10^{-3} = [H^+]$$

$$pH = -\log[1.2 \times 10^{-3}] = 2.92 \quad (1)$$

Practice Problem 16.6 The pH is 4.5 for a 0.20 M solution of a weak acid, HX. What is the percent ionization of the acid?

Practice Problem 16.7 Calculate K_a for propanoic acid ($HC_3H_5O_2$) if the pH is 2.9 for a 0.10 M solution.

Practice Problem 16.8 What is the pH of 0.15 M acetic acid ($HC_2H_3O_2$) solution? $K_a = 1.8 \times 10^{-5}$ at 25°C.

Base Ionization Constants

When a weak base such as ammonia is dissolved in water, ionization occurs to form hydroxide ion and the conjugate acid of the base:

$$NH_3(aq) + H_2O(l) \rightleftarrows NH_4^+(aq) + OH^-(aq)$$

The equilibrium constant for the reaction is called a **base ionization constant,** K_b:

$$K_b = \frac{[NH_4^+]\,[OH^-]}{[NH_3]} = 1.8 \times 10^{-5} \qquad \text{at } 25°C$$

Because there is a large excess of water, its concentration remains essentially constant and is omitted from the K_{eq} expression.* Coincidentally, K_b for ammonia has the same value as K_a for acetic acid. Other base ionization constants are given in Table 16.2.

As with weak acids, the percent ionization of a weak base is very small. When we are working problems involving weak bases, we can assume that the equilibrium concentration of the base is essentially the same as its initial concentration.

*This can be explained by starting with the following complete equilibrium expression for the reaction:

$$K_b' = \frac{[NH_4^+]\,[OH^-]}{[H_2O]\,[NH_3]}$$

Multiplying by $[H_2O]$ gives

$$K_b'\,[H_2O] = \frac{[NH_4^+]\,[OH^-]}{[NH_3]} = K_b$$

Thus, K_b is the product of two constants, $K_b'\,[H_2O]$.

TABLE 16.2 Ionization Constants for Weak Bases at 25°C

Base	*Formula*	K_b
ammonia	NH_3	1.8×10^{-5}
aniline	$C_6H_5NH_2$	4.3×10^{-10}
methylamine	CH_3NH_2	3.7×10^{-4}
ethylamine	$C_2H_5NH_2$	6.5×10^{-4}
morphine	$C_{17}H_{19}O_3N$	1.6×10^{-6}
nicotine	$C_{10}H_{14}N_2$	1.0×10^{-6}

Example 16.11

Calculate K_b for dimethylamine, $(CH_3)_2NH$, a weak base, if the pH is 11.60 for a 0.30 M solution. ■

Solution Begin by writing the chemical equation for the ionization of the base. Find the equilibrium concentrations and calculate the value of K_b.

$$(CH_3)_2NH(aq) + H_2O(l) \rightleftarrows (CH_3)_2NH_2^+(aq) + OH^-(aq)$$

$$pOH = 14.00 - pH = 14.00 - 11.60 = 2.40$$

$$[OH^-] = 4.0 \times 10^{-3}\ M = [(CH_3)_2NH_2^+]$$

Substituting into the K_b expression gives

$$K_b = \frac{[(CH_3)_2NH_2^+]\,[OH^-]}{[(CH_3)_2NH]} = \frac{(4.0 \times 10^{-3})\,(4.0 \times 10^{-3})}{(0.30)}$$

$$= 5.3 \times 10^{-5}$$

Example 16.12

Determine the pH of a 0.12 M aqueous solution of ammonia. $K_b = 1.8 \times 10^{-5}$. ■

Solution

$$pH = 14.00 - pOH \quad (1)$$

$$pOH = -\log[OH^-] \quad (2)$$

$$NH_3(aq) + H_2O(l) \rightleftarrows NH_4^+(aq) + OH^-(aq)$$

Let $x = [OH^-] = [NH_4^+]$. Because ammonia is a weak base, at equilibrium, $[NH_3] \cong 0.12$ M. Then:

$$K_b = \frac{[NH_4^+]\,[OH^-]}{[NH_3]} = \frac{(x)\,(x)}{0.12} = 1.8 \times 10^{-5} \quad (3)$$

$$x^2 = 2.2 \times 10^{-6}$$

$$x = 1.5 \times 10^{-3} = [OH^-]$$

$$pOH = 2.82 \quad (2)$$

$$pH = 14.00 - 2.82 = 11.18 \quad (1)$$

Practice Problem 16.9 A 0.25 M solution of a weak base, B, had a pH of 9.0. What was the percent ionization of the base?

Practice Problem 16.10 Calculate the pH of a 0.012 M solution of morphine ($C_{17}H_{19}O_3N$). $K_b = 1.6 \times 10^{-6}$ at 25°C.

Chapter Review

Directions: From the choices listed at the end of this Chapter Review, choose the word or term that best fits in each blank. Use each choice only once. Answers appear at the end of the chapter.

Section 16.1 The study of reaction rates is called __________ __________ (1). According to the __________ (2) theory of reactions, reactant particles must collide for a chemical reaction to occur. However, products are formed only when collisions are __________ (3). Two factors determine whether a collision is effective. The first is the __________ __________ (4) of the colliding molecules, and the second is their __________ (5) when they collide. The colliding molecules must have sufficient kinetic energy to break bonds of the __________ (6). Furthermore, the colliding molecules must be oriented properly to allow the __________ (7) of the products to form. Large, __________ (8) molecules are less likely to be oriented properly when they collide, and they tend to react more slowly than __________ (9) molecules.

Section 16.2 In an effective collision, the __________ (10) energy of the molecules decreases and their potential energy __________ (11). The reacting substances momentarily exist in a high-energy state called a __________ (12) state in which bonds of the reactants are __________ (13) and bonds of the __________ (14) are forming. When the bond breaking and forming process is completed, the __________ (15) energy of the substances __________ (16) and their kinetic energy increases again.

The difference between the potential energy of the reactants and that of the transition state is called the __________ (17) energy of the reaction. It is the __________ (18) kinetic energy needed by colliding molecules for a reaction to occur. When the kinetic energy of colliding molecules is __________ (19) than E_a, the collision is ineffective. A reaction that has a large E_a is __________ (20) than a reaction that has a __________ (21) E_a.

The __________ (22) of reaction is related to the difference between the potential energy (PE) states of the reactants and products. When the PE of the products is less than the PE of the reactants, the reaction is __________ (23), and energy is __________ (24) to the surroundings. When the PE of the products is greater than the PE of the reactants, the reaction is __________ (25).

Section 16.3 Chemical reactions are always __________ (26) at higher temperatures. Increasing the temperature by 10°C increases the __________ (27) of many reactions two to three times.

A __________ (28) is a substance that increases the rate of a reaction without being __________ (29). Our bodies use catalysts called __________ (30), which are essential to life. When a catalyst is added to a mixture of reactants, the reaction pathway is changed, resulting in a smaller activation energy. A catalyzed reaction is faster because more molecules have kinetic energies greater than __________ (31). Because a catalyst is not consumed in a reaction, only a small __________ (32) is needed.

At higher concentrations, the rate of a reaction is greater because reactants are more __________ (33) and their collision __________ (34) is higher than at lower concentrations.

Section 16.4 __________ (35) equilibrium is a condition in which the rates of a forward and reverse reaction are equal. At equilibrium, the concentrations of reactants and products are __________ (36). Attainment of equilibrium often requires an __________ (37) period of time.

Section 16.5 An equilibrium constant, K_{eq}, is a ratio of the molar concentrations of the products divided by the concentrations of the reactants, each raised to a __________ (38) equal to its balancing coefficient in the chemical equation. The value of K_{eq} is constant even though equilibrium concentrations vary in different experiments. However, K_{eq} varies with __________ (39). If the value of K_{eq} is a very large number (>100), the position of equilibrium lies to the __________ (40).

Section 16.6 The effect of a change in a reaction variable can be predicted using __________ (41) principle. For example, when a reactant or product is added to an equilibrium mixture, the reaction shifts __________ (42) from the substance added. On the other hand, the reaction shifts toward a reactant or product __________ (43) from an equilibrium mixture. In the same way, a reaction shifts away from a gaseous reactant or product when the partial pressure of the substance is __________ (44) and __________ (45) a gaseous reactant or product when its partial pressure is decreased.

Decreasing the __________ (46) occupied by a gaseous equilibrium mixture causes the reaction to shift in the direction that __________ (47) the number of molecules, and vice versa. Raising the temperature of an equilibrium mixture causes the reaction to shift in the direction in which heat is __________ (48). On the other hand, a shift occurs in the direction in which heat is released when the temperature is __________ (49). Changing the temperature also causes the value of __________ (50) to change. For an exothermic reaction, the value of K_{eq} is smaller at a higher temperature, and vice versa. For an endothermic reaction, the value of K_{eq} is larger at a __________ (51) temperature, and vice versa.

A catalyst increases the rates of opposing reactions __________ (52) and has no effect on the position of equilibrium.

Section 16.7 An acid __________ (53) constant, K_a, is an equilibrium constant for the ionization of a weak acid in water. When K_a is very small (10^{-4} or smaller), the acid is almost 100% __________ (54).

A base ionization constant, __________ (55), is an equilibrium constant for the ionization of a weak base in water. As with weak acids, the percent ionization of a weak base is very small.

Choices: absorbed, activation, amount, away, bonds, breaking, catalyst, chemical, collision, complex, consumed, constant, crowded, decreases, E_a, effective, endothermic, enzymes, equally, exothermic, extended, faster, frequency, heat, higher, increased, increases, ionization, ionized, K_b, K_{eq}, kinetic, kinetic energy, Le Chatelier's, less, lowered, minimum, orientation, potential, power, products, rates, reactants, reaction kinetics, reduces, released, removed, right, slower, small, smaller, temperature, toward, transition, volume.

Study Objectives

After studying Chapter 16, you should be able to:

Section 16.1

1. Briefly describe the collision theory of reactions.
2. Explain the criteria for an effective collision.
3. Explain why large, complex molecules tend to react more slowly than small molecules.

Section 16.2

4. Draw potential energy diagrams for exothermic and endothermic reactions showing the reactants, products, transition state, E_a, and ΔH.
5. Explain why a reaction that has a large E_a tends to be slower than a reaction that has a small E_a.

Section 16.3

6. Describe three reaction variables that affect the rate of a reaction.
7. Explain why raising the temperature increases the rate of a reaction.
8. Explain how adding a catalyst increases the rate of a reaction.
9. Explain why increasing the concentrations of reactants generally increases the rate of a reaction.

Section 16.4

10. Describe three important characteristics of chemical equilibrium.

Section 16.5

11. Write equilibrium constant (K_{eq}) expressions for reactions at equilibrium.
12. Calculate the value of K_{eq} from equilibrium concentrations.
13. Predict the position of equilibrium from the value of K_{eq}.

Section 16.6

14. Use Le Chatelier's principle to predict how a change in concentration, a change in pressure or volume, and a change in temperature will affect the position of equilibrium.
15. Describe how an increase or decrease in temperature affects the K_{eq} for a reaction.
16. Describe the effect of a catalyst on a chemical equilibrium.

Section 16.7

17. Write the K_a expression for a weak acid and the K_b expression for a weak base.
18. Correlate the strength of an acid or base with the value of K_a or K_b.
19. Calculate the value of K_a or K_b using the pH of a solution of an acid or base of known concentration.
20. Calculate the percent ionization of a weak acid or weak base from concentration data.
21. Determine the pH of a solution from its concentration and K_a or K_b.

Key Terms

Review the definition of each of the terms listed here. For reference they are listed by chapter section number. You may use the glossary as necessary.

16.1 collision theory, effective collision, ineffective collision, reaction kinetics

16.2 activation energy, heat of reaction, transition state

16.3 catalyst, enzymes

16.4 chemical equilibrium

16.5 equilibrium constant

16.6 Le Chatelier's principle

16.7 acid ionization constant, base ionization constant

Questions and Problems

The answers to questions and problems that have blue numbers appear in Appendix C. Asterisks indicate the more challenging problems.

Reaction Rates, Collision Theory (Section 16.1)

1. What is meant by reaction kinetics?
2. Explain what is meant by the rate of a chemical reaction.
3. What is the basic assumption of the collision theory of reactions?
4. What is meant by an effective collision of reactants?
5. What happens to reactants in an ineffective collision?
6. What two factors determine whether or not a collision is effective?
7. Why do large, complex molecules tend to react more slowly in a chemical reaction than small molecules?

Energy Changes During Reactions (Section 16.2)

8. What is a transition state in a chemical reaction?
9. What is the activation energy of a chemical reaction?
10. What is meant by heat of reaction, ΔH?
11. Explain the relationship of the energy states of reactants and products for an exothermic reaction.
12. During an endothermic reaction, energy is absorbed from the surroundings. Explain the relationship of the energy states of the reactants and products of an endothermic reaction.
13. The reaction of X_2 and Y_2 to give 2XY is exothermic. Draw a potential energy diagram for this reaction. Show the energy states of the reactants, products, and the transition state. Label important energy quantities.
14. The reaction of R_2 and Z_2 to give 2RZ is endothermic. Draw a potential energy diagram for this reaction. Show the energy states of the reactants, products, and the transition state. Label important energy quantities.
15. What is the relationship of E_a and the rate of a chemical reaction?
16. The activation energy for the reaction

    $CH_4(g) + Cl_2(g) \longrightarrow CH_3Cl(g) + HCl(g)$

    is smaller than E_a for the reaction

    $CH_4(g) + Br_2(g) \longrightarrow CH_3Br(g) + HBr(g)$

    Compare the rates of the two reactions at a given temperature.

Variables That Affect Reaction Rate (Section 16.3)

17. How does raising the temperature affect the rate of a reaction?
18. Consider the variables that affect reaction rates and explain why milk (and many other foods) are generally stored under refrigeration rather than at room temperature.
19. In terms of collision theory, explain how raising the temperature increases the rate of a reaction.

20. Explain why it is often difficult to light a charcoal fire in a backyard barbeque on a cold winter day.
21. What is a catalyst?
22. What is the function of an enzyme in the human body?
23. How does adding a catalyst affect the activation energy of a reaction?
24. In general, why is only a small amount of a catalyst needed in a reaction?
25. The decomposition of aqueous hydrogen peroxide, $H_2O_2(aq)$, is slow at room temperature. However, after adding a small amount of KI, the decomposition occurs rapidly. What is the function of KI in this reaction?
26. How does increasing the concentrations of reactants affect the rates of most reactions?
27. Explain, according to collision theory, how increasing the concentrations of reactants increases the rate of a reaction.
28. Explain how fanning a fire causes it to burn faster.

Chemical Equilibrium (Section 16.4)

29. What is meant by chemical equilibrium?
30. Describe three characteristics of chemical equilibrium.
31. Equilbrium is sometimes described as a balanced condition for a reaction. What is in balance—that is, what remains unchanged at equilibrium?
32. Attainment of equilibrium often requires an extended period of time. How can the time for a reaction to reach equilibrium be shortened?
33. The reaction $C_2H_6(g) + Br_2(g) \rightleftarrows C_2H_5Br(g) + HBr(g)$ is quite slow at 20°C in the dark and takes a very long time to reach equilibrium. However, at 250°C, equilibrium is attained quickly. Explain.

Equilibrium Constants (Section 16.5)

34. Write the K_{eq} expression for the reaction

    $H_2(g) + I_2(g) \rightleftarrows 2HI(g)$

35. The oxidation of ammonia by oxygen is described by the following equation. Write the K_{eq} expression for the reaction.

    $4NH_3(g) + 5O_2(g) \rightleftarrows 4NO(g) + 6H_2O(g)$

36. Write the K_{eq} expression for the reaction

    $4H_2(g) + CS_2(g) \rightleftarrows CH_4(g) + 2H_2S(g)$

37. The reaction of sulfur dioxide (SO_2) and chlorine (Cl_2) produces sulfuryl chloride (SO_2Cl_2). At equilibrium at 30°C, the concentrations of both SO_2 and Cl_2 were 0.0239 M, and the concentration of SO_2Cl_2 was 0.952 M. Calculate K_{eq} for the reaction at this temperature.
38. The reaction of acetylene (C_2H_2) with hydrogen produces ethane (C_2H_6).

    $C_2H_2(g) + 2H_2(g) \rightleftarrows C_2H_6(g)$

    After coming to equilibrium, a reaction mixture was analyzed and found to contain 4.9 mol ethane/L, 0.40 mol acetylene/L, and 1.75 mol hydrogen/L. Calculate K_{eq} for the reaction.
39. At a certain temperature, $K_{eq} = 5.0 \times 10^3$ for the following reaction:

    $2CO(g) + O_2(g) \rightleftarrows 2CO_2(g)$

    Write the K_{eq} expression. Is the formation of CO_2 favored at this temperature? Explain.
40. At a certain temperature, $K_{eq} = 0.010$ for the following reaction.

    $2NO_2(g) \rightleftarrows N_2(g) + 2O_2(g)$

    Write the K_{eq} expression. Is this reaction favored at this temperature? Explain.
41. To be at equilibrium, a reaction must be reversible.
    a. Write the chemical equation for the reverse of the reaction described in problem 40.
    b. Write the K_{eq} expression, and calculate the value of K_{eq} for the reverse reaction.
    c. Is the reaction favored in this direction or in the direction as written in problem 40?
42. a. Write the chemical equation for the reverse of the reaction described in problem 39.
    b. Write the K_{eq} expression, and calculate the value of K_{eq} for the reverse reaction.
    c. Is the reaction favored in this direction?

Le Chatelier's Principle (Section 16.6)

43. State Le Chatelier's principle.
44. For the following reaction,

    $2NO(g) + Br_2(g) \rightleftarrows 2NOBr(g)$

    what will be the effect of adding more bromine on
    a. the equilibrium concentration of NOBr?
    b. the equilibrium concentration of NO?
    c. the value of K_{eq}?
45. The reaction described in problem 44 is exothermic. What will be the effect of raising the temperature on
    a. the equilibrium concentration of NOBr?
    b. the value of K_{eq}?
46. The following reaction is endothermic.

    $SO_2Cl_2(g) \rightleftarrows SO_2(g) + Cl_2(g)$

    a. What is the effect of raising the temperature on the equilibrium concentration of SO_2?
    b. What is the effect of raising the temperature on the value of K_{eq}?

47. Is heat absorbed or released when chlorine (Cl_2) is added to the equilibrium mixture described in problem 46? Explain.

48. Sulfur dioxide (SO_2) is a common air pollutant resulting from the burning of coal that contains sulfur. In the atmosphere, SO_2 reacts exothermically with O_2 to form SO_3.

$$2SO_2(g) + O_2(g) \rightleftarrows 2SO_3(g)$$

When this reaction is carried out in a closed reactor, how is the position of equilibrium affected by
a. adding SO_2?
b. increasing the temperature?
c. decreasing the volume?

49. Methyl alcohol (CH_3OH, wood alcohol) is a synthetic fuel that can be made according to the following exothermic reaction:

$$CO(g) + 2H_2(g) \rightleftarrows CH_3OH(g)$$

What is the effect on the position of equilibrium of
a. removing methyl alcohol from the equilibrium mixture?
b. decreasing the temperature?
c. adding CO to the equilibrium mixture?
d. decreasing the volume of the container?

50. How does adding CO to the equilibrium mixture described in problem 49 affect the value of K_{eq}? Explain.

51. What is the effect of adding sodium hydroxide on the color of the following solution? Explain.

$$\underset{\text{orange}}{Cr_2O_7^{2-}(aq)} + 2OH^-(aq) \rightleftarrows \underset{\text{yellow}}{2CrO_4^{2-}(aq)} + H_2O(l)$$

52. What is the effect of adding an acid, such as HCl(aq), to the solution described in problem 51? Describe the color change.

53. Ethylene (C_2H_4) reacts with hydrogen to give ethane (C_2H_6) according to the following equation:

$$C_2H_4(g) + H_2(g) \rightleftarrows C_2H_6(g)$$

What is the effect of the addition of a platinum catalyst on
a. the rate of the forward reaction?
b. the rate of the reverse reaction?
c. the position of equilibrium?

Ionization Constants (Section 16.7)

54. Write the K_a expressions for the following acids:
a. HBr(aq), hydrobromic acid
b. HCN(aq), hydrocyanic acid
c. $HCHO_2$(aq), formic acid (in ant bites and insect stings)

55. Write the K_b expressions for the following bases:
a. NH_3(aq)
b. $C_2H_5NH_2$(aq), ethylamine

56. A 0.10 M solution of a weak acid, HA, had a pH of 4.0. What is the value of K_a?

57. What is the percent ionization of the weak acid in problem 56?

58. A 0.20 M solution of a weak acid had a pH of 3.2. What is the percent ionization of the weak acid?

59. What is the value of the ionization constant of the acid in problem 58?

60. A 0.10 M solution of a weak base (B) had a pH of 9.4. What is the value of K_b?

61. What is the percent ionization of the base described in problem 60?

62. For a certain weak base, $K_b = 1 \times 10^{-10}$. What is the percent ionization of a 0.010 M solution of the base having a pH of 8.63?

63. What is the pH of a 0.11 M solution of lactic acid ($HC_3H_5O_3$)? $K_a = 1.4 \times 10^{-4}$.

64. Calculate the pH of a 0.015 M aqueous solution of aniline ($C_6H_5NH_2$). $K_b = 4.3 \times 10^{-10}$.

Additional Problems

65. Heating isobutane (C_4H_{10}) in the presence of a catalyst produces isobutylene (C_4H_8) and hydrogen (H_2).
a. Write the chemical equation for this reaction and the expression for K_{eq}.
b. After equilibrium was reached at a certain reaction temperature, the concentrations were found to be 1.5 mol/L of isobutane, 4.5 mol/L of isobutylene, and 4.5 mol/L of hydrogen. Calculate the value of K_{eq} at this temperature.

66. Calculate K_{eq} for the following reaction from the equilibrium concentrations given:

$$2SO_3(g) + CO_2(g) \rightleftarrows CS_2(g) + 4O_2(g)$$

$[SO_3] = 0.020$ M, $[CO_2] = 0.0035$ M, $[CS_2] = 0.000\ 45$ M, $[O_2] = 0.000\ 10$ M

67. The reaction of nitrogen with oxygen to give nitrogen dioxide is described by the following equation:

$$N_2(g) + 2O_2(g) \rightleftarrows 2NO_2(g)$$

At equilibrium at a certain temperature, the concentrations were $[N_2] = 1.2 \times 10^{-3}$ M, $[O_2] = 2.3 \times 10^{-4}$ M, and $[NO_2] = 4.5 \times 10^{-2}$ M. Calculate the equilibrium constant for this reaction at this temperature.

68. Salicylic acid ($HC_7H_5O_3$) is a weak acid used to make aspirin. The pH of a 0.10 M solution of the acid was found to be 1.9. What is the percent ionization of salicylic acid in this solution?

69. For the following reaction at equilibrium,

$$4HCl(g) + O_2(g) \rightleftarrows 2Cl_2(g) + 2H_2O(g)$$

predict which direction the reaction will shift when
a. HCl is added to the reaction mixture.
b. water is added to the reaction mixture.
c. O_2 is added to the reaction mixture.
d. the volume of the container is decreased.

70. Consider the following endothermic reaction at equilibrium:

$$3O_2(g) \rightleftarrows 2O_3(g)$$

What is the effect of raising the temperature upon
a. the value of K_{eq}?
b. the concentration of ozone (O_3)?

71. Hydrazine (N_2H_4), a compound used as a rocket fuel, dissolves in water to give a basic solution.
a. Write a chemical equation for the ionization of hydrazine in water.
b. Write the K_b expression.
c. Calculate the percent ionization of hydrazine in a 0.25 M solution having a pH of 9.3.

*72. At a certain temperature, 0.40 mol of N_2, H_2, and NH_3 were mixed in a 1.0-L container, and the reaction was allowed to come to equilibrium.

$$N_2(g) + 3H_2(g) \rightleftarrows 2NH_3(g)$$

The equilibrium concentration of N_2 was 0.35 M.
a. What were the equilibrium concentrations of H_2 and NH_3?
b. Calculate the value of K_{eq} at this temperature.

Solutions to Practice Problems

PP 16.1

First, write the K_{eq} expression:

$$K_{eq} = \frac{[NH_3]^2}{[H_2]^3\,[N_2]}$$

Then, substitute into the equation and calculate K_{eq}:

$$K_{eq} = \frac{(0.150)^2}{(0.575)^3\,(0.425)} = 0.279$$

PP 16.2

The chemical equation is:

$$N_2(g) + O_2(g) \rightleftarrows 2NO(g)$$

$$K_{eq} = \frac{[NO]^2}{[N_2]\,[O_2]} = 1 \times 10^{-30}$$

The formation of NO is not favored because the value of K_{eq} is very small. No appreciable NO will be formed at this temperature.

PP 16.3

a. Addition of H_2O will cause a shift to the right (toward the product), thereby increasing the concentration of H_2.
b. Removing CO will cause a shift to the right resulting in a higher concentration of H_2.

PP 16.4

a. Increasing the partial pressure of O_2 is equivalent to increasing its concentration. A shift to the right will occur, resulting in a higher concentration of H_2O.
b. Decreasing the partial pressure of H_2 is equivalent to decreasing its concentration. The reaction will shift to the left, thereby lowering the concentration of H_2O.

PP 16.5

a. Raising the temperature increases the value of K_{eq} for an endothermic reaction.
b. For an endothermic reaction, raising the temperature causes a shift in the position of equilibrium to the right, resulting in more decomposition of the reactant.

PP 16.6

Find: % ionization = ?

Given: 0.20 M HX, pH = 4.5

Known: $HX(aq) \rightleftarrows H^+(aq) + X^-(aq)$

Solution: When pH = 4.5, $[H^+] = 3 \times 10^{-5}$ M, or 0.000 03 M (1 significant figure). The chemical equation for the ionization of HX shows $[H^+] = [X^-]$, and only 0.000 03 mol/L of the 0.20 mol/L HX is ionized. Now calculate the percent ionization:

$$\%\text{ ionization} = \frac{(0.000\ 03)}{(0.20)} \times 100 = 0.02\%$$

PP 16.7

To calculate K_a, the equilibrium concentrations $[H^+]$, $[C_2H_3O_2^-]$, and $[HC_2H_3O_2]$ are needed. The chemical equation is

$$HC_3H_5O_2(aq) \rightleftarrows H^+(aq) + C_3H_5O_2^-(aq)$$

When pH = 2.9,

$$[H^+] = 1 \times 10^{-3}\text{ M, or } 0.001\text{ M}$$
$$= [C_3H_5O_2^-]$$

Thus,

$$[HC_3H_5O_2] = 0.10\text{ M} - 0.001\text{ M} = 0.10\text{ M}$$

Substituting into the K_a expression gives

$$K_a = \frac{[H^+]\,[C_3H_5O_2^-]}{[HC_3H_5O_2]} = \frac{(1 \times 10^{-3})\,(1 \times 10^{-3})}{(0.10)}$$
$$= 1 \times 10^{-5}$$

PP 16.8

Find: pH = ?

Given: 0.15 M solution $HC_2H_3O_2$, $K_a = 1.8 \times 10^{-5}$

Known: $pH = -\log[H^+]$

$$HC_2H_3O_2(aq) \rightleftarrows H^+(aq) + C_2H_3O_2^-(aq)$$

Solution: To calculate pH, we must find $[H^+]$.

Let $[H^+] = [C_2H_3O_2^-] = x$. Because K_a is smaller than 10^{-4}, the percent ionization of the acid is very small, and the equilibrium concentration of the acid $[HC_2H_3O_2] \cong 0.15$ M.

$$K_a = 1.8 \times 10^{-5} = \frac{[H^+][C_2H_3O_2^-]}{[HC_2H_3O_2]} = \frac{(x)(x)}{(0.15)}$$

$$x^2 = (0.15)(1.8 \times 10^{-5}) = 2.7 \times 10^{-6}$$

$$x = 1.6 \times 10^{-3} = [H^+]$$

$$pH = -\log[1.6 \times 10^{-3}] = 2.80$$

PP 16.9

Find: % ionization = ?

Given: 0.25 M B, pH = 9.0

Known: $B(aq) + H_2O(l) \rightleftarrows BH^+(aq) + OH^-(aq)$

Solution: When pH = 9.0, $[H^+] = 1 \times 10^{-9}$ M, and $[OH^-] = 1 \times 10^{-5}$ M, or 0.000 01 M (1 significant figure). The chemical equation for the ionization of B shows $[BH^+] = [OH^-] = 0.000\ 01$ M. Thus, 0.000 01 mol/L of B is ionized.

Now calculate the percent ionization:

$$\%\ \text{ionization} = \frac{(0.000\ 01)}{(0.25)} \times 100 = 0.004\%$$

PP 16.10

Find: pH = ?

Given: 0.012 M solution of $C_{17}H_{19}O_3N$

$K_b = 1.6 \times 10^{-6}$

Known: $pH + pOH = 14$ (1)

$pOH = -\log[OH^-]$ (2)

Solution: To find $[OH^-]$, use the K_b expression for the ionization of morphine.

$$C_{17}H_{19}O_3N(aq) + H_2O(l) \rightleftarrows C_{17}H_{19}O_3NH^+(aq) + OH^-(aq)$$

Let $x = [OH^-] = [C_{17}H_{19}O_3NH^+]$. Because morphine is a weak base, at equilibrium, $[C_{17}H_{19}O_3N] \cong 0.012$ M.

$$K_b = \frac{[C_{17}H_{19}O_3NH^+][OH^-]}{[C_{17}H_{19}O_3N]} = \frac{(x)(x)}{0.012} = 1.6 \times 10^{-6} \quad (3)$$

$$x^2 = 1.9 \times 10^{-8}$$

$$x = 1.4 \times 10^{-4} = [OH^-]$$

$$pOH = 3.85 \quad (2)$$

$$pH = 14.00 - 3.85 = 10.15 \quad (1)$$

Answers to Chapter Review

1. reaction kinetics
2. collision
3. effective
4. kinetic energy
5. orientation
6. reactants
7. bonds
8. complex
9. small
10. kinetic
11. increases
12. transition
13. breaking
14. products
15. potential
16. decreases
17. activation
18. minimum
19. less
20. slower
21. smaller
22. heat
23. exothermic
24. released
25. endothermic
26. faster
27. rates
28. catalyst
29. consumed
30. enzymes
31. E_a
32. amount
33. crowded
34. frequency
35. chemical
36. constant
37. extended
38. power
39. temperature
40. right
41. Le Chatelier's
42. away
43. removed
44. increased
45. toward
46. volume
47. reduces
48. absorbed
49. lowered
50. K_{eq}
51. higher
52. equally
53. ionization
54. ionized
55. K_b

17

Oxidation–Reduction Reactions

Contents

An oxidation–reduction reaction provides the energy to run this racing car.

Photosynthesis involves oxidation–reduction reactions.

You Probably Know . . .

- Oxygen, an important component of air, is an oxidizing agent (Section 6.8). During combustion, oxygen oxidizes fuels such as gasoline, natural gas, wood, and coal.
- Oxygen is produced naturally during photosynthesis, a plant growth process involving oxidation–reduction (redox) reactions that use carbon dioxide and water.
- Food provides energy for our bodies. Metabolism of compounds from the foods we eat involves many redox reactions.
- Batteries produce electricity using electron-transfer reactions.

Electron-transfer reactions occur in batteries.

Why Should We Study Oxidation–Reduction Reactions?

Many familiar processes involve oxidation–reduction (redox) reactions, including combustion of fuels such as natural gas, gasoline, coal, and wood. Another example is photosynthesis, a process in which solar energy is converted into chemical energy stored in the structure of the molecules that make up plants. When fruits, vegetables, and other foods are eaten, their stored energy is released through redox reactions of metabolism in our bodies. Other examples include batteries for automobiles, flashlights, watches, and calculators. In fact, practically all processes that provide energy for powering vehicles, heating buildings, fueling the human body, or any other use involve redox reactions. Furthermore, redox reactions are used in the manufacture of fertilizers, pesticides, plastics, cosmetics, and a variety of other consumer products.

17.1 Oxidation–Reduction Reactions

Learning Objectives

- Identify elements oxidized and reduced in an oxidation–reduction reaction.
- Identify the oxidizing and reducing agents involved in an oxidation–reduction reaction.

Reduction always *accompanies oxidation.*

Loss of electrons = *oxidation (LEO)*

Gain of electrons = *reduction (GER)*

Remember, LEO *the lion says* GER.

Oxidation–reduction reactions are electron-transfer reactions (Section 6.8). For example, in the reaction of sodium with chlorine, each sodium atom *loses* one electron and is *oxidized* to a sodium ion. Each chlorine atom *gains* one electron and is *reduced* to a chloride ion. Thus, oxidation involves a loss of electrons, and reduction involves a gain of electrons. Note that reduction always accompanies oxidation. One cannot occur without the other.

$$2Na(s) + Cl_2(g) \rightarrow 2NaCl(s)$$

$$Na\cdot + :\ddot{Cl}: \quad Na^+ + [:\ddot{Cl}:]^-$$
$$+ \rightarrow$$
$$Na\cdot + :\ddot{Cl}: \quad Na^+ + [:\ddot{Cl}:]^-$$

Oxidizing agent: Electron acceptor.
Reducing agent: Electron donor.

Sodium, the element oxidized, is called a **reducing agent** because it causes chlorine to be reduced by giving it electrons. Chlorine is called the **oxidizing agent** because it causes sodium to be oxidized by accepting its electrons. Because metals are elements that have a tendency to lose electrons to form cations, they are reducing agents. On the other hand, when nonmetals react with metals (and sometimes with other nonmetals), they gain electrons and act as oxidizing agents.

The term oxidizing agent originated with oxygen (O_2), one of the most abundant elements and by far the most common oxidizing agent. Oxygen reacts with fuels (reducing agents) in combustion reactions, with iron to form rust (Fe_2O_3), and in metabolism where glucose ($C_6H_{12}O_6$) is oxidized to form carbon dioxide and water. Oxygen is an oxidizing agent essential to life.

Note, however, that oxidizing agents are not limited to nonmetallic elements. We find that many compounds, cations, and anions also can act as oxidizing agents. Furthermore, metals are not the only known reducing agents. We also find that many compounds, anions, and even some cations act as reducing agents (Table 17.1).

Oxygen oxidizes fuels such as natural gas (methane, CH_4):

$$CH_4(g) + 2O_2(g) \rightarrow CO_2(g) + 2H_2O(g)$$

TABLE 17.1 Oxidizing and Reducing Agents (Reactions in Aqueous Solution)

$$\text{Oxidizing agent} \underset{\text{oxidized}}{\overset{\text{reduced}}{\rightleftarrows}} \text{Reducing agent}$$

	Oxidizing agent		Reducing agent	
Increasing oxidizing ability (↑)	*Strongly oxidizing*			Increasing reducing ability (↓)
	$F_2(g) + 2e^-$	⇄	$2F^-$	
	$H_2O_2 + 2H^+ + 2e^-$	⇄	$2H_2O(l)$	
	$MnO_4^- + 8H^+ + 5e^-$	⇄	$Mn^{2+} + 4H_2O(l)$	
	$Cl_2(g) + 2e^-$	⇄	$2Cl^-$	
	$Cr_2O_7^{2-} + 14H^+ + 6e^-$	⇄	$2Cr^{3+} + 7H_2O(l)$	
	$O_3(g) + H_2O(l) + 2e^-$	⇄	$O_2(g) + 2OH^-$	
	$O_2(g) + 4H^+ + 4e^-$	⇄	$2H_2O(l)$	
	$2IO_3^- + 12H^+ + 10e^-$	⇄	$I_2(s) + 6H_2O(l)$	
	$NO_3^- + 4H^+ + 3e^-$	⇄	$NO(g) + 2H_2O(l)$	
	$Ag^+ + e^-$	⇄	$Ag(s)$	
	$Fe^{3+} + e^-$	⇄	Fe^{2+}	
	$Cu^{2+} + 2e^-$	⇄	$Cu(s)$	
	$2H^+ + 2e^-$	⇄	$H_2(g)$	
	$Sn^{2+} + 2e^-$	⇄	$Sn(s)$	
	$Zn^{2+} + 2e^-$	⇄	$Zn(s)$	
	$2H_2O(l) + 2e^-$	⇄	$H_2(g) + 2OH^-$	
	$Al^{3+} + 3e^-$	⇄	$Al(s)$	
	$Mg^{2+} + 2e^-$	⇄	$Mg(s)$	
	$Na^+ + e^-$	⇄	$Na(s)$	
			Strongly reducing	

As you consider the information in Table 17.1, note that the strongest oxidizing agents are the best electron acceptors and are reduced easily. Fluorine (F_2), chlorine (Cl_2), and oxygen (O_2) are good oxidizing agents, as expected from their positions in the periodic table. Furthermore, the strongest reducing agents are the best electron donors and are oxidized easily. As expected from their positions in the periodic table, alkali (Group 1A) and alkaline earth (Group 2A) metals are among the best reducing agents.

We can use the information in Table 17.1 to predict the outcome of a redox reaction in the same way that we used an activity series of metals (Section 10.8). When an oxidizing agent is mixed with a reducing agent that lies below it in the table, electrons are transferred to give the respective products of oxidation and reduction. To illustrate, consider what happens when Cl_2 is added to a solution of iron(II) ions:

$$Cl_2(g) + 2Fe^{2+}(aq) \longrightarrow$$

The products of the reaction can be predicted from the table. Because Cl_2 is above Fe^{2+}, Cl_2 will be reduced to Cl^-, and Fe^{2+} will be oxidized to Fe^{3+}.

$$Cl_2(g) + Fe^{2+}(aq) \longrightarrow Fe^{3+}(aq) + Cl^-(aq)$$

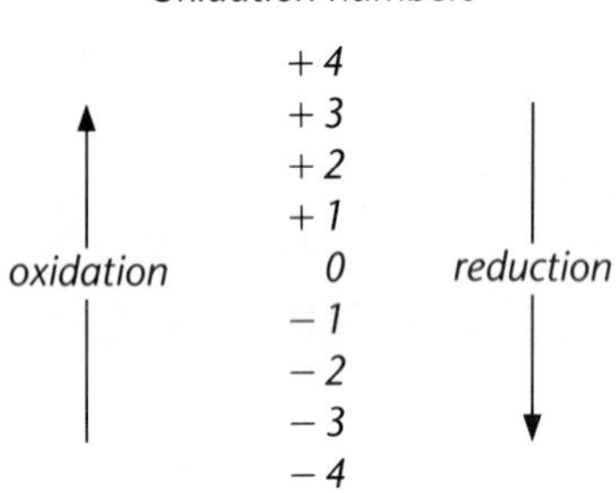

Note that as chlorine is reduced, its oxidation number changes from 0 (in Cl_2) to -1 (for Cl^-), and when iron is oxidized its oxidation number increases from $+2$ (for Fe^{2+}) to $+3$ (for Fe^{3+}). Because the total number of electrons lost during oxidation is always equal to the total number of electrons gained in reduction, no electrons are shown in the balanced equation. The coefficients for Fe^{2+} and Fe^{3+} are needed to balance charge (and electrons lost or gained).

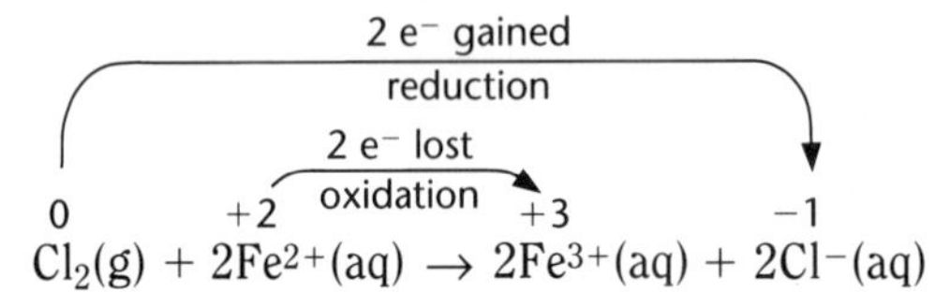

$$Cl_2(g) + 2Fe^{2+}(aq) \longrightarrow 2Fe^{3+}(aq) + 2Cl^-(aq)$$

Remember that during oxidation electrons are lost, and the oxidation number of the oxidized element increases. During reduction, electrons are gained, and the oxidation number of the reduced element decreases (Section 9.2).

Review the rules for assigning oxidation numbers (Section 9.2).

Many redox equations are so complex that balancing by inspection is nearly impossible. Strategies for balancing such equations are presented in Sections 17.2 and 17.3. Before attempting to balance complex redox equations, you must have a working knowledge of oxidation numbers. To review the rules for assigning oxidation numbers, refer to Section 9.2.

Hydrogen gas is formed in the reaction of magnesium with hydrochloric acid.

Example 17.1

What products result from mixing magnesium (Mg) with aqueous acid (H^+)? Write an equation for the reaction. ■

Solution Hydrogen ion is above magnesium in the table, so an oxidation–reduction reaction occurs to give H_2 and Mg^{2+}. The balanced net ionic equation is

$$2H^+(aq) + Mg(s) \longrightarrow H_2(g) + Mg^{2+}(aq)$$

Example 17.2

Assign oxidation numbers to all elements in the reaction described in Example 17.1. ■

Solution You may need to refer to the rules for assigning oxidation numbers in Section 9.2. Oxidation numbers are shown above the formulas:

$$\overset{+1}{2H^+}(aq) + \overset{0}{Mg}(s) \longrightarrow \overset{0}{H_2}(g) + \overset{+2}{Mg^{2+}}(aq)$$

Example 17.3

Assign oxidation numbers to the elements in the following formulas:

a. H_2O_2

b. MnO_4^-

c. $Cr_2O_7^{2-}$

d. O_3 ■

Solutions

a. $\overset{+1}{H_2}\overset{-1}{O_2}$

b. $\overset{+7}{Mn}\overset{-2}{O_4^-}$

c. $\overset{+6}{Cr_2}\overset{-2}{O_7^{2-}}$

d. $\overset{0}{O_3}$

Practice Problem 17.1 Assign oxidation numbers to the elements in the following formulas. (Solutions to practice problems appear at the end of the chapter.)

a. NaClO

b. F_2

c. NO_2

d. HIO_4

17.2

Balancing Redox Equations: The Half-Reaction Method

Learning Objective

■ Balance redox equations using the half-reaction method.

Although many simple redox equations can be balanced by inspection (Chapter 10), a strategy is needed for more complex equations. In the half-reaction (or ion–electron) method, the oxidation and reduction are separated into half-reactions. Balanced equations are written for the half-reactions, and these equations are added to give the balanced equation for the oxidation–reduction reaction.

To illustrate the half-reaction method, consider the reaction of chloride ion with permanganate ion in acidic solution, described by the following unbalanced equation. The symbols indicating physical states are omitted for convenience during the balancing process.

$$Cl^- + MnO_4^- \rightarrow Cl_2 + Mn^{2+}$$

The steps of the half-reaction method for balancing the equation are as follows:

Step 1 Write unbalanced equations for the oxidation and reduction half-reactions.

$$MnO_4^- \rightarrow Mn^{2+}$$

$$Cl^- \rightarrow Cl_2$$

Step 2 Balance the equations for the half-reactions separately.

a. Balance the elements other than oxygen and hydrogen. In this example, Mn is balanced.

$$MnO_4^- \rightarrow Mn^{2+}$$

b. Add H_2O to balance oxygen atoms.

$$MnO_4^- \rightarrow Mn^{2+} + 4H_2O$$

c. Add H^+ ions to balance hydrogen atoms.

$$8H^+ + MnO_4^- \rightarrow Mn^{2+} + 4H_2O$$

d. Add electrons to balance charge. In **Step 2c,** the net charge for the reactants is $+7$, the sum of the charges for the reactants $[8(+1) + 1(-1)]$. The charge for the products is $+2$, the sum of the charges for the products $[1(+2) + 0$ for $H_2O]$. The difference is $+5$. Because electrons are negatively charged, $5e^-$ are added to the more positive side of the equation:

$$5e^- + 8H^+ + MnO_4^- \rightarrow Mn^{2+} + 4H_2O$$

Because electrons are gained, this is the reduction half-reaction. Repeat the process for the other half-reaction. Some steps may be skipped for the oxidation half-reaction because neither oxygen nor hydrogen is involved.

$$2Cl^- \rightarrow Cl_2 + 2e^- \quad \text{(oxidation half-reaction)}$$

Step 3 Combine the half-reactions.

a. Electrons must be canceled when the half-reaction equations are added. Here, multiply the equation for the oxidation half-reaction by 5 and the equation for the reduction half-reaction by 2, which gives a total of $10e^-$ lost and gained.

$$10Cl^- \rightarrow 5Cl_2 + 10e^- \quad \text{(oxidation half-reaction)}$$

$$10e^- + 16H^+ + 2MnO_4^- \rightarrow 2Mn^{2+} + 8H_2O \quad \text{(reduction half-reaction)}$$

b. Cancel electrons and add the equations to give the balanced equation. Also cancel H^+ and H_2O if they appear on both sides of the equation.

$$10Cl^- \rightarrow 5Cl_2 + \cancel{10e^-}$$

$$\cancel{10e^-} + 16H^+ + 2MnO_4^- \rightarrow 2Mn^{2+} + 8H_2O$$

$$10Cl^- + 16H^+ + 2MnO_4^- \rightarrow 2Mn^{2+} + 8H_2O + 5Cl_2$$

Step 4 Check to be sure both the elements and charges are balanced:

10 Cl^- are balanced by 5 Cl_2.
16 H^+ are balanced by 8 H_2O.
2 Mn in $2MnO_4^-$ are balanced by $2Mn^{2+}$.
8 O in $2MnO_4^-$ are balanced by $8H_2O$.

Net charge on the left side (+4) is balanced by net charge on the right side (+4).

Step 5 Add symbols to show whether substances are solids, liquids, gases, or in aqueous solution.

$$10Cl^-(aq) + 16H^+(aq) + 2MnO_4^-(aq) \rightarrow 2Mn^{2+}(aq) + 8H_2O(l) + 5Cl_2(g)$$

A key principle in balancing redox equations is that the number of electrons lost in oxidation is equal to the number gained in reduction. In other words:

Oxidation (number of e^- lost) equals Reduction (number of e^- gained)

Remember that electrons are not shown in the final balanced equation because all electrons lost during oxidation are gained during reduction.

Balancing Redox Equations: The Half-Reaction Method

Step 1 Separate the equation into half-reactions.
Step 2 Balance the equations for the half-reactions.
- **a.** Balance the elements other than oxygen and hydrogen.
- **b.** Add H_2O to balance oxygen atoms.
- **c.** Add H^+ ions to balance hydrogen atoms.
- **d.** Add electrons to balance charge.

Step 3 Combine the equations for the half-reactions, multiplying each by the appropriate factor to give equal numbers of electrons lost and gained.
- **a.** Cancel electrons and add the equations.
- **b.** Also cancel H^+ and H_2O if they appear on both sides of the equation.

Step 4 Check the equation to be sure that both atoms and charges are balanced.
Step 5 Add symbols to show whether substances are solids, liquids, gases, or in aqueous solution.

The half-reaction method of balancing redox equations is used for reactions in aqueous solution involving ions. Most redox reactions occur in acidic solution. For a reaction that occurs in basic solution, a slight variation in the steps is required to balance the equation. However, because reactions in basic solution are less common, they are not included here.

Example 17.4

Balance the following equation (a

$$NO_3^- + Fe^{2+} \rightarrow Fe^{3+} + N$$ ■

Solution

Step 1 Separate into half-reactions:

$$NO_3^- \rightarrow NO$$

$$Fe^{2+} \rightarrow Fe^{3+}$$

Step 2 Balance the equations for the half-reactions separately.

a. For the reduction half-reaction equation, first balance nitrogen (balanced).

$$NO_3^- \rightarrow NO$$

b. Add H_2O to balance oxygen.

$$NO_3^- \rightarrow NO + 2H_2O$$

c. Add H^+ to balance hydrogen.

$$4H^+ + NO_3^- \rightarrow NO + 2H_2O$$

d. Add electrons to balance charge.

$$3e^- + 4H^+ + NO_3^- \rightarrow NO + 2H_2O$$

The oxidation half-reaction equation can be balanced by simply balancing charge by adding one electron on the right side.

$$Fe^{2+} \rightarrow Fe^{3+} + e^-$$

Step 3 Combine the equations for the half-reactions. The oxidation half-reaction must be multiplied by 3 to cancel electrons.

$$\cancel{3e^-} + 4H^+ + NO_3^- \rightarrow NO + 2H_2O$$

$$3Fe^{2+} \rightarrow 3Fe^{3+} + \cancel{3e^-}$$

$$3Fe^{2+} + 4H^+ + NO_3^- \rightarrow 3Fe^{3+} + NO + 2H_2O$$

Step 4 Check the equation to be sure that both atoms and charges are balanced.

Step 5 Add symbols to show whether substances are solids, liquids, gases, or in aqueous solution.

$$3Fe^{2+}(aq) + 4H^+(aq) + NO_3^-(aq) \rightarrow 3Fe^{3+}(aq) + NO(g) + 2H_2O(l)$$

Example 17.5

Write the balanced equation for the following reaction in acidic solution.

$$Sn^{2+} + Cr_2O_7^{2-} \longrightarrow Sn^{4+} + Cr^{3+}$$ ■

Solution

Step 1 Separate into half-reactions:

$$Sn^{2+} \longrightarrow Sn^{4+}$$

$$Cr_2O_7^{2-} \longrightarrow Cr^{3+}$$

Step 2 Balance the equations for the half-reactions. The oxidation half-reaction equation can be balanced by balancing charge.

$$Sn^{2+} \longrightarrow Sn^{4+} + 2e^-$$

To balance the reduction half-reaction equation:

a. Balance Cr.

$$Cr_2O_7^{2-} \longrightarrow 2Cr^{3+}$$

b. Add $7H_2O$ to the right side to balance oxygen.

$$Cr_2O_7^{2-} \longrightarrow 2Cr^{3+} + 7H_2O$$

c. Add 14 H^+ to the left side to balance hydrogen.

$$14H^+ + Cr_2O_7^{2-} \longrightarrow 2Cr^{3+} + 7H_2O$$

d. Add 6 electrons to the left side to balance charge.

$$6e^- + Cr_2O_7^{2-} + 14H^+ \longrightarrow 2Cr^{3+} + 7H_2O$$

Step 3 Combine the equations. To cancel electrons, the oxidation half-reaction equation must be multiplied by 3:

$$3Sn^{2+} \longrightarrow 3Sn^{4+} + \cancel{6e^-}$$

$$\cancel{6e^-} + Cr_2O_7^{2-} + 14H^+ \longrightarrow 2Cr^{3+} + 7H_2O$$

$$3Sn^{2+} + Cr_2O_7^{2-} + 14H^+ \longrightarrow 2Cr^{3+} + 3Sn^{4+} + 7H_2O$$

Step 4 Check the equation. Be sure that charges as well as atoms are balanced.

Step 5. Add symbols to show whether substances are solids, liquids, gases, or in aqueous solution.

$$3Sn^{2+}(aq) + Cr_2O_7^{2-}(aq) + 14H^+(aq) \longrightarrow 2Cr^{3+}(aq) + 3Sn^{4+}(aq) + 7H_2O(l)$$

Practice Problem 17.2 Balance the following equation (acidic solution):

$$U^{4+} + MnO_4^- \longrightarrow UO_2^{2+} + Mn^{2+}$$

17.3 Balancing Redox Equations: The Oxidation-Number Method

Learning Objective

- Balance redox equations using the oxidation-number method.

The half-reaction method is not convenient for balancing redox reactions that do not occur in aqueous solution. One such example is the reaction of iron(III) oxide with carbon monoxide, as described by the following equation (unbalanced):

$$Fe_2O_3(s) + CO(g) \longrightarrow Fe(s) + CO_2(g)$$

Although this equation can be balanced by inspection, we will use it to illustrate the oxidation-number method for balancing redox equations.

We can quickly recognize that this is a redox reaction because elements are involved. However, because this reaction is not in aqueous solution, there are no ions in solution, and the half-reaction (ion–electron) method should not be used in balancing the equation. This is where the oxidation-number method is used.

Before balancing an equation by the oxidation number method, all reactants and products must be given in a skeletal equation. The steps are:

Step 1 Assign oxidation numbers to each element, following the rules discussed earlier (Section 9.2).

$$\overset{+3\,-2}{Fe_2O_3} + \overset{+2\,-2}{CO} \longrightarrow \overset{0}{Fe} + \overset{+4\,-2}{CO_2}$$

Step 2 Diagram the electrons lost (oxidation) by drawing an arrow connecting the reduced and oxidized forms of the element (in this case, C). Write the number of electrons (per atom) next to the arrow. Then diagram the electrons gained (reduction) in a like manner. Balance the electrons gained and lost by multiplying the number of electrons gained and/or lost per atom by the appropriate factor.

3 × (2 e⁻ lost)

$$\overset{+3\,-2}{Fe_2O_3} + \overset{+2\,-2}{CO} \longrightarrow \overset{0}{Fe} + \overset{+4\,-2}{CO_2}$$

2 × (3 e⁻ gained)

Step 3 Determine the balancing coefficients that are needed in the equation from the factor(s) used to balance electrons. In this equation, 3 carbon atoms and 2 iron atoms are needed. On the left side of the equation, 1 Fe_2O_3 is needed to give 2 Fe. However, a balancing coefficient of 1 is understood and is not written in the equation.

3 × (2 e⁻ lost)

$$\overset{+3\,-2}{Fe_2O_3} + 3\overset{+2\,-2}{CO} \longrightarrow 2\overset{0}{Fe} + 3\overset{+4\,-2}{CO_2}$$

2 × (3 e⁻ gained)

Step 4 Balance oxygen (and hydrogen) as necessary. In this equation, oxygen is already balanced.

$Fe_2O_3 + 3CO \longrightarrow 2Fe + 3CO_2$

Step 5 Check the equation.

Step 6 Add symbols to show whether substances are solids, liquids, gases, or in aqueous solution.

$Fe_2O_3(s) + 3CO(g) \longrightarrow 2Fe(s) + 3CO_2(g)$

Remember, the oxidation-number method is particularly useful in balancing equations that do not occur in aqueous solution. However, it also generally works well in balancing equations for solution reactions. Practice the method with the following examples.

Example 17.6

When hydrocarbons (made up only of hydrogen and carbon) are burned, carbon dioxide and water are formed. Write a balanced equation for the combustion of hexene, C_6H_{12}, a liquid hydrocarbon. ■

Solution

Step 1 Write an unbalanced equation and assign oxidation numbers to each element.

$$\overset{-2+1}{C_6H_{12}} + \overset{0}{O_2} \longrightarrow \overset{+4-2}{CO_2} + \overset{+1-2}{H_2O}$$

Step 2 Diagram and balance the electrons lost and the electrons gained per atom. Because there are 6 carbons in C_6H_{12}, the electrons lost per atom must be multiplied by 6.

$6 \times (6\ e^-\ \text{lost}) = 36\ e^-$

$$\overset{-2\ +1}{C_6H_{12}} + \overset{0}{O_2} \longrightarrow \overset{+4\ -2}{CO_2} + \overset{+1\ -2}{H_2O}$$

$18 \times (2\ e^-\ \text{gained}) = 36\ e^-$

Step 3 Add balancing coefficients to the equation. Note that the formula of hexene has 6 carbons, and a total of 36 e^- are lost per C_6H_{12}. Then, 9 O_2 are needed to give 18 oxygen atoms. On the right side of the equation, 6 CO_2 are needed to balance carbon atoms, and 6 H_2O are needed to balance 18 oxygen atoms on the right side.

$6 \times (6\ e^-\ \text{lost})$

$$\overset{-2\ +1}{C_6H_{12}} + 9\overset{0}{O_2} \longrightarrow 6\overset{+4\ -2}{CO_2} + 6\overset{+1\ -2}{H_2O}$$

$18 \times (2\ e^-\ \text{gained})$

Step 4 Inspection shows that hydrogen atoms are now balanced.

Step 5 Check the equation.

Step 6 Add symbols to show whether substances are solids, liquids, or gases.

$C_6H_{12}(l) + 9O_2(g) \longrightarrow 6CO_2(g) + 6H_2O(l)$

Example 17.7

Balance the following equation (acidic solution):

$$CuS(s) + NO_3^-(aq) \longrightarrow Cu^{2+}(aq) + NO_2(g) + S(s)$$ ■

Solution

Step 1 Assign oxidation numbers.

$$\overset{+2-2}{CuS} + \overset{+5-2}{NO_3^-} \longrightarrow \overset{+2}{Cu^{2+}} + \overset{+4-2}{NO_2} + \overset{0}{S}$$

Step 2 Diagram and balance the electrons gained and lost per atom.

(2 e⁻ lost)

$$\overset{+2\ -2}{CuS} + \overset{+5\ -2}{NO_3^-} \longrightarrow \overset{+2}{Cu^{2+}} + \overset{+4\ -2}{NO_2} + \overset{0}{S}$$

2 × (1 e⁻ gained)

Step 3 Add balancing coefficients to the equation.

(2 e⁻ lost)

$$\overset{+2\ -2}{CuS} + 2\overset{+5\ -2}{NO_3^-} \longrightarrow \overset{+2}{Cu^{2+}} + 2\overset{+4\ -2}{NO_2} + \overset{0}{S}$$

2 × (1 e⁻ gained)

Step 4 Add H_2O to balance oxygen atoms. Add H^+ to balance hydrogen atoms. Note that this step is necessary only for reactions in aqueous solution.

(2 e⁻ lost)

$$\overset{+2\ -2}{CuS} + 2\overset{+5\ -2}{NO_3^-} + 4H^+ \longrightarrow \overset{+2}{Cu^{2+}} + 2\overset{+4\ -2}{NO_2} + \overset{0}{S} + 2H_2O$$

2 × (1 e⁻ gained)

Step 5 Check the equation to see if both the elements and the charges are balanced. Do not confuse oxidation numbers with ionic charges.

Step 6 Add symbols to show whether substances are solids, liquids, gases, or in aqueous solution.

$$CuS(s) + 2NO_3^-(aq) + 4H^+(aq) \longrightarrow Cu^{2+}(aq) + 2NO_2(g) + S(s) + 2H_2O(l)$$

Practice Problem 17.3 Balance the equation for the reaction of methane (CH_4) with chlorine (Cl_2) to give carbon tetrachloride (CCl_4):

$$CH_4(g) + Cl_2(g) \longrightarrow CCl_4(l) + HCl(g)$$

Balancing Redox Equations: The Oxidation-Number Method

Step 1 Assign oxidation numbers to the elements.
Step 2 Diagram and balance the electrons lost and the electrons gained.
Step 3 Add balancing coefficients for the substances containing the elements reduced and oxidized.
Step 4 Balance oxygen by adding H_2O to the equation. Balance hydrogen by adding H^+ (for reactions in acidic solution only).
Step 5 Check the equation.
Step 6 Add symbols to show whether substances are solids, liquids, gases, or in aqueous solution.

17.4 Electrochemistry

Chemical reactions (redox) can produce electricity.
Electricity can produce chemical reactions.

Electrochemistry is a branch of chemistry concerned with the relationship of electricity and chemical reactions. Electrochemistry includes the study of batteries, which use chemical reactions to generate electricity, and the use of electricity to cause chemical change. Electrical energy is used to drive chemical reactions for the manufacture of chemicals and chemical products.

Electrochemistry is studied using devices called electrochemical cells. Batteries have become part of our everyday lives. We depend upon battery-powered clocks and watches to keep us on our daily schedules. Our cars are started with the help of a battery. We listen to battery-powered radios and solve chemistry problems with the help of battery-powered calculators.

The connection between electricity and chemical reactions was first noticed by two Italian scientists, Luigi Galvani (1737–1798) and Alessandro Volta (1745–1827). In 1786, Galvani observed that the leg of a dead frog hung by a copper wire twitched when it touched a piece of iron. This observation and others involving two metals in contact with solutions led to the conclusion that an electric current can be produced by a chemical reaction. In 1800, Volta used these ideas to invent the first battery, a group of current-producing devices connected in series. One of these simple devices is often called a *voltaic cell* in his honor, or sometimes a *galvanic cell* in honor of Galvani.

A device in which an electric current results from a redox reaction, or that uses an electric current to cause a redox reaction is called an **electrochemical cell.** The two basic types of electrochemical cells, voltaic and electrolytic, are discussed in the following sections.

The Italian scientists Luigi Galvani and Alessandro Volta (left) first recognized the connection between electricity and chemical reactions. Volta invented the first battery in 1800. The terms galvanic cell and voltaic cell are in honor of their work.

17.5 Voltaic Cells and Batteries

Learning Objectives

- Diagram voltaic cells, labeling the anode, cathode, electrode polarities, current flow, and directions the ions move.
- Write equations for the half-reactions and the total reaction for the Daniell cell, the lead storage battery, the original dry cell, an alkaline cell, and a mercury cell.

Reaction of zinc with copper(II) ions:

$Zn(s) + Cu^{2+}(aq) \longrightarrow Zn^{2+}(aq) + Cu(s)$

Volta's first battery was a series of electrochemical cells that used the reaction of zinc metal with copper(II) ions in aqueous sulfuric acid.

$$Zn(s) + Cu^{2+}(aq) \longrightarrow Zn^{2+}(aq) + Cu(s)$$

In this reaction, Zn is oxidized and is the reducing agent (electron donor) and Cu^{2+} is reduced and is the oxidizing agent (electron acceptor). When a strip of zinc is dipped into a solution containing Cu^{2+}, electrons are transferred directly from zinc atoms to copper(II) ions whenever they collide. Direct transfer of electrons from the reducing agent to the oxidizing agent produces no electric current.

To produce an electric current from a redox reaction, the oxidation and reduction half-reactions are separated, usually in separate containers or compartments called half-cells. Each half-cell contains an electrolyte and an electrode, a conductor at which electrons are either lost by the element oxidized or gained by the element reduced. The electrodes are connected by a wire that allows electrons to flow from the reducing agent to the oxidizing agent. An electrochemical cell similar to one constructed in 1836 by the British chemist John Daniell is illustrated in Figure 17.1.

Electrolyte: A substance that dissolves to give a solution that conducts an electric current.

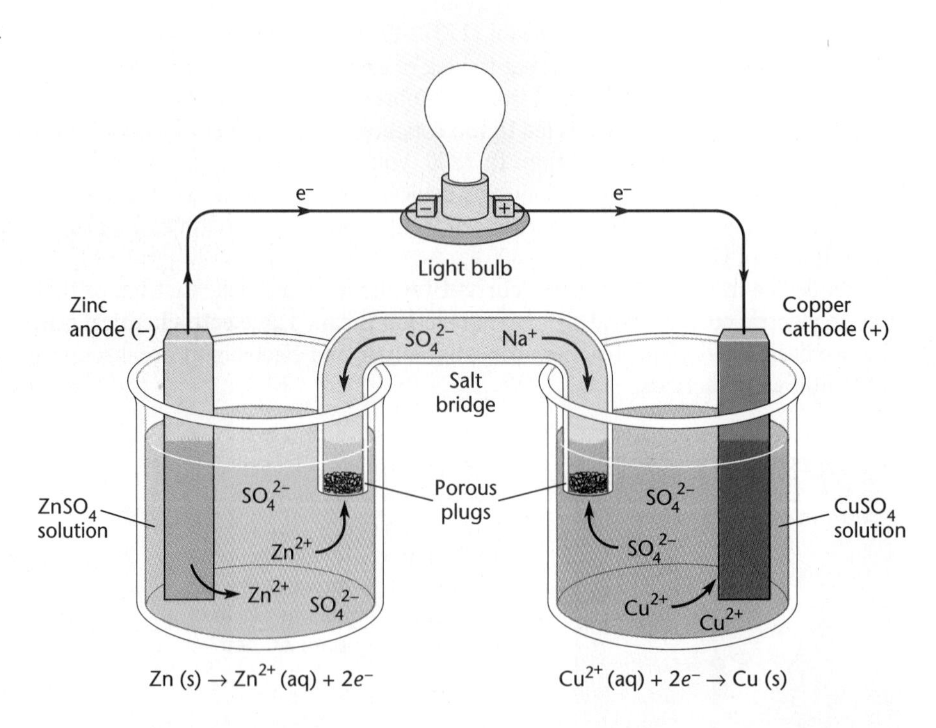

FIGURE 17.1 The Daniell cell. The cell reaction is $Zn(s) + Cu^{2+}(aq) \longrightarrow Cu(s) + Zn^{2+}(aq)$. In this cell, zinc is oxidized at the anode, and copper(II) ions are reduced at the cathode. Electrons flow from the anode (Zn) through the external circuit to the cathode (Cu). Anions move through the salt bridge toward the anode, and cations move toward the cathode to maintain a charge balance in the solutions.

CHEMICAL WORLD

The World's Tiniest Battery

What's left over after you put a battery in a human red blood cell? Answer: Room for 99 more batteries. In 1992, scientists built a battery so small that one hundred of them could fit inside a human red blood cell.

To make the world's tiniest battery, electrochemists used a scanning tunneling microscope (STM), the type of microscope used to create images of surface atoms (such as the opening photograph in Chapter 5). The STM makes these pictures by running the tip of a very fine needle over a surface.[1] Later, scientists discovered that this needle-tip probe could act like a magnet to pick up surface atoms, move them, and then drop them, as an electromagnet moves steel balls. In 1990, scientists at IBM demonstrated this capability by moving 35 xenon atoms around on the face of a nickel crystal to spell "IBM."

Two years later, in 1992, electrochemists at the University of California (Irvine) used an STM to make two piles of atoms on a graphite surface, one of copper atoms and one of zinc atoms a short distance away. When the piles of atoms were covered with an electrolyte, a current flowed between the zinc anode and the copper cathode through the graphite substrate. That device constituted a battery. This battery produced $\frac{1}{50}$ volt for 46 minutes before it died.

There's been a trend in recent years to reduce the sizes of electronic devices, and these batteries will certainly be valuable for miniaturization. Other possible uses for batteries no larger than cold viruses are studies of how human proteins *in vivo* are affected by electric fields and of corrosion (which is an electrochemical reaction) on an atomic scale.

Other researchers have been developing motors the sizes of molecules. Will these Daniell cells supply the power for such motors? What will they drive? This is surely the last possibility for microminiaturization.

[1]The sample to be imaged by STM acts as one electrode and the probe acts as a second electrode. A voltage is applied between them. The amount of electric current flowing between the probe and the sample depends on their separation distance. A feedback mechanism on the STM senses this current and keeps the separation distance constant at about one nanometer. As the probe moves over the surface of the sample, its up-and-down motion is translated by a computer into an image of the surface. The resolution of this surface image is so high that it enables us to distinguish individual surface atoms.

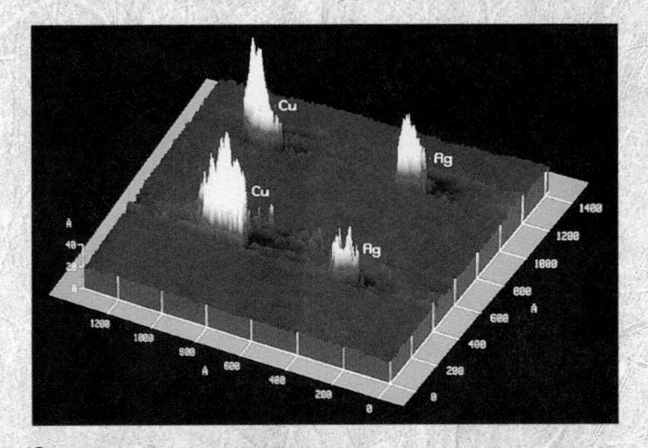

a

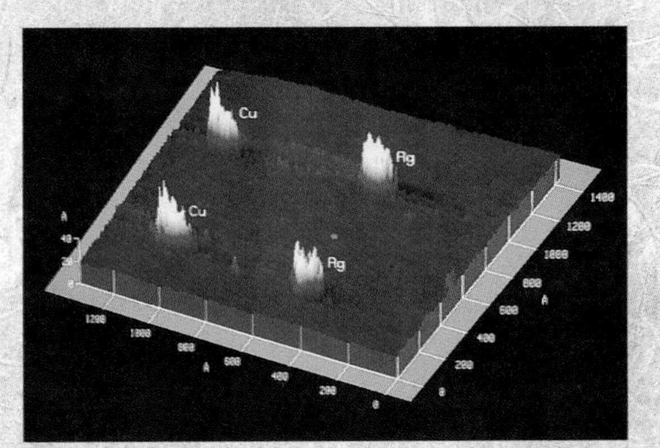

b

Photos taken through a scanning tunneling microscope show the structures of the world's smallest battery. (a) Pillars of silver and copper atoms deposited on a graphite surface one-hundredth the diameter of a red blood cell make up the battery. Numbers along the edges are in Angstrom units (1 Å = 1×10^{-8} cm). The battery generated 20 thousandths of a volt. (b) This photo shows the battery after copper atoms have left the copper pillars and the silver terminals have been coated with a two-atom-thick layer of copper.

In one beaker, a strip of zinc is dipped in a solution of $ZnSO_4$ (contains Zn^{2+}). A wire connects the zinc strip to a copper strip dipped in a solution of $CuSO_4$ (contains Cu^{2+}) in a second beaker. Because zinc is a stronger reducing agent than copper (see Table 17.1), Zn atoms lose electrons to form Zn^{2+}. The oxidation half-reaction is

$$Zn(s) \longrightarrow Zn^{2+} + 2e^{-} \qquad \text{(oxidation half-reaction)}$$

Anode: Electrode where oxidation occurs.

The electrode where oxidation takes place is called the **anode.** In this cell, the zinc strip is the anode. In a voltaic cell, the anode has a negative charge.

The electrons move through the external circuit to the copper strip dipped in a solution of a copper salt, where Cu^{2+} ions accept electrons to form Cu atoms. The cathode is positively charged in a voltaic cell.

$$2e^{-} + Cu^{2+} \longrightarrow Cu(s) \qquad \text{(reduction half-reaction)}$$

Cathode: Electrode where reduction occurs.

The electrode where reduction occurs is called the **cathode.** In this cell, the copper strip is the cathode. The cathode is positively charged in a voltaic cell.

A necessary feature of the cell is the salt bridge that connects the solutions in the half-cells. The salt bridge contains a salt solution such as Na_2SO_4, sometimes in gel form. It allows ions to move between the half-cells without complete mixing of the solutions. When Zn^{2+} ions form at the anode and Cu^{2+} ions are consumed at the cathode, ions move through the salt bridge between the half-cells to maintain a charge balance in the solutions. In other words, it is impossible to build up an excess negative charge in the half-cell toward which the electrons are flowing. Note that the cations move toward the cathode and the anions move toward the anode. Without the salt bridge, the circuit would not be complete, and an electric current would not flow through the wire.

The Lead Storage Battery

The lead storage battery is used in cars and trucks to provide power for starting the engine and for lights and other accessories (Figure 17.2). A 12-volt (V) car battery consists of six 2-volt cells connected together. Each cell is constructed of alternating sheets of lead (Pb), the reducing agent, and lead(IV) oxide (PbO_2), the oxidizing agent. The lead sheets are connected to form the anode, whereas the lead oxide is the cathode. The electrolyte is a 30% solution of sulfuric acid, commonly called battery acid. During discharge of the battery, such as when your car is started, Pb is oxidized and PbO_2 is reduced.

Anode half-reaction (oxidation)

$$Pb(s) + H_2SO_4(aq) \longrightarrow PbSO_4(s) + 2H^{+}(aq) + 2e^{-}$$

Cathode half-reaction (reduction)

$$PbO_2(s) + H_2SO_4(aq) + 2H^{+}(aq) + 2e^{-} \longrightarrow PbSO_4(s) + 2H_2O(l)$$

Overall reaction

$$Pb(s) + PbO_2(s) + 2H_2SO_4(aq) \longrightarrow 2PbSO_4(s) + 2H_2O(l)$$

An important feature of a lead storage battery is that it is rechargeable. During recharging, current is forced through the battery to reverse the half-reactions, thereby replenishing the electrolyte and restoring the electrodes. During recharg-

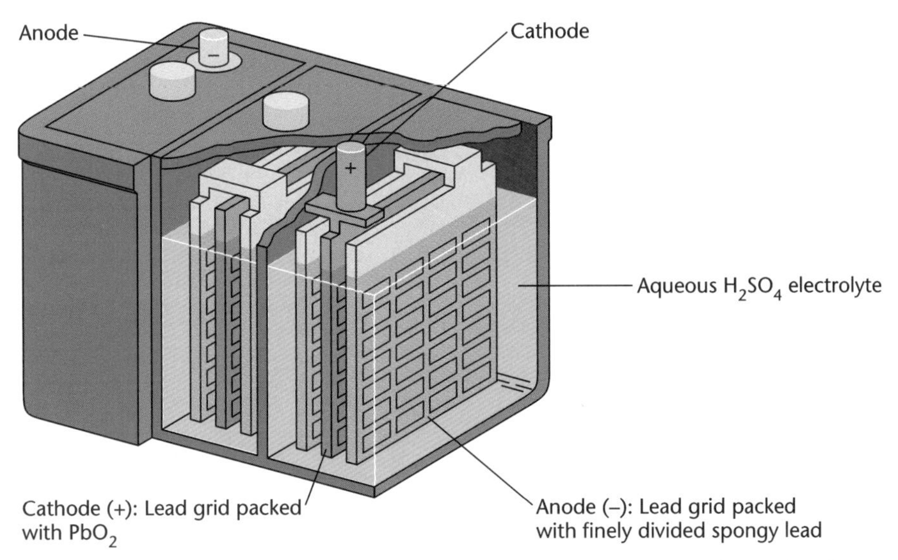

FIGURE 17.2 A lead storage battery. Several lead sheets connected together form the anode. Alternating sheets filled with lead(IV) oxide form the cathode. The electrolyte is 30% sulfuric acid. A 12-volt battery has six 2-volt cells connected inside the battery case.

A lead storage battery can be checked using a hydrometer, a device that indicates the density of the electrolyte.

ing, energy is stored in the battery for later use. A car battery is constantly recharged by the car's generator or alternator as the car is running.

The density of the sulfuric acid in a fully charged battery is 1.28 g/mL. Because discharging consumes sulfuric acid (as illustrated in the equations), the density of the electrolyte is lower for a "weak" battery. On the other hand, when a battery is charging, the half-reactions are reversed, and the concentration and the density of the sulfuric acid can be restored to about their original levels. When a battery is "dead," you can check for a bad cell by testing the electrolyte in each cell with a hydrometer, a simple device that indicates when the density of the electrolyte is low.

Dry Cells

A dry cell is a battery that uses an electrolyte that is a nonflowing paste rather than a liquid. The original "flashlight battery" was invented in 1865 by Georges Leclanche (1839–1882), a French chemist. The zinc anode of this battery also serves as the container for the electrolyte. The cathode is an inert graphite (carbon) rod that is positioned in the center of the cell and surrounded by a paste of MnO_2, $ZnCl_2$, NH_4Cl, powdered carbon, starch, and water (Figure 17.3). The half-reactions are complex but can be described by the following equations:

Anode half-reaction (oxidation)

$$Zn(s) \longrightarrow Zn^{2+}(aq) + 2e^-$$

Cathode half-reaction (reduction)

$$2MnO_2(s) + 2NH_4^+(aq) + 2e^- \longrightarrow Mn_2O_3(s) + 2NH_3(aq) + H_2O(l)$$

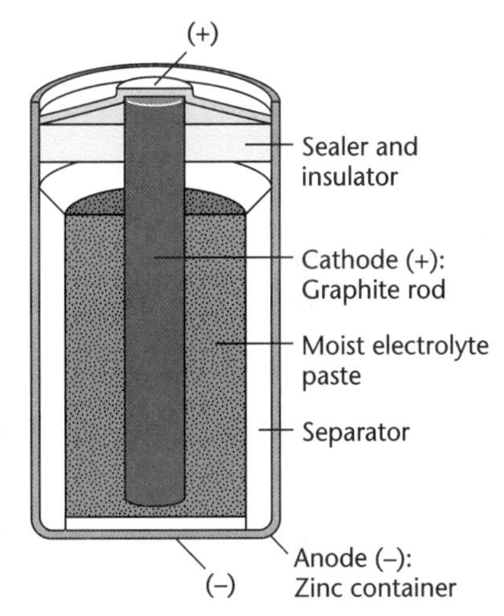

FIGURE 17.3 An ordinary dry cell or flashlight battery. The anode is zinc, which also serves as the container for the electrolyte. The electrolyte paste contains $ZnCl_2$, NH_4Cl, MnO_2, powdered carbon, starch, and water. The cathode in the center of the battery is a graphite rod. A fresh dry cell of this type delivers about 1.5 V.

FIGURE 17.4 An alkaline dry cell. The anode is powdered zinc (in a moist paste mixed with KOH) which donates electrons to a brass collector. Electrons move through the external circuit to the inner steel case (cathode) where they are accepted by the MnO_2 (mixed in a paste with graphite and water), the oxidizing agent.

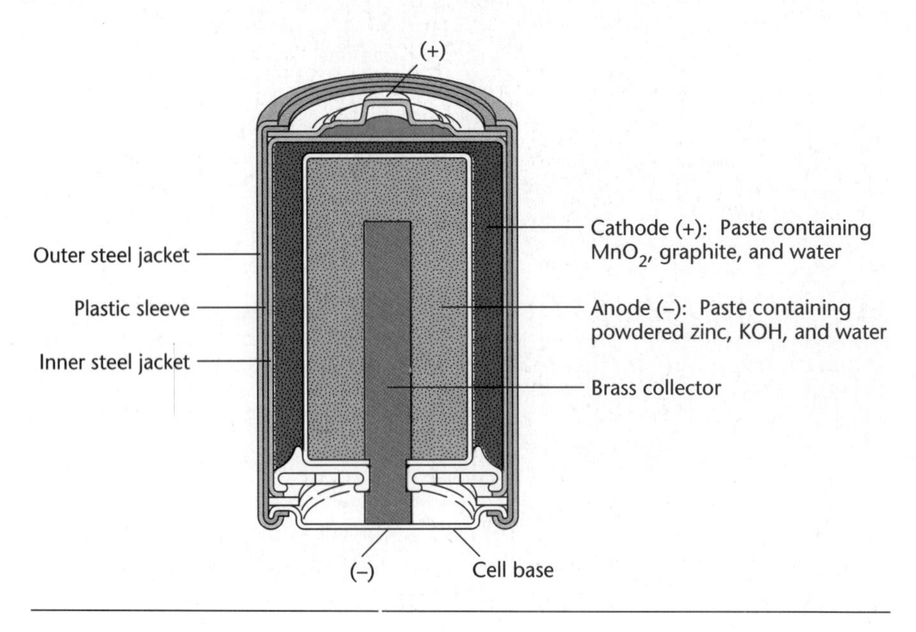

The equation for the overall reaction can be obtained by adding the equations for the half-reactions.

Overall reaction

$$Zn(s) + 2MnO_2(s) + 2NH_4^+(aq) \longrightarrow Zn^{2+}(aq) + Mn_2O_3(s) + 2NH_3(aq) + H_2O(l)$$

The NH_3 reacts with the Zn^{2+} that is migrating toward the cathode to form $Zn(NH_3)_4^{2+}$, a soluble complex ion. Each dry cell delivers about 1.5 V and is not rechargeable. After a period of extended use, a battery may "run down" because movement of the ions through the electrolyte paste is not keeping up with the current demand. A rest period often gives a battery additional life by giving ions time to migrate, thereby restoring charge balance in the electrolyte.

An alkaline dry cell (Figure 17.4) is more efficient than the original dry cell but is also more expensive. The NH_4Cl in the electrolyte is replaced with KOH.

Anode half-reaction (oxidation)

$$Zn(s) + 2OH^-(aq) \longrightarrow ZnO(s) + H_2O(l) + 2e^-$$

Cathode half-reaction (reduction)

$$2MnO_2(s) + H_2O(l) + 2e^- \longrightarrow Mn_2O_3(s) + 2OH^-(aq)$$

Overall reaction

$$2MnO_2(s) + Zn(s) \longrightarrow Mn_2O_3(s) + ZnO(s)$$

The alkaline electrolyte causes less corrosion of the zinc container than the acidic ammonium chloride paste in the ordinary dry cell. A further advantage is that no gaseous ammonia is produced.

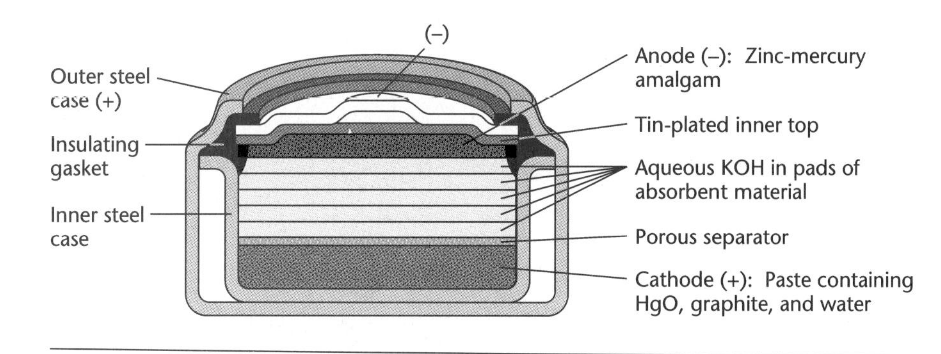

FIGURE 17.5 A mercury cell. The anode is a zinc–mercury amalgam. The oxidizing agent, mercury(II) oxide in a paste of graphite and water, accepts electrons from an inner steel case. A mercury cell delivers about 1.35 V.

The mercury cell (Figure 17.5), sometimes called a "button battery" because of its small size and its shape, is a zinc dry cell commonly used in watches, cameras, and hearing aids. Mercury(II) oxide (HgO) replaces MnO_2 as the oxidizing agent.

Anode half-reaction (oxidation)

$$Zn(s) + 2OH^-(aq) \longrightarrow ZnO(s) + H_2O(l) + 2e^-$$

Cathode half-reaction (reduction)

$$HgO(s) + H_2O(l) + 2e^- \longrightarrow Hg(l) + 2OH^-(aq)$$

Overall reaction

$$Zn(s) + HgO(s) \longrightarrow ZnO(s) + Hg(l)$$

The mercury cell is relatively expensive but is very reliable and has the advantage of maintaining an almost constant voltage of 1.35 V during its entire lifetime. This is essential in watches, which otherwise would slow as the voltage of a more conventional battery drops.

Electrolysis

17.6

Learning Objectives

- Diagram electrolytic cells, labelling the anode, cathode, electrode polarities, current flow, and directions the ions move.
- Write equations for the half-reactions and the total reaction for the electrolysis of water, molten sodium chloride, and aqueous sodium chloride.

A voltaic cell uses a redox reaction to produce an electric current. The reverse of this process, called **electrolysis,** uses an electric current to bring about a redox reaction. A cell used in electrolysis is called an electrolytic cell. A number of elements are prepared commercially using electrolysis.

A voltaic cell produces electricity.

An electrolytic cell consumes electricity.

The electrolysis of water occurs when an electric current is passed through an aqueous solution of an electrolyte. This reaction can be demonstrated using a laboratory apparatus such as that shown in Figure 17.6. The half-reactions take place at inert electrodes such as platinum.

Anode half-reaction (oxidation)

$$2H_2O(l) \rightarrow 4H^+(aq) + O_2(g) + 4e^-$$

Cathode half-reaction (reduction)

$$2H_2O(l) + 2e^- \rightarrow H_2(g) + 2OH^-(aq)$$

Multiplying the cathode half-reaction by 2 and combining the equations gives the balanced total equation. Note that it is necessary to combine $4H^+ + 4OH^-$ to give $4H_2O$ (right side) and then cancel with $4H_2O$ on the left side of the equation.

$$2H_2O(l) \rightarrow 2H_2(g) + O_2(g)$$

Because water is a nonelectrolyte, a small amount of inert ionic solute, such as sodium sulfate, must be added to allow a flow of current. Electrolysis of water may occur to some extent when a lead storage battery is charged or when a car with a dead battery is jump-start. Because of the danger of explosion of H_2–O_2 mixtures, care should be taken to avoid sparks and to vent a battery when it is being charged or when a car is jump-started.

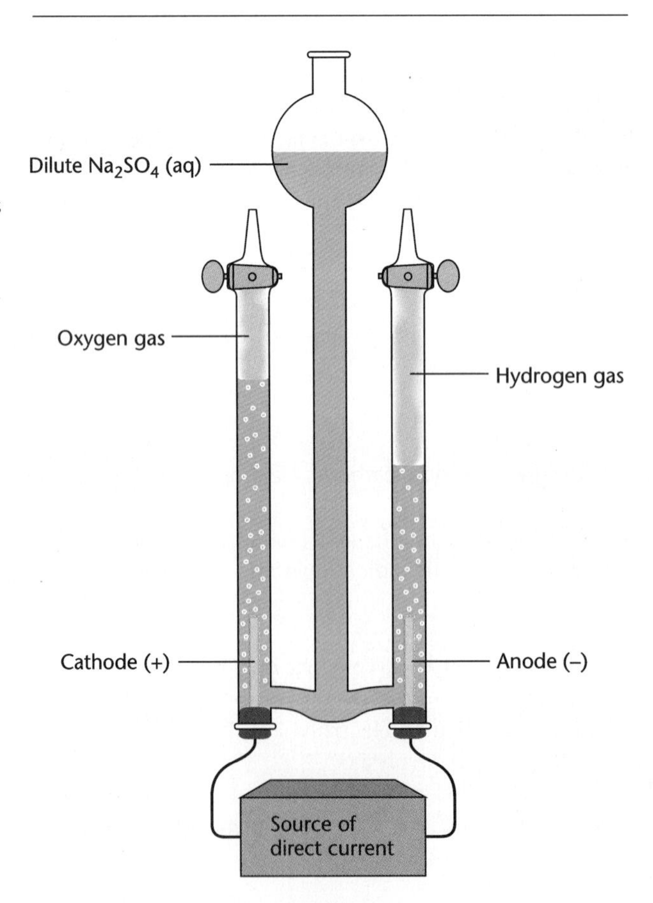

FIGURE 17.6 A laboratory apparatus for the electrolysis of water. When a current is passed through a dilute Na_2SO_4 solution, oxidation occurs at the anode to produce $O_2(g)$. Reduction produces $H_2(g)$ at the cathode. An inert metal, such as platinum, is used for the electrodes. The volume of hydrogen produced is twice the volume of oxygen formed, in agreement with Avogadro's hypothesis (Section 12.8).

The electrolysis of molten sodium chloride (m.p. 801°C) produces sodium and chlorine.

$$2NaCl(l) \longrightarrow 2Na(l) + Cl_2(g)$$

The reaction is carried out in an electrolytic cell (Figure 17.7) consisting of two inert electrodes immersed in the hot liquid sodium chloride. A power supply forces a current through the liquid from the cathode (−) to the anode (+). The half-reactions at the anode (oxidation) and cathode (reduction) are as follows:

Anode half-reaction (oxidation)

$$2Cl^-(l) \longrightarrow Cl_2(g) + 2e^-$$

Cathode half-reaction (reduction)

$$Na^+(l) + e^- \longrightarrow Na(l)$$

Combining the equations for the half-reactions gives the balanced equation shown earlier. Note that electrode polarities for an electrolytic cell are opposite those of a battery. In a battery, the anode is negative and the cathode is positive, whereas for an electrolytic cell, the anode is positive and the cathode is negative. Sodium metal and chlorine gas are produced commercially by this method. Sodium is used in sodium vapor lamps and is a strong reducing agent in chemical reactions.

Electrolysis of aqueous sodium chloride solution (Figure 17.8) produces chlorine and hydrogen rather than sodium. Most of the chlorine produced in the United States is obtained by this method. The anode half-reaction is the same as in the electrolysis of molton NaCl, but reduction of water (instead of Na^+) produces $H_2(g)$ and $OH^-(aq)$. Reduction of Na^+ is very difficult compared with the reduction of H_2O (Table 17.1). Furthermore, oxidation of Cl^- to give Cl_2 occurs more readily than oxidation of H_2O to give O_2.

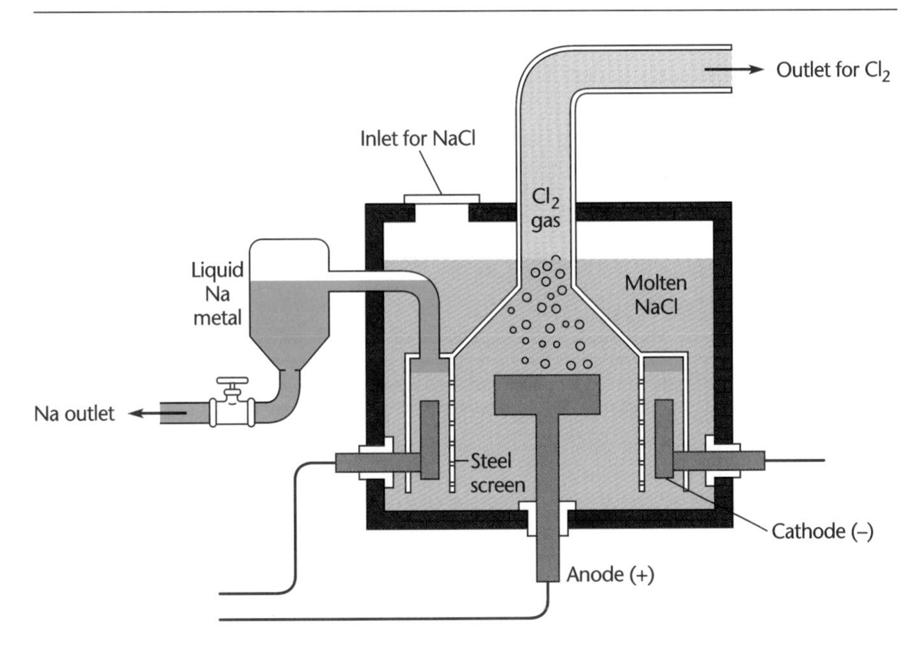

FIGURE 17.7 Electrolysis of molten sodium chloride. A Downs cell utilizes a graphite anode and a ring-shaped steel cathode. A steel grid separates the compartments where sodium and chlorine are formed.

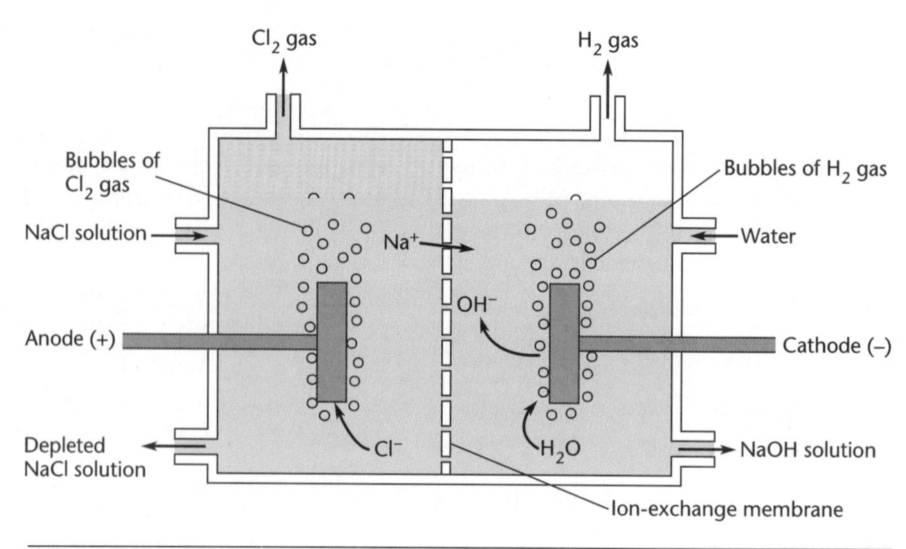

FIGURE 17.8 A commercial cell for the electrolysis of aqueous sodium chloride. An ion-exchange membrane allows only Na^+ ions to pass through and keeps OH^- ions in the cathode compartment to produce a good yield of sodium hydroxide. The membrane also prevents mixing of Cl_2 and H_2 gases.

Anode half-reaction (oxidation)

$$2Cl^-(aq) \longrightarrow Cl_2(g) + 2e^-$$

Cathode half-reaction (reduction)

$$2H_2O(l) + 2e^- \longrightarrow H_2(g) + 2OH^-(aq)$$

The net ionic equation is a result of the combination of the half-reactions:

$$2Cl^-(aq) + 2H_2O(l) \longrightarrow Cl_2(g) + H_2(g) + 2OH^-(aq)$$

The total equation includes sodium ions:

$$2NaCl(aq) + 2H_2O(l) \longrightarrow Cl_2(g) + H_2(g) + 2NaOH(aq)$$

Chlorine is important in water purification and in the manufacture of chemicals, plastics, pesticides, and paper. The design of a commercial electrolysis cell also allows aqueous sodium hydroxide to be removed from the cell. Sodium hydroxide, the most important commercial base, is used in the manufacture of soap and paper and to neutralize acids in industrial processes. Thus, two of the top ten chemicals produced in the United States, chlorine and sodium hydroxide, are manufactured simultaneously in this process.

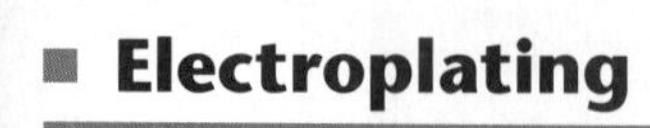

Electroplating

Electroplating is an industrial process in which a metal is electrolytically deposited or "plated out" on the surface of an object. The object to be plated is made the cathode of an electrolytic cell and the anode is often made of the plating metal. Although the object to be plated is usually metal, a plastic object can be plated if it is first coated with graphite to make it electrically conducting. The electrolyte is generally a solution of a salt of the plating metal. Common plating metals include silver, gold, copper, and chromium.

FIGURE 17.9 A chrome part after electroplating.

CHEMICAL WORLD

Gold Is Where You Don't See It

Growlersburg, now Georgetown, is a town in the mother lode of California about fifteen miles from Hangtown, now Placerville. The nuggets in the creeks around Growlersburg were so big you could (it was said) hear them growling in the gold pans.

Now the growlers are gone, along with the vigilantes, and though you can still make 50–100 dollars per very long day if you know the right place, the gold in the ore that's mined commercially today is not visible to the naked eye. Modern gold miners consider ore rich if it contains 1/10 ounce of gold in a ton of rock. Put your shovel away.

Gold dissolves in mercury, and in the days of the gold rush mercury was used to separate the precious stuff from waste. San Francisco Bay is still polluted by mercury that washed down the rivers from concentrators a hundred-odd miles away.

Gold doesn't dissolve in acids, but (as was also known in the 1880s) it can be dissolved in a basic solution of sodium cyanide. In modern concentrators, the gold-bearing rock is crushed fine, exposed to air to oxidize the gold to Au^+ ions, which form a complex with cyanide ions:

$$4Au(s) + 8CN^-(aq) + O_2(g) + 2H_2O(l) \longrightarrow 4Au(CN)_2^-(aq) + 4OH^-$$

Electroplating is used to recover the dissolved gold from the basic sodium cyanide solution. The gold is electroplated onto a cathode when an electric current sends electrons through the cathode and into the solution containing the dissolved gold:

$$Au(CN)_2^-(aq) + e^- \longrightarrow Au(s) + 2CN^-(aq)$$

A prospector pans for gold.

Because gold is not as concentrated in nature now as it was only a few decades ago, as many as ten tons of rock may have been processed for every ounce of gold jewelry you wear.

In chrome plating, a thin layer of chromium is deposited as the final layer on the surface of a metal object such as a steel car bumper (Figure 17.9). Electrolysis of a solution of CrO_3 in H_2SO_4 reduces chromium(VI) to chromium(III) and then to chromium metal. The overall reduction may be described by the equation for the half-reaction:

$$CrO_3(aq) + 6H^+(aq) + 6e^- \longrightarrow Cr(s) + 3H_2O(l)$$

A variety of chrome automobile accessories are plated using this process.

Chapter Review

Directions: From the choices listed at the end of this Chapter Review, choose the word or term that best fits in each blank. Use each choice only once. Answers appear at the end of the chapter.

Section 17.1 Electron-transfer reactions are called __________ (1) reactions. Oxidation is a __________ (2) of electrons, and reduction is a __________ (3) of electrons. An oxidizing agent is an electron __________ (4), and a reducing agent is an electron __________ (5). During oxidation, the oxidation number __________ (6), and in reduction, the oxidation number __________ (7). Oxidation numbers can be assigned by following rules outlined in Section 9.2.

Section 17.2 In the half-reaction (ion–electron) method for balancing redox equations, the oxidation and reduction are divided into __________ (8) reactions. After balancing, half-reactions are combined after canceling __________ (9). A key principle in redox reactions is that the number of electrons lost in reduction __________ (10) the number of electrons gained in oxidation. The __________ __________ (11) method can be used to balance redox equations for reactions that occur in aqueous solution but should not be used to balance equations of nonaqueous redox reactions.

Section 17.3 In the __________ __________ (12) method for balancing redox equations, oxidation numbers are assigned to the elements in the equation, and the electrons lost and gained are balanced and diagrammed above or below the equation. Then balancing __________ (13) are determined from the diagram. This method can be used to balance equations for reactions that are __________ (14) as well as for those that occur in aqueous solution.

Section 17.4 __________ (15) is the branch of chemistry concerned with generation of __________ (16) using a redox reaction and with the use of electricity to cause __________ (17) change. The term __________ (18) cell is in honor of Alessandro Volta who invented the first battery. A voltaic cell is a device that uses a redox reaction to produce an __________ __________ (19).

Section 17.5 In a Daniell cell, constructed of copper and zinc electrodes, zinc is the __________ (20) and copper is the __________ (21). The anode is always the electrode where __________ (22) occurs. __________ (23) occurs at the cathode. __________ (24) migrate toward the cathode, and __________ (25) toward the anode. In a voltaic cell, the anode is __________ (26) charged. A __________ (27) bridge is necessary to maintain a charge balance in the solutions in the two half-cells.

The __________ (28) storage battery is the common car battery. This battery uses lead for the anode and __________ __________ (29) for the cathode. The electrolyte is aqueous __________ __________ (30). The concentration and density of the electrolyte decrease when the battery is discharged. Consequently, a battery can be checked using a __________ (31) to determine if a cell is bad. An important feature of this battery is that it is __________ (32) by the car's alternator or generator.

A battery that uses a nonflowing paste in place of a liquid electrolyte is called a __________ (33) cell. The first flashlight battery used a __________ (34) anode and an electrolyte containing __________ (35), $ZnCl_2$, NH_4Cl, carbon, starch, and water. In the __________ (36) dry cell, ammonium chloride is replaced with a mixture of ZnO and KOH. The mercury cells of the type used in watches and hearing aids use __________ (37) as the oxidizing agent in place of MnO_2.

Section 17.6 A process that uses an electric current to bring about a redox reaction is called __________ (38). In the electrolysis of water, __________ (39) is formed at the anode, and __________ (40) is formed at the cathode. The electrolysis of molten sodium chloride produces sodium and __________ (41). In the electrolysis of aqueous sodium chloride, hydrogen is formed instead of __________ (42). In addition, aqueous __________ (43) is formed in the cathode compartment.

An electrochemical process in which a metal is deposited on the surface of an object is called __________ (44). The object to be plated is made the cathode, and the anode is often made of the __________ (45) metal. In "chrome" plating, the electrolyte is a solution of __________ (46) in sulfuric acid.

Choices: acceptor, alkaline, anions, anode, cathode, cations, chemical, chlorine, coefficients, CrO_3, decreases, donor, dry, electric current, electricity, electrochemistry, electrolysis, electrons, electroplating, equals, gain, half-, half-reaction, HgO, hydrogen, hydrometer, increases, lead, lead(IV) oxide, loss, MnO_2, NaOH, negatively, nonaqueous, oxidation, oxidation-number, oxygen, plating, rechargeable, redox, reduction, salt, sodium, sulfuric acid, voltaic, zinc

Study Objectives

After studying Chapter 17, you should be able to:

Section 17.1

1. Identify elements oxidized and reduced in an oxidation–reduction reaction.
2. Identify the oxidizing and reducing agents involved in an oxidation–reduction reaction.

Section 17.2

3. Balance redox equations using the half-reaction method.

Section 17.3

4. Balance redox equations using the oxidation-number method.

Section 17.5

5. Diagram voltaic cells, labeling the anode, cathode, electrode polarities, current flow, and directions the ions move.
6. Write equations for the half-reactions and the total reaction for the Daniell cell, the lead storage battery, the original dry cell, an alkaline cell, and a mercury cell.

Section 17.6

7. Diagram electrolytic cells, labeling the anode, cathode, electrode polarities, current flow, and directions the ions move.
8. Write equations for the half-reactions and the total reaction for the electrolysis of water, molten sodium chloride, and aqueous sodium chloride.

Key Terms

Review the definition of each of the terms listed here. For reference, they are listed by chapter section number. You may use the glossary as necessary.

17.1 oxidizing agent, reducing agent

17.4 electrochemical cell, electrochemistry

17.5 anode, cathode

17.6 electrolysis, electroplating

Questions and Problems

Answers to questions and problems with blue numbers appear in Appendix C.

Oxidation–Reduction Reactions (Section 17.1)

1. What is meant by oxidation? In general, which elements undergo oxidation readily?
2. What is meant by reduction? Which elements have a tendency to be reduced?
3. What is meant by an oxidizing agent? What is the origin of the term?
4. Which of the elements is the strongest oxidizing agent?
5. What is a reducing agent? In general, which elements are the best reducing agents?
6. Predict the products of the reaction of tin (Sn) with iodate ion (IO_3^-). Refer to Table 17.1.
7. Predict the products of the reaction that occurs (if any) when the following are mixed. Refer to Table 17.1.
   a. $Fe^{2+} + MnO_4^-$
   b. $Sn(s) + NO_3^-$
   c. $Cu^{2+} + Cl_2(g)$
8. Write a balanced equation for the reaction (if any) that occurs when copper (Cu) is mixed with chlorine (Cl_2). Refer to Table 17.1.
9. Identify the oxidizing agent and the reducing agent in problem 8. Which element is oxidized? Reduced?
10. Write a balanced equation for the reaction (if any) that occurs when ozone (O_3) is mixed with fluoride ion (F^-). Refer to Table 17.1.
11. Write a balanced equation for the reaction (if any) that occurs when magnesium (Mg) is added to a solution of copper(II) ion (Cu^{2+}). Refer to Table 17.1.
12. Assign oxidation numbers to the elements that appear in the equation in problem 11.
13. Identify the oxidizing agent and the reducing agent in problem 11. Which element is oxidized? Reduced?
14. Which is the stronger oxidizing agent, MnO_4^- or NO_3^-? Refer to Table 17.1.
15. What are the oxidation numbers of Mn in MnO_4^- and N in NO_3^-?

Balancing Redox Equations:
The Half-Reaction Method (Section 17.2)

16. Using the half-reaction method, write a balanced equation for the reaction of $MnO_4^-(aq)$ with Ag(s) in acidic solution. Refer to Table 17.1.
17. Identify the element oxidized and the element reduced in the reaction described in problem 16. What is the oxidizing agent? The reducing agent?
18. Using the half-reaction method, write a balanced equation for the reaction of $NO_3^-(aq)$ with Cu(s) in acidic solution. Refer to Table 17.1.
19. Identify the element oxidized and the element reduced in the reaction described in problem 18. What is the oxidizing agent? The reducing agent?
20. A piece of copper was added to an acidic solution of $K_2Cr_2O_7$. Write a balanced equation for the reaction using the half-reaction method.
21. Identify the element oxidized and the element reduced in the reaction described in problem 20. What is the oxidizing agent? The reducing agent?
22. Balance the following equations using the half-reaction method (acidic solution):
    a. $Cr_2O_7^{2-}(aq) + Sn^{2+}(aq) \longrightarrow Cr^{3+}(aq) + Sn^{4+}(aq)$
    b. $CdS(s) + I_2(aq) \longrightarrow Cd^{2+}(aq) + I^-(aq) + S(s)$
    c. $C_2O_4^{2-}(aq) + MnO_4^-(aq) \longrightarrow Mn^{2+}(aq) + CO_2(g)$
23. Using the half-reaction method, balance the following equations (acidic solution):
    a. $CuS(s) + NO_3^-(aq) \longrightarrow Cu^{2+}(aq) + S(s) + NO(g)$
    b. $ClO^-(aq) + I^-(aq) \longrightarrow Cl^-(aq) + I_2(aq)$
    c. $Zn(s) + As_2O_3(s) \longrightarrow AsH_3(g) + Zn^{2+}(aq)$
24. Using the half-reaction method, balance the following equations (acidic solution):
    a. $S_2O_3^{2-}(aq) + Ag^+(aq) \longrightarrow Ag(s) + S_4O_6^{2-}(aq)$
    b. $Br_2(aq) + I^-(aq) \longrightarrow I_2(aq) + Br^-(aq)$
    c. $WO_3(s) + Sn^{2+}(aq) + Cl^-(aq) \longrightarrow W_3O_8(s) + SnCl_6^{2-}(aq)$
25. Balance the following equations using the half-reaction method (acidic solution):
    a. $S_2O_3^-(aq) + Cl_2(g) \longrightarrow HSO_4^-(aq) + Cl^-(aq)$
    b. $AsH_3(g) + ClO_3^-(aq) \longrightarrow H_3AsO_4(aq) + Cl^-(aq)$
    c. $P_4(s) + NO_3^-(aq) \longrightarrow HPO_4^{2-}(aq) + NO(g)$

Balancing Redox Equations:
The Oxidation-Number Method (Section 17.3)

26. The Ostwald process for the manufacture of nitric acid occurs in three stages, starting with the high-temperature catalytic oxidation of ammonia.

    $NH_3(g) + O_2(g) \longrightarrow NO(g) + H_2O(g)$

    $NO(g) + O_2(g) \longrightarrow NO_2(g)$

    $NO_2(g) + H_2O(l) \longrightarrow HNO_3(l) + NO(g)$

    Balance these equations using the oxidation-number method.
27. Nitrogen monoxide (nitric oxide, NO) is formed in automobile engines. It then is oxidized in the atmosphere to nitrogen dioxide, a brown gas that gives smog a brown color.
    a. Balance the equations for the formation of NO_2.
    b. Identify the oxidizing agent and the reducing agent in each reaction.

    $N_2(g) + O_2(g) \longrightarrow NO(g)$

    $NO(g) + O_2(g) \longrightarrow NO_2(g)$
28. Balance the following equations using the oxidation-number method.
    a. $CuO(s) + NH_3(g) \longrightarrow N_2(g) + Cu(s) + H_2O(l)$
    b. $Zn(s) + SO_4^{2-}(aq) \longrightarrow Zn^{2+}(aq) + H_2S\ (aq)$ (acidic solution)
    c. $As_2S_5(s) + NO_3^-(aq) \longrightarrow H_3AsO_4(aq) + HSO_4^-(aq) + NO_2(g)$ (acidic solution)
29. Balance the following equations using the oxidation-number method. Assume acidic solutions.
    a. $MnO(s) + PbO_2(s) \longrightarrow MnO_4^-(aq) + Pb^{2+}(aq)$
    b. $SO_2(g) + NO_3^-(aq) \longrightarrow NO(g) + SO_4^{2-}(aq)$
    c. $KBr(aq) + H_2SO_4(aq) \longrightarrow K_2SO_4(aq) + Br_2(aq) + SO_2(g)$
30. Finely divided iron (Fe) reacts with chlorine gas (Cl_2) to form iron(III) chloride. Write a balanced equation for the reaction.
31. Balance the equations in problem 23 using the oxidation-number method.
32. Balance the equations in problem 24 using the oxidation-number method.
33. Complete and balance the following equations (acidic solution). Give the oxidation number of oxygen in the missing product.
    a. $H_2O_2 + I_2 \longrightarrow I^- + ?$
    b. $H_2O_2 + Sn^{2+} \longrightarrow Sn^{4+} + ?$
34. Identify the oxidizing agent and the reducing agent in each reaction in problem 33.
35. Balance the following equations using the oxidation-number method.
    a. $Cr_2O_7^{2-}(aq) + CH_3OH(aq) \longrightarrow Cr^{3+}(aq) + CO_2(g)$
    b. $C_2H_4(g) + MnO_4^-(aq) \longrightarrow C_2H_6O_2(aq) + MnO_2(s)$ (at 20°C)
    c. $C_2H_4(g) + MnO_4^-(aq) \longrightarrow CO_2(g) + Mn^{2+}(aq)$ (at 100°C)
36. In the preparation of uranium, uranium tetrafluoride is heated with magnesium. Write a balanced equation for the reaction. Identify the element oxidized, the element reduced, the oxidizing agent, and the reducing agent.
37. Balance the following equations using the oxidation-number method:
    a. $HNO_3(aq) \longrightarrow NO_2(g) + O_2(g) + H_2O(l)$
    b. $Ce^{4+}(aq) + H_2O_2(aq) \longrightarrow Ce^{3+}(aq) + O_2(g)$ (acidic solution)
    c. $H_2S(aq) + H_2O_2(aq) \longrightarrow S(s)$ (acidic solution)

Voltaic Cells (Sections 17.4, 17.5)

38. A voltaic cell uses aluminum and copper electrodes and aqueous solutions of aluminum sulfate and copper(II) sulfate.
    a. Draw a diagram of the cell and label the anode, cathode, the electrode polarities, direction the current flows, and the directions the cations and anions move.
    b. Write equations for the electrode half-reactions and a balanced equation for the cell reaction.

39. What is the purpose of a salt bridge in the construction of a voltaic cell?

40. A voltaic cell can be constructed based upon the following reaction:

    $Fe^{3+}(aq) + Ag(s) + Cl^{-}(aq) \longrightarrow Fe^{2+}(aq) + AgCl(s)$

    a. Draw a diagram of the cell and label the anode, cathode (platinum, an inert electrode), the electrode polarities, direction the current flows, and the directions the ions move.
    b. Write equations for the electrode half-reactions.

41. A voltaic cell was based on the following reaction:

    $Fe(s) + Sn^{2+}(aq) \longrightarrow Sn(s) + Fe^{2+}(aq)$

    a. Draw a diagram of the cell and label the anode, cathode, the electrode polarities, direction the current flows, and the directions the ions move.
    b. Write the equations for the electrode half-reactions.

42. The following half-reactions occur in an alkaline battery:

    $Zn(s) + 2OH^{-}(aq) \longrightarrow Zn(OH)_2(s) + 2e^{-}$

    $2MnO_2(s) + H_2O(l) + 2e^{-} \longrightarrow Mn_2O_3(s) + 2OH^{-}(aq)$

    a. Which half-reaction occurs at the anode and which occurs at the cathode?
    b. Write the balanced equation for the reaction of the alkaline battery.

Electrolysis (Section 17.6)

43. Diagram a cell for the electrolysis of molten sodium chloride.
    a. Label the anode, cathode, the electrode polarities, direction the current flows, and the directions the ions move.
    b. Write equations for the electrode half-reactions and for the electrolysis reaction.

44. The electrolysis of aqueous sodium bromide produces hydrogen and oxygen.
    a. Diagram the electrolytic cell.
    b. Label the anode, cathode, the electrode polarities, direction the current flows, and the directions the ions move.
    c. Write equations for the electrode half-reactions and for the electrolysis reaction.

45. Silver plating is an important process for the manufacture of silverware and jewelry.
    a. Diagram a cell for silver plating a spoon.
    b. Label the anode, cathode, the electrode polarities, direction the current flows, and the directions the ions move.
    c. Write equations for the electrode half-reactions.

Additional Problems

46. Balance the following equations for redox reactions occurring in acidic or neutral solution. Choose the method that you think is appropriate.
    a. $HgS(s) + Cl^{-}(aq) + NO_3^{-}(aq) \longrightarrow HgCl_4^{2-}(aq) + S(s) + NO(g)$
    b. $I_2(aq) + S_2O_3^{2-}(aq) \longrightarrow S_4O_6^{2-}(aq) + I^{-}(aq)$
    c. $Br^{-}(aq) + H_2SO_4(aq) \longrightarrow Br_2(aq) + SO_2(g)$

47. For the reactions in problem 46, identify
    a. the element oxidized.
    b. the element reduced.
    c. the oxidizing agent.
    d. the reducing agent.

48. Balance the following equations using the oxidation-number method.
    a. $CuO(s) + NH_3(g) \longrightarrow N_2(g) + H_2O(l) + Cu(s)$
    b. $P_2H_4(g) \longrightarrow PH_3(g) + P_4H_2(g)$
    c. $Ca_3(PO_4)_2(s) + SiO_2(s) + C(s) \longrightarrow CaSiO_3(s) + P_4(g) + CO(g)$

49. Balance the following equations. Use the method you think is appropriate.
    a. $HNO_2(aq) + I^{-}(aq) \longrightarrow NO(g) + I_2(aq)$
    b. $Ag(s) + H_2S(aq) + O_2(aq) \longrightarrow Ag_2S(s)$
    c. $MoO_3(s) + H_2(g) \longrightarrow Mo(s) + H_2O(l)$

50. Diagram a voltaic cell constructed using nickel and silver electrodes and aqueous solutions of nickel(II) nitrate and silver nitrate. Show the direction of current flow and migration of the cations and anions. Indicate the charges of the electrodes. Write balanced equations for the half-reactions and the redox reaction.

51. Aluminum is produced commercially by the Hall process which involves electrolysis of a molten mixture of Al_2O_3 in cryolite (Na_3AlF_6), a naturally occurring aluminum compound. The electrolysis is carried out at a temperature above the melting point of the mixture (about 950°C) using graphite (carbon) electrodes. A steel tank lined with graphite serves as the cathode, and molten aluminum is drained from the tank during production. The half-reactions are

    $Al^{3+}(l) + 3e^{-} \longrightarrow Al(l)$

    $2O^{2-}(l) \longrightarrow O_2(g) + 4e^{-}$

    a. Diagram a cell for the Hall process.
    b. Label the anode, cathode, electrode polarities, direction the current flows, and directions the ions move.
    c. Write a balanced equation for the reaction.

Solutions to Practice Problems

PP 17.1

Remember that the oxidation number for O is generally −2, and for H it is generally +1.

a. $\overset{+1}{Na}\overset{+1}{Cl}\overset{-2}{O}$

b. $\overset{0}{F_2}$

c. $\overset{+4}{N}\overset{-2}{O_2}$

d. $\overset{+1}{H}\overset{+7}{I}\overset{-2}{O_4}$

PP 17.2

Step 1 Separate into half-reactions:

$MnO_4^- \longrightarrow Mn^{2+}$

$U^{4+} \longrightarrow UO_2^{2+}$

Step 2 Balance the equations for the half-reactions:

$MnO_4^- + 8H^+ + 5e^- \longrightarrow Mn^{2+} + 4H_2O$

$2H_2O + U^{4+} \longrightarrow UO_2^{2+} + 4H^+ + 2e^-$

Step 3 Combine the half-reaction equations. Note that to cancel electrons, the reduction half-reaction must be multiplied by 2 and the oxidation half-reaction must be multiplied by 5. Some of the H_2O and H^+ can also be canceled.

$$2MnO_4^- + \cancel{16}H^+ + \cancel{10e^-} \longrightarrow 2Mn^{2+} + \cancel{8}H_2O$$

$$\overset{2}{\cancel{10}}H_2O + 5U^{4+} \longrightarrow 5UO_2^{2+} + \overset{4}{\cancel{20}}H^+ + \cancel{10e^-}$$

$$2MnO_4^- + 5U^{4+} + 2H_2O \longrightarrow 2Mn^{2+} + 5UO_2^{2+} + 4H^+$$

Step 4 Check the equation.

Step 5 Add symbols to indicate whether the substances are solid, liquid, gas, or in aqueous solution.

$$2MnO_4^-(aq) + 5U^{4+}(aq) + 2H_2O(l) \longrightarrow 2Mn^{2+}(aq) + 5UO_2^{2+}(aq) + 4H^+(aq)$$

PP 17.3

Step 1 Assign oxidation numbers to each element:

$$\overset{-4+1}{CH_4} + \overset{0}{Cl_2} \longrightarrow \overset{+4-1}{CCl_4} + \overset{+1-1}{HCl}$$

Step 2 Diagram the electrons lost and the electrons gained per atom. Balance the electrons lost and gained.

(8 e⁻ lost)

$$\overset{-4+1}{CH_4} + \overset{0}{Cl_2} \rightarrow \overset{+4\ -1}{CCl_4} + \overset{+1-1}{HCl}$$

8 × (1 e⁻ gained)

Step 3 Determine what balancing coefficients are needed in the equation from the factors used to balance electrons. In this equation, 8 chlorine atoms are needed, requiring $4Cl_2$. On the right side of the equation, 4 HCl are needed to give a total (with CCl_4) of 8 chlorine atoms.

(8 e⁻ lost)

$$\overset{-4+1}{CH_4} + 4\overset{0}{Cl_2} \rightarrow \overset{+4\ -1}{CCl_4} + 4\overset{+1-1}{HCl}$$

8 × (1 e⁻ gained)

Step 4 Balance hydrogen (and oxygen) as necessary. In this equation, hydrogen is already balanced.

Step 5 Check the equation.

Step 6 Add symbols to indicate whether the substances are solid, liquid, gas, or in aqueous solution.

$$CH_4(g) + 4Cl_2(g) \longrightarrow CCl_4(l) + 4HCl(g)$$

Answers to Chapter Review

1. redox
2. loss
3. gain
4. acceptor
5. donor
6. increases
7. decreases
8. half-
9. electrons
10. equals
11. half-reaction
12. oxidation-number
13. coefficients
14. nonaqueous
15. electrochemistry
16. electricity
17. chemical
18. voltaic
19. electric current
20. anode
21. cathode
22. oxidation
23. reduction
24. cations
25. anions
26. negatively
27. salt
28. lead
29. lead(IV) oxide
30. sulfuric acid
31. hydrometer
32. rechargeable
33. dry
34. zinc
35. MnO_2
36. alkaline
37. HgO
38. electrolysis
39. oxygen
40. hydrogen
41. chlorine
42. sodium
43. NaOH
44. electroplating
45. plating
46. CrO_3

18

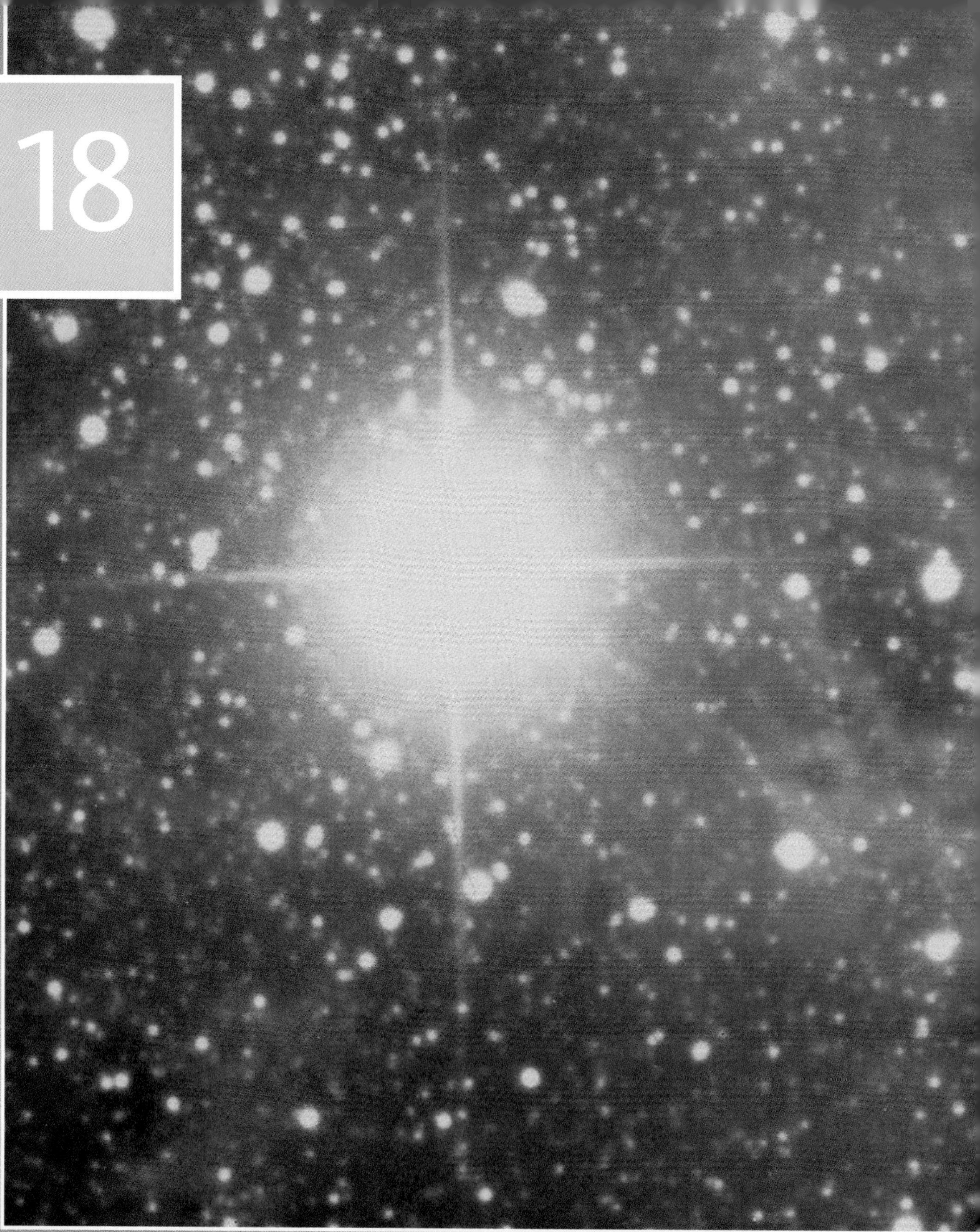

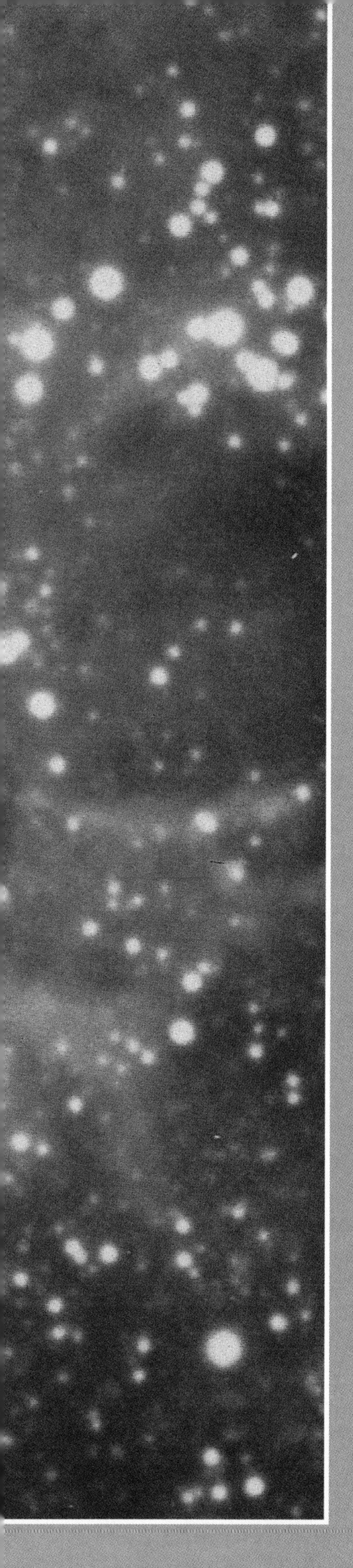

Radioactivity and Nuclear Energy

Contents

The great supernova of 1987. This exploding star is the nearest to earth and the brightest seen in almost 400 years.

You Probably Know . . .

Explosion of a nuclear weapon.

- The energy of the stars and our sun is a result of nuclear reactions.
- Some naturally occurring substances, such as radium and uranium, are radioactive and emit various kinds of radiation (Section 5.3). For example, pitchblende, a uranium-bearing ore, emits radiation that leaves an image on photographic film.
- Nuclear explosions release tremendous amounts of energy and potentially lethal radiation.
- The sun is a gigantic nuclear reactor that produces the energy that makes life on earth possible.
- Nuclear reactors on earth can harness nuclear energy to produce electricity.
- Radiation therapy is a widely used treatment for cancer.

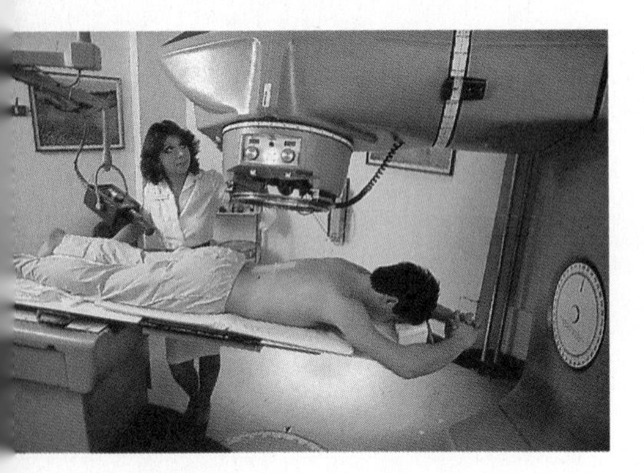

Physicians prepare a cancer patient for radiation treatment.

Why Should We Study Radioactivity and Nuclear Energy?

During the night of February 23, 1987, Canadian astronomer Ian Shelton, using a University of Toronto telescope at Las Campanas Observatory in Chile, discovered a bright *supernova.* This exploding star increased in brightness a thousand times overnight, becoming bright enough to be seen with the naked eye. According to generally accepted theory, nuclear fusion reactions involving hydrogen gas in a star's core provide the energy that makes a star shine. Heavier elements, such as iron, also form, increasing until the star burns out in an enormous nuclear explosion called a supernova.

The explosion of atomic bombs over the Japanese cities of Hiroshima and Nagasaki in August 1945 ended the war with Japan and ushered in the nuclear age. The incredible destructive potential of nuclear energy had been demonstrated during wartime. The potential of peaceful uses of nuclear energy is even more astounding.

About 20% of the electricity generated in the United States is produced in nuclear power plants. Two major issues in the operation of nuclear power plants are safety and nuclear waste. The future of nuclear power plants depends upon resolution of these issues. Increased use of nuclear power could reduce our dependence on fossil fuels for the energy we need in the future.

Numerous applications of radiation have developed in recent decades. Medical use of radiation in diagnosis and therapy is now routine. Radiation therapy is used to treat cancer. Radioisotopes are also widely used to diagnose a number of diseases such as those of the thyroid and the circulatory system.

Chemists use isotopic tracers to study reaction pathways. Furthermore, several methods of modern chemical analysis, such as neutron activation analysis, involve nuclear reactions.

Other applications of nuclear reactions and radiation include preservation of food, radioiosotope dating of objects of archaeological interest, and industrial process control.

Today, the potential of peacetime uses of nuclear energy and radioactive substances appears to be unlimited. At some time in our lives, we all will experience the impact of this constantly advancing technology.

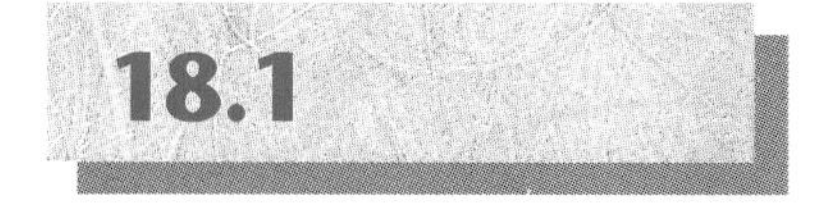

Discovery of Radioactivity

Radioactivity was discovered in 1896 by Antoine Henri Becquerel (1852–1908), a French scientist. He was experimenting with different substances to see if they would emit light after absorbing ultraviolet radiation from the sun. In preparation for one of his experiments, Becquerel put a sample of a uranium salt and some photographic film wrapped in dark paper in a drawer while he waited for a period of bright sunlight. Later, when he examined the photographic film, he discovered that it had been exposed. Since the film had been protected from light by the wrapping, some kind of radiation from the uranium salt must have passed through the protective paper and exposed the film.

Further experiments showed that exposure of Becquerel's photographic film was due to radioactivity from uranium nuclei. **Radioactivity** is the spontaneous emission of rays or particles from the nucleus of an atom. Within two years of Becquerel's discovery, investigations by Marie Curie (1867–1934) and her husband, Pierre Curie (1859–1906), led to the discovery of other radioactive substances. Their experiments with pitchblende, an ore containing uranium, resulted in the discovery of polonium (Po, atomic number 84), a new radioactive element named after Poland, Marie Curie's native country. A short time later, they discovered the highly radioactive element radium (Ra, atomic number 88). The Curies and Henri Becquerel shared the Nobel prize in physics in 1903 for their work with radioactivity. In 1911, Marie Curie was given the Nobel prize in chemistry for the discovery of polonium and radium.

Marie and Pierre Curie discovered the radioactive elements polonium and radium, both hundreds of times more radioactive than uranium.

Natural Radioactivity

Learning Objectives

- Describe and give symbols for alpha, beta, and gamma radiation.
- Compare the penetrating power of alpha, beta, and gamma radiation.
- Describe a Geiger–Müller counter, a scintillation counter, and a film badge.

Natural radioactivity results from the spontaneous decay or disintegration of naturally occurring nuclei. The nature of natural radioactivity can be demonstrated by allowing emissions from a radioactive mineral to pass between charged plates (Figure 18.1). The emissions separate into three kinds of "rays." Alpha (α) particles are attracted to the negative plate because they are positively charged. Beta (β) particles are negatively charged and are attracted to the positive plate. Gamma (γ) rays, which are uncharged, pass between the plates undeflected.

Penetrating effects of alpha, beta, and gamma radiation. Although all types of radiation have a wide range of energies, alpha radiation is considered the least penetrating and is absorbed by several sheets of paper or by clothing. High-energy beta radiation can penetrate paper or clothing but is absorbed by a sheet of plastic. Gamma radiation is the most penetrating and requires the most shielding.

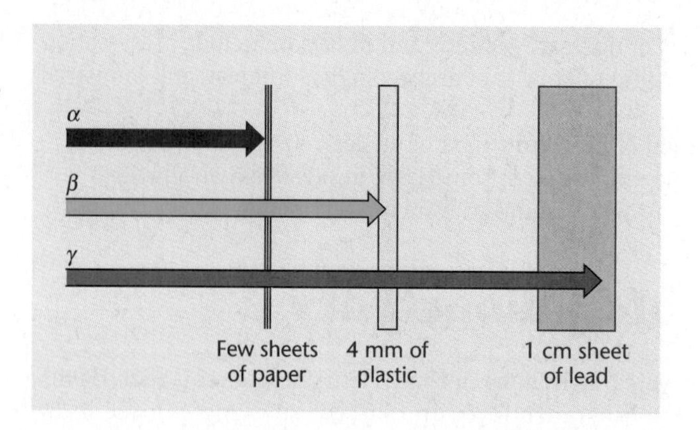

Alpha particles: ^4_2He (positively charged).

Beta particles: $_{-1}^{0}\beta$ (negatively charged).

Gamma rays: $^0_0\gamma$ (uncharged, no mass).

Alpha particles are now known to be helium nuclei (Section 5.3), represented by the nuclear symbol, ^4_2He. Alpha radiation is readily absorbed by the outer layer of skin or even by ordinary clothing. Its penetrating effect is not great.

Beta particles have been identified as fast-moving particles identical to electrons but originating from the nucleus of an atom. Beta particles have a wide range of energies depending upon their source. Higher-energy betas, such as those emitted by phosphorus-32, are considerably more penetrating than alpha particles and will pass through clothing and the skin. The charge and mass number are often included with the symbol, $_{-1}^{0}\beta$.

Gamma rays are high-energy rays similar to x-rays. Gamma radiation is electromagnetic energy, and as such has no charge or mass. Because of its high energy and lack of charge, gamma radiation is very penetrating. Gamma rays can pass through our bodies and even through ordinary concrete walls. Shields to protect against gamma rays are often made of lead, a high-density metal, or high-density concrete. The thickness of the shielding required depends upon the level of the radiation.

Measuring Radiation

Emissions from radioactive samples are called **ionizing radiation** because they have sufficient energy to ionize ordinary atoms and molecules that are in their path. Using the symbol, M, to represent a molecule, ionization can be illustrated by the equation:

$$M \xrightarrow[\text{radiation}]{\text{ionizing}} M^+ + e^-$$

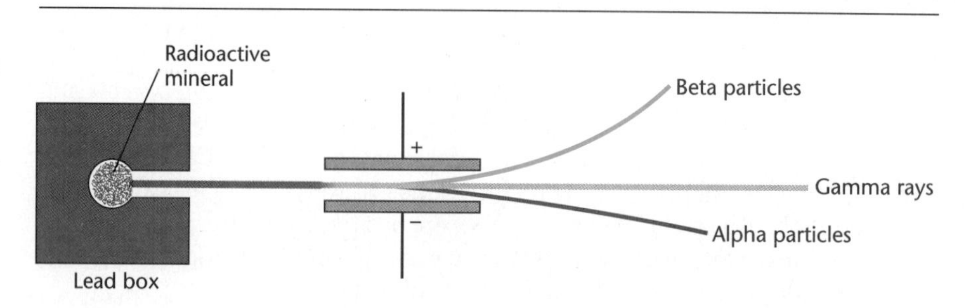

FIGURE 18.1 Emissions from a radioactive mineral separate into three beams when passed between two charged plates. The negatively charged beta (β) particles are deflected toward the positive plate. The positively charged alpha (α) particles are deflected toward the negative plate. Gamma (γ) rays are uncharged and are not deflected.

The ability to ionize matter distinguishes these emissions from nonionizing radiation such as ultraviolet, infrared, and visible light. Ionizing radiation can also cause covalent bonds to break resulting in the degradation of molecular structure. Exposure to high levels of ionizing radiation can be very damaging to living cells and can cause severe illness (radiation sickness) and death. Genetic defects can also result. All of these effects were experienced by people who survived nuclear explosions in Japan and elsewhere.

The **activity** of a radioactive element is related to the number of nuclear disintegrations that occur per second. The greater the number of disintegrations per second, the greater the activity. The SI unit of activity is the **becquerel,** Bq, where one Bq is one disintegration per second. Another widely used unit of activity is the **curie,** Ci, named for Marie Curie who discovered radium. One curie corresponds to 3.7×10^{10} disintegrations per second, the activity of one gram of radium-226. One gram of radium-226 is a "one-curie source" of radioactivity. Thus, the activity of a sample in curies is

1 Bq = 1 disintegration per second.

1 Ci = 3.7×10^{10} disintegrations per second.

$$\text{activity (Ci)} = \frac{\text{measured number of disintegrations per second}}{3.7 \times 10^{10} \text{ disintegrations per second}}$$

For example, if a sample undergoes 4.2×10^7 disintegrations per second, its activity is

$$\text{activity (Ci)} = \frac{4.5 \times 10^7}{3.7 \times 10^{10}} = 1.2 \times 10^{-3} \text{ Ci, or } 1.2 \text{ mCi}$$

The ionizing effect of radioactive emissions allows them to be detected by ionization counters such as the well known Geiger–Müller counter (Figure 18.2). The main component of this instrument consists of a tube filled with a gas such as argon. A wire in the center of the tube is positively charged with respect to the wall of the tube. Radioactive emissions enter the tube through a thin window and cause ionization of argon atoms. The resulting electrons move toward the positive wire

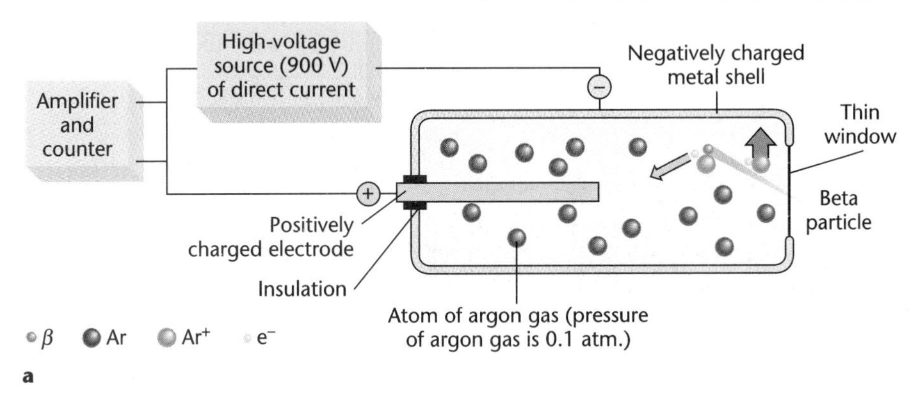

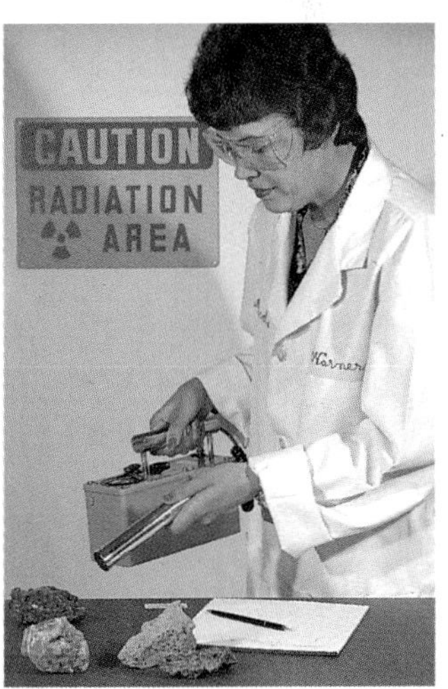

FIGURE 18.2 (a) Diagram of a Geiger–Müller counter. Radioactive emissions enter the tube through the thin window and ionize argon atoms. The resulting electrons move toward the positive electrode, causing an electric pulse which registers on the meter and is heard as a clicking sound. The meter shows the number of particles counted per minute. (b) A Geiger–Müller counter in use.

FIGURE 18.3 A scintillation counter.

causing an electric pulse that is registered on a meter and heard as a clicking sound. Each click indicates the detection of one particle entering the tube.

A scintillation counter (Figure 18.3) contains a substance called a phosphor that emits flashes of light when struck by radioactive emissions (Section 5.3). Each flash indicates one nuclear disintegration. This counter also measures the intensities of the flashes, which are proportional to the energy of the absorbed radiation. From counts per second (proportional to activity) and the energy of the emissions, the source of radiation often can be identified.

A film badge is a detection device that is commonly worn by people working in areas where they might be exposed to radiation. The badge contains photographic film, which registers any ionizing radiation in a worker's environment. After the badge is worn for a specific length of time, the film is removed and developed. The degree of exposure is related to the radiation absorbed during the time the badge was worn.

Effects of Exposure to Radiation

We are exposed to radiation from a number of sources. Some radiation is naturally occurring, such as that from uranium-bearing rock and soil. Radioactive radon gas is continually released from these materials. Other radioactive material was released into the environment during the testing of nuclear weapons. We may have been exposed to radiation when diagnostic x-rays were taken in a dentist's or doctor's office. Some people handle radioactive materials or work near radioactive sources where they are employed. Absorption of large doses of radiation can cause illness or death. Thousands died as a result of the nuclear bombs dropped on Japan during World War II, and more died following the accident at the Chernobyl nuclear power plant in the Ukraine in 1986.

The amount of radiation absorbed by a substance may be expressed in **rads** (*r*adiation *a*bsorbed *d*ose), where one rad is a dose of 10^{-5} J/g of the substance. The SI unit is the *gray,* Gy (named for Harold Gray, a British radiologist); this larger unit corresponds to the absorption of one J/kg of substance.

When human tissue is exposed to radiation, the **rem** is often used as the unit of radiation dosage. The rem takes into consideration the differences in effectiveness of various types of radiation in causing damage to tissues of the human body. The dosage in rems is obtained by multiplying the dose in rads times a biological effectiveness factor (BEF) that is characteristic of the type of radiation:

$$\text{dose (rems)} = \text{dose (rads)} \times \text{BEF}$$

The BEF for alpha radiation is much larger than that for beta and gamma radiation (Table 18.1).

TABLE 18.1 Biological Effectiveness Factors (BEF) for Various Kinds of Radiation

Radiation	*BEF*
β, γ, x-rays	1
slow neutrons	3
fast neutrons	10
α	20

Some tissues are more sensitive than others to radiation. Radiation reduces the rate of cell division, and cells undergoing division are more affected. For example, exposure to radiation poses a much greater risk to a fetus during the third to seventh weeks of pregnancy than to an adult person. Furthermore, radiation produces longer-lasting effects in some tissues in which cells are replaced slowly, such as eyes. The health and age of an exposed person are also factors.

Alpha radiation from an external source is readily absorbed by the surface layer of skin. However, inhalation of alpha-emitting particles can pose significant

CHEMICAL WORLD

Cleaning Up with Ionizing Radiation

Emissions from radioactive samples are called ionizing radiation because they have enough energy to ionize atoms and molecules. It is this ability to ionize matter that makes radioactive substances valuable in a wide range of applications. Few people are aware that the economic impact of these uses is billions of dollars. For example, ionizing nuclear radiation is used to sterilize hospital clothing and equipment such as syringes and to sterilize food to preserve it. It is even used to enhance the properties of natural rubber, which is then used in the manufacture of your automobile tires, and it is a cleaner method than the alternative.

Despite people's fascination with rubber, for a long time one major problem kept it from being widely used: its physical properties changed dramatically with temperature. At warmer temperatures rubber was gooey and sticky, and at colder temperatures it was hard and brittle—until about 1840 when Charles Goodyear had his lucky accident. He spilled sulfur and rubber onto a hot stove. The resulting chemical reaction ruined his indoor air quality, but it improved the rubber, making it less temperature sensitive. Let's see what sulfur does to rubber.

Natural rubber molecules are long chains of carbon and hydrogen atoms. The sulfur changes the properties of natural rubber by crosslinking its molecules, that is, by producing crosswise sulfur-to-sulfur linkages between the long carbon chains. Crosslinking converts natural rubber from many long and independent molecular chains to a state of one large three-dimensional molecular network. As a result, chain movement is restrained, which increases the stability and elasticity of the natural rubber.

Rubber can be crosslinked using ionizing radiation instead of sulfur.

When molecules absorb ionizing radiation, they can lose electrons to form cations. Loss of hydrogen ions (H^+) can leave unpaired electrons on neighboring carbon chains. Sharing of two electrons results in a covalent bond—a crosslink that joins the carbon chains.

The process of heating a mixture of rubber, sulfur, and various toxic chemicals is called vulcanization, and for decades following Goodyear's discovery, rubber was vulcanized to crosslink it. Today, natural rubber is frequently irradiated because ionizing radiation also can crosslink large molecules—without using noxious chemicals.

health risks. Uranium miners and others who may inhale dust containing alpha emitters often wear respirators and must take other precautions to protect themselves from the inhalation hazards in their work environment.

The U.S. government has set limits of 5 rems per year for radiation workers and 0.5 rem for the general population. The average exposure from natural sources is about 0.1 rem/yr. Following the accident at the Chernobyl nuclear power plant, an average of more than 40 rem was received by people living within 15 km of the plant.

18.3 Nuclear Equations

Learning Objectives

- Write symbols for a neutron, positron, and proton.
- Complete and balance nuclear equations.

Nuclear symbol (Section 5.4):

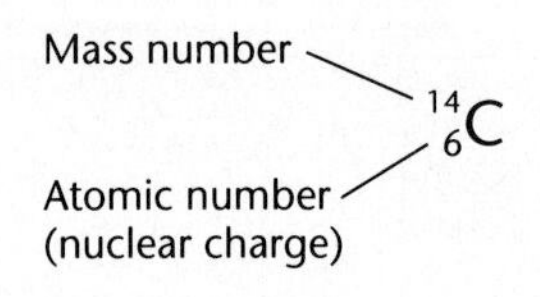

Many different nuclei are radioactive. One example is carbon-14, which decays with the emission of a beta particle:

$$^{14}_{6}C \longrightarrow ^{14}_{7}N + ^{\;\,0}_{-1}\beta$$

The decay of carbon-14 is the basis of the technique of carbon dating of ancient artifacts (Section 18.5). This equation is called a **nuclear equation** because it shows changes in the nuclei as indicated by nuclear symbols. The symbol for the beta particle (an electron) shows its charge and mass number. Most nuclear reactions result in a change of one element into another. When balancing a nuclear equation, both the charge and the mass number must be balanced.

In a nuclear equation, both the charge *and the* mass number *must be balanced.*

As shown in this example, beta decay always results in an increase in atomic number (number of protons and nuclear charge) with no change in mass number (Figure 18.4). In other words, the net result of beta decay is that a neutron is converted into a proton with loss of an electron (or beta particle).

Beta emission: The atomic number increases by 1 unit and the mass number does not change.

$$^{1}_{0}n \longrightarrow ^{1}_{1}H + ^{\;\,0}_{-1}\beta$$

Remember, this electron is lost from the nucleus, not from an electron orbital outside the nucleus.

Alpha decay is observed only for heavy nuclei with atomic numbers greater than 83. A well-known example is uranium-238, the most abundant isotope of uranium.

Alpha emission: The atomic number decreases by 2 units and the mass number decreases by 4 units.

$$^{238}_{\;92}U \longrightarrow ^{234}_{\;90}Th + ^{4}_{2}He$$

Note that loss of an alpha particle reduces the atomic number by 2 units and reduces the mass number by 4 units. The sum of the mass numbers is 238 on both sides of the equation, and the sum of the atomic numbers is 92.

Gamma emission: There is no change in atomic number or mass number.

Many alpha and beta emissions are accompanied by gamma emission. However, because gamma rays have no mass or charge, gamma emission does not affect the mass number or the atomic number, and it is often omitted from nuclear equations.

These examples illustrate two ideas that will guide you in writing nuclear equations:

$$^{3}_{1}H \longrightarrow ^{3}_{2}He + ^{\;\,0}_{-1}\beta$$

FIGURE 18.4 Disintegration of hydrogen-3 (tritium). Beta decay converts a neutron in the nucleus into a proton. The mass number remains the same and the atomic number increases by one unit.

TABLE 18.2 Symbols for Particles in Nuclear Reactions

Particle	*Preferred Symbol*	*Other Symbol*
alpha	${}^{4}_{2}He$	α, He^{2+}
beta	${}^{0}_{-1}\beta$	${}^{0}_{-1}e$, β
gamma	${}^{0}_{0}\gamma$	γ
neutron	${}^{1}_{0}n$	n
positron	${}^{0}_{+1}\beta$	${}^{0}_{+1}e$
proton	${}^{1}_{1}p$	p, ${}^{1}_{1}H$

1. The sum of the atomic numbers is the same on both sides of the equation.
2. The sum of the mass numbers is the same on both sides of the equation.

The symbols for particles commonly involved in nuclear reactions are shown in Table 18.2. Note that a positron is a positively charged particle with a mass about equal to the mass of an electron.

Example 18.1

Write a nuclear equation for the beta decay of thorium-234. ■

Solution The atomic number (nuclear charge) of thorium (Th) is 90, and a beta particle has a charge of -1. Balance the charge to determine the atomic number of the missing nucleus:

$${}^{234}_{90}Th \longrightarrow {}^{0}_{-1}\beta + {}^{-}_{-}?$$

Thus $90 = -1 + ?$ The missing atomic number is 91. Refer to the periodic table to find the symbol for element number 91 (Pa). Also balance the mass numbers to determine that the missing mass number is 234. Complete the nuclear equation:

$${}^{234}_{90}Th \longrightarrow {}^{0}_{-1}\beta + {}^{234}_{91}Pa$$

Check to be sure that atomic numbers and mass numbers are both balanced.

Example 18.2

Write a nuclear equation for the decay of phosphorus-30 by positron emission (see Table 18.2). ■

Solution ${}^{30}_{15}P \longrightarrow {}^{0}_{+1}\beta + {}^{30}_{14}Si$

Note that emission of a positron results in a nucleus that has one less proton and one more neutron. The mass number remains unchanged.

Practice Problem 18.1 Write a nuclear equation for the decay of radon-222 by alpha emission. (Solutions to practice problems appear at the end of the chapter.)

18.4 Natural Radioactive Decay Series

Learning Objectives

- Describe three natural radioactive decay series.
- Given the starting and final nuclei, determine the combination of alpha and beta decay reactions for a decay series.

The alpha decay of uranium-238 is the first reaction of a series of fourteen nuclear reactions that ends with lead-206, a nonradioactive lead isotope (Figure 18.5). This natural decay series has eight steps involving alpha particle emissions and six steps involving beta particle emissions. Remember, the emission of an alpha particle results in an element with an atomic number two units smaller than that of the decaying element. On the other hand, beta decay results in the formation of an element having an atomic number one unit greater than that of the decaying element. Because lead-206 is stable, the decay series stops with this element.

One of the radioactive elements formed in the uranium-238 decay series, radon-222, has caused widespread concern in the United States. Radon-222 is a gas like the other noble gases, and it readily diffuses through rock and soil into homes and other buildings. It can be breathed into the lungs, where some of it decays by emitting alpha and gamma radiation. Furthermore, the decay products are solids that tend to be trapped in the lungs, and some of them decay rapidly: polonium-218 (alpha decay), lead-214 (beta, gamma decay), and polonium-214 (alpha, gamma decay). The radiation absorbed is significantly higher for smokers and others who are exposed simultaneously to smoke and radon-222. The radiation from these decay products when absorbed in the lungs can cause lung cancer. Some estimates suggest that as many as 10% of the deaths from lung cancer are due to exposure to radon-222.

Another natural decay series begins with thorium-232 and ends with lead-208. This series includes six alpha emissions and four beta emissions. A third natural series begins with uranium-235 and ends with lead-207. A fourth series begins with plutonium-241, a synthetic element, and ends with bismuth-209.

Example 18.3

The thorium-232 decay series includes six alpha emissions and four beta emissions. Without writing nuclear equations for all the steps, show by means of calculations that this combination of alpha and beta decay reactions produces lead-208. ■

Solution The overall decay process can be represented as

$$^{232}_{90}\text{Th} \longrightarrow {}^{208}_{82}\text{Pb} + 6\,^{4}_{2}\text{He} + 4\,_{-1}^{0}\beta$$

1. The sum of mass numbers on the right side of the equation is $208 + 6(4) = 232$, which equals the mass number of thorium-232 on the left side.
2. The sum of the atomic numbers on the right side of the equation is $82 + 6(2) + 4(-1) = 90$, which equals the atomic number of thorium on the left side.

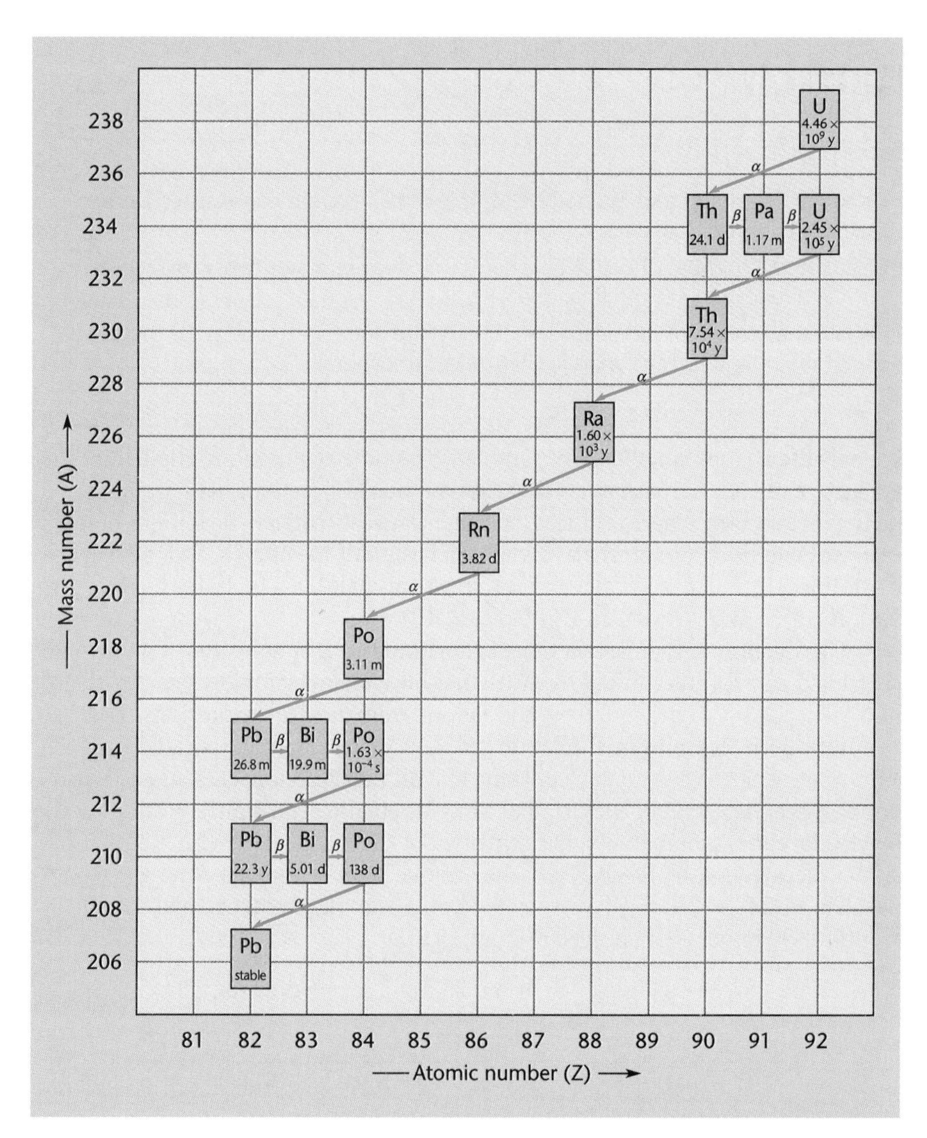

FIGURE 18.5 The uranium-238 decay series consists of eight alpha emissions and six beta emissions. It results in the formation of lead-206 which is stable. The nuclei are shown with their half-lives (y = years, d = days, m = minutes, s = seconds).

Practice Problem 18.2 The uranium-235 series ends with lead-207. What combination of alpha and beta emissions is apparently involved in the series?

18.5 Half-Life

Learning Objectives

- Calculate the quantity of radioactive sample remaining after an elapsed time of 1, 2, 3, 4 (and so on) half-lives.
- Calculate the number of half-lives required for a given quantity of radioactive sample to decay to a given final quantity.
- Explain the process of radiocarbon dating. Write nuclear equations for the decay of carbon-14 and its formation in the upper atmosphere.

The **half-life** ($t_{1/2}$) of a radioactive sample is the time required for one-half of the radioactive nuclei in a sample to decay. Some nuclei have half-lives as short as a fraction of a second, whereas others have half-lives estimated to be billions of years (refer to Table 18.3). For example, carbon-15 has a half-life of 2.4 s, and the half-life of astatine-218 is 1.6 s. Nuclei having very long half-lives include thorium-232 (1.41×10^{10} yr) and uranium-238 (4.46×10^{9} yr).

Fallout is radioactive dust that falls to the earth after it has been carried in the atmosphere from the site of a nuclear explosion.

The fraction of radioactive nuclei remaining in a sample after a given time has passed can be determined from its half-life. As an example, consider strontium-90 ($t_{1/2} = 28$ yr), found in the fallout from nuclear explosions. Because strontium-90 is chemically similar to calcium, it can be incorporated into the bones of our bodies where it can continue to emit radiation for years. After 28 years (one half-life), one-half of the original sample remains. During the next 28 years (56 yr total time), one-half of what remained of the sample after the first half-life has now decayed and $\frac{1}{2} \times \frac{1}{2}$ or $\frac{1}{4}$ of the original sample remains. After three half-lives ($28 + 28 + 28 = 84$ yr), $\frac{1}{2} \times \frac{1}{2} \times \frac{1}{2}$ or $\frac{1}{8}$ of the original sample of strontium-90 remains undecayed (Figure 18.6).

$$1.0 \text{ mg } ^{90}_{38}\text{Sr} \xrightarrow{t_{1/2}} 0.50 \text{ mg } ^{90}_{38}\text{Sr} \xrightarrow{t_{1/2}} 0.25 \text{ mg } ^{90}_{38}\text{Sr} \xrightarrow{t_{1/2}} 0.13 \text{ mg } ^{90}_{38}\text{Sr}$$

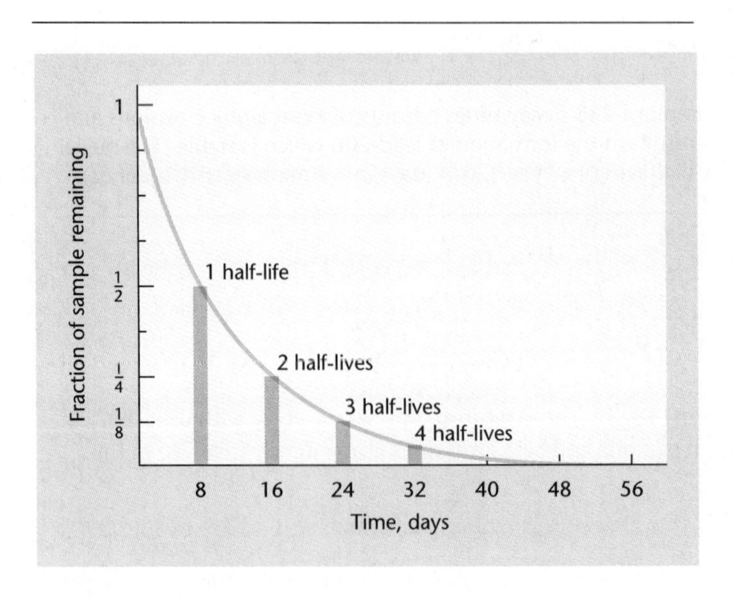

FIGURE 18.6 Radioactive decay. One-half of a radioactive sample decays during each half-life. The half-life of iodine-131 is 8.0 days. In each 8.0-day period, one-half of the I-131 nuclei decays by beta and gamma emission to Xe-131.

TABLE 18.3 Half-Lives of Selected Nuclei		
Nucleus	*Half-Life*	*Type of Decay*
iodine-131	8.0 da	beta, gamma
tritium (hydrogen-3)	12.3 yr	beta
carbon-14	5.73×10^3 yr	beta
cobalt-60	5.26 yr	beta, gamma
strontium-90	28.1 yr	beta
uranium-238	4.46×10^9 yr	alpha, gamma
protactinium-234	6.70 hr	beta, gamma
thorium-232	1.41×10^{10} yr	alpha, gamma

Example 18.4

What percentage of the original quantity of a sample of tritium would remain after 37 yr? The half-life of tritium is 12.3 yr. ■

Solution To calculate the number of half-lives, divide the elapsed time by the half-life.

$$\frac{37}{12.3} = 3.0 \text{ half-lives}$$

Percentage remaining $= 100\% \times 0.5 \times 0.5 \times 0.5 = 13\%$.

Example 18.5

The half-life of bromine-80 is 18 min. Starting with 5.0×10^3 μg, how long would it take for the sample to be reduced to 3.1×10^2 μg? ■

Solution The number of half-lives can be estimated by multiplying by 0.5 until the remaining mass of sample is obtained:

$$5.0 \times 10^3 \,\mu\text{g} \times 0.5 \times 0.5 \times 0.5 \times 0.5 = 0.31 \times 10^3 \mu\text{g}$$
$$= 3.1 \times 10^2 \,\mu\text{g}$$

Four half-lives elapsed, or 4×18 min $= 72$ min.

Practice Problem 18.3 Calculate the fraction of a sample of cesium-137 that would remain after 120. yr. The half-life of cesium-137 is 30.0 yr.

■ Radioisotope Dating

Radioisotope dating is a technique used to estimate the age of an ancient relic or a rock by measuring the activity of a radioisotope. The most important dating technique is **radiocarbon dating,** in which the radioisotope is carbon-14. Radiocarbon dating is used for relics of plant, animal, or human origin, which typically contain carbon.

Cosmic rays are streams of subatomic particles and small nuclei that originate in the stars and travel through space.

Carbon-14 is an isotope of carbon that is formed in the upper atmosphere by the action of cosmic-ray neutrons on nitrogen atoms.

$$^{14}_{7}\text{N} + ^{1}_{0}\text{n} \longrightarrow ^{14}_{6}\text{C} + ^{1}_{1}\text{H}$$

Carbon-14 decays by beta emission (Section 18.3) with a half-life of 5730 years. Because the formation and decay of carbon-14 balance each other, the amount of carbon-14 in the atmosphere remains approximately constant.

Carbon-14 atoms combine with oxygen to form carbon dioxide, which then enters plants through photosynthesis and is absorbed into the bodies of animals that feed on the plants. Through respiration and normal excretion, carbon-14 atoms are returned to the environmental carbon pool. As a result, a fairly constant ratio of one carbon-14 atom to 10^{12} carbon-12 atoms is maintained in the carbon pool. As long as a plant or animal is alive, this ratio is maintained in its molecules. However, when death occurs, intake of carbon-14 stops, and carbon-14 atoms are lost over time due to the constant rate of radioactive decay. By measuring the ratio of carbon-14 to carbon-12 that exists now in a sample, the time lapsed since the death of the plant or animal can be estimated (Figure 18.7).

18.6 Nuclear Transmutations

Learning Objective

- Complete nuclear equations for the formation of synthetic elements.

FIGURE 18.7 The Shroud of Turin was thought possibly to be the burial wrapping of the body of Christ. It has been kept in a chapel in Turin, Italy, since 1578. The claims of its authenticity were apparently settled by radiocarbon dating, which estimated the cloth to be from the twelfth century.

For centuries, alchemists tried in vain to convert lead into gold in their pursuit of wealth. However, an induced nuclear reaction transforming one element into another (a **transmutation**) was not accomplished until 1919 when Ernest Rutherford (Section 5.3) bombarded nitrogen nuclei with alpha particles to produce oxygen-17 and protons.

$$^{4}_{2}\text{He} + ^{14}_{7}\text{N} \longrightarrow ^{17}_{8}\text{O} + ^{1}_{1}\text{p}$$

Since Rutherford's historic experiment, many other transmutations have been accomplished. One of these, the bombardment of beryllium-9 nuclei with alpha particles carried out by James Chadwick in 1932, confirmed the existence of neutrons.

$$^{4}_{2}\text{He} + ^{9}_{4}\text{Be} \longrightarrow ^{12}_{6}\text{C} + ^{1}_{0}\text{n}$$

All the known **transuranium elements,** those beyond uranium in the periodic table, are artificially prepared. For example, bombarding uranium-238 with neutrons produces neptunium-239, which decays by beta emission to plutonium-239.

$$^{238}_{92}\text{U} + ^{1}_{0}\text{n} \longrightarrow ^{239}_{93}\text{Np} + ^{0}_{-1}\beta$$

$$^{239}_{93}\text{Np} \longrightarrow ^{239}_{94}\text{Pu} + ^{0}_{-1}\beta$$

Bombardment with neutrons is easier than with alpha particles because neutrons are not electrostatically repelled by positive nuclei.

The neutron was discovered by James Chadwick in 1932.

Another important neutron-initiated transmutation is the formation of iron-59 from iron-58. Subsequent beta decay gives cobalt-59, which then absorbs another neutron to give cobalt-60. Cobalt-60, a beta and gamma emitter, is used in cancer radiation therapy.

$$^{58}_{26}Fe + ^{1}_{0}n \longrightarrow ^{59}_{26}Fe$$

$$^{59}_{26}Fe \longrightarrow ^{59}_{27}Co + ^{0}_{-1}\beta$$

$$^{59}_{27}Co + ^{1}_{0}n \longrightarrow ^{60}_{27}Co$$

Three elements that appear before uranium in the periodic table do not occur naturally but have been synthesized by nuclear bombardment. Technetium-97 has been produced by bombarding molybdenum-96 with deuterium nuclei.

$$^{96}_{42}Mo + ^{2}_{1}H \longrightarrow ^{97}_{43}Tc + ^{1}_{0}n$$

Astatine-211 has been synthesized by bombarding bismuth-209 with alpha particles.

$$^{209}_{83}Bi + ^{4}_{2}He \longrightarrow ^{211}_{85}At + 2^{1}_{0}n$$

Promethium (atomic number 59) has been found among the fission products formed in nuclear reactors.

Example 18.6

Complete the following equations:

a. $^{241}_{95}Am + ? \longrightarrow ^{243}_{97}Bk + 2^{1}_{0}n$

b. $^{239}_{94}Pu + ^{4}_{2}He \longrightarrow ? + ^{1}_{0}n$ ■

Solution Balance atomic numbers to identify the missing particle. Then balance mass numbers.

a. $^{4}_{2}He$

b. $^{242}_{96}Cm$

Practice Problem 18.4 What nucleus can be bombarded by a proton to produce aluminum-27 (the only product)? Write a nuclear equation for the reaction.

18.7 Nuclear Fission

Learning Objective

- Describe the fission of uranium-235. Write a nuclear equation to illustrate fission.

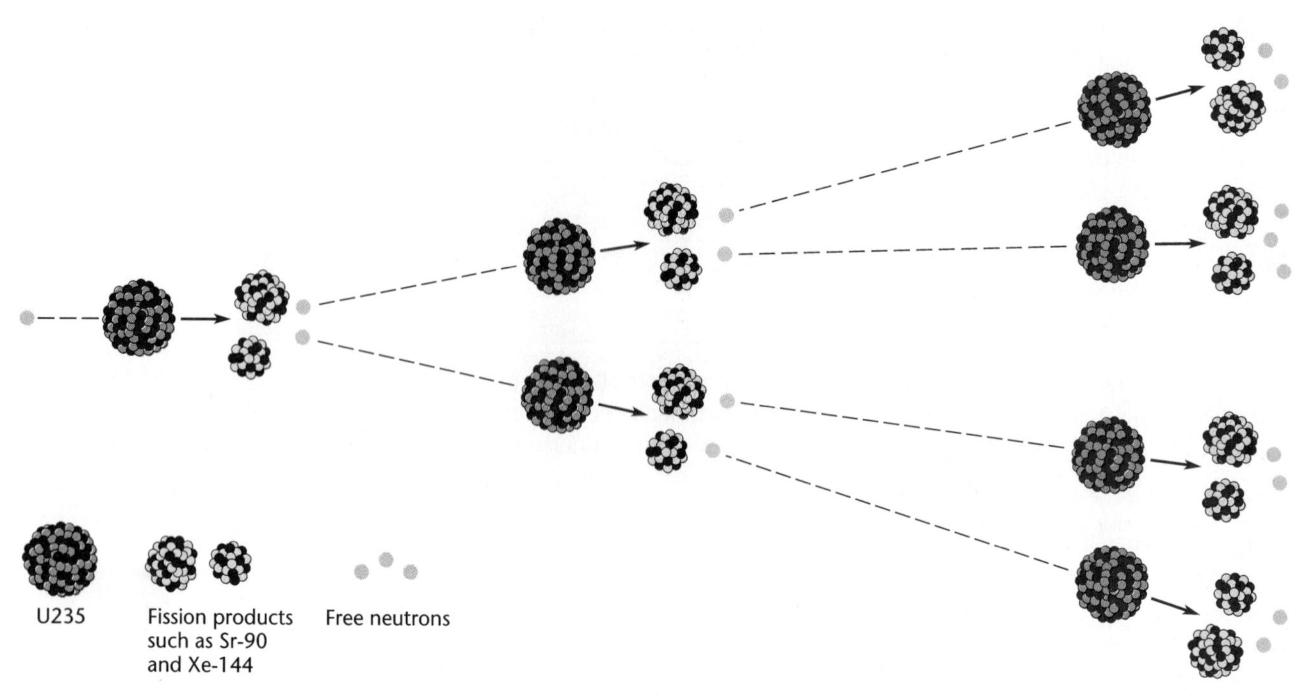

FIGURE 18.8 A nuclear chain reaction. Each uranium fission produces two or more neutrons, which in turn cause the fission of additional uranium nuclei. With a sufficiently large sample of uranium, the chain reaction can go out of control, resulting in a nuclear explosion.

Nuclear fission is the breaking of a nucleus into two smaller nuclei of similar mass. Nuclear fission was discovered in Germany in the late 1930s prior to the outbreak of World War II, in the course of experiments with uranium-235. When bombarded with neutrons, uranium-235 gives many fission products. One possible fission is illustrated by the following equation:

$$^{235}_{92}\text{U} + ^{1}_{0}\text{n} \longrightarrow ^{142}_{56}\text{Ba} + ^{92}_{36}\text{Kr} + 2^{1}_{0}\text{n}$$

If the two neutrons formed strike two other uranium-235 nuclei, four neutrons will be available. These neutrons can cause the fission of four additional uranium nuclei, producing eight neutrons, and so on. The result is a nuclear chain reaction (Figure 18.8). A chain reaction is self-sustaining if it continues after the supply of neutrons from outside is discontinued. To be self-sustaining, at least one neutron from each fission must split another uranium nucleus. Some neutrons escape from the mass of uranium, and to sustain the chain reaction these must be balanced by neutrons formed during fission. When the neutron balance is such that the number of fission events remains constant, the fission process is in a **critical condition.** The mass needed to just maintain the critical condition is called the **critical mass.**

FIGURE 18.9 A replica of the "Little Boy" atomic bomb that destroyed Hiroshima. The 9000-pound bomb was dropped from a B-29 bomber and exploded at about 1800 ft, creating a blast the equivalent of 20 000 tons of TNT.

The atomic bombs dropped on Japan during World War II were fission bombs. A nuclear fission explosion is caused by very rapidly bringing together two subcritical masses of fissionable material so as to exceed its critical mass. The two masses are forced together using an explosive charge. A critical mass of uranium-235 is a 40-kg mass about 16 cm in diameter. However, the uranium used in the bomb dropped on Hiroshima was a mixture of uranium-235 and uranium-238, and a somewhat greater total mass was needed to obtain a critical mass of the fissionable uranium-235 (Figure 18.9).

Nuclear Energy and Nuclear Reactors 18.8

Learning Objectives

- Determine the mass defect for a nucleus.
- Calculate the binding energy of a nucleus.
- Calculate the energy equivalent to a certain mass.
- Write a nuclear equation to describe the formation of fissionable material in a breeder reactor.

Albert Einstein (1879–1955) was awarded the Nobel Prize in physics in 1921.

Protons and neutrons are very tightly packed in the nucleus of an atom. The energy that would be required to decompose a nucleus into its individual protons and neutrons is called the **binding energy** of the nucleus. Some of this energy is released when a nucleus undergoes decay or fission to form more stable nuclei. The energy released in a nuclear reaction is much greater than that of chemical reactions. For example, the energy released per mole of fissioned uranium-235 (235 g) is 1.9×10^{10} kJ, about 2.5 million times the energy released by burning the same mass of coal.

The products of a nuclear reaction always have slightly less mass than the reactants. This change in mass accounts for the tremendous quantities of energy produced in nuclear reactions. Albert Einstein suggested that mass has an energy equivalent that can be calculated using the following equation:

$$E = mc^2$$

where E is the energy equivalent to a mass, m, and c is the speed of light in a vacuum, 2.998×10^8 m/s. When we calculate the energy resulting from the loss of mass in a nuclear reaction, the energy will be in joules if the difference in mass is in kilograms.

$1\ J = 1\ kg \cdot m^2/s^2$

The measured mass of a nucleus is slightly less than the mass calculated by adding the masses of its nucleons (particles in the nucleus—that is, its protons and neutrons). This difference in mass is called the **mass defect** of the nucleus. The binding energy of the nucleus is the energy equivalent to the mass defect and can be calculated using Einstein's equation. This calculation is illustrated in the following example.

Example 18.7

The mass of a hydrogen-3 (tritium) nucleus is 3.015 50 amu. The mass of a proton is 1.007 28 amu, and the mass of a neutron is 1.008 66 amu. Calculate the binding energy of a tritium nucleus. Note that 1 g = 6.0221×10^{23} amu. ■

Solution The binding energy, E, can be calculated from the mass defect, m, using Einstein's equation, $E = mc^2$. The mass defect equals the difference between the mass of the nucleus and the sum of the masses of the nucleons.

1. Sum of the masses of the nucleons:

1 p	1.007 28 amu
2 n 2 (1.008 66 amu) =	2.017 32 amu
total for the nucleons	3.024 60 amu

2. Mass of a tritium nucleus: $-3.015\ 50$ amu

3. Mass defect: 0.009 10 amu

$$(0.009\ 10\ \cancel{\text{amu}})\left(\frac{1\ \text{g}}{6.0221 \times 10^{23}\ \cancel{\text{amu}}}\right) = 1.51 \times 10^{-26}\ \text{g}$$
$$= 1.51 \times 10^{-29}\ \text{kg}$$

4. Binding energy:

$$E = (1.51 \times 10^{-29}\ \text{kg})\ (2.998 \times 10^{8}\ \text{m/s})^2$$
$$= 1.36 \times 10^{-12}\ \text{J}$$

Remember, $1\ \text{J} = 1\ \text{kg} \cdot \text{m}^2/\text{s}^2$.

Practice Problem 18.5 What is the binding energy of a lithium-7 nucleus? Its nuclear mass is 7.014 35 amu, the mass of a proton is 1.007 28 amu, and the mass of a neutron is 1.008 66 amu. $1\ \text{g} = 6.0221 \times 10^{23}$ amu.

Nuclear energy has great potential in meeting the world's demand for energy. Our consumption of fossil fuels—coal, oil, and natural gas—has increased steadily during the last two centuries. At the present rate of consumption, our fossil-fuel reserves could be nearly exhausted within two centuries. Increased use of nuclear energy could reduce dependence on fossil fuels.

In a nuclear power plant, heat produced during a nuclear reaction, rather than from burning a fossil fuel, is used to generate electricity (Figure 18.10). A cooling fluid is circulated through the reactor core and then to a steam generator. The resulting steam is piped to a steam turbine which drives an electric generator. Thus, nuclear energy is converted into mechanical energy in the turbine and then into electrical energy in the generator.

The most abundant isotope of uranium is uranium-238. Natural uranium is only about 0.7% uranium-235, the fissionable isotope. The uranium used in a nuclear fission reactor is enriched to about 3% uranium-235. This is a much lower percentage than is required in an atomic bomb, and an atomic explosion cannot occur in a fission reactor. Furthermore, fission is controlled using rods of neutron-absorbing cadmium, which are inserted into the reactor core as necessary. The control rods reduce the number of neutrons available for fission and thereby prevent overheating of the reactor core.

Two critical issues in the operation of nuclear power plants are safety and nuclear waste. Power plants must be designed to contain radioactive material in

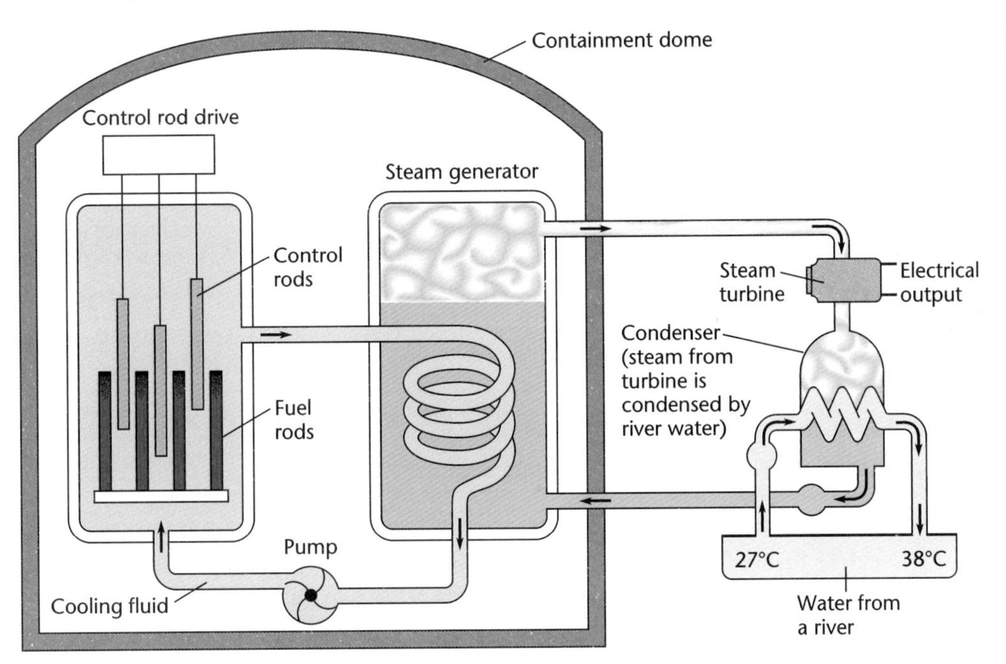

FIGURE 18.10 *Above:* Nuclear fission reactor. Energy from nuclear fission is used to form steam. The steam drives a turbine, which in turn drives an electric generator. *Right:* Three Mile Island nuclear power plant.

the event of an accident. The incident at the Three Mile Island reactor in Pennsylvania in 1979 and the Chernobyl accident in the Ukraine in 1986 have increased public concern about reactor safety. If the reactor temperature goes out of control, "meltdown" of the reactor core can occur. Such an event has the potential of releasing radioactive material into the environment. New reactors have been designed to address this problem, but there still is serious concern about the safety of many older reactors.

Disposal of nuclear waste is another major issue. Some radioactive waste contains materials having half-lives of thousands of years. Clearly, permanent disposal sites will be required for these materials. There is no agreement yet regarding where or how to dispose of nuclear waste.

Because the fissionable isotope, uranium-235, is a small percentage of natural uranium, it is likely that we could face shortages of this reactor fuel in the future. A possible solution to future reactor fuel shortages is the **breeder reactor,** which uses fission to produce additional fissionable fuel. In a breeder reactor, neutrons from fission of uranium-235 convert uranium-238 into plutonium-239:

$$^{238}_{92}\text{U} + ^{1}_{0}\text{n} \longrightarrow ^{239}_{92}\text{U}$$

$$^{239}_{92}\text{U} \longrightarrow ^{0}_{-1}\beta + ^{239}_{93}\text{Np} \longrightarrow ^{0}_{-1}\beta + ^{239}_{94}\text{Pu}$$

After the fuel rods are removed from the reactor, plutonium-239, which is fissionable, can be separated for use as a fuel in a conventional nuclear reactor.

Although breeder reactors could supply sufficient nuclear fuel to meet the world's energy demand for centuries to come, a number of problems need to be solved. One is the extreme toxicity of plutonium. Inhalation of only a few micrograms of the metal can be fatal. Another problem is a matter of economics. Research and development of breeder reactors is very expensive, and much needs to be done to demonstrate that large quantities of plutonium can be produced, transported, and stored safely. Furthermore, because plutonium-239 is used in nuclear weapons, security is a major issue. Many people fear that providing breeder reactor technology to some countries could result in the proliferation of nuclear weapons.

18.9 Nuclear Fusion

Learning Objectives

- Describe nuclear fusion.
- Write a nuclear equation to illustrate fusion.

Nuclear fusion is the process of combining two small nuclei to form a larger one. Fusion accounts for the tremendous energy of the sun. Fusion is illustrated by the following equations:

$${}^{2}_{1}\text{H} + {}^{3}_{1}\text{H} \longrightarrow {}^{4}_{2}\text{He} + {}^{1}_{0}\text{n} + \text{energy}$$

$${}^{2}_{1}\text{H} + {}^{2}_{1}\text{H} \longrightarrow {}^{3}_{2}\text{He} + {}^{1}_{0}\text{n} + \text{energy}$$

$${}^{2}_{1}\text{H} + {}^{2}_{1}\text{H} \longrightarrow {}^{3}_{1}\text{H} + {}^{1}_{1}\text{H} + \text{energy}$$

Fusion reactions are also responsible for the tremendous energy of the hydrogen bomb. Temperatures of millions of degrees are required for fusion. Such temperatures are reached in the interior of the sun but are very difficult to attain on earth except in a fission explosion. Consequently, an explosion of a fission bomb is used to detonate a hydrogen bomb.

Controlled nuclear fusion is an attractive source of energy for several reasons. First, materials needed for fusion are readily available. Fuels for fusion, such as hydrogen and deuterium, are plentiful, but quantities of uranium for fission are limited. Second, large amounts of energy are produced from very small amounts of material. Third, fusion is a cleaner process than fission (which produces large quantities of radioactive waste), and the environmental impact of a fusion reactor would likely be insignificant. Research is continuing, with the hope of developing a controlled method of initiating fusion.

18.10 Uses of Radioisotopes

Learning Objectives

- Describe one or more uses of radioisotopes in medicine.
- Describe one or more uses of radioisotopes in research and industry.

In addition to producing nuclear power, radioisotopes have many uses. In medicine, radioisotopes are used in diagnosis and treatment of a variety of disorders. For

CHEMICAL WORLD

Our Energy Dilemma

The United States is the world's leading consumer of energy. We have only 6% of the world's population, but we consume more than 25% of the energy the world produces each year. However, because of energy conservation and continued technological development in other countries, our share of the energy being consumed is slowly declining.

Industry accounts for about 33% of our energy consumption, in its production of raw materials—such as metals, glass, plastics, and fibers—and finished products. Utilities consume another 25% in generating electricity for lighting, heating, and air conditioning. Transportation uses another 25%, and about 14% is used for heating homes.

Today, about 90% of the energy used in the United States comes from fossil fuels—coal, petroleum, and natural gas. Though 95% of the world's energy was derived from coal 100 years ago, by 1950 petroleum had become the principal fuel because of our increasing use of cars and trucks and the greater convenience of using petroleum. Environmental concerns have further eroded the position of coal in the energy market.

We now face the prospect of greatly diminished reserves of fossil fuels. Although estimates vary, it is likely that less than 50% of the world's original petroleum and natural gas reserves will remain as we enter the twenty-first century. In other words, in one century we have consumed one-half of the fuel that we currently depend upon for most of our energy needs. At our present rate of consumption, petroleum reserves could be at a critically low level by the middle of the next century. Coal reserves are more extensive and could last two to three centuries at the present rate of consumption, a shorter time if consumption increases.

Faced with potential fossil-fuel burnout, we must learn to use other energy sources. Development projects have demonstrated the feasibility of using solar energy, geothermal deposits, wind and ocean currents, and biowaste (garbage and animal waste) to meet our industrial and domestic energy demands. However, as important as these energy sources are, it is unlikely that they will be enough to replace petroleum.

The conclusion of many scientists and national leaders is that we must work toward using nuclear energy as our principal energy source in the next century. However, many problems remain to be solved. Fission reactors raise questions of fuel supply, reactor safety, and nuclear waste disposal. Fusion holds great promise but awaits a breakthrough in technology needed to initiate controlled fusion. The certainty of dwindling fossil-fuel reserves is sure to spur the effort to find answers to the critical questions of our energy dilemma.

A coal strip-mine. Coal is our most abundant fossil fuel. Unfortunately, mining and burning coal cause serious environmental problems, especially because much of the remaining coal contains sulfur, which when burned is converted to sulfur dioxide. Sulfur dioxide is a serious respiratory irritant and is the major contributor to the acid rain problem.

Geothermal energy can contribute to our energy needs. This power plant in Wairakei, New Zealand, uses energy from hot rock below the ground to generate electricity.

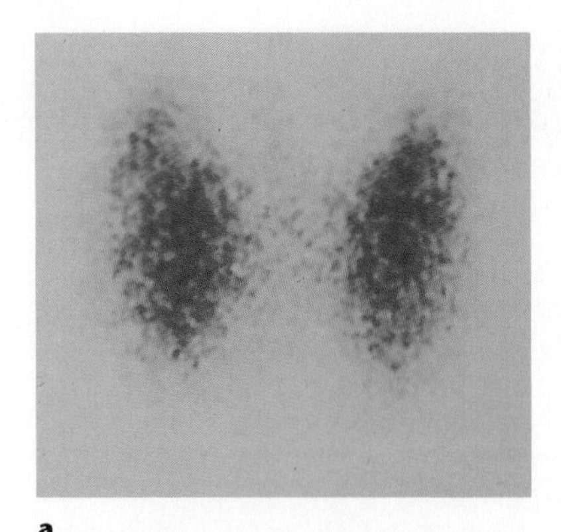

a

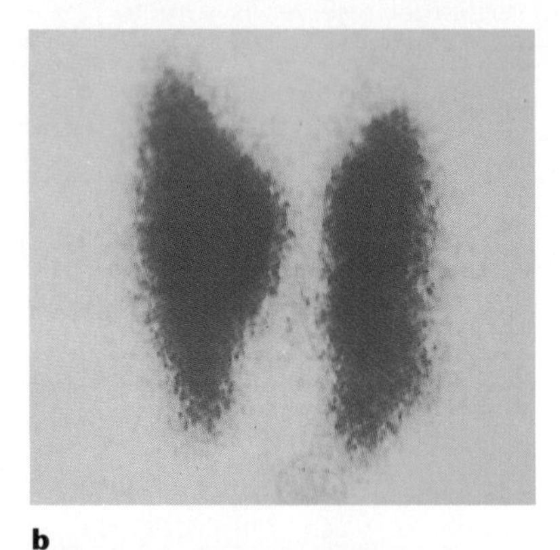

b

FIGURE 18.11 Thyroid scans using radioactive iodide. After drinking a solution of NaI containing radioactive iodide, a patient's thyroid is scanned with a radiation detector coupled with a computer that generates an image of the thyroid gland. (a) A normal thyroid. (b) An overactive thyroid.

TABLE 18.4 Medical Uses of Radioisotopes

Radioisotope	*Half-Life*	*Emission*	*Uses*
chromium-51	27.8 da	gamma	diagnosis: spleen, gastrointestinal disorders
cobalt-60	5.3 yr	beta, gamma	therapy: cancer
gallium-67	78.3 hr	gamma	diagnosis: various tumors
iodine-123	13.3 hr	gamma	diagnosis: thyroid
iodine-131	8.1 da	beta, gamma	diagnosis and therapy: thyroid
iron-59	45.1 da	beta, gamma	diagnosis: anemia, bone-marrow function
krypton-81	2.1×10^5 yr	gamma	diagnosis: lung ventilation
technetium-99m[a]	6 hr	gamma	diagnosis: heart muscle, brain, kidney, liver
thallium-201	73 hr	gamma	diagnosis: coronary function

[a]The *m* stands for "metastable," meaning that this isotope spontaneously loses energy to give a more stable form of the same isotope.

example, absorption of iodine by a patient's thyroid gland can be checked using iodine-123, a gamma emitter, and a radiation detector. The patient drinks a solution of sodium iodide containing iodine-123, and then the thyroid is scanned for levels of radioactivity. An overactive thyroid will absorb more iodide than is normal and will show higher levels of radioactivity. An image of the scan could reveal an overactive thyroid, nodules, or a tumor (Figure 18.11).

Because it is absorbed in the thyroid, the effect of radioactive iodide is localized, making it ideal for treating thyroid cancer. Iodine-131, a beta and a gamma emitter, is the isotope used in thyroid cancer therapy. Other radioisotopes are used in the diagnosis and treatment of other illnesses (Table 18.4).

Radioisotopes have many uses in research, industry, and even in our homes. One is the use of a **radioactive tracer,** a radioisotope that can be followed through a chemical reaction or even a series of reactions. For example, in the study of photosynthesis, carbon dioxide was prepared that contained carbon-14 as a tracer. The

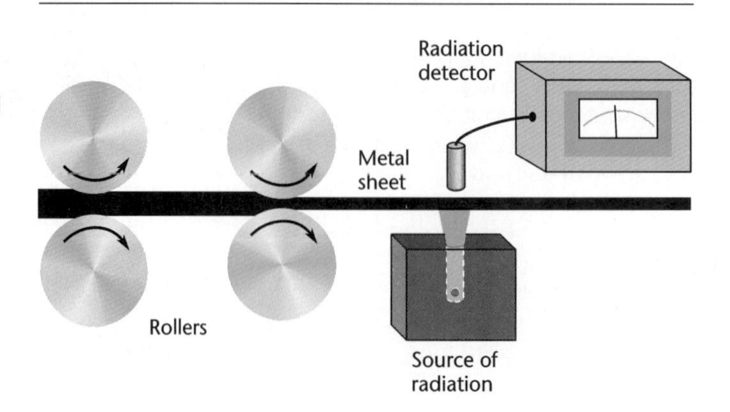

FIGURE 18.12 A thickness gauge uses a radiation source to monitor and control the thickness of a metal sheet or plastic film.

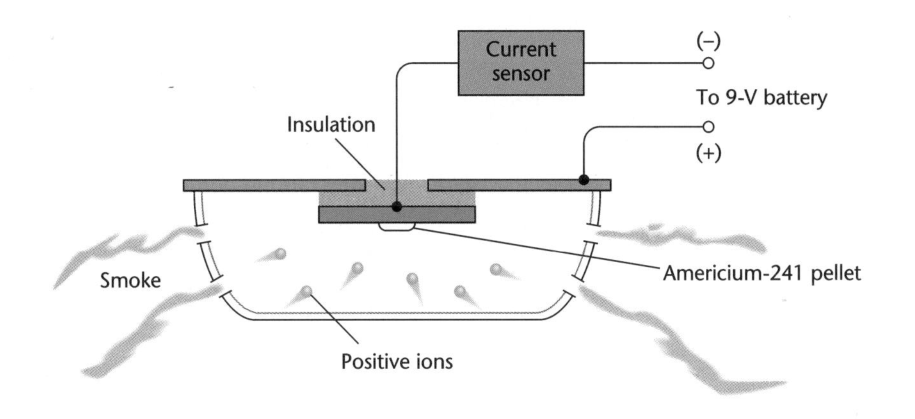

FIGURE 18.13 A smoke detector uses alpha particles from americium-241 to ionize air, causing an electric current. Smoke entering the detector absorbs some of the ions. The current drops, triggering an alarm.

overall process of photosynthesis involves many chemical reactions and is summarized by the equation:

$$6CO_2(g) + 6H_2O(l) \xrightarrow{\text{sunlight}} C_6H_{12}O_6(aq) + 6O_2(g)$$

Using a radiation detector, scientists have been able to follow the radioactive carbon-14 through various intermediates of the process, thereby giving them a better understanding of photosynthesis.

In industry, a thickness gauge containing a radioisotope can be used to control the production of a metal sheet or a plastic film. The level of radiation passing through the material is related to its thickness (Figure 18.12). The petroleum industry uses radioisotopes to monitor the movement of liquids through pipelines. Another application is the use of gamma radiation from cobalt-60 to detect structural flaws in metal castings for aircraft parts, thereby improving aircraft safety.

In one type of smoke detector for home use, alpha particles from a small pellet of americium-241 ionize the air inside the detector (Figure 18.13). Movement of the resulting ions produces a small electric current. When smoke particles enter the detector, they absorb some of the ions, causing a drop in the current. A current sensor then triggers an alarm.

Both medical and nonmedical uses of radioisotopes are expanding rapidly, and the applications seem almost unlimited.

Chapter Review

Directions: From the choices listed at the end of this Chapter Review, choose the word or term that best fits in each blank. Use each choice only once. Answers appear at the end of the chapter.

Section 18.1 Radioactivity was discovered in 1896 by ____________ (1). The spontaneous emission of rays or particles from the nucleus of an atom is called ____________ (2). Shortly after the discovery of radioactivity, Marie and Pierre Curie discovered ____________ (3) and ____________ (4), two radioactive elements.

Section 18.2 Natural radioactivity results from the spontaneous decay or disintegration of ____________ (5) occurring nuclei. Natural radioactive emissions include three kinds of rays: alpha, beta, and ____________ (6) rays. Alpha particles are ____________ (7) nuclei and are ____________ (8) charged. Beta particles are fast-moving ____________ (9) and are ____________ (10) charged. Gamma rays are high-energy rays similar to ____________ (11) and have no charge or ____________ (12). Of the three, gamma radiation is the most ____________ (13). Consequently, ____________ (14) or high-density concrete are often used to shield from gamma radiation.

__________ (15) radiation has sufficient energy to ionize atoms and molecules in its path. The ability to ionize matter distinguishes radioactive emissions from __________ (16) radiation such as ultraviolet, visible, and __________ (17) light. Ionizing radiation can be very damaging to living cells and can cause __________ (18) defects, sickness, and death.

The __________ (19) counter is the most well known ionization counter for detecting and counting ionizing radiation. A __________ (20) counter measures the energy of the radiation as well as the counts per minute. A film badge is used by radiation workers to measure radiation absorbed during a specific period of time.

Radiation absorbed may be expressed in __________ (21), or radiation absorbed dose. The SI unit is the gray (Gy). The unit that takes into consideration the effectiveness of radiation to cause damage to tissue is the __________ (22).

Section 18.3 A __________ (23) equation shows changes in nuclei. Nuclear equations must be balanced with regard to both mass number and __________ (24), or __________ (25) number. Loss of an alpha particle reduces the atomic number by __________ (26) units. Beta decay results in an increase in the atomic number by one unit.

Section 18.4 The uranium-238 decay series involves fourteen steps including eight alpha emissions and six __________ (27) emissions. This decay series ends with __________ (28), a stable isotope of lead. Another decay series begins with thorium-232 and ends with lead-208. The thorium-232 series includes __________ (29) alpha emissions and four beta emissions.

Section 18.5 The time required for one-half of a sample to decay is called its __________ (30). After an elapsed time equal to three half-lives, __________ (31) percent of the original sample remains.

__________ (32) dating is a technique used to estimate the age of an ancient relic by measuring the activity of a radioisotope. The most important dating technique is __________ (33) dating in which the radioisotope is __________ (34). This isotope of carbon is formed in the upper atmosphere by the action of cosmic-ray neutrons on __________ (35) atoms. The ratio of carbon-14 to __________ (36) atoms is fairly constant in the environment and in living plants, animals, and people. However, at the time of death, __________ (37) of carbon-14 stops, and the percentage of carbon-14 atoms __________ (38) due to radioactive decay. The __________ (39) of a sample can be estimated by determining the current ratio of carbon-14 to carbon-12.

Section 18.6 All of the __________ (40) elements have been artificially prepared. For example bombardment of uranium-238 with neutrons produces __________ (41) and a beta particle. In addition to the transuranium elements, three other elements have been prepared artificially.

Section 18.7 The splitting of a nucleus into two smaller nuclei of similar mass is called nuclear __________ (42). The first fission process involved the neutron bombardment of __________ (43). Because each fission generally produces two or more neutrons, a __________ (44) reaction may result. A fission process is in a __________ (45) condition when the number of fission events remains constant. A critical mass is necessary to maintain a critical condition, and exceeding the critical mass may result in a nuclear __________ (46). The atomic bombs dropped on __________ (47) were fission bombs.

Section 18.8 The energy equivalent to the mass defect of a nucleus is called its __________ (48) energy. During __________ (49) or fission of a nucleus, some of this energy is released. The energy from nuclear reactions is millions of times greater than the energy from __________ (50) of an equal mass of a fossil fuel.

In a nuclear power plant, the heat from a nuclear reaction is used to form __________ (51) which drives a turbine which then drives an __________ (52) generator. The fission process is controlled using __________ (53) control rods that are inserted into the reactor core to absorb __________ (54).

Section 18.9 Nuclear __________ (55) is responsible for the energy of the sun. The combining of two __________ (56) nuclei to form a larger one releases a tremendous quantity of energy. Fusion research is focusing upon developing a controlled method of __________ (57) fusion.

Section 18.10 Radioisotopes are used in medicine in both the __________ (58) and treatment of various disorders. In research, scientists can gain a better understanding of chemical reactions by following a particular element through a reaction using a radioactive __________ (59). Industrial processes can be monitored with instruments that use radioisotopes.

Choices: age, atomic, Becquerel, beta, binding, cadmium, carbon-12, carbon-14, chain, charge, combustion, critical, decay, diagnosis, drops, electric, electrons, explosion, fission, fusion, gamma, Geiger–Müller, genetic, half-life, helium, infrared, initiating, intake, ionizing, Japan, lead, lead-206, mass, naturally, negatively, neptunium-239, neutrons, nitrogen-14, nonionizing, nuclear, penetrating, polonium, positively, radioactivity, radiocarbon, radioisotope, radium, rads, rem, scintillation, six, small, steam, 12.5, tracer, transuranium, two, uranium-235, x-rays

Study Objectives

After studying Chapter 18, you should be able to:

Section 18.2

1. Describe and give symbols for alpha, beta, and gamma radiation.
2. Compare the penetrating power of alpha, beta, and gamma radiation.
3. Describe a Geiger–Müller counter, a scintillation counter, and a film badge.

Section 18.3

4. Write symbols for a neutron, positron, and proton.
5. Complete and balance nuclear equations.

Section 18.4

6. Describe three natural radioactive decay series.
7. Given the starting and final nuclei, determine the combination of alpha and beta decay reactions for a decay series.

Section 18.5

8. Calculate the quantity of radioactive sample remaining after an elapsed time of 1, 2, 3, 4 (and so on) half-lives.
9. Calculate the number of half-lives required for a given quantity of radioactive sample to decay to a given final quantity.
10. Explain the process of radiocarbon dating. Write nuclear equations for the decay of carbon-14 and its formation in the upper atmosphere.

Section 18.6

11. Complete nuclear equations for the formation of synthetic elements.

Section 18.7

12. Describe the fission of uranium-235. Write a nuclear equation to illustrate fission.

Section 18.8

13. Determine the mass defect for a nucleus.
14. Calculate the binding energy of a nucleus.
15. Calculate the energy equivalent to a certain mass.
16. Write a nuclear equation to describe the formation of fissionable material in a breeder reactor.

Section 18.9

17. Describe nuclear fusion.
18. Write a nuclear equation to illustrate fusion.

Section 18.10

19. Describe one or more uses of radioisotopes in medicine.
20. Describe one or more uses of radioisotopes in research and industry.

Key Terms

Review the definition of each of the terms listed here. For reference, they are listed by chapter section number. You may use the glossary as necessary.

18.1 radioactivity

18.2 activity, alpha particle, becquerel, beta particle, curie, gamma ray, ionizing radiation, rad, rem

18.3 nuclear equation

18.5 fallout, half-life, radiocarbon dating, radioisotope dating

18.6 transmutation, transuranium element

18.7 critical condition, critical mass, nuclear fission

18.8 binding energy, breeder reactor, mass defect

18.9 nuclear fusion

18.10 radioactive tracer

Questions and Problems

Answers to questions and problems with blue numbers appear in Appendix C.

Natural Radioactivity (Sections 18.1, 18.2)

1. Who is given the credit for discovering radioactivity?
2. Explain what is meant by radioactivity.
3. Describe the three principal types of radioactive emissions. Write a symbol for each including the charge and mass where appropriate.
4. Compare the penetrating power of alpha, beta, and gamma radiation. Which of these requires the greatest shielding to protect the human body?
5. Who discovered the elements polonium and radium?
6. Explain what is meant by ionizing radiation and nonionizing radiation. Give examples of each type.
7. What is the most well known ionization counter for detecting and measuring radioactive emissions?
8. Explain how a Geiger–Müller counter works.
9. What is the activity of a radioactive sample?
10. What is the SI unit used for activity?
11. Explain what is meant by a "one-curie sample?"
12. What is a rad?
13. What is a rem? How is a rem related to a rad?
14. What information is provided by a scintillation counter that permits a source of radiation to be identified?
15. What is a natural decay series? Why does the uranium-238 decay series go no further than lead-206?

Nuclear Equations, Nuclear Decay Series (Sections 18.3, 18.4)

16. Write a nuclear equation for the alpha decay of the following:
    a. uranium-238
    b. thorium-238
17. Write a nuclear equation for the beta decay of the following:
    a. actinium-228
    b. tritium (hydrogen-3)
18. Identify the missing particle and write a balanced nuclear equation for these decays:
    a. beta decay to give sulfur-32
    b. alpha decay to give radium-226
19. Write a balanced nuclear equation for the decay of the following by beta emission:
    a. iron-59
    b. protactinium-234
    c. cobalt-60
20. Write a balanced nuclear equation for the decay of the following by alpha emission:
    a. thorium-230
    b. polonium-210
    c. radium-226
21. The uranium-238 decay series has eight alpha emissions and six beta emissions. Without writing nuclear equations for all steps, show by means of calculations that this combination of alpha and beta emissions produces lead-206.
22. Write nuclear equations for the following:
    a. alpha decay of gold-179
    b. positron decay (refer to Table 18.2) of cobalt-56
    c. beta decay of lithium-8
23. Identify the type of decay and write a nuclear equation for each of the following nuclear transformations:
    a. cesium-142 to barium-142
    b. tin-128 to antimony-128
    c. uranium-229 to thorium-225

Half-Life (Section 18.5)

24. Explain what is meant by the half-life of a radioactive sample.
25. Calculate the percentage of the original quantity of a sample remaining after six half-lives.
26. How many half-lives elapse in the decay of 75% of a radioactive element?
27. Carbon-14 has a half-life of 5730 yr. Starting with 1.0 mol of carbon-14, approximately what mass would remain after 17 200 yr?
28. Write a nuclear equation for the decay of carbon-14.
29. Write a nuclear equation for the formation of carbon-14 from nitrogen in the upper atmosphere.
30. Cobalt-60, used in cancer radiation therapy, has a half-life of 5.26 yr. Starting with 1.0 mg of cobalt-60, calculate what mass would remain after 10.5 yr.
31. Explain why the ratio of carbon-14 to carbon-12 remains constant in a growing tree but decreases steadily after the tree is cut down.

Nuclear Transmutations (Section 18.6)

32. In transmutations, explain why a neutron is often preferred as a bombarding particle rather than an alpha particle.
33. Identify the starting nucleus for the preparation of iron-59 by neutron bombardment. Write a nuclear equation for the reaction.
34. The bombardment of bismuth-209 with an alpha particle gives astatine-211. Identify the other product particle(s) and write a nuclear equation for the process.

35. When bombarded with an alpha particle, americium-241 (atomic number 95) is coverted to berkelium-243 (atomic number 97). Write a nuclear equation for this transmutation.

36. Bombardment of calcium-44 with a proton produces a new element and a neutron. Identify the element and write a nuclear equation for the transmutation.

37. Transmutation of curium-246 (atomic number 96) into nobelium-254 (atomic number 102) also produces four neutrons. What was the bombarding particle? Write a nuclear equation for the reaction.

Nuclear Fission (Section 18.7)

38. A certain nuclear fission of uranium-235 produced strontium-94 and three neutrons. What was the other particle produced? Write a nuclear equation for the reaction.

39. What is meant by a critical condition for a fission process?

40. The critical mass of uranium-235 is about 40 kg. Explain what is meant by critical mass.

41. Write nuclear equations for the fission of uranium-235 by a neutron to give the following fission products:
    a. rubidium-90 and cesium-144
    b. strontium-90 and xenon-144

42. Write a nuclear equation for the fission of plutonium-239 to give cerium-144 and krypton-85.

Nuclear Energy and Nuclear Reactors (Section 18.8)

43. Diagram a nuclear fission reactor for generating electricity.

44. The fission of uranium-235 produces about 2.5 million times the energy produced by burning the same mass of coal. Calculate the mass of coal in metric tons that must be burned to produce the same energy as the fission of 1.5 kg of uranium-235. (one metric ton = 1000 kg)

45. Explain why a nuclear explosion cannot occur in a nuclear fission reactor.

46. What are two major issues in the operation of nuclear power plants?

47. Calculate the mass defect of an oxygen-16 nucleus having a nuclear mass of 15.990 52 amu. The mass of a proton is 1.007 28 amu, and the mass of a neutron is 1.008 66 amu.

48. Determine (a) the loss in mass and (b) the energy released for the fusion of a tritium (hydrogen-3) and a deuterium (hydrogen-2) nucleus:

    $^{3}_{1}H + ^{2}_{1}H \longrightarrow ^{4}_{2}He + ^{1}_{0}n$

    The masses are:

neutron	1.008 66 amu
proton	1.007 28 amu
tritium	3.015 50 amu
deuterium	2.013 55 amu

49. Americium-241, used in smoke detectors, decays by alpha emission to give neptunium-237 with a loss in mass of 0.0061 amu. Calculate the energy in joules released upon decay of 100 americium-241 nuclei.

Nuclear Fusion (Section 18.9)

50. What is meant by nuclear fusion?

51. Where does nuclear fusion occur naturally?

52. Under what circumstances has artificially induced nuclear fusion been successful?

53. Write a nuclear equation for the fusion of two deuterium atoms to form helium-3.

54. Identify the missing particles and write nuclear equations for the following fusion reactions:
    a. tritium plus deuterium to form helium-4
    b. two helium-3 nuclei to form a helium-4 nucleus

55. Why is nuclear fusion more difficult to initiate than fission? Explain how fusion has been artificially induced.

56. What are some advantages of nuclear fusion over fission as a potential source of energy?

Uses of Radioisotopes (Section 18.10)

57. Describe the use of radioactive iodine in medicine.

58. What radioisotope is used in a smoke detector? Explain how a smoke detector works.

59. In the manufacture of plastic film, its thickness can be monitored and controlled using a gauge containing a radioisotope. Describe how a thickness gauge works.

Solutions to Practice Problems

PP 18.1

$^{222}_{86}Rn \longrightarrow ^{4}_{2}He + ^{218}_{84}Po$

PP 18.2

$^{235}_{92}U \longrightarrow ^{207}_{82}Pb + x\,^{4}_{2}He + y\,^{0}_{-1}\beta$

To balance the mass numbers:

$235 = 207 + 4x$

$4x = 28, x = 7$

Therefore, 7 alpha emissions are involved.

To balance the atomic numbers (charge):

$$92 = 82 + (7 \times 2) + y(-1)$$

$$y = 82 + 14 - 92 = 4$$

Therefore, 4 beta emissions are involved.

PP 18.3

$$\frac{120.\text{ yr}}{30.0\text{ yr/half-life}} = 4 \text{ half-lives}$$

Fraction remaining is $0.5 \times 0.5 \times 0.5 \times 0.5 = 0.0625$.

PP 18.4

Write an incomplete nuclear equation, indicating what is unknown:

$${}^{1}_{1}\text{H} + ? \longrightarrow {}^{27}_{13}\text{Al}$$

The missing atomic number is 12, and the missing mass number is 26. The element with atomic number 12 is magnesium. The complete equation is:

$${}^{1}_{1}\text{H} + {}^{26}_{12}\text{Mg} \longrightarrow {}^{27}_{13}\text{Al}$$

PP 18.5

Lithium-7 has 3 protons and 4 neutrons. To calculate the binding energy, we must first calculate the mass defect:

1. Sum of the masses of the nucleons:
   mass of 3 p = 3(1.007 28 amu) = 3.021 84 amu
   mass of 4 n = 4(1.008 66 amu) = 4.034 64 amu
   total mass of nucleons: 7.056 48 amu
2. Minus the nuclear mass: −7.014 35 amu
3. Mass defect: 0.042 13 amu

$$\text{mass defect (g)} = \frac{1\text{ g}}{(6.0221 \times 10^{23}\ \cancel{\text{amu}})}(0.042\ 13\ \cancel{\text{amu}})$$

$$= 6.996 \times 10^{-26}\text{ g}$$

$$= 6.996 \times 10^{-29}\text{ kg}$$

4. Then, to calculate the binding energy:

$$E = mc^2 = (6.996 \times 10^{-29}\text{ kg})(2.998 \times 10^{8}\text{ m/s})^2$$

$$= 6.288 \times 10^{-12}\text{ J}$$

Answers to Chapter Review

1. Becquerel
2. radioactivity
3. radium
4. polonium
5. naturally
6. gamma
7. helium
8. positively
9. electrons
10. negatively
11. x-rays
12. mass
13. penetrating
14. lead
15. ionizing
16. nonionizing
17. infrared
18. genetic
19. Geiger–Müller
20. scintillation
21. rads
22. rem
23. nuclear
24. charge
25. atomic
26. two
27. beta
28. lead-206
29. six
30. half-life
31. 12.5
32. radioisotope
33. radiocarbon
34. carbon-14
35. nitrogen-14
36. carbon-12
37. intake
38. drops
39. age
40. transuranium
41. neptunium-239
42. fission
43. uranium-235
44. chain
45. critical
46. explosion
47. Japan
48. binding
49. decay
50. combustion
51. steam
52. electric
53. cadmium
54. neutrons
55. fusion
56. small
57. initiating
58. diagnosis
59. tracer

19

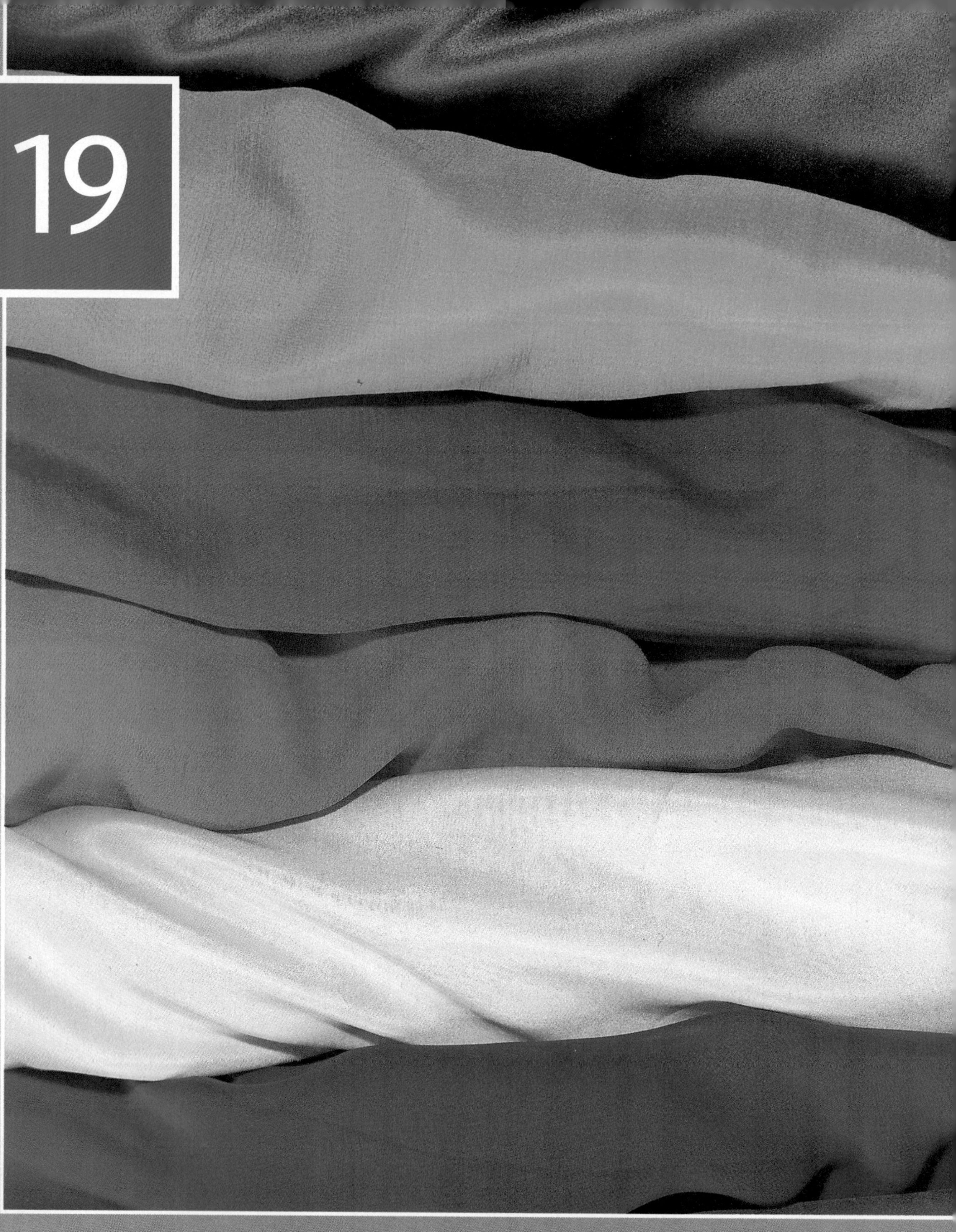

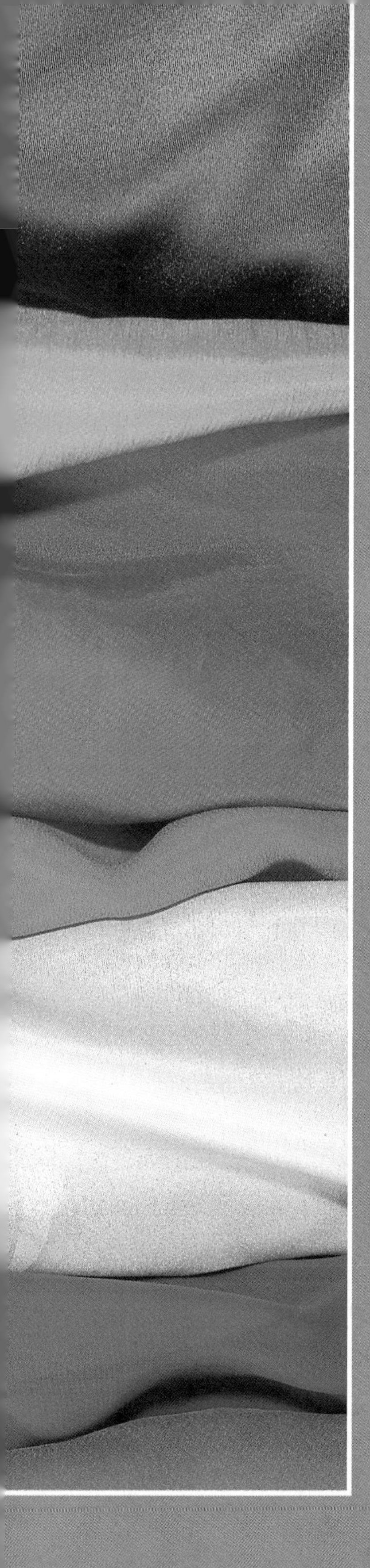

Introduction to Organic Chemistry

Contents

Silk, a natural organic fiber, can be woven to produce beautiful fabric.

Gasoline and other important fuels are obtained from petroleum.

You Probably Know . . .

- Organic compounds are compounds of carbon. They are found in living organisms and are left behind after they die. Petroleum and coal are sources of many organic compounds.
- Most organic compounds burn. Natural gas, gasoline, and other fuels obtained from petroleum are indispensable to us. Ordinary combustible materials are organic substances.
- Cotton, wool, and silk fibers and fabrics are organic substances of natural origin. Moreover, many useful synthetic fibers are used in the manufacture of a wide variety of textile products.
- Most of the common drugs are organic compounds, including aspirin, Tylenol®, morphine, and codeine.

Aspirin, a common pain reliever.

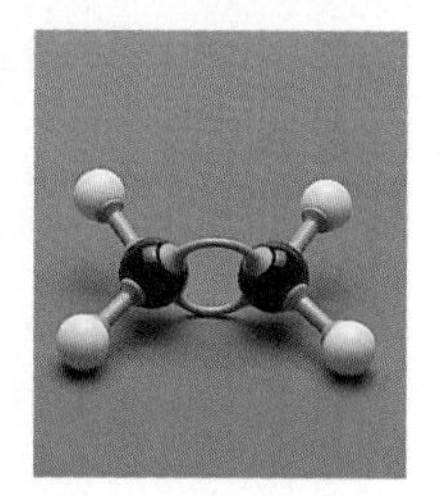
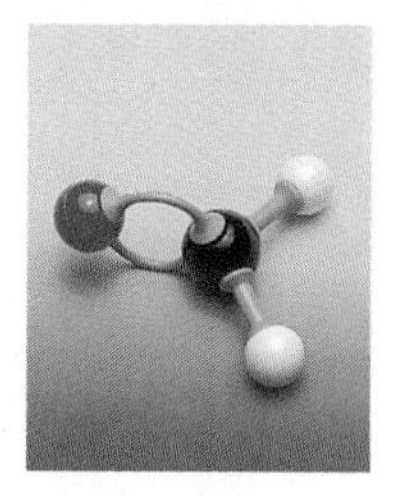

Models of some common organic compounds: propane, ethene, formaldehyde, and ethyne.

Why Should We Study Organic Chemistry?

Organic compounds: Compounds containing carbon.

Inorganic compounds: Compounds of mineral origin (Section 1.1).

Chemical compounds fall into two categories: organic and inorganic. Most organic compounds are made up of relatively few elements—carbon, hydrogen, oxygen, and a few other nonmetals. Yet the number of known organic compounds is estimated in the millions, far exceeding the number of inorganic compounds. Organic compounds have a carbon structure. Most are found in living organisms (plants, animals, and the human body) and in natural deposits such as petroleum and coal that are thought to have resulted from plant and animal matter. However, in the last century, many have been synthesized by chemists. Synthetic organic chemistry has opened the door to countless numbers of new products and new uses of organic compounds.

Organic compounds are very much a part of our daily lives. Fuels such as natural gas (mostly methane), liquified petroleum gas (LPG, propane), and gasoline are used in homes and cars. After water, our bodies are made mostly of organic compounds. Our food is made of carbohydrates, fats, and proteins—all organic substances, as are the spices and flavorings that make our food more enjoyable. We take vitamins such as vitamin C and niacin to maintain good health. We relieve headache and arthritis pain with aspirin and ibuprofen, and we treat bacterial infections with penicillin and sulfa drugs. Plastics, fabrics, and

Propane, a liquified petroleum gas, is used as a fuel for barbeque grills.

rubber are materials used to make eyeglasses and clothing, as well as many other products that we consider indispensable.

An understanding of organic chemistry can give us a better appreciation of the necessities and the conveniences of our lives. We will then be better prepared to make good choices about our food, our health, and the products we buy.

In this chapter, we consider the common classes of organic compounds. Their names, molecular structures, syntheses, and common reactions are all important topics. An understanding of organic chemistry is essential to a study of biochemistry, a topic that is introduced in Chapter 20.

Nature of Organic Compounds

19.1

Learning Objectives

- Write Lewis formulas and condensed formulas of compounds from their molecular formulas.
- Describe the bond angles and shapes of organic molecules.

Carbon is a unique element. It is the only element that bonds with itself to form chains of atoms, branched structures, and rings. A carbon framework is the structural basis for millions of organic compounds. The variety of structures is almost unlimited, and chemists are discovering or synthesizing new structures every day.

Organic compounds are generally molecular and have physical properties that are typical of other molecular compounds. In general, they are low-melting solids, liquids, or gases at ordinary temperatures. Most are nonpolar or only slightly polar and are relatively insoluble in water and other polar solvents. Their physical properties are quite different from those of inorganic compounds such as metals, inorganic salts, and network solids, which are typically high-melting substances. Most inorganic compounds are insoluble in nonpolar solvents, and many salts are water soluble.

Carbon generally forms four covalent bonds.

VSEPR model: Valence Shell Electron Pair Repulsion model (Section 8.7).

Carbon structures are very stable because of strong carbon–carbon bonds. Both C—C and C—H bonds, which are also common in organic molecules, are relatively nonpolar and lack the reactivity of more polar bonds. Although polar bonds of organic molecules are quite reactive, the carbon framework often is undisturbed in these reactions.

When we examine the structure of organic molecules, we find that carbon forms four covalent bonds. This is exactly what is predicted for carbon because of its position in Group 4A in the periodic table and from the octet rule. With its four valence electrons, a carbon atom tends to share electrons four times with other atoms to complete its octet of electrons. The bonding in some common organic compounds is shown in Table 19.1.

```
    H
    |
H—C—   is abbreviated  CH3—
    |
    H

    H
    |
 —C—   is abbreviated  —CH2—
    |
    H

    |                    |
 —C—H  is abbreviated  —CH
    |                    |
```

Remember that Lewis formulas are structural formulas, but they do not show the shapes of molecules. Shapes can be predicted using the VSEPR model. The spatial arrangement of the atoms surrounding a carbon atom can be predicted by counting the number of pairs (or sets) of electrons around the carbon atom. When

TABLE 19.1 Bonding in Some Common Organic Compounds

Compound	*Molecular Formula*	*Lewis Formula*	*Abbreviated Formula*
methane (in natural gas)	CH_4	H—C(H)(H)—H	CH_4
propane (liquified petroleum gas)	C_3H_8	H—C(H)(H)—C(H)(H)—C(H)(H)—H	$CH_3CH_2CH_3$
ethene (ethylene)	C_2H_4	$H_2C{=}CH_2$	$CH_2{=}CH_2$
ethyne (acetylene)	C_2H_2	H—C≡C—H	CH≡CH
ethanol (ethyl alcohol)	C_2H_6O	H—C(H)(H)—C(H)(H)—O—H	CH_3CH_2OH
formaldehyde (a tissue preservative)	CH_2O	H—C(=O)—H	HCHO
formic acid (in ant bites)	CH_2O_2	H—C(=O)—O—H	HCO_2H

a carbon atom has four single bonds (as in CH_4 and C_3H_8), the bond angles are about 109° resulting in a tetrahedral shape (Figure 19.1).

The electrons of a double bond are counted as one set, and a carbon atom that has a double bond is surrounded by only three sets of electrons. As a result, the bond angles in C_2H_4 and CH_2O are about 120°, and the carbons are trigonal. A triple bond is considered one set of electrons, and C_2H_2 has bond angles of 180° and is a linear molecule.

It is common practice to represent organic compounds using abbreviated formulas. They are easier to write than detailed Lewis formulas. Note that the H's bonded to a carbon are generally written following the C. The lines for double and triple bonds are usually shown in the condensed formulas, but single bonds are not drawn except when groups need to be spaced to avoid crowding.

Isomers: Compounds that have the same molecular formula but different structural formulas.

Frequently, two or more Lewis formulas having atoms in different positions can be drawn for a given molecular formula. These compounds are called **isomers.** For example, there are two isomers with the molecular formula C_2H_6O:

$$CH_3CH_2OH \qquad CH_3OCH_3$$

Isomers are very common among organic compounds.

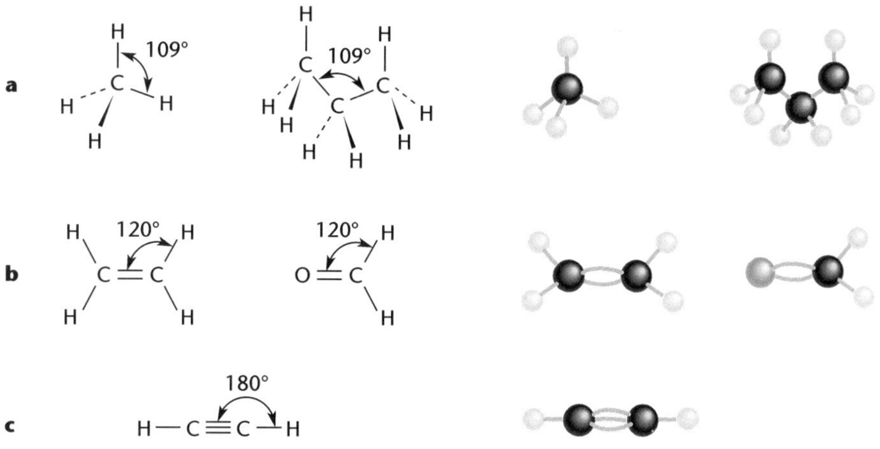

FIGURE 19.1 Shapes of organic molecules. (a) The bond angles in methane (CH_4) and propane (C_3H_8) are about 109°, and carbon is at the center of a tetrahedron. (b) In ethene (C_2H_4) and formaldehyde (CH_2O), the bond angles are about 120°, and the carbons are trigonal. (c) Ethyne (C_2H_2) has bond angles of 180° and is a linear molecule.

Example 19.1

Write structural formulas for two isomers that have the molecular formula C_2H_4O. Label the approximate bond angles. ■

Solution

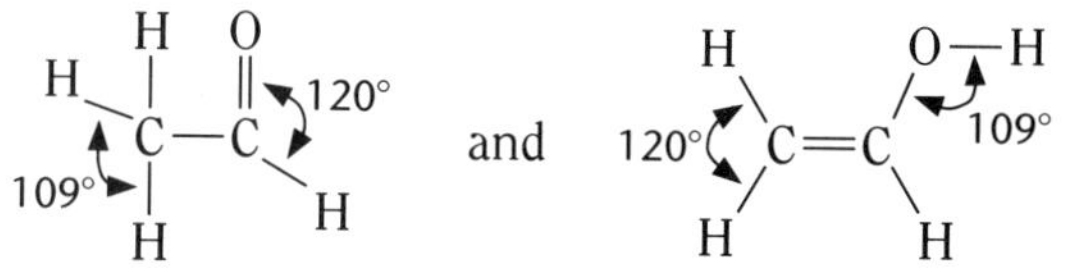

Practice Problem 19.1 Three isomers have the molecular formula C_3H_8O. Write their structural formulas. (Solutions to practice problems appear at the end of the chapter.)

Alkanes

19.2

Learning Objectives

- Write the molecular formula of an alkane having a specific number of carbons.
- Name alkanes using IUPAC and common names.
- Name and write formulas of alkyl groups with one to four carbons.
- Write equations for the combustion of alkanes and their reactions with halogens.

Alkanes are hydrocarbons that contain only single bonds. They are said to be **saturated** compounds because the carbon structure has the maximum number of

Hydrocarbon: A compound composed of only hydrogen and carbon.

TABLE 19.2 Alkanes (C_nH_{2n+2})			
Name	*Formula*	*Name*	*Formula*
methane	CH_4	hexane	C_6H_{14}
ethane	C_2H_6	heptane	C_7H_{16}
propane	C_3H_8	octane	C_8H_{18}
butane	C_4H_{10}	nonane	C_9H_{20}
pentane	C_5H_{12}	decane	$C_{10}H_{22}$

Alkane: A saturated hydrocarbon.

Saturated: Referring to a carbon structure having the maximum number of H's possible; C_nH_{2n+2}.

hydrogen atoms possible. The general formula of an alkane is C_nH_{2n+2} (refer to Table 19.2). Note that this formula refers only to noncyclic alkanes, those that do not have ring structures. The simplest alkane is methane, CH_4 ($n = 1$), followed by ethane, C_2H_6 ($n = 2$), and propane, C_3H_8 ($n = 3$). The number of H's in the formula of an alkane having a particular number of C's is easily calculated using the general formula. For example, octane, having eight carbons, has $(2 \times 8) + 2 = 18$ H's, and its molecular formula is C_8H_{18}.

Naming Alkanes

The name of an alkane has two parts: a prefix to indicate the number of carbons and the suffix *-ane* to indicate that it is an alkane (Table 19.2). In general, each type of compound has its own suffix. Two kinds of names are used for organic compounds. The International Union of Pure and Applied Chemistry (IUPAC) has devised systematic names, those that follow a set of rules. Those that do not follow the rules are called common or trivial names. They have been passed from one generation of chemists to another, and many are still widely used. Some common names have been accepted by IUPAC as official names.

Normal alkane: a continuous-chain alkane.
Butane: $CH_3CH_2CH_2CH_3$

Alkanes that have a continuous chain of carbons are called normal alkanes. Examples include butane and pentane:

butane: $CH_3CH_2CH_2CH_3$
pentane: $CH_3CH_2CH_2CH_2CH_3$

Other alkanes have branched structures. In addition to their IUPAC names, the small branched alkanes have common names that are still widely used. Although a pattern can be recognized for some of these common names, they are not part of the IUPAC system. Note that the total number of carbon atoms, not the number in the longest chain, determines the choice of the common name.

isobutane: CH_3CHCH_3 (with CH_3 branch on the central C)

isopentane: $CH_3CHCH_2CH_3$ (with CH_3 branch on the second C)

In the IUPAC names, the branches, referred to as alkyl groups, are treated as substituents on the parent carbon chain. An **alkyl group** has the general formula

TABLE 19.3 Alkyl Groups

Name	*Formula*[a]	*Name*	*Formula*
methyl	$CH_3—$	butyl	$CH_3CH_2CH_2CH_2—$
ethyl	$CH_3CH_2—$	*sec*-butyl	$CH_3CH_2CHCH_3$ (with bond from CH)
propyl	$CH_3CH_2CH_2—$	isobutyl	$(CH_3)_2CHCH_2—$
isopropyl	$(CH_3)_2CH—$	*tert*-butyl	$(CH_3)_3C—$

[a]The dash shows where another atom of the molecule is attached to the alkyl group.

C_nH_{2n+1} and has one less hydrogen atom than an alkane. The name of an alkyl group, such as methyl, $CH_3—$, is derived from the name of the corresponding alkane by changing the suffix to *-yl* (Table 19.3).

When two or more alkyl groups have the same molecular formula, they must be given different names. For example, there are two groups with the formula C_3H_7:

propyl: $CH_3CH_2CH_2—$

isopropyl:
```
CH3CH—
   |
   CH3
```

Remember that alkyl groups are parts of molecules, not molecules themselves. Their names are used to derive names of compounds of which they are a part.

A few simple rules serve as the basis for IUPAC names. These are illustrated in naming the following compound:

```
           CH3
           |
CH3CHCH2CCH3
     |     |
    CH3   CH3
```

1. Identify the kind of compound and choose the appropriate suffix. We can see that this is a hydrocarbon having only single bonds (an alkane), and its suffix is *-ane*.
2. Count the carbons of the *longest continuous chain*. Combine the appropriate prefix (see Table 19.2) with the suffix. The longest continuous chain has five carbons, and the name for this chain is pent*ane*.
3. Number the longest carbon chain, giving the carbons bearing substituents the smallest combination of numbers. For this compound, we must number from right to left.

```
           CH3
 5   4  3  2| 1
CH3CHCH2CCH3
     |     |
    CH3   CH3
```

4. Arrange the substituents in the name alphabetically, using a number to indicate the position of each on the chain. Note that counting prefixes are used when

Counting prefixes (Section 9.5):

mono	*1*
di	*2*
tri	*3*
tetra	*4*
penta	*5*
etc.	

there are two or more of the same substituent. In this example, the substituents are methyl groups, three of them, so the prefix *tri-* is used. Numbers are separated by commas, and numbers and words are separated by hyphens.

name: 2,2,4-trimethylpentane

Cycloalkane: An alkane having a ring structure.

Carbon atoms can bond together to form a ring, resulting in a **cycloalkane.** These are named using the prefix *cyclo-*. Because of the ring structure, a cycloalkane has two fewer H's than other alkanes, as indicated by the general formula C_nH_{2n}.

cyclopropane: CH_2 bonded to $CH_2—CH_2$ (three-membered ring) or △

cyclobutane: $CH_2—CH_2$ / $CH_2—CH_2$ (four-membered ring) or □

The formula of a cycloalkane is commonly represented by drawing a polygon in which each corner represents a carbon atom and each side is a C—C bond. The H's on each carbon are understood and are not drawn.

An oil refinery. A catalytic cracking unit is shown at the right.

Source of Alkanes: Petroleum

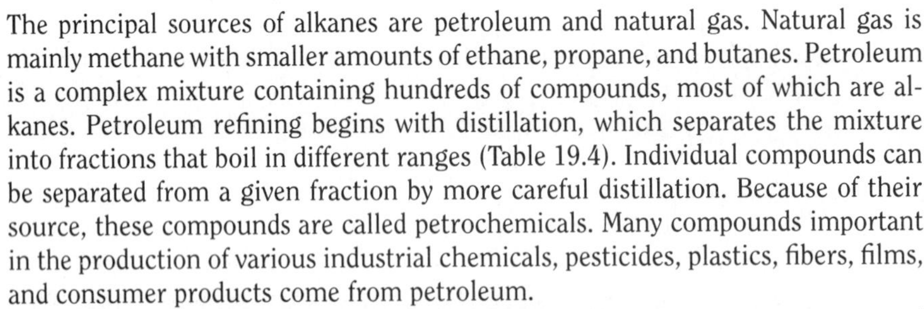

The principal sources of alkanes are petroleum and natural gas. Natural gas is mainly methane with smaller amounts of ethane, propane, and butanes. Petroleum is a complex mixture containing hundreds of compounds, most of which are alkanes. Petroleum refining begins with distillation, which separates the mixture into fractions that boil in different ranges (Table 19.4). Individual compounds can be separated from a given fraction by more careful distillation. Because of their source, these compounds are called petrochemicals. Many compounds important in the production of various industrial chemicals, pesticides, plastics, fibers, films, and consumer products come from petroleum.

Cracking: large molecules → small molecules

Reforming: unbranched molecules → branched molecules

Alkylation: small molecules → large molecules

After petroleum is separated into various fractions, three processes are used to change the structure and composition of the compounds involved. **Cracking** changes large molecules into smaller ones. **Reforming** changes unbranched molecules into branched ones. **Alkylation** uses small molecules to make larger molecules. Hydrogen, used as an alternative fuel and as a reducing agent, is a

TABLE 19.4 Typical Distillation Fractions from Petroleum Refining

Fraction	*Carbon Number*	*Boiling Range*
gasoline	C_4–C_{12}	20°–200°C
kerosene	C_{10}–C_{14}	200°–275°C
fuel oil, diesel fuel	C_{14}–C_{18}	275°–350°C
lubricating oil	C_{18}–C_{20}	higher than 350°C
residue (used for greases and asphalt)	greater than C_{20}	—

byproduct of cracking and reforming. Petroleum refining is a major source of this important gas. The three processes require the use of different catalysts.

Chemical Properties of Alkanes

More than 95% of the oil and gas produced is burned, and combustion is the most important reaction of alkanes. Carbon dioxide and water are the products of complete combustion:

$$\underset{\text{methane}}{CH_4} + 2O_2 \rightarrow CO_2 + 2H_2O$$

$$\underset{\text{propane}}{C_3H_8} + 5O_2 \rightarrow 3CO_2 + 4H_2O$$

Because their structures involve only single bonds, and because of their lack of polarity, alkanes are typically unreactive. One of the few reactions is halogenation (reaction with a halogen), which is used to make halogenated solvents, chlorofluorocarbons (CFC's), and organic intermediates of importance in the synthesis of other chemicals. These reactions are frequently initiated with ultraviolet light.

$$CH_4 + 3Cl_2 \xrightarrow{\text{UV light}} \underset{\substack{\text{trichloromethane}\\ \text{(chloroform)}}}{CHCl_3} + 3HCl$$

$$CH_3CH_3 + Br_2 \xrightarrow{\text{UV light}} \underset{\text{bromoethane}}{CH_3CH_2Br} + HBr$$

Halogenation of an alkane is an example of a **substitution** reaction, one in which one atom is replaced by another. One or more of the hydrogen atoms can be replaced by halogens, and mixtures of products may result. However, the reactions often can be controlled to obtain mostly one desired product.

Substitution: A reaction in which one atom is replaced by another.

Example 19.2

Write structural formulas for the following alkanes:

a. butane

b. 3-methylpentane

c. isobutane

d. 4-ethyl-2,2-dimethylhexane ■

Solutions

a. $CH_3CH_2CH_2CH_3$

b.
$$CH_3CH_2\underset{\substack{|\\ CH_3}}{CH}CH_2CH_3$$

c.
$$CH_3\overset{\substack{CH_3\\ |}}{C}HCH_3$$

d.
$$CH_3\overset{\substack{CH_3\\ |}}{\underset{\substack{|\\ CH_3}}{C}}CH_2\underset{\substack{|\\ CH_2CH_3}}{C}HCH_2CH_3$$

Example 19.3

Name the following compounds:

a. $CH_3CH_2CH_3$

b. 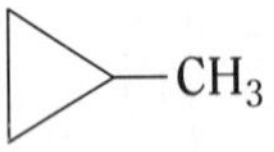 (cyclopropane ring)$-CH_3$

c. 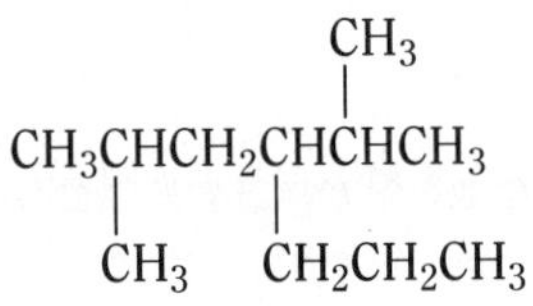
$$\begin{array}{l} \qquad\qquad\qquad\quad CH_3 \\ \qquad\qquad\qquad\quad | \\ CH_3\underset{|}{C}HCH_2\underset{|}{C}HCHCH_3 \\ \quad\;\; CH_3 \quad\; CH_2CH_2CH_3 \end{array}$$

d. 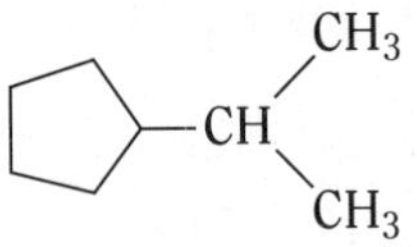 (cyclopentane ring)$-CH(CH_3)_2$ ■

Solutions

a. propane

b. methylcyclopropane

c. 4-isopropyl-2-methylheptane

d. isopropylcyclopentane

Example 19.4

Write equations for the following reactions:

a. the combustion of butane

b. the chlorination of ethane to give chloroethane ■

Solutions

a. Combustion involves reaction of the hydrocarbon with oxygen to give carbon dioxide and water. The equation can be balanced by inspection.

$$2C_4H_{10} + 13O_2 \longrightarrow 8CO_2 + 10H_2O$$

b. Halogenation of an alkane is a substitution reaction.

$$CH_3CH_3 + Cl_2 \xrightarrow{\text{UV light}} \underset{\text{chloroethane}}{CH_3CH_2Cl} + HCl$$

Practice Problem 19.2 Write structural formulas for the following compounds:

a. methylbutane

b. 3,3-dimethylpentane

c. ethylcyclobutane

d. 3-ethyl-4-methylhexane

Practice Problem 19.3 Write equations for the following reactions:

a. the preparation of bromocyclopentane from cyclopentane

b. the combustion of pentane, C_5H_{12}

Alkenes and Alkynes

19.3

Learning Objectives

- Write the molecular formula of an alkene or alkyne having a specific number of carbons.
- Name (using IUPAC and common names) and write formulas of alkenes and alkynes.
- Write equations for the preparation of ethylene, propylene, and acetylene.
- Write equations for reactions of alkenes with hydrogen, halogens, and water.
- Write equations for the formation of polymers from monomers.
- Write equations for reactions of acetylene with HCl, HCN, and halogens.

Alkenes contain C=C bonds.

Alkenes are hydrocarbons that contain a C=C bond. They are called **unsaturated** compounds because they have fewer H's than saturated compounds, as indicated by the general formula C_nH_{2n}. Ethene (ethylene) and propene (propylene) are the simplest alkenes:

ethene: $CH_2{=}CH_2$ (C_2H_4)
propene: $CH_2{=}CHCH_3$ (C_3H_6)

Alkynes contain C≡C bonds

Alkynes are unsaturated hydrocarbons that have a C≡C bond. They are represented by the general formula C_nH_{2n-2}. The most important alkyne is ethyne, commonly called acetylene. It is the fuel of the welder's torch.

ethyne (acetylene): $HC{\equiv}CH$

-ene: Suffix for alkenes.

-yne: Suffix for alkynes.

The IUPAC names of alkenes and their derivatives use the suffix *-ene*. Although the rules for naming alkanes generally apply, two things must be noted. First, we select the longest carbon chain that contains the C=C bond as the basis for the prefix. Second, we begin numbering the chain at the end closest to the C=C bond. The position of the C=C bond in the carbon chain is indicated using the smaller of the numbers for the C=C bond. The following example illustrates these rules:

```
1      2  3  4
CH2=CHCHCH3
        |
       CH3
```

3-methyl-1-but*ene*

The IUPAC names of alkynes follow the same rules except that the suffix *-yne* is used.

```
5   4  3    2 1
CH3CHC≡CCH3
    |
   CH3
```

4-methyl-2-pent*yne*

Common names are widely used for some of the smaller alkenes (Table 19.5).

Some isomeric alkenes have the same carbon structure and have the C=C bond in the same position, but they have groups that are oriented differently in

Worker using an acetylene torch.

TABLE 19.5 Common Names of Some Alkenes

Formula	*Common Name*	*IUPAC Name*
$CH_2{=}CH_2$	ethylene	ethene
$CH_2{=}CHCH_3$	propylene	propene
$CH_2{=}C(CH_3)CH_3$	isobutylene	methylpropene

Stereo: *Referring to three dimensions.*

Cis: *Two like groups on the same side of a C=C bonds.*

Trans: *Two like groups on opposite sides of a C=C bonds.*

space. These isomers are called **stereoisomers.** Note that stereoisomers differ only in the arrangement of groups in space and not in the basic carbon structure or the position of a C=C bond or substituent on a carbon chain. When naming alkenes that are stereoisomers, the prefix *cis* means that two H's or like groups are positioned on the same side of the C=C bond. Groups or H's that are positioned on opposite sides of the C=C bond are indicated by the prefix *trans.*

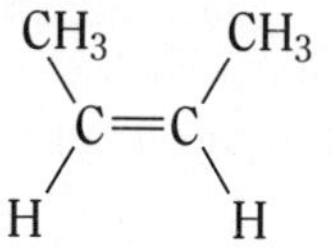

$$\begin{array}{ccc} CH_3 & & CH_3 \\ & C{=}C & \\ H & & H \end{array} \qquad \begin{array}{ccc} CH_3 & & H \\ & C{=}C & \\ H & & CH_3 \end{array}$$

cis-2-butene 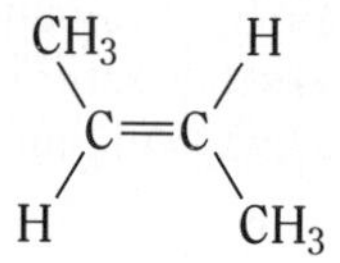   *trans*-2-butene

Remember that the bonds surrounding the carbons of a C=C bond form angles of 120° and lie in one plane (VSEPR, Section 8.7).

Example 19.5

Name the following compounds:

a. $CH_3CH(CH_3)CH{=}C(CH_3)CH_3$

b. $CH_3CH(CH_3)CH_2CH_2C{\equiv}CCH_3$

c. $$\begin{array}{ccc} CH_3 & & H \\ & C{=}C & \\ H & & CH_2CH_3 \end{array}$$ ■

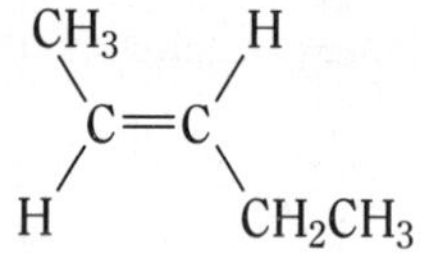

Solutions

a. 2,4-dimethyl-2-pentene
b. 6-methyl-2-heptyne
c. *trans*-2-pentene

Only small amounts of alkenes are found in petroleum. However, catalytic cracking produces a number of industrially important ones.

$$CH_3CH_3 \xrightarrow{\text{catalyst/heat}} CH_2{=}CH_2 + H_2$$

ethylene

$$CH_3CH(CH_3)CH_3 \xrightarrow{\text{catalyst/heat}} CH_2{=}C(CH_3)CH_3 + H_2$$

isobutylene

Alkynes are not found in petroleum in significant quantities and are generally obtained by synthesis. Acetylene can be prepared by the reaction of calcium carbide with water.

$$CaC_2 + 2H_2O \longrightarrow HC{\equiv}CH + Ca(OH)_2$$

Reactions of Alkenes and Alkynes

Addition reaction: A reaction in which one compound adds to another.

Alkenes typically undergo **addition** reactions, or reactions in which one compound combines with another. Note that adding the formulas of the two reactants gives the formula of the product. Addition reactions of alkenes lead to saturated compounds. Hydrogenation, or the addition of hydrogen, is usually carried out using catalysts such as platinum (Pt), palladium (Pd), or nickel (Ni). The following are examples of addition reactions:

$$CH_2{=}CH_3 + H_2 \xrightarrow{Pt} CH_3CH_3$$

$$CH_2{=}CH_2 + Br_2 \longrightarrow CH_2BrCH_2Br$$

1,2-dibromoethane
(additive in leaded gasoline)

$$CH_2{=}CH_2 + H_2O \xrightarrow{H^+} CH_3CH_2OH$$

ethanol
(grain alcohol)

The reactions of alkynes are similar except two additions are possible. The reactions can be controlled to obtain a product that has a C=C bond.

$$HC{\equiv}CH + 2Br_2 \longrightarrow CHBr_2CHBr_2$$

1,1,2,2-tetrabromoethane

$$HC{\equiv}CH + HCl \longrightarrow CH_2{=}CHCl$$

vinyl chloride

$$HC{\equiv}CH + HCN \longrightarrow CH_2{=}CHCN$$

acrylonitrile

Polymer: from poly-, *meaning "many," and* meros, *meaning "part" (Greek).*

Monomer: One part.

One of the most important reactions of alkenes is polymerization, or polymer formation. A **polymer** is a very large molecule composed of a repeating structural unit. The reactant in a polymerization reaction is called a **monomer.** These so-called addition polymers result from the adding together of many monomer units.

$$\underset{\text{ethylene}}{CH_2{=}CH_2} \xrightarrow{\text{catalyst}} \underset{\text{polyethylene}}{\sim CH_2CH_2(CH_2CH_2)_n\sim}$$

$$\underset{\text{propylene}}{CH_2{=}CHCH_3} \xrightarrow{\text{catalyst}} \underset{\text{polypropylene}}{\sim CH_2\underset{|\atop CH_3}{C}H(CH_2\underset{|\atop CH_3}{C}H)_n\sim}$$

The n in the formula of a polymer indicates that the structural unit repeats a large number of times. The ~ at the ends of the formula indicates that the carbon chain

TABLE 19.6 Some Common Polymers

Monomer	*Polymer*	*Trade Names*	*Uses*
$CH_2{=}CH_2$ ethylene	$\sim(CH_2CH_2)_n\sim$ polyethylene	Polyfilm[a]	plastic articles, such as bottles, dishpans, tubing
$CH_2{=}CH(CH_3)$ propylene	$\sim(CH_2CH(CH_3))_n\sim$ polypropylene	Herculon[b]	carpeting, upholstery fabric, artificial turf
$CH_2{=}CH(Cl)$ vinyl chloride	$\sim(CH_2CH(Cl))_n\sim$ polyvinyl chloride (PVC)	Tygon[c]	pipes and tubing, wire insulation, vinyl upholstery, shower curtains, baby bibs
$CH_2{=}CH(CN)$ acrylonitrile	$\sim(CH_2CH(CN))_n\sim$ polyacrylonitrile	Orlon[d] Acrilan[e]	fibers for clothing, carpeting
$CH_2{=}CH(C_6H_5)$ styrene	$\sim(CH_2CH(C_6H_5))_n\sim$ polystyrene	Styrofoam[a]	plastic toys, combs, brushes; foam insulation, flotation devices
$CF_2{=}CF_2$ tetrafluoroethylene (TFE)	$\sim(CF_2CF_2)_n\sim$ polytetrafluoroethylene	Teflon[d]	coatings for pans, valves, gaskets, engine wear treatment

[a]Dow Chemical Co.
[b]Hercules, Inc.
[c]U.S. Stoneware Co.
[d]E.I. du Pont de Nemours & Co.
[e]Monsanto Industrial Chemicals, Inc.

All these plastic containers and materials are made of polymers.

continues in both directions. Polymer molecules are typically very large, and molar masses often exceed 100 000 g/mol. Polymers are very important materials, and many different products are manufactured from polymers of this type (Table 19.6).

Example 19.6

Write structural formulas for the products of the following reactions:

a. $CH_2{=}CHCH_3 + H_2/Pt \longrightarrow$
b. $CH_3CH{=}CHCH_3 + Cl_2 \longrightarrow$
c. acetylene + HBr $\longrightarrow$
(controlled addition)
d. $CH_2{=}CHBr$ + polymerization catalyst $\longrightarrow$ ■

Solutions The products of addition reactions of alkenes are saturated compounds. However, controlled addition to an alkyne gives a product with a C=C bond.

a. $CH_3CH_2CH_3$

b. $CH_3CH(Cl){-}CH(Cl)CH_3$

c. $CH_2{=}CHBr$

d. $\sim(CH_2CH(Br))_n\sim$

Example 19.7

Write formulas for the following polymers:

a. polyisobutylene

b. polyvinyl chloride

c. polytetrafluoroethylene ■

Solutions The formula of a polymer can be derived from the formula of its monomer.

a. monomer: $CH_2{=}C(CH_3)_2$ polymer: $\sim(CH_2C(CH_3)_2)_n\sim$

b. monomer: $CH_2{=}CHCl$ polymer: $\sim(CH_2CH(Cl))_n\sim$

c. monomer: $CF_2{=}CF_2$ polymer: $\sim(CF_2CF_2)_n\sim$

Practice Problem 19.4 Write equations for the following reactions:

a. addition of bromine to 1-butene

b. addition of hydrogen to isobutylene

c. addition of water to 2-butene

d. addition of HCl (1:1 mole ratio) to 2-butyne

Practice Problem 19.5 Name the polymers that are prepared from the following monomers:

a. $CH_2{=}CH_2$, ethylene

b. $CH_2{=}CHCN$, acrylonitrile

c. $CH_2{=}CHC_6H_5$, styrene

CHEMICAL WORLD

Perfect Packaging

Energy shortage, acid rain, pollution, poverty—everywhere in this modern world we face problems and hard choices.

But for many people, life is good at present in our affluent nation, and convenient. Food comes prepared (pre-prepared?), our tables have fruits and vegetables in winter, fish a thousand miles from the ocean, ice cream in the desert.

All kept by energy-using refrigeration and packaged in styrofoam, cardboard, paper, plastic, and metals. All headed for the landfill, where there is any land left. A few years ago, a New York city garbage barge went on an odyssey down the east coast of America looking for a place to dump its cargo. It was turned back from New Orleans.

The perfect food package should protect its contents from moisture, from drying, and from contaminants (including bacteria, oxygen, and carbon dioxide), and it should be easy to open. When the contents are eaten, it should simply disappear. Mad dreams of a mad scientist?

Wait: The Japanese, who inhabit crowded islands where there is little room for landfills, have for years used a kind of rice paper to wrap some foods. It's not a perfect packaging material, but it's edible—it disappears with its contents.

Edible packaging. Edible plastic? Plastic is a polymer. Couldn't chemists make a polymer that we can digest? Well, they're working on it.

It would be a spray. If you have a product that has dry parts and moist parts, such as ice cream cake, or dry parts and fatty parts, such as fish and chips, you can separate them with edible plastic so that the moisture or fat doesn't migrate. Frozen pizza. It comes to you with nothing to obscure its delectable look, heat it up, and eat it—the whole thing. Fruit and cereal—treat them with edible packaging, no dry fruit, no soggy cereal. Pie with flaky crust. And it could be used at home. Dad makes his lunch, sprays the bread before applying mayonnaise: fresh sandwich hours later. Fresh fruit at home keeps for days or weeks—after it's been sliced. The shelf life of perishables is extended without additives, and some of it without refrigeration.

An edible plastic coating could keep ice cream cones crisp and crunchy.

And no wandering garbage barges. We ourselves would be the answer to some of the landfill problems: we would be our own individual waste disposal units. And we'd enjoy more healthful eating because food could be sprayed and eaten as finger food: vegetables, raw, cooked au gratin, sliced peaches, oranges, everything would be as easy to eat in its own package as grapes are.

And the quality and taste of prepared foods would improve, too, because producers would have to make their product taste better than its packaging.

19.4 Aromatic Compounds

Learning Objectives

- Name and draw formulas of benzene and other common aromatic compounds.
- Write equations for the reactions of benzene with Cl_2 or Br_2 and with HNO_3/H_2SO_4.
- Describe the octane rating scale for gasoline and note what types of compounds have high octane numbers.

An **aromatic compound** has a six-carbon ring containing three C═C bonds. Benzene is the simplest aromatic compound.

Aromatic: Containing a benzene ring.

benzene: H H H H H H (Lewis structure) or (resonance structures) or (hexagon with circle)

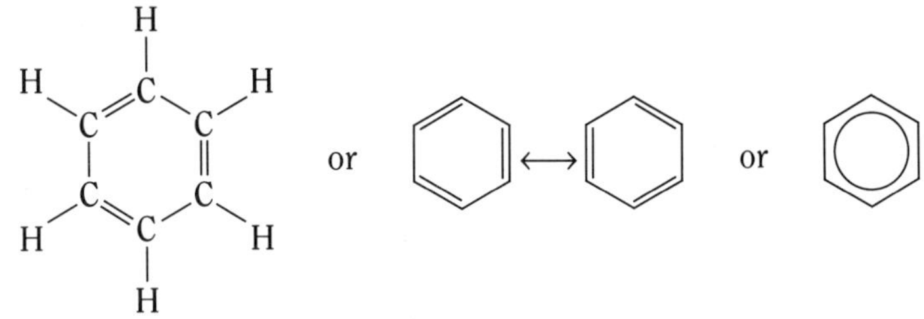

Resonance can be used when drawing the formulas of aromatic compounds. However, it is common practice to show only one of the resonance structures or to use a hexagon with a circle instead of three double bonds. Note that the circle represents six electrons spread out among the six carbons of the ring.

Resonance occurs when two or more Lewis formulas can be drawn for a substance that differ only in the position of electrons (Section 8.4).

Using the IUPAC system, we usually name aromatic compounds as derivatives of benzene. However, common names are also widely used.

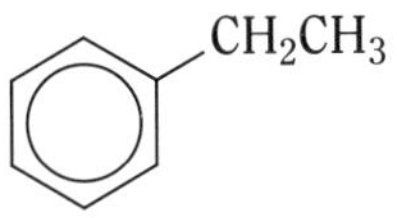

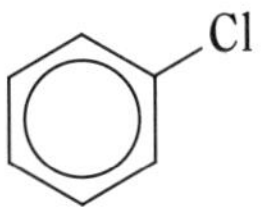

CH_2CH_3 — ethylbenzene

Cl — chlorobenzene

CH_3 — methylbenzene (toluene)

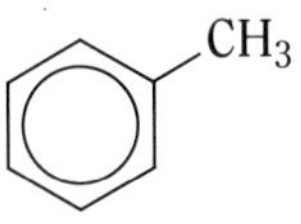

Phenyl can be used to name a benzene ring attached to a carbon chain.

Phenyl (C_6H_5):

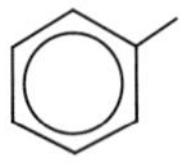

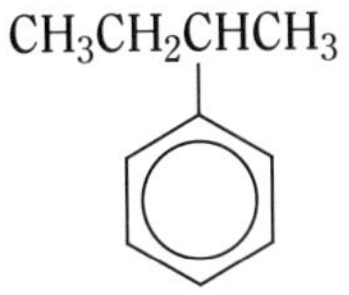

$CH_3CH_2CHCH_3$

2-phenylbutane

With an aromatic compound that has two or more substituents, the ring is numbered to indicate their positions. We begin numbering with the carbon bearing the substituent mentioned or indicated last in the name. The common names of disubstituted compounds use the prefixes *ortho-*, *meta-*, and *para-* (abbreviated *o-*, *m-*, and *p-*) instead of numbers.

ortho: *A 1,2 disubstituted benzene.*

meta: *A 1,3 disubstituted benzene.*

para: *A 1,4 disubstituted benzene.*

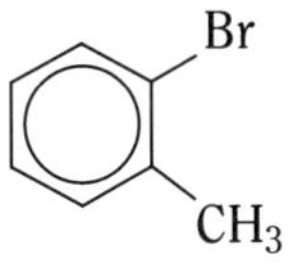

Br, CH_3 — 1-bromo-2-methylbenzene (*o*-bromotoluene)

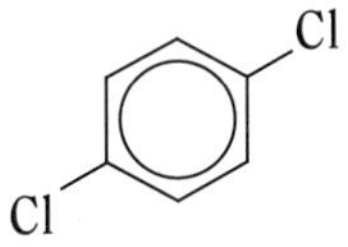

Cl, Cl — 1,4-dichlorobenzene (*p*-dichlorobenzene)

Some aromatic hydrocarbons are found in petroleum. For example, toluene (methylbenzene) is a high-octane compound in gasoline. Petroleum companies adjust the octane rating of a fuel (a measure of antiknock characteristics) by adding a compound having a high octane number. The reference compound for the octane rating scale is 2,2,4-trimethylpentane (isooctane) with an octane number of 100. Branched alkanes, as well as aromatic compounds, have high octane numbers (Table 19.7).

TABLE 19.7 Octane Numbers of Some Hydrocarbons

Name	*Formula*	*Octane Number*
heptane	$CH_3CH_2CH_2CH_2CH_2CH_2CH_3$	0
2,4-dimethylhexane	$CH_3CHCH_2CHCH_2CH_3$ CH_3 CH_3	65
2,2,4-trimethylpentane	CH_3 $CH_3CCH_2CHCH_3$ CH_3 CH_3	100
meta-xylene (1,3-dimethylbenzene)	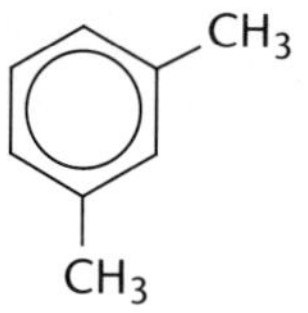 CH_3 CH_3	118
toluene	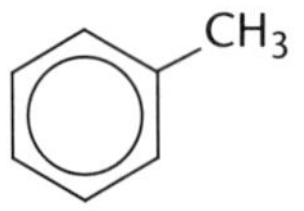 CH_3	118

Reactions of Aromatic Compounds

An aromatic ring is a very stable structure. Consequently, addition reactions, which consume unsaturation, are not usually observed with aromatic compounds. Instead, aromatic compounds undergo substitution reactions. Because of the stability of an aromatic ring, catalysts are usually required for these reactions.

Alkenes, addition reactions:

$$CH_2{=}CH_2 + Br_2 \rightarrow BrCH_2CH_2Br$$

ethylene

Aromatic compounds, substitution reactions:

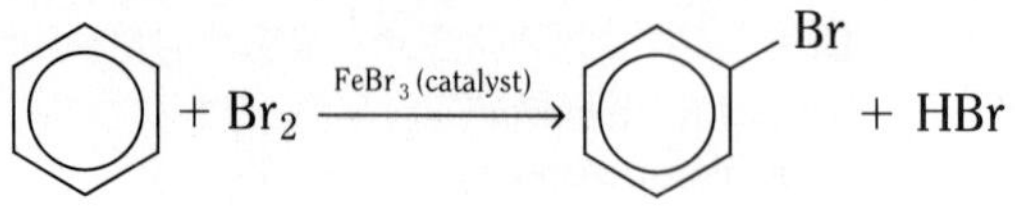

$$\text{benzene} + Br_2 \xrightarrow{FeBr_3\ \text{(catalyst)}} \text{bromobenzene (Br)} + HBr$$

benzene bromobenzene

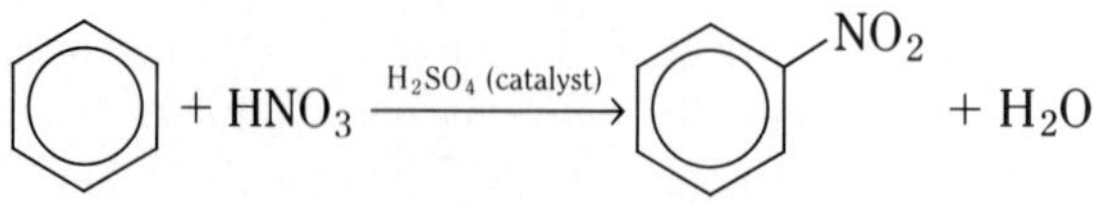

$$\text{benzene} + HNO_3 \xrightarrow{H_2SO_4\ \text{(catalyst)}} \text{nitrobenzene } (NO_2) + H_2O$$

nitrobenzene

This last reaction is called nitration and results in the introduction of a nitro group (NO_2).

Example 19.8

Write formulas for the following compounds:

a. *m*-bromonitrobenzene
b. isopropylbenzene
c. *p*-chlorotoluene ■

Solutions

a. (benzene ring with NO_2 and Br in meta positions)

b. (benzene ring with $CH(CH_3)-CH_3$ substituent)

c. (benzene ring with CH_3 and Cl in para positions)

1-chloro-4-methylbenzene

Example 19.9

Which compound has the higher octane number, octane or toluene? ■

Solution Toluene. Octane numbers of normal alkanes are very small, whereas aromatic hydrocarbons generally have high octane numbers.

Example 19.10

Write equations for the following reactions:

a. benzene + $Cl_2/FeCl_3$
b. 1,4-dimethylbenzene + HNO_3/H_2SO_4 ■

Solutions

a. $C_6H_6 + Cl_2/FeCl_3 \rightarrow C_6H_5Cl + HCl$

b. $CH_3-C_6H_4-CH_3 + HNO_3/H_2SO_4 \rightarrow CH_3-C_6H_3(NO_2)-CH_3 + H_2O$

Practice Problem 19.6 Name the following compounds:

a. (benzene ring with Cl and Cl on adjacent positions)

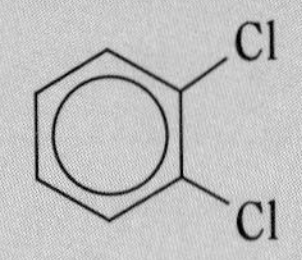

b. (benzene ring with Br and $CHCH_3$ with CH_3 substituent)

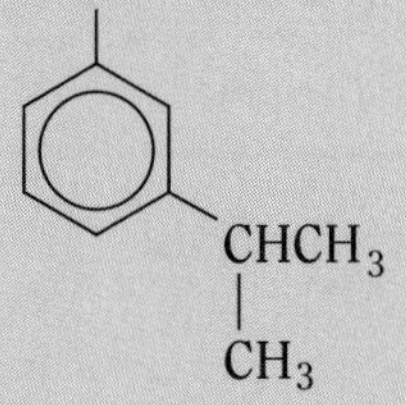

c. (benzene ring with NO_2 and CH_2CH_3 on adjacent positions)

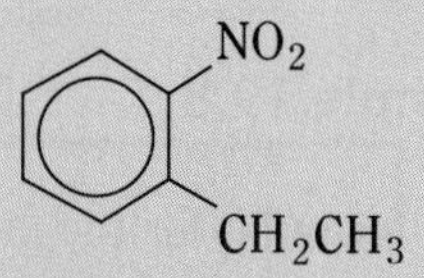

Practice Problem 19.7 Write equations for the reaction of 1,3,5-trimethylbenzene with

a. $Br_2/FeBr_3$
b. HNO_3/H_2SO_4

19.5 Functional Groups in Organic Compounds

Learning Objectives

- Give the functional group of an alkene, alkyne, alkyl halide, alcohol, ether, aldehyde, ketone, amine, carboxylic acid, ester, and amide.
- Identify the class of a compound from its formula.

The physical and chemical properties of many organic compounds are in large part due to the characteristics of one atom or a group of atoms called a **functional group.** Organic compounds are classified and named according to their functional groups. For example, a compound in which a halogen is bonded to a carbon having four single bonds is called an **alkyl halide.** An alkyl halide is sometimes represented by the general formula RX, where R represents an alkyl group (refer to Table 19.3) and X is the general symbol for a halogen, the functional group. The functional group of an alkene is C═C, and the functional group of an alkyne is C≡C. Many of the important functional groups are shown in Table 19.8.

Example 19.11

To what class of compound do each of the following belong?

a. CH_3CH_2Br
b. CH_3OH
c. $CH_3CH_2\overset{\overset{O}{\|}}{C}OH$ ■

TABLE 19.8 Important Functional Groups in Organic Compounds

Class of Compound	*Functional Group*	*Example*
alkene	$>C=C<$	$CH_2{=}CHCH_3$
alkyne	$-C{\equiv}C-$	$HC{\equiv}CH$
alkyl halide	$-X$	CH_3CH_2Cl
alcohol	$-OH$	CH_3CH_2OH
ether	$-\overset{\vert}{\underset{\vert}{C}}-O-\overset{\vert}{\underset{\vert}{C}}-$	$CH_3CH_2OCH_2CH_3$
aldehyde	$-\overset{O}{\overset{\Vert}{C}}-H$	$CH_3\overset{O}{\overset{\Vert}{C}}H$
ketone	$-\overset{O}{\overset{\Vert}{C}}-$	$CH_3\overset{O}{\overset{\Vert}{C}}CH_3$
amine	$-NH_2$	$CH_3CH_2NH_2$
carboxylic acid	$-\overset{O}{\overset{\Vert}{C}}-OH$	$CH_3\overset{O}{\overset{\Vert}{C}}OH$
ester	$-\overset{O}{\overset{\Vert}{C}}-O-$	$CH_3\overset{O}{\overset{\Vert}{C}}OCH_2CH_3$
amide	$-\overset{O}{\overset{\Vert}{C}}-NH_2$	$CH_3\overset{O}{\overset{\Vert}{C}}NH_2$

Solutions

a. alkyl halide
b. alcohol
c. carboxylic acid

Practice Problem 19.8 Write formulas for the following:
a. an ether that has two carbon atoms
b. a ketone that has three carbon atoms
c. an ester that has three carbon atoms

19.6 Alcohols and Ethers

Learning Objectives

- Name and write formulas of alcohols and ethers.
- Write equations for the preparation of ethanol, 2-propanol, and diethyl ether.

An alcohol: R—OH.

-ol: Suffix for alcohols.

An **alcohol** is a compound that has an OH group bonded to a carbon that has four single bonds. The IUPAC names of alcohols have the suffix *-ol.* When numbering the carbon chain of an alcohol, we begin at the end closest to the OH group. The position of the OH group on the carbon chain is indicated by a number in the name. Note, however, that a number is necessary only when there are two possible positions for the OH.

CH_3OH	CH_3CH_2OH	$CH_3CH_2CH_2OH$
methanol (methyl alcohol)	ethanol (ethyl alcohol)	1-propanol (propyl alcohol)

$$CH_3\underset{OH}{\underset{|}{C}}HCH_2\underset{CH_2CH_3}{\underset{|}{C}}HCH_2CH_3$$

4-ethyl-2-hexanol

Rubbing alcohol is an aqueous solution of 2-propanol.

Common names continue to be used for the smaller alcohols. The name of the alkyl group is followed by *alcohol* as shown in the preceding examples.

Ethanol, or grain alcohol, found in alcoholic beverages, is the best known alcohol. It also is an ingredient of gasohol, which is typically 10% ethanol and 90% gasoline. It is a common solvent for flavorings, perfumes, and medicines. Methanol, known as wood alcohol, is a synthetic fuel that is being used experimentally in some buses and fleet vehicles. It is also used as an antifreeze in windshield washer fluid. Methanol is highly toxic, and ingestion of as little as one tablespoon can be fatal. Rubbing alcohol is a 70% solution of 2-propanol (isopropyl alcohol).

Ethanol can be prepared by fermentation of grains or by hydration of ethylene using an acid catalyst.

$$CH_2{=}CH_2 + H_2O \xrightarrow{H^+} CH_3CH_2OH$$

The same method is used to prepare 2-propanol.

$$CH_2{=}CHCH_3 + H_2O \xrightarrow{H^+} CH_3\underset{OH}{\underset{|}{C}}HCH_3$$

A glycol is a 1,2-diol.

Some alcohols have two or more OH groups. Ethylene glycol (1,2-ethanediol) is the common automobile antifreeze. It is also used to make polyesters (Section 19.9) and polyurethane foam. Glycerol, or glycerin (1,2,3-propanetriol), is a lubricant used in laboratories and for medicinal purposes.

$HOCH_2CH_2OH$	$HOCH_2\overset{OH}{\overset{|}{C}}HCH_2OH$
ethylene glycol (1,2-ethanediol)	glycerol (1,2,3-propanetriol)

Alcohols are frequently used in the preparation of other organic compounds. Some of these reactions will be described later in this chapter.

Automobile antifreeze is ethylene glycol.

Example 19.12

Name the following alcohols:

a. $CH_3CH_2\underset{\displaystyle OH}{\underset{|}{C}}HCH_3$

b. $CH_3\underset{\displaystyle OH}{\underset{|}{C}}HCH_2CH_2\underset{\displaystyle CH_2CH_3}{\underset{|}{C}}HCH_3$ ■

Solutions

a. 2-butanol
b. 5-methyl-2-heptanol

Example 19.13

Write an equation for the preparation of 2-butanol from 2-butene. ■

Solution Alcohols can be made by hydration of alkenes using an acid catalyst such as sulfuric acid.

$$CH_3CH{=}CHCH_3 + H_2O \xrightarrow{H_2SO_4} CH_3\underset{\displaystyle OH}{\underset{|}{C}}HCH_2CH_3$$

■ Ethers

Ether: R—O—R, Ar—O—R, Ar—O—Ar, where Ar represents an aromatic (aryl) group.

An **ether** has two alkyl groups bonded to an oxygen, although in some ethers an alkyl group is replaced by an aromatic ring. The most common ether is diethyl ether, often simply called ether. It once was widely used as an anesthetic, but it is highly flammable and tends to cause nausea during a patient's period of recovery. Although no longer used for medical purposes, ether continues to be used as a solvent in laboratory experiments. Ethers are generally named by naming the two alkyl groups followed by *ether:*

$CH_3CH_2OCH_2CH_3$ — diethyl ether

$CH_3OCH_2CH_2CH_3$ — methyl propyl ether

Diethyl ether is prepared by the dehydration of ethanol using an acid catalyst:

$$2CH_3CH_2OH \xrightarrow{H_2SO_4} CH_3CH_2OCH_2CH_3 + H_2O$$

Example 19.14

Write structural formulas for the following:

a. dimethyl ether
b. ethyl isopropyl ether ■

Solutions

a. CH_3OCH_3
b. $CH_3CH_2O\underset{\underset{\displaystyle CH_3}{|}}{C}HCH_3$

Practice Problem 19.9 Write structural formulas for the following compounds:

a. methyl-2-propanol
b. diisopropyl ether
c. 3-methyl-2-butanol
d. 1,2-propanediol (propylene glycol)

Practice Problem 19.10 Write structural formulas for the missing compounds:

a. $\underline{\quad ? \quad} \xrightarrow{H_2SO_4} CH_3CH_2OCH_2CH_3 + H_2O$
b. $CH_2{=}CHCH_3 + \underline{\quad ? \quad} \xrightarrow{H^+} CH_3\underset{\underset{\displaystyle OH}{|}}{C}HCH_3$

19.7 Aldehydes and Ketones

Learning Objectives

- Name and write structural formulas for aldehydes, ketones, and their acetals and ketals.
- Write equations for the preparation of aldehydes and ketones from alcohols.
- Write equations for the oxidation and reduction of aldehydes, the reduction of ketones, and the preparation of an acetal.

Carbonyl is pronounced "car-bon-eel."

Aldehyde: $R{-}\overset{\overset{\displaystyle O}{\|}}{C}{-}H$

Ketone: $R{-}\overset{\overset{\displaystyle O}{\|}}{C}{-}R$

-al: Suffix for aldehydes.

Aldehydes and ketones are both carbonyl compounds. Carbonyl compounds contain the C=O, or **carbonyl group.** An **aldehyde** has a hydrogen bonded to the carbonyl group, whereas a **ketone** has two alkyl (or aryl) groups.

The IUPAC names of aldehydes use the suffix *-al.* When numbering the carbon chain of an aldehyde, we always start with the C=O. Because this is always the first carbon, it is not necessary to include its number in the name.

$$\mathrm{CH_3CHCH_2CHCH_2CH}$$
(with $=$O on the terminal C, $\mathrm{CH_3}$ on C-5 and $\mathrm{CH_2CH_3}$ on C-3)

3-ethyl-5-methylhexanal

Common names of aldehydes are derived from the names of the related carboxylic acids (refer to Table 19.9, Section 19.9). The word "acid" is dropped and the *-ic* suffix is changed to *-aldehyde.*

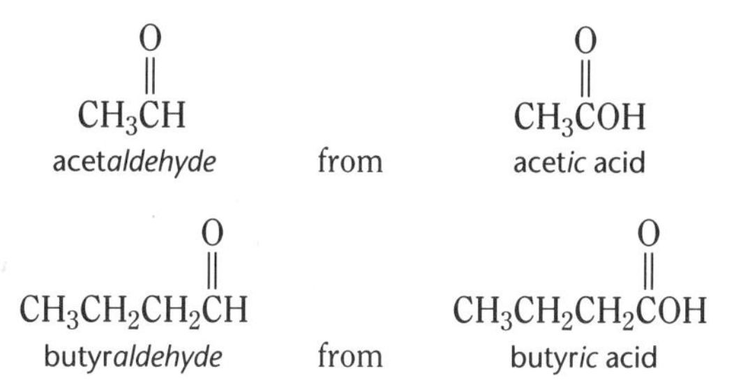

-one: suffix for ketones

For ketones, the suffix is *-one* in both IUPAC and common names. Some common names are derived by naming the groups bonded to the C═O, but others must be memorized. In the IUPAC names, the chain is numbered starting with the end closest to the C═O.

$\mathrm{CH_3\overset{O}{\overset{\|\|}{C}}CH_3}$	$\mathrm{CH_3\overset{O}{\overset{\|\|}{C}}\underset{CH_3}{\underset{\|}{C}H}CH_3}$
propanone (acetone)	3-methylbutanone (isopropyl methyl ketone)

Aldehydes and ketones are often prepared by oxidizing alcohols. A variety of oxidizing agents are used, including CrO_3 and $KMnO_4$. In preparing an aldehyde, the oxidation must be carefully controlled to prevent further oxidation to a carboxylic acid.

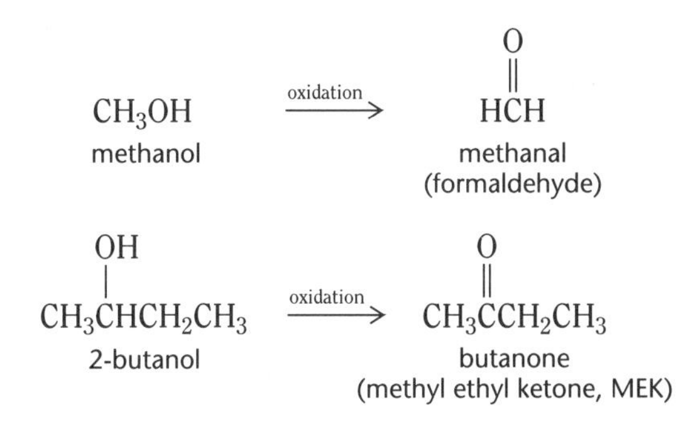

These examples illustrate the common practice of not balancing equations for organic reactions. Because of the complexity of many organic reactions, equations frequently show only the organic reactants and products and the reagent used to carry out the reaction.

A frog preserved in formalin, an aqueous solution of formaldehyde.

Vanilla flavoring is a solution of vanillin, isolated from vanilla beans.

The simplest aldehyde, formaldehyde, is used in aqueous solution (called formalin) as a tissue preservative by biologists and as an embalming fluid by morticians. Large quantities of formaldehyde are consumed in the manufacture of insulating plastics (Bakelite®), plastic dishes, thermal insulation, and plywood adhesive. A number of aldehydes and ketones are common spices and flavorings.

$C_6H_5CH{=}CHCHO$ (structure: $-CH{=}CH\overset{O}{\overset{\|}{C}}H$ on a benzene ring)
cinnamaldehyde
(oil of cinnamon)

CH_3O, HO, $\overset{O}{\overset{\|}{C}}H$ on a benzene ring
vanillin
(vanilla flavor)

CH_3, O on a cyclohexane ring; $C(CH_3){=}CH_2$
carvone
(spearmint)

Acetal: A geminal diether, $RCH(OR')_2$ (geminal, from "gemini," meaning twin).

Aldehydes and ketones readily undergo addition reactions, generally using an acid catalyst. One important example is the addition of an alcohol to give an **acetal or ketal,** geminal diethers related to aldehydes and ketones. A hemiacetal is an intermediate formed in this reaction. These structures will be important in our discussion of carbohydrates (Section 20.2). Hemiacetals usually react further to give acetals.

$$C_6H_5\overset{O}{\overset{\|}{C}}H \xrightarrow{CH_3CH_2OH/H^+} C_6H_5\overset{OH}{\overset{|}{C}}HOCH_2CH_3 \xrightarrow{CH_3CH_2OH/H^+} C_6H_5CH(OCH_2CH_3)_2$$

benzaldehyde — a hemiacetal — benzaldehyde diethyl acetal

Aldehydes are readily oxidized, a chemical property that distinguishes them from ketones. Almost any oxidizing agent, even air (O_2), will oxidize an aldehyde to a carboxylic acid. Consequently, aldehydes are often contaminated with carboxylic acids that form during storage.

$$C_6H_5\overset{O}{\overset{\|}{C}}H + CrO_3 \rightarrow C_6H_5\overset{O}{\overset{\|}{C}}OH + Cr^{3+}$$

benzaldehyde — benzoic acid

Both aldehydes and ketones are readily reduced to alcohols. Sodium borohydride ($NaBH_4$) is a good reducing agent for this reaction.

$$CH_3\overset{O}{\overset{\|}{C}}CH_3 + NaBH_4 \rightarrow CH_3\overset{OH}{\overset{|}{C}}HCH_3$$

Example 19.15

Write formulas for the following compounds:

a. 3-methylpentanal
b. 3-pentanone
c. 3-ethyl-2-pentanone
d. phenylpropanone ■

Solutions

a. $CH_3CH_2\underset{\displaystyle CH_3}{\underset{|}{C}}HCH_2\overset{\displaystyle O}{\overset{\|}{C}}H$

b. $CH_3CH_2\overset{\displaystyle O}{\overset{\|}{C}}CH_2CH_3$

c. $CH_3CH_2\underset{\displaystyle CH_3CH_2}{\underset{|}{C}}H\overset{\displaystyle O}{\overset{\|}{C}}CH_3$

d. $C_6H_5CH_2\overset{\displaystyle O}{\overset{\|}{C}}CH_3$

Example 19.16

Write formulas for the missing compounds:

a.

$$\underline{\quad ? \quad} + NaBH_4 \longrightarrow CH_3\overset{\displaystyle OH}{\overset{|}{C}}HCH_3$$

b.

$$CH_3\overset{\displaystyle OH}{\overset{|}{C}}HCH_2CH_3 \xrightarrow{\text{oxidation}} \underline{\quad ? \quad}$$

c.

$$C_6H_5CH_2\overset{\displaystyle O}{\overset{\|}{C}}H + \underline{\quad ? \quad} \xrightarrow{H^+} C_6H_5CH_2CH(OCH_3)_2$$ ■

Solutions

a. $CH_3\overset{\displaystyle O}{\overset{\|}{C}}CH_3$

b. $CH_3\overset{\displaystyle O}{\overset{\|}{C}}CH_2CH_3$

c. CH_3OH. An acetal is formed by the reaction of an alcohol with an aldehyde using an acid catalyst.

Practice Problem 19.11 Write structural formulas for the following compounds:
a. benzaldehyde
b. propanal
c. 3-ethyl-2-methyl-4-heptanone
d. 2-methylpropanal

Practice Problem 19.12 Write equations for the following reactions:
a. the reduction of butanal with $NaBH_4$
b. the oxidation of 1-phenylethanol

19.8 Amines

Learning Objectives

- Name and write formulas for amines.
- Write equations for the preparation of amines from alkyl halides, ammonia, and other amines.
- Recognize a formula as that of a heterocyclic amine.
- Write equations for the reactions of amines with acids.

Amines: RNH_2, R_2NH, and R_3N.

An **amine** is a compound with one, two, or three alkyl or aryl groups bonded to nitrogen. The simplest amines contain an NH_2 or **amino group.**

$CH_3CH_2NH_2$ — ethylamine

$CH_3CH(CH_3)NHCH_3$ — isopropylmethylamine

$CH_3N(CH_3)CH_3$ — trimethylamine

Amines are generally named by naming the groups bonded to nitrogen followed by *-amine.* Many amines are known by their common names.

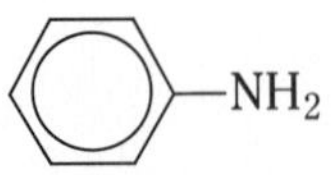

$C_6H_5-NH_2$ — aniline

$C_6H_5-CH_2CH(NH_2)CH_3$ — amphetamine (benzedrine)

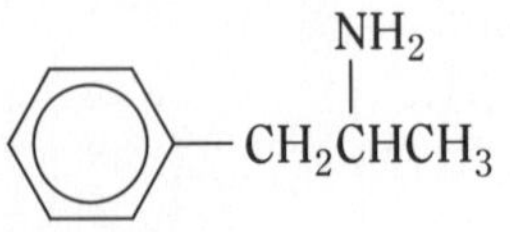

Amines usually have strong odors. The odors of some amines resemble the odor of raw fish, which is known to contain volatile amines. The odors of a few amines are particularly foul, as indicated by the common names of two examples:

$NH_2CH_2CH_2CH_2CH_2NH_2$
putrescine
(1,4-diaminobutane)

$NH_2CH_2CH_2CH_2CH_2CH_2NH_2$
cadaverine
(1,5-diaminopentane)

These compounds are end products of decaying flesh.

A **heterocyclic** amine has one or more nitrogen atoms forming a ring with carbon atoms. Heterocyclic amine structures are common in biologically active compounds and are found in some drugs.

Heterocyclic: "Hetero" means containing more than one kind of atom; "cyclic" means a ring.

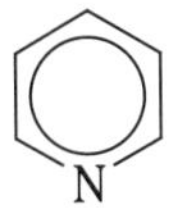

pyridine 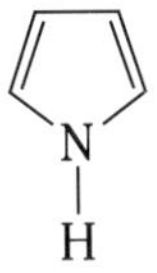   pyrrole 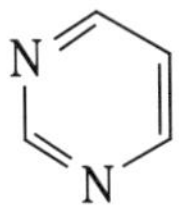   pyrimidine 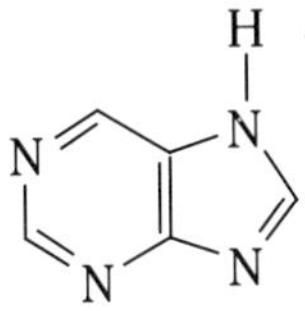   purine

Simple alkyl amines are prepared by the reaction of ammonia or an amine with an alkyl halide. The alkyl group replaces one of the hydrogens of ammonia to form the amine. The hydrogen chloride that forms is neutralized by excess ammonia:

$$\underset{\text{chloroethane}}{CH_3CH_2Cl} + \underset{\text{ammonia}}{NH_3} \longrightarrow \underset{\text{ethylamine}}{CH_3CH_2NH_2} + HCl$$

$$NH_3 + HCl \longrightarrow NH_4Cl$$

Amines are basic, and their most characteristic reaction is with an acid to form a salt.

$$CH_3NH_2 + HCl \longrightarrow \underset{\text{methylammonium chloride}}{CH_3NH_3^+Cl^- \text{ or } CH_3NH_3Cl}$$

$$CH_3CH_2NH_2 + CH_3CO_2H \longrightarrow \underset{\text{ethylammonium acetate}}{CH_3CH_2NH_3^+ + CH_3CO_2^-}$$

Amines are used in the manufacture of insecticides, dyes, and drugs. Some amines, called **alkaloids** (Figure 19.2), are isolated from plants such as tobacco (nicotine), coffee beans (caffeine), opium (morphine), and coca (cocaine).

Alkaloids: Naturally occurring amines found in plants (from "alkaline," meaning "basic").

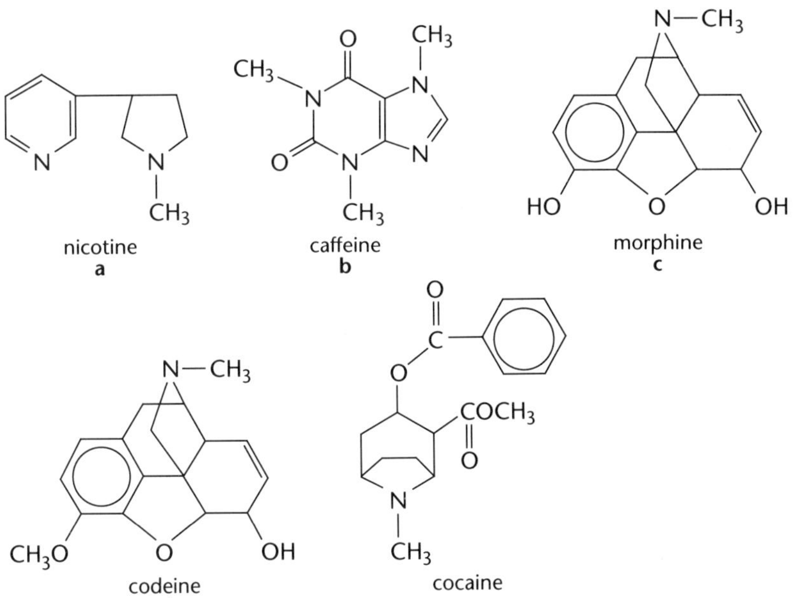

nicotine a   caffeine b   morphine c   codeine d   cocaine e

FIGURE 19.2 Some common alkaloids. (a) Nicotine, found in tobacco leaves, is a highly addictive alkaloid. (b) Sources of caffeine include coffee, tea, cocoa, and chocolate, and many soft drinks. Caffeine is an effective stimulant and is used in NoDoz® pills. (c) Morphine, an opium alkaloid, is one of the most effective pain killers. (d) Codeine, used in cough syrups, is closely related in structure. (e) Cocaine, found in the leaves of the coca bush common in Peru and Bolivia, is a powerful stimulant and widely abused drug.

Example 19.17

Name the following compounds:

a. $(CH_3CH_2)_2NH$

b. $C_6H_5{-}NH_2$ (benzene ring with NH_2) ■

Solutions

a. diethylamine

b. aniline

Example 19.18

Write equations for the following reactions:

a. preparation of diethylamine from ethylamine

b. reaction of diethylamine with hydrobromic acid ■

Solutions

a. $CH_3CH_2NH_2 + CH_3CH_2Cl \rightarrow (CH_3CH_2)_2NH + HCl$

b. $(CH_3CH_2)_2NH + HBr \rightarrow (CH_3CH_2)_2NH_2{}^+Br^-$

Practice Problem 19.13 Write formulas for the missing compounds:

a. aniline + ___?___ $\rightarrow$ $C_6H_5{-}NH_3{}^+ + CH_3\overset{\overset{\large O}{\|}}{C}O^-$

b. $CH_3\underset{\underset{\large CH_3}{|}}{C}HCH_2NH_2 + CH_3Br \rightarrow$ ___?___ + HBr

19.9 Carboxylic Acids, Esters, and Amides

Learning Objectives

- Name and write formulas for carboxylic acids, esters, and amides.
- Write equations for the preparation of carboxylic acids, esters, and amides.
- Write equations for the neutralization of carboxylic acids and for the reaction of an ester with a base.
- Recognize and write formulas for triglycerides, nitrate and phosphate esters, polyesters, and polyamides (nylon).

TABLE 19.9 Some Carboxylic Acids, Esters, and Amides

Acids	*Esters*	*Amides*
HCO_2H formic acid (methanoic acid)	HCO_2CH_3 methyl formate (methyl methanoate)	$HCONH_2$ formamide (methanamide)
CH_3CO_2H acetic acid (ethanoic acid)	$CH_3CO_2CH_3$ methyl acetate (methyl ethanoate)	CH_3CONH_2 acetamide (ethanamide)
$CH_3CH_2CO_2H$ propionic acid (propanoic acid)	$CH_3CH_2CO_2CH_3$ methyl propionate (methyl propanoate)	$CH_3CH_2CONH_2$ propionamide (propanamide)
$HO_2C(CH_2)_4CO_2H$ adipic acid (hexanedioic acid)	$CH_3O_2C(CH_2)_4CO_2CH_3$ methyl adipate (methyl hexanedioate)	$NH_2CO(CH_2)_4CONH_2$ adipamide (hexanediamide)
$C_6H_5CO_2H$ benzoic acid	$C_6H_5CO_2CH_3$ methyl benzoate	$C_6H_5CONH_2$ benzamide

A **carboxylic acid** is a compound that contains a **carboxyl group,** an OH group coupled with a C=O group. The most common carboxylic acids are formic acid, found in ant bites and insect stings, and acetic acid, the acid in vinegar.

Carboxylic acids: $R\overset{O}{\overset{\|}{C}}OH$

Esters: $R\overset{O}{\overset{\|}{C}}OR'$

Amides: $R\overset{O}{\overset{\|}{C}}NH_2$

$$H\overset{O}{\overset{\|}{C}}OH \text{ or } HCO_2H \quad \text{(formic acid)} \qquad CH_3\overset{O}{\overset{\|}{C}}OH \text{ or } CH_3CO_2H \quad \text{(acetic acid)}$$

The IUPAC names of acids use the *-oic* suffix and the word *acid.* The carbon chain is numbered beginning with the carboxyl carbon. Common names of acids are also widely used (Table 19.9).

Carboxylic acids can be made by oxidation of alcohols. Thus, oxidation of ethanol gives acetic acid.

$$CH_3CH_2OH + CrO_3/H^+ \longrightarrow CH_3CO_2H + Cr^{3+}$$

During fermentation of juices, the ethanol that is formed can be oxidized further, resulting in the formation of vinegar. The sour taste of vinegar is due to acetic acid, which is present in a concentration of about 5%.

$$C_6H_{12}O_6 \xrightarrow{\text{fermentation}} CH_3CH_2OH \xrightarrow{\text{further oxidation}} CH_3CO_2H$$

Carboxylic acids typically have acid ionization constants (K_a) of about 1×10^{-5} and are weaker acids than strong inorganic acids such as hydrochloric, nitric, and sulfuric acids. Neutralization produces a salt and water.

$$C_6H_5CO_2H \text{ (benzoic acid)} + NaOH \longrightarrow C_6H_5CO_2^-Na^+ \text{ (sodium benzoate)} + H_2O$$

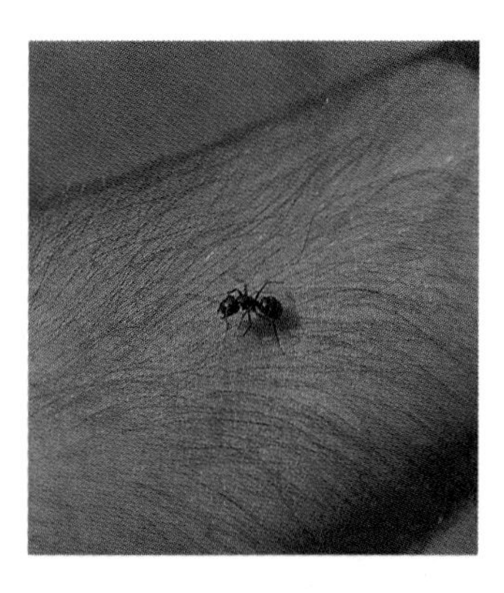

The stinging pain of an ant bite is caused by formic acid.

Name of an ester: Isopropyl acetate

$$CH_3\overset{\overset{\displaystyle O}{\|}}{C}O\overset{\overset{\displaystyle CH_3}{|}}{C}HCH_3$$

Name of an amide: Acetamide

$$CH_3\overset{\overset{\displaystyle O}{\|}}{C}NH_2$$

N-*methylbenz*amide

$$C_6H_5-\overset{\overset{\displaystyle O}{\|}}{C}-NHCH_3$$

Esters and amides are derivatives of carboxylic acids. The formula of an **ester** has an alkyl or aryl group in place of the H of a carboxyl group. An **amide** has NH_2 in place of the OH of a carboxylic acid. Some amides have one or both of the NH's replaced by alkyl or aryl groups. In naming an ester, the group bonded to oxygen is named first, and the *-ate* suffix is used with the root of the name of the related acid. The names of amides generally use the *-amide* suffix with the root of the name of the related acid. With nitrogen substituted amides, the position of the substituent is indicated in the name with $N-$.

Esters can be prepared by the reaction of a carboxylic acid with an alcohol using an acid catalyst. The reaction, called esterification, is reversible, but good yields of the ester can be obtained by removing water from the reaction (Le Chatelier's principle, Section 16.6).

$$CH_3\overset{\overset{\displaystyle O}{\|}}{C}OH + HOCH_2CH_3 \overset{H^+}{\rightleftarrows} CH_3\overset{\overset{\displaystyle O}{\|}}{C}OCH_2CH_3 + H_2O$$

Amides can be prepared in similar reactions from carboxylic acids, esters, or other acid derivatives.

$$CH_3\overset{\overset{\displaystyle O}{\|}}{C}OCH_2CH_3 + NH_3 \rightarrow CH_3\overset{\overset{\displaystyle O}{\|}}{C}NH_2 + CH_3CH_2OH$$

Nitroglycerin: CH_2-ONO_2, $CH-ONO_2$, CH_2-ONO_2 (each bonded in a chain)

Phosphate ester: $RO-\overset{\overset{\displaystyle O}{\|}}{\underset{\underset{\displaystyle OR}{|}}{P}}-OR$

Triglycerides: RCO_2CH_2, RCO_2CH, RCO_2CH_2 (each bonded in a chain)

where R is a chain of 11 to 19 carbons.

Esters, which usually have sweet odors, give fruits their characteristic flavors (Table 19.10). Some esters are related to inorganic oxyacids rather than carboxylic acids. Nitroglycerin, used in making dynamite, is an example. It is also used as a vasodilator to increase blood flow to the heart in the treatment of angina pectoris. Phosphate esters are used as insecticides, and a phosphate ester linkage is part of the structure of RNA and DNA (Section 20.4).

Animal fats and vegetable oils are esters of glycerol and long-chain carboxylic acids and are called **triglycerides.** The fats, which are semisolid, have saturated carbon chains, whereas the carbon chains of oils generally contain C=C bonds.

Soaps are prepared by the reaction of triglycerides with an aqueous base such as sodium hydroxide. Past generations made soap at home from lard and lye (NaOH) leached from wood ashes.

$$\begin{array}{l} CH_3(CH_2)_{16}\overset{\overset{\displaystyle O}{\|}}{C}OCH_2 \\ CH_3(CH_2)_{16}\overset{\overset{\displaystyle O}{\|}}{C}OCH \\ CH_3(CH_2)_{16}\overset{\overset{\displaystyle O}{\|}}{C}OCH_2 \end{array} + 3NaOH \longrightarrow 3CH_3(CH_2)_{16}CO_2Na + HOCH_2\overset{\overset{\displaystyle OH}{|}}{C}HCH_2OH$$

stearin (from lard) — sodium stearate (a soap)

Nitroglycerin is the explosive used to make dynamite.

Polyesters and polyamides, important synthetic fibers and plastics, are prepared from dicarboxylic acids or their derivatives and diols and diamines. These are sometimes called **condensation polymers** because a small molecule, usually water,

TABLE 19.10 Some Common Esters (Fruit Flavors)

Ester	Flavor
$CH_3C(=O)OCH_2CH_2CH(CH_3)CH_3$ isopentyl acetate	banana
$CH_3C(=O)OCH_2(CH_2)_6CH_3$ octyl acetate	orange
$CH_3CH_2CH_2C(=O)OCH_2CH_3$ ethyl butyrate	pineapple
$CH_3CH_2CH_2C(=O)OCH_3$ methyl butyrate	apple
$HC(=O)OCH_2CH(CH_3)CH_3$ isobutyl formate	raspberry
$C_6H_4(CO_2CH_3)(NH_2)$ methyl anthranilate	grape

The flavor of bananas is due to isopentyl acetate.

is lost as the monomer molecules are joined together. The most important polyester is polyethylene terephthalate, widely known by the trade name Dacron®.

$$\underset{\text{terephthalic acid}}{HOC(=O)-C_6H_4-C(=O)OH} \quad \underset{\text{ethylene glycol}}{HOCH_2CH_2OH} \xrightarrow{\text{acid catalyst}}$$

$$\underset{\text{Dacron® (a polyester)}}{(-C(=O)-C_6H_4-C(=O)OCH_2CH_2O-)_n}$$

The repeating structural units of the polymers are joined by ester and amide linkages. The amide or peptide linkage joins the amino acid units of proteins, which can be viewed as naturally occurring polyamides (Section 20.1).

$$\underset{\text{nylon-6,6 (a polyamide)}}{(-NHCH_2CH_2CH_2CH_2CH_2CH_2NHC(=O)CH_2CH_2CH_2CH_2C(=O)-)_n}$$

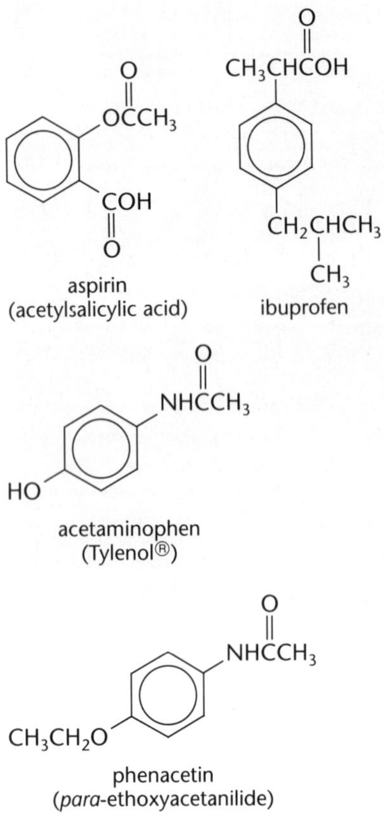

O
||
CH_3CHCOH

O
||
$OCCH_3$

COH
||
O

CH_2CHCH_3
|
CH_3

aspirin
(acetylsalicylic acid)

ibuprofen

O
||
$NHCCH_3$

HO

acetaminophen
(Tylenol®)

O
||
$NHCCH_3$

CH_3CH_2O

phenacetin
(*para*-ethoxyacetanilide)

FIGURE 19.3 Some common pain relievers.

A number of important pharmaceuticals are carboxylic acids, esters, or amides. Aspirin, or acetylsalicylic acid, is both an acid and an ester. Other common pain relievers include acetaminophen, phenacetin, and ibuprofen (Figure 19.3).

Example 19.19

Name the following compounds, using either common or IUPAC names:

a. HCO_2H

b. $HCNH_2$ (with C=O)

c. $CH_3CH_2CO_2CH_2CH_3$

d. $CH_3CH_2OCCH_3$ (with C=O) ■

Solutions

a. formic acid (methanoic acid)
b. formamide (methanamide)
c. ethyl propionate (ethyl propanoate)
d. ethyl acetate (ethyl ethanoate)

Example 19.20

Write the formula of the major product of each reaction:

a. $CH_3CH_2CH_2OH + CrO_3/H^+ \longrightarrow$
b. acetic acid + NaOH $\longrightarrow$ ___?___ + H_2O
c. acetic acid + isopropyl alcohol/H^+ $\longrightarrow H_2O$ + ___?___ ■

Solutions

a. Oxidation of an alcohol produces a carboxylic acid: $CH_3CH_2CO_2H$.
b. Neutralization of an acid produces a salt and water: $CH_3CO_2{}^-Na^+$.
c. An ester is formed: $CH_3CHOCCH_3$ (with CH_3 on CH and C=O) (isopropyl acetate.)

Practice Problem 19.14 Write formulas for the following compounds:

a. butanoic acid
b. benzamide
c. butyl formate
d. methyl benzoate

Practice Problem 19.15 Write formulas for the missing compounds:

a. octanoic acid + KOH $\longrightarrow$ ___?___ + H_2O
b. octyl acetate + NaOH $\longrightarrow$ sodium acetate + ___?___

Chapter Review

Directions: From the list of choices at the end of this Chapter Review, choose the word or term that best fits in each blank. Use each choice only once. Answers appear at the end of the chapter.

Section 19.1 The only element that bonds with itself to form chains of atoms, branched structures, and rings is __________ (1). Organic compounds are generally __________ (2), and they typically are low-melting solids, liquids, or __________ (3) like other molecular compounds. In agreement with its position in Group 4A in the periodic table, carbon generally forms __________ (4) covalent bonds. The bonding is a combination of single, double, and __________ (5) bonds. The bond angles are a result of the number of sets of __________ (6) surrounding a given carbon atom. For example, a carbon atom having four single bonds has bond angles of about __________ (7), resulting in a __________ (8) shape. Compounds that have the same molecular formula but different structural formulas are called __________ (9).

Section 19.2 Hydrocarbons that contain only single bonds are called __________ (10), also called __________ (11) hydrocarbons. The general formula of an alkane is __________ (12). Alkanes are named using the suffix __________ (13). Those that have a continuous chain of carbons are called __________ (14) alkanes. In addition to their IUPAC names, some small branched alkanes have common names.

In IUPAC names, the branches are named as __________ (15) groups. These groups have one less hydrogen than an alkane and have the general formula __________ (16). There are two alkyl groups with the formula C_3H_7, propyl and __________ (17). The prefix in an IUPAC name indicates the number of carbons in the longest continuous carbon chain. For example, an alkane having a chain of five carbons is named __________ (18). Chains having substituents are numbered so that the carbons bearing the substituents have the __________ (19) combination of numbers. Substituents are arranged in the name __________ (20). An alkane having a ring structure is called a __________ (21).

The principal sources of alkanes are natural gas and __________ (22). After petroleum is separated by distillation into several fractions, other processes are used to change the structure of the compounds involved. Large molecules are changed into smaller ones by __________ (23). Unbranched molecules are changed into branched ones by __________ (24). Small molecules are used to make larger ones in a process called __________ (25).

The most important reaction of alkanes is __________ (26). Alkanes are typically unreactive compounds, and the only other important reaction is __________ (27). For example, chloroform, $CHCl_3$, is made by the reaction of chlorine with __________ (28).

Section 19.3 Hydrocarbons that contain a C=C bond are called __________ (29). They have the general formula __________ (30). Alkynes contain a __________ (31) bond. The most important alkyne is __________ (32). The IUPAC names of alkenes use the suffix __________ (33) and follow the same general rules used for alkanes. However, we must remember to select the carbon chain that contains the C=C bond and to begin numbering at the end __________ (34) to it. The names of alkynes follow the same rules except the suffix __________ (35) is used. Isomers that differ only in the arrangement of groups in space and not in the basic carbon structure or the position of a C=C bond or substituent on a carbon chain are called __________ (36). Examples include *cis*- and __________ (37) 2-butene.

Alkenes are sometimes prepared by cracking. Acetylene can be prepared by the reaction of __________ __________ (38) with water. Alkenes and alkynes typically undergo __________ (39) reactions. For example, addition of water to ethylene produces __________ (40). Controlled addition of HCl to acetylene results in __________ (41). One of the most important addition reactions of alkenes and other similar __________ (42) is __________ (43). For example, treatment of propylene with the appropriate catalyst produces __________ (44), a polymer used in the manufacture of upholstery fabric and artificial turf.

Section 19.4 A compound that has a six-carbon ring containing three C=C bonds is said to be __________ (45). The simplest aromatic compound is __________ (46), and the names of aromatic compounds are generally named as derivatives of this compound. Aromatic hydrocarbons often have high __________ (47) numbers, and toluene is added to gasoline for this reason.

Because of the stability of an aromatic ring, aromatic compounds generally undergo __________ (48) reactions rather than addition. For example, reaction of benzene with bromine and a catalyst gives ________ (49).

Section 19.5 An atom or group of atoms in a molecule that is responsible for its characteristic chemical properties is called its ________ (50) group. For example, an aldehyde is represented by the general formula ________ (51), and a carboxylic acid has the general formula ________ (52).

Section 19.6 An alcohol is represented by the formula ________ (53). Ethanol is sometimes called ________ (54) alcohol because it can be prepared by fermentation of various grains. Rubbing alcohol is an aqueous solution of ________ (55). Glycols are 1,2-diols. The most common glycol is ethylene glycol, used as ________ (56) antifreeze.

An ________ (57) is represented by the formula R—O—R. The most common ether is ________ (58) ether.

Section 19.7 Aldehydes and ketones are examples of ________ (59) compounds. The suffixes *-al* and *-one* are used in their ________ (60) names. Common names continue to be widely used, for example, ________ (61) for propanone. The simplest aldehyde is ________ (62), used in aqueous solution as a tissue preservative and an embalming fluid. It is also used in the manufacture of plastics and plywood ________ (63).

Aldehydes react with alcohols to form ________ (64) or geminal diethers. Furthermore, aldehydes are readily ________ (65) to carboxylic acids.

Section 19.8 A compound with one, two, or three alkyl groups bonded to nitrogen is called an ________ (66). The simplest amines contain $—NH_2$ or the ________ (67) group. An amine that has one or more nitrogen atoms forming a ring with carbon atoms is called a ________ (68) amine. The most characteristic reaction of an amine is reaction with an ________ (69) to form a salt. Amines isolated from plants are called ________ (70). A highly addictive alkaloid found in tobacco is ________ (71).

Section 19.9 The carboxylic acid found in ant bites is ________ (72) acid. Vinegar contains ________ (73) acid. Their IUPAC names use the suffix ________ (74) and *acid*. A compound in which an alkyl or aryl group has replaced the H of a carboxyl group is called an ________ (75). Some esters are ________ (76) flavors. An ester can be prepared by the reaction of an alcohol with an acid. An example of a nitrate ester is ________ (77), a high explosive. Phosphate esters are part of the structure of RNA and ________ (78). Esters called triglycerides react with sodium hydroxide to form ________ (79).

Polyesters and polyamides are very important ________ (80) polymers. They include Dacron® and nylon-6,6. A number of pharmaceutical products such as aspirin, acetaminophen, and ibuprofen are carboxylic acids, esters, or amides.

Choices: acetals, acetic, acetone, acetylene, acid, adhesive, addition, alkaloids, alkanes, alkenes, alkyl, alkylation, alphabetically, amine, amino, *-ane,* aromatic, automobile, benzene, bromobenzene, C≡C, calcium carbide, carbon, carbonyl, closest, C_nH_{2n}, C_nH_{2n+1}, C_nH_{2n+2}, combustion, condensation, cracking, cycloalkane, DNA, electrons, *-ene,* ester, ethanol, ether, ethyl, formaldehyde, formic, four, fruit, functional, gases, grain, halogenation, heterocyclic, isomers, isopropyl, IUPAC, methane, molecular, monomers, normal, nicotine, nitroglycerin, octane, *-oic,* 109°, oxidized, pentane, petroleum, polymerization, polypropylene, 2-propanol, RCHO, RCO_2H, reforming, ROH, saturated, smallest, soaps, stereoisomers, substitution, tetrahedral, *trans-,* triple, vinyl chloride, *-yne*

Study Objectives

After studying Chapter 19, you should be able to:

Section 19.1

1. Write Lewis formulas and condensed formulas of compounds from their molecular formulas.
2. Describe the bond angles and shapes of organic molecules.

Section 19.2

3. Write the molecular formula of an alkane having a specific number of carbons.

4. Name alkanes using IUPAC and common names.
5. Name and write formulas of alkyl groups with one to four carbons.
6. Write equations for the combustion of alkanes and their reactions with halogens.

Section 19.3

7. Write the molecular formula of an alkene or alkyne having a specific number of carbons.
8. Name (using IUPAC and common names) and write formulas of alkenes and alkynes.
9. Write equations for the preparation of ethylene, propylene, and acetylene.
10. Write equations for reactions of alkenes with hydrogen, halogens, and water.
11. Write equations for the formation of polymers from monomers.
12. Write equations for reactions of acetylene with HCl, HCN, and halogens.

Section 19.4

13. Name and draw formulas of benzene and other common aromatic compounds.
14. Write equations for the reactions of benzene with Cl_2 or Br_2 and with HNO_3/H_2SO_4.
15. Describe the octane rating scale for gasoline and note what types of compounds have high octane numbers.

Section 19.5

16. Give the functional group of an alkene, alkyne, alkyl halide, alcohol, ether, aldehyde, ketone, amine, carboxylic acid, ester, and amide.
17. Identify the class of a compound from its formula.

Section 19.6

18. Name and write formulas of alcohols, glycols, and ethers.
19. Write equations for the preparation of ethanol, 2-propanol, and diethyl ether.

Section 19.7

20. Name and write structural formulas for aldehydes, ketones, and their acetals and ketals.
21. Write equations for the preparation of aldehydes and ketones from alcohols.
22. Write equations for the oxidation and reduction of aldehydes, the reduction of ketones, and the preparation of an acetal.

Section 19.8

23. Name and write formulas for amines.
24. Write equations for the preparation of amines from alkyl halides, ammonia, and other amines.
25. Recognize a formula as that of a heterocyclic amine.
26. Write equations for the reactions of amines with acids.

Section 19.9

27. Name and write formulas for carboxylic acids, esters, and amides.
28. Write equations for the preparation of carboxylic acids, esters, and amides.
29. Write equations for the neutralization of carboxylic acids and for the reaction of an ester with a base.
30. Recognize and write formulas for triglycerides, nitrate and phosphate esters, polyesters, and polyamides (nylon).

Key Terms

Review the definition of each of the terms listed here. For reference, they are listed by chapter section number. You may use the glossary as necessary.

19.1 isomers

19.2 alkanes, alkylation, alkyl group, cracking, cycloalkane, reforming, saturated, substitution

19.3 addition, alkenes, alkynes, monomer, polymer, stereoisomers, unsaturated

19.4 aromatic

19.5 alkyl halide, functional group

19.6 alcohol, ether

19.7 acetal, aldehyde, carbonyl group, ketal, ketone

19.8 alkaloid, amine, amino group, heterocyclic

19.9 amide, carboxylic acid, carboxyl group, condensation polymer, ester, triglyceride

Questions and Problems

The answers to questions and problems with blue numbers appear in Appendix C.

Nature of Organic Compounds. Alkanes, Alkenes, and Alkynes. (Sections 19.1, 19.2, 19.3)

1. Draw detailed Lewis formulas for two isomers with the molecular formula C_3H_4. Note that one isomer has some bond angles of 109°, but the other does not. Label the bond angles.
2. Draw condensed formulas for the isomers having the formula C_3H_6. Name the compounds.
3. There are two isomers with the formula C_3H_7Cl. Write structural formulas and the names for these compounds.
4. Write structural formulas for all isomers with the formula C_5H_{12}. Name the compounds.
5. Write structural formulas for and name the alkenes with the formula C_4H_8.
6. Name and write structural formulas for the cycloalkanes with the formula C_4H_8.
7. There are five isomers with the formula C_6H_{14}. Write their structural formulas.
8. Name the compounds in problem 7.
9. Write formulas for the following compounds:
   a. 2-methylhexane
   b. 4-isopropylheptane
   c. 3-ethylhexane
   d. 4-*t*-butyloctane
10. Write formulas for:
   a. 2-methyl-2-pentene
   b. *trans*-2-pentene
   c. 3-methyl-1-pentyne
   d. *cis*-4-methyl-2-pentene
11. Write structural formulas for the following compounds:
   a. heptane
   b. 4-ethyl-2-hexyne
   c. 2,3,3-trimethylhexane
   d. 1,1-dimethylcyclopentane
12. Name the following compounds: 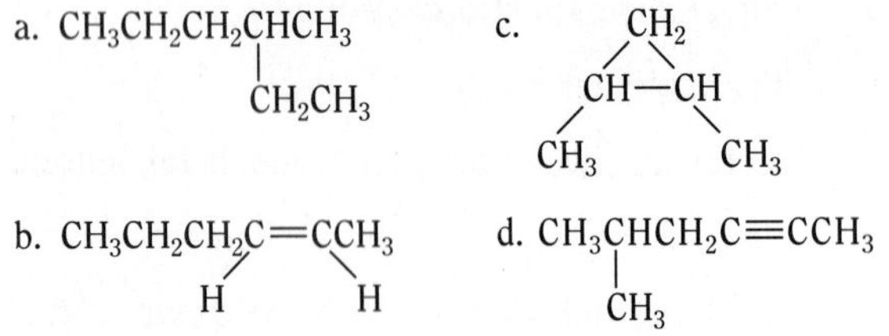
   a. $CH_3CH_2CH_2CHCH_3$ with CH_2CH_3 on the fourth carbon
   b. $CH_3CH_2CH_2C{=}CCH_3$ with H on each double-bonded carbon
   c. CH_2 ring: $CH—CH$ with CH_3 on each CH
   d. $CH_3CHCH_2C{\equiv}CCH_3$ with CH_3 on the second carbon
13. Write structural formulas for the missing compounds:
   a. propane + oxygen $\longrightarrow$ water + ___?___
   b. ___?___ + bromine/light $\longrightarrow$ HBr + bromoethane
   c. propylene + H_2/Pt $\longrightarrow$ ___?___
14. Write the structural formula for the organic product of each of the following reactions:
   a. calcium carbide + $H_2O \longrightarrow$
   b. propylene + $Br_2 \longrightarrow$
   c. 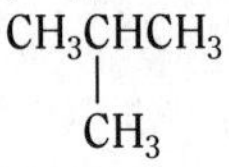 CH_3CHCH_3 (with CH_3 on the middle carbon) catalyst/heat $\longrightarrow$
15. Write the formula of the missing compound in each reaction: 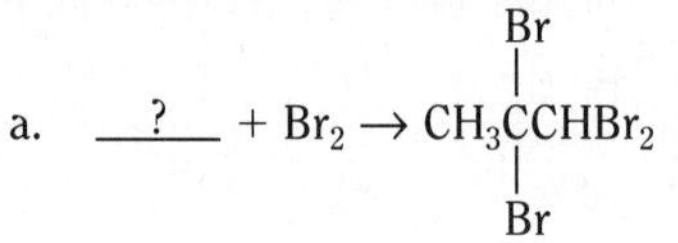 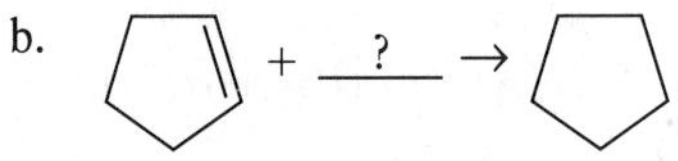
   a. ___?___ + $Br_2 \rightarrow CH_3CBr_2CHBr_2$ (drawn as CH_3CCHBr_2 with Br above and below the second carbon)
   b. cyclopentene + ___?___ $\rightarrow$ cyclopentane
   c. $CH_2{=}CCl_2$ + polymerization catalyst $\longrightarrow$ ___?___
16. Write the formula of the monomer used to make each polymer:
   a. $\sim(CH_2CH)_n\sim$ with CO_2H on CH
   b. $\sim(CH_2CH)_n\sim$ with $OCCH_3$ (C=O) on CH
   c. $\sim(CH_2CH)_n\sim$ with CCH_3 (C=O) on CH

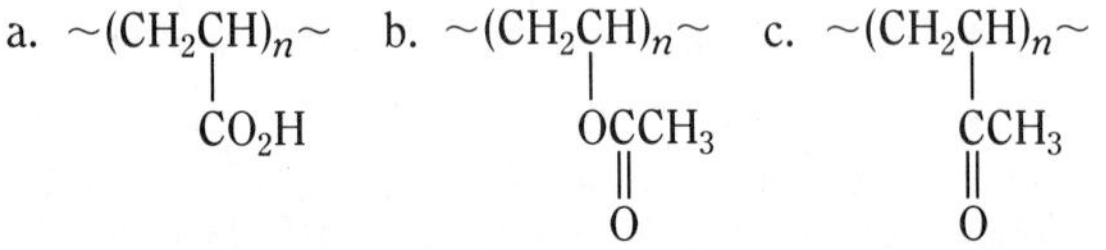

Aromatic Compounds (Section 19.4)

17. Draw resonance structures and a resonance hybrid formula for benzene. What is the meaning of a circle drawn inside a six carbon ring?
18. Draw formulas for the following compounds:
   a. *t*-butylbenzene
   b. 3-nitrobenzoic acid
   c. iodobenzene
   d. *o*-chloroethylbenzene
19. Briefly explain the significance of the octane rating of a gasoline. What kinds of hydrocarbons have high octane numbers?
20. Which of the following compounds has the highest octane number?
   a. heptane
   b. 2-methylhexane
   c. 2,2,4-trimethylpentane
21. Which of the following compounds has the highest octane number?
   a. toluene
   b. octane
   c. 3-methylhexane

22. Write formulas for the missing compounds:
   a. 1,3,5-trimethylbenzene + $Cl_2/FeCl_3 \rightarrow$ ___?___
   b. $CH_3CH_2-C_6H_4-CH_2CH_3$ + ___?___ $\rightarrow$ $CH_3CH_2-C_6H_3(NO_2)-CH_2CH_3$

Alcohols and Ethers (Section 19.6)

23. Write structural formulas for the following compounds:
   a. 2-chloroethanol
   b. 3,3-dimethyl-2-butanol
   c. phenylmethanol
   d. 2-methyl-2-pentanol

24. Name the following compounds:
   a. $CH_3CH_2OCH(CH_3)CH_3$
   b. $CH_3CH_2CH_2OCH_2CH_2CH_3$
   c. $CH_3CH_2CH(CH_3)CH(OH)CH_2CH_3$

25. Write structural formulas for the following compounds:
   a. ethyl phenyl ether
   b. 3-methyl-1-butanol
   c. 2-cyclopentyl-2-propanol
   d. isobutyl alcohol

26. Write equations for the following reactions:
   a. preparation of diisopropyl ether from isopropyl alcohol
   b. reaction of 2-butene with aqueous acid
   c. reaction of cyclopentene with aqueous acid

27. Name the following compounds:
   a. $HOCH_2CH_2OH$
   b. $C_6H_5-O-C_6H_5$
   c. C_5H_9-OH (cyclopentane ring bearing OH)

28. Write formulas for the missing compounds:
   a. ___?___ + $H_2SO_4 \rightarrow CH_3OCH_3 + H_2O$
   b. $CH_3CH_2CH{=}CHCH_2CH_3 + H_2O/H^+ \rightarrow$ ___?___

Aldehydes and Ketones (Section 19.7)

29. Write formulas for the following compounds:
   a. 2-methyl-3-pentanone
   b. cyclopentanone
   c. propionaldehyde
   d. methanal

30. Name the following compounds:
   a. $CH_3CH(CH_3)CHO$
   b. $CH_3C(=O)CH_2CH(CH_3)CH_3$
   c. $CH_3CH_2C(=O)CH_2C(CH_3)_2CH_3$

31. Write equations for the following reactions:
   a. oxidation of 2-pentanol
   b. reaction of sodium borohydride with butanone
   c. oxidation of butyraldehyde

32. Write formulas for the missing compounds:
   a. benzaldehyde + $CH_3OH/H^+ \rightarrow$ ___?___
   b. benzaldehyde + ___?___ $\rightarrow$ phenylmethanol
   c. 1-phenylethanol + ___?___ $\rightarrow$ $C_6H_5C(=O)CH_3$

33. Write structural formulas for the following:
   a. the ethyl acetal of propanal
   b. the methyl ketal of acetone

Amines (Section 19.8)

34. Name the following amines:
   a. CH_3NHCH_3
   b. $CH_3CH_2CH_2CH_2NH_2$
   c. $CH_3CH(CH_3)NH_2$
   d. $(CH_3)_3CNH_2$

35. Write structural formulas for the following amines:
   a. *p*-chloroaniline
   b. trimethylamine
   c. dipropylamine
   d. 2-phenylethylamine

36. Morphine, an alkaloid found in poppy seeds, is a highly effective pain killer. Explain what is meant by an alkaloid. What alkaloid is found in tobacco? In coffee beans?

37. Pyridine, C_5H_5N, is a heterocyclic amine. Write a structural formula for pyridine. Explain what is meant by a heterocyclic compound.

38. Write equations for the following reactions:
   a. reaction of ammonia with propyl bromide
   b. reaction of butylamine with hydrochloric acid

39. Write formulas for the missing compounds:
   a. H_2SO_4 + ___?___ → $(CH_3)_2NH_2^+$ + HSO_4^-
   b. aniline + ___?___ → (benzene ring)–$NHCH_3$
   c. 2-phenylethyl chloride + ___?___ → (benzene ring)–$CH_2CH_2NH_2$

Carboxylic Acids, Esters, and Amides (Section 19.9)

40. Name the following compounds:
   a. $CH_3CH_2CO_2H$
   b. $CH_3CH_2C(=O)NH_2$
   c. $CH_3CH_2CO_2CH_2CH_3$
   d. (benzene ring)–$C(=O)NH_2$
41. Write formulas for the following compounds:
   a. butyl butyrate
   b. 2-bromobenzoic acid
   c. isopropyl acetate
   d. *N*-methylhexanamide
42. Write equations for the following reactions:
   a. oxidation of 1-propanol with CrO_3/H^+
   b. reaction of acetic acid with potassium hydroxide
   c. reaction of a triglyceride (use a general formula) with sodium hydroxide
43. Butyric acid forms when butter spoils or becomes rancid. Write an equation for the reaction of butyric acid with sodium hydroxide.
44. Aspirin, or acetylsalicylic acid (see Figure 19.3), is sometimes taken with milk of magnesia, $Mg(OH)_2$. Write an equation for the reaction of aspirin with milk of magnesia.
45. Write an equation for the reaction of isobutyl formate, the raspberry flavor, with sodium hydroxide.
46. Ammonia reacts with methyl benzoate to give benzamide. Write an equation for the reaction.
47. Write the formula for ethyl phosphate, a phosphate ester.
48. Alfred Nobel, for whom the Nobel prizes are named, invented dynamite, which contains nitroglycerin, a high explosive. Write the formula of nitroglycerin, a nitrate ester of glycerol.
49. Nylon-6,6 is a polyamide of 1,6-hexanediamine, $H_2N(CH_2)_6NH_2$, and hexanedioic acid, $HO_2C(CH_2)_4CO_2H$. Write a formula for nylon-6,6.
50. The pain caused by ant bites and many insect stings is due to formic acid. A common first-aid remedy involves putting moist sodium bicarbonate, $NaHCO_3$, on the area of skin that was bitten. Write an equation for the reaction of formic acid with sodium bicarbonate. (See Section 10.9.)

Additional Questions and Problems

51. Which of the following compounds are unsaturated?
   a. isobutane
   b. 1-butene
   c. 2-butyne
52. Cycloalkanes and noncyclic alkenes that have the same number of carbons are isomers. Draw the structural formulas for all isomers of C_5H_{10}. Five of the isomers are cycloalkanes.
53. Write structural formulas for all the aldehydes having the formula $C_5H_{10}O$.
54. Write structural formulas for all the aldehydes and ketones having the formula C_4H_8O.
55. Three dichlorobenzenes are known. Write their structural formulas. The para isomer is used as moth crystals.
56. Naphthalene, $C_{10}H_8$, has two aromatic rings joined together. Write a structural formula for naphthalene.
57. Four alcohols are known that have the molecular formula $C_4H_{10}O$. Write the structural formulas and names of these compounds.
58. Alcohols and ethers that have the same number of carbons are isomers. Write structural formulas for the isomers of C_3H_8O. Name these compounds.
59. An ester is generally made from an alcohol and an acid. Write the structural formula and name of the alcohol and the acid that could be used to make methyl salicylate, oil of wintergreen.

   (benzene ring with OH and CO_2CH_3 on adjacent carbons)

60. Sodium benzoate is a preservative used in soft drinks. Write an equation for its formation from benzoic acid.
61. An alkyne and a diene (two C═C bonds) that have the same number of carbons are isomers. Write structural formulas for four isomers having the formula C_4H_6.
62. Write structural formulas for two esters that have the molecular formula $C_3H_6O_2$. Name these compounds.
63. Write equations for the preparation of the esters in problem 62.
64. Write an equation for the oxidation of cinnamaldehyde ($C_6H_5CH{=}CHCHO$), oil of cinnamon, with CrO_3. What is the common name of the acid that is formed?
65. Oxalic acid ($H_2C_2O_4$), also called ethanedioic acid, is a poisonous substance found in rhubarb leaves. Write its structural formula.
66. Lactic acid, also named 2-hydroxypropanoic acid, is found in sour milk and sourdough. Write its structural formula.
67. The flavor of apricots is a result of pentyl propanoate. Write the structural formula of this ester.

68. Acetaminophen, or *N*-(4-hydroxyphenyl)acetamide, is a pain reliever known as Tylenol®. Write its structural formula.

69. Carboxylic acids that contain an amino (NH_2) group are called amino acids. Write the formula of 2-aminopropanoic acid.

70. Sugars are aldehydes and ketones that generally have many hydroxyl (OH) groups. One of the simplest sugars is glyceraldehyde, or 2,3-dihydroxypropanal. Write a structural formula for glyceraldehyde.

Solutions to Practice Problems

PP 19.1

The O may be placed between C and H or between C and C.

$CH_3CH_2CH_2OH$ $\quad$ CH_3CHCH_3 (OH on the middle C) $\quad$ $CH_3CH_2OCH_3$

$$CH_3CH_2CH_2OH \qquad \underset{\displaystyle OH}{CH_3\overset{}{C}HCH_3} \qquad CH_3CH_2OCH_3$$

PP 19.2

a. $CH_3\underset{\displaystyle CH_3}{\underset{|}{C}}HCH_2CH_3$

b. $CH_3CH_2\overset{\displaystyle CH_3}{\overset{|}{\underset{\displaystyle CH_3}{\underset{|}{C}}}}CH_2CH_3$

c. CH_2CH_3 (on a cyclobutane ring)

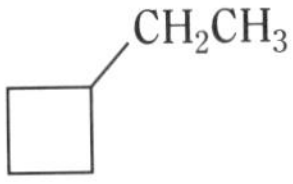

d. $CH_3CH_2\overset{\displaystyle CH_3}{\overset{|}{C}}H\underset{\displaystyle CH_2CH_3}{\underset{|}{C}}HCH_2CH_3$

PP 19.3

a. 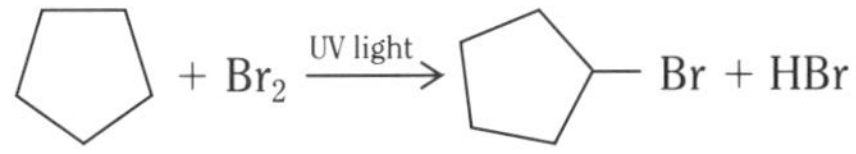 cyclopentane $+ Br_2 \xrightarrow{\text{UV light}}$ cyclopentyl—Br $+ HBr$

b. $C_5H_{12} + 8O_2 \longrightarrow 5CO_2 + 6H_2O$

PP 19.4

a. $CH_2{=}CHCH_2CH_3 + Br_2 \rightarrow BrCH_2\underset{\displaystyle Br}{\underset{|}{C}}HCH_2CH_3$

b. $CH_3\underset{\displaystyle CH_3}{\underset{|}{C}}{=}CH_2 + H_2/Pt \rightarrow CH_3\underset{\displaystyle CH_3}{\underset{|}{C}}HCH_3$

c. $CH_3CH{=}CHCH_3 + H_2O/H^+ \rightarrow CH_3CH_2\underset{\displaystyle OH}{\underset{|}{C}}HCH_3$

d. $CH_3C{\equiv}CCH_3 + HCl \rightarrow CH_3CH{=}\underset{\displaystyle Cl}{\underset{|}{C}}CH_3$

PP 19.5

Addition polymers are named by using the prefix *poly-* with the name of the monomer.

a. polyethylene
b. polyacrylonitrile
c. polystyrene

PP 19.6

a. *o*-dichlorobenzene (1,2-dichlorobenzene)
b. *m*-bromoisopropylbenzene (1-bromo-3-isopropylbenzene)
c. *o*-ethylnitrobenzene (1-ethyl-2-nitrobenzene)

PP 19.7

a. An H on a ring carbon is replaced with Br.

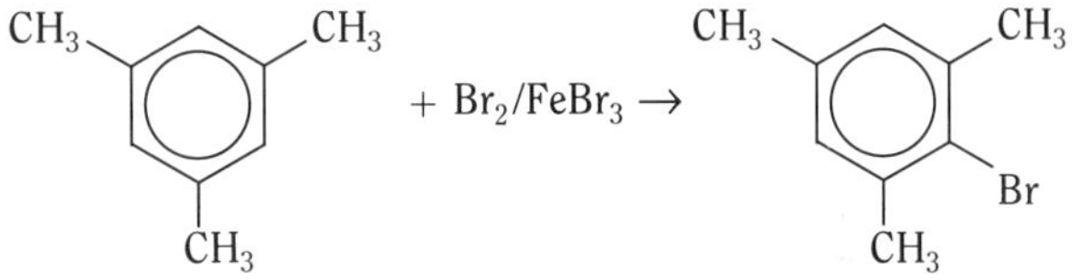

1,3,5-(CH_3)$_3$ benzene $+ Br_2/FeBr_3 \rightarrow$ ring with CH_3, CH_3, CH_3, Br

b. An H on a ring carbon is replaced with NO_2.

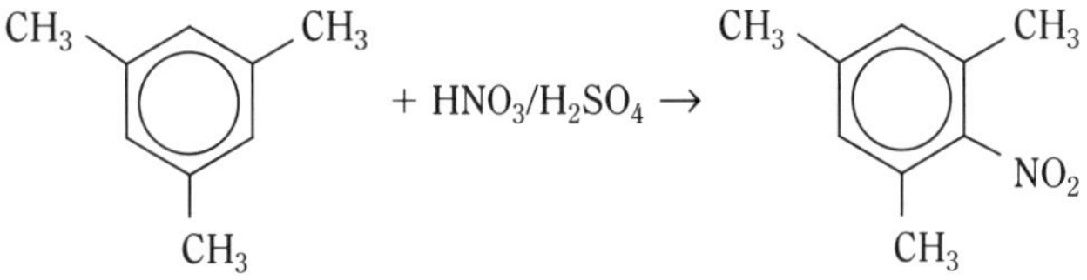

1,3,5-(CH_3)$_3$ benzene $+ HNO_3/H_2SO_4 \rightarrow$ ring with CH_3, CH_3, CH_3, NO_2

PP 19.8

a. CH_3OCH_3

b. $CH_3\overset{\displaystyle O}{\overset{\|}{C}}CH_3$

c. $CH_3\overset{\displaystyle O}{\overset{\|}{C}}OCH_3$ and $H\overset{\displaystyle O}{\overset{\|}{C}}OCH_2CH_3$

PP 19.9

a. $CH_3\overset{\displaystyle CH_3}{\overset{|}{\underset{\displaystyle OH}{\underset{|}{C}}}}CH_3$

b. $CH_3\underset{\displaystyle CH_3}{\underset{|}{C}}HO\overset{\displaystyle CH_3}{\overset{|}{C}}HCH_3$

c. $CH_3\overset{\displaystyle CH_3}{\overset{|}{C}}H\underset{\displaystyle OH}{\underset{|}{C}}HCH_3$

d. $CH_3\underset{\displaystyle OH}{\underset{|}{C}}HCH_2OH$

PP 19.10

a. CH_3CH_2OH. A coefficient of 2 is needed to balance the equation.
b. H_2O. Hydration of an alkene can be used to prepare alcohols.

PP 19.11

a. 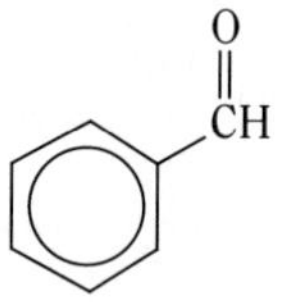 $\overset{\overset{\displaystyle O}{\|}}{C}H$

b. $CH_3CH_2\overset{\overset{\displaystyle O}{\|}}{C}H$

c. $CH_3\underset{\underset{\displaystyle CH_2CH_3}{|}}{\overset{\overset{\displaystyle CH_3}{|}}{C}H}CH\overset{\overset{\displaystyle O}{\|}}{C}CH_2CH_2CH_3$

d. $CH_3\underset{\underset{\displaystyle CH_3}{|}}{C}H\overset{\overset{\displaystyle O}{\|}}{C}H$

PP 19.12

a. $CH_3CH_2CH_2\overset{\overset{\displaystyle O}{\|}}{C}H + NaBH_4 \rightarrow CH_3CH_2CH_2CH_2OH$

b. 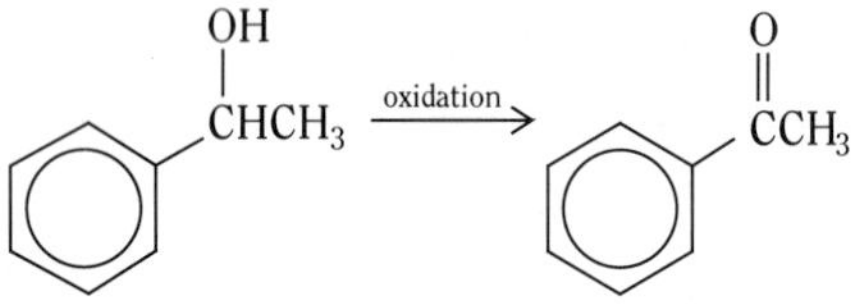 $\overset{\overset{\displaystyle OH}{|}}{C}HCH_3 \xrightarrow{\text{oxidation}} \overset{\overset{\displaystyle O}{\|}}{C}CH_3$

PP 19.13

a. CH_3CO_2H
b. $CH_3\underset{\underset{\displaystyle CH_3}{|}}{C}HCH_2NHCH_3$

PP 19.14

a. $CH_3CH_2CH_2CO_2H$

b. 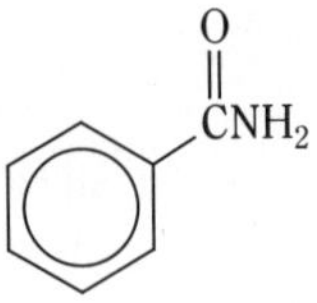 $\overset{\overset{\displaystyle O}{\|}}{C}NH_2$

c. $CH_3CH_2CH_2CH_2O\overset{\overset{\displaystyle O}{\|}}{C}H$

d. 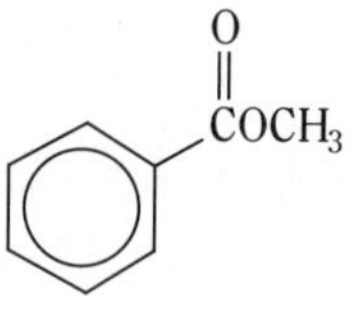 $\overset{\overset{\displaystyle O}{\|}}{C}OCH_3$

PP 19.15

a. $CH_3(CH_2)_6CO_2^-\ K^+$
b. $CH_3(CH_2)_7OH$

Answers to Chapter Review

1. carbon
2. molecular
3. gases
4. four
5. triple
6. electrons
7. 109°
8. tetrahedral
9. isomers
10. alkanes
11. saturated
12. C_nH_{2n+2}
13. *-ane*
14. normal
15. alkyl
16. C_nH_{2n+1}
17. isopropyl
18. pentane
19. smallest
20. alphabetically
21. cycloalkane
22. petroleum
23. cracking
24. reforming
25. alkylation
26. combustion
27. halogenation
28. methane
29. alkenes
30. C_nH_{2n}
31. $C\equiv C$
32. acetylene
33. *-ene*
34. closest
35. *-yne*
36. stereoisomers
37. *trans-*
38. calcium carbide
39. addition
40. ethanol
41. vinyl chloride
42. monomers
43. polymerization
44. polypropylene
45. aromatic
46. benzene
47. octane
48. substitution
49. bromobenzene
50. functional
51. RCHO
52. RCO_2H
53. ROH
54. grain
55. 2-propanol
56. automobile
57. ether
58. ethyl
59. carbonyl
60. IUPAC
61. acetone
62. formaldehyde
63. adhesive
64. acetals
65. oxidized
66. amine
67. amino
68. heterocyclic
69. acid
70. alkaloids
71. nicotine
72. formic
73. acetic
74. *-oic*
75. ester
76. fruit
77. nitroglycerin
78. DNA
79. soaps
80. condensation

20

Biochemistry: The Chemistry of Life

Contents

Golden aspen cover many slopes of the Rocky Mountains in early fall. Cellulose is the principal structural component of trees and other plants.

A termite can utilize cellulose as an energy source.

Muscles contain proteins. Steroids help control muscle growth.

You Probably Know . . .

- Proteins are a major component of muscles. Moreover, some proteins called enzymes are catalysts that enable chemical reactions within cells to occur very rapidly. (Section 16.3)
- Sugars and starches are carbohydrates that can be used by most organisms as sources of energy. Cellulose is also a carbohydrate, but humans and most animals cannot use it for energy. Cellulose is a structural material for plants.
- Fats and oils are examples of lipids. Steroids are also lipids. One of the best known steroids is cholesterol, which can be a contributor to heart disease.
- DNA (*d*eoxyribo*n*ucleic *a*cid) is a very complex molecule that carries a code for the information related to the characteristics and functions of an organism. Specific sections of DNA molecules are called genes.

These animals all have different DNA.

Why Should We Study Biochemistry?

Chemistry is everywhere. We see evidence of it in the weather, in the aging and corrosion of buildings and machines, in minerals, and in the burning of fossil fuels. However, one of the most direct examples of chemistry in action is in our own bodies and in the other living organisms around us.

A knowledge of biochemistry gives us a better appreciation for the nature of life in general. Moreover, those who study biology are better able to understand biological principles if they know something about biochemistry. Cellular structure and function are a direct result of biochemistry. This knowledge also helps us to understand the medical problems we personally encounter.

Biochemistry is the chemistry of life processes (Section 1.1), and it includes both inorganic and organic chemistry. Inorganic chemistry affects living systems in several ways. For example, inorganic compounds help control the acidity of body fluids within narrow pH ranges. The proper pH is maintained by bicarbonate (HCO_3^-) and phosphate (PO_4^{3-}) buffers (Section 15.7) in the blood and in cellular fluids. In addition, a proper concentration of salts is very

1A																	8A
1 H	2A											3A	4A	5A	6A	7A	2 He
3 Li	4 Be											5 B	6 C	7 N	8 O	9 F	10 Ne
11 Na	12 Mg											13 Al	14 Si	15 P	16 S	17 Cl	18 Ar
19 K	20 Ca	21 Sc	22 Ti	23 V	24 Cr	25 Mn	26 Fe	27 Co	28 Ni	29 Cu	30 Zn	31 Ga	32 Ge	33 As	34 Se	35 Br	36 Kr
37 Rb	38 Sr	39 Y	40 Zr	41 Nb	42 Mo	43 Tc	44 Ru	45 Rh	46 Pd	47 Ag	48 Cd	49 In	50 Sn	51 Sb	52 Te	53 I	54 Xe
55 Cs	56 Ba	71 Lu	72 Hf	73 Ta	74 W	75 Re	76 Os	77 Ir	78 Pt	79 Au	80 Hg	81 Tl	82 Pb	83 Bi	84 Po	85 At	86 Rn
87 Fr	88 Ra	103 Lr	104 Unq	105 Unp	106 Unh	107 Ns	108 Hs	109 Mt									

57 La	58 Ce	59 Pr	60 Nd	61 Pm	62 Sm	63 Eu	64 Gd	65 Tb	66 Dy	67 Ho	68 Er	69 Tm	70 Yb
89 Ac	90 Th	91 Pa	92 U	93 Np	94 Pu	95 Am	96 Cm	97 Bk	98 Cf	99 Es	100 Fm	101 Md	102 No

FIGURE 20.1 The chemical elements necessary for life. The essential elements are shown in blue and the trace elements are shown in red.

A restored dining room, ancient Pompeii. The Romans are thought to have poisoned themselves as a society by using drinking vessels that contained lead and by using lead pipe in their plumbing.

important to control flow of water through membranes (osmosis) and to provide ions for body needs. A number of *essential* elements (present in a concentration of 0.05% of more) are found in most organisms (Figure 20.1). *Trace* elements are those present in a concentration of less than 0.05%. You may be surprised to learn that elements such as lead, arsenic, and selenium, which are generally considered to be poisonous in trace amounts, are actually essential nutrients.

Of the many organic compounds that are important in biochemistry, we will focus in this chapter upon four biologically important types: proteins, carbohydrates, lipids, and nucleic acids. Although the chemistry occurring in living organisms involves reactions and interactions that can be demonstrated outside of biological systems, there is one significant difference. Biochemical reactions in living cells require special catalysts called enzymes. These are discussed in the next section.

The chemistry of living cells is similar to chemistry that can be demonstrated in a laboratory flask, but it takes place in a vastly more complex system.

20.1 Proteins

Learning Objectives

- Describe fibrous and globular proteins.
- Identify and explain the four levels of protein structure.
- Explain denaturation and how it occurs.
- Explain the general structure of enzymes and a model for the mechanism of enzyme activity.

Proteins are natural polyamides.

Proteins are natural polyamides (Section 19.9). They make up about 15% of our bodies and about two-thirds of the total dry weight of cells. Proteins occur in the blood, in muscles, in bone, in hair, and in the brain. In fact, no living part of the human body is without proteins.

Fibrous proteins: Rod-shaped, structural proteins.

The name protein *comes from the Greek word "protos," meaning "first importance."*

Globular proteins: Generally spherical proteins that facilitate processes.

Proteins are commonly considered to be either fibrous or globular, according to their physical characteristics. **Fibrous proteins** are insoluble in water and tend to be rod-shaped with great mechanical strength. They are structural proteins and are the main components of hair, muscle, and cartilage. **Globular proteins** tend to be more compatible with water than fibrous proteins, and their shapes are more spherical than linear. They are proteins that enable processes to occur, such as transporting and storing oxygen, acting as catalysts, and aiding in regulating body processes.

The structure of protein molecules is described as primary, secondary, tertiary, or quaternary. We now discuss each of these and the chemical forces that contribute to the structure.

Primary Structure

The fundamental building blocks of all proteins are α-amino acids. An amino acid is both an amine and a carboxylic acid, as illustrated by the following general formula:

$$\underbrace{H_2N}_{\text{amino group}}-\underset{\substack{| \\ H}}{\overset{\substack{R \\ |}}{C}}-\underbrace{\overset{\substack{O \\ \|}}{C}OH}_{\text{carboxyl group}}$$

(labels: side chain → R; α-carbon → C)

Some amino acids have nonpolar R groups, which are hydrophobic. Others have polar R groups, which are hydrophilic. Figure 20.2 gives examples of each type along with their three-letter codes. There are 20 amino acids that are found in almost all proteins. Those in Figure 20.2 show some of the variety that exists in the R groups. The sizes of the R groups are important in determining the type of protein structure, as we shall soon discover.

In forming proteins, amino acids react head to tail in the following manner:

$$\underset{\text{tail}}{H_2N}-\underset{\substack{| \\ R}}{\overset{\substack{H \\ |}}{C}}-\underset{\text{head}}{\overset{\substack{O \\ \|}}{C}-OH} + \underset{\text{tail}}{H-\underset{H}{N}}-\underset{\substack{| \\ R'}}{\overset{\substack{H \\ |}}{C}}-\underset{\text{head}}{\overset{\substack{O \\ \|}}{C}-OH} \rightarrow H_2N-\underset{\substack{| \\ R}}{\overset{\substack{H \\ |}}{C}}-\underbrace{\overset{\substack{O \\ \|}}{C}-\overset{\substack{H \\ |}}{N}}_{\text{amide or peptide linkage}}-\underset{\substack{| \\ R'}}{\overset{\substack{H \\ |}}{C}}-\overset{\substack{O \\ \|}}{C}-OH + H_2O$$

The new bond formed is an **amide bond,** frequently called a **peptide linkage** by biologists and some biochemists. The product in the preceding reaction is called *di*peptide because it is made up of two amino acid units.

Amides can be prepared in the laboratory as well. An amine reacts with a carboxylic acid to form a salt, which must be heated to 100°–200°C to form an amide according to the following equations:

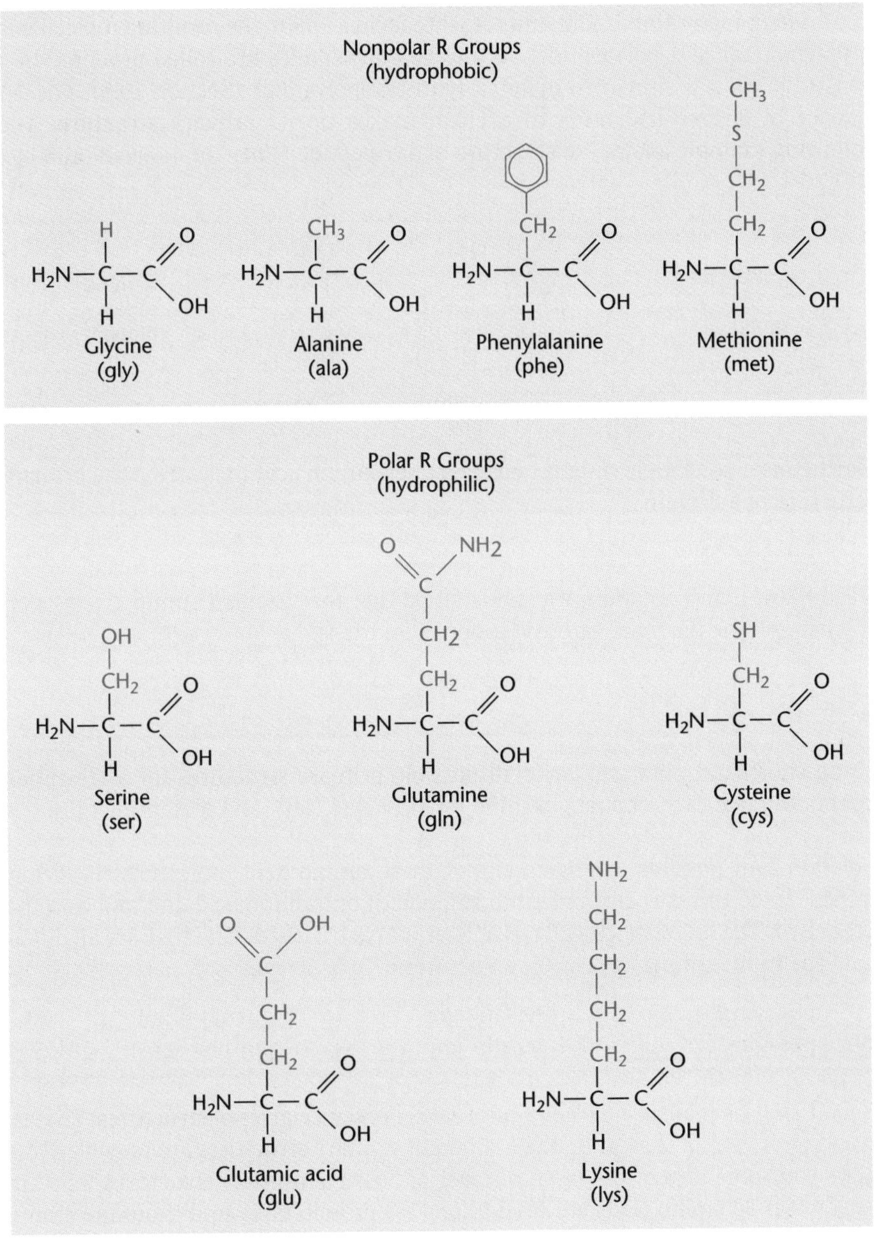

Nonpolar R Groups (hydrophobic)

Glycine (gly) · Alanine (ala) · Phenylalanine (phe) · Methionine (met)

Polar R Groups (hydrophilic)

Serine (ser) · Glutamine (gln) · Cysteine (cys) · Glutamic acid (glu) · Lysine (lys)

FIGURE 20.2 Some of the 20 α-amino acids found in most proteins. The R group is shown in red.

$$\underset{\text{carboxylic acid}}{R-\overset{O}{\overset{\|}{C}}-OH} + \underset{\text{amine}}{H_2N-R'} \rightarrow \underset{\text{a salt}}{R-\overset{O}{\overset{\|}{C}}-O^- \;{}^+H_3N-R'} \xrightarrow{100°-200°C} \underset{\text{an amide}}{R-\overset{O}{\overset{\|}{C}}-\underset{H}{\underset{|}{N}}-R'} + H_2O$$

However, these high temperatures are much more severe than cells can tolerate, and the efficiency is low (about 75%). Thus, at normal cellular temperature, an enzyme is necessary to catalyze the biochemical formation of amides.

A spider's web, made of protein (nature's nylon), is very strong.

When many amino acid units react to form a chain, the resulting molecule is a polymer called a polypeptide. Very large polypeptides are called proteins. In a sense a protein is similar to nylon, a synthetic polyamide (Section 19.9). The sequence of amino acid units in a chain makes up its **primary structure.** The following example shows the structure of a tripeptide (*three* amino acid units).

```
                                                 SH
                                                 |
                  CH3       H   H          H    CH2
terminal           |        |   |          |     |                     terminal
amino  → H2N—C—C—N—C—C—N—C—C—OH ←  carboxyl
group              |   ||       |   ||           |   ||                 group
                   H   O        H   O            H   O
            alanine        glycine           cysteine
             (ala)          (gly)              (cys)
```

Primary structure: The amino acid sequence in a protein chain.

Biochemists use the three-letter code for each amino acid to describe the primary structure of a protein:

ala-gly-cys

Proteins are very long polyamides (polypeptides) made from amino acids.

In this shorthand notation, it is understood that the terminal amino group is on the left and the terminal carboxyl group is on the right.

Example 20.1

Using shorthand notation, write all possible primary structures for a tripeptide containing one each of phenylalanine (phe), serine (ser), and glutamine (gln). ■

Solution All possible arrangements of these amino acid units must be determined. To do this it is convenient to begin with one amino acid unit and vary the other two. The following are all possible primary structures, and they are all different. The total number of possible structures is 3! or $3 \times 2 \times 1 = 6$.

phe-ser-gln	ser-phe-gln	gln-ser-phe
phe-gln-ser	ser-gln-phe	gln-phe-ser

Using each of six different amino acids only once gives 6! structures. That is, $6 \times 5 \times 4 \times 3 \times 2 \times 1$, or 2580, different primary structures are possible. This gives you some idea of the large number of possible primary structures that can exist when 20 amino acids are used to make a protein of several thousand amino acid units.

The importance of primary structure is illustrated in the differences between vasopressin and oxytocin. Each of these polypeptides contains nine amino acid units. Although they differ by only two amino acid units, these polypeptides perform completely different functions in the human body. Vasopressin raises blood pressure, and oxytocin triggers contraction in the uterus and secretion of milk.

■ Secondary Structure

Secondary structure: The arrangement of a protein chain about an axis.

Secondary structure of a protein refers to the general shape of a protein chain, or the arrangement of a chain about an axis. It is maintained by hydrogen bonding due to the presence of carbonyl groups and N—H groups in the molecule.

amide group in a protein chain

C=O······H—N

H—N

C=O

amide group in a protein chain

a hydrogen bond

Hydrogen bond: A strong attraction between the hydrogen atom of O—H or N—H and another nitrogen or oxygen atom (Section 13.2).

The size of the R group in the protein determines which type of secondary structure will predominate. If the R groups are all small—such as H, CH_3, or CH_2OH—hydrogen bonding between side-by-side chains is possible. This results in a shape that is commonly referred to as a **pleated sheet** (Figure 20.3). These sheets can then be stacked on one another.

Pleated sheet secondary structure is found in some fibrous proteins. It is present in some muscle fiber, but silk is one of the best examples of a protein with pleated sheet structure. Silk is known for its flexibility, strength, and resistance to stretching. The presence of small side chains allows for great flexibility. However, because the chains are always fully extended, the fiber cannot be easily stretched.

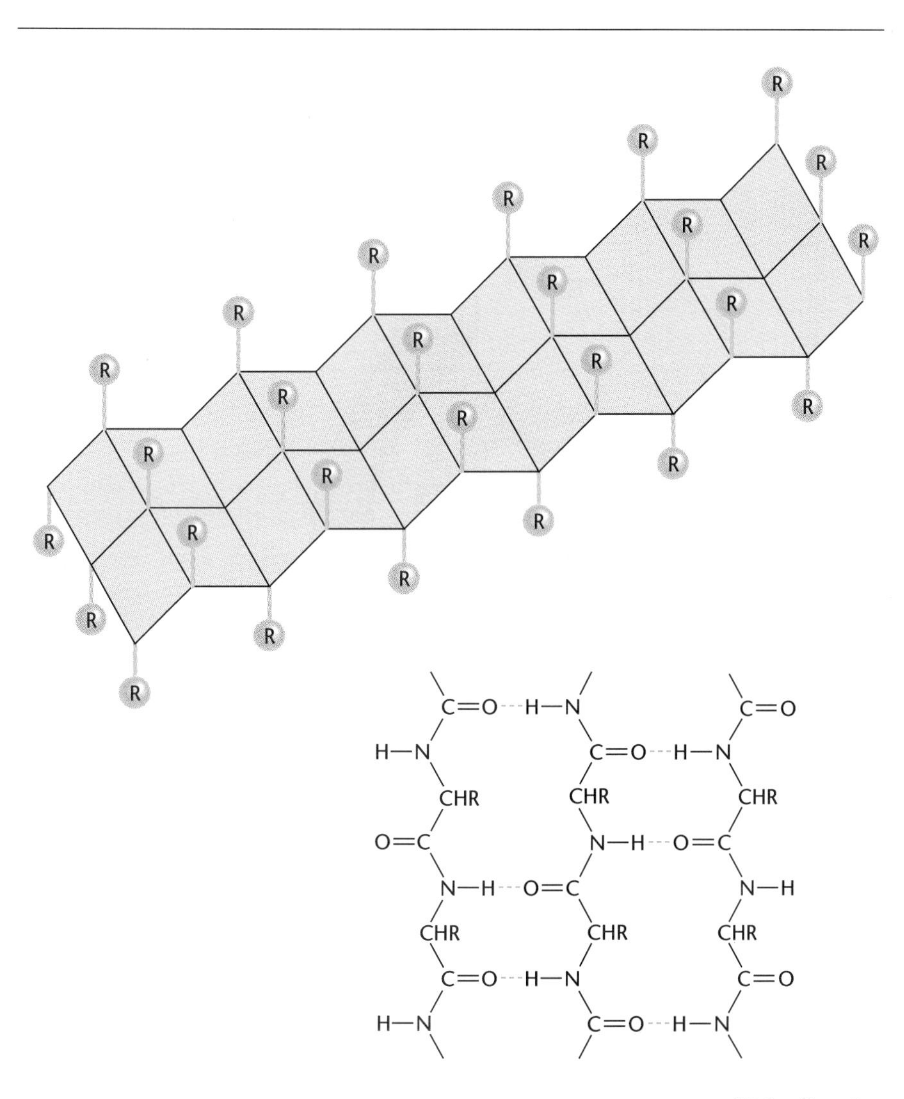

FIGURE 20.3 Pleated sheet secondary structure of proteins.

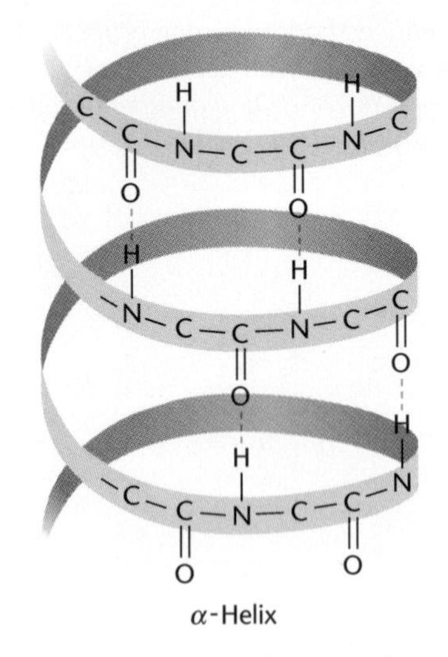

FIGURE 20.4 α-Helix secondary structure of proteins.

Breaking silk fiber requires either rupturing thousands of hydrogen bonds or breaking covalent bonds. Each requires a large amount of energy.

When R groups of the side chains are large, hydrogen bonding at regular intervals between protein chains is difficult. In these proteins hydrogen bonding occurs within a single chain, resulting in a coil, or **α- helix** (Figure 20.4). Wool, hair, and globular proteins frequently have an α-helix secondary structure.

Proteins with the α-helix secondary structure are more easily stretched because the coil can be elongated by breaking only a few hydrogen bonds. Furthermore, the coil can be interrupted over a short segment to allow for sharp bends, frequently found in globular proteins.

Tertiary Structure

Tertiary structure refers to the folding of a protein chain back on itself to form its overall shape (Figure 20.5). This is analogous to a tangled wad of yarn or a coiled (secondary structure) telephone cord that has become tangled with itself. To maintain its tertiary structure, the protein chain must be "tied" or attached at points

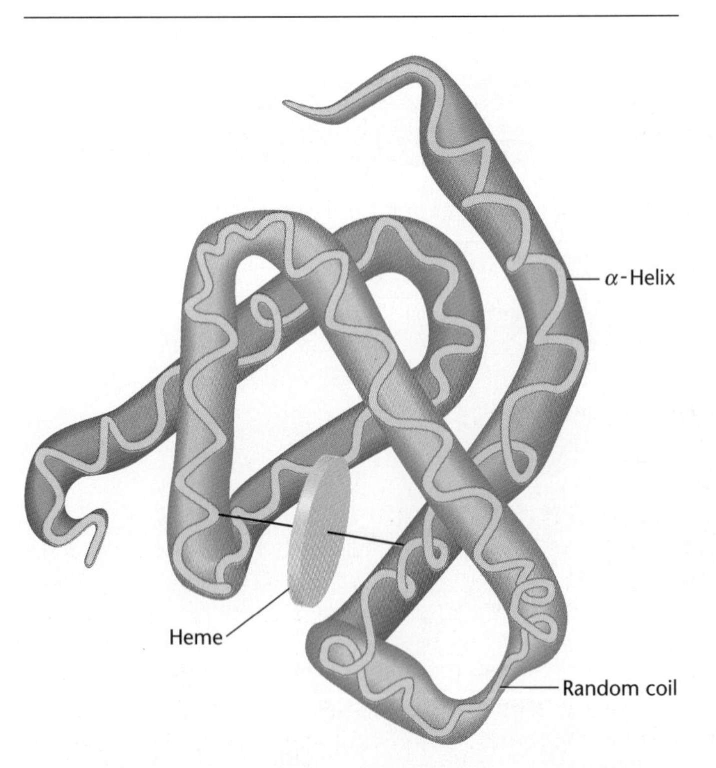

FIGURE 20.5 Myoglobin is a globular protein whose tertiary structure is formed by the chain folding back on itself. Its secondary structure is mainly helical except at the bends.

FIGURE 20.6 Types of interactions that hold a protein in its tertiary structure.

where the loops "touch." The attachments at these points are the result of several types of attractions (Figure 20.6):

Tertiary structure: The three-dimensional shape resulting from the folding and bending of a protein.

1. hydrogen bonding of the general nature described for holding secondary structure
2. **salt bridges,** $—CO_2^- \ H_3N^+—$ which are formed by the interactions of basic amino groups with carboxyl groups
3. **disulfide linkages,** —S—S— which are covalent bonds formed between close —SH groups.
4. **hydrophobic interactions,** which are the association of nonpolar (hydrophobic) groups with each other by dispersion forces (Section 13.2)

Disulfide linkages (covalent bonds) and salt bridges (ionic attractions) are generally stronger interactions than hydrogen bonding. Hydrophobic interactions are the weakest.

Tertiary structure is important to the function of proteins, especially enzymes. These molecules must be precisely shaped to function properly. This is explained in the discussion of enzymes.

Quaternary Structure

Some molecules contain several protein chains that fit together to perform a specific function. Hemoglobin is probably the most familiar example. Each hemoglobin molecule contains two pairs of protein chains. Each chain is about the size and shape of myoglobin (see Figure 20.5). These chains fit together in a precise pattern, which also includes four heme groups containing iron to transport oxygen. The manner in which the four protein chains fit together in hemoglobin is its **quaternary structure.**

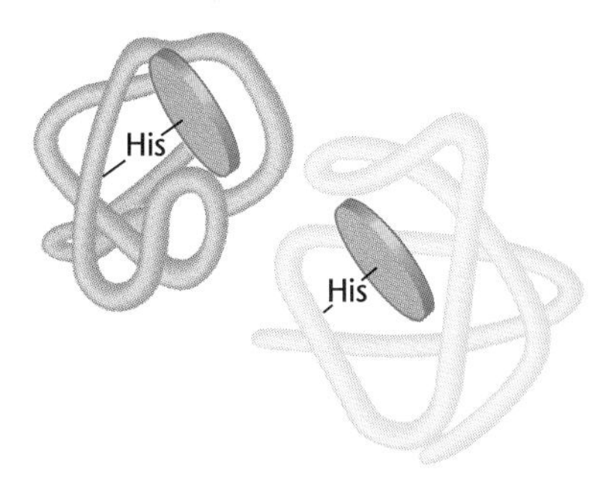

Two pairs of protein strands, like the pair shown here, interlock to give the quaternary structure of hemoglobin.

Quaternary structure: The way two or more protein chains fit together to form specific three-dimensional shapes.

Quaternary structure is important to the function of a protein such as hemoglobin. Two of the four protein chains are identical and contain 146 amino acid units each. People with sickle cell anemia have hemoglobin that differs from normal hemoglobin by only one amino acid unit in each of the 146-unit proteins. This one different amino acid causes the hemoglobin to be misshapen. These sickled

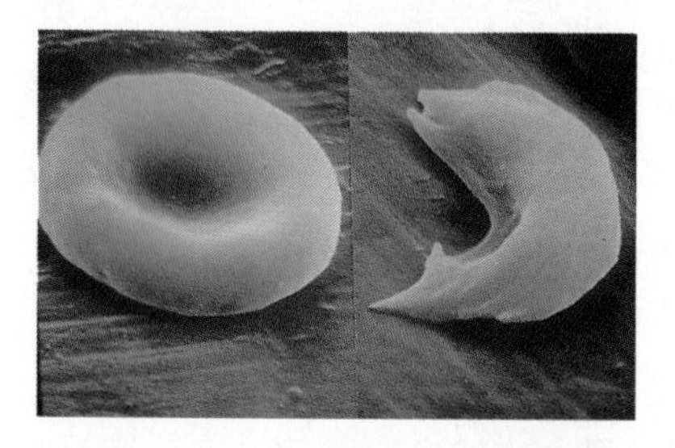

Left: A normal red blood cell. *Right:* A sickled cell. A difference of only one amino acid unit in a protein can have a dramatic effect upon tertiary and quaternary structure.

cells survive only half as long as normal blood cells, causing a deficiency of oxygen in the blood. Thus, a small change in primary structure results in a significant difference in tertiary and quaternary structure, with usually fatal consequences.

Denaturation

We have discussed in some detail the levels of structure that exist in proteins. However, it should be pointed out that protein structure is very delicate and can be destroyed easily. Altering the three-dimensional structure of a protein is called **denaturation.** Heating egg white is an example of denaturing protein. The protein in egg white is a more or less transparent liquid. However, this protein structure is altered by heat so that it becomes a white opaque solid. Another example of de-

CHEMICAL WORLD

Soap

Pioneer families struggled to survive by living off the land. The men worked the fields and tended the animals, and the women worked at the house. Thus, the women raised chickens, grew gardens, supervised the children, fixed meals, cleaned the house, and did the laundry. However, before they could do the laundry, they needed soap. Because soap was not readily available, they made their own.

Everything that went into the soap-making process was obtained from materials at hand. One necessity was fat, either the skillet residue after frying meat or the lard produced by heating raw animal fat, especially pig fat. The other necessity was alkali, mainly metal hydroxides and carbonates, obtained by leaching wood ashes. The alkali and fat were then heated together in an iron kettle. Today, we understand that the pioneer's soap formed in a reaction of triglycerides with sodium hydroxide:

$$\begin{array}{ccccccc} \overset{\displaystyle O}{\|} & & & & & & \\ H_2COCR & & & & H_2COH & & \overset{\displaystyle O}{\|} \\ \overset{\displaystyle O}{\|} & & & & | & & \\ HCOCR & + & 3NaOH & \rightarrow & HCOH & + & 3RCO^-Na^+ \\ \overset{\displaystyle O}{\|} & & & & | & & \\ H_2COCR & & & & H_2COH & & \\ \text{triglyceride} & & \text{found in alkali} & & \text{glycerol} & & \text{soap} \end{array}$$

After about 1900, lye (NaOH) could be obtained at a store instead of by processing wood ashes. The soap made with lye was often called lye soap. If too much lye was used, the soap could produce chemical burns; and if not enough lye was used, the soap was greasy. However, some soapmakers were skilled enough to avoid both extremes and could produce skin soaps as well as laundry soaps.

It is doubtful that the frontier folks knew why the soap worked, but the theory is now quite well de-

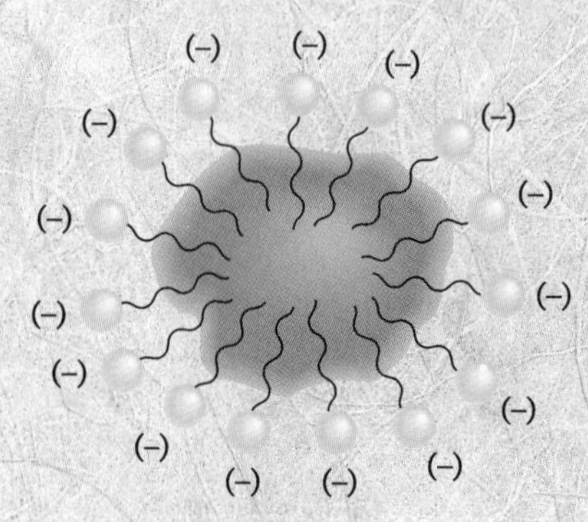

FIGURE 1 Soap forms a micelle with a grease particle. Micelles are hydrophilic and are easily washed away with water.

naturing is the curdling of milk when lemon juice is added. The protein structure in the milk is changed by the acid in the juice so that the protein precipitates.

Because protein structure involves several kinds of attractions, there are several factors that can cause denaturation. Primary structure is considerably more durable than the other levels of structure because primary structure is held in place by strong covalent bonds. Thus, proteins are usually denatured by breaking the other types of attractions. These can be disrupted by heating, changing the pH, or introducing an organic solvent, such as alcohol, or heavy metals. Heating or exposing to alcohol are common ways of killing bacteria by denaturing their proteins. Heavy metals are poisonous because they denature vital proteins by reacting with sulfur-containing groups. Egg whites can be denatured not only by heating, but also by changing the pH. For example, pouring vinegar (containing acetic acid) on a broken fresh egg causes it to appear to be cooked at the vinegar–protein interface.

Heating an egg denatures the protein so that it changes to an opaque solid.

veloped. The R group is a long nonpolar hydrocarbon chain, typically containing 15–17 carbon atoms. This tail is hydrophobic and thus is compatible with nonpolar oil and grease. On the head end of the soap is an ionic group ($—CO_2^- Na^+$) that is hydrophilic. It is not compatible with oil and grease but is very compatible with water. The attraction between soap and oil results in very small particles called micelles that attract water molecules through ion–dipole attractions. However, because micelles are negatively charged, they repel each other and stay suspended rather than merging to form a large drop. The soap is acting as an emulsifier, that is, it enables oil to form an emulsion in water (Figure 1).

Soaps have one serious drawback. They form an insoluble material by reacting with Mg^{2+} and Ca^{2+} ions in hard water:

$$\underset{\substack{\text{soluble soap} \\ \text{(easily suspended in water)}}}{2RC(=O)ONa} + Ca^{2+} \rightarrow \underset{\substack{\text{insoluble soap} \\ \text{(precipitates)}}}{(RC(=O)O)_2Ca} + 2Na^+$$

The precipitated soap is difficult to remove from clothing and is the tattletale-gray in white garments. Moreover, it is the principal component of bathtub-ring. Furthermore, it leaves a residue on hair after shampooing, resulting in a dull rather than a bright and shiny appearance. The same is true for dishes.

To overcome the problem of insoluble soap, chemists developed synthetic detergents. Detergents have structural features similar to those of soap: a long hydrophobic hydrocarbon tail and a hydrophilic polar or ionic head. However, the calcium and magnesium salts of detergents do not precipitate like those of soap, so they do not leave objectionable residues (Figure 2). Out with the tattletale-gray!

Hydrophilic head

$SO_3^- Na^+$

$CHCH_3$
CH_2
CH_2
CH_2
CH_3

Hydrocarbon tail

FIGURE 2 A typical synthetic detergent molecule.

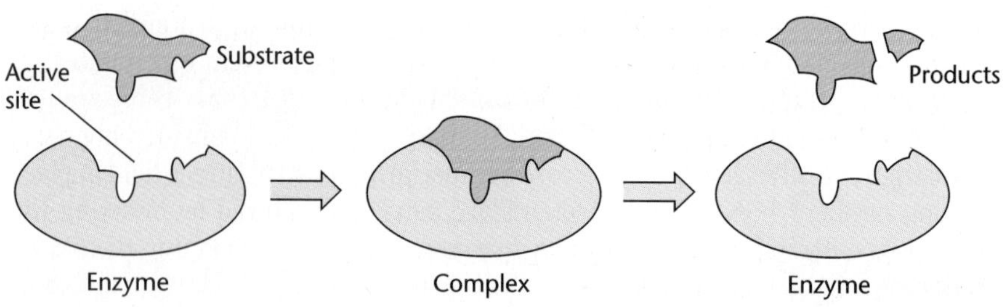

FIGURE 20.7 The lock-and-key model of enzyme action. After the reaction occurs, the products separate and the enzyme is ready for a new substrate molecule.

Enzymes

Enzymes are a highly specialized class of proteins that catalyze biochemical reactions due to their specific tertiary structure. Without enzymes, biochemical reactions would be so slow and inefficient that life could not exist.

The actual structure of an enzyme is complex and difficult to illustrate in a realistic drawing. For this reason we use a schematic (Figure 20.7) to represent the lock-and-key model for enzyme mechanism. Because of its specific tertiary structure, an enzyme often can fit the shape of only one reacting substance, or **substrate,** like a key in a lock. In fitting a substrate closely, an enzyme is thought to function by orienting the reacting groups for formation of bonds, by assisting the making (or breaking) of bonds by influencing polarities, or both. The position in the structure of an enzyme that influences the reaction is called the **active site.** There are many types of enzymes, and each catalyzes a specific reaction. Enzymes are so efficient that they can interact with several hundred thousand substrate molecules per second. Thus, only a very small amount of an enzyme is necessary for a reaction.

Enzymes can be deactivated by several means. They can be inhibited by binding to a substrate imposter, that is, a molecule of the same general shape as the substrate. Enzymes can also be inhibited by chemicals that react at their active sites or at locations that are vital to achieving a "fit" with a substrate. As with other proteins, enzymes can be denatured by heating, changing the pH, or reacting with heavy metals.

20.2 Carbohydrates

Learning Objectives

- Describe the general structure of sugars.
- Describe the principal differences in the structure of starch and cellulose.
- Describe the structure and function of glycogen.

Carbohydrates are polyhydroxy aldehydes and ketones and their derivatives. Because many carbohydrates have the empirical formula CH_2O (or $C \cdot H_2O$), early chemists thought these compounds must be hydrated carbon (hence their name). Carbohydrates are widely distributed in nature and include such familiar substances as sugars, starches, and cellulose. The latter two substances are polymers, as we shall see. Sugars and starches are used by humans and many animals as food sources.

Carbohydrates are named using the suffix *-ose.* Common names are widely used. For example, glucose and fructose are isomeric sugars that have the molecular formula $C_6H_{12}O_6$ (Figure 20.8). A prefix may be used to indicate the number of carbons. Thus, because glucose and fructose contain *six* carbons, they are called *hex*oses. Ribose, with *five* carbons, is a *pent*ose. Most sugars found in living organisms are pentoses and hexoses. You will also note that both glucose and ribose are aldehydes (Section 19.7). Therefore, glucose and ribose may also be called aldoses (*ald* for aldehyde). On the other hand, because fructose is a ketone, it is a ketose (*ket* for ketone). Frequently two prefixes are combined to indicate both classifications. Thus, glucose is an aldohexose; fructose, a ketohexose; and ribose, an aldopentose.

Glucose is also called dextrose.

While sugars may be represented by continuous-chain formulas (refer to Figure 20.8), most sugars are found mainly in cyclic form. The cyclic structure of glucose results when the hydroxyl group on C-5 reacts with the carbonyl group, as shown in the following equation.

Chemists often use the continuous-chain formula of sugars for convenience in distinguishing structural differences.

$$
\begin{array}{c}
\text{H}\diagdown\;\diagup\!\!\!\diagup\text{O} \\
{}_{1}\text{C} \\
\text{H}-{}_{2}\text{C}-\text{OH} \\
\text{HO}-{}_{3}\text{C}-\text{H} \\
\text{H}-{}_{4}\text{C}-\text{OH} \\
\text{H}-{}_{5}\text{C}-\text{OH} \\
{}_{6}\text{CH}_2\text{OH}
\end{array}
\;\rightarrow\;
\begin{array}{c}
\text{OH} \\
\text{H}-{}_{1}\text{C}- \\
\text{H}-{}_{2}\text{C}-\text{OH} \\
\text{HO}-{}_{3}\text{C}-\text{H} \\
\text{H}-{}_{4}\text{C}-\text{OH} \\
\text{H}-{}_{5}\text{C}-\text{O}- \\
{}_{6}\text{CH}_2\text{OH}
\end{array}
\;+\;
\begin{array}{c}
\text{H} \\
\text{HO}-{}_{1}\text{C}- \\
\text{H}-{}_{2}\text{C}-\text{OH} \\
\text{HO}-{}_{3}\text{C}-\text{H} \\
\text{H}-{}_{4}\text{C}-\text{OH} \\
\text{H}-{}_{5}\text{C}-\text{O}- \\
{}_{6}\text{CH}_2\text{OH}
\end{array}
$$

Hemiacetal isomers

This product of the addition of an alcohol to an aldehyde is called a hemiacetal (Section 19.7). Because the OH of C-5 can react at either side of the carbonyl group, two stereoisomers are formed. One is called α-glucose and the other is called β-glucose (Figure 20.9).

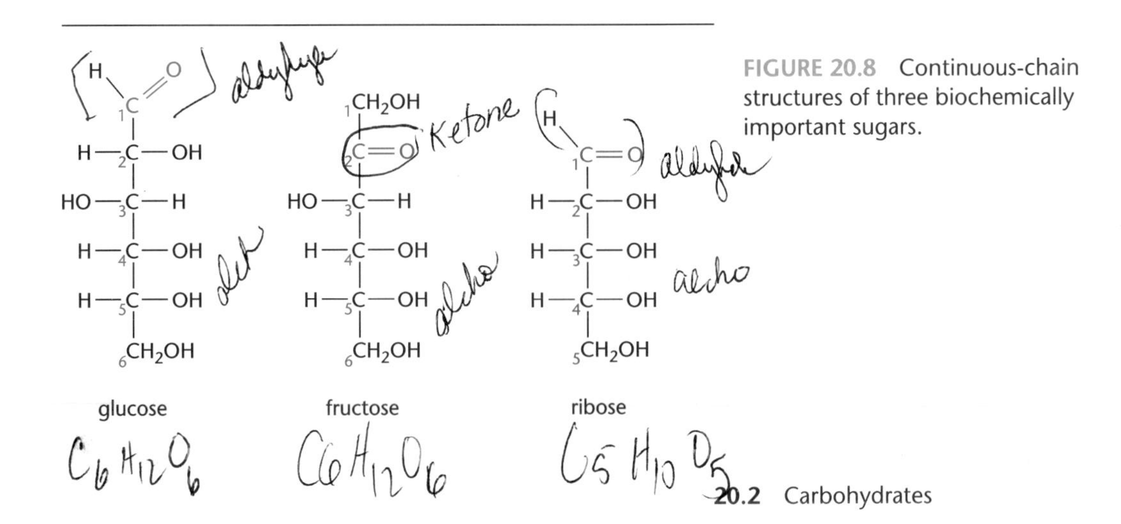

FIGURE 20.8 Continuous-chain structures of three biochemically important sugars.

FIGURE 20.9 Structural representation of the cyclic forms of α-glucose and β-glucose. Note that on C-1 the OH is down in α-glucose, whereas the OH is up in β-glucose.

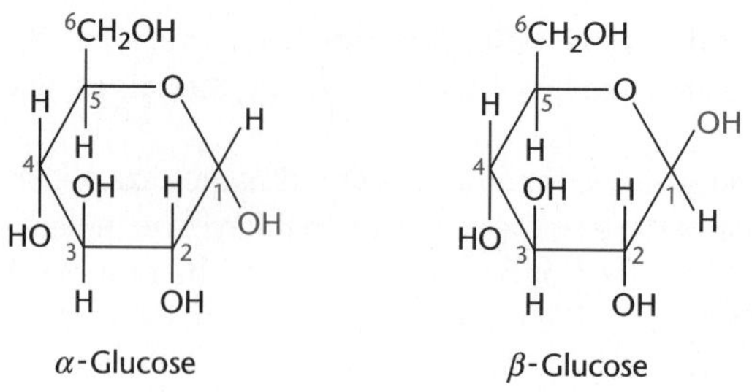

At this point the distinction between the two products may seem trivial because the only difference is the orientation of the OH and the H on C-1. However, the structures shown in Figure 20.9 more accurately show the difference in the structure of α- glucose and β-glucose. Moreover, this difference in structure at C-1 accounts for some of the differences in the properties of starch and cellulose. For example, we can digest starch, composed of α-glucose units, but not cellulose, composed of β-glucose units.

So far we have considered only **monosaccharides,** that is molecules with only one sugar unit. Combining two monosaccharides gives a **disaccharide,** as shown in Figure 20.10. Sucrose, common table sugar, is a disaccharide formed by the combination of glucose, the sugar found commonly in the cells of our bodies, and fructose, which is found in fruits and honey. (See Table 20.1 for the relative sweetness of some compounds.)

FIGURE 20.10 Sucrose (c) is formed by the loss of water in a reaction of glucose (a) with fructose (b).

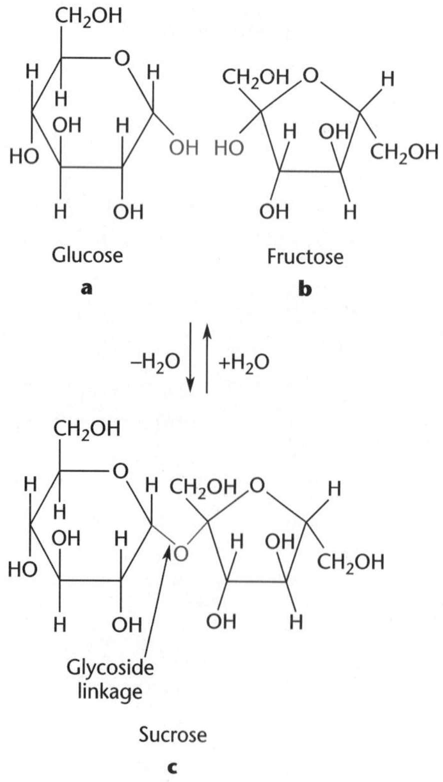

TABLE 20.1 Sweetness of Some Compounds Relative to Sucrose as 100

Compound	*Class*	*Relative Sweetness*
Glucose	Monosaccharide	74
Fructose	Monosaccharide	173
Galactose	Monosaccharide	70
Lactose	Disaccharide	16
Sucrose	Disaccharide	100
Maltose	Disaccharide	33
Aspartame*	Dipeptide	16 000
Saccharin*	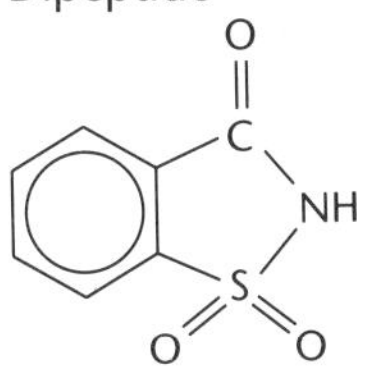	30 000

*Note that aspartame and saccharin are *not* saccharides.

Larger molecules called **polysaccharides** are formed when many monosaccharide units are joined. The two most common polysaccharides are starch and cellulose. Both are polymers formed by loss of water between glucose molecules in a manner similar to the formation of sucrose from glucose and fructose (Figure 20.10). The primary structures of starch and cellulose differ because of the difference in the orientation of the OH group on C-1 of α- and β- glucose. Compare the two structural formulas in Figure 20.11. You will note that the oxygen of C-1 is down for starch (α-linkage), and the oxygen of C-1 for celluose is up (β-linkage).

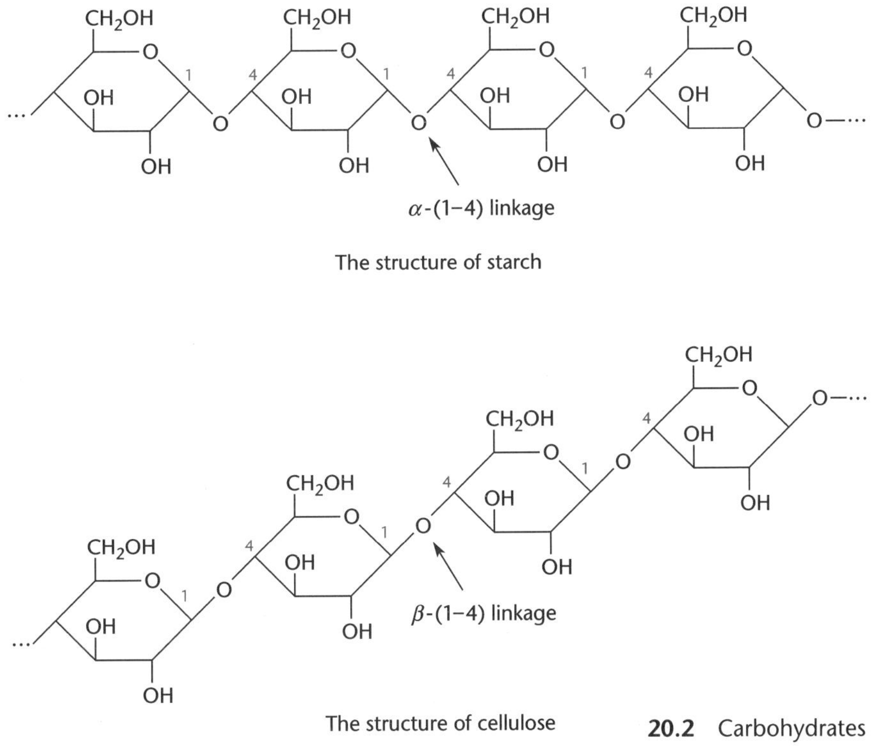

α-(1–4) linkage

The structure of starch

β-(1–4) linkage

The structure of cellulose

FIGURE 20.11 Starch and cellulose differ only in the orientation of the oxygen on C-1.

Cells store energy in the form of glycogen, which has a highly branched structure as shown in this drawing. Enzymes cause glucose units to be released from the ends of the branches when the cell calls for more energy.

Starch occurs in many plants where it functions to store the glucose manufactured by the plants. There are two types of starch. *Amylose* is starch that is made up of long unbranched chains. It has a helical secondary structure and forms a colloid (Section 14.7) in hot water. The second type, *amylopectin,* is highly branched and precipitates in water.

Humans and many animals store glucose in a polymeric form called *glycogen.* It is highly branched and is structurally similar to amylopectin. Glycogen is stored in liver and muscle cells and serves as a source of reserve energy when the glucose level in the blood is low. Enzymes that break off glucose units can work only on the ends of chains. So, when a cell needs energy in an emergency, the ends of the many branches of glycogen can release glucose rapidly.

Storing glucose in the form of glycogen is an important design feature of the cell. To store a sufficient amount of glucose as the monosaccharide would produce a highly concentrated glucose solution in the cell. Because of osmosis (Section 14.6), the cell would swell to a very large size and eventually burst. Thus, glycogen is a clever way to store glucose, avoiding a high solution concentration that would result in rupture of cells.

Cellulose, like starch, is produced in plants from glucose. However, the secondary structure of cellulose is rod-shaped, which allows the chains to closely align with one another and to hydrogen bond. This gives cellulose more structural stability than starch. Thus, it is suitable for supporting plant structures. When cellulose fibers are bonded together with lignin, a noncarbohydrate natural glue, a very strong structure results.

The enzymes that can break down starch to glucose are not effective with cellulose. Thus, many organisms that do not have these enzymes cannot utilize cellulose directly. Grass-eating animals that obtain nutrition from cellulose have microorganisms in their digestive systems that will break down cellulose molecules into glucose. This is also true for wood-eating termites. When we consider all the plants, including trees, the amount of glucose on the earth is amazing. If it were easily accessible for human nutrition, starvation would be rare.

This giant redwood tree is supported by cellulose, a material made from β-glucose.

Lipids

20.3

Learning Objectives

- Write a general structural formula of a triglyceride.
- Describe the differences in triglycerides in fats and oils.
- Recognize a compound as a steroid from its structural formula.

Water-insoluble substances in cells are called **lipids.** Simple lipids that are esters of glycerol are called **triglycerides** (Section 19.9) and are represented by the formula

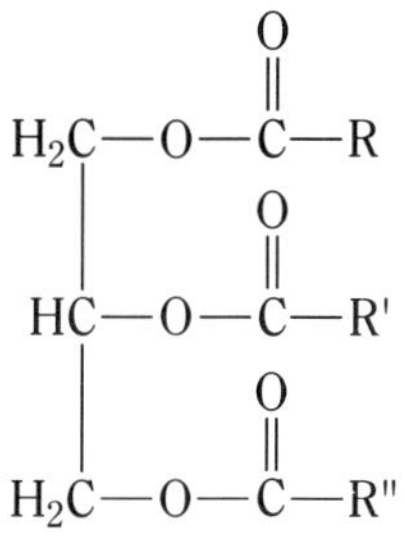

$$\begin{array}{l} H_2C-O-\overset{\displaystyle O}{\overset{\|}{C}}-R \\ \;\;| \\ HC-O-\overset{\displaystyle O}{\overset{\|}{C}}-R' \\ \;\;| \\ H_2C-O-\overset{\displaystyle O}{\overset{\|}{C}}-R'' \end{array}$$

where R, R′, and R″ are usually hydrocarbon chains 15–17 carbons long. These chains can be saturated or unsaturated. Those that are saturated are called **fats** and are solids at room temperature. Most natural fats are produced by animals rather than plants. Triglycerides that have two or more C=C bonds are said to be **polyunsaturated** and are oils at room temperature. Plants are the major source of oils. Vegetable oils can be converted to fats by hydrogenation of some of the double bonds (Section 19.3). These hydrogenated materials are found in grocery stores as shortening or as margarine. Polyunsaturated fats and oils are reported to be more healthful than saturated fats because they do not contribute to the production of cholesterol.

Glycerol can be purchased as "glycerin" in many grocery stores. In a dilute aqueous solution it acts as a skin moisturizer.

Triglycerides can be hydrolyzed just like any other ester. The products of **hydrolysis** are glycerol and **fatty acids** (Figure 20.12). The reverse reaction, called esterification (Section 19.9), is the process by which cells produce triglycerides with the help of enzymes.

Hydrolysis: A chemical reaction in which a water molecule is split.

Triglycerides occur many places in the body. The most common locations are in cell membranes and in body fat, which serves as a storehouse of excess energy. Fats store almost 2.5 times more energy per gram than carbohydrates, so they

Fatty acid: A carboxylic acid containing 16–18 carbons.

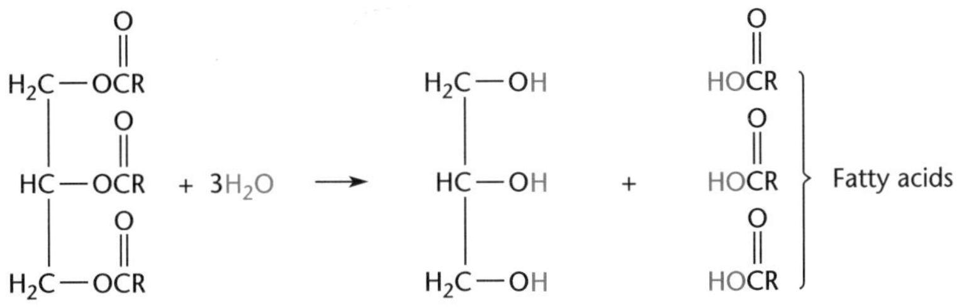

$$\begin{array}{l} H_2C-O\overset{\displaystyle O}{\overset{\|}{C}}R \\ \;\;| \\ HC-O\overset{\displaystyle O}{\overset{\|}{C}}R \\ \;\;| \\ H_2C-O\overset{\displaystyle O}{\overset{\|}{C}}R \end{array} + 3H_2O \longrightarrow \begin{array}{l} H_2C-OH \\ \;\;| \\ HC-OH \\ \;\;| \\ H_2C-OH \end{array} + \left.\begin{array}{l} HO\overset{\displaystyle O}{\overset{\|}{C}}R \\ HO\overset{\displaystyle O}{\overset{\|}{C}}R \\ HO\overset{\displaystyle O}{\overset{\|}{C}}R \end{array}\right\} \text{Fatty acids}$$

Triglyceride — Glycerol

FIGURE 20.12 Hydrolysis of a triglyceride.

CHEMICAL WORLD

Reforming Your Hair

Some people with straight hair want it curly, or at least less straight. Some people with curly hair want it straight, or at least less curly. It's fortunate that the structure of hair is easy to modify.

Hair is a fibrous protein. Whether it's naturally straight or curly depends on its protein structure, which depends on hydrogen bonding and disulfide linkages between various sites on the long protein molecules, which depends on the person's genetic makeup. Hydrogen bonds can be broken by moisture and heat, and disulfide linkages can be broken by certain chemicals. In other words, the structure of hair can be modified just as other proteins can be denatured.

So, wetting hair breaks many of the hydrogen bonds, which reform as the hair dries. If you bend or straighten the hair while it's wet, then the bonds reform to maintain that form. Heating wet hair breaks more of the bonds and allows more "curling" bonds to form as it dries. Of course, if you go out in the rain or fog, all is lost (Figure 1).

The disulfide linkages follow the same curl-or-straight rules as the hydrogen bonds, but they're covalent bonds and are harder to break. If you can rearrange those, your curl or noncurl is less temporary. First, we treat the hair with a reducing agent, which breaks the S—S bonds and forms two SH groups. Then we reform the hair in some way, plus or minus curl, whichever you desire, and finally treat it with an oxidizing agent (neutralizer), which "sets" the hair by forming new S—S bonds. These covalent bonds are "permanent," and your restructured hair will remain more or less permanent (Figure 2).

A permanent can give a curl to straight hair.

However, like human nature, your hair will revert to old habits as it grows out. Curly or straight—it's in the genes!

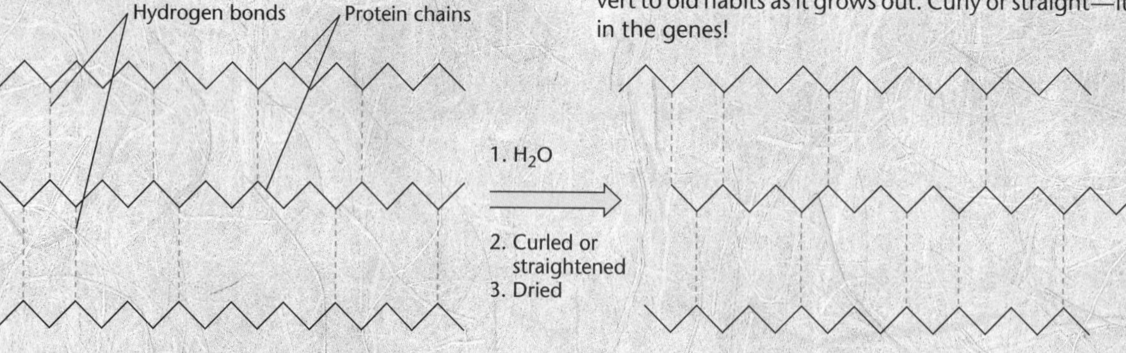

FIGURE 1 Wetting hair breaks hydrogen bonds, which form again when the hair dries.

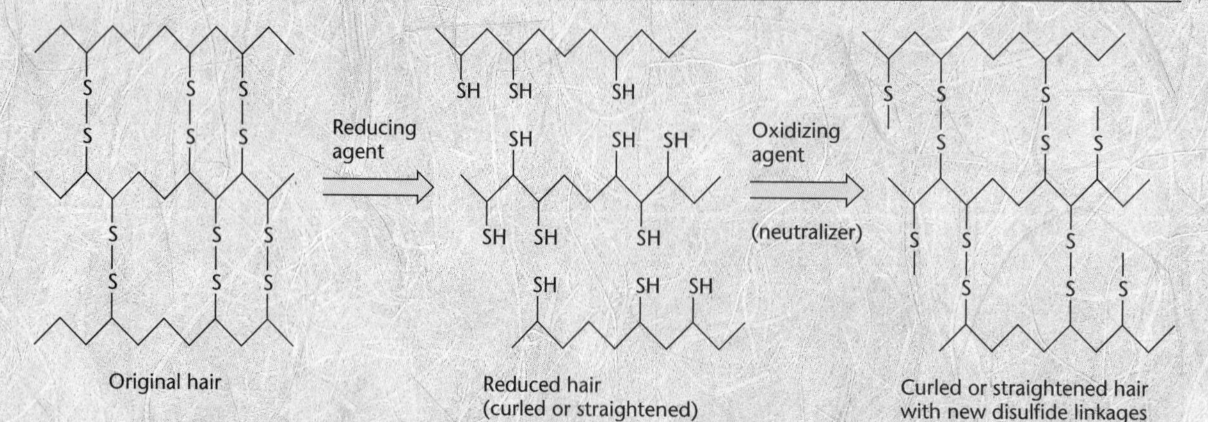

FIGURE 2 Hair structure is changed with a permanent.

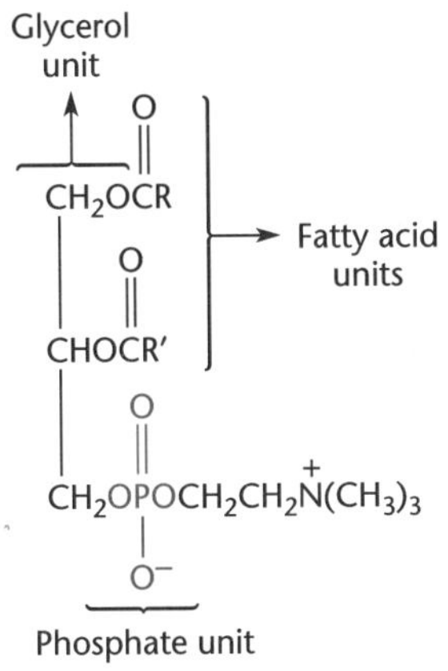

Lecithin, which is used as an emulsifier in some candy bars, is a phospholipid.

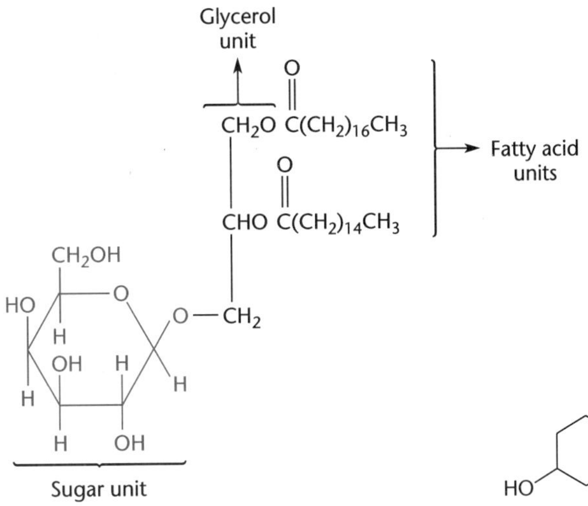

A glycolipid.

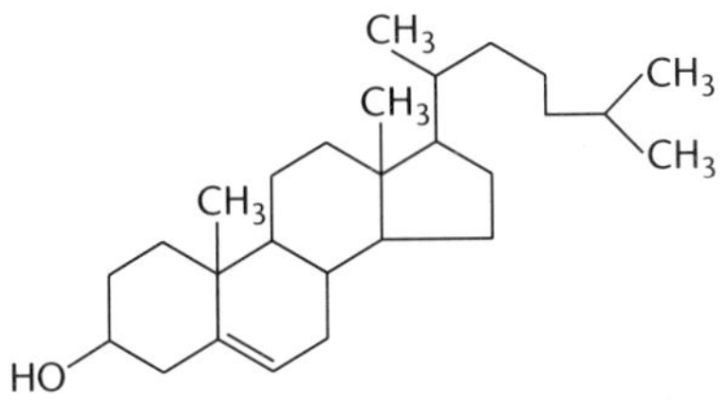

Cholesterol.

provide an efficient way to store energy. The average person would be huge if excess energy had to be stored as carbohydrates.

Other types of lipids are variations of triglycerides. **Phospholipids** have both a fatty acid ester part and a phosphate ester group. **Glycolipids** contain a sugar unit instead of the phosphate ester.

There are other lipids that have no chemical relationship to glycerides except that they are water insoluble. These are the **steroids,** which have the polycyclic ring structure shown in Figure 20.13. Cholesterol is the principal steroid in the body. It is required by the body to form other steroids, including sex hormones. However, cholesterol is probably best known for its role in the formation of the plaque present in "hardened" arteries. It can also precipitate in the gall bladder to form gall stones.

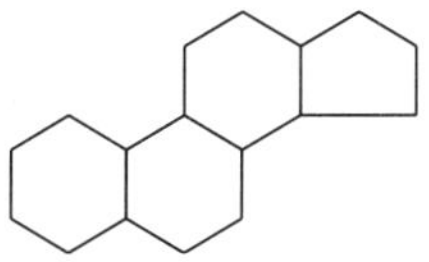

FIGURE 20.13 Skeletal structure of steroids.

Nucleic Acids

20.4

Learning Objectives

- Explain the general structure of nucleic acids.
- Describe the structural differences between DNA and RNA.
- Describe the functions of messenger RNA, ribosomal RNA, and transfer RNA.

Nucleic acids are responsible for preserving the blueprint for all the characteristics and functions of an organism. They are also responsible for converting that blueprint into biological structures necessary to achieve the characteristics and perform the functions of a particular organism. To accomplish this, nucleic acids have an ingeneous design that is composed of three types of chemical groups: sugars (Section 20.2), phosphate, and organic bases. We will systematically build

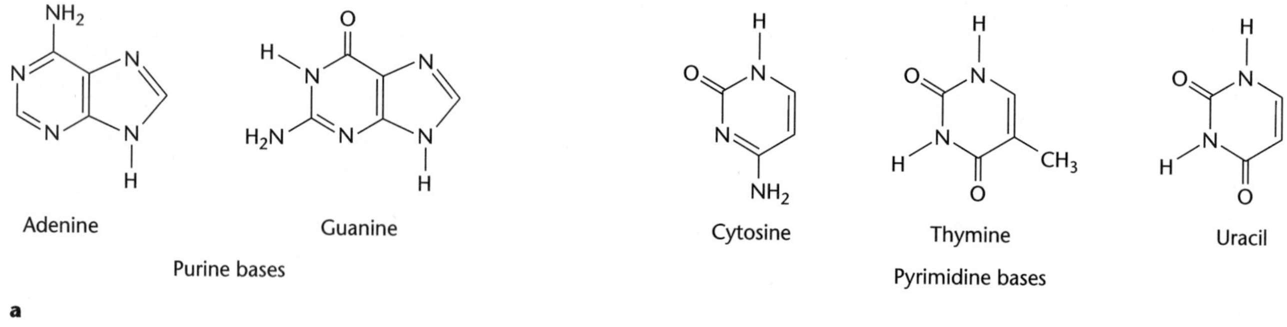

Adenine

Guanine

Purine bases

Cytosine

Thymine

Uracil

Pyrimidine bases

a

FIGURE 20.14 (a) The principal organic bases, and (b) the two sugars found in nucleic acids.

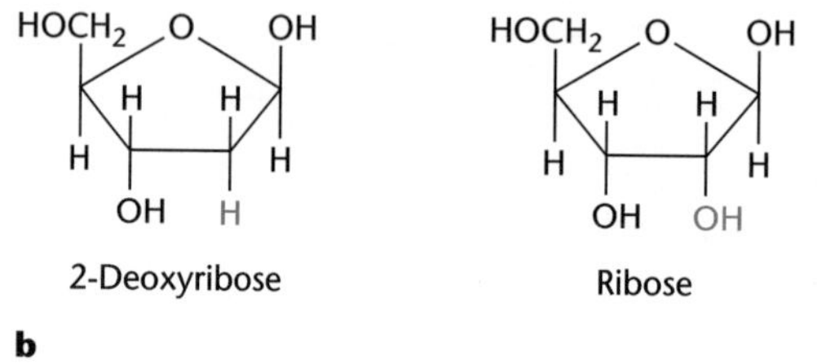

2-Deoxyribose

Ribose

b

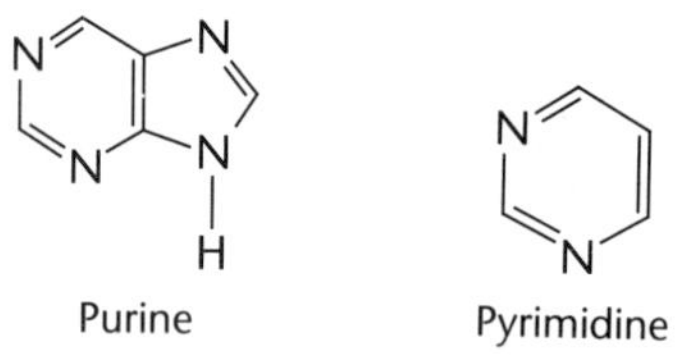

Purine

Pyrimidine

the primary structure from these groups to demonstrate the composition of nucleic acids.

Organic bases and sugars are shown in Figure 20.14. The organic bases are heterocyclic compounds (Section 19.8) classified as purines and pyrimidines. Note that pyrimidines contain one ring and purines contain two fused rings. The sugars that are found in nucleic acids are ribose and 2-deoxyribose. Both are pentoses, but 2-deoxyribose has no OH group at C-2. When a purine or pyrimidine combines with one of the sugars, they form a **nucleoside** (Figure 20.15).

When phosphate reacts with an OH group on the sugar of a nucleoside, a **nucleotide** is formed (Figure 20.16). The phosphate group on adenosine mono-

FIGURE 20.15 Ribose combines with adenine to produce adenosine, a nucleoside.

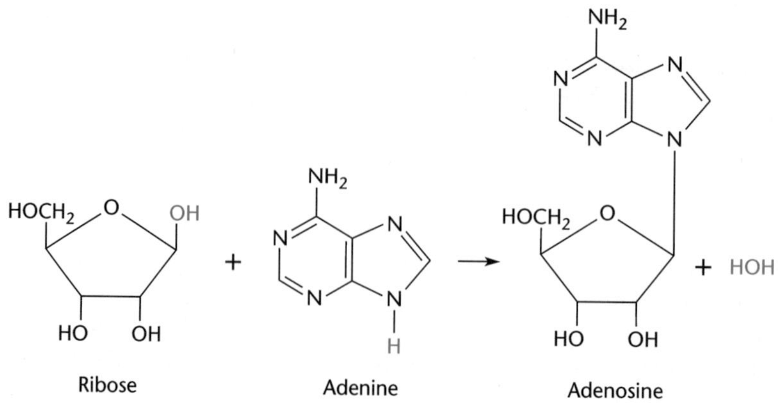

Ribose + Adenine → Adenosine + HOH

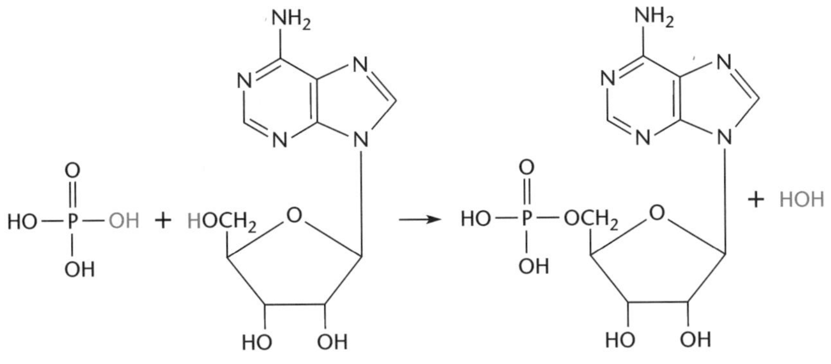

Phosphate · Adenosine · Adenosine monophosphate

FIGURE 20.16 Adenosine combines with phosphate to produce adenosine monophosphate, a nucleotide.

phosphate is capable of reacting with a sugar molecule of another nucleotide, and so on, producing the polymeric structure of a nucleic acid (Figure 20.17). This structure can be represented schematically as

```
        base               base               base
         |                  |                  |
... sugar—phosphate—sugar—phosphate—sugar—phosphate ...
```

Thus, a nucleic acid has a polymer backbone composed of alternating phosphate and sugar units with pendant base groups. We will see that these pendant groups are the key to information storage. When the sugar is 2-deoxyribose, the nucleic acid is deoxyribonucleic acid (DNA). When the sugar is ribose, the nucleic acid is ribonucleic acid (RNA).

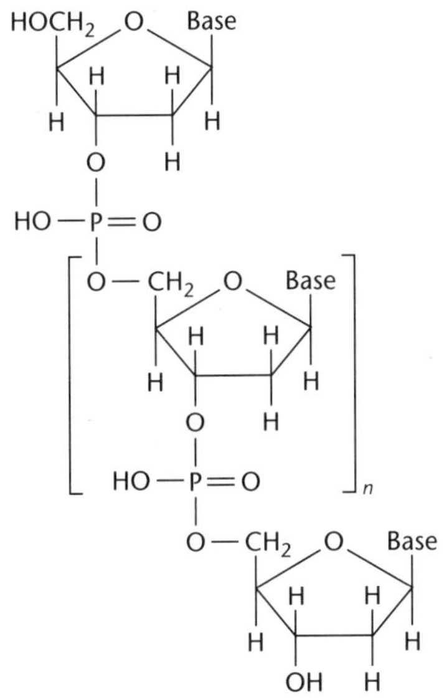

FIGURE 20.17 The general structure of a nucleic acid, a polynucleotide.

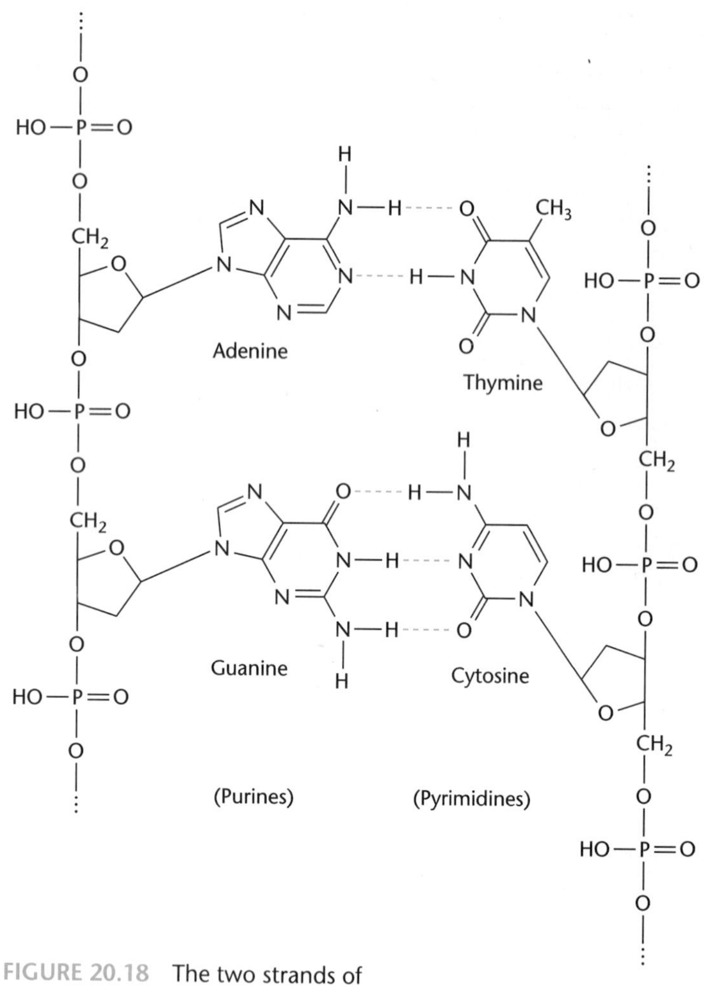

FIGURE 20.18 The two strands of DNA are held together by hydrogen bonding between complementary base pairs.

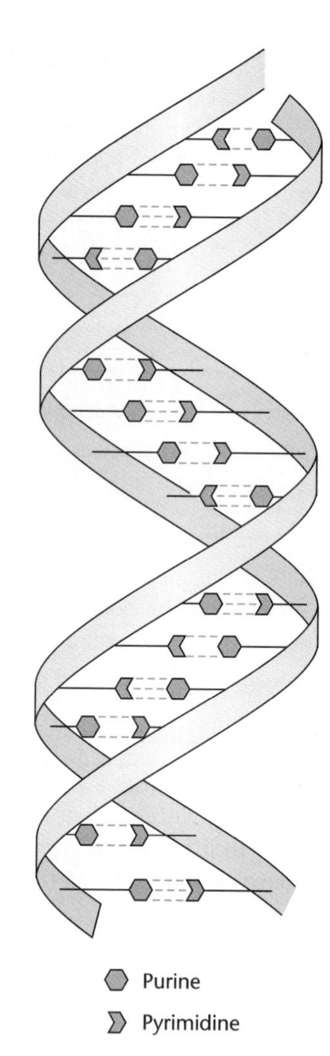

The quaternary structure of DNA is a double helix.

Nucleic acids, like proteins, can be described in terms of their levels of structure (Section 20.1). A major difference in the primary structure of DNA and RNA is the sugar unit. The differences in the secondary, tertiary, and quaternary structures of DNA and RNA are also significant. DNA has a helical secondary structure held in place by quaternary structure with a second helix (refer to Figure 20.19). Thus, DNA is called a **double helix.** However, DNA has no tertiary structure because the chain does not usually fold back on itself. RNA, on the other hand, occasionally has a helical secondary structure, but often the secondary structure is disordered, or random. This randomness and the absence of quaternary structure allow RNA to have tertiary structure; that is, it can fold back on itself. The tertiary structure is especially important to the function of transfer RNA, which is discussed later.

The double helix of DNA is held together by hydrogen bonds between a purine on one chain and a pyrimidine on the other chain (Figures 20.18 and 20.19). Thus, each double helix has the same number of pyrimidine and purine groups, and each chain is a complement of the other. That is, when one chain has a purine at a given location, the other has a pyrimidine.

One DNA double helix carries all of the information for the characteristics and functions of an organism.

The information that gives an organism its specific characteristics and functions is carried as a code by DNA. The code is the sequential arrangement of the four purine and pyrimidine bases on the chains. Human DNA contains about 5.5 billion bases along the chain. A segment of DNA that carries information about the primary structure of a specific protein is called a **gene.** About one million genes are carried on human DNA. Every cell in an organism, except mature red blood cells, contains DNA. It is essential that new DNA with exactly the same code is made every time a cell divides.

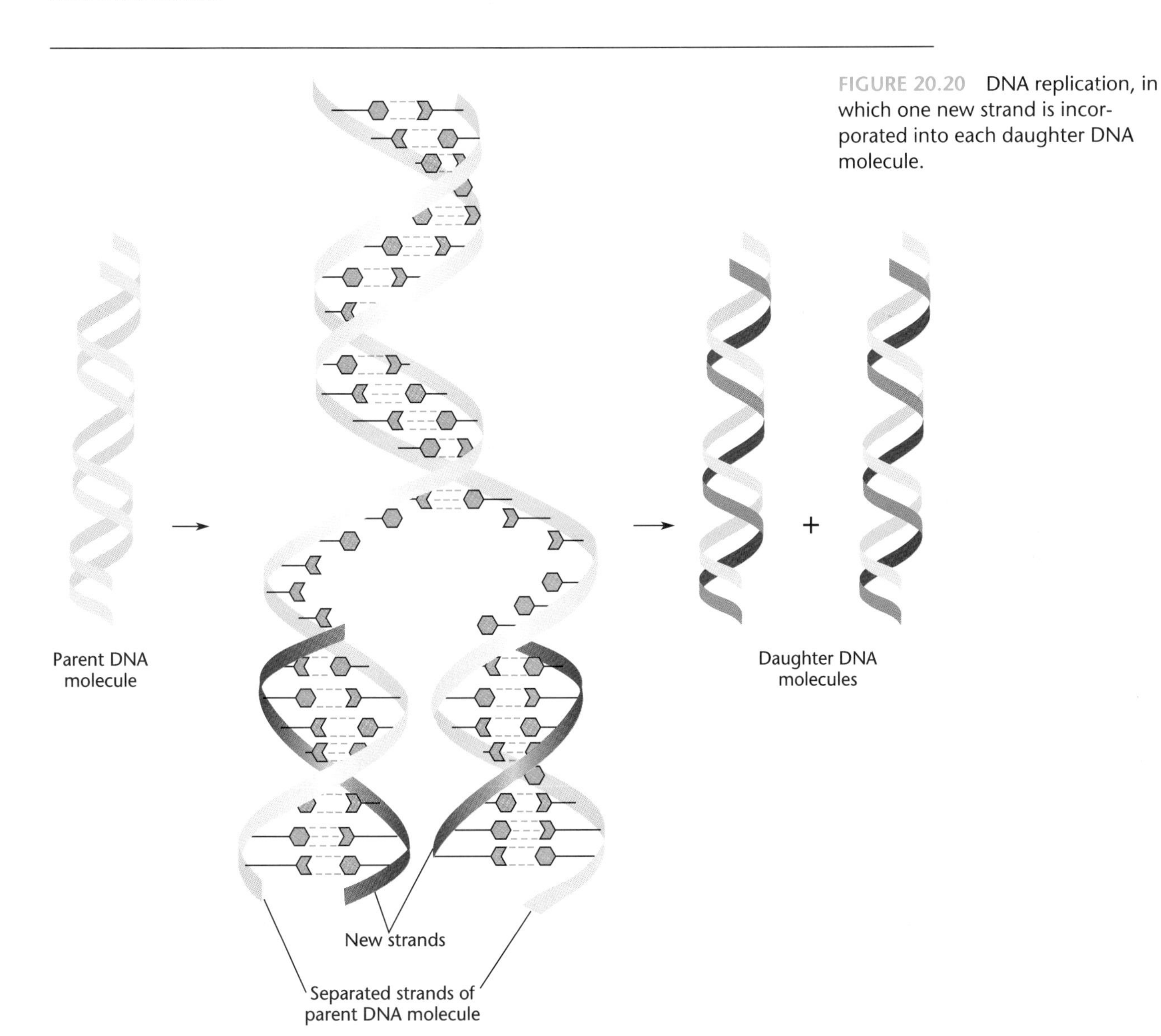

FIGURE 20.20 DNA replication, in which one new strand is incorporated into each daughter DNA molecule.

FIGURE 20.21 A tRNA molecule has tertiary structure that places the three-base code in a position to interact with mRNA, enabling the amino acid to be placed in the proper position in a newly synthesized protein.

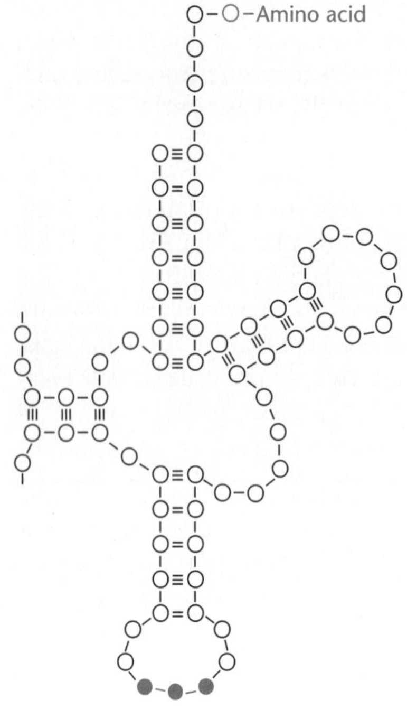

Proteins are made at ribosomes, using the code transcribed from DNA by mRNA.

The process of producing an exact replica of DNA when cells divide is called **replication.** In this process one end of the double helix separates into two strands. Each of the strands acts as a template, or pattern, on which a new complementary strand is synthesized. Thus, each double helix that results from replication contains an old strand and one new strand (Figure 20.20).

RNA copies the code from a gene and carries the information to **ribosomes** (cellular substructures) where proteins are made. The process of copying the code is called **transcription.** This is done by messenger RNA (mRNA). Ribosomal RNA (rRNA) makes up the ribosomes where the mRNA will be attached. Amino acids are brought to the ribosomes by transfer RNA (tRNA) to be incorporated into a protein in the specific order required by mRNA. A tRNA molecule has tertiary structure that places a three-base code in a position to interact with mRNA, enabling the amino acid to be placed in the proper position in a newly synthesized protein (Figure 20.21). An enzyme promotes the rapid formation of the amide bond in the reaction of an amino group of one amino acid with a carboxyl group of another amino acid (Section 20.1).

This discussion of nucleic acids illustrates the incredible order required to preserve the blueprint for all the characteristics and functions of an organism. A remarkable amount of order is also necessary to convert the code into biological structures. Not only are highly structured nucleic acids required in this process, but very specific enzymes are necessary to catalyze reactions under the relatively mild conditions existing within a cell.

Chapter Review

Directions: From the choices listed at the end of this Chapter Review, choose the word or term that best fits in each blank. Use each choice only once. Answers appear at the end of the chapter.

Section 20.1 A major component of muscle tissue is ___________ (1). Its functional group, sometimes called a peptide group, is also called an ___________ (2) group. It is formed when ___________ (3) acids are condensed to form polymers. These polymers are in some ways similar to ___________ (4), a synthetic polyamide.

Proteins have several levels of structure. The primary structure is that type of structure that involves amino acid ___________ (5). When a protein chain bends around on itself, the shape of the molecule is called ___________ (6) structure. On the other hand, the shape of the chain itself is called ___________ (7) structure. In some cases several protein chains fit together, causing ___________ (8) structure. Two types of secondary structure at the α-helix and ___________ (9) sheet. Tertiary structure in proteins is maintained by ___________ (10) bonding, ___________ (11) bridges (ionic), ___________ (12) linkages (covalent bonds), and ___________ (13) interactions (dispersion forces). Considering the strength of these forces, disulfide linkages and salt bridges are the ___________ (14) and hydrophobic interactions are the ___________ (15). Altering three-dimensional protein structure is called ___________ (16). This can happen by heating, treating with an organic solvent, or reacting with a heavy ___________ (17) or an acid.

Biochemical catalysts are called ___________ (18). They fit the ___________ (19), the reacting molecule, like a ___________ (20) fits a lock. Enzymes that have been deactivated because they have reacted with a substrate imposter are said to be ___________ (21).

Section 20.2 Carbohydrates contain ___________ (22) or hydroxyl groups. Carbohydrates containing only one sugar unit are called ___________ (23). These molecules are either aldehydes or ketones, depending upon the position of the ___________ (24) group. When two monosaccharides combine, a ___________ (25) results. The building block that makes up both starch and cellulose is ___________ (26). Starch and cellulose differ because of the difference in the orientation of the OH on ___________ (27) of α- and β-glucose. Most animals cannot use ___________ (28) for nutrition without help from the enzymes of other organisms.

Section 20.3 Lipids are all ___________ (29) in water. ___________ (30) are lipids that are esters of glycerol. Those that contain C═C bonds are said to be ___________ (31) and are usually liquids at room temperature. When hydrolyzed, triglycerides produce glycerol and ___________ (32) acids. Lipids that contain phosphate ester groups are called ___________ (33). ___________ (34) are lipids that contain a polycyclic ring structure. An example that is involved in hardening of arteries is ___________ (35).

Section 20.4 ___________ (36) acids are biologically important molecules that are responsible for preserving genetic information. These materials are polymers that have a backbone containing phosphate and ___________ (37) units with ___________ (38) base groups. The bases are called purines and ___________ (39), and the sugars are ribose and ___________ (40). A compound formed from a base and a sugar is called a ___________ (41). When a phosphate group reacts with a nucleoside, a ___________ (42) forms.

The specific type of nucleic acid that contains the genetic code of an organism is ___________ (43). Its quarternary structure is a double ___________ (44) and is held together by hydrogen bonding of ___________ (45) pairs. Making another DNA molecule that is an exact duplicate of the first is called ___________ (46). A segment of DNA that contains a code for a specific characteristic or function is called a ___________ (47).

The nucleic acid responsible for implementing the genetic code is ___________ (48). Copying the message is called ___________ (49) and is carried out by ___________ (50). This molecule moves to a ribosome composed of ___________ (51) where protein is synthesized from amino acids carried to the ribosome by ___________ (52).

Choices: amide, amino, base, C-1, carbonyl, cellulose, cholesterol, denaturation, 2-deoxyribose, disaccharide, disulfide, DNA, enzymes, fatty, gene, glucose, helix, hydrogen, hydrophobic, inhibited, insoluble, key, mRNA, metal, monosaccharides, nucleic, nucleoside, nucleotide, nylon, OH, pendant, phospholipids, pleated, protein, pyrimidines, quaternary, rRNA, replication, RNA, salt, secondary, sequence, steroids, strongest, substrate, sugar, triglycerides, tRNA, tertiary, transcription, unsaturated, weakest

Study Objectives

After studying Chapter 20, you should be able to:

Section 20.1

1. Describe fibrous and globular proteins.
2. Identify and explain the four levels of protein structure.
3. Explain denaturation and how it occurs.
4. Explain the general structure of enzymes and a model for the mechanism of enzyme activity.

Section 20.2

5. Describe the general structure of sugars.
6. Describe the principal differences in the structure of starch and cellulose.
7. Describe the structure and function of glycogen.

Section 20.3

8. Write a general structural formula of a triglyceride.
9. Describe the differences in triglycerides in fats and oils.
10. Recognize a compound as a steroid from its structural formula.

Section 20.4

11. Explain the general structure of nucleic acids.
12. Describe the structural differences between DNA and RNA.
13. Describe the functions of messenger RNA, ribosomal RNA, and transfer RNA.

Key Terms

Review the definition of each of the terms listed here. For reference, they are listed by chapter section number. You may use the glossary as necessary.

20.1 active site, α-helix, denaturation, disulfide linkage, enzyme, fibrous protein, globular protein, hydrophobic interaction, peptide linkage, pleated sheet, primary structure, protein, quarternary structure, salt bridge, secondary structure, substrate, tertiary structure

20.2 carbohydrate, disaccharide, monosaccharide, polysaccharide

20.3 fat, fatty acid, glycolipid, hydrolysis, lipid, phospholipid, polyunsaturated, steroid, triglyceride

20.4 double helix, gene, nucleic acid, nucleoside, nucleotide, replication, ribosome, transcription

Questions and Problems

Answers to questions and problems with blue numbers appear in Appendix C.

Proteins (Section 20.1)

1. Describe the differences in structure between globular and fibrous proteins.
2. Draw the general formula for an α-amino acid. Indicate the α-carbon atom, the amino group, and the carboxyl group.
3. Explain how the R groups influence the compatibility of an amino acid with water.
4. From the formulas of amino acids in Figure 20.2, draw the structures of the following dipeptides. Circle the amide groups in the structures.
   a. lys-met b. met-lys c. ala-gly d. gly-ala
5. From the formulas of amino acids in Figure 20.2, draw structures of the six different tripeptides possible containing serine, glutamine, and cysteine (each used only once). Circle the amide groups in your structures.
6. Explain what is meant by the primary structure of proteins.
7. How many possible structures exist for a polypeptide composed of five different amino acids if each is used only once?
8. Explain what is meant by the secondary structure of proteins.
9. Describe the secondary structure of proteins that is known as the pleated sheet. List two materials that contain proteins with this structure.
10. How does the size of the R group of an amino acid affect the secondary structure of a protein?
11. In general terms, explain what is meant by the tertiary structure of a protein.
12. List four types of bonds or attractive forces that hold a protein in its tertiary structure and describe each. Which of these are strongest and which are the weakest?
13. Explain what is meant by the quaternary structure of a protein. Give an example of a protein that has quaternary structure.

14. What is meant by denaturation of proteins? List four ways that proteins can be denatured.
15. What is the term given to a protein that catalyzes biochemical reactions?
16. What is the term given to the specific portion of an enzyme where catalysis actually occurs?
17. What is the term for the molecule acted upon by an enzyme?
18. Explain what is meant by inhibition of an enzyme.

Carbohydrates (Section 20.2)

19. What functional groups are present in simple sugars (monosaccharides)?
20. What general name is given to sugars containing six carbon atoms and a ketone group?
21. What would be an appropriate general term for a molecule composed of three sugar units?
22. What is meant by the term polysaccharide?
23. What is the building block of both starch and cellulose? How do the primary structures of starch and cellulose differ?
24. Describe the secondary structure of cellulose.
25. Explain how the structure of glycogen facilitates the release of glucose.

Lipids (Section 20.3)

26. What characteristic is common to all lipids?
27. Draw the general structure of a triglyceride. What are the two components that make up a typical triglyceride?
28. Draw the ring structure common to all steroids.
29. Describe a beneficial use for cholesterol in the body. How can cholesterol be harmful to the body?

Nucleic Acids (Section 20.4)

30. What substance contains the information for all the characteristics and functions of an organism?
31. Draw the general ring structure of a purine and of a pyrimidine.
32. Draw the structures of ribose and 2-deoxyribose.
33. Draw the general structure of a typical nucleoside.
34. Draw the general structure of a typical nucleotide. Label the structural components.
35. Describe how the DNA double helix is held together.
36. What is a major difference between the primary structures of DNA and RNA?
37. What name is given to a section of DNA that contains the code for construction of a specific protein?
38. Explain the difference between the functions of mRNA and tRNA.

Answers to Chapter Review

1. protein
2. amide
3. amino
4. nylon
5. sequence
6. tertiary
7. secondary
8. quaternary
9. pleated
10. hydrogen
11. salt
12. disulfide
13. hydrophobic
14. strongest
15. weakest
16. denaturation
17. metal
18. enzymes
19. substrate
20. key
21. inhibited
22. OH
23. monosaccharides
24. carbonyl
25. disaccharide
26. glucose
27. C-1
28. cellulose
29. insoluble
30. triglycerides
31. unsaturated
32. fatty
33. phospholipids
34. steroids
35. cholesterol
36. nucleic
37. sugar
38. pendant
39. pyrimidines
40. 2-deoxyribose
41. nucleoside
42. nucleotide
43. DNA
44. helix
45. base
46. replication
47. gene
48. RNA
49. transcription
50. mRNA
51. rRNA
52. tRNA

The Scientific Calculator

A hand calculator is necessary for most of the calculations in this book. When selecting a calculator, a scientific calculator will be more useful than a business calculator.

Addition, Subtraction, Multiplication, and Division

These basic operations are illustrated in the following examples:

	Example	*Enter*	*Display*	*Rounded*
a.	14.61 + 3.18	14.61 [+] 3.18 [=]	17.79	17.79
b.	65.73 − 13.2	65.73 [−] 13.2 [=]	52.53	52.5
c.	2.79 × 5.410	2.79 [×] 5.410 [=]	15.0939	15.1
d.	44.5 ÷ 1.25	45.5 [÷] 1.25 [=]	35.6000	35.6

Remember, your calculator does not necessarily display the correct number of significant figures for the answer. The rules for significant figures are given in Chapter 3.

Second Function Key

Scientific calculators have more functions than there are keys, so some of the keys are used for more than one function. The first function usually appears on the key, and the second function is shown above the key. To use the second function, first press the [2nd] key and then press the function key. The second function key may be called a [SHIFT] key or an [INV] key on some calculators.

Change Sign

This key, [+/−], will change the sign of a number to the opposite sign. If a number is positive, pressing this key will change it to a negative sign, and vice versa.

Exponential Operations

Very large and very small numbers are conveniently expressed using exponential or scientific notation (Section 3.3). A scientific calculator performs exponential operations very well.

To enter an exponential number such as 4.3×10^3, first enter the nonexponential number (4.3), and then press the [EXP] key followed by the exponent (3). When the exponent is negative, complete the entry by pressing the [+/−] key. Never enter the × *10* part of the exponential number. This is included in the function of the [EXP] key.

Example	Enter	Display
a. 4.3×10^3	4.3 [EXP] 3	4.3 03
b. 6.13×10^{-4}	6.13 [EXP] 4 [+/−]	6.13 −04
c. 1.18×10^{-8}	1.18 [EXP] 8 [+/−]	1.18 −08
d. 7.65×10^{13}	7.65 [EXP] 13	7.65 13

Chain Calculations

Some problems require multiple operations. A chain calculation can save time and avoid rounding errors that occur when calculations are performed in successive steps. The following examples illustrate some chain calculations using a scientific calculator.

Example A.1

Use your calculator to solve the following problems. Express the answers in scientific notation.

a. $(3.21 \times 10^3)(6.72 \times 10^{-6})$

b. $\dfrac{7.63 \times 10^4}{2.13 \times 10^{-3}}$

c. $\dfrac{(106)(2.4 \times 10^6)}{(7.49 \times 10^{-5})}$

d. $\dfrac{(1.15 \times 10^{-4})(6.23)}{(4.33 \times 10^{-8})}$ ■

Solutions

	Enter	Display
a.	3.21 [EXP] 3 [×] 6.72 [EXP] 6 [+/−] [=]	2.1571 −02
b.	7.63 [EXP] 4 [÷] 2.13 [EXP] 3 [+/−] [=]	3.5822 07
c.	106 [×] 2.4 [EXP] 6 [÷] 7.49 [EXP] 5 [+/−] [=]	3.3965 12
d.	1.15 [EXP] 4 [+/−] [×] 6.23 [÷] 4.33 [EXP] 8 [+/−] [=]	1.6546 04

Answers

a. 2.16×10^{-2}

b. 3.58×10^7

c. 3.4×10^{12}

d. 1.65×10^4

Logarithms

A common logarithm, or logarithm to the base 10, is calculated using the [LOG] key. In this book, logarithms are used to calculate the pH from the molarity of the hydrogen ion, [H+], in aqueous solution (Section 15.6).

To calculate a logarithm of a number, enter the number and press the [LOG] key. For example, to calculate log(156), enter 156 and press [LOG]. The display is 2.1931246. Because the number (156) has three significant figures, the logarithm is rounded off to three decimal places, or 2.193.

An antilogarithm is calculated using the [10^x] key. For example, to calculate the number whose logarithm is 2.193, enter the logarithm (2.193), then press the

2nd key, followed by the 10^x key. The display is 155.95525, which rounds off to 156.

Example A.2

Calculate the logarithms of the following numbers.

a. 6.02×10^{23}
b. 1.83×10^{-11}
c. 67.9 ■

Solutions

	Enter	*Display*	*Answer*
a.	6.02 EXP 23 LOG	2.378 01	23.780
b.	1.83 EXP 11 +/− LOG	−1.0738 01	−10.738
c.	67.9 LOG	1.8318698	1.832

Example A.3

Find the antilogarithm of each of the following.

a. 4.15
b. 2.6
c. −3.53 ■

Solutions

	Enter	*Display*	*Answer*
a.	4.15 2nd 10^x	14125.375	1.4×10^4
b.	2.6 2nd 10^x	398.10717	4×10^2
c.	3.53 +/− 2nd 10^x	0.0002951	3.0×10^{-4}

Mathematical Operations

Exponential Notation

Because many of the numbers used in science are either very large or very small, chemists and other scientists often use exponential notation. In exponential notation, a number is expressed as a number (usually a number between 1 and 10) times a power of 10, for example,

$$2425.3 = 2.4253 \times 10^3$$

where 10^3 means $10 \times 10 \times 10$. If no power of 10 is given, the power of 10 is assumed to be zero. Any number to the zeroth power equals 1

$$10^0 = 1$$

and therefore

$$6.82 = 6.82 \times 10^0$$

Interconverting Ordinary Decimal Notation and Exponential Notation

When you go back and forth between ordinary decimal notation and exponential notation, you are simply writing the same number in a different way; therefore, the *value of the number must remain the same.* When converting a decimal to exponential notation, for each place the decimal point is moved to the left, the number must be multiplied by 10:

$$2425.3 = 2.4253 \times 10 \times 10 \times 10 = 2.4253 \times 10^3$$

If the number is made smaller by moving the decimal point to the left, the power of 10 must be made larger (more positive or less negative). For each place the decimal point is moved to the right, the number must be multiplied by one tenth.

$$0.0136 = 1.36 \times 0.1 \times 0.1 = 1.36 \times \frac{1}{10} \times \frac{1}{10} = 1.36 \times \frac{1}{10^2}$$
$$= 1.36 \times 10^{-2}.$$

If a number is made larger by moving the decimal point to the right, the power of 10 must be made smaller (more negative or less positive). Examples B.1 and B.2 provide more illustrations of the concept.

Example B.1

Write each of the following numbers in exponential notation:

a. 7 653 192

b. 0.000 000 96 ■

Solutions

a. To write 7 653 192 as a number between 1 and 10, the decimal point (which is not shown in the original number) must be moved six places to the left:

$$7\ 653\ 192 = 7\ 653\ 192. = 7.653\ 192 \times 10 \times 10 \times 10 \times 10 \times 10 \times 10$$
$$= 7.653\ 192 \times 10^6$$

Notice that the exponent of 10 is equal to the number of places the decimal point is moved. When a number ≥ 10 is written in exponential notation, the power of 10 is positive.

b. To write 0.000 000 96 as a number between 1 and 10, the decimal point must be moved seven places to the right:

$$0.000\ 000\ 96 = 9.6 \times 0.1 \times 0.1 \times 0.1 \times 0.1 \times 0.1 \times 0.1 \times 0.1$$
$$= 9.6 \times 10^{-7}$$

Again, the exponent of 10 is equal to the number of places the decimal point is moved; however, when a number <1 is written in exponential notation, the power of 10 is negative.

Example B.2

Write each of the following numbers in ordinary decimal notation:

a. 6.83×10^{-4}

b. 5.4×10^5 ■

Solutions

a. The power of 10, -4, means multiply by 0.1 four times,

$$6.83 \times 10^{-4} = 6.83 \times 0.1 \times 0.1 \times 0.1 \times 0.1 = 0.000\ 683$$

or, in other words, move the decimal point four places to the left:

$$6.83 \times 10^{-4} = 0.000\ 683$$

b. The power of 10, 5, means multiply by 10 five times,

$$5.4 \times 10^5 = 5.4 \times 10 \times 10 \times 10 \times 10 \times 10 = 540\ 000,$$

or, in other words, move the decimal point five places to the right:

$$5.4 \times 10^5 = 540\ 000$$

Study these examples carefully so that you see the relationship between the number of places the decimal point is moved and the power of 10. Then do Practice

Problems B.1 and B.2 both by hand and with your calculator. Answers to the practice problems are given at the end of Appendix B.

Practice Problem B.1 Write each of the following numbers in exponential notation.
a. 365.2
b. 0.0007
c. 0.0142
d. 23 652
e. 7.5

Practice Problem B.2 Write each of the following numbers in ordinary decimal notation.
a. 7.4×10^{-7}
b. 3.982×10^{1}
c. 3.27×10^{0}
d. 5.81×10^{-3}
e. 7.456×10^{8}

The number of significant figures in numbers such as 300 and 528 000 is uncertain. Exponential notation can be used to show the number of significant figures. To show that there are three significant figures in the number 300, write 3.00×10^2; to show two significant figures, write 3.0×10^2, and to show one significant figure, write 3×10^2. The number 528 000 may have three, four, five, or six significant figures:

$5.280\ 00 \times 10^5$ has six significant figures

5.2800×10^5 has five significant figures

5.280×10^5 has four significant figures

5.28×10^5 has three significant figures

If you have trouble telling how many significant figures there are in numbers like 0.002 50, write the number in exponential notation:

$$0.002\ 50 = 2.50 \times 10^{-3}$$

Thus, there are three significant figures in 0.002 50. The zeros on the left-hand side are not significant; they are necessary to show the magnitude of the number. The zero on the right-hand side is significant; if it were not, it would be omitted and the number would be written 0.0025.

Multiplication of Exponential Numbers

To multiply exponential numbers, multiply the numbers between 1 and 10 and add the exponents to find the power of 10 in the product:

$$(2.3 \times 10^5)(3.42 \times 10^8)(6.1 \times 10^{-4}) = (2.3 \times 3.42 \times 6.1) \times \{10^{[5 + 8 + (-4)]}\}$$
$$= 48 \times 10^9 = 4.8 \times 10^{10}$$

Notice that the product of the numbers, 47.9826, is rounded to two significant figures because there are only two significant figures in 2.3 and 6.1. In multiplication and division, the number of significant figures in the answer is limited by the number or numbers that have the smallest number of significant figures. The product is written 4.8×10^{10}—not 48×10^9—because, in exponential notation, the number should usually be between 1 and 10.

To convert a number from one power of 10 to another, remember that *the value of the number must not be changed when the number is rewritten.* If the number is made smaller, the power of 10 must be made larger by the same factor. If the number is made larger, the power of 10 must be made smaller by the same factor. For example, to convert 48 to 4.8, the number 48 must be multiplied by 0.1, or 10^{-1}; to keep the value of the number the same, the power of 10 must be multiplied by 10, or 10^{+1}. Any number to the zeroth power is equal to 1:

$$(10^{-1}) \times (10^{+1}) = 10^0 = 1$$

Multiplication by 1 does not change the value of a number.

Example B.3

The number 6.34×10^{-6} is equal to what number times 10^{-3}? ■

Solution To convert 10^{-6} to 10^{-3}, multiply by 10^{+3}; to keep the value of the number the same, multiply 6.34 by 10^{-3}:

$$6.34 \times 10^{-3} = 0.006\ 34$$

Therefore

$$6.34 \times 10^{-6} = 0.006\ 34 \times 10^{-3}$$

Check: Remember to check your answer:

$6.34 \times 10^{-6} = 0.000\ 006\ 34$, and $0.006\ 34 \times 10^{-3}$ is also equal to 0.000 006 34.

Practice Problem B.3 The number 4.9×10^8 is equal to what number times 10^6?

Practice Problem B.4 What is the value of the following product?

$(8.22 \times 10^{7})(3.874 \times 10^{-9})(7.629 \times 10^{3})$

Division of Exponential Numbers

To divide exponential numbers, divide the numbers between 1 and 10, and subtract the sum of the powers of 10 in the denominator from the sum of the powers of 10 in the numerator to find the power of 10 in the quotient:

$$\frac{(4.5 \times 10^{8})(6.32 \times 10^{6})}{(8.41 \times 10^{10})(9.63 \times 10^{-4})} = \frac{(4.5 \times 6.32)}{(8.41 \times 9.63)} \times 10^{\{(8 + 6) - [10 + (-4)]\}}$$

$$= 0.35 \times 10^{8} = 3.5 \times 10^{7}$$

Practice Problem B.5 What is the value of each of the following quotients?

a. $\dfrac{(9.6 \times 10^{8})}{(3 \times 10^{3})}$

b. $\dfrac{(9.42 \times 10^{7})}{(2.8 \times 10^{-6})}$

c. $\dfrac{(8.6 \times 10^{-4})}{(2.3 \times 10^{9})}$

d. $\dfrac{(4.7 \times 10^{-6})}{(7.5 \times 10^{-10})}$

e. $\dfrac{(8.76 \times 10^{14})(5.8 \times 10^{9})}{(2.63 \times 10^{22})(4.54 \times 10^{12})}$

Addition and Subtraction of Exponential Numbers

Exponential numbers that are to be added or subtracted must first be written with the same power of 10. Example B.4 illustrates the addition of exponential numbers.

To find the sum of one dollar and two cents, you would first either convert two cents to \$0.02 and then add—\$1.00 + \$0.02 = \$1.02—or you would convert the one dollar to 100 cents and add—100 cents + 2 cents = 102 cents. You would not simply add one dollar to two cents and get three dollars or three cents.

Example B.4

What is the sum of (3.2×10^{8}), (2.4×10^{9}), and (5.63×10^{10})? ■

Solution Convert the other number or numbers to the same power of 10 as the largest number to be added or subtracted, 5.63×10^{10} in this example.

$$\begin{aligned} 3.2 \times 10^{8} &= 0.032 \times 10^{10} \\ 2.4 \times 10^{9} &= 0.24 \times 10^{10} \\ & \underline{5.63 \times 10^{10}} \\ & 5.902 \times 10^{10} = 5.90 \times 10^{10} \end{aligned}$$

Because the last two numbers have only two decimal places, the answer can have only two decimal places.

Practice Problem B.6 Write the answer to each of the following problems in exponential notation.

a. $(4.91 \times 10^{7}) + (3.62 \times 10^{6})$
b. $(5.14 \times 10^{23}) - (8.29 \times 10^{22})$
c. $(6.9 \times 10^{-13}) - (7.2 \times 10^{-14})$

Powers and Roots

Raising an exponential number to a power is similar to multiplication: Find the power of the number between 1 and 10, and multiply the exponent of 10 by the power. For example,

$$(3.8 \times 10^{7})^{3} = (3.8)^{3} \times 10^{7(3)}$$
$$= 55 \times 10^{21} = 5.5 \times 10^{22}$$

To find a root of an exponential number by hand, first rewrite the number so that the power of 10 is divisible by the root. For example, to take the square root of 4.9×10^{9}, first rewrite 4.9×10^{9} so that the power of 10 is divisible by 2:

$$\sqrt{4.9 \times 10^{9}} = \sqrt{49 \times 10^{8}}$$

Then, take the square root of the number and divide the exponent by 2:

$$\sqrt{49 \times 10^{8}} = 7.0 \times 10^{4}$$

Be careful to show the correct number of significant figures in the root.

Practice Problem B.7

a. What is the fourth power of 2.3×10^{2}?
b. Find the square root of 8.10×10^{13}.
c. Find the cube root of 6.4×10^{-11}.

Logarithms

Logarithms are exponents. Two kinds of logarithms are used in chemistry: the **common logarithm,** which is referred to as *log,* and the **natural logarithm,** which is referred to as *ln.* The logarithms used in pH are common logarithms, or logarithms to the base 10; that is, they are the power to which 10 must be raised to give a number. For example, $3 = \log(1000)$ because $10^{3} = 1000$, and $-3 = \log(0.001)$ because $10^{-3} = 0.001$. The **antilogarithm** of a number x is the number that has x as its logarithm; that is, the antilog of 6 is 10^{6} and the antilog of -6 is 10^{-6}. The antilogarithm is referred to as *antilog.*

Practice Problem B.8 Find the log of

a. 10^{5}
b. 10^{-4}

Practice Problem B.9 Find the antilog of
a. -8
b. 15

For most numbers the power of 10 is not a whole number. Consult the instruction book for your calculator to learn how to find the logs of numbers such as 2762 and 0.002 762 and the antilogs of numbers such as 6.348 and -6.348 ($\log 2762 = 3.4412$, $\log 0.002\ 762 = -2.5588$, antilog $6.348 = 2.23 \times 10^6$, and antilog $-6.348 = 4.49 \times 10^{-7}$). The number of digits to the right of the decimal point in a logarithm should be equal to the number of significant figures as shown in the preceding examples.

Practice Problem B.10 Use your calculator to find the value of each of the following expressions.
a. $\log (3 \times 10^{-12})$
b. $\log (8.42 \times 10^{16})$
c. $\log 1$
d. antilog (-22.498)
e. antilog (5.7)

Natural logarithms are logarithms to the base e ($e = 2.7183\ldots$):

$$\ln x = e^x$$

The number e occurs often in the mathematical relationships used in chemistry and physics. Consult the instruction book for your calculator to learn how to find the ln of numbers such as 2762 and 0.002 762 and the antiln of numbers such as 6.348 and -6.348 ($\ln 2762 = 7.9237$, $\ln 0.002\ 762 = -5.8918$, antiln $6.348 = 571$, antiln $-6.438 = 0.001\ 60$). To convert a logarithm to the base 10 to a natural logarithm, multiply by 2.303:

$$\ln x = 2.303 \log x$$

Practice Problem B.11 Use your calculator to find the value of each of the following expressions.
a. $\ln (3 \times 10^{-12})$
b. $\ln (8.42 \times 10^{16})$
c. $\ln 1$
d. antiln (-22.498)
e. antiln (5.7)

Solving Equations Using Algebra

Equations that contain only one unknown can be solved for the unknown by rearranging the equation so that the unknown is on one side and the knowns are on the other side. When an equation is rearranged, *whatever is done to one side of the equation must be done to the other side.* For example, to solve the equation

$$2.43 = \frac{x}{6.97}$$

multiply each side of the equation by 6.97 and simplify:

$$6.97 \times 2.43 = \cancel{6.97} \times \frac{x}{\cancel{6.97}}$$

or

$$6.97 \times 2.43 = x = 16.9$$

Example B.5

Solve the following equations for x.

a. $3.54 = \frac{7.08}{x}$

b. $4.65 = x + 8.19$ ■

Solutions

a. To solve the equation $3.54 = \frac{7.08}{x}$ for x, multiply each side by x,

$$x \cdot 3.54 = x \cdot \frac{7.08}{x}$$

divide each side by 3.54, and simplify:

$$\frac{x \cdot \cancel{3.54}}{\cancel{3.54}} = \frac{\cancel{x} \cdot 7.08}{\cancel{x} \cdot 3.54}$$

$$x = \frac{7.08}{3.54} = 2.00$$

b. To solve the equation $4.65 = x + 8.19$ for x, subtract 8.19 from each side:

$$(4.65 - 8.19) = (x + \cancel{8.19}) - \cancel{8.19}$$

$$-3.54 = x$$

Practice Problem B.12 Solve the following equations for x.

a. $6.9 = \frac{2.0}{x}$

b. $2.84 = \frac{x}{7.82}$

c. $5.82 = x - 6.43$

Answers to Practice Problems

PP B.1

a. 3.652×10^{2} b. 7×10^{-4} c. 1.42×10^{-2} d. 2.3652×10^{4}
e. 7.5×10^{0}

PP B.2

a. 0.000 000 74 b. 39.82 c. 3.27 d. 0.005 81 e. 745 600 000

PP B.3

490

PP B.4

2.43×10^{2}

PP B.5

a. 3×10^{5} b. 3.4×10^{13} c. 3.7×10^{-13} d. 6.3×10^{3} e. 4.3×10^{-11}

PP B.6

a. 5.27×10^{7} b. 4.31×10^{23} c. 6.2×10^{-13}

PP B.7

a. 2.8×10^{9} b. 9.00×10^{6} c. 4.0×10^{-4}

PP B.8

a. 5 b. -4

PP B.9

a. 10^{-8} b. 10^{15}

PP B.10

a. -11.5 b. 16.925 c. 0.0 d. 3.18×10^{-23} e. 5×10^{5}

PP B.11

a. -26.5 b. 38.972 c. 0.0 d. 1.70×10^{-10} e. 3×10^{2}

PP B.12

a. 0.29 b. 22.2 c. 12.25

Answers to Selected Problems

Chapter 1

1.1 (b) sunlight (d) electricity (g) heat (h) love (i) fear
1.3 analytical chemist
1.5 environmental chemistry
1.6 organic: (a) sugar from sugar cane
(c) a piece of wood
(d) human hair
inorganic: (b) table salt
(e) sand
(f) water
1.10 A precise statement of a problem is important to a scientist so that he or she has a clear focus in choosing a direction to follow in solving a problem.
1.11 (a) hypothesis (b) law (c) law (d) theory (e) hypothesis (f) theory (g) hypothesis

Chapter 2

2.5 (b) air (c) motor oil (e) gasoline
2.7 homogeneous: (a) salt water
(e) gasoline
(f) vegetable oil
heterogeneous: (b) Italian salad dressing
(c) iced tea (assuming there are actual ice cubes in the tea)
(d) orange juice
(g) grape jam
(h) concrete
2.9 (a) hydrogen (c) aluminum (h) oxygen
2.11 The observations suggest that rubbing alcohol is a mixture. The remaining substance was clearly different. It had no odor and was not flammable. The flammable alcohol burned, leaving behind nonflammable water.
2.13 Baking soda is a single substance. Single substances have definite composition, while mixtures have variable composition.
2.15 (a) nitrogen (b) phosphorus (c) silicon (d) lithium (e) chlorine (f) helium
2.17 (a) 2 (b) 3 (c) 6
2.19 (a) sodium hydride (b) carbon tetrachloride (c) calcium sulfide
2.21 (a) magnesium nitride (b) lithium chloride (c) aluminum oxide
2.23 (b) exploding a firecracker (g) lighting a fire
2.25 mixture; no chemical change; yes, a chemical change occurs; water, H_2O
2.27 (b), (c), and (e) are not balanced
2.29 (b) noodles and water in a pan
(c) water and sawdust
(e) mixture of broken glass and water
2.31 chromatography
2.33 positive
2.35 exothermic: (a) a burning candle
(e) a burning marshmallow
endothermic: (b) pancakes cooking on a griddle
(c) digesting a meal
(d) a roasting marshmallow
2.37 Potential energy decreases as the apple falls out of the tree, since potential energy depends on position. As it falls, the stored or potential energy of the apple is converted into kinetic energy. Kinetic energy is the energy of motion, and it increases as the apple falls. Since energy is conserved, there is no change in the total energy of the apple when it falls.
2.39 Inverting a test tube and placing it in water traps a clear colorless gas. Air resists being compressed by the water as you push down on the test tube. Since the air takes up space, it must be matter.
2.41 chemical: (a) Cooking oil burns.
(c) A black coating forms on copper when it is heated in air.
physical: (b) Gold is yellow in color.
2.43 (a) calcium fluoride (b) dinitrogen oxide (c) xenon tetrafluoride
2.45 filtration: (a) cooked carrots and water
(c) ice and vinegar (d) cereal and milk
distillation: (b) a solution of paraffin wax and gasoline

Chapter 3

3.1 qualitative: (a) red hair (b) a long rope
quantitative: (c) a ten-story building
3.3 (a) m (meter) (b) s (second) (c) K (kelvin) (d) kg (kilogram)
3.5 (a) 1 Mm (b) 1 mm (c) 1 hm (d) 1 nm (e) 1 cm (f) 1 μm

3.7 (a) 3.45×10^{-3} (b) 1.032×10^2
(c) 4.5623×10^4

3.9 602 000 000 000 000 000 000 000

3.11 (a) 5.40×10^{-2} (rounded to three significant figures)
(b) 9.45×10^{12}

3.13 (a) 3.47×10^{12} (b) 2.41×10^{10}
(c) 9.76×10^{-8}

3.15 (a) 9.25 (b) 1.03×10^{-4}

3.17 Mass expresses the quantity of matter in a sample. Mass is independent of location and/or gravity. Weight measures the force exerted by a mass under the influence of gravity. The mass of an astronaut's suit will not change when taken from the earth to the moon, but the weight of the suit will be much less on the moon because the force of gravity is less on the surface of the moon than it is on the surface of the earth.

3.19 (a) 1 000 000 g (b) 0.001 g (c) 0.01 g
(d) 10 g (e) 0.000 000 001 g (f) 1 000 g

3.21 "Two-dimensional" quantitites are those requiring two dimensions in their calculation. Area is an example of a two-dimensional quantity, as the area of an office could be 2.5 m by 3.0 m = 7.5 m^2. "Three-dimensional" quantities are those requiring three dimensions in their calculation. Volume is the most common example of a three-dimensional quantity: the volume of a box is calculated 25.0 cm × 25.0 cm × 20.0 cm = 1.25×10^4 cm^3.

3.23 (a) a pipet

3.25 liquids—g/mL or g/cm^3; solids—g/cm^3; gases—g/L or g/mL

3.27 The density of water (0.998 g/mL or 998 g/L) is about 1000 times greater than the density of air (1.2 g/L). Fluid must be pushed out of the way when running through it. Since a volume of water is more massive than an equal volume of air, more work must be done (more exertion) when running in water than when running in air.

3.29 (a) A calorie is a quantity of energy; specifically, the amount of energy required to raise the temperature of 1.0 gram of water by one degree Celsius. (b) A food calorie is also a quantity of energy. It is the unit used by nutritionists to describe the energy value of foods. One Calorie contains the same amount of energy as one kilocalorie or 1000 calories. (c) Absolute zero is the theoretically lowest temperature possible. It is equal to −273.15°C or 0 K.

3.31 (a) a pound of gold (b) a gallon of water
(c) ten bricks

3.33 K (kelvin)

3.35
(a) $T_F = 1.8T_C + 32$ or $T_C = \dfrac{T_F - 32}{1.8}$

(b) $T_K = T_C + 273$ or $T_C = T_K - 273$

3.37 (a) 200 K (b) 90°C

3.39 Numbers from measurements differ from exact numbers in that measured numbers always contain uncertainty. The last digit recorded is an estimated value. There are no estimates in exact numbers.

3.41 The accuracy of a measurement refers to how well the measurement agrees with the true value. Since precision refers to the repeatability of a measurement without reference to accuracy, it is possible to have measurements with good precision and poor accuracy. This condition occurs when the measuring device is incorrectly calibrated, but correctly used.

3.43 The estimated digit is determined by judging the distance between the smallest calibration marks on the measuring device. Typically, it is first determined whether the measurement is more than half way, about half way, or less than half way between the calibration marks. If more than half way, how much more? In this way, an estimated value for the last digit is obtained. If it is estimated that the measurement falls on a calibration mark, this is reported by estimating a zero as the last digit.

3.45 (c) a zero between two nonzero digits

3.47 (a) 3 (b) 2 (c) 1 (d) 2 (e) 4 (f) 3

3.49 (a) 0.102 (b) 10.8 (c) 213 (d) 1.34×10^4

3.51 (a) 3.61×10^1 (b) 2.03×10^5
(c) 4.40×10^{-4} (d) 7.65×10^2

3.53 (a) 79 m^2 (b) 1.62 g (c) 57 mi/hr

3.55 5.00 g

3.57 181 lb

3.59 0.095 oz per person per day

3.61 You should check the freezer with a Fahrenheit thermometer. Since it is desirable for the meat to be solidly frozen, a temperature below the freezing point of water is needed. 0°C is at the freezing point of water, and water coexists with ice at this temperature. 0°F is equal to (0°F − 32°F)/1.8 = −18°C and is therefore colder than the freezing point of water. The meat will be solidly frozen at this temperature.

3.63 (a) 4 (b) 3 (c) 4 (d) exact number

3.65 (a) 16.0 mL (b) 36.5°C (c) 12.00 cm

Chapter 4

4.9 (a) $\dfrac{1 \text{ lb}}{16 \text{ oz}}, \dfrac{16 \text{ oz}}{1 \text{ lb}}$

(b) $\dfrac{2.20 \text{ lb}}{1 \text{ kg}}, \dfrac{1 \text{ kg}}{2.20 \text{ lb}}$

(c) $\dfrac{10^6 \ \mu\text{g}}{1 \text{ g}}, \dfrac{1 \text{ g}}{10^6 \ \mu\text{g}}$

(d) $\dfrac{1 \text{ tsp}}{5 \text{ mL}}, \dfrac{5 \text{ mL}}{1 \text{ tsp}}$

(e) $\dfrac{1760 \text{ yd}}{1 \text{ mi}}, \dfrac{1 \text{ mi}}{1760 \text{ yd}}$

4.11 1.5×10^6 mm

4.13 1.19×10^3 g

4.15 3.6 g/cup

4.17 6.8×10^3 g

4.19 45 L

4.21 0.75 L/pkg

4.23 25 doses

4.25 3×10^1 cups

4.27 $\frac{\$2.3 \times 10^4}{\text{acre}}$

4.29 \$0.0483/shave or 4.83¢/shave

4.31 1.5×10^7 wheels

4.33 5.9×10^9 ft^3

4.35 Tokyo: $\frac{1.48 \times 10^4 \text{ people}}{\text{sq mi}}$, Japan: $\frac{898 \text{ people}}{\text{sq mi}}$

4.37 3×10^2 lb/week

4.39 40.0 m^3

4.41 12 qt

4.43 1.04¢/page

4.45 2×10^2 gal

4.47 $\$1.10 \times 10^3$/hr

4.49 0.869 g/cm^3

4.51 0.354 cm^3

4.53 6.80×10^2 g

4.55 198 g

4.57 1.1×10^2 g

4.59 49 g

4.61 102.9°F

4.63 53°C

4.65 7.2×10^2 J

4.67 60. g

4.69 0.014 g or 1.4×10^{-2} g

4.71 4.0×10^5 particles

4.73 465 K, 378°F

4.75 1.6×10^8 g

4.77 6.5×10^2 trucks

4.79 \$335.01

Chapter 5

5.13 atomic number

5.15 The numbers are mass numbers (sum of protons plus neutrons). Atomic number = 8.

5.17 (a) $^4_2$He (b) $^{14}_6$C (c) $^{244}_{94}$Pu (d) $^{10}_4$Be (e) $^{39}_{19}$K (f) $^{37}_{17}$Cl

5.19

Element	Nuclear Symbol	Atomic Number	Mass Number	Number of Protons	Number of Electrons	Number of Neutrons
helium	$^3_2$He	2	3	2	2	1
sodium	$^{22}_{11}$Na	11	22	11	11	11
chlorine	$^{35}_{17}$Cl	17	35	17	17	18
copper	$^{65}_{29}$Cu	29	65	29	29	36
iron	$^{57}_{26}$Fe	26	57	26	26	31
uranium	$^{235}_{92}$U	92	235	92	92	143
aluminum	$^{27}_{13}$Al	13	27	13	13	14
fluorine	$^{19}_9$F	9	19	9	9	10
arsenic	$^{70}_{33}$As	33	70	33	33	37
gold	$^{197}_{79}$Au	79	197	79	79	118

5.21 The atomic mass of an element is the mass of the naturally occurring isotopes. It is different from the mass number of the most abundant isotope because the element is composed of several isotopes, each of which is present in a different amount and each of which contributes to the average mass of the element. Furthermore, it is possible that the most abundant isotope might contribute little more than 50% to the average mass. Thus the mass number of the most abundant isotope would not reflect the average composition of the element.

5.23 107.87 amu, silver

5.25 12.01 amu

5.27 2.009×10^{-23} g; mass decreases when subatomic particles are combined to make a nucleus

5.29 5.309×10^{-23} g

5.31 Mendeleev and Meyer

5.33 The periodic law states that similar chemical and physical properties of the elements recur periodically when the elements are arranged according to their atomic numbers.

5.35 (a) Group 7A (b) Group 8A (c) Group 2A (d) period 7

5.37 Be, Mg, Ca, Sr, Ba, Ra

5.39 Cl, F, I, Br

5.41 (a) 7A (b) 8A (c) 1A (d) 2A

5.43 shiny or lustrous, good heat conductor, good electrical conductor, malleable, and ductile

5.45 tendency to gain electrons in a chemical reaction

5.47 ionic compound, e.g., sodium chloride

5.49 (a) N (b) Br (c) F

5.51 (a) Ge (b) Sb (c) As

5.53 Ca, Ba, Mg

5.55 (a) O^{2-} (b) N^{3-} (c) Al^{3+}

5.57 (a) $2K + Br_2 \longrightarrow 2KBr$ (made up of K^+ and Br^-)
(b) $4Li + O_2 \longrightarrow 2Li_2O$ (made up of Li^+ and O^{2-})
(c) $Ca + S \longrightarrow CaS$ (made up of Ca^{2+} and S^{2-})

5.59 different chemical properties: (a) N/Ne (d) K/Kr
similar chemical properties: (b) O/S (c) P/As

5.61 (a) $2Na + S \longrightarrow Na_2S$
(b) $2Ca + O_2 \longrightarrow 2CaO$
(c) $Mg + I_2 \longrightarrow MgI_2$

5.63 (a) $^{70}_{30}$Zn (b) $^{197}_{79}$Au (c) $^{57}_{26}$Fe

5.65 (a) 8 protons, 10 electrons
(b) 20 protons, 18 electrons
(c) 35 protons, 36 electrons
(d) 13 protons, 10 electrons

5.67 Sn or Pb

5.69 B

Chapter 6

6.3 The difference between a continuous spectrum and a bright-line emission spectrum is that a continuous spectrum is observed when so many colors of light are being emitted at the same time that they all blend together to form a rainbow. If only a few colors of light are being emitted, they are seen as individual lines of color in a bright-line emission spectrum.

6.10 The notation for principal energy levels is $n = x$, where $x = 1, 2, 3, \ldots$

6.13 $n = 1$, 1 sublevel; $n = 2$, 2 sublevels; $n = 3$, 3 sublevels; $n = 4$, 4 sublevels

6.16 An atomic orbital is a three-dimensional region about the nucleus in which there is a high probability of locating an electron. An orbit is a circular path for an electron.

6.20 *s* sublevel, 1 orbital; *p* sublevel, 3 orbitals; *d* sublevel, 5 orbitals; *f* sublevel, 7 orbitals

6.24 two spin states possible

6.32 (a) Al: $1s^22s^22p^63s^23p^1$ (b) P: $1s^22s^22p^63s^23p^3$
(c) Cl: $1s^22s^22p^63s^23p^5$

6.34 (a) Li: $1s^22s^1$ (b) Na: $1s^22s^22p^63s^1$
(c) K: $1s^22s^22p^63s^23p^64s^1$

These elements all have one valence electron and are all found in Group 1A of the periodic table.

6.35 (a) F: $1s^22s^22p^5$ (b) Cl: $1s^22s^22p^63s^23p^5$
(c) Br: $1s^22s^22p^63s^23p^64s^23d^{10}4p^5$

The similarity is that all three elements have 5 electrons in the outermost occupied *p* sublevel and 7 electrons in the outermost occupied principal energy level, which is known as the valence shell. All of the elements are found in Group 7A in the periodic table.

6.39 The electron configurations of transition metals are similar in that they have from 1 to 10 electrons in the highest-occupied *d* subshell and no electrons in the next higher *p* subshell.

6.42 (a) Cl: $3s^23p^5$ (b) As: $4s^24p^3$ (c) Si: $3s^23p^2$
(d) Rb: $5s^1$ (e) Sb: $5s^25p^3$

6.43 (a) Sr: [Kr]$5s^2$ (b) Ga: [Ar]$4s^23d^{10}4p^1$
(c) Se: [Ar]$4s^23d^{10}4p^4$ (d) I: [Kr]$5s^24d^{10}5p^5$
(e) Sb: [Kr]$5s^24d^{10}5p^3$

6.45 (a) $4s^1$ (b) $3s^23p^5$ (c) $2s^2$ (d) $4s^24p^6$

6.47 (a) Sr: (b) K· (c) :$\dot{\text{S}}$: (d) :$\dot{\underset{\cdot}{\text{As}}}$·

6.49 (a) 2 unpaired electrons (b) 1 unpaired electron (c) 3 unpaired electrons (d) 1 unpaired electron (e) 0 unpaired electrons (f) 3 unpaired electrons

6.54 (a) +1, K^+ (b) +2, Mg^{2+} (c) −1, F^-
(d) −2, O^{2-} (e) −3, P^{3-} (f) +2, Ba^{2+}

6.57 (a) Ca^{2+} (b) $[:\ddot{\underset{..}{Cl}}:]^-$ (c) $[:\ddot{\underset{..}{S}}:]^{2-}$ (d) K^+
(e) Al^{3+}

6.58 (a) Ca^{2+}: $1s^22s^22p^63s^23p^6$ (b) Cl^-: $1s^22s^22p^63s^23p^6$
(c) S^{2-}: $1s^22s^22p^63s^23p^6$ (d) K^+: $1s^22s^22p^63s^23p^6$
(e) Al^{3+}: $1s^22s^22p^6$

6.61 (a) Ne (b) Ne (c) Ar (d) Ar (e) Ar

6.64 Rb^+, Sr^{2+}, Y^{3+}

6.66 Li· → $Li^+ + e^-$

6.68 $:\ddot{\underset{..}{Cl}}\cdot + e^- \rightarrow [:\ddot{\underset{..}{Cl}}:]^-$

6.69 (a) Li· + $:\ddot{\underset{..}{Cl}}\cdot \rightarrow Li^+ + [:\ddot{\underset{..}{Cl}}:]^-$ or LiCl

Li·: oxidized, reducing agent; Cl: reduced, oxidizing agent

(b) Ca: + $:\dot{\underset{\cdot}{S}}\cdot \rightarrow Ca^{2+} + [:\ddot{\underset{..}{S}}:]^{2-}$ or CaS

Ca: oxidized, reducing agent; S: reduced, oxidizing agent

6.74 (a) Cs > K (b) I > Cl (c) Ba > Mg
(d) K > Br (e) Be > O

6.76 (a) $Cl^- > Cl$ (b) $Cl^- > Li^+$ (c) $S^{2-} > O^{2-}$

6.81 (a) Na (b) I (c) Cl (d) Mg

6.86 (a) F (b) Cl (c) O (d) O (e) Cl

6.87 (a) F (b) Cl (c) O (d) S

6.89 (a) 7 valence electrons (b) 5 (c) 3
(d) 6 (e) 5 (f) 7

6.91 (a) K (b) Mg (c) Al (d) Ca

6.94 (a) Sn: $1s^22s^22p^63s^23p^64s^23d^{10}4p^65s^24d^{10}5p^2$
(b) Rb^+: $1s^22s^22p^63s^23p^64s^23d^{10}4p^6$
(c) Se^{2-}: $1s^22s^22p^63s^23p^64s^23d^{10}4p^6$

6.96 (a) Mg (b) As (c) Se

6.99 (a) O (b) O (c) O* (d) Po (e) O

Chapter 7

7.9 (a) 358 g/mol (b) 138 g/mol (c) 74.5 g/mol
(d) 80.1 g/mol (e) 58.3 g/mol

7.11 (a) 58.1 g/mol (b) 46.1 g/mol (c) 98.0 g/mol
(d) 123.1 g/mol (e) 284 g/mol (f) 44.0 g/mol
(g) 58.1 g/mol

7.13 (a) 2.86 mol (b) 0.646 mol (c) 0.1024 mol
(d) 1.385 mol

7.15 (a) 225 g (b) 353 g (c) 61.7 g (d) 80.0 g

7.17 (a) 206 g (b) 0.155 g (c) 21.7 g (d) 23.1 g

7.19 (a) 1.41×10^{22} molecules (b) 3.0×10^{21} molecules
(c) 3.9×10^{23} molecules (d) 7.53×10^{23} molecules

7.21 (a) 1.00×10^{25} molecules
(b) 2.33×10^{22} molecules
(c) 1.2×10^{19} molecules

7.23 (a) 9.93×10^{19} atoms
(b) 3.8×10^{16} atoms
(c) 1.7×10^{19} atoms
(d) 4.88×10^{25} atoms

7.25 (a) 5.66×10^{-23} g (b) 7.31×10^{-23} g
(c) 7.33×10^{-23} g (d) 2.23×10^{-22} g

7.27 (a) 6.66×10^{-23} g (b) 9.75×10^{-23} g
(c) 1.98×10^{-22} g (d) 1.15×10^{-23} g
(e) 4.67×10^{-23} g (f) 1.24×10^{-22} g

7.29 4.2×10^{-18} mol

7.31 (a) 39.3% Na, 60.7% Cl (b) 7.79% C, 92.2% Cl
(c) 23.6% K, 76.5% I
(d) 27.4% Na, 1.20% H, 14.3% C, 57.1% O

7.33 (a) 81.6% (b) 47.7% (c) 83.5% (d) 50.0%
(e) 44.1%

7.35 46.7% N

7.37 (a) 93.1% Ag (b) 63.5% Fe (c) 24.7% K
(d) 28.5% Cu (e) 38.0% Co (f) 68.4% Cr

7.39 (d) $NaHCO_3$ (e) CH_2O (f) C_4H_9

7.41 (b) HO (d) CH_2O

7.43 Fe_2S_3

7.45 $NaMnO_4$

7.47 $C_7H_7NO_2$

7.49 NO_2

7.51 N_2O_3
7.53 FeS_2
7.55 CH_3
7.57 empirical formula and molar mass
7.59 $C_6H_8O_7$
7.61 $C_4H_{12}N_2$
7.63 $C_{10}H_8$
7.65 (a) 55.9 g Fe (b) 63.6 g Cu (c) 108 g Ag
(d) 24.3 g Mg (e) 52.0 g Cr (f) 24.3 g Mg
7.67 (a) 0.816 g (b) 0.835 g (c) 0.500 g
(d) 0.441 g
7.69 $Na_2Al_2Si_3O_{10} \cdot 2H_2O$
7.71 142 g Cl
7.73 1.64×10^7g
7.75

Compound	*Mass, g*	*Number of Moles*	*Number of Molecules*
$C_{14}H_9Cl_5$	88.8	0.250	1.51×10^{23}
$C_{10}H_8$	10.0	0.0781	4.70×10^{22}
NO_2	95.7	2.08	1.25×10^{24}
C_2H_6O	2.68×10^{-21}	5.81×10^{-23}	35

7.77 1.50 g C
7.79 (a) 40.1% Ca, 12.0% C, 48.0% O
(b) 41.7% Mg, 54.9% O, 3.46% H

Chapter 8

8.9 covalent bonding
8.11 ionic: (a) KBr (c) BaS (f) FeO covalent: (b) N_2 (d) PH_3 (e) $SOCl_2$
8.13 endothermic
8.15 (a) CCl_4 (b) CCl_4
8.17 Lewis formula
8.19 (a) $[H_4N]^+$ (H—N—H with H above and below) $[:\ddot{Cl}:]^-$ (b) H—As: (with H above and below)
(c) :Cl—N: with :Cl: above and :Cl: below (d) $K^+[:\ddot{O}—H]^-$
(e) :Ö—Se=Ö: (f) $Mg^{2+}\ 2[:\ddot{Br}:]^-$
8.21 Na^+ [:Ö=C(—O—H)—Ö: ⟷ :Ö—C(—O—H)=Ö:]$^-$
8.23 :Ö—S=Ö: ⟷ :Ö=S—Ö:
8.25 O_2 has unpaired electrons, which make it paramagnetic. N_2 does not have unpaired electrons.

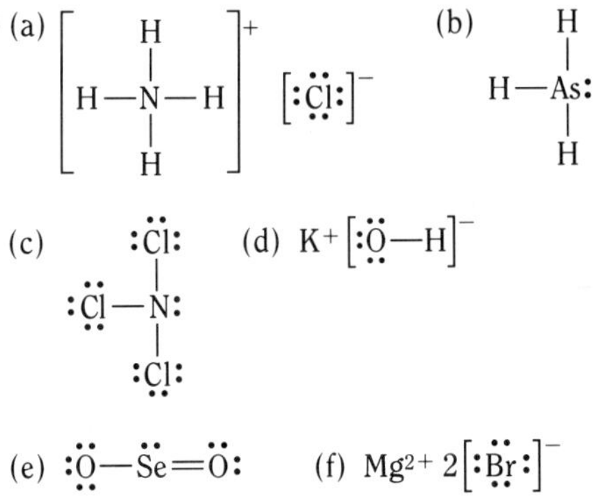

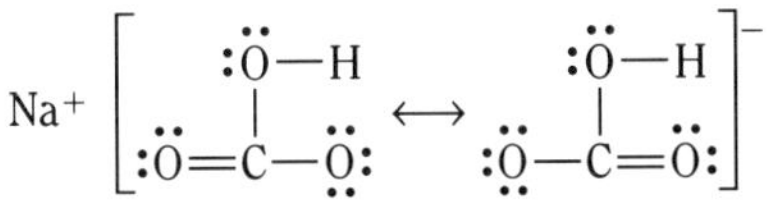

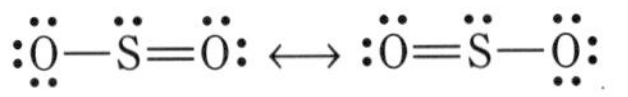

8.27 fluorine
8.29 (a) polar (b) nonpolar (c) polar (d) nonpolar
8.31 Because the two different atoms joined together have almost the same electronegativity (difference in electronegativity = 0.2), and therefore essentially the same attraction for shared electrons, the bond is practically nonpolar.
8.33 (a) H—F (b) equally polar (c) O—H
(d) C—O (e) P—Cl
8.35 (a) tetrahedral (b) linear (c) linear
(d) trigonal
8.37 H—C≡N:; bond angle is 180°; shape is linear
8.39 nonpolar; carbon and sulfur have the same electronegativity, so there are no polar bonds; a molecule cannot be polar if there are no polar bonds, regardless of its shape
8.41 The Lewis formulas, below, show that the S in sulfur dioxide has three sets of electrons, whereas the C in carbon dioxide has two sets of electrons. In carbon dioxide, there are two equal dipoles that exactly cancel because of the linear shape of the molecule. Therefore, CO_2 is nonpolar. The extra set of electrons on the S in sulfur dioxide yields a shape with bond angles of 120°. There are two dipoles, but they do not cancel one another because of the bond angles. Therefore, SO_2 is polar.

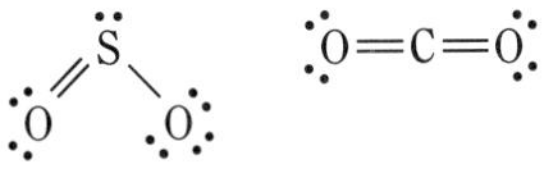

O=S̈—O (bent) :Ö=C=Ö:

8.43 Propane is not polar because it does not contain polar bonds.

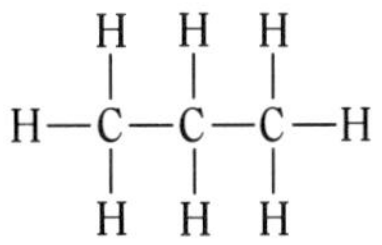

CH_3—CH_2—CH_3 (H—C—C—C—H with H above and below each C)

8.45 The Lewis formulas drawn below are not resonance structures because they do differ in the placement of atoms as well as the placement of electrons.

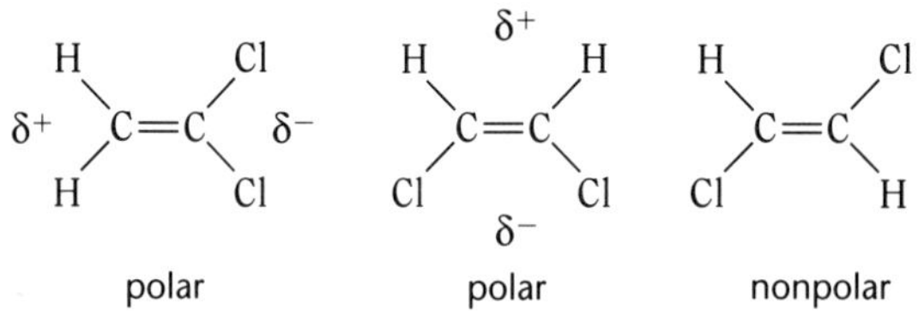

$H_2C=CCl_2$ (δ+ on H side, δ− on Cl side)	HClC=CClH, cis (δ+ on H side, δ− on Cl side)	HClC=CClH, trans
polar	polar	nonpolar

8.47 (a) H—C(H)(H)—C(H)(H)—Ö—H and H—C(H)(H)—Ö—C(H)(H)—H
(b) $H_2C=CH—CH_3$ (C=C—C—H) and cyclopropane (H—C—C—H ring with CH_2)

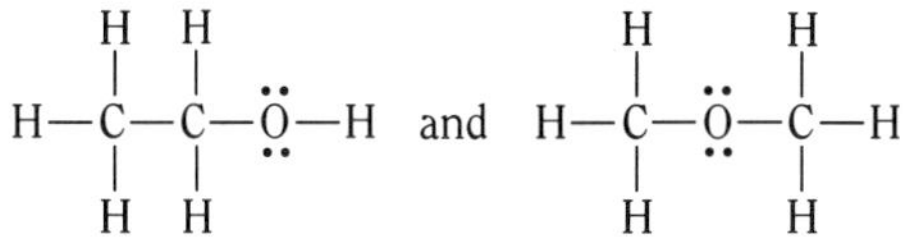

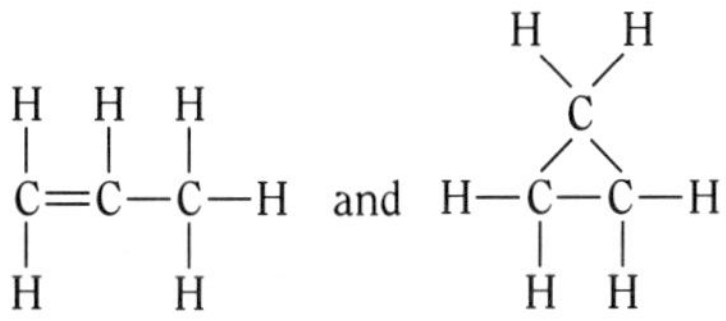

8.49

```
   :Cl:
      \
:Cl—Si—Cl:
      /
   :Cl:
```

bond angles = 109°; molecule is tetrahedral; nonpolar because the bond dipoles cancel one another

8.51

```
     :Cl:              :Br:
       |                 |
:Cl—C—Cl:       :Br—C—Br:
       |                 |
     :Cl:              :Br:
```

Both compounds are nonpolar because the tetrahedral shape results in the cancellation of the bond dipoles.

Chapter 9

9.5 (a) −3 (b) +4 (c) +3 (d) +3
9.7 (a) +7 (b) +5 (c) +3 (d) +1
9.9 (a) −2 (b) −1 (c) −2 (d) −2
9.11 (a) S = +4 (b) P = −3 (c) Cl = +7 (d) Se = −2
9.13 (a) I = +7, O = −2 (b) N = −3, H = +1 (c) As = +5, O = −2 (d) Mn = +7, O = −2
9.15 (a) potassium iodide (b) tetraphosphorus decoxide (c) iron(II) oxide (d) tin(IV) sulfide (e) magnesium oxide
9.17 (a) iron(III) ion (b) mercury(I) ion (c) tin(IV) ion (d) copper(I) ion
9.19 (a) copper(II) chloride (b) iron(II) nitride (c) tin(IV) iodide (d) copper(I) sulfide
9.21 (a) Al_2S_3 (b) $PbBr_4$ (c) $Cu(CN)_2$ (d) $SnCl_2$ (e) Mg_3N_2
9.23 (a) potassium cyanide (b) magnesium hydroxide (c) copper(II) hydroxide (d) iron(III) hydroxide
9.25 (a) $Sn(OH)_4$ (b) NH_4Br (c) $Fe(OH)_2$
9.27 binary acids: (b) HBr(aq) (d) H_2S(aq)
9.29 (a) HI(aq) (b) HBr(aq) (c) H_2Se(aq)
9.31 (a) arsenic acid (b) hypobromous acid (c) selenic acid (d) selenous acid (e) iodic acid (f) arsenous acid
9.33 (a) HNO_3 (b) $HClO_2$ (c) HIO_4 (d) H_2SO_3
9.35 (a) sodium hypochlorite (b) calcium sulfite (c) potassium nitrate (d) sodium nitrite
9.37 (a) copper(II) nitrate (b) tin(II) sulfate (c) copper(II) phosphate
9.39 nitrous acid, HNO_2
9.41 (a) potassium bisulfite (b) copper(II) bisulfate (c) sodium biselenate
9.43 (a) NO_3^- (b) ClO_4^- (c) NO_2^- (d) IO_3^- (e) SO_4^{2-} (f) ClO_2^- (g) ClO^- (h) BrO_3^-
9.45 (a) MnO_4^- (b) ClO_3^- (c) $Cr_2O_7^{2-}$ (d) CrO_4^{2-} (e) PO_4^{3-}
9.47 (a) $Fe_2(SO_4)_3$ (b) Cu_3PO_4 (c) $Cr_2(SO_4)_3$ (d) $Sn(ClO_2)_2$ (e) Na_2CO_3
9.49 (a) $KMnO_4$ (b) $K_2Cr_2O_7$ (c) Na_2CrO_4 (d) $NaClO_3$
9.51 NH_4NO_3
9.53 NaClO

Chapter 10

10.3 (a) $C_5H_{12} + 8O_2 \longrightarrow 5CO_2 + 6H_2O$
(b) $Ca(OH)_2 + 2HCl \longrightarrow CaCl_2 + 2H_2O$
(c) $CaCl_2 + 2AgNO_3 \longrightarrow 2AgCl + Ca(NO_3)_2$
(d) $Na_2CO_3 + 2HBr \longrightarrow 2NaBr + H_2O + CO_2$
(e) $4Al + 3O_2 \longrightarrow 2Al_2O_3$
10.5 (a) $4NH_3 + 5O_2 \longrightarrow 4NO + 6H_2O$
(b) $Ni + 2HCl \longrightarrow NiCl_2 + H_2$
(c) $3Ca(OH)_2 + 2H_3PO_4 \longrightarrow 6H_2O + Ca_3(PO_4)_2$
(d) $2NaI + Cl_2 \longrightarrow 2NaCl + I_2$
10.7 (a) iron(II) acetate + potassium sulfide ⟶ iron(II) sulfide + potassium acetate
(b) dinitrogen tetroxide $\xrightarrow{\Delta}$ nitrogen dioxide
(c) mercury(I) nitrate + sodium chloride ⟶ mercury(I) chloride + sodium nitrate
10.9 (a) $Fe(C_2H_3O_2)_2(aq) + K_2S(aq) \longrightarrow FeS(s) + 2KC_2H_3O_2(aq)$
(b) $N_2O_4(g) \xrightarrow{\Delta} 2NO_2(g)$
(c) $Hg_2(NO_3)_2(aq) + 2NaCl(aq) \longrightarrow Hg_2Cl_2(s) + 2NaNO_3(aq)$
10.11 (a) sulfuric acid + iron ⟶ iron(II) sulfate + hydrogen
(b) sodium chloride $\xrightarrow{\text{electrolysis}}$ sodium + chlorine
(c) copper(I) sulfide + oxygen ⟶ copper + sulfur dioxide
10.13 $NaOCl(aq) + NaCl(aq) + 2HC_2H_3O_2(aq) \longrightarrow Cl_2(g) + H_2O(l) + 2NaC_2H_3O_2(aq)$
10.15 (a) $3KOH(aq) + H_3PO_4(aq) \longrightarrow 3H_2O(l) + K_3PO_4(aq)$
(b) $Mg(s) + NiCl_2(aq) \longrightarrow MgCl_2(aq) + Ni(s)$
(c) $MgCO_3(s) + 2HCl(aq) \longrightarrow H_2O(l) + MgCl_2(aq) + CO_2(g)$
10.17 (a) $2K(s) + 2H_2O(l) \longrightarrow 2KOH(aq) + H_2(g)$
(b) $2Al(s) + 3H_2SO_4(aq) \longrightarrow 3H_2(g) + Al_2(SO_4)_3(aq)$
(c) $MgCO_3(s) \longrightarrow MgO(s) + CO_2(g)$
(d) $Na_2SO_4(aq) + BaCl_2(aq) \longrightarrow 2NaCl(aq) + BaSO_4(s)$
10.19 (a) $N_2(g) + O_2(g) \longrightarrow 2NO(g)$
(b) $SO_3(g) + H_2O(l) \longrightarrow H_2SO_4(aq)$
(c) $P_4(s) + 6Cl_2(g) \longrightarrow 4PCl_3(s)$
10.21 (a) $2Na_2O_2(s) \longrightarrow 2Na_2O(s) + O_2(g)$
(b) $H_2CO_3(aq) \longrightarrow CO_2(g) + H_2O(l)$
10.23 (a) $C_6H_{12}O_6(s) + 6O_2(g) \longrightarrow 6CO_2(g) + 6H_2O(g)$
(b) $2C_8H_{18}(l) + 25O_2(g) \longrightarrow 16CO_2(g) + 18H_2O(g)$
10.25 $C_7H_8(l) + 9O_2(g) \longrightarrow 7CO_2(g) + 4H_2O(g)$
10.27 (a) $Zn(s) + H_2SO_4(aq) \longrightarrow ZnSO_4(aq) + H_2(g)$
(b) $Ca(s) + 2H_2O(l) \longrightarrow Ca(OH)_2(aq) + H_2(g)$
(c) $Al(s) + H_2O(l) \longrightarrow$ no reaction
(d) $Mg(s) + 2AgNO_3(aq) \longrightarrow Mg(NO_3)_2(aq) + 2Ag(s)$

10.29 (a) $AgNO_3(aq) + HCl(aq) \longrightarrow AgCl(s) + HNO_3(aq)$
(b) $CuCl_2(aq) + Na_2S(aq) \longrightarrow CuS(s) + 2NaCl(aq)$
(c) $Fe(NO_3)_3(aq) + NaCl(aq) \longrightarrow$ no reaction
(d) $Pb(NO_3)_2(aq) + K_2CrO_4(aq) \longrightarrow PbCrO_4(s) + 2KNO_3(aq)$

10.31 (a) $2Na_3PO_4(aq) + 3CaCl_2(aq) \longrightarrow Ca_3(PO_4)_2(s) + 6NaCl(aq)$
(b) $K_2SO_4(aq) + Ba(NO_3)_2(aq) \longrightarrow BaSO_4(s) + 2KNO_3(aq)$
(c) $Hg(NO_3)_2(aq) + NaCl(aq) \longrightarrow$ no reaction
(d) $2Al(NO_3)_3(aq) + 3Na_2S(aq) \longrightarrow Al_2S_3(s) + 6NaNO_3(aq)$

10.33 (a) $Ag^+(aq) + Cl^-(aq) \longrightarrow AgCl(s)$
(b) $Cu^{2+}(aq) + S^{2-}(aq) \longrightarrow CuS(s)$
(c) no net reaction
(d) $Pb^{2+}(aq) + CrO_4^{2-}(aq) \longrightarrow PbCrO_4(s)$

10.35 $2Na(s) + 2H_2O(l) \longrightarrow 2Na^+(aq) + 2OH^-(aq) + H_2(g)$

10.37 $NH_4^+(aq) + OH^-(aq) \longrightarrow NH_3(g) + H_2O(l)$

10.39 (a) $2Ca(s) + O_2(g) \longrightarrow 2CaO(s)$
(b) $2HClO_3(aq) + Ba(OH)_2(aq) \longrightarrow Ba(ClO_3)_2(aq) + 2H_2O(l)$
(c) $2HgO(s) \longrightarrow O_2(g) + 2Hg(l)$
(d) $Ca(s) + 2H_2O(l) \longrightarrow H_2(g) + 2Ca(OH)_2(aq)$

10.41 (a) combustion
(b) double replacement
(c) double replacement
(d) double replacement/decomposition
(e) combination

10.43 (a) $Mg(s) + 2HCl(aq) \longrightarrow MgCl_2(aq) + H_2(g)$
(b) $2C_6H_6(l) + 15O_2(g) \longrightarrow 12CO_2(g) + 6H_2O(g)$
(c) $Cd(NO_3)_2(aq) + H_2S(aq) \longrightarrow CdS(s) + 2HNO_3(aq)$
(d) $FeCl_3(aq) + Na_3PO_4(aq) \longrightarrow FePO_4(s) + 3NaCl(aq)$

10.45 (a) oxidized: Mg; reduced: H in HCl
(b) oxidized: C in C_6H_6; reduced: O

10.47 (a) $Al(NO_3)_3(aq) + 3NaOH(aq) \longrightarrow Al(OH)_3(s) + 3NaNO_3(aq)$
(b) $Ni(s) + 2HBr(aq) \longrightarrow NiBr_2(aq) + H_2(g)$
(c) $C_4H_{10}O(l) + 6O_2(g) \longrightarrow 4CO_2(g) + 5H_2O(g)$

10.49 Nothing happened to the ring because gold is below hydrogen in the activity series. The ring could be recovered by carefully pouring off the acid into another beaker, leaving the ring in the original beaker. Then the ring should be thoroughly rinsed with water before handling.

Chapter 11

11.1 6 mol H_2
11.3 3 mol O_2
11.5 2.6 mol $NaHCO_3$
11.7 (a) 3.28 mol O_2 (b) 1.50 mol P_4O_{10}
11.9 (a) 0.250 mol Na_2SO_3 (b) 0.500 mol HCl
11.11 (a) 12.7 mol C_2H_5OH (b) 12.7 mol CO_2
11.13 928 g CO_2
11.15 6.30×10^2 g $NaHCO_3$
11.17 0.405 g AgCl
11.19 0.22 g MgO
11.21 21.9 g CuO
11.23 8.42 kg CaO
11.25 1.27×10^3 g P_4
11.27 88.5 g Na and 137 g Cl_2
11.29 43.7 g Fe_2O_3
11.31 4.19 g Ag
11.33 485 g CCl_4
11.35 theoretical yield: 1.34 g Hg; percent yield: 92.5%
11.37 (a) 95.8 g Ti (b) 74.1%
11.39 limiting reactant, O_2; 5.15 g CO_2 theoretical yield
11.41 (a) NH_3 (b) 149 g $(NH_4)_3PO_4$
11.43 3.9×10^3 g or 3.9 kg
11.45 (a) CCl_4 (b) 39.3 g CCl_2F_2 (c) 11 g SbF_3 excess
11.47 1.46 g C_2H_5Cl
11.49 (a) 1.02×10^3 g C_2H_5OH (b) 61.3% yield
11.51 (a) H_2 (b) 51.5 g CH_3OH
11.53 (a) 23.4 mol HNO_3 (b) 287 g H_2SO_4

Chapter 12

12.7 The sense of smell depends on the tendency of a gas to fill its container. Gaseous molecules move rapidly through space until they encounter an obstacle. When an animal is swimming underwater, its sense of smell is practically ineffective because the molecules having odor are stopped by the water before they can encounter the animal's nose.

12.9 A volume of gas cannot be measured by pouring it into a graduated cylinder because the gaseous molecules would completely fill the cylinder and then escape from the cylinder to fill the room.

12.11 molecular velocity: $CO_2 < Ar < O_2 < N_2 < CH_4$

12.13 Gasoline molecules evaporated from the pan and moved around in the garage to fill the garage. When gas molecules in appropriate concentration reached the pilot light of the space heater, ignition occurred.

12.15 increasing average molecular velocity: $CO_2 < Ar < O_2 < N_2$

12.17 Pressure is force per unit area. Pressure in a gas is caused by collisions of gas particles with the walls of the container.

12.19 574 torr
12.21 152.8 atm, 1.161×10^5 torr
12.23 0.149 atm
12.25 0.933 atm
12.27 63.8 mL
12.29 7.59 L
12.31 36 mL
12.33 13.0 atm
12.35 24.3 atm
12.37 STP refers to a standard temperature and pressure, which has been defined as 273 K and 1 atm.
12.39 1.37 L
12.41 5.01×10^3 torr

12.43 7.73×10^{-3} L
12.45 1.43 mol
12.47 1.98 L
12.49 5.15 L
12.51 12.6 L
12.53 1.67×10^5 mol
12.55 4.4 L
12.57 13 L
12.59 3.11 atm
12.61 0.370 mol
12.63 1.4 g
12.65 4.69 L
12.67 44.1 g/mol
12.69 6.24 L
12.71 *D:* $CH_4 < O_2 < HCl < SO_2 < HI$
12.73 17.0 g/mol
12.75 1.16 g/mL
12.77 (a) 619 torr (b) 0.001 38 mol
12.79 9.9 L
12.81 0.0181 mol H_2O_2
12.83 194 g HgO
12.85 12 g NH_4Cl
12.87 1.42×10^4 g Cu_2S
12.89 2.66 g $BaCO_3$
12.91 119 L
12.93 29.3 L/mol
12.95 0.0931 g
12.97 2.4 L
12.99 0.156 g $KClO_3$
12.101 2.46×10^{22} molecules
12.103 16.6 L

Chapter 13

13.5 A liquid diffuses more slowly than a gas because its molecules are so close together that they inhibit the rapid movement possible for molecules in a gas. Many more collisions occur per unit time in a liquid, and many of these collisions inhibit forward motion.
13.7 dipole–dipole attractions, hydrogen bonds, and dispersion forces
13.9 Since HBr is a polar molecule, it has both dipole–dipole attractions and dispersion forces.
13.11 Dipole–dipole attractions are practically ineffective with gases because they are weak forces that are not significant at the great distances between gas molecules.
13.13 Polar liquids are generally volatile because dipole–dipole attractions are a weak force of attraction.
13.15 an H—F, H—O, or H—N bond
13.17

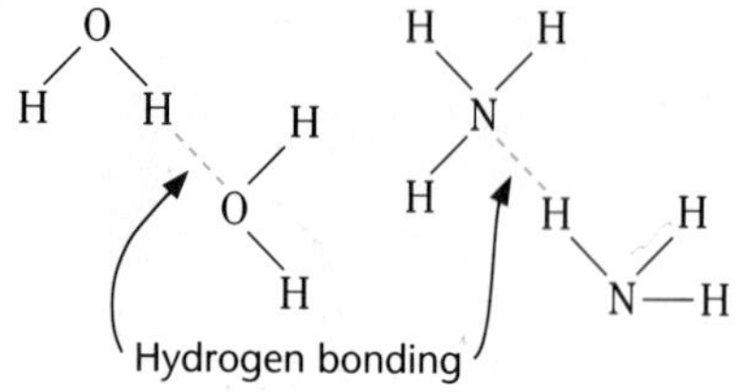

O, H, H, H, O, H, H, H, N, H, H, H, N—H

Hydrogen bonding

13.19 hydrophilic: (a) ethyl alcohol, (b) methylamine, (c) ammonia. These molecules are hydrophilic because each has either an O—H or an N—H bond and therefore can form hydrogen bonds with water molecules.
13.21 Nonpolar molecules such as C_5H_{12} are attracted to one another through temporary induced dipoles that result from the movement of electrons within the molecule. These intermolecular forces are called dispersion or London forces.
13.23 Since dispersion forces increase with size and molar mass, large molecules have strong dispersion forces. An example is octane, C_8H_{18}.
13.25 The rate of vaporization of a liquid increases with increasing temperature because a larger fraction of molecules has enough kinetic energy to escape to the vapor phase.
13.27 Dynamic equilibrium of a liquid and its vapor means that molecules are continually evaporating and condensing even though the total number of molecules in the vapor phase remains constant. Equilibrium vapor pressure of a liquid refers to the vapor pressure exerted by the liquid at equilibrium with its vapor.
13.29 Sublimation is the process of going directly from the solid state to the vapor state. Some substances that sublime are iodine, *p*-dichlorobenzene (moth crystals), and dry ice (solid carbon dioxide).
13.31 (a) benzene (b) ethyl ether (c) chloroform (d) toluene
13.33 (a) toluene (b) butyl alcohol (c) carbon tetrachloride (d) chlorobenzene
13.35 2.3×10^3 kJ
13.37 The electrostatic attractions between oppositely charged ions are much greater than the attractions between molecules. Therefore, ionic compounds have low vapor pressures and high boiling points.
13.39 NaCl has a higher normal boiling point than CH_3Cl because ionic compounds have stronger attractions between particles (ions) than do molecular compounds.
13.41 All three molecules are polar and have dipole–dipole attractions. In addition, there are dispersion forces. Because of its molar mass (128 g/mol), HI has stronger dispersion forces than either HBr (*MM* = 81 g/mol) or HCl (*MM* = 36.5 g/mol). As intermolecular forces increase in strength, it becomes more difficult to separate the molecules to form a vapor. Thus, the temperature required to produce a vapor pressure of one atmosphere is greater as the dispersion forces increase.
13.43 A crystalline solid is one that has a regular geometric pattern. Some examples are sodium chloride, diamond, quartz, and other gemstones.
13.45 A crystalline solid is distinguished from an amorphous solid by the physical properties of hardness and brittleness. Amorphous solids are flexible and elastic.
13.47 types of crystalline solids: ionic, molecular, network, and metallic
13.49 Sucrose forms a molecular crystal. The intermolecular forces in a sugar crystal are dipole–dipole attractions, hydrogen bonding, and dispersion forces.

13.51 Network solids have very high melting points compared with molecular crystals because the interparticle attractions are covalent bonds, which are much stronger than dipole–dipole attractions, hydrogen bonding, or dispersion forces, and which, therefore, require more energy to disrupt for melting.
13.53 An aqueous sugar solution is a nonconductor of electricity because sugar is molecular. Without ions, there can be no conductivity in aqueous solutions.
13.55 (a) molecular (b) ionic (c) network (d) molecular (e) molecular (f) ionic
13.57 3.35×10^5 J or 335 kJ
13.59 636 kJ
13.61 4.00×10^2 g
13.63

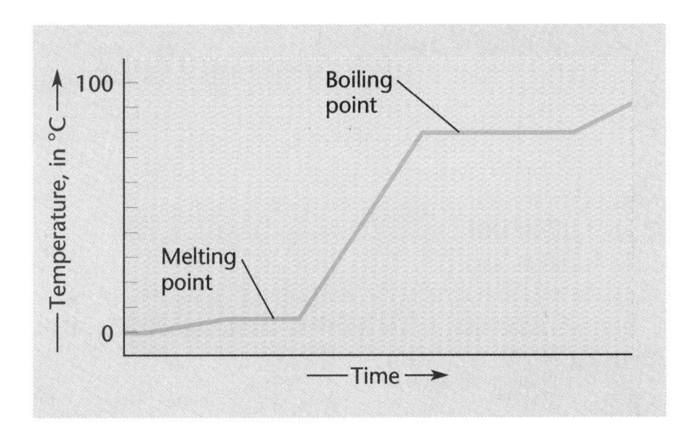

13.65 Surface tension is a result of intermolecular attractive forces. Increased temperature results in increased energy and velocity for the molecules of a liquid. As the energy of the molecules increases, intermolecular forces are relatively weaker—the molecules have too much energy to stay close to one another. Surface tension is the result of attractive forces pulling molecules together, but increased molecular velocity tends to pull molecules apart. The result is that surface tension decreases as temperature increases.
13.67 There are two reasons for the high viscosity of vegetable oils. There are very strong dispersion forces because the molecules are large. Additionally, the long carbon chains get tangled together and cannot flow freely over one another.
13.69 its high specific heat and its large heat of vaporization
13.71 Ice would form at the bottom of lakes and rivers and freeze up to the top. This would kill marine life.
13.73 Hydration means the attraction or bonding of water molecules to an ion or molecule.
13.75 A hydrate is a substance that contains water molecules in its crystalline structure. Hydrates have definite proportions of water as shown by their formulas and are therefore compounds. Examples include $CuSO_4 \cdot 5H_2O$, $MgSO_4 \cdot 7H_2O$, and $CaCl_2 \cdot 2H_2O$.
13.77 (a) $Na_2CO_3 \cdot 10H_2O$ (b) $MgSO_4 \cdot 7H_2O$ (c) $Ba(OH)_2 \cdot 8H_2O$
13.79 A deliquescent compound is a compound that absorbs enough moisture from the air to dissolve in it. Calcium chloride, $CaCl_2$, is an example of a deliquescent compound.
13.81 The boiling point of hydrogen peroxide, H_2O_2, is higher than the boiling point of water because there are two oxygen atoms in hydrogen peroxide that can participate in hydrogen bonding. The more hydrogen bonds that can form, the stronger the intermolecular forces holding the molecules together in the liquid state. Additionally, hydrogen peroxide molecules have a greater molar mass than water molecules, with accordingly larger dispersion forces. Therefore, more energy, as represented by a higher boiling point, is required to reach the boiling point.
13.83 Water in a narrow glass tube has a meniscus because of attractive forces acting between the liquid and the surface of the glass. Since pentane (C_5H_{12}) is a nonpolar molecule, it is not attracted to the same kinds of substances as water, and no meniscus is formed in a narrow glass tube.
13.85 Ethyl alcohol has a much lower equilibrium vapor pressure at 25°C than does ethyl ether because ethyl alcohol has an OH group that can participate in hydrogen bonding. Hydrogen bonding is not possible in ether. In the absence of this strong attractive force, it is easier for ether molecules to vaporize than it is for alcohol molecules to vaporize. Thus, ethyl ether has a higher vapor pressure than ethyl alcohol.
13.87 100°C
13.89 The fire might have started from the vapors that were produced when the motor oil was heated. Raising the temperature increases the vapor pressure of a liquid. The vapors would burn when they encountered the gas flame. The investigator was not able to ignite a small volume of motor oil with a match because it was such a cold day that there was only a small amount of vapor in the absence of heating.

Chapter 14

14.7 factors in solubility: the nature of intermolecular forces and temperature
14.9 Immiscible means substances do not dissolve in one another. Water and vegetable oil do not mix.
14.11 When KI dissolves, the ions are surrounded, or hydrated, by water molecules. The attractions between the ions and the water dipole are strong enough to counter the ion–ion attractions in the solid.
14.13 Ammonia gas is very soluble in water because it can form hydrogen bonds with water molecules. Nitrogen molecules are nonpolar and are not attracted to water molecules. Therefore, nitrogen is practically insoluble in water.
14.15 Hydrogen chloride, HCl(g), is very soluble in water as a result of ion–dipole attractions. Polar water molecules are attracted to the polar HCl molecules. The attractions are strong enough to ionize the HCl molecules, producing H^+(aq) and Cl^-(aq) ions surrounded by water molecules. Thus, even without hydrogen bonding, hydrogen chloride is able to dissolve in water.
14.17 36.4 g NaCl/100.0 g water
14.19 The solubility of most solids in water increases when the temperature is increased.

14.21 A supersaturated solution is one that contains more solute than a saturated solution at that temperature. Supersaturated solutions are not stable.
14.23 When a supersaturated solution is allowed to stand for a period of time, the excess solute usually crystallizes until the remaining solution is saturated and therefore stable.
14.25 Since boiling decreases the solubility of gases, there is less oxygen in water that has been boiled and then cooled. Over time, oxygen diffuses back into the water, but if goldfish are placed in the cooled water right away, they will die for lack of oxygen.
14.27 A finely divided solid dissolves more rapidly than one composed of coarse particles because dissolving occurs at the surface. Finely divided solids have a larger surface area where water molecules can collide with solute particles and take them into solution.
14.29 5.0 g dextrose
14.31 (a) 3.8 g (b) 1.88 g (c) 0.14 g (d) 81.0
14.33 5.0 g NaOH and 95 g water
14.35 0.50 g $KMnO_4$ and 24.5 g water
14.37 8.50 g HCl
14.39 0.25 g acetic acid
14.41 0.025 g NO_3^-
14.43 4.0 ppm F^-
14.45 0.060 ppm Cr
14.47 13.8 M H_2SO_4
14.49 7.89 M HCl
14.51 5.00 M NaOH
14.53 (a) 0.900 M KCl (b) 0.517 M $Ca(NO_3)_2$ (c) 0.0673 M $BaCl_2$ (d) 3.39 M KOH
14.55 (a) 50. g KI (b) 21.3 g Na_2SO_4 (c) 23 g K_3PO_4 (d) 26.0 g $BaCl_2$
14.57 0.347 M $C_6H_{12}O_6$
14.59 0.0488 mol sucrose
14.61 (a) 19.7 g $Ca(NO_3)_2$ (b) 20.0 g NaOH (c) 7.31 g NaCl (d) 567 g Na_2SO_3
14.63 0.142 L (142 mL)
14.65 2.40 M H_2SO_4
14.67 0.60 M NH_3
14.69 2.00×10^{-3} L (2.00 mL)
14.71 0.25 L
14.73 17.1 g $CaSO_4$
14.75 0.0622 g $CaCl_2$
14.77 654 mL
14.79 0.252 L (252 mL)
14.81 5.25 mL (5.25×10^{-3} L)
14.83 0.491 g $NaHCO_3$
14.85 102.9°C
14.87 100.45°C
14.89 6.2 m
14.91 −119°C
14.93 (a) 10.0 g urea in 100. g water
14.95 The Tyndall effect is the scattering of light by the particles in a colloid, which results in a foggy or cloudy appearance to the part of the solution through which the light is passing.
14.97 emulsions
14.99 0.838 M $HC_2H_3O_2$
14.101 3.39 M NH_3
14.103 20.% HCl
14.105 1.1 g/mL
14.107 0.005 53 mol H_2O

Chapter 15

15.11 (a) KOH and (c) $Ba(OH)_2$
15.13 (a) OH^- (b) H_2O (c) Cl^- and (d) NH_3
15.15 (a) Br^- (b) HSO_4^- (c) CH_3NH^- (d) ClO_3^-
15.17 (a) $H^-(aq) + H_2O(l) \longrightarrow H_2(g) + OH^-(aq)$
(b) $NH_2^-(aq) + H_2O(l) \longrightarrow OH^-(aq) + NH_3(aq)$
(c) $O^{2-}(aq) + H_2O(l) \longrightarrow OH^-(aq) + OH^-(aq)$
$[O^{2-}(aq) + H_2O(l) \longrightarrow 2OH^-(aq)]$
(d) $NH_3(aq) + H_2O(l) \longrightarrow OH^-(aq) + NH_4^+(aq)$
(e) $CO_3^{2-}(aq) + H_2O(l) \longrightarrow OH^-(aq) + HCO_3^-(aq)$
(f) $CN^-(aq) + H_2O(l) \longrightarrow OH^-(aq) + HCN(aq)$
15.19 (a), (c), and (d) are amphoteric. Conjugate acids: (a) H_2O, (c) H_2S, (d) H_3PO_4; conjugate bases: (a) O^{2-}, (c) S^{2-}, (d) HPO_4^{2-}
15.21 0.126 M HCl
15.23 0.160 M NaOH
15.25 An acid anhydride is a compound that reacts with water to form an acid. SO_2 is an example involving an element in Group 6A: $SO_2(g) + H_2O(l) \longrightarrow H_2SO_3(aq)$.
15.27 $H_2SO_4(aq)$
15.29 $Cl_2O(g) + H_2O(l) \longrightarrow 2HClO(aq)$. The oxidation number of chlorine is +1 in each compound.
15.31 (a) NaOH (b) $Ba(OH)_2$ (d) $Cu(OH)_2$
15.33 $ZnO(s) + 2HCl(aq) \longrightarrow ZnCl_2(aq) + H_2O(l)$
15.35 $3H_2SO_4(aq) + Al_2O_3(s) \longrightarrow Al_2(SO_4)_3(aq) + 3H_2O(l)$
15.37 A weak acid is one that is only slightly ionized in dilute solution. The notation using unequal arrows in the equation below indicates that most of the acetic acid molecules are not ionized.

$$HC_2H_3O_2\,(aq) + H_2O(l) \rightleftarrows C_2H_3O_2^-\,(aq) + H_3O^+(aq)$$

15.39 F^-
15.41 $OH^- > NH_3 > F^- > Cl^- > ClO_4^-$
15.43 neutral solution: $[H_3O^+] = [OH^-] = 1.0 \times 10^{-7}$ mol/L
15.45 neutral solution: pH = pOH = 7
15.47 pH = 1.0
15.49 Sourness depends on the acidity of the solution. The pH of orange juice is higher (3.7) than that of lemon juice (2.3). The higher the pH, the less acidic the solution. Therefore, orange juice is less acidic and less sour than lemon juice. Additionally, orange juice may have more natural sugar than lemon juice, with the sweetness masking some of the sour taste.
15.51 (a) pOH = 9.0, $[OH^-] = 1 \times 10^{-9}$ M
(b) pOH = 6.0, $[OH^-] = 1 \times 10^{-6}$ M
(c) pOH = 11.0, $[OH^-] = 1 \times 10^{-11}$ M
(d) pOH = 2.0, $[OH^-] = 1 \times 10^{-2}$ M
15.53 pH meter

15.55 most common type of buffer solution: solution consisting of a mixture of a weak acid and its conjugate base
15.57 $H_2PO_4^-/HPO_4^{2-}$ (acid/conjugate base)
15.59 An acetic acid buffer resists a change in pH when a strong acid is added by converting some of the acetate ion into its conjugate acid, acetic acid. In the process, the hydronium ions from the strong acid are consumed, and there is very little change in pH.

$$C_2H_3O_2^-(aq) + H_3O^+(aq) \longrightarrow HC_2H_3O_2(aq) + H_2O(l)$$

15.61 When strong acid is added to the HCO_3^-/CO_3^{2-} buffer, the carbonate ion reacts with the hydronium ion to form the bicarbonate ion. Thus, there is no significant increase in the concentration of hydronium ion, and the pH is unchanged.

$$CO_3^{2-}(aq) + H_3O^+(aq) \longrightarrow HCO_3^-(aq) + H_2O(l)$$

15.63 0.156 M HCO_2H
15.65 first step: $H_2SO_4(aq) + H_2O(l) \longrightarrow HSO_4^-(aq) + H_3O^+(aq)$
second step: $HSO_4^-(aq) + H_2O(l) \longrightarrow SO_4^{2-}(aq) + H_3O^+(aq)$
HSO_4^- is a weaker acid than H_2SO_4 because the negatively charged ion holds onto its positively charged proton more strongly than does the neutral molecule.
15.67 pH = 2.82
15.69 (a) base anhydride, $Zn(OH)_2$
(b) base anhydride, $Al(OH)_3$
(c) acid anhydride, H_3AsO_4
(d) acid anhydride, H_2SeO_3
15.71 4×10^{-3} M

Chapter 16

16.13

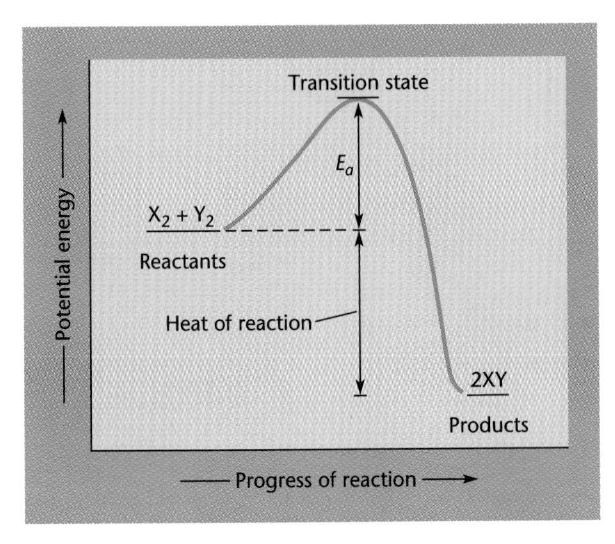

16.15 The larger the value of E_a, the slower the rate of a chemical reaction.
16.17 Raising the temperature increases the rate of a chemical reaction.
16.19 Raising the temperature increases the rate of a reaction by increasing the number of molecules that have kinetic energies greater than E_a. Therefore, more collisions will be effective at higher temperatures. Additionally, as the particles move faster, more collisions will occur in a given time.
16.21 A catalyst is a substance that speeds up the rate of a reaction without being consumed by the reaction.
16.23 Adding a catalyst provides an additional pathway that has a lower activation energy. The reaction will primarily occur by this alternate pathway, since a larger number of particles will have sufficient energy to react by the catalyzed pathway.
16.25 KI serves as a catalyst for the decomposition of aqueous hydrogen peroxide.
16.27 Increasing the concentrations of reactants increases the number of collisions that occur in a given period of time. If there are more frequent collisions, there will be more frequent effective collisions. As a result, the rate of the reaction will increase.
16.29 Chemical equilibrium is the state in which the rate of the forward reaction equals the rate of the reverse reaction.
16.31 The concentrations of reactants and products remains unchanged at equilibrium.
16.33 Equilibrium is achieved rapidly at 250°C because the rates of reactions increase with increasing temperature. Since equilibrium is the condition in which the rates of forward and reverse reactions are equal, the faster the reactions occur, the sooner will equilibrium be reached.
16.35

$$K_{eq} = \frac{[NO]^4[H_2O]^6}{[NH_3]^4[O_2]^5}$$

16.37 1.67×10^3
16.39 $K_{eq} = \dfrac{[CO_2]^2}{[CO]^2[O_2]} = 5.0 \times 10^3$. The formation of CO_2 is favored. The large value of K_{eq} indicates a large concentration of product (CO_2) and a small concentration of reactants at equilibrium.
16.41 (a) $N_2(g) + 2O_2(g) \rightleftarrows 2NO_2(g)$
(b) $K_{eq} = \dfrac{[NO_2]^2}{[N_2][O_2]^2} = \dfrac{1}{0.010} = 1.0 \times 10^2$
(c) The reaction is favored in the direction represented in this problem, as indicated by the value of K_{eq} greater than 1.
16.43 Le Chatelier's principle states that if a reaction at equilibrium is changed, the reaction will adjust to counteract the change and establish a new equilibrium.
16.45 (a) decreases (b) decreases
16.47 Adding chlorine, a product, to the equilibrium mixture in problem 46 will cause the reaction to shift to the left. Reaction to the left is the exothermic direction. Therefore, heat will be released when chlorine is added.
16.49 (a) shift right (b) shift right (c) shift right
(d) shift right
16.51 Adding NaOH causes the equilibrium to shift away from the reactant side, increasing the concentration of $CrO_4^{2-}(aq)$, which is yellow. Therefore, the solution will turn yellow.

16.53 (a) increases (b) increases (c) no change

16.55 (a) $K_b = \frac{[NH_4^+][OH^-]}{[NH_3]}$

(b) $K_b = \frac{[C_2H_5NH_3^+][OH^-]}{[C_2H_5NH_2]}$

16.57 0.1% ionized

16.59 $K_a = 2 \times 10^{-6}$

16.61 0.03%

16.63 pH = 2.41

16.65 (a) $C_4H_{10}(aq) \overset{\text{catalyst}}{\rightleftarrows} C_4H_8(g) + H_2(g)$; $K_{eq} = \frac{[C_4H_8][H_2]}{[C_4H_{10}]}$

(b) $K_{eq} = 14$

16.67 $K_{eq} = 3.2 \times 10^7$

16.69 (a) shift right (b) shift left (c) shift right (d) shift right

16.71 (a) $N_2H_4(aq) + H_2O(l) \rightleftarrows N_2H_5^+(aq) + OH^-(aq)$

(b) $K_{eq} = \frac{[N_2H_5^+][OH^-]}{[N_2H_4]}$

(c) 0.008%

Chapter 17

17.7 (a) Mn^{2+} and Fe^{3+} (b) NO(g) and Sn^{2+} (c) no reaction

17.9 oxidizing agent is reduced: $Cl_2(g)$; reducing agent is oxidized: Cu

17.11 $Mg(s) + Cu^{2+}(aq) \longrightarrow Mg^{2+}(aq) + Cu(s)$

17.13 oxidizing agent is reduced: $Cu^{2+}(aq)$; reducing agent is oxidized: Mg(s)

17.15 Mn is +7 in MnO_4^- and N is +5 in NO_3^-

17.17 element oxidized: Ag; element reduced: Mn; oxidizing agent: $MnO_4^-(aq)$; reducing agent: Ag(s)

17.19 element oxidized: Cu; element reduced: N; oxidizing agent: $NO_3^-(aq)$; reducing agent: Cu(s)

17.21 element oxidized: Cu; element reduced: Cr; oxidizing agent: $Cr_2O_7^{2-}(aq)$; reducing agent: Cu(s)

17.23 (a) $3CuS(s) + 2NO_3^-(aq) + 8H^+(aq) \longrightarrow 3Cu^{2+}(aq) + 3S(s) + 2NO(g) + 4H_2O(l)$

(b) $ClO^-(aq) + 2I^-(aq) + 2H^+(aq) \longrightarrow Cl^-(aq) + I_2(aq) + H_2O(l)$

(c) $6Zn(s) + As_2O_3(s) + 12H^+(aq) \longrightarrow 6Zn^{2+}(aq) + 2AsH_3(g) + 3H_2O(l)$

17.25 (a) $S_2O_3^{2-}(aq) + 4Cl_2(g) + 5H_2O(l) \longrightarrow 2HSO_4^-(aq) + 8Cl^-(aq) + 8H^+(aq)$

(b) $3AsH_3(g) + 4ClO_3^-(aq) \longrightarrow 3H_3AsO_4(aq) + 4Cl^-(aq)$

(c) $3P_4(s) + 20NO_3^-(aq) + 8H_2O(l) \longrightarrow 12HPO_4^{2-}(aq) + 20NO(g) + 4H^+(aq)$

17.27 (a) $N_2(g) + O_2(g) \longrightarrow 2NO(g)$ and $2NO(g) + O_2(g) \longrightarrow 2NO_2(g)$

(b) oxidizing agent in each reaction is O_2; reducing agents are N_2 and NO

17.29 (a) $2MnO(s) + 5PbO_2(s) + 8H^+(aq) \longrightarrow 2MnO_4^-(aq) + 5Pb^{2+}(aq) + 4H_2O(l)$

(b) $3SO_2(g) + 2NO_3^-(aq) + 2H_2O(l) \longrightarrow 2NO(g) + 3SO_4^{2-}(aq) + 4H^+(aq)$

(c) $2KBr(aq) + 2H_2SO_4(aq) \longrightarrow K_2SO_4(aq) + Br_2(aq) + SO_2(g) + 2H_2O(l)$

17.31 (a) $3CuS(s) + 2NO_3^-(aq) + 8H^+(aq) \longrightarrow 3Cu^{2+}(aq) + 3S(s) + 2NO(g) + 4H_2O(l)$

(b) $ClO^-(aq) + 2I^-(aq) + 2H^+(aq) \longrightarrow Cl^-(aq) + I_2(aq) + H_2O(l)$

(c) $6Zn(s) + As_2O_3(s) + 12H^+(aq) \longrightarrow 6Zn^{2+}(aq) + 2AsH_3(g) + 3H_2O(l)$

17.33 (a) $H_2O_2(aq) + I_2(aq) \longrightarrow 2I^-(aq) + O_2(g) + 2H^+(aq)$; oxidation number of O in product is 0

(b) $H_2O_2(aq) + Sn^{2+}(aq) + 2H^+(aq) \longrightarrow Sn^{4+}(aq) + 2H_2O(l)$; oxidation number of O in product is −2

17.35 (a) $Cr_2O_7^{2-}(aq) + CH_3OH(aq) + 8H^+(aq) \longrightarrow 2Cr^{3+}(aq) + CO_2(g) + 6H_2O(l)$

(b) $3C_2H_4(g) + 2MnO_4^-(aq) + 2H_2O(l) + 2H^+(aq) \longrightarrow 3C_2H_6O_2(aq) + 2MnO_2(s)$

(c) $5C_2H_4(g) + 12MnO_4^-(aq) + 36H^+(aq) \longrightarrow 10CO_2(g) + 12Mn^{2+}(aq) + 28H_2O(l)$

17.37 (a) $4HNO_3(aq) \longrightarrow 4NO_2(g) + O_2(g) + 2H_2O(l)$

(b) $2Ce^{4+}(aq) + H_2O_2(aq) \longrightarrow 2Ce^{3+}(aq) + O_2(g) + 2H^+(aq)$

(c) $H_2S(aq) + H_2O_2(aq) \longrightarrow S(s) + 2H_2O(l)$

17.39 The purpose of a salt bridge in the construction of a voltaic cell is to permit the flow of ions without complete mixing of the solutions.

17.41 (a)

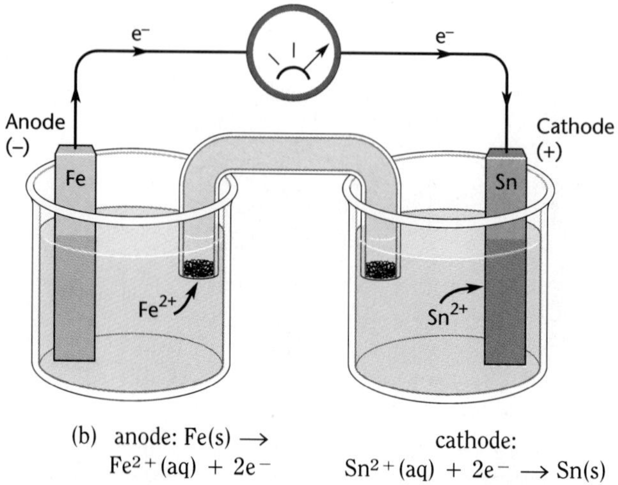

(b) anode: $Fe(s) \longrightarrow Fe^{2+}(aq) + 2e^-$ cathode: $Sn^{2+}(aq) + 2e^- \longrightarrow Sn(s)$

17.43 (a) See Figure 17.7 in the text. In your diagram, the anode will be the electrode at which chlorine gas is produced. The cathode will be the electrode at which sodium ions react to produce liquid sodium. The cathode will be negative and the anode will be positive. The current flows through the external circuit

from the anode to the cathode. The sodium ions are attracted to the cathode, and the chloride ions are attracted to the anode.
(b) anode: $2Cl^-(l) \longrightarrow Cl_2(g) + 2e^-$;
cathode: $Na^+(l) + e^- \longrightarrow Na(l)$
electrolysis reaction: $2NaCl(l) \longrightarrow 2Na(l) + Cl_2(g)$
17.45 (a, b)

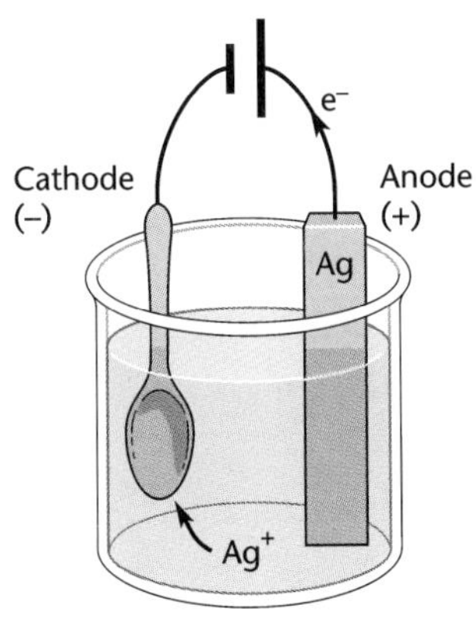

(c) anode: $Ag(s) \longrightarrow Ag^+(aq) + e^-$;
cathode: $Ag^+(aq) + e^- \longrightarrow Ag(s)$
17.47 (a) element oxidized: (a) S, (b) S, (c) Br
(b) element reduced: (a) N, (b) I, (c) S
(c) oxidizing agent: (a) NO_3^-, (b) I_2, (c) H_2SO_4
(d) reducing agent: (a) HgS, (b) $S_2O_3^{2-}$, (c) Br^-
17.49 (a) $2HNO_2(aq) + 2I^-(aq) + 2H^+(aq) \longrightarrow 2NO(g) + I_2(aq) + 2H_2O(l)$
(b) $4Ag(s) + 2H_2S(aq) + O_2(aq) \longrightarrow 2Ag_2S(s) + 2H_2O(l)$
(c) $MoO_3(s) + 3H_2(g) \longrightarrow Mo(s) + 3H_2O(l)$
17.51 (a, b)

e⁻
Cathode
(−)
Anode
(+)
Al_2O_3 (l) in Na_3AlF_6 (l)
O_2 (g)
Al^{3+}
O^{2-}
Al (l)

(c) $2Al_2O_3(l) \longrightarrow 4Al(l) + 3O_2(g)$

Chapter 18

18.7 Geiger–Müller counter
18.9 Activity is equal to the number of radioactive disintegrations per second.
18.11 A "one-curie source" is equivalent to a one-gram sample of radium-226. It yields 3.7×10^{10} disintegrations per second.
18.13 The rem is a unit of radiation dosage that takes into consideration the difference in effectiveness of various types of radiation in causing damage. A rem is related to a rad by the following equation: dose (rems) = dose (rads) × BEF, where BEF is selected for the type of radiation.
18.15 A natural decay series is a series of nuclear transformations that occurs in nature and that converts a radioactive isotope into a stable nonradioactive isotope of a different element. The uranium-238 decay series goes no further than lead-206 because the nucleus of lead-206 is stable and therefore not radioactive.
18.17 (a) ${}^{228}_{89}Ac \longrightarrow {}^{0}_{-1}\beta + {}^{228}_{90}Th$
(b) ${}^{3}_{1}H \longrightarrow {}^{0}_{-1}\beta + {}^{3}_{2}He$
18.19 (a) ${}^{59}_{26}Fe \longrightarrow {}^{0}_{-1}\beta + {}^{59}_{27}Co$
(b) ${}^{234}_{91}Pa \longrightarrow {}^{0}_{-1}\beta + {}^{234}_{92}U$
(c) ${}^{60}_{27}Co \longrightarrow {}^{0}_{-1}\beta + {}^{60}_{28}Ni$
18.21 ${}^{238}_{92}U \longrightarrow 8{}^{4}_{2}He + 6{}^{0}_{-1}\beta + {}^{?}_{?}?$
mass number = 238 − 8(4) − 6(0) = 238 − 32 = 206
atomic number = 92 − 8(2) − 6(−1) = 92 − 16 + 6 = 82 ∴ ${}^{206}_{82}Pb$ is the product
18.23 (a) ${}^{142}_{55}Cs \longrightarrow {}^{142}_{56}Ba + {}^{0}_{-1}\beta$; beta decay
(b) ${}^{128}_{50}Sn \longrightarrow {}^{128}_{51}Sb + {}^{0}_{-1}\beta$; beta decay
(c) ${}^{229}_{92}U \longrightarrow {}^{225}_{90}Th + {}^{4}_{2}He$; alpha beta decay
18.25 1.56%
18.27 1.8 g
18.29 ${}^{14}_{7}N + {}^{1}_{0}n \longrightarrow {}^{14}_{6}C + {}^{1}_{1}H$
18.31 The ratio of carbon-14 to carbon-12 remains constant in a growing tree because the tree is incorporating atmospheric carbon-14 into its structure as fast as the carbon-14 already in its structure is decaying. The ratio decreases steadily after the tree is cut down because the processes that would incorporate carbon-14 have stopped (e.g., growth), while the decay process continues.
18.33 starting nucleus is iron-58: ${}^{58}_{26}Fe + {}^{1}_{0}n \longrightarrow {}^{59}_{26}Fe$
18.35 ${}^{241}_{95}Am + {}^{4}_{2}He \longrightarrow {}^{243}_{97}Bk + 2{}^{1}_{0}n$
18.37 bombarding particle is the nucleus of a carbon-12 atom: ${}^{246}_{96}Cm + {}^{12}_{6}C \longrightarrow {}^{254}_{102}No + 4{}^{1}_{0}n$
18.39 A critical condition for a fission process is the condition in which the number of fission events per unit time remains constant.
18.41 (a) ${}^{235}_{92}U + {}^{1}_{0}n \longrightarrow {}^{90}_{37}Rb + {}^{144}_{55}Cs + 2{}^{1}_{0}n$
(b) ${}^{235}_{92}U + {}^{1}_{0}n \longrightarrow {}^{90}_{38}Sr + {}^{144}_{54}Xe + 2{}^{1}_{0}n$
18.43 See Figure 18.10 in the text. Be sure to include a core of radioactive material with control rods made of a moderating material, cooling fluid passing through and around the nuclear core, and a place for the now-heated cooling fluid to exchange heat with water to make the steam to run a steam turbine.
18.45 An explosion cannot occur in a nuclear fission reactor because the uranium used in a nuclear reactor is only 3% uranium-235. This percentage of the fissionable isotope of uranium is too low for a nuclear explosion.
18.47 0.137 00 amu = 2.2750×10^{-28} kg

18.49 9.0×10^{-11} J

18.51 Nuclear fusion occurs naturally in stars such as the sun.

18.53 ${}^{2}_{1}H + {}^{2}_{1}H \rightarrow {}^{3}_{2}He + {}^{1}_{0}n$

18.55 Nuclear fusion is more difficult to initiate than fission because extremely high temperatures are required for fusion. Fission has been used to provide the necessary energy to initiate fusion.

18.57 Radioactive iodine is used in two ways in medicine. It can be used for the diagnosis of thyroid problems and for the treatment of thyroid cancer. Different radioisotopes are used for the two purposes. In the diagnosis of thyroid problems, enough radioactive iodine is administered in a solution for a scan of the thyroid gland. In the treatment of thyroid cancer, radiation can be localized in the thyroid gland via radioactive iodine, and this radiation then kills the cancer cells.

18.59 A thickness gauge works by measuring the amount of radiation that passes through a plastic film. The amount of radiation that passes through the film is inversely related to the thickness of the film; i.e., the thicker the film, the less radiation passes through the film to reach a detector on the other side from the radioactive source.

Chapter 19

19.1 $H_3C{-}C(=O){-}C{\equiv}C{-}H$ (109°, 180°) H₂C=C=CH₂ (120°)

19.3

$H{-}CH_2{-}CH_2{-}CH_2Cl$
1-chloropropane

$CH_3{-}CHCl{-}CH_3$
2-chloropropane

19.5 $CH_3CH_2CH{=}CH_2$
1-butene

$H(CH_3)C{=}C(CH_3)H$
trans-2-butene

$(CH_3)HC{=}CH(CH_3)$
cis-2-butene

$CH_2{=}C(CH_3)_2$
methylpropene

19.7 $CH_3CH_2CH_2CH_2CH_2CH_3$ $CH_3CH_2CH_2CH(CH_3)CH_3$ $CH_3CH_2CH(CH_3)CH_2CH_3$ $CH_3CH_2C(CH_3)_2CH_3$ $CH_3CH(CH_3)CH(CH_3)CH_3$

19.9 (a) $CH_3CH(CH_3)CH_2CH_2CH_2CH_3$

(b) $CH_3CH_2CH_2CH(CH(CH_3)_2)CH_2CH_3$

(c) $CH_3CH_2CH(CH_2CH_3)CH_2CH_2CH_3$

(d) $CH_3CH_2CH_2CH(C(CH_3)_3)CH_2CH_2CH_2CH_3$

19.11 (a) $CH_3CH_2CH_2CH_2CH_2CH_2CH_3$

(b) $CH_3C{\equiv}CCH(CH_2CH_3)CH_2CH_3$

(c) $CH_3CH(CH_3)C(CH_3)_2CH_2CH_2CH_3$

(d) 1,1-dimethylcyclopentane (CH_3, CH_3 on the same ring carbon)

19.13 (a) $O{=}C{=}O$ (b) CH_3CH_3 (c) $CH_3CH_2CH_3$

19.15 (a) $CH_3C{\equiv}CH$ (b) H_2
(c) $\sim CH_2CCl_2(CH_2CCl_2)_n \sim$

19.17 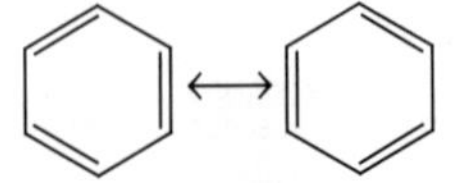 or 

The circle drawn inside a six carbon ring means six electrons spread out over the six carbon atoms by resonance.

19.19 The octane rating of a gasoline compares the antiknock performance of the gasoline to the performance of isooctane. Branched alkanes and aromatic hydrocarbons have the highest octane numbers.

19.21 highest octane number: (a) toluene

19.23 (a) $ClCH_2CH_2OH$ (b) $CH_3CH(OH)C(CH_3)_3$

(c) $C_6H_5{-}CH_2OH$ (d) $CH_3C(OH)(CH_3)CH_2CH_2CH_3$

19.25 (a) C_6H_5–OCH_2CH_3 (b) $CH_3CH(CH_3)CH_2CH_2OH$

(c) $CH_3C(OH)(CH_3)$–cyclopentyl (d) $CH_3CH(CH_3)CH_2OH$

19.27 (a) 1,2-ethanediol (ethylene glycol)
(b) diphenyl ether
(c) cyclopentanol

19.29 (a) $CH_3CH(CH_3)C(=O)CH_2CH_3$ (b) cyclopentanone

(c) $CH_3CH_2C(=O)H$ (d) $HC(=O)H$

19.31 (a) $CH_3CH(OH)CH_2CH_2CH_3 \xrightarrow{\text{oxidation}} CH_3C(=O)CH_2CH_2CH_3$

(b) $CH_3C(=O)CH_2CH_3 + NaBH_4 \rightarrow CH_3CH(OH)CH_2CH_3$

(c) $CH_3CH_2CH_2C(=O)H + CrO_3 \rightarrow CH_3CH_2CH_3C(=O)OH$

19.33 (a) $CH_3CH_2C(OCH_2CH_3)_2$–H (b) $CH_3C(OCH_3)_2$–CH_3

19.35 (a) NH_2–C_6H_4–Cl (para) (b) $CH_3N(CH_3)CH_3$

(c) $CH_3CH_2CH_2NHCH_2CH_2CH_3$

(d) $C_6H_5CH_2CH_2NH_2$

19.37 (pyridine, N in ring) A heterocyclic compound has a noncarbon atom as part of a ring.

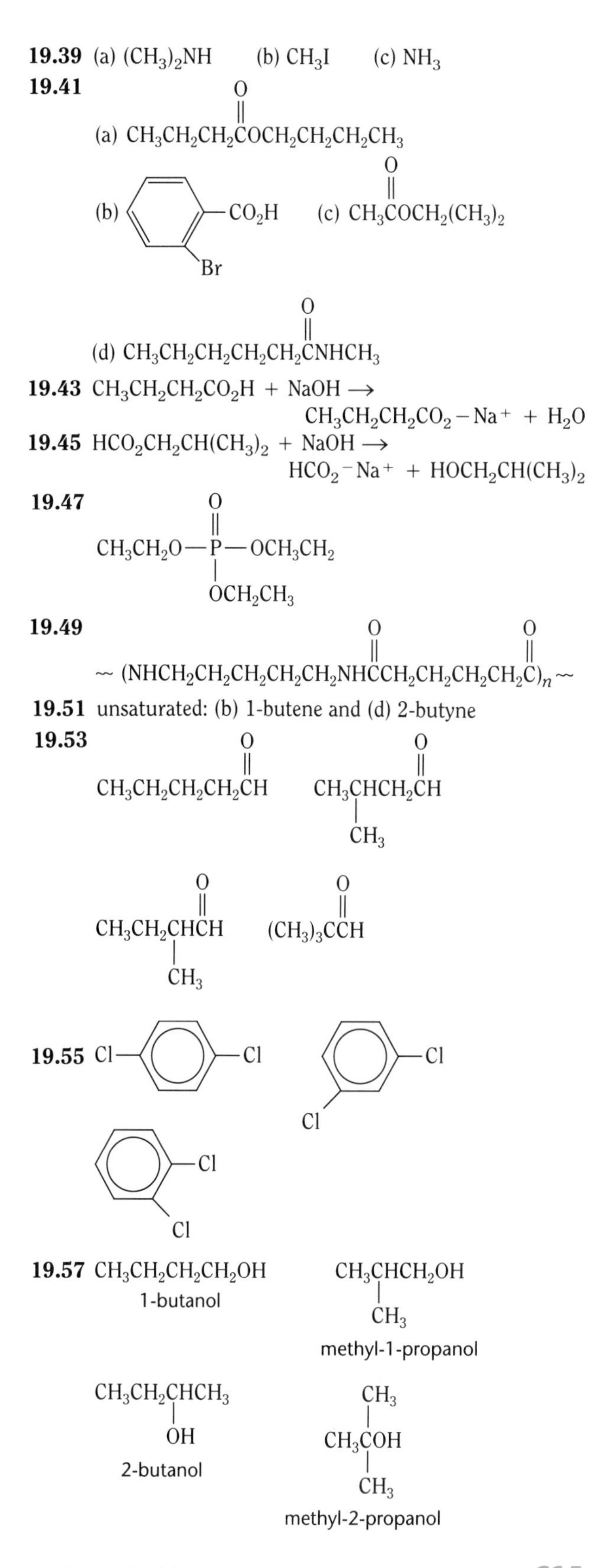

19.39 (a) $(CH_3)_2NH$ (b) CH_3I (c) NH_3

19.41 (a) $CH_3CH_2CH_2C(=O)OCH_2CH_2CH_2CH_3$

(b) 2-bromobenzoic acid (C_6H_4(Br)–CO_2H) (c) $CH_3C(=O)OCH_2(CH_3)_2$

(d) $CH_3CH_2CH_2CH_2CH_2C(=O)NHCH_3$

19.43 $CH_3CH_2CH_2CO_2H + NaOH \longrightarrow CH_3CH_2CH_2CO_2{}^-Na^+ + H_2O$

19.45 $HCO_2CH_2CH(CH_3)_2 + NaOH \longrightarrow HCO_2{}^-Na^+ + HOCH_2CH(CH_3)_2$

19.47 CH_3CH_2O–$P(=O)(OCH_2CH_3)$–OCH_3CH_2

19.49 $\sim(NHCH_2CH_2CH_2CH_2CH_2CH_2NHC(=O)CH_2CH_2CH_2CH_2C(=O))_n\sim$

19.51 unsaturated: (b) 1-butene and (d) 2-butyne

19.53 $CH_3CH_2CH_2CH_2C(=O)H$ $CH_3CH(CH_3)CH_2C(=O)H$

$CH_3CH_2CH(CH_3)C(=O)H$ $(CH_3)_3CC(=O)H$

19.55 Cl–C_6H_4–Cl (para), Cl–C_6H_4–Cl (meta), Cl–C_6H_4–Cl (ortho)

19.57 $CH_3CH_2CH_2CH_2OH$ 1-butanol; $CH_3CH(CH_3)CH_2OH$ methyl-1-propanol

$CH_3CH_2CH(OH)CH_3$ 2-butanol; $CH_3C(CH_3)_2OH$ methyl-2-propanol

19.59

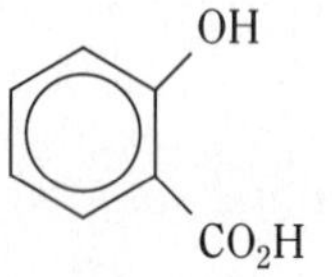

```
        OH
   [benzene ring]
        CO2H            CH3OH
2-hydroxybenzoic acid   methanol
```

19.61 $HC{\equiv}CCH_2CH_3$ $CH_3C{\equiv}CCH_3$
$CH_2{=}C{=}CHCH_3$ $CH_2{=}CHCH{=}CH_2$

19.63

$$\overset{O}{\overset{\|}{HCOH}} + HOCH_2CH_3 \xrightarrow{H^+} \overset{O}{\overset{\|}{HCOCH_2CH_3}} + H_2O$$

$$CH_3\overset{O}{\overset{\|}{COH}} + HOCH_3 \xrightarrow{H^+} CH_3\overset{O}{\overset{\|}{COCH_3}} + H_2O$$

19.65

$$\overset{O}{\overset{\|}{HOC}}-\overset{O}{\overset{\|}{COH}}$$

19.67 $CH_3CH_2CO_2CH_2CH_2CH_2CH_2CH3$

19.69

```
CH3CHCO2H
    |
    NH2
```

Chapter 20

20.1 Fibrous proteins are rod-shaped, whereas globular proteins are more spherical in shape.

20.3 The R groups in amino acids can be polar or nonpolar. Polar R groups increase the compatibility of an amino acid with water, since water is polar. Nonpolar R groups decrease the compatibility of an amino acid with water.

20.5

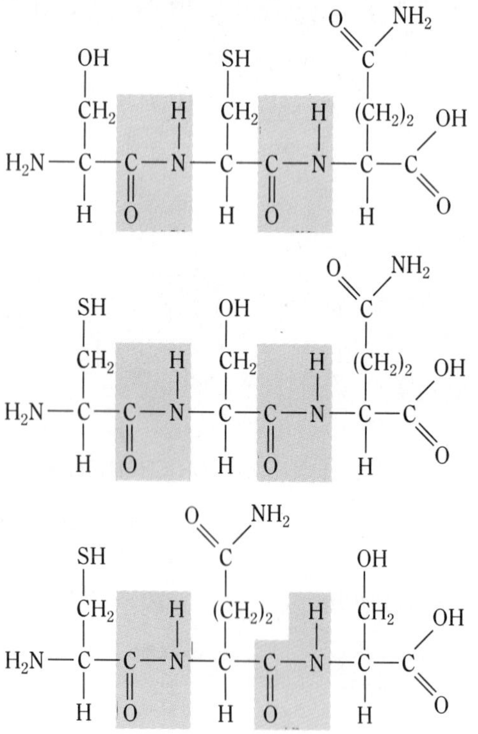

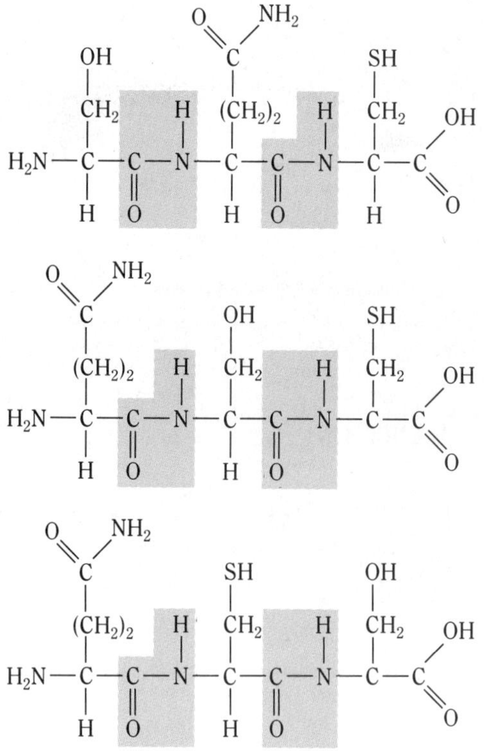

```
                                  O   NH2
                                   \\ /
     OH          SH                 C
     |           |                  |
     CH2    H    CH2    H         (CH2)2  OH
     |      |    |      |           |    /
H2N—C—C—N—C—C—N—C—C
     |   ||      |  ||              |    \\
     H   O       H  O               H     O
```

```
                                  O   NH2
                                   \\ /
     SH          OH                 C
     |           |                  |
     CH2    H    CH2    H         (CH2)2  OH
     |      |    |      |           |    /
H2N—C—C—N—C—C—N—C—C
     |   ||      |  ||              |    \\
     H   O       H  O               H     O
```

```
              O   NH2
               \\ /
     SH         C                 OH
     |          |                 |
     CH2    H  (CH2)2   H         CH2    OH
     |      |   |       |         |     /
H2N—C—C—N—C—C—N—C—C
     |   ||     |   ||            |     \\
     H   O      H   O             H      O
```

```
                     O   NH2
                      \\ /
     OH                C              SH
     |                 |              |
     CH2    H        (CH2)2   H       CH2    OH
     |      |          |      |       |     /
H2N—C—C—N—C—C—N—C—C
     |   ||            |  ||          |     \\
     H   O             H  O           H      O
```

```
   O   NH2
    \\ /
     C           OH               SH
     |           |                |
   (CH2)2   H    CH2    H         CH2    OH
     |      |    |      |         |     /
H2N—C—C—N—C—C—N—C—C
     |   ||      |  ||            |     \\
     H   O       H  O             H      O
```

```
   O   NH2
    \\ /
     C           SH               OH
     |           |                |
   (CH2)2   H    CH2    H         CH2    OH
     |      |    |      |         |     /
H2N—C—C—N—C—C—N—C—C
     |   ||      |  ||            |     \\
     H   O       H  O             H      O
```

20.7 120

20.9 The pleated sheet form of protein secondary structure involves long strands of protein running parallel with hydrogen bonding between the side-by-side chains. The result is a sheet-like structure that is pleated along the lines of the protein chains. Silk and some muscle fiber contain this structure.

20.11 Tertiary structure of a protein refers to the overall shape of a protein that results from the protein chain folding back on itself.

20.13 Hemoglobin is an example of a protein that has quaternary structure. Quaternary structure is the result of combining two or more proteins in order to perform a specific function, such as the transport of oxygen by hemoglobin.

20.15 enzyme

20.17 substrate

20.19 carbonyl groups (aldehydes and ketones) and hydroxyl groups

20.21 trisaccharide

20.23 Glucose is the building block. Starch and cellulose differ in the position of the OH group on C-1. In starch, this OH group is down (α) and in cellulose it is up (β).

20.25 The structure of glycogen facilitates the release of glucose by having many branches. Glucose is easily released only from the ends of chains. Therefore, the more ends, the more rapidly glucose can be released in the body upon demand.

20.27 Two components of a typical triglyceride are glycerol and fatty acids.

$$\begin{array}{l} H_2C-O-\overset{\displaystyle O}{\overset{\|}{C}}-R \\ \;| \\ HC-O-\overset{\displaystyle O}{\overset{\|}{C}}-R' \\ \;| \\ H_2C-O-\overset{\displaystyle O}{\overset{\|}{C}}-R'' \end{array}$$

20.29 A beneficial use for cholesterol in the body is to serve as the basis for the formation of other steroids, e.g., sex hormones. Cholesterol can be harmful to the body by the formation of plaque, which deposits in arteries, and by precipitating as gall stones.

20.31

N N N N H

a purine

N N

a pyrimidine

20.33

NH_2 N N N N

$HO-CH_2$ H O H H H

HO OH

20.35 The DNA double helix is held together by hydrogen bonding between complementary bases on the antiparallel strands of the double helix.

20.37 gene

Glossary

α-helix Secondary structure of a protein, in which hydrogen bonding occurs within a single chain, resulting in a coil

absolute zero Equivalent to $-273.15°C$; the basis for the Kelvin scale and theoretically the lowest possible temperature

accuracy How close a measured value is to a true value

acetal Geminal diether related to an aldehyde; $RCH(OR)_2$

acid anhydride Substance that combines with water to form an acid; a nonmetal oxide

acid–base indicator Compound that changes color when a solution reaches a certain level of acidity

acid ionization constant Equilibrium constant for the ionization of an acid, HA:

$$K_a = \frac{[H^+][A^-]}{[HA]}$$

acid salt Salt formed from a di- or triprotic acid that has an anion capable of giving up a proton

activation energy Minimum kinetic energy needed for colliding molecules to react

active site Position in the structure of an enzyme that influences a reaction

activity Number of nuclear disintegrations per second of a radioactive element

actual yield Measured mass of product in a reaction

addition Reaction in which one compound combines with another

adhesion Attractive force between a liquid and a surface

adsorb To stick to a surface

aerosol Colloid in which colloidal particles are dispersed in a gas

alcohol Compound that has an OH group bonded to a carbon that has four single bonds

aldehyde Carbonyl compound to which a hydrogen is bonded;

$$\overset{\displaystyle O}{\overset{\|}{RCH}}$$

alkali metals Elements of Group 1A in the periodic table

alkaline earth metals Elements of Group 2A in the periodic table

alkaloid Naturally occurring amine found in plants

alkane Saturated hydrocarbon represented by the formula C_nH_{2n+2}; a hydrocarbon that has only single bonds

alkene Hydrocarbon that contains a carbon–carbon double bond ($C{=}C$)

alkylation Process of petroleum refining that uses small molecules to make larger molecules

alkyl group Group represented by the formula C_nH_{2n+1}

alkyl halide Compound in which a halogen is bonded to a carbon having four single bonds

alkyne Hydrocarbon that contains a carbon–carbon triple bond ($C{\equiv}C$)

alpha particle Nucleus of a helium atom, ${}^4_2He$ or He^{2+}

amide Derivative of a carboxylic acid in which NH_2 is in place of the OH; $R\underset{\underset{\displaystyle O}{\|}}{C}NH_2$

amine Compound with one, two, or three alkyl or aryl groups bonded to nitrogen

amino group $—NH_2$

amorphous solid Solid that has no identifiable repeating geometric pattern

amphoteric Referring to a substance that can act as either an acid or a base

analytical chemistry Study of matter to determine the identity and quantity of its components

anhydride Compound derived from another by the removal of water

anhydrous Free from water

anion A negative ion

anode Electrode where oxidation occurs in an electrochemical cell

area Measure of a surface bounded by a set of lines

aromatic Compound containing a benzene ring

Arrhenius acid Substance that forms hydrogen ions (H^+) when dissolved in water

Arrhenius base Substance that dissolves in water to liberate hydroxide ions (OH^-)

atmospheric pressure Pressure of the earth's atmosphere

atom Smallest particle of an element that can combine with other elements to form compounds

atomic mass Average mass of an element's atoms calculated from the masses of the isotopes and their abundances in a naturally occurring sample

atomic mass unit (amu) $\frac{1}{12}$ the mass of a carbon-12 atom; equal to a mass of 1.6605×10^{-24} g

atomic number Number of protons in an atom's nucleus

atomic orbital Region in space around the nucleus of an atom where there is a high probability of finding an electron

atomic weight *See atomic mass*

Avogadro's hypothesis Hypothesis that equal volumes of different gases at the same temperature and pressure contain the same number of molecules

Avogadro's number Number of carbon atoms in 12 grams of carbon-12; 6.022×10^{23}

barometer Instrument used to measure atmospheric pressure

barometric pressure *See atmospheric pressure*

base Compound containing hydroxide ion

base anhydride Metal oxide that combines with water to form a base

base ionization constant Equilibrium constant for the ionization of a weak base, B:

$$K_b = \frac{[BH^+][OH^-]}{[B]}$$

becquerel (Bq) SI unit of the activity of a radioactive sample; 1 Bq is one disintegration per second

beta particle Fast-moving electron resulting from radioactive decay of a nucleus

binary acid Acid composed of two elements: hydrogen and an electronegative element

binary compound Compound made up of two elements

binding energy Energy that would be required to decompose a nucleus into its protons and neutrons

biochemistry Study of life processes

biological sciences Sciences involved with the study of life forms

boiling point Temperature at which the vapor pressure of a liquid is equal to amtospheric pressure

bond dissociation energy Energy needed to break the bond of a molecule

Boyle's law Law that the pressure and volume of a sample of a gas are inversely proportional at constant temperature

breeder reactor Nuclear reactor that uses fission to produce more fissionable fuel

Brønsted–Lowry acid A proton (H^+) donor

Brønsted–Lowry base A proton (H^+) acceptor

buffer solution Solution that resists changes in pH even when a strong acid or a strong base is added

calibration Process of adjusting an instrument to agree with the standard instrument

calorie Quantity of energy required to raise the temperature of one gram of water one Celsius degree; equivalent to 4.184 J

carbohydrates Compounds composed of C, H, and O, including sugars, starches, and cellulose

carbonyl group $>C{=}O$

carboxyl group $-C(=O)OH$

carboxylic acid Compound that contains a carboxyl group; $RC(=O)OH$

catalyst Substance that increases the rate of a reaction without being consumed

cathode Electrode where reduction occurs in an electrochemical cell

cation A positive ion

Charles' law Law that volume and absolute temperature of a sample of a gas are directly proportional at constant pressure

chemical bond Attractive force that keeps the oppositely charged ions of an ionic compound together and holds two atoms of a molecule together

chemical change Conversion of one substance into another

chemical energy Energy stored in chemical substances; a form of potential energy

chemical equation Equation representing a chemical reaction, in which the symbols or formulas of reactants are written on the left side of an arrow and those of the products are written on the right side

chemical equilibrium State in which the rates of the forward and reverse reactions are equal

chemical property Characteristic of a substance that establishes its chemical identity and distinguishes it from other substances

chemical reaction *See chemical change*

chemistry Study of the composition of materials and how their structure affects their properties and behavior

chromatography Separation method that depends upon differences in the ability of the compounds of a mixture to stick to surfaces

colligative property Physical property of a solution that depends upon the concentration of solute particles but is independent of the type of solute

collision theory Theory that reactant particles must collide for a chemical reaction to occur

colloid Heterogeneous mixture containing suspended, dispersed particles that are aggregates of molecules, ions, or atoms

combination reaction Reaction in which two or more substances combine to form a single compound

combined gas law Law that the volume of a fixed quantity of gas is directly proportional to its temperature and inversely proportional to its pressure

combustion Chemical reaction that releases both heat and light

compound Substance that is made up of two or more elements joined together and with a definite composition

concentrated Referring to a solution that contains a relatively large amount of solute per quantity of solution

concentration Ratio of the amount of solute to a given amount of solution

condensation Process of a vapor becoming a liquid

conjugate acid–base pair An acid and the base that results when the acid loses one proton

conversion factor Ratio that expresses the relationship of one quantity to another

covalent bond Attractive force between two atoms of a molecule that results when the atoms share a pair of electrons

cracking Process of petroleum refining that breaks down large molecules into smaller ones

critical condition Referring to a fission process in which the neutron balance is such that the number of fission events remains constant

critical mass Minimum mass needed to maintain the critical condition

crystalline solid Solid with a structure that has a regular, geometric pattern

crystallization Formation of crystals from a solution

curie (Ci) Activity of 3.7×10^{10} disintegrations per second, the activity of one gram of radium-226

cycloalkane Alkane having a ring structure

Dalton's law of partial pressures For a mixture of gases, the total pressure is the sum of the partial pressures of the component gases

decomposition reaction Reaction in which a single compound decomposes or breaks down to give two or more elements or compounds

deliquescent substance Substance that absorbs enough water from the air to form a solution

denaturation Altering the three-dimensional structure of a protein

density Ratio of the mass of a substance to its volume

diatomic Composed of two atoms

diffusion Process of spontaneous mixing of gases or liquids that occurs because of the random movement of the molecules

dilute Referring to a solution that contains a relatively small amount of solute per quantity of solution

dimensional analysis Method of writing an equation by examining the units of given and known quantities in a problem

dipole A covalent bond or a molecule which has ends that are oppositely charged

dipole–dipole attractions Electrostatic attractive forces between polar molecules

diprotic Describes an acid that has two acidic hydrogens (protons) in each formula unit

disaccharide Molecule composed of two monosaccharide units

dispersion forces Attractions between molecules produced by momentarily induced dipoles; proportional to the molar mass

distillation Method for separating solid–liquid and liquid–liquid solutions in which the components of the solution differ in their volatility

disulfide linkages S—S bonds formed from two SH groups

double bond Bond in which two atoms share two pairs of electrons

double replacement reaction Reaction in which the cation of each reactant combines with the anion of the other reactant to give two new compounds

dynamic equilibrium State in which opposing changes are occurring at equal rates, resulting in no net change over time

effective collision Collision that causes bonds of the reactants to break, allowing new bonds of the products to form

effusion Passage of a gas through a tiny hole or porous membrane from a region of high pressure into a region of low pressure

electrochemical cell Device in which an electric current results from a redox reaction, or that uses an electric current to cause a redox reaction

electrochemistry Branch of chemistry concerned with the relationship of electricity and chemical reactions

electrolysis Process whereby an electric current brings about a redox reaction

electrolyte Compound, either as a liquid or dissolved in water, that conducts electricity

electron Low-mass, negatively charged particle found outside the nucleus of an atom

electron affinity Energy change resulting when a gaseous atom gains an electron into its valence shell

electron configuration Shorthand notation that shows the energy states of an atom's or ion's electrons

electronegativity Capacity of an atom to attract its bonding electrons

electroplating Process in which a metal is electrolytically deposited on the surface of an object

electrostatic force Attractive or repulsive force between electrically charged objects

element Substance that cannot be broken down into simpler substances by chemical means

empirical formula Formula that shows the simplest mole ratio of the elements making up the compound

emulsion Colloid in which a liquid is dispersed in a liquid

endothermic Referring to a change in which heat is absorbed from the surroundings

endpoint Point at which an acid–base indicator changes color during a titration

energy Ability to do work

enzyme Protein that catalyzes a biochemical reaction

equilibrium constant The constant

$$K_{eq} = \frac{[C]^c[D]^d}{[A]^a[B]^b}$$

where C and D are products, A and B are reactants, and *a*, *b*, *c*, and *d* are coefficients in the balanced equation for a reaction at equilibrium

equivalence point Point in an acid–base titration at which the stoichiometric quantity of reactant indicated in the chemical equation has been added

ester Derivative of a carboxylic acid in which an alkyl or aryl group replaces the H of a carboxyl group; $\mathrm{R\underset{\underset{\displaystyle O}{\|}}{C}OR}$

ether Compound that has two alkyl groups bonded to an oxygen; R—O—R

exact numbers Numbers that have no uncertainty, including small numbers obtained by counting and numbers that define quantities

exothermic Referring to a change in which heat is released into the surroundings

exponential notation Expression of a number as a power of 10

families *See groups*

fat Triglyceride with saturated hydrocarbon chains

fatty acid Carboxylic acid containing 16–18 carbons

fibrous protein Protein that is insoluble in water and tends to be rod-shaped

filter Device used in filtration of a mixture to collect the solid and allow the liquid to pass through

filtration Method for separating a solid from a liquid of a heterogeneous mixture

fluid A material that flows

formula Chemical representation that includes the symbols of the elements making up a substance and the numbers, as subscripts, that indicate the number of atoms of each element

formula unit Atom or group of atoms indicated by the formula of the substance

formula weight *See molar mass*

functional group Atom or group of atoms that is primarily responsible for determining the physical and chemical properties of a compound

gamma ray High-energy electromagnetic radiation resulting from radioactive decay

gas Physical state in which matter has no definite shape or volume

Gay-Lussac's law Law that the pressure and absolute temperature of a sample of a gas are directly proportional when a constant volume is maintained

gel Sol that has a semirigid structure

gene Segment of DNA that carries information about the primary structure of a specific protein

globular protein A protein that is more spherical than linear

glycolipid Lipid with a fatty acid ester part and a sugar unit

ground state Lowest energy state of an atom

groups Columns of elements as arranged in the periodic table

half-life Time required for one-half of a radioactive sample to decay

halogens Elements of Group 7A in the periodic table

heat Form of energy that flows spontaneously from a warm object to one at a lower temperature

heat of reaction Heat absorbed or released to the surroundings during a chemical reaction

Heisenberg uncertainty principle Principle that it is impossible to precisely determine the exact position and speed of an electron at the same time

heterocyclic A compound that has a ring formed from more than one kind of atom

heterogeneous Referring to mixtures that are nonuniform

homogeneous Referring to mixtures that are uniform in composition, properties, and appearance

Hund's rule Rule that electrons are evenly distributed among orbitals of a subshell

hydrate Substance containing water molecules in its crystal structure

hydration Attraction or bonding of water molecules to an ion or molecule

hydrocarbon Compounds made up only of hydrogen and carbon

hydrogen bonding Attractive forces between molecules that have H—F, H—O, and H—N bonds; the strongest dipole–dipole attractions

hydrolysis Reaction in which the water molecule is split

hydrophilic Attracted to water

hydrophobic Repelled by water

hydrophobic interactions Association of hydrophobic groups with each other by dispersion forces

hygroscopic substance Substance that readily absorbs moisture from the air

hypothesis Tentative answer to a question or an explanation of observations supported by a limited amount of data

ideal gas Hypothetical gas that behaves exactly according to the kinetic-molecular theory

ideal gas equation Relationship of quantity, volume, pressure, and temperature of a gas: $PV = nRT$

immiscible Referring to liquids that do not mix (dissolve) in all proportions

induced dipole Temporary dipole in a molecule induced by the dipole of another molecule

ineffective collision Collision in which the colliding molecules bounce apart unchanged; no bonds are broken and no products are formed

inorganic chemistry Study of substances that are primarily of mineral origin

ion Positively or negatively charged atom or group of atoms

ionic bond Attractive force between oppositely charged ions in a compound

ionic crystal Crystalline solid composed of cations and anions

ionic equation Equation for a reaction that shows the ions present in a solution

ionization energy Amount of energy required to remove an element's outermost electron

ionizing radiation Emissions from radioactive samples

isoelectronic Atoms and ions that have the same electron configuration

isomers Compounds that have the same molecular formula but different structural formulas

isotopes Atoms of an element that have different numbers of neutrons, and hence different masses

IUPAC International Union of Pure and Applied Chemistry, the organization that sets the rules for the names and formulas of compounds

joule SI unit of energy; the kinetic energy (KE) of a 2-kg mass moving with a velocity of 1 m/s: $KE = \frac{1}{2} mv^2$

kelvin SI unit of temperature equivalent to one Celsius degree

ketal Geminal diether related to a ketone; $R_2C(OR)_2$

ketone $\overset{\displaystyle O}{\underset{}{\overset{\|}{RCR}}}$

kinetic Referring to motion

kinetic energy Energy of motion

kinetic-molecular theory Theory that all matter is made up of molecules in constant motion

law Statement of natural events that occur with consistency under the same conditions

law of combining volumes Law that the volumes of gases reacting with each other, when measured at the same temperature and pressure, are in ratios of small, whole numbers

law of conservation of energy Law that energy can neither be created nor destroyed in an ordinary chemical or physical change; energy is conserved

law of conservation of matter (mass) Law that matter is neither created nor destroyed during a chemical reaction

law of definite composition Law that every pure substance has a definite and fixed composition

law of multiple proportions Law that when two elements combine to form more than one compound, the masses of one element combined with a given mass of the other element are in a ratio of small, whole numbers

Le Chatelier's principle Principle that if a change is imposed on a reaction at equilibrium, the reaction will adjust to counteract the effect of the change and establish a new equilibrium state

Lewis acid Substance that accepts a pair of electrons in forming a covalent bond

Lewis base Substance that donates a pair of electrons in forming a covalent bond

Lewis formula Structural formula in which a line represents a covalent bond (two electrons) and dots represent nonbonding valence electrons

Lewis (electron-dot) symbol Symbol for an element surrounded by dots that represent its valence electrons

limiting reactant Reactant that is totally consumed in a reaction; the reactant that is not in excess

lipid Water-insoluble substance found in cells

liquid Physical state in which matter has a definite volume and takes the shape of the part of the container it occupies

main-group elements *See representative elements*

manometer Instrument used to measure the pressure of a confined gas

mass Quantity of matter in a sample or an object

mass defect Difference between the measured mass of a nucleus and the mass calculated by adding the masses of its protons and neutrons

mass number Sum of the protons and neutrons in the nucleus of an atom

mass percent Grams of solute per 100 grams of solution

matter Anything that has mass and occupies space

melting point Temperature at which a solid melts and the solid and liquid are in equilibrium

meniscus Curved surface of water or other liquid

metal Element that has a tendency to lose one or more electrons in a chemical reaction to form a positive ion

metallic bonding Attraction characterized by outer electrons that are able to freely flow through the crystal

metallic crystal Crystalline solid in which metal atoms occupy regular positions in the crystal

metalloid Elements whose properties are between those of metals and nonmetals

miscible Referring to liquids that mix (dissolve) in all proportions

mixture Composed of two or more substances

molality Number of moles of solute per kilogram of solvent

molar heat of fusion Heat that is absorbed in melting one mole of a solid

molar heat of vaporization Energy needed to vaporize one mole of liquid

molarity Number of moles of solute per liter of solution

molar mass Mass in grams of one mole of any substance

molar volume Volume of one mole of gas; 22.4 L at standard conditions

mole Amount of substance that contains the same number of formula units (atoms, molecules, etc.) as there are carbon atoms in exactly 12 g of carbon-12

molecular crystal Crystalline solid made up of molecules

molecular formula Formula that shows the actual numbers of atoms in a molecule

molecular weight *See molar mass*

molecule Neutral particle composed of two or more atoms joined together

monomer Reactant in a polymerization reaction

monoprotic Referring to an acid that has one acidic hydrogen (proton) in each formula unit

monosaccharide Molecule with one sugar unit

net ionic equation Equation that results when spectator ions are canceled from an ionic equation

network solid Crystalline solid in which atoms are covalently bonded to other atoms in an extended network

neutralization Combination of H^+ and OH^- to form H_2O

neutron Subatomic particle with a mass approximately the same as the mass of a proton, and with no electrical charge

noble gas Element of Group 8A in the periodic table

noble gas core Electron configuration for one of the noble gases

noble gas rule *See octet rule*

nonmetal Element that has a tendency to gain one or more electrons in a chemical reaction

nonpolar covalent bond Bond that results when two atoms having the same electronegativity share bonding electrons equally

nonpolar molecule Molecule that has no net dipole

normal boiling point Boiling temperature of a liquid at one atmosphere pressure

nuclear fission Breaking of a nucleus into two smaller nuclei of similar mass

nuclear fusion Process of combining two small nuclei to form a larger one

nuclear symbol Symbol of an atom that includes its atomic number and its mass number

nucleic acid Polynucleotide composed of sugars, phosphate, and nitrogen bases

nucleoside Compound composed of a sugar and a nitrogen base

nucleotide Phosphate ester of a nucleoside

nucleus Tiny, very dense region at the center of an atom containing all the positive charge and most of the mass of the atom

octet Set of eight electrons

octet rule Rule that when forming compounds, an atom has a tendency to lose, gain or share electrons resulting in eight electrons in its last occupied shell

orbit Specific path on which the electron of an atom travels, according to the Bohr model

organic chemistry Study of carbon compounds

oxidation Chemical change in which an element loses electrons

oxidation number Number that expresses the oxidation state of an element

oxidizing agent Substance that causes the oxidation of another substance; the electron acceptor in an oxidation–reduction (redox) reaction

oxyacid Ternary acids made up of hydrogen, oxygen, and a third element

oxyanion Anion formed when an oxyacid loses one or more hydrogen ions (protons)

paramagnetic Referring to a substance that is attracted by a magnetic field

partial pressure Pressure exerted by each gas making up a mixture

parts per million Grams of solute per one million grams of solution

Pauli exclusion principle Principle that no more than two electrons can occupy a given orbital, and that when two electrons occupy an orbital, their spins must be opposite

peptide linkage Bonding that joins amino acid units in a protein;

```
      H
      |
—C—N—
 ||
 O
```

percentage composition Mass percentage of each element making up a compound

percent yield Ratio of actual yield to theoretical yield expressed as a percent

periodic law Law that similar chemical and physical properties of the elements recur periodically when elements are arranged according to their atomic numbers

periods Rows of elements as arranged in the periodic table

pH Negative logarithm of $[H_3O^+]$; $-\log[H_3O^+]$

phospholipid Lipid with both a fatty acid ester part and a phosphate ester group

physical change Change in a substance that alters its physical properties without changing its identity

physical chemistry Study of the fundamental structure of matter and the influence of energy upon it

physical property Characteristic of a substance that can be determined without changing its composition

physical sciences Sciences concerned with the materials in the universe and how energy affects them

pleated sheet Secondary structure of a protein with small R groups in which hydrogen bonding between side-by-side chains is possible

pOH Negative logarithm of $[OH^-]$; $-\log[OH^-]$

polar covalent bond Bond in which there is unequal sharing of electrons

polar molecule Molecule that has a dipole

polymer Very large molecule composed of a repeating structural unit

polysaccharide Molecule with many monosaccharide units joined together

polyunsaturated Referring to a substance that has two or more C=C double bonds

potential energy Energy of position, or stored energy

precipitate Insoluble solid

precision Agreement among repeated measurements

pressure Force exerted per unit area of surface

primary structure Amino acid sequence in a protein chain

principal energy level Major energy level for electrons of an atom

product Result of a chemical reaction, written to the right of the arrow of a chemical equation

property Characteristic used to identify a substance

protein Natural polyamide

proton Positively charged particle within the nucleus of an atom

qualitative observations Observations that are not numerical and refer to what, what kind, or to a quality observed
quantitative observations Observations that are expressed numerically and result from making measurements of various kinds
quantum of energy Definite amount of energy emitted in the form of light when an electron drops to a lower energy level
quaternary structure The way two or more protein chains fit together to form specific three-dimensional shapes
rad Unit of radiation dosage: 10^{-5} J/g
radioactive tracer Radioisotope that can be followed through a chemical reaction or a series of reactions
radioactivity Spontaneous emission of rays or particles from the nucleus of an atom
radiocarbon dating Dating technique in which the radioisotope used is carbon-14; used for relics of plant, animal, or human origin, which typically contain carbon
radioisotope dating Technique used to estimate the age of an ancient relic or a rock by measuring the activity of a radioisotope
reactant Substance initially present in a chemical reaction, written to the left of the arrow in a chemical equation
reaction kinetics Study of reaction rates
redox reaction Oxidation–reduction or electron-transfer reaction
reducing agent Substance that causes the reduction of another substance; electron donor in an oxidation–reduction (redox) reaction
reduction Chemical change in which an element gains electrons
reforming Process of petroleum refining that changes unbranched molecules into branched ones
rem Unit of radiation dosage that takes into consideration the differences in effectiveness of different types of radiation in causing tissue damage
replication Process of producing an exact replica of DNA when cells divide
representative elements Elements in Groups 1A–8A in the periodic table
resonance Concept described by drawing two or more Lewis formulas for a substance that differ only in the position of the electrons
resonance structures Lewis formulas used to show resonance
ribosome Cellular substructure that receives genetic information from RNA
salt Ionic compound that results when a metal ion or ammonium ion replaces a hydrogen of an acid
salt bridges Ionic attractions resulting from the interaction of amine groups with carboxyl groups
saturated **(a)** Referring to a carbon structure that has the maximum number of hydrogen atoms possible; **(b)** referring to a solution that contains the most solute normally possible at a given temperature; dissolved and undissolved solute are in equilibrium
scientific method The logical, systematic process that scientists follow in solving problems
scientific notation Form of exponential notation in which a number is written as a product of a number between 1 and 10, and a power of 10, i.e., $A \times 10^{B}$
secondary structure Arrangement of a protein chain about an axis
shell *See principal energy level*
SI International System of Units; system of units based on the metric system and used by scientists everywhere
significant figures All the digits that are known in a given measurement plus the estimated (uncertain) digit
single bond Covalent bond in which two atoms share one pair of electrons
single replacement reaction Reaction in which an element reacting with a compound in aqueous solution replaces one of the elements making up the compound
sol Liquid colloid in which a solid is dispersed in a liquid
solid Physical state in which matter has a definite volume and shape
solubility The concentration of a substance in a saturated solution
solute Substance dispersed in the solvent; usually the minor component of a solution
solvent Substance that determines the physical state of a solution; usually the major component
specific heat Quantity of heat required to raise the temperature of one gram of a substance one Celsius degree
spectator ion Ion in solution that does not take part in a reaction
spectrum Photograph or chart of the display resulting when light is separated into its various colors
standard conditions *See standard temperature and pressure*
standard temperature and pressure (STP) Reference conditions established for gases; standard temperature is 273 K (0°C), and standard pressure is 1 atm (760 torr)
stereoisomers Isomers that differ only in the orientation of groups in space
steroid Lipid that has a characteristic polycyclic ring structure
strong acid Acid that ionizes completely, or almost so
strong base Good proton acceptor
structural formula Formula that shows how atoms are joined together in a molecule
subatomic particles Particles that make up atoms; electrons, protons, and neutrons
sublevel Subdivisions of the principal energy levels for electrons of an atom
sublimation Vaporization of a solid without first melting
subshell *See sublevel*
substance One kind of matter; element or a compound
substitution Reaction in which one atom is replaced by another

substrate Substance with which an enzyme reacts
supersaturated Referring to a solution that has a concentration greater than that of a saturated solution
surface tension Tendency of the surface of a liquid to resist stretching and behave like an elastic skin
temperature Measure of intensity of heat
ternary compound Compound made up of three elements
tertiary structure Three-dimensional shape resulting from the folding and bending of a protein
theoretical yield Calculated mass of product that will form when a reactant completely reacts
theory A good explanation of observed phenomena supported by a significant amount of data
titration Experimental procedure that involves the measurement of the volume of one solution required to react with another
transcription Process of copying genetic information
transition elements Block of elements separating the representative elements on the left and right in the periodic table
transition state High-energy complex in which reactant bonds are breaking and/or product bonds are forming
transmutation Change of one element into another
transuranium elements Those beyond uranium in the periodic table; all have been artificially prepared
triglyceride A simple lipid; an ester of glycerol and long-chain carboxylic acids
triple bond Covalent bond in which two atoms share three pairs of electrons
triprotic Referring to an acid that has three acidic hydrogens (protons) in each formula unit
Tyndall effect Light scattering by a colloid
unit Defining quantity for a measurement
unsaturated **(a)** Referring to a carbon structure that has fewer hydrogen atoms than a saturated structure; **(b)** a solution that is less than saturated
valence Combining capacity of an element
valence shell Outermost occupied principal energy level for the electrons of an atom
vapor Substance in the gaseous state above its liquid
vaporization Formation of a vapor from a liquid
vapor pressure Pressure exerted by vapor molecules
viscosity Resistance of a liquid to flow
viscous Referring to a liquid that flows slowly and does not spread easily when spilled
volume Amount of space occupied by matter
water of crystallization Water in a hydrate
water of hydration *See water of crystallization*
weight Gravitational attractive force exerted upon an object
work Moving an object through a distance

Index

References to figures and tables are printed in italic type.

Photo Credits

Contents
p. iv top Bill Kamin/Visuals Unlimited; **middle** Joel Gordon; **bottom** Randall J. Davis; **p. v top** Science VU-IBMRL/Visuals Unlimited; **middle** Randall J. Davis; **bottom** Warren Stone/Visuals Unlimited; **p. vi top** Anna Teetsov/McCrone Associates, Inc.; **middle** E. R. Degginger; **bottom** Reuters/Bettmann; **p. vii top** E. R. Degginger; **middle** Peticolas-Megna/Fundamental Photographs; **bottom** Bruce Gaylord/Visuals Unlimited; **p. viii left** NASA; **right** E. R. Degginger; **p. ix top** E. R. Degginger; **middle** Simon Bruty/Tony Stone Images; **bottom** Arnold J. Karpoff/Visuals Unlimited; **p. x top** Phil Degginger; **middle** Anglo-Australian Telescope Board; **bottom** Randall J. Davis.

1
pp. xvi–1 Bill Kamin/Visuals Unlimited; **p. 2 top** E. R. Degginger; **p. 2 bottom** Tom McCarthy/The Stock Market; **p. 3 top** Peticolas-Megna/Fundamental Photographs; **p. 3 bottom** Les Christman/Visuals Unlimited; **p. 4** Tom Pantages; **p. 5 top** Science VU/Visuals Unlimited; **p. 5 bottom** The Granger Collection; **p. 6** The Granger Collection; **p. 7 top, bottom** The Granger Collection; **p. 8** E. R. Degginger; **p. 9** Tom Pantages; **p. 10** E. R. Degginger

2
pp. 14–15 Randall J. Davis; **p. 16 top** Steve McCutcheon/Visuals Unlimited; **p. 16 bottom** Phil Degginger; **p. 17 left, right** E. R. Degginger; **p. 27 top** E. R. Degginger; **p. 27 bottom** Bruce Roberts/Photo Researchers; **p. 32 left, right** E. R. Degginger; **p. 33 top** Tom Pantages; **p. 33 bottom** Phil Degginger; **p. 34** Richard Megna/Fundamental Photographs; **p. 35** Michael Mathers/Peter Arnold, Inc.; **p. 36 top, bottom** Phil Degginger

3
pp. 42–43 Joel Gordon; **p. 44 top** E. R. Degginger; **p. 44 bottom** John D. Cunningham/Visuals Unlimited; **p. 45** Albert Copley/Visuals Unlimited; **p. 51** Science VU/Visuals Unlimited; **p. 52 top** Richard Hutchings/Photo Researchers; **p. 52 bottom** E. R. Degginger; **p. 54 top, bottom** Phil Degginger; **p. 58** Randall J. Davis; **p. 62** Robert E. Daemmrich/Tony Stone Images

4
pp. 72–73 Randall J. Davis; **p. 74 top** Randall J. Davis; **p. 74 bottom** Jean-Michel Labat/Photo Researchers; **p. 75 top** Phil Degginger; **p. 75 bottom** Joseph P. Sinnot/Fundamental Photographs; **p. 82** Doug Sokell/Visuals Unlimited; **p. 84** Tom Pantages; **p. 87** E. R. Degginger; **p. 89** NASA; **p. 91** Phil Degginger; **p. 92** Des and Jen Bartlett/Bruce Coleman, Inc.

5
pp. 100–101 Science VU-IBMRL/Visuals Unlimited; **p. 102 top, bottom left** Joel Gordon; **p. 102 bottom right** John D. Cunningham/Visuals Unlimited; **p. 103 top** The Granger Collection; **p. 103 bottom** Science VU-IBMRL/Visuals Unlimited; **p. 109** VU-Cabisco/Visuals Unlimited; **p. 116** Phil Degginger; **p. 117 text left, middle, right** Joel Gordon; **p. 117 margin** E. R. Degginger; **p. 121 left** Phil Degginger; **p. 121 right** Science VU/Visuals Unlimited

6
pp. 128–129 Warren Stone/Visuals Unlimited; **p. 130** Science VU/Visuals Unlimited; **p. 134** AIP Niels Bohr Library; **p. 138** Tom Pantages; **p. 157** Phil Degginger

7
pp. 166–167 E. R. Degginger; **p. 168 top, bottom** E. R. Degginger; **p. 168 middle** Phil Degginger; **p. 169** E. R. Degginger; **p. 170** E. R. Degginger; **p. 171 top** Phil Degginger; **p. 171 bottom** E. R. Degginger; **p. 172** E. R. Degginger; **p. 173** Tom Pantages; **p. 177** Matt Meadows/Peter Arnold, Inc.; **p. 179** E. R. Degginger; **p. 180** E. R. Degginger; **p. 181** Phil Degginger; **p. 185** E. R. Degginger; **p. 187** Science VU-NRCC/Visuals Unlimited

8
pp. 196–197 Anna Teetsov/McCrone Associates, Inc.; **p. 198 top** E. R. Degginger; **p. 198 bottom** Tom Pantages; **p. 201 top, bottom** E. R. Degginger; **p. 205** The Bancroft Library/University of California, Berkeley; **p. 212** Thomas Hollyman/Photo Researchers; **p. 220** Kirtley-Perkins/Visuals Unlimited

9
pp. 228–229 Reuters/Bettmann; **p. 230 bottom** E. R. Degginger; **p. 231 top** E. R. Degginger; **p. 231 bottom** Joel Gordon; **p. 232** Lester V. Bergman & Associates, Inc.; **p. 233** Phillip Wright/Visuals Unlimited; **p. 242 top, bottom left and right** E. R. Degginger; **p. 243 left** Phil Degginger; **p. 243 right** E. R. Degginger; **p. 247** St. Anthony Hospital Central/Denver, Colorado; **p. 259** Phil Degginger

10
p. 266 Peticolas-Megna/Fundamental Photographs; **p. 268 top, bottom** E. R. Degginger; **p. 269** Science VU-AISI/Visuals Unlimited; **p. 270 top, bottom** E. R. Degginger; **p. 271** UPI/Bettmann; **p. 274** Phil Degginger; **p. 277** Ron Spomer/Visuals Unlimited; **p. 278** Tom Pantages; **p. 279** John Sohlden/Visuals Unlimited; **p. 281 top** E. R. Degginger; **p. 281 bottom** Richard Megna/Fundamental Photographs; **p. 282** E. R. Degginger; **p. 283** U.S. Army Natick Research, Development and Engineering Center; **p. 285** Phil Degginger

11
pp. 298–299 E. R. Degginger; **p. 300 top** Phil Degginger; **p. 300 bottom** Joseph P. Sinnot/Fundamental Photographs; **p. 304** Dora Lambrecht/Visuals Unlimited; **p. 306** E. R. Degginger; **p. 307** Phil Degginger; **p. 308**

The Denver Post; **p. 311 top** Phil Degginger; **p. 311 bottom** Randall J. Davis; **p. 314** Blair Seitz/Photo Researchers

12

pp. 322–323 Bruce Gaylord/Visuals Unlimited; **p. 324 top** Mark A. Jacobsen; **p. 324 bottom** Randall J. Davis; **p. 325 top, bottom** Tom Pantages; **p. 328** Linda H. Hopson/Visuals Unlimited; **p. 329** Randall J. Davis; **p. 330** Phil Degginger; **p. 331** The Granger Collection; **p. 334** The Granger Collection; **p. 337** The Granger Collection; **p. 347** Clayton/Visuals Unlimited; **p. 348** Science VU/Visuals Unlimited

13

pp. 364–365 NASA; **p. 366, from top to bottom** E. R. Degginger, Phil Degginger, E. R. Degginger, Tom Pantages; **p. 373** Joel Gordon; **p. 377 top** P. Starborn/Visuals Unlimited; **p. 377 bottom** Phil Degginger; **p. 378** E. R. Degginger; **p. 379** E. R. Degginger; **p. 380 top, middle, bottom** E. R. Degginger; **p. 381 top left** E. R. Degginger; **p. 381 top right** Richard C. Walters/Visuals Unlimited; **p. 381 middle** Cabisco/Visuals Unlimited; **p. 381 bottom** John Cancalosi/Peter Arnold, Inc.; **p. 382** Adrian P. Davies/Bruce Coleman Inc.; **p. 384** Tom Pantages; **p. 389** Will and Deni McIntyre/Visuals Unlimited; **p. 390 left** Diana Schiumo/Fundamental Photographs; **p. 390 right** Philip Wright/Visuals Unlimited; **p. 391 left, middle** E. R. Degginger; **p. 391 right** Paul Silverman/Fundamental Photographs

14

p. 400 E. R. Degginger; **p. 402 top** E. R. Degginger; **p. 402 bottom** Steve McCutcheon/Visuals Unlimited; **p. 404** Tom Pantages; **p. 408 left** E. R. Degginger; **p. 408 right** Daniel W. Gotshall/Visuals Unlimited; **p. 409** Tom Pantages; **p. 412** Lester V. Bergman/Lester V. Bergman & Associates, Inc.; **p. 413 top, middle, bottom** E. R. Degginger; **p. 414 top, bottom** E. R. Degginger; **p. 417** E. R. Degginger; **p. 422 top left, right** E. R. Degginger; **p. 422 bottom left, right** Richard Megna/Fundamental Photographs; **p. 423** Phil Degginger

15

pp. 432–433 E. R. Degginger; **p. 434 top, bottom** E. R. Degginger; **p. 435 top, bottom left, right** E. R. Degginger; **p. 436 left, right** E. R. Degginger; **p. 438** E. R. Degginger; **p. 439** E. R. Degginger; **p. 440** E. R. Degginger; **p. 441** Richard Megna/Fundamental Photographs; **p. 452 left, right** E. R. Degginger

16

pp. 460–461 Arnold J. Karpoff/Visuals Unlimited; **p. 463 left, right** Richard Megna/Fundamental Photographs; **p. 468 top** AC-GM/Peter Arnold, Inc.; **p. 468 middle** Phil Degginger; **p. 468 bottom left, right** Richard Megna/Fundamental Photographs; **p. 470** Phil Degginger; **p. 471 top, bottom** Joel Gordon; **p. 472 left, right** Richard Megna/Fundamental Photographs; **p. 479** Tom Pantages

17

pp. 494–495 Simon Bruty/Tony Stone Images; **p. 496 top** E. R. Degginger; **p. 496 bottom** Phil Degginger; **p. 497** Ken Kay/Fundamental Photographs; **p. 498** E. R. Degginger; **p. 507** The Granger Collection; **p. 508** E. R. Degginger; **p. 509 top, bottom** Reginald Penner/University of California, Irvine; **p. 511** Phil Degginger; **p. 516** Tom Pantages; **p. 517** Les Christman/Visuals Unlimited

18

pp. 524–525 Anglo-Australian Telescope Board; **p. 526 top** U.S. Navy, Science Photo Library/Photo Researchers; **p. 526 bottom** Martin Dohrn, Science Photo Library/Photo Researchers; **p. 527** The Granger Collection; **p. 529** E. R. Degginger; **p. 530** Science VU, Technical Associates/Visuals Unlimited; **p. 531** Phil Degginger; **p. 538** Gianni Tortoli, Science Source/Photo Researchers; **p. 540** John D. Cunningham/Visuals Unlimited; **p. 541** The Granger Collection; **p. 543** Phil Degginger; **p. 545 top** T. Havill/Visuals Unlimited; **p. 545 bottom** E. R. Degginger; **p. 546 top, bottom** SIU/Visuals Unlimited

19

pp. 554–555 Phil Degginger; **p. 556 top** Phil Degginger; **p. 556 middle, left to right** E. R. Degginger, Joel Gordon, Joel Gordon, Joel Gordon, Joel Gordon; **p. 556 bottom** Phil Degginger; **p. 562** Joseph P. Sinnot/Fundamental Photographs; **p. 565** Blair Seitz/Photo Researchers; **p. 568** E. R. Degginger; **p. 570** John Serrao/Visuals Unlimited; **p. 576** E. R. Degginger; **p. 577** Phil Degginger; **p. 579** Phil Degginger; **p. 580** David S. Addison/Visuals Unlimited; **p. 585** Tom Pantages; **p. 586** Tom Pantages; **p. 587** Walt Anderson/Visuals Unlimited

20

p. 598 Randall J. Davis; **p. 600 top** George Loun/Visuals Unlimited; **p. 600 middle, left to right** W. Omerod/Visuals Unlimited, E. R. Degginger, E. R. Degginger; **p. 601** The Granger Collection; **p. 604** Gregory Gorel/Visuals Unlimited; **p. 608** Stanley Flegler/Visuals Unlimited; **p. 609** Phil Degginger; **p. 614** Randall J. Davis; **p. 616** Michael Long/Visuals Unlimited

Formulas and Names of Some Common Ions

Cations

1+		2+		3+	
Li^{+}	lithium	Be^{2+}	beryllium	Al^{3+}	aluminum
Na^{+}	sodium	Mg^{2+}	magnesium	Ga^{3+}	gallium
K^{+}	potassium	Ca^{2+}	calcium		
Rb^{+}	rubidium	Sr^{2+}	strontium		
Cs^{+}	cesium	Ba^{2+}	barium		
		Cr^{2+}	chromium(II)	Cr^{3+}	chromium(III)
		Mn^{2+}	manganese(II)	Mn^{3+}	manganese(III)
		Fe^{2+}	iron(II)	Fe^{3+}	iron(III)
		Co^{2+}	cobalt(II)	Co^{3+}	cobalt(III)
		Ni^{2+}	nickel(II)		
Cu^{+}	copper(I)	Cu^{2+}	copper(II)		
Ag^{+}	silver				
Au^{+}	gold(I)			Au^{3+}	gold(III)
		Zn^{2+}	zinc		
		Cd^{2+}	cadmium		
		Hg_2^{2+}	mercury(I)		
		Hg^{2+}	mercury(II)		
		Sn^{2+}	tin(II)		
		Pb^{2+}	lead(II)		
NH_4^{+}	ammonium				
H_3O^{+}	hydronium				

Anions

1−		2−		3−	
F^{-}	fluoride	O^{2-}	oxide	N^{3-}	nitride
Cl^{-}	chloride	S^{2-}	sulfide	P^{3-}	phosphide
Br^{-}	bromide	O_2^{2-}	peroxide		
I^{-}	iodide				
H^{-}	hydride				
OH^{-}	hydroxide				
CN^{-}	cyanide				
NO_3^{-}	nitrate	SO_4^{2-}	sulfate		
NO_2^{-}	nitrite	SO_3^{2-}	sulfite		
ClO_4^{-}	perchlorate	HPO_4^{2-}	hydrogen phosphate	PO_4^{3-}	phosphate
ClO_3^{-}	chlorate	$HAsO_4^{2-}$	hydrogen arsenate	AsO_4^{3-}	arsenate
ClO_2^{-}	chlorite				
ClO^{-}	hypochlorite				
HCO_3^{-}	hydrogen carbonate (bicarbonate)	CO_3^{2-}	carbonate		
		SiO_3^{2-}	silicate		
$C_2H_3O_2^{-}$	acetate	CrO_4^{2-}	chromate		
MnO_4^{-}	permanganate	$Cr_2O_7^{2-}$	dichromate		